HELICOPTER FLIGHT DYNAMICS

HELICOPTER FLIGHT DYNAMICS

The Theory and Application
of
Flying Qualities
and
Simulation Modelling

Second Edition

Gareth D. Padfield

BSc, PhD, C Eng, FRAeS

Blackwell
Publishing

© 1996, 2007 by G.D. Padfield

Blackwell Publishing editorial offices:
Blackwell Publishing Ltd, 9600 Garsington Road, Oxford OX4 2DQ, UK
 Tel: +44 (0)1865 776868
Blackwell Publishing Asia Pty Ltd, 550 Swanston Street, Carlton, Victoria 3053, Australia
 Tel: +61 (0)3 8359 1011

ISBN: 978-14051-1817-0

Published in North America by American Institute of Aeronautics and Astronautics, Inc.
370 L' Enfant Promenade, SW, Washington DC 20024-2518

First published 1996
Second edition published 2007

Library of Congress Cataloging-in-Publication Data:
A catalogue record for this title is available from the Library of Congress

British Library Cataloguing-in-Publication Data:
Padfield, G. D.
Helicopter flight dynamics : the theory and application of flying qualities and simulation modelling/Gareth D. Padfield. – 2nd ed.
        p. cm.
    Includes bibliographical references and index.
    ISBN-13: 978-1-4051-1817-0 (hardback : alk. paper)
    ISBN-10: 1-4051-1817-2 (hardback : alk. paper)
    1. Helicopters–Aerodynamics.   2. Helicopters–Handling characteristics.   I. Title.
    TL716.P23 2007
    629.132'5252–dc22

                                                        2007004737

Set in 9.5/12 pt Times
by Techbooks, New Delhi, India
Printed and bound in Singapore
by Markono Print Media Pte Ltd

The publisher's policy is to use permanent paper from mills that operate a sustainable forestry policy, and which has been manufactured from pulp processed using acid-free and elementary chlorine-free practices. Furthermore, the publisher ensures that the text paper and cover board used have met acceptable environmental accreditation standards.

For further information on Blackwell Publishing, visit our website:
www.blackwellpublishing.com

To my family
Joey, Jude and George

Contents

Preface to first edition   xiii
Preface to second edition   xvii
Copyright acknowledgements   xxi
Notation   xxiii
List of abbreviations   xxxiii

Chapter 1   Introduction
1.1 Simulation modelling   1
1.2 Flying qualities   3
1.3 Missing topics   4
1.4 Simple guide to the book   5

Chapter 2   Helicopter flight dynamics – an introductory tour
2.1 Introduction   9
2.2 Four reference points   10
   2.2.1 The mission and piloting tasks   11
   2.2.2 The operational environment   14
   2.2.3 The vehicle configuration, dynamics and flight envelope   15
      Rotor controls   15
      Two distinct flight regimes   17
      Rotor stall boundaries   20
   2.2.4 The pilot and pilot–vehicle interface   22
   2.2.5 Résumé of the four reference points   24
2.3 Modelling helicopter flight dynamics   25
      The problem domain   25
      Multiple interacting subsystems   26
      Trim, stability and response   28
      The flapping rotor *in vacuo*   30
      The flapping rotor in air – aerodynamic damping   33
      Flapping derivatives   36
      The fundamental 90° phase shift   36
      Hub moments and rotor/fuselage coupling   38
      Linearization in general   41
      Stability and control résumé   42
      The static stability derivative M_w 43
      Rotor thrust, inflow, Z_w and vertical gust response in hover   46
      Gust response in forward flight   48
      Vector-differential form of equations of motion   50

Validation   52
Inverse simulation   57
Modelling review   58
2.4 Flying qualities   59
Pilot opinion   60
Quantifying quality objectively   61
Frequency and amplitude – exposing the natural dimensions   62
Stability – early surprises compared with aeroplanes   63
Pilot-in-the-loop control; attacking a manoeuvre   66
Bandwidth – a parameter for all seasons?   67
Flying a mission task element   70
The cliff edge and carefree handling   71
Agility factor   72
Pilot's workload   73
Inceptors and displays   75
Operational benefits of flying qualities   75
Flying qualities review   77
2.5 Design for flying qualities; stability and control augmentation 78
Impurity of primary response   79
Strong cross-couplings   79
Response degradation at flight envelope limits   80
Poor stability   80
The rotor as a control filter   81
Artificial stability   81
2.6 Chapter review   84

Chapter 3   Modelling helicopter flight dynamics: building a simulation model
3.1 Introduction and scope   87
3.2 The formulation of helicopter forces and moments in level 1 modelling   91
3.2.1 Main rotor   93
Blade flapping dynamics – introduction   93
The centre-spring equivalent rotor   96
Multi-blade coordinates   102
Rotor forces and moments   108
Rotor torque   114
Rotor inflow   115
Momentum theory for axial flight   116
Momentum theory in forward flight   119
Local-differential momentum theory and dynamic inflow   125
Rotor flapping–further considerations of the centre-spring
  approximation   128
Rotor in-plane motion – lead–lag   135
Rotor blade pitch   138
Ground effect on inflow and induced power   139
3.2.2 The tail rotor   142
3.2.3 Fuselage and empennage   146
The fuselage aerodynamic forces and moments   146
The empennage aerodynamic forces and moments   149

3.2.4 Powerplant and rotor governor 152
3.2.5 Flight control system 154
   Pitch and roll control 154
   Yaw control 158
   Heave control 158
3.3 Integrated equations of motion of the helicopter 159
3.4 Beyond level 1 modelling 162
3.4.1 Rotor aerodynamics and dynamics 163
   Rotor aerodynamics 163
   Modelling section lift, drag and pitching moment 164
   Modelling local incidence 167
   Rotor dynamics 168
3.4.2 Interactional aerodynamics 171
Appendix 3A   Frames of reference and coordinate transformations 175
3A.1 The inertial motion of the aircraft 175
3A.2 The orientation problem – angular coordinates of the aircraft 180
3A.3 Components of gravitational acceleration along the aircraft axes 181
3A.4 The rotor system – kinematics of a blade element 182
3A.5 Rotor reference planes – hub, tip path and no-feathering 184

Chapter 4   Modelling helicopter flight dynamics: trim and stability analysis

4.1 Introduction and scope 187
4.2 Trim analysis 192
4.2.1 The general trim problem 194
4.2.2 Longitudinal partial trim 196
4.2.3 Lateral/directional partial trim 201
4.2.4 Rotorspeed/torque partial trim 203
4.2.5 Balance of forces and moments 204
4.2.6 Control angles to support the forces and moments 204
4.3 Stability analysis 208
4.3.1 Linearization 209
4.3.2 The derivatives 214
   The translational velocity derivatives 215
   The angular velocity derivatives 224
   The control derivatives 231
   The effects of non-uniform rotor inflow on damping and control
      derivatives 234
   Some reflections on derivatives 235
4.3.3 The natural modes of motion 236
   The longitudinal modes 241
   The lateral/directional modes 247
   Comparison with flight 250
Appendix 4A   The analysis of linear dynamic systems (with special reference to
      6 DoF helicopter flight) 252
Appendix 4B   The three case helicopters: Lynx, Bo105 and Puma 261
4B.1 Aircraft configuration parameters 261
   The DRA (RAE) research Lynx, ZD559 261
   The DLR research Bo105, S123 261

The DRA (RAE) research Puma, SA330   263
Fuselage aerodynamic characteristics   264
Empennage aerodynamic characteristics   268
4B.2 Stability and control derivatives   269
4B.3 Tables of stability and control derivatives and system eigenvalues   277
Appendix 4C   The trim orientation problem   293

**Chapter 5   Modelling helicopter flight dynamics: stability under constraint
                and response analysis**
5.1 Introduction and scope   297
5.2 Stability under constraint   298
     5.2.1 Attitude constraint   299
     5.2.2 Flight-path constraint   306
          Longitudinal motion   306
          Lateral motion   310
5.3 Analysis of response to controls   315
     5.3.1 General   315
     5.3.2 Heave response to collective control inputs   317
          Response to collective in hover   317
          Response to collective in forward flight   323
     5.3.3 Pitch and roll response to cyclic pitch control inputs   325
          Response to step inputs in hover – general features   325
          Effects of rotor dynamics   327
          Step responses in hover – effect of key rotor parameters   327
          Response variations with forward speed   330
          Stability versus agility – contribution of the horizontal tailplane   331
          Comparison with flight   332
     5.3.4 Yaw/roll response to pedal control inputs   338
5.4 Response to atmospheric disturbances   344
          Modelling atmospheric disturbances   346
          Modelling helicopter response   348
          Ride qualities   350

Chapter 6   Flying qualities: objective assessment and criteria development
6.1 General introduction to flying qualities   355
6.2 Introduction and scope: the objective measurement of quality   360
6.3 Roll axis response criteria   364
     6.3.1 Task margin and manoeuvre quickness   364
     6.3.2 Moderate to large amplitude/low to moderate frequency: quickness
            and control power   371
     6.3.3 Small amplitude/moderate to high frequency: bandwidth   378
          Early efforts in the time domain   378
          Bandwidth   381
          Phase delay   386
          Bandwidth/phase delay boundaries   387
          Civil applications   389
          The measurement of bandwidth   391

Estimating ω_{bw} and τ_p 397
Control sensitivity 399
6.3.4 Small amplitude/low to moderate frequency: dynamic stability 401
6.3.5 Trim and quasi-static stability 402
6.4 Pitch axis response criteria 404
   6.4.1 Moderate to large amplitude/low to moderate frequency: quickness
     and control power 404
   6.4.2 Small amplitude/moderate to high frequency: bandwidth 408
   6.4.3 Small amplitude/low to moderate frequency: dynamic stability 410
   6.4.4 Trim and quasi-static stability 413
6.5 Heave axis response criteria 417
   6.5.1 Criteria for hover and low speed flight 420
   6.5.2 Criteria for torque and rotorspeed during vertical axis manoeuvres 424
   6.5.3 Heave response criteria in forward flight 424
   6.5.4 Heave response characteristics in steep descent 427
6.6 Yaw axis response criteria 429
   6.6.1 Moderate to large amplitude/low to moderate frequency: quickness
     and control power 430
   6.6.2 Small amplitude/moderate to high frequency: bandwidth 433
   6.6.3 Small amplitude/low to moderate frequency: dynamic stability 433
   6.6.4 Trim and quasi-static stability 436
6.7 Cross-coupling criteria 437
   6.7.1 Pitch-to-roll and roll-to-pitch couplings 437
   6.7.2 Collective to pitch coupling 440
   6.7.3 Collective to yaw coupling 440
   6.7.4 Sideslip to pitch and roll coupling 440
6.8 Multi-axis response criteria and novel-response types 442
   6.8.1 Multi-axis response criteria 442
   6.8.2 Novel response types 444
6.9 Objective criteria revisited 447

Chapter 7   Flying qualities: subjective assessment and other topics
7.1 Introduction and scope 455
7.2 The subjective assessment of flying quality 456
   7.2.1 Pilot handling qualities ratings – HQRs 457
   7.2.2 Conducting a handling qualities experiment 464
     Designing a mission task element 464
     Evaluating roll axis handling characteristics 466
7.3 Special flying qualities 478
   7.3.1 Agility 478
     Agility as a military attribute 478
     The agility factor 481
     Relating agility to handling qualities parameters 484
   7.3.2 The integration of controls and displays for flight in degraded visual
     environments 487
     Flight in DVE 487
     Pilotage functions 488
     Flying in DVE 489

The usable cue environment   490
UCE augmentation with overlaid symbology   496
7.3.3 Carefree flying qualities   500
7.4 Pilot's controllers   508
7.5 The contribution of flying qualities to operational effectiveness and the safety
of flight   511

Chapter 8   Flying qualities: forms of degradation
8.1 Introduction and scope   517
8.2 Flight in degraded visual environments   519
8.2.1 Recapping the usable cue environment   520
8.2.2 Visual perception in flight control – optical flow and motion parallax 523
8.2.3 Time to contact; optical tau, τ 532
8.2.4 τ control in the deceleration-to-stop manoeuvre   536
8.2.5 Tau-coupling – a paradigm for safety in action   538
8.2.6 Terrain-following flight in degraded visibility   545
τ on the rising curve   548
8.3 Handling qualities degradation through flight system failures   559
8.3.1 Methodology for quantifying flying qualities following flight function
failures   562
8.3.2 Loss of control function   564
Tail rotor failures   564
8.3.3 Malfunction of control – hard-over failures   568
8.3.4 Degradation of control function – actuator rate limiting   574
8.4 Encounters with atmospheric disturbances   576
8.4.1 Helicopter response to aircraft vortex wakes   578
The wake vortex   578
Hazard severity criteria   579
Analysis of encounters – attitude response   587
Analysis of encounters – vertical response   588
8.4.2 Severity of transient response   593
8.5 Chapter Review   597
Appendix 8A   HELIFLIGHT and FLIGHTLAB at the University of Liverpool   599
FLIGHTLAB   601
Immersive cockpit environment   602

References   608
Index   633

Preface to first edition

In this preface, I want to communicate three things. First, I would like to share with the reader my motivation for taking on this project. Second, I want to try to identify my intended audience and, third, I want to record some special acknowledgements to colleagues who have helped me.

When I decided to pursue a career as an aeronautical engineer, my motivation stemmed from an aesthetic delight in flight and things that flew, combined with an uncanny interest in tackling, and sometimes solving, difficult technical problems. Both held a mystery for me and together, unbeknown to me at the time, helped me to 'escape' the Welsh mining community in which I had been sculptured, on to the roads of learning and earning. Long before that, in the late 1940s, when I was taking my first gasps of Welsh air, the Royal Aircraft Establishment (RAE) had been conducting the first research flight trials to understand helicopter stability and control. It should be remembered that at that time, practical helicopters had been around for less than a decade. From reading the technical reports and talking with engineers who worked in those days, I have an image of an exciting and productive era, with test and theory continuously wrestling to provide first-time answers to the many puzzles of helicopter flight dynamics.

Although there have been quiet periods since then, the RAE sustained its helicopter research programme through the 1950s, 1960s and 1970s and by the time I took charge of the activities at Bedford in the mid-1980s, it had established itself at the leading edge of research into rotor aerodynamics and helicopter flight dynamics. My own helicopter journey began in the Research Department at Westland Helicopters in the early 1970s. At that time, Westland were engaged with the flight testing of the prototype Lynx, a helicopter full of innovation for a 1960s design. This was also an exciting era, when the foundations of my understanding of helicopter flight dynamics were laid down. Working with a small and enthusiastic group of research engineers, the mysteries began to unfold, but at times it felt as if the more I learned, the less I understood. I do not want to use the word enthusiastic lightly in this context; a great number of helicopter engineers that I have known have a degree of enthusiasm that goes way beyond the call of duty, so to speak, and I do believe that this is a special characteristic of people in this relatively small community. While it is inevitable that our endeavours are fuelled by the needs of others – the ubiquitous customer, for example – enthusiasm for the helicopter and all of the attendant technologies is a powerful and dynamic force. In writing this book I have tried to share some of my enthusiasm and knowledge of helicopter flight dynamics with as large an audience as possible, and that was probably sufficient personal motivation to undertake the task. This motivation is augmented by a feeling that my own experience in theory and test has given me insight into, and a somewhat unique way of looking at, the

subject of flight dynamics that I hope will appeal to the reader in search of understanding.

There are, however, more pragmatic reasons for writing this book. While fixed-
wing flight dynamics, stability and control have been covered from a number of perspectives in more than a dozen treatise over the years, there has never been a helicopter
textbook dedicated to the subject; so there is, at least, a perceived gap in the available
literature, and, perhaps more importantly, the time is ripe to fill that gap. The last 10–20
years has seen a significant amount of research in flight simulation and flying qualities
for helicopters, much of which has appeared in the open literature but is scattered in
scores of individual references. This book attempts to capture the essence of this work
from the author's perspective, as a practitioner involved in the DRA (RAE) research
in national and international programmes. It has been a busy and productive period,
indeed it is still continuing, and I hope that this book conveys the impression of a living
and mature subject, to which many contributions are yet to be made.

The book is written mainly for practising flight dynamics engineers. In some
organizations, such a person may be described as a flying qualities engineer, a flight
simulation engineer or even a flight controls engineer, but my personal view is that these
titles reflect subdisciplines within the larger field of flight dynamics. Key activities of the
flight dynamics engineer are simulation modelling, flying qualities and flight control.
Simulation brings the engineer into a special and intimate relationship with the system
he or she is modelling and the helicopter is a classic example. The present era appears to
be characterized by fast-disappearing computational constraints on our ability to model
and simulate the complex aeroelastic interactions involved in helicopter flight. Keeping
step with these advances, the flight dynamics engineer must, at the same time, preserve
an understanding of the link between cause and effect. After all, the very objectives of
modelling and simulation are to gain an understanding of the effects of various design
features and insight into the sensitivity of flight behaviour to changes in configuration
and flight condition. In the modelling task, the flight dynamics engineer will need to
address all the underlying assumptions, and test them against experimental data, in a
way that provides as complete a calibration as possible. The flight dynamics engineer
will also have a good understanding of flying qualities and the piloting task, and he
or she will appreciate the importance of the external and internal influences on these
qualities and the need for mission-oriented criteria. Good flying qualities underpin
safe flight, and this book attempts to make the essence of the theoretical developments
and test database, assembled over the period from the early 1980s through to the
present time, accessible to practising engineers. Flight testing is an important part of
flight dynamics, supporting both simulation validation and the development of flying
qualities criteria. In this book I have attempted to provide the tools for building and
analysing simulation models of helicopter flight, and to present an up-to-date treatment
of flying qualities criteria and flight test techniques.

While this is primarily a specialist's book, it is also written for those with empathy
for the broader vision, within which flight dynamics plays its part. It is hoped that the
book, or parts of the book, will appeal to test pilots and flight test engineers and
offer something useful to engineers without aeronautical backgrounds, or those who
have specialized in the aerodynamic or controls disciplines and wish to gain a broader
perspective of the functionality of the total aircraft. In writing Chapters 2, 6 and 7, I
have tried to avoid a dependence on 'difficult' mathematics. Chapters 3, 4 and 5, on
the other hand, require a reasonable grasp of analytical and vectorial mechanics as

would, for example, be taught in the more extensive engineering courses at first and higher degree levels. With regard to education programmes, I have had in mind that different parts of the book could well form the subject of one or two term courses at graduate or even advanced undergraduate level. I would strongly recommend Chapter 2 to all who have embarked on a learning programme with this book. Taught well, I have always believed that flight dynamics is inspirational and, hence, a motivating subject at university level, dealing with whole aircraft and the way they fly, and, at the same time, the integration of the parts that make the whole. I have personally gained much from the subject and perhaps this book also serves as an attempt to return my own personal understandings into the well of knowledge.

In the sense that this book is an offering, it also reflects the great deal of gratitude I feel towards many colleagues over the years, who have helped to make the business enjoyable, challenging and stimulating for me. I have been fortunate to be part of several endeavours, both nationally and internationally, that have achieved significant progress, compared with the sometimes more limited progress possible by individuals working on their own. International collaboration has always held a special interest for me and I am grateful to AGARD, Garteur, TTCP and other, less formal, ties with European and North American agencies, for providing the auspices for collaboration. Once again, this book is full of the fruits of these activities. I genuinely believe that helicopters of the future will perform better, be safer and be easier to fly because of the efforts of the various research groups working together in the field of flight dynamics, feeding the results into the acquisition processes in the form of the requirements specifications, and into the manufacturing process, through improved tools and technologies.

In the preparation of this book several colleagues have given me specific support which I would like to acknowledge. For assistance in the generation and presentation of key results I would like to acknowledge the Rotorcraft Group at DRA Bedford. But my gratitude to the Bedford team goes far beyond the specific support activities, and I resist identifying individual contributions for that reason. As a team we have pushed forward in many directions over the last 10 years, sometimes at the exciting but lonely leading edge, at other times filling in the gaps left by others pushing forward with greater pace and urgency. I want to record that this book very much reflects these team efforts, as indicated by the many cited references. I was anxious to have the book reviewed in a critical light before signing it off for publication, and my thanks go to colleagues and friends Ronald Milne, Ronald DuVal, Alan Simpson, Ian Simons and David Key for being kind enough to read individual chapters and for providing me with important critical reviews. A special thanks to Roy Bradley for reviewing the book in its entirety and for offering many valuable ideas which have been implemented to make the book better.

I first had the serious idea of writing this book about 4 years ago. I was familiar with the Blackwell Science series and I liked their productions, so I approached them first. From the beginning, my publisher at Blackwell's, Julia Burden, was helpful and encouraging. Later, during the preparation, the support from Julia and her team was sustained and all negotiations have been both positive and constructive; I would like to express my gratitude for this important contribution. I would like also to acknowledge the vital support of my employer, the Defence Research Agency, for allowing me to use material from my research activities at RAE and DRA over the past 18 years. My particular thanks to my boss, Peter England, Manager, Flight Dynamics and Simulation Department at DRA Bedford, who has been continually supportive with a positive attitude that has freed me from any feelings of conflict of interest. Acknowledgements for DRA material used and figures or quotes from other sources are included elsewhere in this book. The figures in this book were produced by two artists, those in Chapter 2 by Peter Wells and the rest by Mark Straker. Both worked from often very rough drafts and have, I believe, done an excellent job – thank you both.

All these people have helped me along the road in a variety of different ways, as I have tried to indicate, but I am fully accountable for what is written in this book. I am responsible for the variations in style and 'colour', inevitable and perhaps even desirable in a book of this scope and size. There have been moments when I have been guided by some kind of inspiration and others where I have had to be more concerned with making sure the mathematics was correct. I have done my best in this second area and apologise in advance for the inevitable errors that will have crept in. My final thanks go to you, the reader, for at least starting the journey through this work. I hope that you enjoy the learning and I wish you good fortune with the application of your own ideas, some of which may germinate as a result of reading this book. It might help to know that this book will continue to be my guide to flight dynamics and I will be looking for ways in which the presentation can be improved.

Gareth D. Padfield

Sharnbrook, England

Preface to second edition

In the preface to the first edition of my book I talked about flight dynamics as a '*living and mature subject, to which many contributions are yet to be made*'; I believe this statement is still true and every new generation of engineers has something new to add to the store of knowledge. During the 10 years since its publication, the disciplines of flight dynamics and handling/flying qualities engineering have matured into a systems approach to the design and development of those functions and technologies required to support the piloting task. At the same time, as pilot-centred operational attributes, flying qualities are recognised as the product of a continual tension between performance and safety. These two descriptions and the interplay between them highlight the importance of the subject to continuing helicopter development. The most obvious contributors to flying qualities are the air vehicle dynamics – the stability and control characteristics – and these aspects were treated in some depth in the first edition. Flying qualities are much more, however, and this has also been emphasized. They are a product of the four elements: the aircraft, the pilot, the task and the environment, and it is this broader, holistic view of the subject which is both a technical discipline and an operational attribute, which emphasizes the importance to flight safety and operational effectiveness. I have tried to draw out this emphasis in the new material presented in Chapter 8, Degraded Flying Qualities, which constitutes the bulk of the new content in this second edition.

During the preparation of the first edition, ADS-33C was being used extensively in a range of military aircraft programmes. The handling qualities (HQs) criteria represented key performance drivers for the RAH-66 Comanche, and although this aircraft programme would eventually be cancelled, Industry and the surrounding helicopter 'community' would learn about the technology required to deliver Level 1 HQs across a range of operational requirements. The last decade has seen ADS-33 applied to aircraft such as NH-90 and the UK's attack helicopter, and also to new operations including maritime rotorcraft and helicopters carrying external loads, and used as a design guide for civil tilt rotor aircraft. It is now common at annual European and American Helicopter Fora to hear presentations on new applications of ADS-33 or extensions to its theoretical basis. The Standard has also been refined over this period and currently exists in the ADS-33E-PRF (performance) version, emphasizing its status as a performance requirement. A brief resume of developments is added to Chapter 6.

Significant advances have also been made on the modelling and simulation front, and it is very satisfying to see the considerable pace at which the modelling of complex helicopter aerodynamics is moving. It surely will not be very long before the results of accurate physical flow modelling will be fully embodied into efficient, whole aircraft design codes and real-time simulation. A combination of high-quality computer tools for comprehensive synthesis and analysis and robust design criteria pave the way for massive reductions in timescales and costs for design, development and certification. The modelling and simulation material in Chapters 3, 4 and 5 is largely unchanged in this second edition. This is simply a result of the author needing to put limits on what is achievable within the timescale available.

In August 1999, I left government 'service' to join The University of Liverpool with a mandate to lead the aerospace activity, both on the research and the learning and teaching (L&T) axes. I was confident that my 30 years of experience would enable me to transition fairly naturally into academia on the research axis. I had very little experience on the L&T side however, but have developed undergraduate modules in rotorcraft flight, aircraft performance and flight handling qualities. I confirm the old adage – to learn something properly, you need to teach it – and it has been very satisfying to 'plough' some of my experience back into the formative 'soil' of future careers.

As with the first edition, while this work is a consolidation of my knowledge and understanding, much has been drawn from the efforts and results of others, and not only is acknowledging this fact appropriate but it also feels satisfying to record these thanks, particularly to the very special and highly motivated group of individuals in the Flight Science and Technology Research Group at the University of Liverpool. This group has formed and grown organically, as any university research group might, over the period since 2000 and, hopefully, will continue to develop capabilities and contribute to the universal pool of knowledge and understanding. Those, in academe, who have had the pleasure and privilege to 'lead' a group of young post-graduate students and post-doctoral researchers will perhaps understand the sense in which I derive satisfaction from witnessing the development of independent researchers, and adding my mite to the process. Thanks to Ben Lawrence and Binoy Manimala who have become experts in FLIGHTLAB and other computational flight dynamics analyses and helped me in numerous ways, but particularly related to investigating the effects of trailing wake vortices on helicopters. Neil Cameron derived the results presented in Chapter 8 on the effects of control system failures on the handing qualities of tilt rotor aircraft. Gary Clark worked closely with me to produce the results in Chapter 8, relating to terrain following flight in degraded visibility. Immeasurable gratitude to Mark White, the simulation laboratory manager in FS&T, who has worked with me on most of the research projects initiated over the last 5 years. The support of Advanced Rotorcraft Technology, particularly Ronald Du Val and Chengian Ho, with various FLIGHTLAB issues and the development of the HELIFLIGHT simulator has been huge and is gratefully acknowledged.

Those involved in flight dynamics and handling qualities research will understand the significant contribution that test pilots make to the subject, and at Liverpool we have been very fortunate indeed to have the sustained and consistently excellent support from a number of ex-military test pilots, and this is the place to acknowledge their contribution to my developing knowledge captured in this book. Sincere thanks to Andy Berryman, Nigel Talbot, Martin Mayer and Steve Cheyne; they should hopefully know how important I consider their contributions to be.

Thanks to Roger Hoh and colleagues at Hoh Aeronautics, whose continuous commitment to handling qualities excellence has been inspirational to me. Roger has also made contributions to the research activities in FS&T particularly related to the development of handling criteria in degraded conditions and the attendant design of displays for flight in degraded visual environments. The whole subject of visual perception in flight control has been illuminated to me through close collaboration with David Lee, Professor of Perception in Action at The University of Edinburgh. David's contributions to my understanding of the role of optical flow and optical tau in the control of motion has been significant and is gratefully acknowledged.

Over the last 10 years I have received paper and electronic communications from colleagues and readers of the first edition worldwide who have been complementary and have politely identified various errors or misprints, which have been corrected. These communications have been rather too numerous to identify and mention individually here but it is hoped that a collective thanks will be appreciated.

Mark Straker produced the figures in the form they appear in this book to his usual very high standard; thanks again Mark for your creative support.

Finally, grateful thanks to Julia Burden at Blackwell Publishing who has been unrelenting in her encouragement, dare I say persistence, with me to produce material for this second edition. Any Head of a fairly large academic department (at Liverpool I am currently Head of Engineering with 900 students and 250 staff) will know what a challenging and rather absorbing business it can be, especially when one takes it on to direct and increase the pace of change. So, I was reluctant to commit to this second edition until I felt that I had sufficient new research completed to 'justify' a new edition; the reader will now find a consolidation of much of that new work in the new Chapter 8. Only the authors who have worked under the pressures of a tight schedule, whilst at the same time having a busy day job, will know how and where I found the time.

So this book is offered to both a new and old readership, who might also find some light-hearted relief in a 'refreshed' version of my poem, or sky-song as I call it, *Helicopter Blues*, which can also be sung in a 12-bar blues arrangement (*normally in Emaj but in Am if you're feeling cool*)

I got the helicopter blues
They're going round in my head
I got the helicopter blues
They're still going round in my head
brother please tell me what to do about these helicopter blues

My engine she's failing
Gotta reduce my torque
My engine she keeps failing
Gotta pull back on my power
seems like I'm autorotating from all these helicopter blues

My tail rotor ain't working
Ain't got no place to go
My tail rotor she ain't working
Ain't got no place to turn
These helicopter blues brother
They're driving me insane

My humms are a humming
Feel all fatigued, used and abused
My humms are humming
I'm worn out from all this aerofoil toil
If I don't get some maintenance
sister I've had it with all these helicopter blues

My gearbox is whining
Must need more lubrication

I said I can't stand this whining
please ease my pain with boiling oil
If I don't get that stuff right now
I'm gonna lock up with those helicopter blues

Dark blue or light
The blues got a strong hold on me
It really don't matter which it is
The blues got no respect for me
Well, if only I could change to green
Maybe I could shake off these helicopter blues

I've designed a new helicopter
It'll be free of the blues
I've used special techniques and powerful computers
I'm sure I know what I'm doing
now I gotta find someone to help me chase away these helicopter blues

I went to see Boeing
Said I got this new blues-free design
I went up to see Boeing, told them my story and it sounded fine
But they said why blue's our favourite colour
Besides which, you're European

So I took my design to Eurocopter
I should have thought of them first
If I'd only gone to Eurocopter
I wouldn't be standing here dying of thirst
They said 'ces la vie mon frere' you can't make a sans bleu helicoptre

I went to see Sikorsky
I thought – They'll fix the blues
They sent for Nick Lappos
To fix the helicopter blues
Nick said don't be such a baby Gareth
(besides, I don't work here anymore)
Just enjoy those helicopter blues

I'll go see Ray Prouty
People say, Ray – he ain't got no blues
Please help me Ray – how much more aerodynamics do I need – I'll clean your shoes
Ray said, wake up and smell the coffee fella
Learn how to hide those helicopter blues

I've learned to live with them now
I'm talking about the helicopter blues
Even got to enjoy them
Those sweet, soothing helicopter blues
I'm as weary as hell but please don't take away my helicopter blues

Gareth D. Padfield

Caldy, England

The cover photograph is reproduced with permission from AgustaWestland.

Copyright acknowledgements

The following people and organizations are gratefully acknowledged for granting permission for the use of copyright material.

The UK MoD and Defence Research Agency for Figs 2.31, 2.43, 2.44, 2.50, 3.15, 3.28, 3.29, 3.35, 3.37, 3.38, 5.7–5.9, 5.28–5.31, 5.34, 6.7, 6.8, 6.9, 6.10, 6.18, 6.19, 6.35, 6.36, 6.38, 6.39, 6.47–6.52, 6.59, 7.10–7.24, 7.38, 7.44, 7.45 and 7.46.* The US Army for Figs 6.15, 6.17, 6.20, 6.25, 6.30, 6.33, 6.40–6.45, 6.56, 6.61, 6.64, 6.65, 6.70 and 7.28 and Table 7.4. The American Helicopter Society (AHS) for Figs 3.16 and 7.5 (with the US Army). Bob Heffley for Figs 6.6 and 6.11. Cambridge University Press for the quote from Duncan's book at the beginning of Chapter 3. Chengjian He and the AHS for Fig. 5.27. Chris Blanken, the US Army and the AHS for Figs 7.29 and 7.30. Courtland Bivens, the AHS and the US Army for Fig. 6.63. David Key and the Royal Aeronautical Society for Figs 6.3 and 6.31. David Key for the quote at the beginning of Chapter 7. DLR Braunschweig for Figs 6.21, 6.23 (with RAeSoc), 6.32, 6.37, 6.58 (with the AHS), 6.68 (with the US Army) and 7.4 (with AGARD). Eurocopter Deutschland for Figs 6.46 and 6.66. Ian Cheeseman and MoD for Figs 3.28 and 3.29. Jeff Schroeder and the AHS for Figs 7.32–7.36. Jeremy Howitt and the DRA for Figs 7.39, 7.40 and 7.41. Knute Hanson and the Royal Aeronautical Society for Fig. 6.69. Lt Cdr Sandy Ellin and the DRA for Figs 2.7, 3.44 and 3.45. Mark Tischler and AGARD for Figs 5.25, 5.26, 6.34 and 6.57. McDonnell Douglas Helicopters, AGARD and the US Army for Fig. 6.71. NASA for Figs 4.12 and 6.2. Institute for Aerospace Research, Ottawa, for Figs 6.54 and 7.7 (with the AHS). Pat Curtiss for Figs 3.46, 3.47 and 5.4. Roger Hoh for Figs 6.24, 6.26 (with the AHS), 6.29 (with the RAeSoc) and 7.27 (with the AHS). Sikorsky Aircraft, the US Army and the AHS for Fig. 6.72. Stewart Houston and the DRA for Figs 5.10–5.13. Tom Beddoes for Fig. 3.42. Jan Drees for Fig. 2.8. AGARD for selected text from References 6.72 and 7.25. Westland Helicopters for granting permission to use configuration data and flight test data for the Lynx helicopter. Eurocopter Deutschland for granting permission to use configuration data and flight test data for the Bo105 helicopter. Eurocopter France for granting permission to use configuration data and flight test data for the SA330 Puma helicopter.

In this second edition, once again the author has drawn from the vast store of knowledge and understanding gained and documented by others and the following people and organizations are gratefully acknowledged for the use of copyright material.

Philippe Rollet and Eurocopter for the use of Table 8.9. John Perrone at the University of Waikato for Figs 8.4, 8.6 and 8.11. James Cutting at Cornell University and MIT Press for Figs 8.7, 8.8 and the basis of Fig 8.10. NASA for Fig. 8.14. David Lee for Figs 8.18 and 8.19. The US Army Aviation Engineering Directorate for the use of Table 6.6 and Figs 6.74, 6.75 and 6.77 and general reference to ADS33. AgustaWestland

Helicopters for the use of the photographs of the EH101 at the start of Chapter 8 and also on the book cover. Roger Hoh and the American Helicopter Society for Fig. 8.2. The American Helicopter Society for a variety of the author's own figures published in Ref 8.31, 8.33 and 8.55. The Institution of Mechanical Engineers for Fig. 8.45 from the author's own paper. The Royal Aeronautical Society for the use of the author's own figures from Ref 8.53. J. Weakly and the American Helicopter Society for Fig. 8.43. Franklin Harris for Fig. 8.62.

Notation

a_0	main rotor blade lift curve slope (1/rad)
a_g	constant acceleration of the τ guide
a_{0T}	tail rotor blade lift curve slope (1/rad)
$a_{n-1}, a_{n-2}, \ldots$	coefficients of characteristic (eigenvalue) equation
$\mathbf{a}_p$	acceleration of P relative to fixed earth (components a_x, a_y, a_z) (m/s², ft/s²)
$\mathbf{a}_{p/g}$	acceleration vector of P relative to G (m/s², ft/s²)
a_{xb}, a_{yb}, a_{zb}	acceleration components of a blade element in rotating blade axes system (m/s², ft/s²)
a_{zpk}	peak normal acceleration (m/s², ft/s²)
c	rotor blade chord (m, ft)
c	constant τ motion
$d(\psi, r_b)$	local drag force per unit span acting on blade element (N/m, lbf/ft)
eR	flap hinge offset (m, ft)
$e_\zeta R$	lag hinge offset (m, ft)
$\mathbf{f}(t)$	forcing function vector
$f_\beta(\psi), f_\lambda(\psi)$	coefficients in blade flapping equation
$f_y(r_b), f_z(r_b)$	in-plane and out-of-plane aerodynamic loads on rotor blade at radial station r_b
g	acceleration due to gravity (m/s², ft/s²)
g_{1c0}, g_{1c1}	lateral cyclic stick–blade angle gearing constants
g_{1s0}, g_{1s1}	longitudinal cyclic stick–blade angle gearing constants
g_{cc0}, g_{cc1}	collective lever–lateral cyclic blade angle gearing constants
g_{cT0}	pedal/collective lever–tail rotor control run gearing constant
g_θ, g_ϕ	nonlinear trim functions
g_{sc0}, g_{sc1}	collective lever–longitudinal cyclic blade angle gearing constants
g_{T0}, g_{T1}	pedal–tail rotor collective blade angle gearing constant
g_T	tail rotor gearing
h	height above ground (m(ft))
h_e	eye-height
$h, \dot{h}$	height (m, ft), height rate (m/s, ft/s)
h_{fn}	height of fin centre of pressure above fuselage reference point along negative z-axis (m, ft)
h_R	height of main rotor hub above fuselage reference point (m, ft)
h_T	height of tail rotor hub above fuselage reference point (m, ft)
$\mathbf{i}, \mathbf{j}, \mathbf{k}$	unit vectors along x-, y- and z-axes
k	τ coupling constant
k_1, k_2, k_3	inertia coupling parameters

k_{1s}, k_{1c}	feedforward gains (rad/unit stick movement)
k_3	$= \tan$ (rad / m^2) tail rotor delta 3 angle
k_ϕ, k_p	feedback gains in roll axis control system (rad/rad, rad/(rad/s))
k_g	feedback gain in collective – normal acceleration loop (rad /m^2)
$k_{\lambda f}$	main rotor downwash factor at fuselage
$k_{\lambda fn}$	main rotor downwash factor at fin
$k_{\lambda T}$	main rotor downwash factor at tail rotor
$k_{\lambda tp}$	main rotor downwash factor at tailplane
k_0, k_q	feedback gains in pitch axis control system (rad/rad, rad/(rad s))
$k_{\theta i}, k_{\phi i}$	trim damping factors
$\ell(\psi, r)$	lift per unit span (N/m, Ibf/ft)
l_f	fuselage reference length (m, ft)
l_{fn}	distance of fin centre of pressure aft of fuselage reference point along negative x-axis (m, ft)
l_T	distance of tail rotor hub aft of fuselage reference point (m, ft)
l_{tp}	distance of tailplane centre of pressure aft of fuselage reference point (m, ft)
$m(r)$	blade mass distribution
m_{am}	apparent mass of air displaced by rotor in vertical motion
n, n_{zpk}	load factor (g)
p, q, r	angular velocity components of helicopter about fuselage x-, y- and z-axes (rad/s)
$p_{pk}/\Delta\phi$	attitude quickness parameter (1/s)
p_{ss}, p_s	steady state roll rate (rad/s)
r, r_b $(^-)$	blade radial distance (with overbar – normalized by radius R) (m, ft)
r, r_c	radial distance from vortex core and vortex core radius
$\mathbf{r}_{p/g}$	position vector of P relative to G (components x, y, z) (m, ft)
s	Laplace transform variable
s	rotor solidity $= N_b c/\pi R$
s_T	tail rotor solidity
t	time (s)
$\bar{t}$	normalized time (t/T)
t_r	time in a manoeuvre when the reversal occurs (s)
t_w	heave time constant $(-1/Z_w)$ (s)
$\bar{t}_w$	t_w normalized by T
t_1	manoeuvre time (s)
$t_{r10,50,90}$	time constants – time to 10%, 50%, 90% of steady-state response (s)
$\mathbf{u}(t)$	control vector
u, v, w	translational velocity components of helicopter along fuselage x-, y- and z-axes (δw $\equiv$ w, etc.) (m/s, ft/s)
v_i	induced velocity at disc (m/s, ft/s)
v_{ihover}	induced velocity at disc in hover (m/s, ft/s)
$v_{i\infty}$	induced velocity in the far field below rotor (m/s, ft/s)
$\mathbf{v}_j$	eigenvectors of $\mathbf{A}^T$
$\mathbf{v}_g, \mathbf{v}_p$	velocity vector of G, P relative to fixed Earth

$\mathbf{v}_{p/g}$	velocity vector of P relative to G (components $u_{p/g}$, $v_{p/g}$, $w_{p/g}$)
v_g	velocity of motion guide (m/s, ft/s)
v_{g0}	initial velocity of motion guide (m/s, ft/s)
w	velocity along aircraft z-axis (ms, fts)
w_{ss}	steady-state velocity along aircraft z-axis (m/s, ft/s)
$w(r, t)$	blade out-of-plane bending displacement (m, ft)
w_0	vertical velocity (m/s, ft/s)
$w_g(t)$	gust velocity component along z-axis (m/s, ft/s)
w_{gm}	maximum value of velocity in ramp gust (m/s, ft/s)
$\mathbf{w}_i$	eigenvectors of $\mathbf{A}$
w_λ	$w - k_{\lambda\mathrm{f}}\Omega R\lambda_0$ total downwash over fuselage (m/s, ft/s)
w_{ss}	steady-state normal velocity (m/s, ft/s)
w_{ss}	steady state velocity along aircraft z axis (m/s, ft/s)
$\mathbf{x}(t)$	state vector
x, x_{cmd}	position and position command in pilot/vehicle system
x, z	distance along x- and z-directions
x, $\overline{x}$	distance (normalized distance) to go in manoeuvre (m, ft)
$\overline{x}'$, $\overline{x}''$	normalized velocity and acceleration in menoeuvre
x, y, z	mutually orthogonal directions of fuselage axes – x forward, y to starboard, z down; centred at the helicopter's centre of mass
$\mathbf{x}_0$	initial condition vector $\mathbf{x}(0)$
x_{cg}	centre of gravity (centre of mass) location forward of fuselage reference point (m, ft)
$\mathbf{x}_e$	equilibrium value of state vector
x_e	distance in eye-height/s
$\dot{x}_e$	velocity in eye-heights/s
x_{g0}	initial displacement of motion guide (m(ft))
x_g	distance to go in motion guide (m(ft))
x_m	distance to go in manoeuvre (m(ft))
x_r	edge rate (1/s)
$\mathbf{x}_f$, $\mathbf{x}_r$, $\mathbf{x}_p$, $\mathbf{x}_c$	elemental state vectors (f – fuselage, r – rotor, p – powerplant, c – control)
z_g	distance of ground below rotor (m, ft)
$\mathbf{A}$, $\mathbf{B}$	system and control matrices
$\mathbf{A}_{ff}$, $\mathbf{A}_{fr}$, etc.	system matrices; ff – fuselage subsystem, fr – rotor to fuselage coupling
$\mathbf{A}_{11}$, $\mathbf{A}_{12}$ ...	submatrices in partitioned form of $\mathbf{A}$
A_b	blade area (m^2, ft^2)
A_d	rotor disc area (m^2, ft^2)
A_f	agility factor – ratio of ideal to actual manoeuvre time
A_x, A_y	x- and y-axes acceleration components of aircraft relative to Earth (m/s^2, ft/s^2)
$\mathbf{B}_{ff}$, $\mathbf{B}_{fr}$, etc.	control matrices; ff fuselage subsystem, fr rotor to fuselage coupling
C'_1	$= \dfrac{1}{1 + a_0 s/16\lambda_0}$ lift deficiency factor
C'_2	$= \dfrac{a_0 s}{16\lambda_0}$
$\mathbf{C}_I(\psi)$	time–dependent damping matrix in individual blade flapping equations

C_{if}	normalized fuselage force and moment coefficients, $i = x, y, z, l, m, n$
C_{La}	aerodynamic flap moment coefficient about roll axis
C_{Lmax}	maximum aerofoil lift coefficient
$\mathbf{C}_M(\psi)$	time-dependent damping matrix in multi-blade flapping equations
$\mathbf{C}_{M0}(\psi)$	constant damping matrix in multi-blade flapping equations
C_{Ma}	aerodynamic flap moment about pitch axis
C_{nfa}, C_{nfb}	fuselage aerodynamic yawing moment coefficients
C_Q	main rotor torque coefficient
C_{Qi}, C_{Qp}	induced and profile torque coefficients
C_{QT}	tail rotor torque coefficient
C_T	rotor thrust coefficient
C_{T_T}	tail rotor thrust coefficient
C_W	weight coefficient
$C_x, \mathbf{C}_y, \mathbf{C}_z$	main rotor force coefficients
$C_{yf\eta}$	normalized sideforce on fin
C_ζ	lag damping
C_{ztp}	normalized tailplane force
D	aircraft drag (N, lbf)
$D(s)$	denominator of closed-loop transfer function
$\mathbf{D}_I(\psi)$	time-dependent stiffness matrix in individual blade flapping equations
$\mathbf{D}_M(\psi)$	time-dependent stiffness matrix in multi-blade flapping equations
$\mathbf{D}_{M0}(\psi)$	constant stiffness matrix in multi-blade flapping equations
$E(r)I(r)$	distributed blade stiffness
$F^{(1)}$	out-of-plane rotor blade force
$F^{(2)}$	in-plane rotor blade force
$F(r, t)$	distributed aerodynamic load normal to blade surface
$\mathbf{F}(\mathbf{x}, \mathbf{u}, t)$	nonlinear vector function of aircraft motion
$F_0^{(1)}$	main rotor force component
$F_0^{(1)}$	one-per-rev cosine component of $F^{(1)}$
$F_{1s}^{(1)}$	one-per-rev sine component of $F^{(1)}$
$F_{2c}^{(1)}$	two-per-rev cosine component of $F^{(1)}$
$F_{2s}^{(1)}$	two-per-rev sine component of $F^{(1)}$
$F_{1c}^{(2)}$	one-per-rev cosine component of $F^{(2)}$
$F_{1s}^{(2)}$	one-per-rev sine component of $F^{(2)}$
$\mathbf{F}_g$	vector of external forces acting at centre of mass (components X, Y, Z)
F_T	tail rotor-fin blockage factor
F_{vi}, F_w, etc.	flap derivatives in heave/coning/inflow rotor model
$G_e(s), H_e(s)$	engine/rotorspeed governor transfer function
$G_{\eta 1cp}(\omega)$	cross-spectral density function between lateral cyclic and roll rate
$H_{\eta 1cp}(\omega)$	frequency response function between lateral cyclic and roll rate
$\mathbf{H}_I(\psi)$	time-dependent forcing function matrix in individual blade flapping equations
$\mathbf{H}_M(\psi)$	time-dependent forcing function matrix in multi-blade flapping equations
$\mathbf{H}_{M0}(\psi)$	forcing function matrix in multi-blade flapping equations
I_β	flap moment of inertia (kg m^2, slug ft^2)
I_n	moment of inertia of nth bending mode (kg m^2, slug ft^2)

I_R	moment of inertia of rotor and transmission system (kg m^2; slug ft^2)
I_{vi}, I_w, etc.	inflow derivatives in heave/coning/inflow rotor model
I_{xx}, I_{yy}, I_{zz}	moments of inertia of the helicopter about the x-, y- and z-axes (kg m^2; slug ft^2)
I_{xz}	product of inertia of the helicopter about the x- and z-axes (kg m^2; slug ft^2)
K_3	rotorspeed droop factor
K_β	centre-spring rotor stiffness (N m/rad, ft Ib/rad)
K_p, K_x	pilot and display scaling gains
L, M, N	external aerodynamic moments about the x-, y- and z-axes (N m, ft lb)
$\mathbf{L}_\beta$	transformation matrix from multi-blade to individual blade coordinates
L_f, M_f, N_f	fuselage aerodynamic moments about centre of gravity (N m, ft Ib)
L_{fn}, N_{fn}	fin aerodynamic moments about centre of gravity (N m, ft Ib)
$L_{\theta_0}, M_{\theta_{1s}}$	control derivatives normalized by moments of inertia (1/s^2)
L_T, N_T, M_T	tail rotor moments about centre of gravity (N m, ft Ib)
L_v, M_q, etc.	moment derivatives normalized by moments of inertia (e.g., $\partial L/\partial v$) (rad/(m s), rad/(ft s), 1/s)
L_w	turbulence scale for vertical velocity component (m, ft)
M, M_d	Mach number, drag divergence Mach number
M_a	mass of helicopter (kg, Ib)
M_β	first moment of mass of rotor blade (kg m; slug ft)
$\mathbf{M}_g$	vector of external moments acting at centre of mass (components L, M, N)
$M_h^{(r)}(0, t)$	rotor hub moment (N m, ft Ib)
M_h, L_h	main rotor hub pitch and roll moments (N m, ft Ib)
M_R, L_R	main rotor pitch and roll moments (N m, ft Ib)
M_{tp}	tail plane pitching moment (N m, ft Ib)
N_b	number of blades on main rotor
N_H	yawing moment due to rotor about rotor hub (N m, ft Ib)
$N_{r_{effective}}$	effective yaw damping in Dutch roll motion (1/s)
P_e, Q_e, R_e	trim angular velocities in fuselage axes system (rad/s)
P_i	rotor induced power (kW, HP)
$P_n(t)$	blade generalized coordinate for out-of-plane bending
P_R	main rotor power (kW, HP)
P_T	tail rotor power (kW, HP)
P_x, P_y	position of aircraft from hover box (m, ft)
Q_{acc}	accessories torque (N m, ft Ib)
Q_e, Q_{eng}	engine torque (N m, ft Ib)
Q_{emax}	maximum continuous engine torque (N m, ft Ib)
Q_R	main rotor torque (N m, ft Ib)
Q_T	tail rotor torque (N m, ft Ib)
Q_w	quickness for aircraft vertical gust response (1/s)
R	rotor radius (m, ft)
$R(s)$	numerator of closed-loop transfer function
R_T	tail rotor radius (m, ft)
S_β	Stiffness number $\dfrac{\lambda_\beta^2 - 1}{\gamma/8}$

S_{fn}	fin area (m², ft²)
$S_n(r)$	blade mode shape for out-of-plane bending
S_p, S_s	fuselage plan and side areas (m², ft²)
S_{tp}	tail plane area (m², ft²)
$S_z(0, t)$	shear force at rotor hub (N, Ibf)
T	main rotor thrust (N, Ibf)
T	manoeuvre duration (s)
$T_{h_{eq}}$	time constant in heave axis first-order equivalent system (s)
T_{ige}	rotor thrust in-ground effect (N, Ibf)
T_{oge}	rotor thrust out-of-ground effect (N, Ibf)
T_x	distance between edges on surface (m(ft))
T_T	tail rotor thrust (N, Ibf)
U_e, V_e, W_e	trim velocities in fuselage axes system (m/s, ft/s, knot)
U_P, U_T	normal and in-plane rotor velocities (m/s, ft/s)
V, V_x	aircraft forward velocity (m/s, ft/s)
V_c	rotor climb velocity (m/s, ft/s)
V_c	tangential velocity at the edge of the vortex core (m/s, ft/s)
V_d	rotor descent velocity (m/s, ft/s)
V_f	total velocity incident on fuselage (m/s, ft/s)
V_{fe}	total velocity in trim (m/s, ft/s, knot)
V_{fn}	total velocity incident on fin (m/s, ft/s)
$V_h^{(r)}(0, t)$	rotor hub shear force (N, Ibf)
V_{res}	resultant velocity at rotor disc (m/s, ft/s)
V_{tp}	total velocity incident on tailplane (m/s, ft/s)
$V_T(r)$	tangential velocity in vortex as a function of distance from core r (m/s, ft/s)
V_x, V_y	velocity components of aircraft relative to Earth
W	eigenvector matrix associated with **A**
X, Y, Z	external aerodynamic forces acting along the x-, y- and z-axes (N, Ibf)
X_f, Y_f, Z_f	components of X, Y, Z from fuselage (N, Ibf)
X_{hw}, Y_{hw}	rotor forces in hub/wind axis system (N, Ibf)
X_R, X_T	components of X from main and tail rotors (N, Ibf)
X_{tp}, X_{fn}	components of X from empennage (tp – horizontal tailplane, fn – vertical fin) (N, Ibf)
$X_u, X_p,$ etc.	X force derivatives normalized by aircraft mass (1/s, m/(s rad), etc.)
Y(t)	principal matrix solution of dynamic equations of motion in vector form
Y_{fn}	aerodynamic sideforce acting on fin (N, Ibf)
$Y_p, Y_a(s)$	transfer function of pilot and aircraft
Y_T	component of Y force from tail rotor (N, Ibf)
$Y_v, Y_r,$ etc.	Y force derivatives normalized by aircraft mass (1/s, m/(s rad), etc.)
Z_w	heave damping derivative (1/s)
$Z_{\theta 0}$	heave control sensitivity derivative (m/(s² rad), ft/(s² rad))
Z_{tp}	component of Z force from tailplane (N, Ibf)
$Z_w, Z_q,$ etc.	Z force derivatives normalized by aircraft mass (1/s, m/(s rad), etc.)
$\alpha(\psi, r, t)$	total incidence at local blade station (rad)

α_1, α_2	incidence break points in Beddoes theory (rad)
α_{1cw}	effective cosine component of one-per-rev rotor blade incidence (rad)
α_{1sw}	effective sine component of one-per-rev rotor blade incidence (rad)
α_d	disc incidence (rad)
α_f	incidence of resultant velocity to fuselage (rad)
$\alpha_{flap}, \alpha_{wh}$	components of local blade incidence (rad)
α_{inflow}	component of local blade incidence (rad)
$\alpha_{pitch}, \alpha_{twist}$	components of local blade incidence (rad)
α_{tp}	incidence of resultant velocity to tailplane
α_{tp0}	zero-lift incidence angle on tailplane (rad)
$\beta(t)$	rotor flap angle (positive up) (rad)
$\beta(t)$	sideslip velocity (rad)
β_f	sideslip angle at fuselage (rad)
β_{fn}	sideslip angle at fin (rad)
$\beta_{1c_{\theta 1s}}$	$= \partial \beta_{1c} / \partial \theta_{1s}$, flapping derivative with respect to cyclic pitch
$\beta_0, \beta_{1c}, \beta_{1s}$	rotor blade coning, longitudinal and lateral flapping angles (subscript w denotes hub/wind axes) – in multi-blade coordinates (rad)
β_{0T}	tail rotor coning angle (rad)
β_{1cT}	tail rotor cyclic (fore – aft) flapping angle (rad)
β_{1cwT}	tail rotor cyclic (fore – aft) flapping angle in tail rotor hub/wind axes (rad)
β_d	differential coning multi-blade flap coordinate (rad)
β_{fn0}	zero-lift sideslip angle on fin (rad)
$\boldsymbol{\beta}_1$	vector of individual blade coordinates
$\beta_i(t)$	flap angle of ith blade (rad)
β_{jc}, β_{js}	cyclic multi-blade flap coordinates (rad)
$\boldsymbol{\beta}_M$	vector of multi-blade coordinates
δ	ratio of instantaneous normal velocity to steady state value $\delta = \frac{w}{w_{ss}}$
δ_0	main rotor profile drag coefficient
δ_2	main rotor lift dependent profile drag coefficient
δ_3	tail rotor delta 3 angle ($\tan^{-1} k_3$)
$\delta_a, \delta_b, \delta_x, \delta_y$	pilot cyclic control displacements
δ_c	collective lever displacement
δ_{T0}	tail rotor profile drag coefficient
δ_{T2}	tail rotor lift dependent profile drag coefficient
$\delta u, \delta w$, etc.	perturbations in velocity components (m/s, ft/s)
γ	flight path angle (rad or deg)
$\dot{\gamma}$	rate of change of γ with time (rad/s or deg/s)
γ_a	$\gamma - \gamma_f$ (rad or deg)
$\overline{\gamma}_a$	γ_a normalized by final value γ'_f
$\overline{\gamma}'_a$	rate of change with normalized time $\overline{t}$
γ_f	final value of flight path angle (rad or deg)
$\underset{\sim}{\gamma}$	tuned aircraft response
γ	Lock number $= \frac{\rho c a_0 R^4}{I_\beta}$
γ^*	$= C'_1 \gamma$; equivalent Lock number
γ_{fe}	flight path angle in trim (rad)
γ_s	shaft angle (positive forward, rad)

γ_T	tail rotor Lock number
$\gamma_{\eta 1cp}$	coherence function associated with frequency response fit between lateral cyclic and roll rate
$\eta_c, \eta_{1s}, \eta_{1c}$	pilot's collective lever and cyclic stick positions (positive up, aft and to port)
η_{1s0}, η_{1c0}	cyclic gearing constants
η_{ct}	tail rotor control run variable
η_p	pedal position
$\lambda_0, \lambda_{1c}, \lambda_{1s}$	rotor uniform and first harmonic inflow velocities in hub/shaft axes (normalized by ΩR)
λ_{0T}	tail rotor uniform inflow component
λ_{C_T}	inflow gain
λ_i	eigenvalue
λ_i	main rotor inflow
λ_{ih}	hover inflow
λ_r	roll subsidence eigenvalue
λ_s	spiral eigenvalue
λ_β	flap frequency ratio; $\lambda_\beta^2 = 1 + \frac{K_\beta}{I_\beta \Omega^2}$
χ	main rotor wake angle (rad)
χ_ε	track angle in equilibrium flight (rad)
χ_1, χ_2	wake angle limits for downwash on tail (rad)
$\lambda_{\beta T}$	tail rotor flap frequency ratio
λ_n	flap frequency ratio for nth bending mode
λ_θ	blade pitch frequency ratio
λ_{tp}	normalized downwash at tailplane
λ_ζ	blade lag frequency ratio
μ	advance ratio $V/\Omega R$
μ	real part of eigenvalue or damping (1/s)
μ_c	normalized climb velocity
μ_d	normalized descent velocity
μ_T	normalized velocity at tail rotor
μ_{tp}	normalized velocity at tailplane
μ_x, μ_y, μ_z	velocities of the rotor hub in hub/shaft axes (normalized by ΩR)
μ_{zT}	total normalized tail rotor inflow velocity
v	lateral acceleration (normalized sideforce) on helicopter (m/s^2, ft/s^2)
v	turbulence component wavenumber = frequency/airspeed
θ	optical flow angle (rad)
θ_0	collective pitch angle (rad)
$\bar{\theta}_0$	collective pitch normalized by θ_{0f}
θ_{0f}	final value of collective (rad)
θ, ϕ, ψ	Euler angles defining the orientation of the aircraft relative to the Earth (rad)
θ_0, θ_{0T}	main and tail rotor collective pitch angles (rad)
θ_{0T}^*	tail rotor collective pitch angle after delta 3 correction (rad)
$\theta_{0.75R}$	blade pitch at 3/4 radius (rad)
θ_{1s}, θ_{1c}	longitudinal and lateral cyclic pitch (subscript w denotes hub/wind axes) (rad)

θ_{1sT}	tail rotor cyclic pitch applied through δ_3 angle (rad)
θ_{tw}	main rotor blade linear twist (rad)
ρ	air density (kg/m³, slug/ft³)
σ	rms turbulence intensity
τ	time to contact surface or object or time to close a gap in a state(s)
$\dot{\tau}$	rate of change of τ with time
τ_g	guide (constant accel or decel) (s)
$\tau_{surface}$	to the surface during climb manoeuvre (s)
τ_x	of the motion variable x, defined as $\frac{x}{\dot{x}}$ where x is the distance or gap to a surface, object or new state and $\dot{x}$ is the instantaneous velocity (s)
τ_1, τ_2	time constants in Beddoes dynamic stall model (s)
τ_β	time constant of rotor flap motion (s)
$\tau_{c1} - \tau_{c4}$	actuator time constants (s)
$\tau_{e1}, \tau_{e2}, \tau_{e3}$	engine time constants (s)
τ_{heq}	time delay in heave axis equivalent system
τ_λ	inflow time constant (s)
τ_{lat}	estimated time delay between lateral cyclic input and aircraft response (s)
τ_p	roll time constant $(= -1/Lp)$ (s)
τ_p	phase delay between attitude response and control input at high frequency (s)
τ_{ped}	estimated time delay between pedal input and aircraft response (s)
ω_{bw}	bandwidth frequency for attitude response (rad/s)
ω_m	natural frequency of low-order equivalent system for roll response (rad/s)
ω_c	crossover frequency defined by point of neutral stability (rad/s)
ω_d	Dutch roll frequency (rad/s)
ω_f	fuel flow variable
ω_ϕ	natural frequency of roll regressing flap mode (rad/s)
$\omega_{fmax}, \omega_{fidle}$	fuel flow variable at maximum contingency and flight idle
$\boldsymbol{\omega}_g$	angular velocity vector of aircraft with components p, q, r
ω_p	phugoid frequency (rad/s)
ω_θ	frequency associated with control system stiffness (rad/s)
ω_{sp}	pitch short period frequency (rad/s)
ω_t	task bandwidth (rad/s)
ω_x	angular velocity in blade axes $= p_{hw} \cos \psi - q_{hw} \sin \psi$ (rad/s)
ω_y	angular velocity in blade axes $= p_{hw} \sin \psi - q_{hw} \cos \psi$ (rad/s)
ψ	heading angle, positive to starboard (rad)
ψ	rotor blade azimuth angle, positive in direction of rotor rotation (rad)
ψ_w	rotor sideslip angle (rad)
ψ_i	azimuth angle of ith rotor blade (rad)
ζ	blade lag angle (rad)
ζ_d	Dutch roll damping factor
ζ_p	phugoid damping factor
ζ_{sp}	pitch short period damping factor
Φ_m	phase margin (degrees)
$\Phi_{wg}(v)$	power spectrum of w component of turbulence

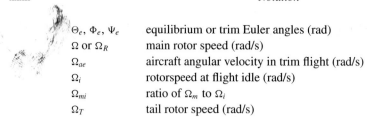

Θ_e, Φ_e, Ψ_e	equilibrium or trim Euler angles (rad)
Ω or Ω_R	main rotor speed (rad/s)
Ω_{ae}	aircraft angular velocity in trim flight (rad/s)
Ω_i	rotorspeed at flight idle (rad/s)
Ω_{mi}	ratio of Ω_m to Ω_i
Ω_T	tail rotor speed (rad/s)

SUBSCRIPTS

$1c$	first harmonic cosine component
$1s$	first harmonic sine component
d	Dutch roll
e	equilibrium or trim condition
g	gravity component or centre of mass G
h	hub axes
hw	hub/wind axes
nf	no-feathering (plane/axes)
p	phugoid
p, a	in-control system, relating to pilot and autostabilizer inputs
s	spiral
s, ss	steady state
sp	short period
tp	tip path (plane/axes)
R, T, f, fn, tp	main rotor, tail rotor, fuselage, fin, tailplane

DRESSINGS

$\dot{u} = \dfrac{\mathrm{d}u}{\mathrm{d}t}$	differentiation with respect to time t
$\beta' = \dfrac{\mathrm{d}\beta}{\mathrm{d}\psi}$	differentiation with respect to azimuth angle ψ
$-$	Laplace transformed variable

List of abbreviations

AC	attitude command
ACAH	attitude command attitude hold
ACS	active control system
ACT	active control technology
ACVH	attitude command velocity hold
AD	attentional demands
ADFCS	advanced digital flight control system
ADS	Aeronautical Design Standard
AEO	Air Engineering Officer
AFCS	automatic flight control system
AFS	advanced flight simulator
AGARD	Advisory Group for Aeronautical Research and Development
AH	attack helicopter
AHS	American Helicopter Society
AIAA	American Institute of Aeronautics and Astronautics
AS	Aerospatiale
ATA	air-to-air
CAA	Civil Aviation Authority
CAP	control anticipation parameter
CGI	computer-generated imagery
CH	cargo helicopter
CHR	Cooper–Harper Rating (as in HQR)
CSM	conceptual simulation model
DERA	Defence Evaluation and Research Agency
DLR	Deutsche Forschungs- und Versuchsantalt fuer Luft- und Raumfahrt
DoF	degree of freedom
DRA	Defence Research Agency
DVE	degraded visual environment
ECD	Eurocopter Deutschland
ECF	Eurocopter France
FAA	Federal Aviation Authority
FoV	field of view
FPVS	flight path vector system
FRL	Flight Research Laboratory (Canada)
FSAA	flight simulator for advanced aircraft
FUMS	fatigue usage monitoring system
GVE	good visual environment
HMD	helmet-mounted display

HP	horse power
HQR	handling qualities rating
HUMS	health and usage monitoring system
IHST	International Helicopter Safety Team
IMC	instrument meteorological conditions
LOES	low-order-equivalent system
MBB	Messerschmit–Bolkow–Blohm
MTE	mission task element
NACA	National Advisory Committee for Aeronautics
NAE	National Aeronautical Establishment
NASA	National Aeronautics and Space Administration
NoE	nap of the earth
NRC	National Research Council (Canada)
OFE	operational flight envelope
OH	observation helicopter
OVC	outside visual cues
PIO	pilot-induced oscillation
PSD	power spectral density
RAE	Royal Aircraft Establishment
RAeSoc	Royal Aeronautical Society
RC	rate command
RCAH	rate command attitude hold
RT	response type
SA	situation awareness
SAE	Society of Automotive Engineers
SA	Sud Aviation
SCAS	stability and control augmentation system
SDG	statistical discrete gust
SFE	safe flight envelope
SHOL	ship-helicopter operating limits
SNIOPs	simultaneous, non-interfering operations
SS	sea state
TC	turn coordination
TQM	total quality management
TRC	translational rate command
TRCPH	translational rate command position hold
TTCP	The Technical Cooperation Programme (United Kingdom, United States, Canada, Australia, New Zealand)
T/W	thrust/weight ratio
UCE	usable cue environment
UH	utility helicopter
VCR	visual cue ratings
VMC	visual meteorological conditions
VMS	vertical motion simulator
VNE	never-exceed velocity
VSTOL	vertical/short take-off and landing
WG	Working Group (AGARD)
agl	above ground level

cg	centre of gravity
ige	in-ground effect
oge	out-of-ground effect
rms	root mean square
rpm	revs per minute
rrpm	rotor revs per minute

*The DRA research Lynx ALYCAT (Aeromechanics LYnx
Control and Agility Testbed) shown flying by the large motion
system of the DRA advanced flight simulator
(Photograph courtesy of Simon Pighills)*

1 Introduction

The underlying premise of this book is that flight dynamics and control is a central discipline, at the heart of aeronautics, linking the aerodynamic and structural sciences with the applied technologies of systems and avionics and, above all, with the pilot. Flight dynamics engineers need to have breadth and depth in their domain of interest, and often hold a special responsibility in design and research departments. It is asserted that more than any other aerospace discipline, flight dynamics offers a viewpoint on, and is connected to, the key rotorcraft attributes and technologies – from the detailed fluid dynamics associated with the interaction of the main rotor wake with the empennage, to the servo-aeroelastic couplings between the rotor and control system, through to the evaluation of enhanced safety, operational advantage and mission effectiveness of good flying qualities. It is further asserted that the multidisciplinary nature of rotorcraft flight dynamics places it in a unique position to hold the key to concurrency in requirements capture and design, i.e., the ability to optimize under the influence of multiple constraints. In the author's view, the role of the practising flight dynamics engineer is therefore an important one and there is a need for guidebooks and practitioner's manuals to the subject to assist in the development of the required skills and knowledge. This book is an attempt at such a manual, and it discusses flight dynamics under two main headings – simulation modelling and flying qualities. The importance of good simulation fidelity and robust flying qualities criteria in the requirements capture and design phases of a new project cannot be overstated, and this theme will be expanded upon later in this chapter and throughout the book. Together, these attributes underpin confidence in decision making during the high-risk early design phase and are directed towards the twin goals of achieving super-safe flying qualities and getting designs right, first time. These goals have motivated much of the research conducted in government research laboratories, industry and universities for several decades.

In this short general Introduction, the aim is to give the reader a qualitative appreciation of the two main subjects – simulation modelling and flying qualities. The topics that come within the scope of flight dynamics are also addressed briefly, but are not covered in the book for various reasons. Finally, a brief 'roadmap' to the seven technical chapters is presented.

1.1 Simulation Modelling

It is beyond dispute that the observed behaviour of aircraft is so complex and puzzling that, without a well developed theory, the subject could not be treated intelligently.

We use this quotation from Duncan (Ref. 1.1) in expanded form as a guiding light at the beginning of Chapter 3, the discourse on building simulation models. Duncan wrote these words in relation to fixed-wing aircraft over 50 years ago and they still hold a profound truth today. However, while it may be 'beyond dispute' that well-developed theories of flight are vital, a measure of the development level at any one time can be gauged by the ability of Industry to predict behaviour correctly before first flight, and rotorcraft experience to date is not good. In the 1989 AHS Nikolsky Lecture (Ref. 1.2), Crawford promotes a 'back to basics' approach to improving rotorcraft modelling in order to avoid major redesign effort resulting from poor predictive capability. Crawford cites examples of the redesign required to improve, or simply 'put right', flight performance, vibration levels and flying qualities for a number of contemporary US military helicopters. A similar story could be told for European helicopters. In Ref. 1.3, the author presents data on the percentage of development test flying devoted to handling and control, with values between 25 and 50% being quite typical. The message is that helicopters take a considerable length of time to qualify to operational standard, usually much longer than originally planned, and a principal reason lies with the deficiencies in analytical design methods.

Underlying the failure to model flight behaviour adequately are three aspects. First, there is no escaping that the rotorcraft is an extremely complex dynamic system and the modelling task requires extensive skill and effort. Second, such complexity needs significant investment in analytical methods and specialist modelling skills and the recognition by programme managers that these are most effectively applied in the formative stages of design. The channelling of these investments towards the critically deficient areas is also clearly very important. Third, there is still a serious shortage of high-quality, validation test data, both at model scale and from flight test. There is an old adage in the world of flight dynamics relating to the merits of test versus theory, which goes something like – 'everyone believes the test results, except the person who made the measurements, and nobody believes the theoretical results, except the person who calculated them'. This saying stems from the knowledge that it is much easier, for example, to tell the computer to output rotor blade incidence at 3/4 radius on the retreating side of the disc than it is to measure it. What are required, in the author's opinion, are research and development programmes that integrate the test and modelling activities so that the requirements of the one drive the other.

There are some signs that the importance of modelling and modelling skills is recognized at the right levels, but the problem will require constant attention to guard against the attitude that the 'big' resources should be reserved for production, when the user and manufacturer, in theory, receive their greatest rewards. Chapters 3, 4 and 5 of this book are concerned with modelling, but we shall not dwell on the deficiencies of the acquisition process, but rather on where the modelling deficiencies lie. The author has taken the opportunity in this Introduction to reinforce the philosophy promoted in Crawford's Nikolsky Lecture with the thought that the reader may well be concerned as much with the engineering 'values' as with the technical detail.

No matter how good the modelling capability, without criteria as a guide, helicopter designers cannot even start on the optimization process; with respect to flying qualities, a completely new approach has been developed and this forms a significant content of this book.

1.2 FLYING QUALITIES

> *Experience has shown that a large percentage, perhaps as much as 65%, of the life-cycle cost of an aircraft is committed during the early design and definition phases of a new development program. It is clear, furthermore, that the handling qualities of military helicopters are also largely committed in these early definition phases and, with them, much of the mission capability of the vehicle. For these reasons, sound design standards are of paramount importance both in achieving desired performance and avoiding unnecessary program cost.*

This quotation, extracted from Ref. 1.4, states the underlying motivation for the development of flying qualities criteria – they give the best chance of having mission performance designed in, whether through safety and economics with civil helicopters or through military effectiveness. But flying quality is an elusive topic and it has two equally important facets that can easily get mixed up – the objective and the subjective. Only recently has enough effort been directed towards establishing a valid set of flying qualities criteria and test techniques for rotorcraft that has enabled both the subjective and objective aspects to be addressed in a complementary way. That effort has been orchestrated under the auspices of several different collaborative programmes to harness the use of flight and ground-based simulation facilities and key skills in North America and Europe. The result was Aeronautical Design Standard (ADS)-33, which has changed the way the helicopter community thinks, talks and acts about flying quality. Although the primary target for ADS-33 was the LHX and later the RAH-66 Comanche programme, other nations have used or developed the standard to meet their own needs for requirements capture and design. Chapters 6, 7 and 8 of this book will refer extensively to ADS-33, with the aim of giving the reader some insight into its development. The reader should note, however, that these chapters, like ADS-33 itself, address how a helicopter with good flying qualities should behave, rather than how to construct a helicopter with good flying qualities.

In search of the meaning of Flying Quality, the author has come across many different interpretations, from Pirsig's somewhat abstract but appealing 'at the moment of pure quality, subject and object are identical' (Ref. 1.5), to a point of view put forward by one flight dynamics engineer: 'flying qualities are what you get when you've done all the other things'. Unfortunately, the second interpretation has a certain ring of truth because until ADS-33, there was very little coherent guidance on what constituted good flying qualities. The first breakthrough for the flying qualities discipline came with the recognition that criteria needed to be related to task. The subjective rating scale, developed by Cooper and Harper (Ref. 1.6) in the late 1960s, was already task and mission oriented. In the conduct of a handling qualities experiment, the Cooper–Harper approach forces the engineer to define a task with performance standards and to agree with the pilot on what constitutes minimal or extensive levels of workload. But the objective criteria at that time were more oriented to the stability and control characteristics of aircraft than to their ability to perform tasks well. The relationship clearly is important but the lack of task-oriented test data meant that early attempts to define criteria boundaries involved a large degree of guesswork and hypothesis. Once the two ingredients essential for success in the development of new criteria, task-orientation and test data, were recognized and resources were channelled effectively, the combined expertise of several agencies focused their efforts, and during the 1980s and 1990s, a completely new approach was developed. With the advent of digital flight control systems, which provide the capability to confer different mission flying qualities in the same aircraft, this new approach can now be exploited to the full.

One of the aspects of the new approach is the relationship between the internal attributes of the air-vehicle and the external influences. The same aircraft might have perfectly good handling qualities for nap-of-the-earth operations in the day environment, but degrade severely at night; obviously, the visual cues available to the pilot play a fundamental role in the perception of flying qualities. This is a fact of operational life, but the emphasis on the relationship between the internal attributes and the external influences encourages design teams to think more synergistically, e.g., the quality of the vision aids, and what the symbology should do, becomes part of the same flying qualities problem as what goes into the control system, and, more importantly, the issues need to be integrated in the same solution. We try to emphasize the importance of this synergy first in Chapter 2, then later in Chapters 6 and 7.

The point is made on several occasions in this book, for emphasis, that good flying qualities make for safe and effective operations; all else being equal, less accidents will occur with an aircraft with good handling qualities compared with an aircraft with merely acceptable handling, and operations will be more productive. This statement may be intuitive, but there is very little supporting data to quantify this. Later, in Chapter 7, the potential benefits of handling to flight safety and effectiveness through a probabilistic analysis are examined, considering the pilot as a component with failure characteristics similar to any other critical aircraft component. The results may appear controversial and they are certainly tentative, but they point to one way in which the question, 'How valuable are flying qualities?', may be answered. This theme is continued in Chapter 8 where the author presents an analysis of the effects of degraded handling qualities on safety and operations, looking in detail at the impact of degraded visual conditions, flight system failures and strong atmospheric disturbances.

1.3 MISSING TOPICS

It seems to be a common feature of book writing that the end product turns out quite different than originally intended and *Helicopter Flight Dynamics* is no exception. It was planned to be much shorter and to cover a wider range of subjects! In hindsight, the initial plan was perhaps too ambitious, although the extent of the final product, cut back considerably in scope, has surprised even the author. There are three 'major' topic areas, originally intended as separate chapters, that have virtually disappeared – 'Stability and control augmentation (including active control)', 'Design for flying qualities' and 'Simulation validation (including system identification tools)'. All three are referred to as required, usually briefly, throughout the book, but there have been such advances in recent years that to give these topics appropriate coverage would have extended the book considerably. They remain topics for future treatment, particularly the progress with digital flight control and the use of simulators in design, development and certification. In the context of both these topics, we appear to be in an era of rapid development, suggesting that a better time to document the state of the art may well be in a few years from now. The absence of a chapter or section on simulation model validation techniques may appear to be particularly surprising, but is compensated for by the availability of the AGARD (Advisory Report on Rotorcraft System Identification), which gives a fairly detailed coverage of the state of the art in this subject up to the early 1990s (Ref. 1.7). Since the publication of the first edition, significant strides have been made in the development of simulation models for use in design and also training simulators. Reference 1.8 reviews some of these developments but we are somewhat in mid-stream with this new push to increase fidelity and the author has resisted the temptation to bring this topic into the second edition.

The book says very little about the internal hardware of flight dynamics – the pilot's controls and the mechanical components of the control system including the hydraulic actuators. The pilot's displays and instruments and their importance for flight in poor visibility are briefly treated in Chapter 7 and the associated perceptual issues are treated in some depth in Chapter 8, but the author is conscious of the many missing elements here. In Chapter 3, the emphasis has been on modelling the main rotor, and many other elements, such as the engine and transmission systems, are given limited coverage.

It is hoped that the book will be judged more on what it contains than on what it doesn't.

1.4 Simple Guide to the Book

This book contains seven technical chapters. For an overview of the subject of helicopter flight dynamics, the reader is referred to the Introductory Tour in Chapter 2. Engineers familiar with flight dynamics, but new to rotorcraft, may find this a useful starting point for developing an understanding of how and why helicopters work. Chapters 3, 4 and 5 are a self-contained group concerned with modelling helicopter flight dynamics. To derive benefit from these chapters requires a working knowledge of the mathematical analysis tools of dynamic systems. Chapter 3 aims to provide sufficient knowledge and understanding to enable a basic flight simulation of a helicopter to be created.

Chapter 4 discusses the problems of trim and stability, providing a range of analytical tools necessary to work at these two facets of helicopter flight mechanics. Chapter 5 extends the analysis of stability to considerations of constrained motion and completes the 'working with models' theme of Chapters 4 and 5 with a discussion on helicopter response characteristics. In Chapters 4 and 5, flight test data from the DRA's research Puma and Lynx and the DLR's Bo105 are used extensively to provide a measure of validation to the modelling. Chapters 6 and 7 deal with helicopter flying qualities from objective and subjective standpoints respectively, although Chapter 7 also covers a number of what we have described as 'other topics', including agility and flight in degraded visual conditions. Chapters 6 and 7 are also self-contained and do not require the same background mathematical knowledge as that required for the modelling chapters. A unified framework for discussing the response characteristics of flying qualities is laid out in Chapter 6, where each of the four 'control' axes are discussed in turn. Quality criteria are described, drawing heavily on ADS-33 and the associated publications in the open literature. Chapter 8 is new in the second edition and contains a detailed treatment of the sources of degraded flying qualities, particularly flight in degraded visual conditions, the effects of failures in flight system functions and the impact of severe atmospheric disturbances. These subjects are also discussed within the framework of quantitative handling qualities engineering, linking with ADS-33, where appropriate. The idea here is that degraded flying qualities should be taken into consideration in design with appropriate mitigation technologies.

Chapters 3 and 4 are complemented and supported by appendices. Herein lie the tables of configuration data and stability and control derivative charts and Tables for the three case study aircraft.

The author has found it convenient to use both metric and British systems of units as appropriate throughout the book, although with a preference for metric where an option was available. Although the metric system is strictly the primary world system of units of measurements, many helicopters are designed to the older British system. Publications, particularly those from the United States, often contain data and charts using the British system, and it has seemed inappropriate to change units for the sake of unification. This does not apply, of course, to cases where data from different sources are compared. Helicopter engineers are used to working in mixed units; for example, it is not uncommon to find, in the same European paper, references to height in feet, distance in metres and speed in knots – such is the rich variety of the world of the helicopter engineer.

*An EH101 Merlin approaching a Type 23 Frigate during
development flight trials
(Photograph courtesy of Westland Helicopters)*

2 Helicopter flight dynamics – an introductory tour

In aviation history the nineteenth century is characterized by man's re-lentless search for a practical flying machine. The 1860s saw a peculiar burst of enthusiasm for helicopters in Europe and the above picture, show-ing an 1863 'design' by Gabrielle de la Landelle, reflects the fascination with aerial tour-boats at that time. The present chapter takes the form of a 'tour of flight dynamics' on which the innovative, and more practical, European designs from the 1960s – the Lynx, Puma and Bo105 – will be introduced as the principal reference aircraft of this book. These splendid designs are significant in the evolution of the modern helicopter and an understanding of their behaviour provides important learning material on this tour and throughout the book.

2.1 INTRODUCTION

This chapter is intended to guide the reader on a Tour of the subject of this book with the aim of instilling increased motivation by sampling and linking the wide range of subtopics that make the whole. The chapter is likely to raise more questions than it will answer and it will point to later chapters of the book where these are picked up and addressed in more detail. The Tour topics will range from relatively simple concepts such as how the helicopter's controls work, through to more complex effects such as the influence of rotor design on dynamic stability, the conflict between stability and controllability and the specialized handling qualities required for military and civil mission task elements. All these topics lie within the domain of the flight dynamics engineer and within the scope of this book. This chapter is required reading for the reader who wishes to benefit most from the book as a whole. Many concepts are introduced and developed in fundamental form here in this chapter, and the material in later chapters will draw on the resulting understanding gained by the reader.

One feature is re-emphasized here. This book is concerned with modelling flight dynamics and developing criteria for flying qualities, rather than how to design and build helicopters to achieve defined levels of quality. We cannot, nor do we wish to, ignore design issues; requirements can be credible only if they are achievable with the available hardware. However, largely because of the author's own background and experience, design will not be a central topic in this book and there will be no chapter dedicated to it. Design issues will be discussed in context throughout the later chapters and some of the principal considerations will be summarized on this Tour, in Section 2.5.

2.2 FOUR REFERENCE POINTS

We begin by introducing four useful reference points for developing an appreciation of flying qualities and flight dynamics; these are summarized in Fig. 2.1 and comprise the following:

(1) the mission and the associated piloting tasks;
(2) the operational environment;
(3) the vehicle configuration, dynamics and the flight envelope;
(4) the pilot and pilot–vehicle interface.

With this perspective, the vehicle dynamics can be regarded as internal attributes, the mission and environment as the external influences and the pilot and pilot–vehicle interface (pvi) as the connecting human factors. While these initially need to be discussed separately, it is their interaction and interdependence that widen the scope of the subject of flight dynamics to reveal its considerable scale. The influences of the

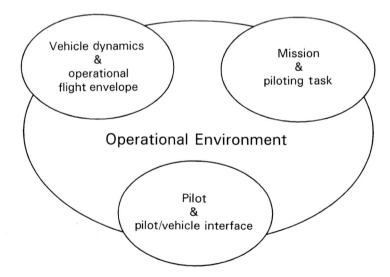

Fig. 2.1 The four reference points of helicopter flight dynamics

mission task on the pilot's workload, in terms of precision and urgency, and the external environment, in terms of visibility and gustiness, and hence the scope for exploiting the aircraft's internal attributes, are profound, and in many ways are key concerns and primary drivers in helicopter technology development. Flying qualities are determined at the confluence of these references.

2.2.1 The mission and piloting tasks

Flying qualities change with the weather or, more generally, with the severity of the environment in which the helicopter operates; they also change with flight condition, mission type and phase and individual mission tasks. This variability will be emphasized repeatedly and in many guises throughout this book to emphasize that we are not just talking about an aircraft's stability and control characteristics, but more about the synergy between the internals and the externals referred to above. In later sections, the need for a systematic flying qualities structure that provides a framework for describing criteria will be addressed, but we need to do the same with the mission and the associated flying tasks. For our purposes it is convenient to describe the flying tasks within a hierarchy as shown in Fig. 2.2. An operation is made up of many missions which, in turn, are composed of a series of contiguous mission task elements (MTE). An MTE is a collection of individual manoeuvres and will have a definite start and finish and prescribed temporal and spatial performance requirements. The manoeuvre sample is the smallest flying element, often relating to a single flying axis, e.g., change in pitch or roll attitude. Objective flying qualities criteria are normally defined for, and tested with, manoeuvre samples; subjective pilot assessments are normally conducted by flying MTEs. The flying qualities requirements in the current US Army's handling qualities requirements, ADS-33C (Ref. 2.1), are related directly to the required MTEs. Hence, while missions, and correspondingly aircraft type, may be quite different, MTEs are often common and are a key discriminator of flying qualities. For example, both

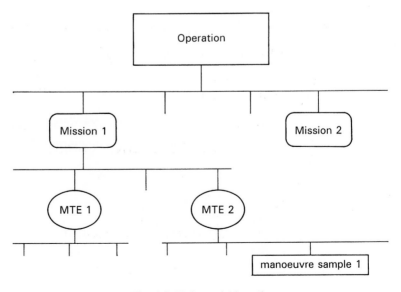

Fig. 2.2 Flying task hierarchy

utility transports in the 30-ton weight category and anti-armour helicopters in the 10-ton weight category may need to fly slaloms and precision hovers in their nap-of-the-earth (NoE) missions. This is one of the many areas where ADS-33C departs significantly from its predecessor, Mil Spec 8501A (Ref. 2.2), where aircraft weight and size served as the key defining parameters. The MTE basis of ADS-33C also contrasts with the fixed-wing requirements, MIL-F-8785C (Ref. 2.3), where flight phases are defined as the discriminating mission elements. Thus, the non-terminal flight phases in Category A (distinguished by rapid manoeuvring and precision tracking) include air-to-air combat, in-flight refuelling (receiver) and terrain following, while Category B (gradual manoeuvres) includes climb, in-flight refuelling (tanker) and emergency deceleration. Terminal flight phases (accurate flight path control, gradual manoeuvres) are classified under Category C, including take-off, approach and landing. Through the MTE and Flight Phase, current rotary and fixed-wing flying qualities requirements are described as mission oriented.

To understand better how this relates to helicopter flight dynamics, we shall now briefly discuss two typical reference missions. Figure 2.3 illustrates a civil mission, described as the offshore supply mission; Fig. 2.4 illustrates the military mission, described as the armed reconnaissance mission. On each figure a selected phase has been expanded and shown to comprise a sequence of MTEs (Figs 2.3(b), 2.4(b)). A typical MTE is extracted and defined in more detail (Figs 2.3(c), 2.4(c)). In the case of the civil mission, we have selected the landing onto the helideck; for the military mission, the 'mask–unmask–mask' sidestep is the selected MTE. It is difficult to break the MTEs down further; they are normally multi-axis tasks and, as such, contain a

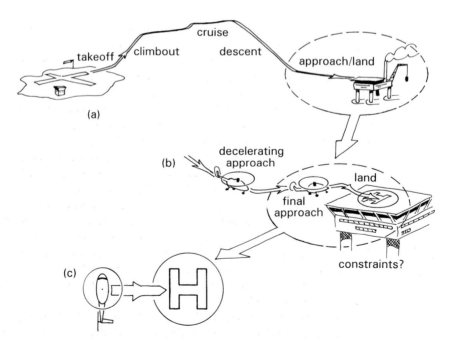

Fig. 2.3 Elements of a civil mission – offshore supply: (a) offshore supply mission;
(b) mission phase: approach and land; (c) mission task element: landing

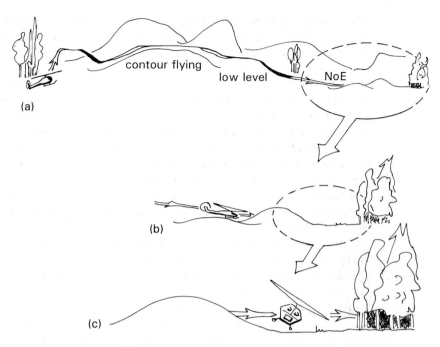

Fig. 2.4 Elements of a military mission – armed reconnaissance: (a) armed reconnaissance mission; (b) mission phase – NoE; (c) mission task element – sidestep

number of concurrent manoeuvre samples. The accompanying MTE text defines the constraints and performance requirements, which are likely to be dependent on a range of factors. For the civil mission, for example, the spatial constraints will be dictated by the size of the helideck and the touchdown velocity by the strength of the undercarriage. The military MTE will be influenced by weapon performance characteristics and any spatial constraints imposed by the need to remain concealed from the radar systems of threats. Further discussion on the design of flight test manoeuvres as stylized MTEs for the evaluation of flying qualities is contained in Chapter 7.

Ultimately, the MTE performance will determine the flying qualities requirements of the helicopter. This is a fundamental point. If all that helicopters had to do was to fly from one airport to another in daylight and good weather, it is unlikely that flying qualities would ever be a design challenge; taking what comes from meeting other performance requirements would probably be quite sufficient. But if a helicopter is required to land on the back of a ship in sea state 6 or to be used to fight at night, then conferring satisfactory flying qualities that minimize the probability of mission or even flight failure is a major design challenge. Criteria that adequately address the developing missions are the cornerstones of design, and the associated MTEs are the data source for the criteria.

The reference to weather and flying at night suggests that the purely 'kinematic' definition of the MTE concept is insufficient for defining the full operating context; the environment, in terms of weather, temperature and visibility, are equally important and bring us to the second reference point.

2.2.2 The operational environment

A typical operational requirement will include a definition of the environmental conditions in which the helicopter needs to work in terms of temperature, density altitude,
wind strength and visibility. These will then be reflected in an aircraft's flight manual.
The requirements wording may take the form: 'this helicopter must be able to operate
(i.e., conduct its intended mission, including start-up and shut-down) in the following
conditions – 5000 ft altitude, 15°C, wind speeds of 40 knots gusting to 50 knots, from
any direction, in day or night'. This description defines the limits to the operational
capability in the form of a multidimensional envelope.

   Throughout the history of aviation, the need to extend operations into poor weather
and at night has been a dominant driver for both economic and military effectiveness.
Fifty years ago, helicopters were fair weather machines with marginal performance;
now they regularly operate in conditions from hot and dry to cold, wet and windy,
and in low visibility. One of the unique operational capabilities of the helicopter is its
ability to operate in the NoE or, more generally, in near-earth conditions defined in
Ref. 2.1 as 'operations sufficiently close to the ground or fixed objects on the ground,
or near water and in the vicinity of ships, oil derricks, etc., that flying is primarily
accomplished with reference to outside objects'. In near-earth operations, avoiding
the ground and obstacles clearly dominates the pilot's attention and, in poor visibility, the pilot is forced to fly more slowly to maintain the same workload. During the
formative years of ADS-33, it was recognized that the classification of the quality of
the visual cues in terms of instrument or visual flight conditions was inadequate to
describe the conditions in the NoE. To quote from Hoh (Ref. 2.4), 'The most critical contributor to the total pilot workload appears to be the quality of the out-of-
the-window cues for detecting aircraft attitudes, and, to a lesser extent, position and
velocity. Currently, these cues are categorized in a very gross way by designating the
environment as either VMC (visual meteorological conditions) or IMC (instrument
meteorological conditions). A more discriminating approach is to classify visibility
in terms of the detailed attitude and position cues available during the experiment or
proposed mission and to associate handling qualities requirements with these finer
grained classifications.' The concept of the outside visual cues (OVC) was introduced,
along with an OVC pilot rating that provided a subjective measure of the visual cue
quality. The stimulus for the development of this concept was the recognition that
handling qualities are particularly affected by the visual cues in the NoE, yet there
was no process or methodology to quantify this contribution. One problem is that the
cue is a dynamic variable and can be judged only when used in its intended role.
Eventually, out of the confusion surrounding this subject emerged the usable cue environment (UCE), which was to become established as one of the key innovations
of ADS-33. In its developed form, the UCE embraces not only the OVC, but also
any artificial vision aids provided to the pilot, and is determined from an aggregate
of pilot visual cue ratings (VCR) relating to the pilot's ability to perceive changes
in, and make adjustments to, aircraft attitude and velocity. Handling qualities in degraded visual conditions, the OVC and the UCE will be discussed in more detail in
Chapter 7.

   The MTE and the UCE are two important building blocks in the new parlance of
flying qualities; a third relates to the aircraft's response characteristics and provides a
vital link between the MTE and UCE.

2.2.3 The vehicle configuration, dynamics and flight envelope

The helicopter is required to perform as a dynamic system within the user-defined operational flight envelope (OFE), or that combination of airspeed, altitude, rate of climb/descent, sideslip, turn rate, load factor and other limiting parameters that bound the vehicle dynamics, required to fulfil the user's function. Beyond this lies the manufacturer-defined safe flight envelope (SFE), which sets the limits to safe flight, normally in terms of the same parameters as the OFE, but represents the physical limits of structural, aerodynamic, powerplant, transmission or flight control capabilities. The margin between the OFE and the SFE needs to be large enough so that inadvertent transient excursions beyond the OFE are tolerable. Within the OFE, the flight mechanics of a helicopter can be discussed in terms of three characteristics – *trim, stability* and *response*, a classification covered in more detail in Chapters 4 and 5.

Trim is concerned with the ability to maintain flight equilibrium with controls fixed; the most general trim condition is a turning (about the vertical axis), descending or climbing (assuming constant air density and temperature), sideslipping manoeuvre at constant speed. More conventional flight conditions such as hover, cruise, autorotation or sustained turns are also trims, of course, but the general case is distinguished by the four 'outer' flight-path states, and this is simply a consequence of having four independent helicopter controls – three for the main rotor and one for the tail rotor. The rotorspeed is not normally controllable by the pilot, but is set to lie within the automatically governed range. For a helicopter, the so-called inner states – the fuselage attitudes and rates – are uniquely defined by the flight path states in a trim condition. For tilt rotors and other compound rotorcraft, the additional controls provide more flexibility in trim, but such vehicles will not be covered in this book.

Stability is concerned with the behaviour of the aircraft when disturbed from its trim condition; will it return or will it depart from its equilibrium point? The initial tendency has been called the static stability, while the longer term characteristics, the dynamic stability. These are useful physical concepts, though rather crude, but the keys to developing a deeper understanding and quantification of helicopter stability comes from theoretical modelling of the interacting forces and moments. From there come the concepts of small perturbation theory and linearization, of stability and control derivatives and the natural modes of motion and their stability characteristics. The insight value gained from theoretical modelling is particularly high when considering the *response* to pilot controls and external disturbances. Typically, a helicopter responds to a single-axis control input with multi-axis behaviour; cross-coupling is almost synonymous with helicopters. In this book we shall be dealing with direct and coupled responses, sometimes described as on-axis and off-axis responses. On-axis responses will be discussed within a framework of response types – rate, attitude and translational-rate responses will feature as types that characterize the initial response following a step control input. Further discussion is deferred until the modelling section within this Tour and later in Chapters 3, 4 and 5. Some qualitative appreciation of vehicle dynamics can be gained, however, without recourse to detailed modelling.

Rotor controls

Figure 2.5 illustrates the conventional main rotor collective and cyclic controls applied through a swash plate. Collective applies the same pitch angle to all blades and is the primary mechanism for direct lift or thrust control on the rotor. Cyclic is more

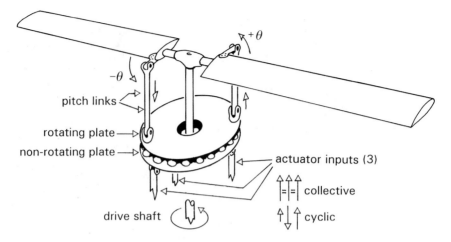

Fig. 2.5 Rotor control through a swash plate

complicated and can be fully appreciated only when the rotor is rotating. The cyclic operates through a swash plate or similar device (see Fig. 2.5), which has non-rotating and rotating halves, the latter attached to the blades with pitch link rods, and the former to the control actuators. Tilting the swash plate gives rise to a one-per-rev sinusoidal variation in blade pitch with the maximum/minimum axis normal to the tilt direction. The rotor responds to collective and cyclic inputs by flapping as a disc, in coning and tilting modes. In hover the responses are uncoupled with collective pitch resulting in coning and cyclic pitch resulting in rotor disc tilting. The concept of the rotor as a coning and tilting disc (defined by the rotor blade tip path plane) will be further developed in the modelling chapters. The sequence of sketches in Fig. 2.6 illustrates how the pilot would need to apply cockpit main rotor controls to transition into forward flight from an out-of-ground-effect (oge) hover. Points of interest in this sequence are:

(1) forward cyclic (η_{1s}) tilts the rotor disc forward through the application of cyclic pitch with a maximum/minimum axis laterally – pitching the blade down on the advancing side and pitching up on the retreating side of the disc; this 90° phase shift between pitch and flap is *the most fundamental facet of rotor behaviour* and will be revisited later on this Tour and in the modelling chapters;

(2) forward tilt of the rotor directs the thrust vector forward and applies a pitching moment to the helicopter fuselage, hence tilting the thrust vector further forward and accelerating the aircraft into forward flight;

(3) as the helicopter accelerates, the pilot first raises his collective (η_c) to maintain height, then lowers it as the rotor thrust increases through so-called 'translational lift' – the dynamic pressure increasing more rapidly on the advancing side of the disc than it decreases on the retreating side; cyclic needs to be moved increasingly forward and to the left (η_{1c}) (for anticlockwise rotors) as forward speed is increased. The cyclic requirements are determined by the asymmetric fore–aft and lateral aerodynamic loadings induced in the rotor by forward flight.

The main rotor combines the primary mechanisms for propulsive force and control, aspects that are clearly demonstrated in the simple manoeuvre described above. Typical

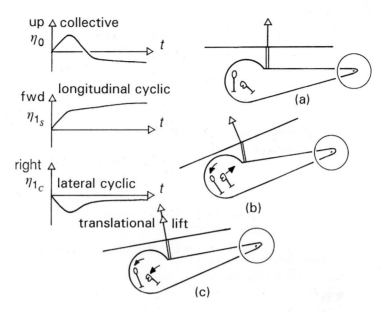

up collective
η_0

fwd longitudinal cyclic
η_{1_s}

right lateral cyclic
η_{1_c}

translational lift

(a)

(b)

(c)

Fig. 2.6 Control actions as helicopter transitions into forward flight: (a) hover; (b) forward acceleration; (c) translational lift

control ranges for main rotor controls are 15° for collective, more than 20° for longitudinal cyclic and 15° for lateral cyclic, which requires that each individual blade has a pitch range of more than 30°. At the same time, the tail rotor provides the antitorque reaction (due to the powerplant) in hover and forward flight, while serving as a yaw control device in manoeuvres. Tail rotors, or other such controllers on single main rotor helicopters, e.g., fenestron/fantail or Notar (Refs 2.5, 2.6), are normally fitted only with collective control applied through the pilot's pedals on the cockpit floor, often with a range of more than 40°; such a large range is required to counteract the negative pitch applied by the built-in pitch/flap coupling normally found on tail rotors to alleviate transient flapping.

Two distinct flight regimes
It is convenient for descriptive purposes to consider the flight of the helicopter in two distinct regimes – hover/low speed (up to about 45 knots), including vertical manoeuvring, and mid/high speed flight (up to V_{ne} – never-exceed velocity). The low-speed regime is very much unique to the helicopter as an operationally useful regime; no other flight vehicles are so flexible and efficient at manoeuvring slowly, close to the ground and obstacles, with the pilot able to manoeuvre the aircraft almost with disregard for flight direction. The pilot has direct control of thrust with collective and the response is fairly immediate (time constant to maximum acceleration $O(0.1 \text{ s})$); the vertical rate time constant is much greater, $O(3 \text{ s})$, giving the pilot the impression of an acceleration command response type (see Section 2.3). Typical hover thrust margins at operational weights are between 5 and 10% providing an initial horizontal acceleration capability of about 0.3–0.5 g. This margin increases through the low-speed regime as the (induced rotor) power required reduces (see Chapter 3). Pitch and roll manoeuvring are accomplished through tilting the rotor disc and hence rotating the

fuselage and rotor thrust (time constant for rate response types $O(0.5 \text{ s})$), yaw through tail rotor collective (yaw rate time constant $O(2 \text{ s})$) and vertical through collective, as described above. Flight in the low-speed regime can be gentle and docile or aggressive and agile, depending on aircraft performance and the urgency with which the pilot 'attacks' a particular manoeuvre. The pilot cannot adopt a carefree handling approach, however. Apart from the need to monitor and respect flight envelope limits, a pilot has to be wary of a number of behavioural quirks of the conventional helicopter in its privileged low-speed regime. Many of these are not fully understood and similar physical mechanisms appear to lead to quite different handling behaviour depending on the aircraft configuration. A descriptive parlance has built up over the years, some of which has developed in an almost mythical fashion as pilots relate anecdotes of their experiences 'close to the edge'. These include ground horseshoe effect, pitch-up, vortex ring and power settling, fishtailing and inflow roll. Later, in Chapter 3, some of these effects will be explained through modelling, but it is worth noting that such phenomena are difficult to model accurately, often being the result of strongly interacting, nonlinear and time-dependent forces. A brief glimpse of just two will suffice for the moment.

Figure 2.7 illustrates the tail rotor control requirements for early Marks (Mks 1–5) of Lynx at high all-up-weight, in the low-speed regime corresponding to winds from

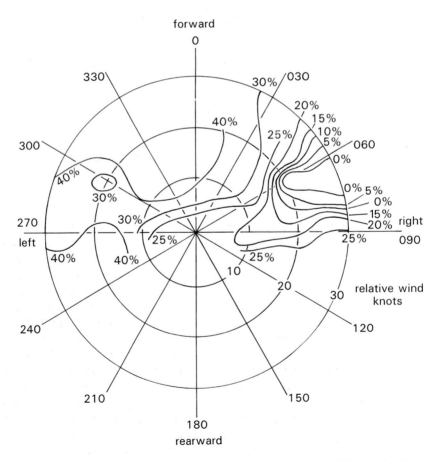

Fig. 2.7 Lynx Mk 5 tail rotor control limits in hover with winds from different directions

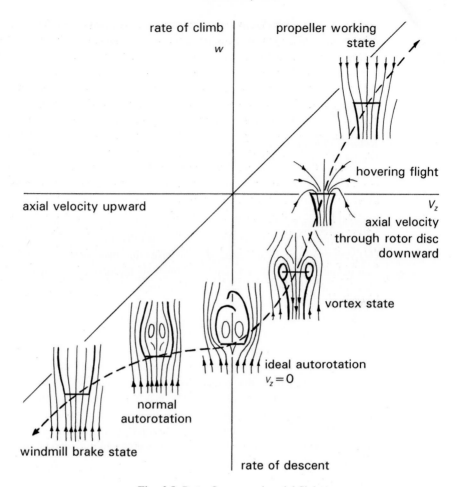

Fig. 2.8 Rotor flow states in axial flight

different 'forward' azimuths (for pedal positions <40%). The asymmetry is striking, and the 'hole' in the envelope with winds from 'green 060–075' (green winds from starboard in directions between 60° and 75° from aircraft nose) is clearly shown. This has been attributed to main rotor wake/tail rotor interactions, which lead to a loss of tail rotor effectiveness when the main rotor wake becomes entrained in the tail rotor. The loss of control and high power requirements threatening at this particular edge of the envelope provide for very little margin between the OFE and SFE.

A second example is the so-called vortex-ring condition, which occurs in near-vertical descent conditions at moderate rates of descent (O(500–800 ft/min)) on the main rotor and corresponding conditions in sideways motion on the tail rotor. Figure 2.8, derived from Drees (Ref. 2.7), illustrates the flow patterns through a rotor operating in vertical flight. At the two extremes of helicopter (propeller) and windmill states, the flow is relatively uniform. Before the ideal autorotation condition is reached, where the induced downwash is equal and opposite to the upflow, a state of irregular and strong vorticity develops, where the upflow/downwash becomes entrained together in a doughnut-shaped vortex. The downwash increases as the vortex grows in strength, leading to large reductions in rotor blade incidences spanwise. Entering the vortex-ring state, the helicopter will increase its rate of descent very rapidly as the lift is lost; any further application of collective by the pilot will tend to reduce the rotor efficiency even further – rates of descent of more than 3000 ft/min can build up very rapidly. The consequences of entering a vortex ring when close to the ground are extremely hazardous.

Rotor stall boundaries

While aeroplanes stall boundaries in level flight occur at low speed, helicopter stall boundaries occur typically at the high-speed end of the OFE. Figure 2.9 shows the aerodynamic mechanisms at work at the boundary. As the helicopter flies faster, forward cyclic is increased to counteract the lateral lift asymmetry due to cyclical dynamic pressure variations. This increases retreating blade incidences and reduces advancing blade incidences (α); at the same time forward flight brings cyclical Mach number (M) variations and the α versus M locus takes the shape sketched in Fig. 2.10. The stall boundary is also drawn, showing how both advancing and retreating blades are close to the limit at high speed. The low-speed, trailing edge-type, high incidence ($O(15°)$) stall on the retreating blade is usually triggered first, often by the sharp local incidence perturbations induced by the trailing tip vortex from previous blades. Shock-induced boundary layer separation will stall the advancing blade at very low incidence ($O(1–2°)$). Both retreating and advancing blade stall are initially local, transient effects and self-limiting on account of the decreasing incidence and increasing velocities in the fourth quadrant of the disc and the decreasing Mach number in the second quadrant. The overall effect on rotor lift will not be nearly as dramatic as when an aeroplane stalls at low speed. However, the rotor blade lift stall is usually accompanied by a large change in blade chordwise pitching moment, which in turn induces a strong, potentially more sustained, torsional oscillation and fluctuating stall, increasing vibration levels and inducing strong aircraft pitch and roll motions.

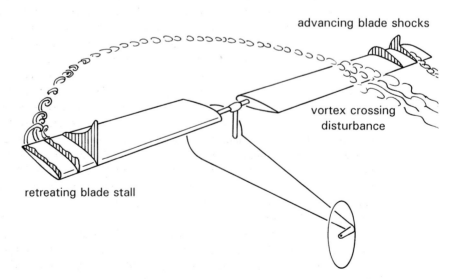

Fig. 2.9 Features limiting rotor performance in high-speed flight

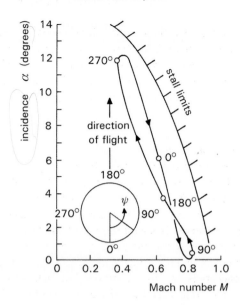

Fig. 2.10 Variation of incidence and Mach number encountered by the rotor blade tip in forward flight

Rotor stall and the attendant increase in loads therefore determine the limits to forward speed for helicopters. This and other effects can be illustrated on a plot of rotor lift (or thrust T) limits against forward speed V. It is more general to normalize these quantities as thrust coefficient C_T and advance ratio μ, where

$$C_T = \frac{T}{\rho(\Omega R)^2 \pi R^2}, \qquad \mu = \frac{V}{\Omega R}$$

where Ω is the rotorspeed, R the rotor radius and ρ the air density. Figure 2.11 shows how the thrust limits vary with advance ratio and includes the sustained or power limit boundary, the retreating and advancing blade lines, the maximum thrust line and the structural boundary. The parameter s is the solidity defined as the ratio of blade area to disc area. The retreating and advancing blade thrust lines in the figure correspond to both level and manoeuvring flight. At a given speed, the thrust coefficient can be increased in level flight, by increasing weight or height flown or by increasing the load factor in a manoeuvre. The manoeuvre can be sustained or transient and the limits will be different for the two cases, the loading peak moving inboard and ahead of the retreating side of the disc in the transient case. The retreating/advancing blade limits define the onset of increased vibration caused by local stall, and flight beyond these limits is accompanied by a marked increase in the fatigue life usage. These are soft limits, in that they are contained within the OFE and the pilot can fly through them. However, the usage spectrum for the aircraft will, in turn, define the amount of time the aircraft is likely (designed) to spend at different C_T or load factors, which, in turn, will define the service life of stressed components. The maximum thrust line defines the potential limit of the rotor, before local stall spreads so wide that the total lift reduces. The other imposed limits are defined by the capability of the powerplant and structural strength of critical components in the rotor and fuselage. The latter is an

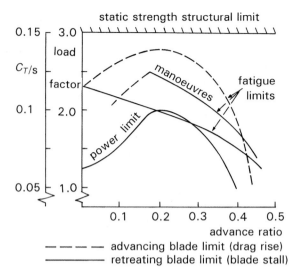

Fig. 2.11 Rotor thrust limits as a function of advance ratio

SFE design limit, set well outside the OFE. However, rotors at high speed, just like the wings on fixed-wing aeroplanes, are sometimes aerodynamically capable of exceeding this.

Having dwelt on aspects of rotor physics and the importance of rotor thrust limits, it needs to be emphasized that the pilot does not normally know what the rotor thrust is; he or she can infer it from a load factor or '*g*' meter, and from a knowledge of take-off weight and fuel burn, but the rotor limits of more immediate and critical interest to the pilot will be torque (more correctly a coupled rotor/transmission limit) and rotorspeed. Rotorspeed is automatically governed on turbine-powered helicopters, and controlled to remain within a fairly narrow range, dropping only about 5% between autorotation and full power climb, for example. Overtorquing and overspeeding are potential hazards for the rotor at the two extremes and are particularly dangerous when the pilot tries to demand full performance in emergency situations, e.g., evasive hard turn or pop-up to avoid an obstacle.

Rotor limits, whether thrust, torque or rotorspeed in nature, play a major role in the flight dynamics of helicopters, in the changing aeroelastic behaviour through to the handling qualities experienced by the pilot. Understanding the mechanisms at work near the flight envelope boundary is important in the provision of carefree handling, a subject we shall return to in Chapter 7.

2.2.4   The pilot and pilot–vehicle interface

This aspect of the subject draws its conceptual and application boundaries from the engineering and psychological facets of the human factors discipline. We are concerned in this book with the piloting task and hence with only that function in the crew station; the crew have other, perhaps more important, mission-related duties, but the degree of spare capacity which the pilot has to share these will depend critically on his flying workload. The flying task can be visualized as a closed-loop feedback

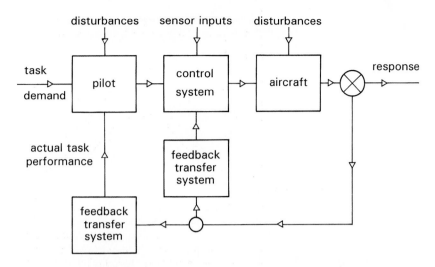

Fig. 2.12 The pilot as sensor and motivator in the feedback loop

system with the pilot as the key sensor and motivator (Fig. 2.12). The elements of Fig. 2.12 form this fourth reference point. The pilot will be well trained and highly adaptive (this is particularly true of helicopter pilots), and ultimately his or her skills and experience will determine how well a mission is performed. Pilots gather information visually from the outside world and instrument displays, from motion cues and tactile sensory organs. They continuously make judgements of the quality of their flight path management and apply any required corrections through their controllers. The pilot's acceptance of any new function or new method of achieving an existing function that assists the piloting task is so important that it is vital that prototypes are evaluated with test pilots prior to delivery into service. This fairly obvious statement is emphasized at this point because of its profound impact on the flying qualities 'process', e.g., the development of new handling criteria, new helmet-mounted display formats or multi-axis sidesticks. Pilot-subjective opinion of quality, its measurement, interpretation and correlation with objective measures, underpins all substantiated data and hence needs to be central to all new developments. Here lies a small catch; most pilots learn to live with and love their aircraft and to compensate for deficiencies. They will almost certainly have invested some of their ego in their high level of skill and ability to perform well in difficult situations. Any developments that call for changes in the way they fly can be met by resistance. To a large extent, this reflects a natural caution and needs to be heeded; test pilots are trained to be critical and to challenge the engineer's assumptions because ultimately they will have to work with the new developments.

Later in this book, in Chapter 6 and, more particularly, Chapter 7, the key role that test pilots have played in the development of flying qualities and flight control technology over the last 10 years will be addressed. In Chapter 8 the treatment of the topic of degraded handling qualities will expose some of the dangerous conditions pilots can experience. Lessons learnt through the author's personal experience of working with test pilots will be covered.

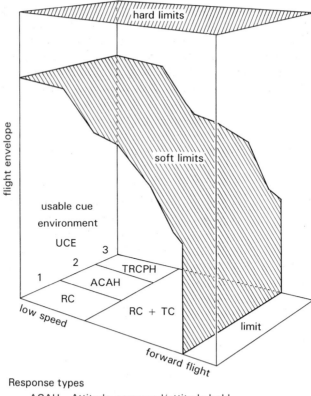

Fig. 2.13 Response types required to achieve Level 1 handing qualities in different UCEs

2.2.5 Résumé of the four reference points

Figure 2.13 illustrates in composite form the interactional nature of the flight dynamics process as reflected by the four reference points. The figure, drawing from the parlance of ADS-33, tells us that to achieve Level 1 handling qualities in a UCE of 1, a rate response type is adequate; to achieve the same in UCEs of 2 and 3 require AC (attitude command) or TRC (translational rate command) response types respectively. This classification represents a fundamental development in helicopter handling qualities that lifts the veil off a very complex and confused matter. The figure also shows that if the UCE can be upgraded from a 3 to a 2, then reduced augmentation will be required. A major trade-off between the quality of the visual cues and the quality of the control augmentation emerges. This will be a focus of attention in later chapters. Figure 2.13 also reflects the requirement that the optimum vehicle dynamic characteristics may need to change for different MTEs and at the edges of the OFE; terminology borrowed from fixed-wing parlance serves to describe these features – task-tailored or mission-oriented flying qualities and carefree handling. Above all else, the quality requirements for flying are driven by the performance and piloting workload

demands in the MTEs, which are themselves regularly changing user-defined requirements. The whole subject is thus evolving from the four reference points – the mission,
the environment, the vehicle and the pilot; they support the flight dynamics discipline
and provide an application framework for understanding and interpreting the modelling and criteria of task-oriented flying qualities. Continuing on the Tour, we address
the first of three key technical areas with stronger analytical content – theoretical
modelling.

2.3 MODELLING HELICOPTER FLIGHT DYNAMICS

A mathematical description or simulation of a helicopter's flight dynamics needs to
embody the important aerodynamic, structural and other internal dynamic effects (e.g.,
engine, actuation) that combine to influence the response of the aircraft to pilot's
controls (handling qualities) and external atmospheric disturbances (ride qualities).
The problem is highly complex and the dynamic behaviour of the helicopter is often
limited by local effects that rapidly grow in their influence to inhibit larger or faster
motion, e.g., blade stall. The helicopter behaviour is naturally dominated by the main
and tail rotors, and these will receive primary attention in this stage of the Tour; we
need a framework to place the modelling in context.

The problem domain
A convenient and intuitive framework for introducing this important topic is illustrated
in Fig. 2.14, where the natural modelling dimensions of frequency and amplitude are
used to characterize the range of problems within the OFE. The three fundamentals of
flight dynamics – trim, stability and response – can be seen delineated, with the latter
expressed in terms of the manoeuvre envelope from normal to maximum at the OFE
boundary. The figure also serves as a guide to the scope of flight dynamics as covered
in this book. At small amplitudes and high frequency, the problem domain merges with
that of the loads and vibration engineer. The separating frequency is not distinct. The
flight dynamicist is principally interested in the loads that can displace the aircraft's

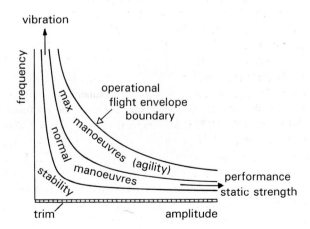

Fig. 2.14 Frequency and amplitude – the natural modelling dimensions for flight mechanics

flight path, and over which the human or automatic pilot has some direct control. On the rotor, these reduce to the zeroth and first harmonic motions and loads – all higher harmonics transmit zero mean vibrations to the fuselage; so the distinction would appear deceptively simple. The first harmonic loads will be transmitted through the various load paths to the fuselage at a frequency depending on the number of blades. Perhaps the only general statement that can be made regarding the extent of the flight dynamicists' domain is that they must be cognisant of all loads and motions that are of primary (generally speaking, controlled) and secondary (generally speaking, uncontrolled) interest in the achievement of good flying qualities. So, for example, the forced response of the first elastic torsion mode of the rotor blades (natural frequency $O(20$ Hz)) at one-per-rev could be critical to modelling the rotor cyclic pitch requirements correctly (Ref. 2.8); including a model of the lead/lag blade dynamics could be critical to establishing the limits on rate stabilization gain in an automatic flight control system (Ref. 2.9); modelling the fuselage bending frequencies and mode shapes could be critical to the flight control system sensor design and layout (Ref. 2.10).

At the other extreme, the discipline merges with that of the performance and structural engineers, although both will be generally concerned with behaviour across the OFE boundary. Power requirements and trim efficiency (range and payload issues) are part of the flight dynamicist's remit. The aircraft's static and dynamic (fatigue) structural strength presents constraints on what can be achieved from the point of view of flight path control. These need to be well understood by the flight dynamicist.

In summary, vibration, structural loads and steady-state performance traditionally define the edges of the OFE within the framework of Fig. 2.14. Good flying qualities then ensure that the OFE can be used safely, in particular that there will always be sufficient control margin to enable recovery in emergency situations. But control margin can be interpreted in a dynamic context, including concepts such as pilot-induced oscillations and agility. Just as with high-performance fixed-wing aircraft, the dynamic OFE can be limited, and hence defined, by flying qualities for rotorcraft. In practice, a balanced design will embrace these in harmony with the central flight dynamics issues, drawing on concurrent engineering techniques (Ref. 2.11) to quantify the trade-offs and to identify any critical conflicts.

Multiple interacting subsystems

The behaviour of a helicopter in flight can be modelled as the combination of a large number of interacting subsystems. Figure 2.15 highlights the main rotor element, the fuselage, powerplant, flight control system, empennage and tail rotor elements and the resulting forces and moments. Shown in simplified form in Fig. 2.16 is the orthogonal body axes system, fixed at the centre of gravity/mass (cg/cm) of the whole aircraft, about which the aircraft dynamics are referred. Strictly speaking, the cg will move as the rotor blades flap, but we shall assume that the cg is located at the mean position, relative to a particular trim state. The equations governing the behaviour of these interactions are developed from the application of physical laws, e.g., conservation of energy and Newton's laws of motion, to the individual components, and commonly take the form of nonlinear differential equations written in the first-order vector form

$$\frac{d\mathbf{x}}{dt} = \mathbf{f}(\mathbf{x}, \mathbf{u}, t) \tag{2.1}$$

with initial conditions $\mathbf{x}(0) = \mathbf{x}_0$.

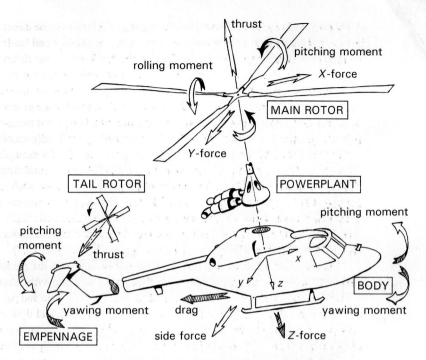

Fig. 2.15 The modelling components of a helicopter

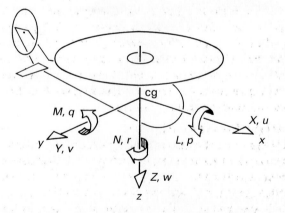

Fig. 2.16 The orthogonal axes system for helicopter flight dynamics

$\mathbf{x}(t)$ is the column vector of state variables; $\mathbf{u}(t)$ is the vector of control variables and $\mathbf{f}$ is a nonlinear function of the aircraft motion, control inputs and external disturbances. The reader is directed to Appendix 4A for a brief exposition on the matrix–vector theory used in this and later chapters. For the special case where only the six rigid-body degrees of freedom (DoFs) are considered, the state vector $\mathbf{x}$ comprises the three translational velocity components u, v and w, the three rotational velocity components p, q and r and the Euler angles ϕ, θ and ψ. The three Euler attitude angles augment the equations of motion through the kinematic relationship between the fuselage rates p, q and r and the rates of change of the Euler angles. The velocities are referred to an axes system fixed at the cg as shown in Fig. 2.16 and the Euler angles define the orientation of the fuselage with respect to an earth fixed axes system.

The DoFs are usually arranged in the state vector as longitudinal and lateral motion subsets, as

$$\mathbf{x} = \{u, w, q, \theta, v, p, \phi, r, \psi\}$$

The function $\mathbf{f}$ then contains the applied forces and moments, again referred to the aircraft cg, from aerodynamic, structural, gravitational and inertial sources. Strictly speaking, the inertial and gravitational forces are not 'applied', but it is convenient to label them so and place them on the right-hand side of the describing equation. The derivation of these equations from Newton's laws of motion will be carried out later in Chapter 3 and its appendix. It is important to note that this six DoF model, while itself complex and widely used, is still an approximation to the aircraft behaviour; all higher DoFs, associated with the rotors (including aeroelastic effects), powerplant/transmission, control system and the disturbed airflow, are embodied in a quasi-steady manner in the equations, having lost their own individual dynamics and independence as DoFs in the model reduction. This process of approximation is a common feature of flight dynamics, in the search for simplicity to enhance physical understanding and ease the computational burden, and will feature extensively throughout Chapters 4 and 5.

Trim, stability and response

Continuing the discussion of the 6 DoF model, the solutions to the three fundamental problems of flight dynamics can be written as

Trim: $$\mathbf{f}(\mathbf{x}_e, \mathbf{u}_e) = \mathbf{0} \tag{2.2}$$

Stability: $$\det\left[\lambda\mathbf{I} - \left(\frac{\partial\mathbf{f}}{\partial\mathbf{x}}\right)_{\mathbf{x}_e}\right] = 0 \tag{2.3}$$

Response: $$\mathbf{x}(t) = \mathbf{x}(0) + \int_0^t \mathbf{f}(\mathbf{x}(\tau), \mathbf{u}(\tau), \tau)\,\mathrm{d}\tau \tag{2.4}$$

The *trim* solution is represented by the zero of a nonlinear algebraic function, where the controls $\mathbf{u}_e$ required to hold a defined state $\mathbf{x}_e$ (subscript e refers to equilibrium) are computed. With four controls, only four states can be prescribed in trim, the remaining set forming into the additional unknowns in eqn 2.1. A trimmed flight condition is defined as one in which the rate of change (of magnitude) of the aircraft's state vector is zero and the resultant of the applied forces and moments is zero. In a trimmed manoeuvre, the aircraft will be accelerating under the action of non-zero resultant aerodynamic and gravitational forces and moments, but these will then be balanced by effects such as centrifugal and gyroscopic inertial forces and moments. The trim equations and associated problems, e.g., predicting performance and control margins, will be further developed in Chapter 4.

The solution of the *stability* problem is found by linearizing the equations about a particular trim condition and computing the eigenvalues of the aircraft system matrix, written in eqn 2.3 as the partial derivative of the forcing vector with respect to the system states. After linearization of eqn 2.1, the resulting first-order, constant coefficient differential equations have solutions of the form $e^{\lambda t}$, the stability of which is determined by the signs of the real parts of the eigenvalues λ. The stability thus found refers to small motions about the trim point; will the aircraft return to – or depart from – the trim point if disturbed by, say, a gust? For larger motions, nonlinearities can alter the behaviour and recourse to the full equations is usually necessary.

The *response* solution given by eqn 2.4 is found from the time integral of the forcing function and allows the evolution of the aircraft states, forces and moments to be computed following disturbed initial conditions $\mathbf{x}(0)$, and/or prescribed control inputs and atmospheric disturbances. The nonlinear equations are usually solved numerically; analytical solutions generally do not exist. Sometimes, narrow-range approximate solutions can be found to describe special large-amplitude nonlinear motion, e.g., limit cycles, but these are exceptional and usually developed to support the diagnosis of behaviour unaccounted for in the original design.

The sketches in Fig. 2.17 illustrate typical ways in which trim, stability and response results are presented; the key variable in the trim and stability sketches is the helicopter's forward speed. The trim control positions are shown with their characteristic shapes; the stability characteristics are shown as loci of eigenvalues plotted on the complex plane; the short-term responses to step inputs, or the step responses, are shown as a function of time. This form of presentation will be revisited later on this Tour and in later chapters.

The reader of this Tour may feel too quickly plunged into abstraction with the above equations and their descriptions; the intention is to give some exposure to mathematical concepts which are part of the toolkit of the flight dynamicist. Fluency in the

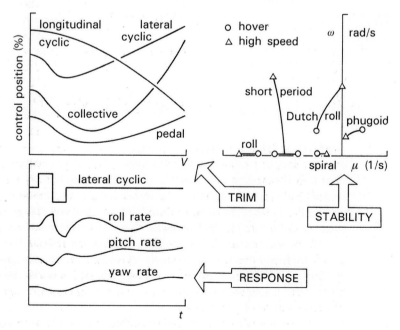

Fig. 2.17 Typical presentation of flight mechanics results for trim, stability and response

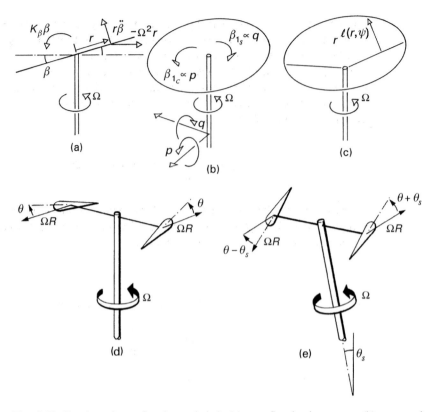

Fig. 2.18 Sketches of rotor flapping and pitch: (a) rotor flapping in vacuum; (b) gyroscopic moments in vacuum; (c) rotor coning in air; (d) before shaft tilt; (e) after shaft tilt showing effective cyclic path

parlance of this mathematics is essential for the serious practitioner. Perhaps even more essential is a thorough understanding of the fundamentals of rotor flapping behaviour, which is the next stop on this Tour; here we shall need to rely extensively on theoretical analysis. A full derivation of the results will be given later in Chapters 3, 4 and 5.

The flapping rotor in vacuo

The equations of motion of a flapping rotor will be developed in a series of steps (Figs 2.18(a)–(e)), designed to highlight a number of key features of rotor behaviour. Figure 2.18(a) shows a rotating blade (Ω, rad/s) free to flap (β, rad) about a hinge at the centre of rotation; to add some generality we shall add a flapping spring at the hinge (K_β, N m/rad). The flapping angle β is referred to the rotor shaft; other reference systems, e.g., relative to the control axis, are discussed in Appendix 3A. It will be shown later in Chapter 3 that this simple centre-spring representation is quite adequate for describing the flapping behaviour of teetering, articulated and hingeless or bearingless rotors, under a wide range of conditions. Initially, we consider the case of flapping in a vacuum, i.e., no aerodynamics, and we neglect the effects of gravity. The first qualitative point to grasp concerns what happens to the rotor when the rotor shaft is suddenly tilted to a new angle. For the case of the zero spring stiffness, the rotor disc will remain aligned in its original position, there being no mechanism to generate a turning moment on the blade. With a spring added, the blade will develop a persistent oscillation about the new shaft orientation, with the inertial moment due to out-of-plane flapping and the centrifugal moment continually in balance.

The dynamic equation of flapping can be derived by taking moments about the flap hinge during accelerated motion, so that the hinge moment $K_\beta \beta$ is balanced by the inertial moments, thus

$$K_\beta \beta = -\int_0^R rm(r)\,\{r\ddot{\beta} + r\Omega^2\beta\}\,dr \tag{2.5}$$

where $m(r)$ is the blade mass distribution (kg/m) and $(\cdot)$ indicates differentiation with respect to time t. Setting $(')$ as differentiation with respect to $\psi = \Omega t$, the blade azimuth angle, eqn 2.5 can be rearranged and written as

$$\beta'' + \lambda_\beta^2 \beta = 0 \tag{2.6}$$

where the flapping frequency ratio λ_β is given by the expression

$$\lambda_\beta^2 = 1 + \frac{K_\beta}{I_\beta \Omega^2} \tag{2.7}$$

and where the flap moment of inertia is

$$I_\beta = \int_0^R m(r) r^2 \, dr \tag{2.8}$$

The two inertial terms in eqn 2.5 represent the contributions from accelerated flapping out of the plane of rotation, $r\ddot{\beta}$, and the in-plane centrifugal acceleration arising from the blade displacement acting towards the centre of the axis of rotation, $r\Omega^2\beta$. Here, as will be the case throughout this book, we make the assumption that β is small, so that $\sin \beta \sim \beta$ and $\cos \beta \sim 1$.

For the special case where $K_\beta = 0$, the solution to eqn 2.6 is simple harmonic motion with a natural frequency of one-per-rev, i.e., $\lambda_\beta^2 = 1$. If the blade is disturbed in flap, the motion will take the form of a persistent, undamped, oscillation with frequency Ω; the disc cut by the blade in space will take up a new tilt angle equal to the angle of the initial disturbance. Again, with K_β set to zero, there will be no tendency for the shaft to tilt in response to the flapping, since no moments can be transmitted through the flapping hinge. For the case with non-zero K_β, the frequency ratio is greater than unity and the natural frequency of disturbed motion is faster than one-per-rev, disturbed flapping taking the form of a disc precessing against the rotor rotation, if the shaft is fixed. With the shaft free to rotate, the hub moment generated by the spring will cause the shaft to rotate into the direction normal to the disc. Typically, the stiffness of a hingeless rotor blade can be represented by a spring giving an equivalent λ_β^2 of between 1.1 and 1.3. The higher values are typical of the first generation of hingeless rotor helicopters, e.g., Lynx and Bo105, the lower more typical of modern bearingless designs. The overall stiffness is therefore dominated by the centrifugal force field.

Before including the effects of blade aerodynamics, we consider the case where the shaft is rotated in pitch and roll, p and q (see Fig. 2.18(b)). The blade now experiences additional gyroscopic accelerations caused by mutually perpendicular angular velocities, p, q and Ω. If we neglect the small effects of shaft angular accelerations, the equation of motion can be written as

$$\beta'' + \lambda_\beta^2 \beta = \frac{2}{\Omega}(p \cos \psi - q \sin \psi) \tag{2.9}$$

The conventional zero reference for blade azimuth is at the rear of the disc and ψ is positive in the direction of rotor rotation; in eqn 2.9 the rotor is rotating anticlockwise when viewed from above. For clockwise rotors, the roll rate term would be negative. The steady-state solution to the 'forced' motion takes the form

$$\beta = \beta_{1c} \cos \psi + \beta_{1s} \sin \psi \tag{2.10}$$

where

$$\beta_{1c} = \frac{2}{\Omega(\lambda_\beta^2 - 1)} \, p, \qquad \beta_{1s} = \frac{-2}{\Omega(\lambda_\beta^2 - 1)} \, q \tag{2.11}$$

These solutions represent the classic gyroscopic motions experienced when any rotating mass is rotated out of plane; the resulting motion is orthogonal to the applied rotation. β_{1c} is a longitudinal disc tilt in response to a roll rate; β_{1s} a lateral tilt in response to a pitch rate. The moment transmitted by the single blade to the shaft, in the rotating axes system, is simply $K_\beta \beta$; in the non-rotating shaft axes, the moment can be written as pitch (positive nose up) and roll (positive to starboard) components

$$M = -K_\beta \beta (\cos \psi) = -\frac{K_\beta}{2}(\beta_{1c}(1 + \cos 2\psi) + \beta_{1s} \sin 2\psi) \tag{2.12}$$

$$L = -K_\beta \beta (\sin \psi) = -\frac{K_\beta}{2}(\beta_{1s}(1 - \cos 2\psi) + \beta_{1c} \sin 2\psi) \tag{2.13}$$

Each component therefore has a steady value plus an equally large wobble at two-per-rev. For a rotor with N_b evenly spaced blades, it can be shown that the oscillatory moments cancel, leaving the steady values

$$M = -N_b \frac{K_\beta}{2} \beta_{1c} \tag{2.14}$$

$$L = -N_b \frac{K_\beta}{2} \beta_{1s} \tag{2.15}$$

This is a general result that will carry through to the situation when the rotor is working in air, i.e., the zeroth harmonic hub moments that displace the flight path of the aircraft are proportional to the tilt of the rotor disc. It is appropriate to highlight that we have neglected the moment of the in-plane rotor loads in forming these hub moment expressions. They are therefore strictly approximations to a more complex effect, which we shall discuss in more detail in Chapter 3. We shall see, however, that the aerodynamic loads are not only one-per-rev, but also two and higher, giving rise to vibratory moments. Before considering the effects of aerodynamics, there are two points that need to be made about the solution given by eqn 2.11. First, what happens when $\lambda_\beta^2 = 1$? This is the classic case of resonance, when according to theory, the response becomes infinite; clearly, the assumption of small flap angles would break down well before this and the nonlinearity in the centrifugal stiffening with amplitude would limit the motion. The second point is that the solution given by eqn 2.11 is only part of the complete solution. Unless the initial conditions of the blade motion were very carefully set up, the response would actually be the sum of two undamped motions, one with the one-per-rev forcing frequency, and the other with the natural frequency λ_β. A complex response would develop, with the combination of two close frequencies leading to a beating response or, in special cases, non-periodic 'chaotic' behaviour. Such situations are somewhat academic for the helicopter, as the aerodynamic forces distort the response described above in a dramatic way.

The flapping rotor in air – aerodynamic damping

Figure 2.18(c) shows the blade in air, with the distributed aerodynamic lift $\ell(r, \psi)$ acting normal to the resultant velocity; we are neglecting the drag forces in this case. If the shaft is now tilted to a new reference position, the blades will realign with the shaft, even with zero spring stiffness. Figures 2.18(d) and (e) illustrate what happens. When the shaft is tilted, say, in pitch by angle θ_s, the blades experience an effective cyclic pitch change with maximum and minimum at the lateral positions ($\psi = 90°$ and $180°$). The blades will then flap to restore the zero hub moment condition.

For small flap angles, the equation of flap motion can now be written in the approximate form

$$\beta'' + \lambda_\beta^2 \beta = \frac{2}{\Omega}(p \cos \psi - q \sin \psi) + \frac{1}{I_\beta \Omega^2} \int_0^R \ell(r, \psi) r \, dr \qquad (2.16)$$

A simple expression for the aerodynamic loading can be formulated with reference to Fig. 2.19, with the assumptions of two-dimensional, steady aerofoil theory, i.e.,

$$\ell(r, \psi) = \frac{1}{2}\rho V^2 c a_0 \alpha \qquad (2.17)$$

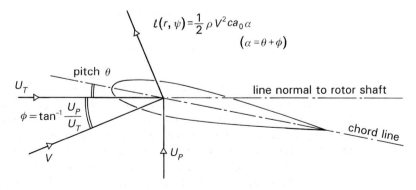

Fig. 2.19 Components of rotor blade incidence

where V is the resultant velocity of the airflow, ρ the air density and c the blade chord. The lift is assumed to be proportional to the incidence of the airflow to the chord line, α, up to stalling incidence, with lift curve slope a_0. In Fig. 2.19 the incidence is shown to comprise two components, one from the applied blade pitch angle θ and one from the induced inflow ϕ, given by

$$\phi = \tan^{-1} \frac{\overline{U}_P}{\overline{U}_T} \approx \frac{\overline{U}_P}{\overline{U}_T} \tag{2.18}$$

where U_T and U_P are the in-plane and normal velocity components respectively (the bar signifies non-dimensionalization with ΩR). Using the simplification that $\overline{U}_P \ll \overline{U}_T$, eqn 2.16 can be written as

$$\beta'' + \lambda_\beta^2 \beta = \frac{2}{\Omega}(p \cos \psi - q \sin \psi) + \frac{\gamma}{2} \int_0^1 (\overline{U}_T^2 \theta + \overline{U}_T \overline{U}_P) \bar{r} \, d\bar{r} \tag{2.19}$$

where $\bar{r} = r/R$ and the Lock number, γ, is defined as (Ref. 2.12)

$$\gamma = \frac{\rho c a_0 R^4}{I_\beta} \tag{2.20}$$

The Lock number is an important non-dimensional scaling coefficient, giving the ratio of aerodynamic to inertia forces acting on a rotor blade.

To develop the present analysis further, we consider the hovering rotor and a constant inflow velocity v_i over the rotor disc, so that the velocities at station r along the blade are given by

$$\overline{U}_T = \bar{r}, \qquad \overline{U}_P = -\lambda_i + \frac{\bar{r}}{\Omega}(p \sin \psi + q \cos \psi) - \bar{r}\beta' \tag{2.21}$$

where

$$\lambda_i = \frac{v_i}{\Omega R}$$

We defer the discussion on rotor downwash until later in this Chapter and Chapter 3; for the present purposes, we merely state that a uniform distribution over the disc is a reasonable approximation to support the arguments developed in this chapter.

Equation 2.19 can then be expanded and rearranged as

$$\beta'' + \frac{\gamma}{8}\beta' + \lambda_\beta^2 \beta = \frac{2}{\Omega}(p \cos \psi - q \sin \psi) + \frac{\gamma}{8}\left(\theta - \frac{4}{3}\lambda_i + \frac{p}{\Omega} \sin \psi\right.$$

$$\left. + \frac{q}{\Omega} \cos \psi \right) \tag{2.22}$$

The flapping eqn 2.22 can tell us a great deal about the behaviour of a rotor in response to aerodynamic loads; in particular the presence of the flap damping β' alters the response characteristics significantly. We can write the applied blade pitch in the form (cf. Fig. 2.5 and the early discussion on rotor controls)

$$\theta = \theta_0 + \theta_{1c} \cos \psi + \theta_{1s} \sin \psi \tag{2.23}$$

where θ_0 is the collective pitch and θ_{1s} and θ_{1c} the longitudinal and lateral cyclic pitch respectively. The forcing function on the right-hand side of eqn 2.22 is therefore made up of constant and first harmonic terms. In the general flight case, with the pilot active on his controls, the rotor controls θ_0, θ_{1c} and θ_{1s} and the fuselage rates p and q will vary continuously with time. As a first approximation we shall assume that these variations are slow compared with the rotor blade transient flapping. We can quantify this approximation by noting that the aerodynamic damping in eqn 2.22, $\gamma/8$, varies between about 0.7 and 1.3. In terms of the response to a step input, this corresponds to rise times (to 63% of steady-state flapping) between 60 and 112° azimuth ($\psi_{63\%} = 16 \ln(2)/\gamma$). Rotorspeeds vary from about 27 rad/s on the AS330 Puma to about 44 rad/s on the MBB Bo105, giving flap time constants between 0.02 and 0.07 s at the extremes. Provided that the time constants associated with the control activity and fuselage angular motion are an order of magnitude greater than this, the assumption of rotor quasi-steadiness during aircraft motions will be valid. We shall return to this assumption a little later on this Tour, but, for now, we assume that the rotor flapping has time to achieve a new steady-state, one-per-rev motion following each incremental change in control and fuselage angular velocity. We write the rotor flapping motion in the quasi-steady-state form

$$\beta = \beta_0 + \beta_{1c} \cos \psi + \beta_{1s} \sin \psi \qquad (2.24)$$

β_0 is the rotor coning and β_{1c} and β_{1s} the longitudinal and lateral flapping respectively. The cyclic flapping can be interpreted as a tilt of the rotor disc in the longitudinal (forward) β_{1c} and lateral (port) β_{1s} planes. The coning has an obvious physical interpretation (see Fig. 2.20).

The quasi-steady coning and first harmonic flapping solution to eqn 2.22 can be obtained by substituting eqns 2.23 and 2.24 into eqn 2.22 and equating constant and first harmonic coefficients. Collecting terms, we can write

$$\beta_0 = \frac{\gamma}{8\lambda_\beta^2} \left(\theta_0 - \frac{4}{3}\lambda_i \right) \qquad (2.25)$$

$$\beta_{1c} = \frac{1}{1 + S_\beta^2} \left\{ S_\beta\theta_{1c} - \theta_{1s} + \left(S_\beta \frac{16}{\gamma} - 1 \right) \bar{p} + \left(S_\beta + \frac{16}{\gamma} \right) \bar{q} \right\} \qquad (2.26)$$

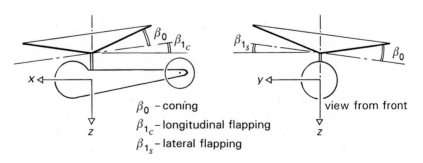

β_0 – coning
β_{1c} – longitudinal flapping
β_{1s} – lateral flapping

view from front

Fig. 2.20 The three rotor disc degrees of freedom

$$\beta_{1s} = \frac{1}{1 + S_\beta^2} \left\{ S_\beta \theta_{1s} + \theta_{1c} + \left(S_\beta + \frac{16}{\gamma} \right) \bar{p} - \left(S_\beta \frac{16}{\gamma} - 1 \right) \bar{q} \right\} \qquad (2.27)$$

where the Stiffness number

$$S_\beta = \frac{8 \left(\lambda_\beta^2 - 1 \right)}{\gamma} \qquad (2.28)$$

and

$$\bar{p} = \frac{p}{\Omega}, \qquad \bar{q} = \frac{q}{\Omega}$$

The Stiffness number S_β is a useful non-dimensional parameter in that it provides a measure of the ratio of hub stiffness to aerodynamic moments.

Flapping derivatives
The coefficients in eqns 2.26 and 2.27 can be interpreted as partial derivatives of flapping with respect to the controls and aircraft motion; hence we can write

$$\frac{\partial \beta_{1c}}{\partial \theta_{1s}} = -\frac{\partial \beta_{1s}}{\partial \theta_{1c}} = -\frac{1}{1 + S_\beta^2} \qquad (2.29)$$

$$\frac{\partial \beta_{1c}}{\partial \theta_{1c}} = \frac{\partial \beta_{1s}}{\partial \theta_{1s}} = \frac{S_\beta}{1 + S_\beta^2} \qquad (2.30)$$

$$\frac{\partial \beta_{1c}}{\partial \bar{q}} = \frac{\partial \beta_{1s}}{\partial \bar{p}} = \frac{1}{1 + S_\beta^2} \left(S_\beta + \frac{16}{\gamma} \right) \qquad (2.31)$$

$$\frac{\partial \beta_{1c}}{\partial \bar{p}} = -\frac{\partial \beta_{1s}}{\partial \bar{q}} = \frac{1}{1 + S_\beta^2} \left(S_\beta \frac{16}{\gamma} - 1 \right) \qquad (2.32)$$

The partial derivatives in eqns 2.29–2.32 represent the changes in flapping with changes in cyclic pitch and shaft rotation and are shown plotted against Stiffness number for different values of γ in Figs 2.21(a)–(c). Although S_β is shown plotted up to unity, a maximum realistic value for current hingeless rotors with heavy blades (small value of γ) is about 0.5, with more typical values between 0.05 and 0.3. The control derivatives illustrated in Fig. 2.21(a) show that the direct flapping response, $\partial \beta_{1c}/\partial \theta_{1s}$, is approximately unity up to typical maximum values of stiffness, i.e., a hingeless rotor blade flaps by about the same amount as a teetering or articulated rotor. However, the variation of the coupled flap response, $\partial \beta_{1c}/\partial \theta_{1s}$, is much more significant, being as much as 30% of the primary response at an S_β of 0.3. When this level of flap cross-coupling is transmitted through the hub to the fuselage, an even larger ratio of pitch/roll response coupling can result due the relative magnitudes of the aircraft inertias.

The fundamental 90° phase shift
A fundamental result of rotor dynamics emerges from the above analysis, that the flapping response is approximately 90° out of phase with the applied cyclic pitch, i.e., θ_{1s} gives β_{1c}, and θ_{1c} gives β_{1s}. For blades freely articulated at the centre of rotation, or teetering rotors, the response is lagged by exactly 90° in hover; for hingeless rotors,

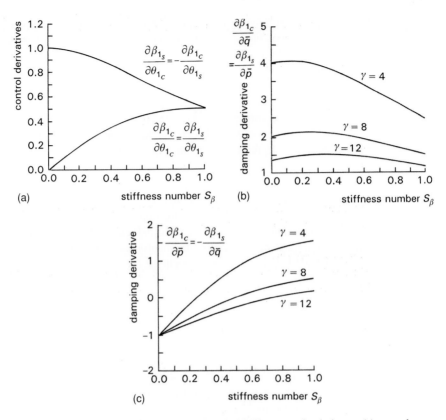

Fig. 2.21 Variation of flap derivatives with Stiffness number in hover: (a) control;
(b) damping; (c) cross-coupling

such as the Lynx and Bo105, the phase angle is about 75–80°. The phase delay is a result
of the rotor being aerodynamically forced, through cyclic pitch, close to resonance,
i.e., one-per-rev. The second-order character of eqn 2.22 results in a low-frequency
response in-phase with inputs and a high-frequency response with a 180° phase lag.
The innovation of cyclic pitch, forcing the rotor close to its natural flapping frequency,
is amazingly simple and effective – practically no energy is required and a degree of
pitch results in a degree of flapping. A degree of flapping can generate between 0 (for
teetering rotors), 500 (for articulated rotors) and greater than 2000 N m (for hingeless
rotors) of hub moment, depending on the rotor stiffness.

The flap damping derivatives, given by eqns 2.31 and 2.32, are illustrated in
Figs 2.21(b) and (c). The direct flap damping, $\partial \beta_{1c}/\partial \bar{q}$, is practically independent
of stiffness up to $S_\beta = 0.5$; the cross-damping, $\partial \beta_{1c}/\partial \bar{p}$, varies linearly with S_β and
actually changes sign at high values of S_β. In contrast with the *in vacuo* case, the direct
flapping response now opposes the shaft motion. The disc follows the rotating shaft,
lagged by an angle given by the ratio of the flap derivatives in the figures. For very heavy
blades (e.g., $\gamma = 4$), the direct flap response is about four times the coupled motion; for
very light blades, the disc tilt angles are more equal. This rather complex response stems
from the two components on the right-hand side of the flapping equation, eqn 2.22,
one aerodynamic due to the distribution of airloads from the angular motion, the other

from the gyroscopic flapping motion. The resultant effect of these competing forces on the helicopter motion is also complex and needs to be revisited for further discussion in Chapters 3 and 4. Nevertheless, it should be clear to the reader that the calculation of the correct Lock number for a rotor is critical to the accurate prediction of both primary and coupled responses. Complicating factors are that most blades have strongly non-uniform mass distributions and aerodynamic loadings and any blade deformation will further effect the ratio of aerodynamic to inertia forces. The concept of the equivalent Lock number is often used in helicopter flight dynamics to encapsulate a number of these effects. The degree to which this approach is valid will be discussed later in Chapter 3.

Hub moments and rotor/fuselage coupling

From the above discussion, we can see the importance of the two key parameters, λ_β and γ, in determining the flapping behaviour and hence hub moment. The hub moments due to the out-of-plane rotor loads are proportional to the rotor stiffness, as given by eqns 2.14 and 2.15; these can be written in the form

Pitch moment:
$$M = -N_b \frac{K_\beta}{2} \beta_{1c} = -\frac{N_b}{2} \Omega^2 I_\beta \left(\lambda_\beta^2 - 1 \right) \beta_{1c} \tag{2.33}$$

Roll moment:
$$L = -N_b \frac{K_\beta}{2} \beta_{1s} = -\frac{N_b}{2} \Omega^2 I_\beta \left(\lambda_\beta^2 - 1 \right) \beta_{1s} \tag{2.34}$$

To this point in the analysis we have described rotor motions with fixed or prescribed shaft rotations to bring out the partial effects of control effectiveness and flap damping. We can now extend the analysis to shaft-free motion. To simplify the analysis we consider only the roll motion and assume that the centre of mass of the rotor and shaft lies at the hub centre. The motion of the shaft is described by the simple equation relating the rate of change of angular momentum to the applied moment:

$$I_{xx}\dot{p} = L \tag{2.35}$$

where I_{xx} is the roll moment of inertia of the helicopter. By combining eqn 2.27 with eqn 2.34, the equation describing the 1 DoF roll motion of the helicopter, with quasi-steady rotor, can be written in the first-order differential form of a rate response type:

$$\dot{p} - L_p p = L_{\theta_{1c}} \theta_{1c} \tag{2.36}$$

where the rolling moment 'derivatives' are given by

$$L_p \approx -\frac{N_b S_\beta I_\beta \Omega}{I_{xx}}, \quad L_{\theta_{1c}} \approx -\frac{N_b S_\beta \gamma I_\beta \Omega^2}{16 I_{xx}} \tag{2.37}$$

where the approximation that $S_\beta^2 \ll 1$ has been made. Non-dimensionalizing by the roll moment of inertia I_{xx} transforms these into angular acceleration derivatives.

These are the most primitive forms of the roll damping and cyclic control derivatives for a helicopter, but they contain most of the first-order effects, as will be observed later in Chapters 4 and 5. The solution to eqn 2.36 is a simple exponential transient

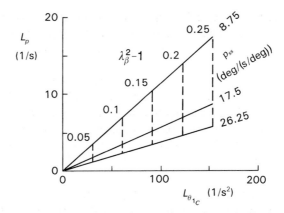

Fig. 2.22 Linear variation of rotor damping with control sensitivity in hover

superimposed on the steady state solution. For a simple step input in lateral cyclic, this takes the form

$$p = -(1 - e^{L_p t}) \frac{L_{\theta_{1c}}}{L_p} \theta_{1c} \tag{2.38}$$

The time constant (time to reach 63% of steady state) of the motion, τ_p, is given by $-(1/L_p)$, the control sensitivity (initial acceleration) by $L_{\theta_{1c}}$ and the rate sensitivity (steady-state rate response per degree of cyclic) by

$$p_{ss}(\text{deg}/(\text{s deg})) = -\frac{L_{\theta_{1c}}}{L_p} = -\frac{\gamma \Omega}{16} \tag{2.39}$$

These are the three handling qualities parameters associated with the time response of eqn 2.36, and Fig. 2.22 illustrates the effects of the primary rotor parameters. The fixed parameters for this test case are $\Omega = 35$ rad/s, $N_b = 4$, $I_\beta/I_{xx} = 0.25$.

Four points are worth highlighting:

(1) contrary to 'popular' understanding, the steady-state roll rate response to a step lateral cyclic is independent of rotor flapping stiffness; teetering and hingeless rotors have effectively the same rate sensitivity;
(2) the rate sensitivity varies linearly with Lock number;
(3) both control sensitivity and damping increase linearly with rotor stiffness;
(4) the response time constant is inversely proportional to rotor stiffness.

These points are further brought out in the generalized sketches in Figs 2.23(a) and (b), illustrating the first-order time response in roll rate from a step lateral cyclic input. These time response characteristics were used to describe short-term handling qualities until the early 1980s when the revision to Mil Spec 8501A (Ref. 2.2) introduced the frequency domain as a more meaningful format, at least for non-classical short-term response. One of the reasons for this is that the approximation of quasi-steady flapping motion begins to break down when the separation between the frequency of rotor flap modes and fuselage attitude modes decreases. The full derivation of the equations of flap motion will be covered in Chapter 3, but to complete this analysis of rotor/fuselage

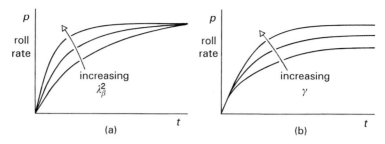

Fig. 2.23 Effects of rotor parameters on roll rate response: (a) rotor stiffness; (b) Lock number

coupling in hover, we shall briefly examine the next, improved, level of approximation. Equations 2.40 and 2.41 describe the coupled motion when only first-order lateral flapping (the so-called flap regressive mode) and fuselage roll are considered. The other rotor modes – the coning and advancing flap mode – and coupling into pitch, are neglected at this stage.

$$\dot{\beta}_{1s} + \frac{\beta_{1s}}{\tau_{\beta_{1s}}} = p + \frac{\theta_{1c}}{\tau_{\beta_{1s}}} \tag{2.40}$$

$$\dot{p} - L_{\beta_{1s}}\beta_{1s} = 0 \tag{2.41}$$

where

$$L_{\beta_{1s}} = L_{\theta_{1c}} = -\frac{N_b}{2}\Omega^2\frac{I_\beta}{I_{xx}}\left(\lambda_\beta^2 - 1\right) = -\frac{1}{\tau_{\beta_{1s}}\tau_p} \tag{2.42}$$

and

$$\tau_{\beta_{1s}} = \frac{16}{\gamma\Omega}, \qquad \tau_p = -\frac{1}{L_p} \tag{2.43}$$

The time constants $\tau_{\beta_{1s}}$ and τ_p are associated with the disc and fuselage (shaft) response respectively. The modes of motion are now coupled roll/flap with eigenvalues given by the characteristic equation

$$\lambda^2 + \frac{1}{\tau_{\beta_{1s}}}\lambda + \frac{1}{\tau_{\beta_{1s}}\tau_p} = 0 \tag{2.44}$$

The roots of eqn 2.44 can be approximated by the 'uncoupled' values only for small values of stiffness and relatively high values of Lock number. Figure 2.24 shows the variation of the exact and uncoupled approximate roots with $(\lambda_B^2 - 1)$ for the case when $\gamma = 8$. The approximation of quasi-steady rotor behaviour will be valid for small offset articulated rotors and soft bearingless designs, but for hingeless rotors with λ_β^2 much above 1.1, the fuselage response is fast enough to be influenced by the rotor transient response, and the resultant motion is a coupled roll/flap oscillation. Note again that the rotor disc time constant is independent of stiffness and is a function only of rotorspeed and Lock number (eqn 2.43).

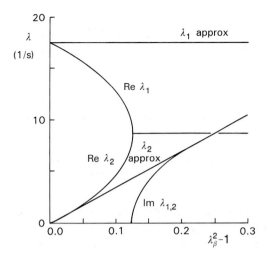

Fig. 2.24 Variation of roll/flap exact and approximate mode eigenvalues with rotor stiffness

Linearization in general

The assumptions made to establish the above approximate results have not been discussed; we have neglected detailed blade aerodynamic and deformation effects and we have assumed the rotorspeed to be constant; these are important effects that will need to be considered later in Chapter 3, but would have detracted from the main points we have tried to establish in the foregoing analysis. One of these is the concept of the motion derivative, or partial change in the rotor forces and moments with rotor motion. If the rotor were an entirely linear system, then the total force and moment could be formulated as the sum of individual effects each written as a derivative times a motion.

This approach, which will normally be valid for small enough motion, has been established in both fixed- and rotary-wing flight dynamics since the early days of flying (Ref. 2.13) and enables the stability characteristics of an aircraft to be determined. The assumption is made that the aerodynamic forces and moments can be expressed as a multi-dimensional analytic function of the motion of the aircraft about the trim condition; hence the rolling moment, for example, can be written as

$$L = L_{trim} + \frac{\partial L}{\partial u}u + \frac{\partial^2 L}{\partial u^2}u^2 + \cdots + \frac{\partial L}{\partial v}v + \cdots + \frac{\partial L}{\partial w}w + \cdots + \frac{\partial L}{\partial p}p + \cdots + \frac{\partial L}{\partial q}q$$

$$+ \text{ terms due to higher motion derivatives (e.g., } \dot{p}) \text{ and controls} \qquad (2.45)$$

For small motions, the linear terms will normally dominate and the approximation can be written in the form

$$L = L_{trim} + L_u u + L_v v + L_w w + L_p p + L_q q + L_r r$$

$$+ \text{ acceleration and control terms} \qquad (2.46)$$

In this 6 DoF approximation, each component of the helicopter will contribute to each derivative; hence, for example, there will be an X_u and an N_p for the rotor, fuselage, empennage and even the tail rotor, although many of these components, while dominating some derivatives, will have a negligible contribution to others. Dynamic effects beyond the motion in the six rigid-body DoFs will be folded into the latter in quasi-steady form, e.g., rotor, air mass dynamics and engine/transmission. For example, if the rotor DoFs were represented by the vector $\mathbf{x}_r$ and the fuselage by $\mathbf{x}_r$, then the linearized, coupled equations can be written in the form

$$
\begin{bmatrix} \dot{\mathbf{x}}_f \\ \dot{\mathbf{x}}_r \end{bmatrix} - \begin{bmatrix} \mathbf{A}_{ff} & \mathbf{A}_{fr} \\ \mathbf{A}_{rf} & \mathbf{A}_{rr} \end{bmatrix} \begin{bmatrix} \mathbf{x}_f \\ \mathbf{x}_r \end{bmatrix} = \begin{bmatrix} \mathbf{B}_{ff} & \mathbf{B}_{fr} \\ \mathbf{B}_{rf} & \mathbf{B}_{rr} \end{bmatrix} \begin{bmatrix} \mathbf{u}_f \\ \mathbf{u}_r \end{bmatrix}
\tag{2.47}
$$

We have included, for completeness, fuselage and rotor controls. Folding the rotor DoFs into the fuselage as quasi-steady motions will be valid if the characteristic frequencies of the two elements are widely separate and the resultant approximation for the fuselage motion can then be written as

$$
\dot{\mathbf{x}}_f - \left[\mathbf{A}_{ff} - \mathbf{A}_{fr} \mathbf{A}_{rr}^{-1} \mathbf{A}_{rf} \right] \mathbf{x}_f = \left[\mathbf{B}_{ff} - \mathbf{A}_{fr} \mathbf{A}_{rr}^{-1} \mathbf{B}_{rf} \right] \mathbf{u}_f + \left[\mathbf{B}_{fr} - \mathbf{A}_{fr} \mathbf{A}_{rr}^{-1} \mathbf{B}_{rr} \right] \mathbf{u}_r
$$

$$
\tag{2.48}
$$

In the above, we have employed the weakly coupled approximation theory of Milne (Ref. 2.14), an approach used extensively in Chapters 4 and 5. The technique will serve us well in reducing and hence isolating the dynamics to single DoFs in some cases, hence maximizing the potential physical insight gained from such analysis. The real strength in linearization comes from the ability to derive stability properties of the dynamic motions.

Stability and control résumé

This Tour would be incomplete without a short discussion on 'stability and control derivatives' and a description of typical helicopter stability characteristics. To do this we need to introduce the helicopter model configurations we shall be working with in this book and some basic principles of building the aircraft equations of motion. The three baseline simulation configurations are described in Appendix 4B and represent the Aerospatiale (ECF) SA330 Puma, Westland Lynx and MBB (ECD) Bo105 helicopters. The Puma is a transport helicopter in the 6-ton class, the Lynx is a utility transport/anti-armour helicopter in the 4-ton class and the Bo105 is a light utility/anti-armour helicopter in the 2.5-ton class. Both the Puma and Bo105 operate in civil and military variants throughout the world; the military Lynx operates with both land and sea forces throughout the world. All three helicopters were designed in the 1960s and have been continuously improved in a series of new Marks since that time. The Bo105 and Lynx were the first hingeless rotor helicopters to enter production and service. On these aircraft, both flap and lead–lag blade motion are achieved through elastic bending, with blade pitch varied through rotations at a bearing near the blade root. On the Puma, the blade flap and lead–lag motions largely occur through articulation with the hinges close to the hub centre. The distance of the hinges from the hub centre is a critical parameter in determining the magnitude of the hub moment induced by blade flapping and lagging; the moments are approximately proportional to the hinge offset, up to values of about 10% of the blade radius. Typical values of the flap hinge offset are found between 3 and 5% of the blade radius. Hingeless rotors are often quoted as having an effective hinge offset, to describe their moment-producing capability, compared with articulated rotor helicopters. The Puma has a flap hinge offset of 3.8%, while the Lynx and Bo105 have effective offsets of about 12.5 and 14% respectively. We can expect the moment capability of the two hingeless rotor aircraft to be about three times that of the Puma. This translates into higher values of λ_β and S_β, and hence higher rotor moment derivatives with respect to all variables, not only rates and controls as described in the above analysis.

The simulation model of the three aircraft will be described in Chapter 3 and is based on the DRA *Helisim* model (Ref. 2.15). The model is generic in form, with two input files, one describing the aircraft configuration data (e.g., geometry, mass properties, aerodynamic and structural characteristics, control system parameters), the other the flight condition parameters (e.g., airspeed, climb/descent rate, sideslip and turn rate) and atmospheric conditions. The datasets for the three Helisim aircraft are located in Chapter 4, Section 4B.1, while Section 4B.2 contains charts of the stability and control derivatives. The derivatives are computed using a numerical perturbation technique applied to the full nonlinear equations of motion and are not generally derived in explicit analytic form. Chapters 3 and 4 will include some analytic formulations to illustrate the physics at work; it should be possible to gain insight into the primary aerodynamic effects for all the important derivatives in this way. The static stability derivative M_w is a good example and allows us to highlight some of the differences between fixed- and rotary-wing aircraft.

The static stability derivative M_w

In simple physical terms the derivative M_w represents the change in pitching moment about the aircraft's centre of mass when the aircraft is subjected to a perturbation in normal velocity w or, effectively, incidence. If the perturbation leads to a positive, pitch-up, moment, then M_w is positive and the aircraft is said to be statically unstable in pitch; if M_w is negative then the aircraft is statically stable. Static stability refers to the initial tendency only and the M_w effect is analogous to the spring in a simple spring/mass/damper dynamic system. In fixed-wing aircraft flight dynamics, the derivative is proportional to the distance between the aircraft's centre of mass and the overall aerodynamic centre, i.e., the point about which the resultant lift force acts when the incidence is changed. This distance metric, in normalized form referred to as the static margin, does not carry directly across to helicopters, because as the incidence changes, not only does the aerodynamic lift on the rotor change, but it also rotates (as the rotor disc tilts). So, while we can consider an effective static margin for helicopters, this is not commonly used because the parameter is very configuration dependent and is also a function of perturbation amplitude. There is another reason why the static margin concept has not been adopted in helicopter flight dynamics. Prior to the deliberate design of fixed-wing aircraft with negative static margins to improve performance, fundamental configuration and layout parameters were defined to achieve a positive static margin. Most helicopters are inherently unstable in pitch and very little can be achieved with layout and configuration parameters to change this, other than through the stabilizing effect of a large tailplane at high-speed (e.g., UH-60). When the rotor is subjected to a positive incidence change in forward flight, the advancing blade experiences a greater lift increment than does the retreating blade (see Fig. 2.25). The $90°$ phase shift in response means that the rotor disc flaps back and cones up and hence applies a positive pitching moment to the aircraft. The rotor contribution to M_w will tend to increase with forward speed; the contributions from the fuselage and horizontal stabilizer will also increase with airspeed but tend to cancel each other, leaving the rotor contribution as the primary contribution. Figure 2.26 illustrates the variation in M_w for the three baseline aircraft in forward flight. The effect of the hingeless rotors on M_w is quite striking, leading to large destabilizing moments at high speed. It is

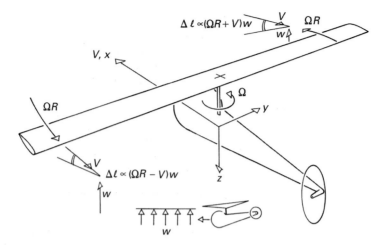

Fig. 2.25 Incidence perturbation on advancing and retreating blades during encounter with
vertical gust

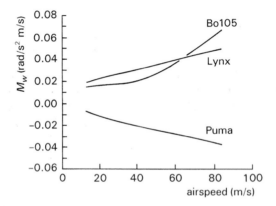

Fig. 2.26 Variation of static stability derivative, M_w, with forward speed for Bo105, Lynx
and Puma

interesting to consider the effect of this static instability on the dynamic, or longer
term, stability of the aircraft.

A standard approximation to the short-term dynamic response of a fixed-wing
aircraft can be derived by considering the coupled pitch/heave motions, assuming
that the airspeed is constant. This is a gross approximation for helicopters but can be
used to approximate high-speed flight in certain circumstances (Ref. 2.16). Figure 2.27
illustrates generalized longitudinal motion, distinguishing between pitch and incidence.
For the present, we postulate that the assumption of constant speed applies, and that the
perturbations in heave velocity w, and pitch rate q, can be described by the linearized
equations:

$$I_{yy}\dot{q} = \delta M$$

$$M_a \dot{w} = M_a U_e q + \delta Z \qquad (2.49)$$

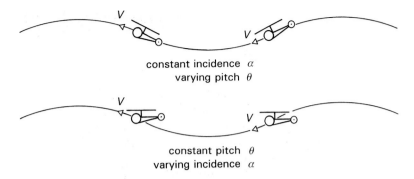

constant incidence α
varying pitch θ

constant pitch θ
varying incidence α

Fig. 2.27 Constant pitch and heave motions

where I_{yy} is the pitch moment of inertia of the helicopter about the reference axes and M_a is the mass. U_e is the trim or equilibrium forward velocity and δZ and δM are the perturbation Z force and pitching moment. Expanding the perturbed force and moment into derivative form, we can write the perturbation equations of motion in matrix form:

$$\frac{d}{dt}\begin{bmatrix} w \\ q \end{bmatrix} - \begin{bmatrix} Z_w & Z_q + U_e \\ M_w & M_q \end{bmatrix}\begin{bmatrix} w \\ q \end{bmatrix} = \begin{bmatrix} Z_{\theta 1s} & Z_{\theta 0} \\ M_{\theta 1s} & M_{\theta 0} \end{bmatrix}\begin{bmatrix} \theta_{1s} \\ \theta_0 \end{bmatrix} \tag{2.50}$$

The derivatives Z_w, M_q, etc., correspond to the linear terms in the expansion of the normal force and pitch moment, as described in eqn 2.45. It is more convenient to discuss these derivatives in semi-normalized form, and we therefore write these in eqn 2.50, and throughout the book, without any distinguishing dressings, as

$$M_w \equiv \frac{M_w}{I_{yy}}, \qquad Z_w \equiv \frac{Z_w}{M_a}, \qquad \text{etc.} \tag{2.51}$$

The solution to eqn 2.50 is given by a combination of transient and steady-state components, the former having an exponential character, with the exponents, the stability discriminants, as the solutions to the characteristic equation

$$\lambda^2 - (Z_w + M_q)\lambda + Z_w M_q - M_w(Z_q + U_e) = 0 \tag{2.52}$$

According to eqn 2.52, when the static stability derivative M_w is zero, then the pitch and heave motions are uncoupled giving two first-order transients (decay rates given by Z_w and M_q). As M_w becomes increasingly positive, the aircraft will not experience dynamic instability until the manoeuvre margin, the stiffness term in eqn 2.52, becomes zero. Long before this however, the above approximation breaks down.

One of the chief reasons why this short period approximation has a limited application range with helicopters is the strong coupling with speed variations, reflected in the speed derivatives, particularly M_u. This speed stability derivative is normally zero for fixed-wing aircraft at subsonic speeds, on account of the moments from all aerodynamic surfaces being proportional to dynamic pressure and hence perturbations tend to cancel one another. For the helicopter, the derivative M_u is significant even in the hover, again caused by differential effects on advancing and retreating blades leading to flapback; so while this positive derivative can be described as statically stable, it actually contributes to the dynamic instability of the pitch phugoid. This effect will be further explored in Chapter 4, along with the second reason why low-order approximations are less widely applicable for helicopters, namely cross-coupling. Practically all helicopter motions are coupled, but some couplings are more significant than others, in terms of their effect on the direct response on the one hand, and the degree of pilot off-axis compensation required, on the other.

Alongside the fundamentals of flapping, the rotor thrust and torque response to normal velocity changes are key rotor aeromechanics effects that need some attention on this Tour.

Rotor thrust, inflow, Z_w and vertical gust response in hover

The rotor thrust T in hover can be determined from the integration of the lift forces on the blades

$$T = \sum_{i=1}^{N_b} \int_0^R \ell(\psi, r)\, dr \tag{2.53}$$

Using eqns 2.17–2.21, the thrust coefficient in hover and vertical flight can be written as

$$C_T = \frac{a_0 s}{2} \left(\frac{\theta_0}{3} + \frac{\mu_z - \lambda_i}{2} \right) \tag{2.54}$$

Again, we have assumed that the induced downwash λ_i is constant over the rotor disc; μ_z is the normal velocity of the rotor, positive down and approximates to the aircraft velocity component w. Before we can calculate the vertical damping derivative Z_w, we need an expression for the uniform downwash. The induced rotor downwash is one of the most important individual components of helicopter flight dynamics; it can also be the most complex. The downwash, representing the discharged energy from the lifting rotor, actually takes the form of a spiralling vortex wake with velocities that vary in space and time. We shall give a more comprehensive treatment in Chapter 3, but in this introduction to the topic we make some major simplifications. Assuming that the rotor takes the form of an actuator disc (Ref. 2.17) supporting a pressure change and accelerating the air mass, the induced velocity can be derived by equating the work done by the integrated pressures with the change in air-mass momentum. The hover downwash over the rotor disc can then be written as

$$v_{i_{hover}} = \sqrt{\frac{T}{2\rho A_d}} \tag{2.55}$$

where A_d is the rotor disc area and ρ is the air density.

Or, in normalized form

$$\lambda_i = \frac{v_i}{\Omega R} = \sqrt{\left(\frac{C_T}{2} \right)} \tag{2.56}$$

The rotor thrust coefficient C_T will typically vary between 0.005 and 0.01 for helicopters in 1 g flight, depending on the tip speed, density altitude and aircraft weight. Hover downwash λ_i then varies between 0.05 and 0.07. The physical downwash is proportional to the square root of the rotor disc loading, L_d, and at sea level is given by

$$v_{i_{hover}} = 14.5\sqrt{L_d} \tag{2.57}$$

For low disc loading rotors ($L_d = 6 \, \text{lb/ft}^2$, $280 \, \text{N/m}^2$), the downwash is about 35 ft/s (10 m/s); for high disc loading rotors ($L_d = 12 \, \text{lb/ft}^2$, $560 \, \text{N/m}^2$), the downwash rises to over 50 ft/s (15 m/s).

The simple momentum considerations that led to eqn 2.55 can be extended to the energy and hence power required in the hover

$$P_i = Tv_i = \frac{T^{3/2}}{\sqrt{(2\rho A_d)}} \tag{2.58}$$

The subscript i refers to the induced power which accounts for about 70% of the power required in hover; for a 10000-lb (4540 kg) helicopter developing a downwash of 40 ft/s (typical of a Lynx), the induced power comes to nearly 730 HP (545 kW).

Equations 2.54 and 2.56 can be used to derive the heave damping derivative

$$Z_w = -\frac{\rho(\Omega R)\pi R^2}{M_a} \frac{\partial C_T}{\partial \mu_z} \tag{2.59}$$

where

$$\frac{\partial C_T}{\partial \mu_z} = \frac{2a_0 s \lambda_i}{16\lambda_i + a_0 s} \tag{2.60}$$

and hence

$$Z_w = -\frac{2a_0 A_b \rho(\Omega R)\lambda_i}{(16\lambda_i + a_0 s)M_a} \tag{2.61}$$

where A_b is the blade area and s the solidity, or ratio of blade area to disc area. For our reference Helisim Lynx configuration, the value of Z_w is about -0.33/s in hover, giving a heave motion time constant of about 3 s (rise time to 63% of steady state). This is typical of heave time constants for most helicopters in hover. With such a long time constant, the vertical response would seem more like an acceleration than a velocity type to the pilot. The response to vertical gusts, w_g, can be derived from the first-order approximation to the heave dynamics

$$\frac{dw}{dt} - Z_w w = Z_w w_g \tag{2.62}$$

The initial acceleration response to a sharp-edge vertical gust provides a useful measure of the ride qualities of the helicopter, in terms of vertical bumpiness

$$\frac{dw}{dt}_{t=0} = Z_w w_g \tag{2.63}$$

A gust of magnitude 30 ft/s (10 m/s) would therefore produce an acceleration bump in Helisim Lynx of about 0.3 g. Additional effects such as the blade flapping, downwash lag and rotor penetration will modify the response. Vertical gusts of this magnitude are rare in the hovering regime close to the ground, and, generally speaking, the low values of Z_w and the typical gust strengths make the vertical gust response in hovering flight fairly insignificant. There are some important exceptions to this general result, e.g., helicopters operating close to structures or obstacles with large downdrafts (e.g., approaching oil rigs), that make the vertical performance and handling qualities, such as power margin and heave sensitivity, particularly critical. We shall return to gust response as a special topic in Chapter 5.

Gust response in forward flight

A similar analysis can be conducted for the rotor in forward flight, leading to the following set of approximate equations for the induced downwash and heave damping; V is the flight speed and V' is the total velocity at the disc

$$v_{i\mu} = \frac{T}{2\rho A_d V'} \tag{2.64}$$

$$\frac{\partial C_T}{\partial \mu_z} = \frac{2a_0 s \mu}{8\mu + a_0 s} \tag{2.65}$$

$$\mu = \frac{V}{\Omega R} \tag{2.66}$$

$$Z_w = -\frac{\rho a_0 V A_b}{2M_a} \left(\frac{4}{8\mu + a_0 s} \right) \tag{2.67}$$

The coefficient outside the parenthesis in eqn 2.67 is the expression for the corresponding value of heave damping for a fixed-wing aircraft with wing area A_w.

$$Z_{w_{FW}} = -\frac{\rho a_0 V A_w}{2M_a} \tag{2.68}$$

The key parameter is again blade/wing loading. The factor in parenthesis in eqn 2.67 indicates that the helicopter heave damping or gust response parameter flattens off at high-speed while the fixed-wing gust sensitivity continues to increase linearly. At lower speeds, the rotary-wing factor in eqn 2.67 increases to greater than one. Typical values of lift curve slope for a helicopter blade can be as much as 50% higher than a moderate aspect-ratio aeroplane wing. It would seem therefore that all else being equal, the helicopter will be more sensitive to gusts at low-speed. In reality, typical blade loadings are considerably higher than wing loadings for the same aircraft weight; values of 100 lb/ft² (4800 N/m²) are typical for helicopters, while fixed-wing executive transports have wing loadings around 40 lb/ft² (1900 N/m²). Military jets have higher wing loadings, up to 70 lb/ft² (3350 N/m²) for an aircraft like the Harrier, but this is still quite a bit lower than typical blade loadings. Figure 2.28 shows a comparison of heave damping for our Helisim Puma helicopter ($a_0 = 6$, blade area = 144 ft² (13.4 m²)) with a similar class of fixed-wing transport ($a_0 = 4$, wing area = 350 ft² (32.6 m²)), both weighing in at 13 500 lb (6130 kg). Only the curve for the rotary-wing aircraft has been extended to zero speed, the Puma point corresponding to the value

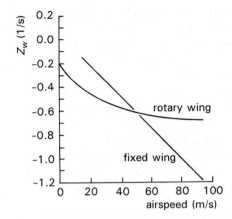

Fig. 2.28 Variation of heave damping, Z_w, with airspeed for rotary- and fixed-wing aircraft

of Z_w given by eqn 2.61. The helicopter is seen to be more sensitive to gusts below about 50 m/s (150 ft/s); above this speed, the helicopter value remains constant, while the aeroplane response continues to increase. Three points are worth developing about this result for the helicopter:

(1) The alleviation due to blade flapping is often cited as a major cause of the lower gust sensitivity of helicopters. In fact, this effect is fairly insignificant as far as the vertical gust response is concerned. The rotor coning response, which determines the way that the vertical load is transmitted to the fuselage, reaches its steady state very quickly, typically in about 100 ms. While this delay will take the edge off a truly sharp gust, in reality, the gust front is usually of ramp form, extending over several of the blade response time constants.

(2) The Z_w derivative reflects the initial response only; a full assessment of ride qualities will need to take into account the short-term transient response of the helicopter and, of course, the shape of the gust. We shall see later in Chapter 5 that there is a key relationship between gust shape and aircraft short-term response that leads to the concept of the worst case gust, when there is 'tuning' or 'resonance' between the aircraft response and the gust scale/amplitude.

(3) The third point concerns the insensitivity of the response with speed for the helicopter at higher speeds. It is not obvious why this should be the case, but the result is clearly connected with the rotation of the rotor. To explore this point further, it will help to revisit the thrust equation, thus exploiting the modelling approach to the full:

$$T = \sum_{i=1}^{N_b} \int_0^R \ell(\psi, r) \, dr$$

or

$$\frac{2C_T}{a_0 s} = \int_0^1 (\overline{U}_T^2 \theta + \overline{U}_P \overline{U}_T) \, d\overline{r} \qquad (2.69)$$

where

$$\overline{U}_T \approx \overline{r} + \mu \sin \psi, \qquad \overline{U}_P = \mu_z - \lambda_i - \mu \beta \cos \psi - \overline{r} \beta' \qquad (2.70)$$

The vertical gust response stems from the product of velocities $\overline{U}_P \, \overline{U}_T$ in eqn 2.69. It can be seen from eqn 2.70 that the forward velocity term in $\overline{U}_T$ varies one-per-rev, therefore contributing nothing to the quasi-steady hub loading. The most significant contribution to the gust response in the fuselage comes through as an N_b-per-rev vibration superimposed on the steady component represented by the derivative Z_w. The ride bumpiness of a helicopter therefore has quite a different character from that of a fixed-wing aircraft where the lift component proportional to velocity dominates the response.

Vector-differential form of equations of motion

Returning now to the general linear problem, we shall find it convenient to use the vector–matrix shorthand form of the equations of motion, written in the form

$$\frac{d\mathbf{x}}{dt} - \mathbf{A}\mathbf{x} = \mathbf{B}\mathbf{u} + \mathbf{f}(t) \qquad (2.71)$$

where

$$\mathbf{x} = \{u, w, q, \theta, v, p, \phi, r, \psi\}$$

$\mathbf{A}$ and $\mathbf{B}$ are the matrices of stability and control derivatives, and we have included a forcing function $\mathbf{f}(t)$ to represent external disturbances, e.g., gusts. Equation 2.71 is a linear differential equation with constant coefficients that has an exact solution with analytic form

$$\mathbf{x}(t) = \mathbf{Y}(t)\,\mathbf{x}(0) + \int_0^t \mathbf{Y}(t - \tau)\,(\mathbf{B}\mathbf{u} + \mathbf{f}(\tau))\,d\tau$$

$$\mathbf{Y}(t) = 0, \qquad\qquad\qquad t < 0$$

$$\mathbf{Y}(t) = \mathbf{U}\,\mathrm{diag}[\exp(\lambda_i t)]\mathbf{U}^{-1}, \quad t \geq 0 \qquad (2.72)$$

The response behaviour is uniquely determined by the principal matrix solution $\mathbf{Y}(t)$ (Ref. 2.18), which is itself derived from the eigenvalues λ_i and eigenvectors $\mathbf{u}_i$ (arranged as columns in the matrix $\mathbf{U}$) of the matrix $\mathbf{A}$. The stability of small motions about the trim condition is determined by the real parts of the eigenvalues and the complete response to controls $\mathbf{u}$ or disturbances $\mathbf{f}$ is a linear combination of the eigenvectors. Figures 2.29(a) and (b) show how the eigenvalues for the Helisim Lynx and Helisim Puma configurations vary with speed from hover to 160 knots; at the higher speeds, the conventional fixed-wing parlance for naming the modes associated with the eigenvalues is appropriate. The pitch instability at high speed for the hingeless rotor Helisim Lynx has already been discussed in terms of the loss of manoeuvre stability. At lower speeds the modes change character, until at the hover they take on shapes peculiar to the helicopter, e.g., heave/yaw oscillation, pitch/roll pendulum mode. The heave/yaw mode tends to be coupled, due to the fuselage yaw reaction to changes in rotor torque, induced by perturbations in the rotor heave/inflow velocity. The eigenvectors represent the mode shapes, or the ratio of the response contributions in the various DoFs. The

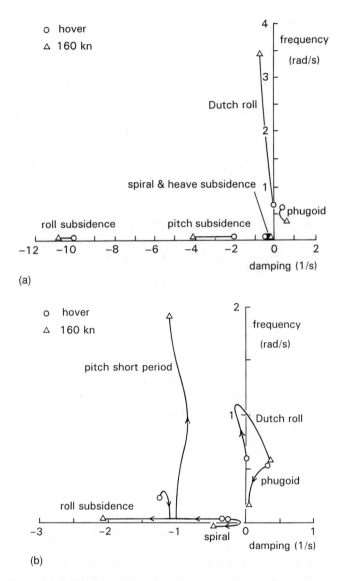

Fig. 2.29 (a) Variation of Lynx eigenvalues with forward speed; (b) variation of Puma eigenvalues with forward speed

modes are linearly independent, meaning that no one can be made up as a collection of the others. If the initial conditions, control inputs or gust disturbance have their energy distributed throughout the DoFs with the same ratio as a particular eigenvector, then the response will be restricted to that mode only. More discussion on the physics of the modes can be found in Chapter 4.

The key value of the linearized equations of motion is in the analysis of stability; they also form the basic model for control system design. Both uses draw on the considerable range of mathematical techniques developed for linear systems analysis. We shall return to these later in Chapters 4 and 5, but we need to say a little more about the two inherently nonlinear problems of flight mechanics – trim and response.

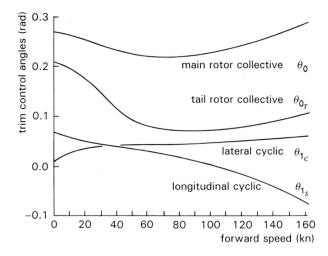

Fig. 2.30 Variation of trim control angles with forward speed for Puma

The former is obtained as the solution to the algebraic eqn 2.2 and generally takes the form of the controls required to hold a steady flight condition. The general form of control variations with forward speed is illustrated in Fig. 2.30. The longitudinal cyclic moves forward as speed increases to counteract the flapback caused by forward speed effects (M_u effect). The lateral cyclic has to compensate for the rolling moment due to the tail rotor thrust and also the lateral flapping induced in response to coning and longitudinal variations in rotor inflow. The collective follows the shape of the power required, decreasing to the minimum power speed at around 70 knots then increasing again sharply at higher speeds. The tail rotor collective follows the general shape of the main rotor collective; at high-speed the pedal required decreases as some of the anti-torque yawing moment is typically produced by the vertical stabilizer.

While it is true that the response problem is inherently nonlinear, it is also true that for small perturbations, the linearized equations developed for stability analysis can be used to predict the dynamic behaviour. Figure 2.31 illustrates and compares the pitch response of the Helisim Lynx fitted with a standard and soft rotor as a function of control input size; the response is normalized by the input size to indicate the degree of nonlinearity present. Also shown in the figure is the normal acceleration response; clearly, for the larger inputs the assumptions of constant speed implicit in any linearization would break down for the standard stiffer rotor. Also the rotor thrust would have changed significantly in the manoeuvre and, together with the larger speed excursions for the stiffer rotor, produce the nonlinear response shown.

Validation

How well a theoretical simulation needs to model the helicopter behaviour depends very much on the application; in the simulation world the measure of quality is described as the fidelity or validation level. Fidelity is normally judged by comparison with test data, both model and full scale. The validation process can be described in terms of two kinds of fidelity–functional and physical (Ref. 2.19), defined as follows:

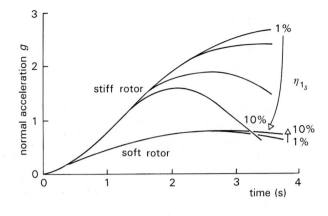

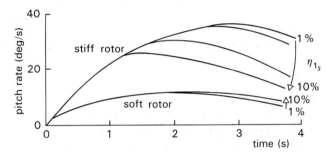

Fig. 2.31 Nonlinear pitch response for Lynx at 100 knots

Functional fidelity is the level of fidelity of the overall model to achieve compliance
with some functional requirement, e.g., for our application, can the model be used
to predict flying qualities parameters?

Physical fidelity is the level of fidelity of the individual modelling assumptions in the
model components, in terms of their ability to represent the underlying physics,
e.g., does the rotor aerodynamic inflow formulation capture the fluid mechanics of
the wake correctly?

It is convenient and also useful to distinguish between these two approaches because
they focus attention on the two ends of the problem – have we modelled the physics
correctly and does the pilot perceive that the simulation 'feels' right? It might be
imagined that the one would follow from the other and while this is true to an extent,
it is also true that simulation models will continue to be characterized by a collection
of aerodynamic and structural approximations, patched together and each correct over
a limited range, for the foreseeable future. It is also something of a paradox that
the conceptual product of complexity and physical understanding can effectively be
constant in simulation. The more complex the model becomes, then while the model
fidelity may be increasing, the ability to interpret cause and effect and hence gain
physical understanding of the model behaviour diminishes. Against this stands the
argument that, in general, only through adding complexity can fidelity be improved.
A general rule of thumb is that the model needs to be only as complex as the fidelity
requirements dictate; improvements beyond this are generally not cost-effective. The
problem is that we typically do not know how far to go at the initial stages of a model

development, and we need to be guided by the results of validation studies reported in the literature. The last few years have seen a surge of activity in this research area, with the techniques of system identification underpinning practically all the progress (Refs 2.20–2.22). System identification is essentially a process of reconstructing a simulation model structure and associated parameters from experimental data. The techniques range from simple curve fitting to complex statistical error analysis, but have been used in aeronautics in various guises from the early days of data analysis (see the work of Shinbrot in Ref. 2.23). The helicopter presents special problems to system identification, but these are nowadays fairly well understood, if not always accounted for, and recent experience has made these techniques much more accessible to the helicopter flight dynamicist.

An example illustrating the essence of system identification can be drawn from the roll response dynamics described earlier in this chapter; if we assume the first-order model structure, then the equation of motion and measurement equation take the form

$$\dot{p} - L_p p = L_{\theta_{1c}} \theta_{1c} + \varepsilon_p$$
$$p = f(p_m) + \varepsilon_m \tag{2.73}$$

The second equation is included to show that in most cases, we shall be considering problems where the variable or state of interest is not the same as that measured; there will generally be some measurement error function ε_m and some calibration function f involved. Also, the equation of motion will not fully model the situation and we introduce the process error function ε_p. Ironically, it is the estimation of the characteristics of these error or noise functions that has motivated the development of a significant amount of the system identification methodology.

The solution for roll rate can be written in either a form suitable for forward (numerical) integration

$$p = p_0 + \int_0^t (L_p p(\tau) + L_{\theta_{1c}} \theta_{1c}(\tau)) \, d\tau \tag{2.74}$$

or an analytic form

$$p = p_0 e^{L_p t} + \int_0^t e^{L_p(t-\tau)} L_{\theta_{1c}} \theta_{1c}(\tau) \, d\tau \tag{2.75}$$

The identification problem associated with eqns 2.73 becomes, 'from flight test measurements of roll rate response to a measured lateral cyclic input, estimate values of the damping and control sensitivity derivatives L_p and $L_{\theta_{1c}}$'. In starting at this point, we are actually skipping over two of the three subprocesses of system identification – state estimation and model structure estimation, processes that aim to quantify better the measurement and process noise. There are two general approaches to solving the identification problem – equation error and output error. With the equation error method, we work with the first equation of 2.73, but we need measurements of both roll rate and roll acceleration, and rewrite the equation in the form

$$\dot{p}_e = L_p p_m + L_{\theta_{1c}} \theta_{1c_m} \tag{2.76}$$

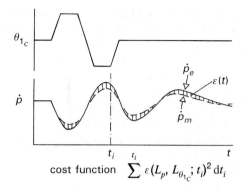

cost function $\sum \varepsilon (L_p, L_{\theta_{1c}}; t_i)^2 \, dt_i$

Fig. 2.32 Equation error identification process

Subscripts m and e denote measurements and estimated states respectively. The identification process now involves achieving the best fit between the estimated roll acceleration $\dot{p}_e$ from eqn 2.76 and the measured roll acceleration $\dot{p}_m$, varying the parameters L_p and $L_{\theta_{1c}}$ to achieve the fit (Fig. 2.32). Equation 2.76 will yield one-fit equation for each measurement point, and hence with n measurement times we have two unknowns and n equations – the classic overdetermined problem. In matrix form, the n equations can be combined in the form

$$\mathbf{x} = \mathbf{By} + \boldsymbol{\varepsilon} \tag{2.77}$$

where $\mathbf{x}$ is the vector of acceleration measurements, $\mathbf{B}$ is the ($n \times 2$) matrix of roll rate and lateral cyclic measurements and $\mathbf{y}$ is the vector of unknown derivatives L_p and $L_{\theta_{1c}}$; $\boldsymbol{\varepsilon}$ is the error vector function. Equation 2.77 cannot be inverted in the conventional manner because the matrix $\mathbf{B}$ is not square. However, a pseudo-inverse can be defined that will provide the so-called least-squares solution to the fitting process, i.e., the error function is minimized so that the sum of the squares of the error between, measured and estimated acceleration is minimized over a defined time interval. The least-squares solution is given by

$$\mathbf{y} = (\mathbf{B}^{\mathrm{T}}\mathbf{B})^{-1}\mathbf{B}^{\mathrm{T}}\mathbf{x} \tag{2.78}$$

Provided that the errors are randomly distributed with a normal distribution and zero mean, the derivatives so estimated from eqn 2.78 will be unbiased and have high confidence factors.

The second approach to system identification is the output error method, where the starting equation is the solution or 'output' of the equation of motion. In the present example, either the analytic (eqn 2.74) or numerical (eqn 2.73) solution can be used; it is usually more convenient to work with the latter, giving the estimated roll rate in this case as

$$p_e = p_{0m} + \int_0^t (L_p p_m(\tau) + L_{\theta_{1c}} \theta_{1c_m}(\tau)) \, d\tau \tag{2.79}$$

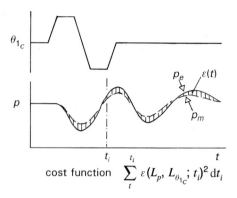

Fig. 2.33 Output error identification process

The error function is then formed from the difference between the measured and estimated roll rate, which can once again be minimized in a least-squares sense across the time history to yield the best estimates of the damping and control derivatives (Fig. 2.33).

Provided that the model structures are correct, the processes we have described will always yield 'good' derivative estimates in the absence of 'noise', assuming that enough measurements are available to cover the frequencies of interest; in fact, the two methods are equivalent in this simple case. Most identification work with simulated data falls into this category, and new variants of the two basic methods are often tested with simulation data prior to being applied to test data. Without contamination with a realistic level of noise, simulation data can give a very misleading impression of the robustness level of system identification methods applied to helicopters. Expanding on the above, we can classify noise into two sources for the purposes of the discussion:

(1) measurement noise, appearing on the measured signals;
(2) process noise, appearing on the response outputs, reflecting unmodelled effects.

It can be shown (Ref. 2.24, Klein) that results from equation error methods are susceptible to measurement noise, while those from output error analysis suffer from process noise. Both can go terribly wrong if the error sources are deterministic and cannot therefore be modelled as random noise. An approach that purports to account for both error sources is the so-called maximum likelihood technique, whereby the output error method is used in conjunction with a filtering process, that calculates the error functions iteratively with the model parameters.

Identifying stability and control derivatives from flight test data can be used to provide accurate linear models for control law design or in the estimation of handling qualities parameters. Our principal interest in this Tour is the application to simulation model validation. How can we use the estimated parameters to quantify the levels of modelling fidelity? The difficulty is that the estimated parameters are made up of contributions from many different elements, e.g., main rotor and empennage, and the process of isolating the source of a deficient force or moment prediction is not obvious. Two approaches to tackling this problem are described in Ref. 2.25; one where the model parameters are physically based and where the modelling element of interest is isolated from the other components through prescribed dynamics – the so-called open-loop or constrained method. The second method involves establishing the relationship between the derivatives and the physical rotorcraft parameters, hence enabling the degree of distortion of the physical parameters required to match the test data. Both these methods are useful and have been used in several different applications over the last few years.

Large parameter distortions most commonly result from one of two sources in helicopter flight dynamics, both related to model structure deficiencies – missing DoFs or missing nonlinearities, or a combination of both. A certain degree of model structure mismatch will always be present and will be reflected in the confidence values in the estimated parameters. Large errors can, however, lead to unrealistic values of some parameters that are effectively being used to compensate for the missing parts. Knowing when this is happening in a particular application is part of the 'art' of system identification. One of the keys to success involves designing an appropriate test input that ensures that the model structure of interest remains valid in terms of frequency and amplitude, bringing us back to the two characteristic dimensions of modelling. A relatively new technique that has considerable potential in this area is the method of *inverse simulation*.

Inverse simulation

The process of validation and fidelity assessment is concerned ultimately with understanding the accuracy and range of application of the various assumptions distributed throughout the modelling. At the heart of this lies the prediction of the external forces and moments, particularly the aerodynamic loads. One of the problems with direct or forward simulation, where the simulation model is driven by prescribed control inputs and the motion time histories derived from the integration of the forces and moments, is that the comparisons of simulation and flight can very quickly depart with even the smallest modelling errors. The value of the comparison in providing validation insight then becomes very dubious, as the simulation and flight are soon engaged in very different manoeuvres. The concept behind inverse simulation is to prescribe, using flight test data, the motion of the helicopter in the simulation and hence derive the required forces and moments for comparison with those predicted by theory. One form of the process can be conceived in closed-loop form with the error between the model and flight forming the function to be minimized by a feedback controller (Fig. 2.34). If we assume that the model structure is linear with n DoFs $\mathbf{x}$, for which we also have flight

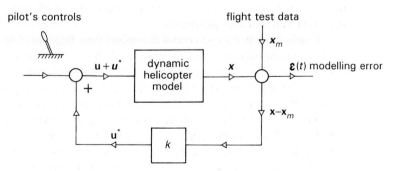

Fig. 2.34 Inverse simulation as a feedback process

measurements $\mathbf{x}_m$, then the process can be written as

$$\frac{d\mathbf{x}}{dt} - \mathbf{Ax} = \mathbf{B}(\mathbf{u} + \mathbf{u}^*) \qquad (2.80)$$

$$\mathbf{u}^* = \mathbf{k}(\mathbf{x} - \mathbf{x}_m) \qquad (2.81)$$

The modelling errors have been embodied into a dummy control variable $\mathbf{u}^*$ in eqns 2.80 and 2.81. The gain matrix $\mathbf{k}$ can be determined using a variety of minimization algorithms to achieve the optimum match between flight and theory; the example given in Ref. 2.26 uses the conventional quadratic least-squares performance index

$$P_I = \int (\mathbf{x} - \mathbf{x}_m)^{\mathrm{T}} \mathbf{R}(\mathbf{x} - \mathbf{x}_m)\, dt \qquad (2.82)$$

The elements of the weighting matrix $\mathbf{R}$ can be selected to achieve distributed fits over the different motion variables. The method is actually a special form of system identification, with the unmodelled effects being estimated as effective controls. The latter can then be converted into residual forces and moments that can be analysed to describe the unmodelled loads or DoFs. A special form of the inverse simulation method that has received greatest attention (Refs 2.27, 2.28) corresponds to the case where the feedback control in Fig. 2.34 and eqn 2.81 has infinite gain. Effectively, four of the helicopter's DoFs can now be prescribed exactly, and the remaining DoFs and the four controls are then estimated. The technique was originally developed to provide an assessment tool for flying qualities; the kinematics of MTEs could be prescribed and the ability of different aircraft configurations to fly through the manoeuvres compared (Ref. 2.29). Later, the technique was used to support validation work and has now become fairly well established (Ref. 2.26).

Modelling review

Assuming that the reader has made it this far, he or she may feel somewhat daunted at the scale of the modelling task described on this Tour; if so, then Chapters 3, 4 and 5 will offer little respite, as the subject becomes even deeper and broader. If, on the other hand, the reader is motivated by this facet of flight dynamics, then the later chapters should bring further delights, as well as the tools and knowledge that are essential for practising the flight dynamics discipline. The modelling activity has been conveniently characterized in terms of frequency and amplitude; we refer back to Fig. 2.14 for setting the framework and highlight again the merging with the loads and vibration disciplines. Later chapters will discuss this overlap in greater detail, emphasizing that while there is a conceptual boundary defined by the pilot-controllable frequencies, in practice the problems actually begin to overlap at the edges of the flight envelope and where high gain active control is employed.

Much of the ground covered in this part of the Tour has utilized analytic approximations to aircraft and rotor dynamics; this approach is always required to provide physical insight and will be employed to a great extent in the later modelling chapters. The general approach will be to search among the coupled-interacting components for combinations of motion that are, in some sense, weakly coupled; if they can be found, there lies the key to analytic approximations. However, we cannot escape the complexity of both the aerodynamic and structural modelling, and Chapter 3 will formulate expressions for the loads from first principles; analytic approximations can then be validated against the more comprehensive theories to establish their range of application. With today's computing performance and new functionality, the approach to modelling is developing rapidly. For example, there are now far more papers published that compare numerical rather than analytic results from comprehensive models with test data. Analytic approximations tend, nowadays, to be a rarity. The comprehensive models are expected to be more accurate and have higher fidelity, but the cost is sometimes the loss of physical understanding, and the author is particularly sensitive to this, having lived through the transition from a previous era, characterized by analytic modelling, to the present, more numerical one. Chapters 3 and 4 will reflect this and will be packed full with the author's well-established prejudices.

We have touched on the vast topic of validation and the question of how good a model has to be. This topic will be revisited in Chapter 5; the answer is actually quite simple – it depends! The author likens the question, 'How good is your model?' to 'what's the weather like on Earth?' It depends on where you are and the time of year, etc. So while the initial, somewhat defensive answer may be simple, to address the question seriously is a major task. This book will take a snapshot of the scene in 1995, but things are moving fast in this field, and new validation criteria along with test data from individual components, all matched to more comprehensive models, are likely to change the 'weather' considerably within the next 5 years.

In the modelling of helicopter flight dynamics, of principal concern are the flying qualities. The last 10 years has seen extensive development of quality criteria, and the accurate prediction of the associated handling and ride qualities parameters is now at the forefront of all functional validation which conveniently leads us to the next stage of the Tour.

2.4 FLYING QUALITIES

In this book we loosely divide flying qualities into two categories – handling qualities, reflecting the aircraft's behaviour in response to pilot controls, and ride qualities, reflecting the response to external disturbances. Agreement on definitions is not widespread and we shall return to some of the debating points later in Chapter 6. A most useful definition of handling qualities has been provided by Cooper and Harper (Ref 2.30) as 'those qualities or characteristics of an aircraft that govern the ease and precision with which a pilot is able to perform the tasks required in support of an aircraft role'. We shall expand on this definition later, but as a starting point it has stood the test of time and is in widespread use today. It is worth elaborating on the key words in this definition. Quantifying an aircraft's characteristics or its internal attributes, while complex and selective, can be achieved on a rational and systematic basis; after all, an aircraft's response is largely predictable and repeatable. Defining a useful task or mission is also relatively straightforward, although we have to be very careful to recognize the importance of the task performance levels required. Quantifying the pilot's abilities is considerably more difficult and elusive. To this end, the Cooper–Harper pilot subjective rating scale (Ref. 2.30) was introduced and has now achieved almost universal acceptance as a measure of handling qualities.

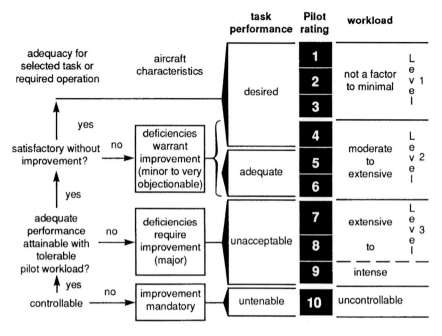

Fig. 2.35 The Cooper – Harper handling qualities rating scale – summarized form

Pilot opinion

The scale, shown in summarized form in Fig. 2.35, is divided into three 'levels'; the crucial discriminators are task performance and pilot workload. Pilot handling quality ratings (HQR) are given for a particular aircraft configuration, flying a particular task under particular environmental conditions; these points cannot be overemphasized. Some projection from the 'simulated' experimental test situation to the operational situation will be required of the test pilot, but extrapolation of handling qualities from known to new conditions is generally unacceptable, which explains why compliance testing needs to be comprehensive and can be so time consuming.

The rating scale is structured as a decision tree; requiring the pilot to arrive at his or her ratings following a sequence of questions/answers, thoughtful considerations and, possibly, dialogue with the test engineer. A Level 1 aircraft is satisfactory without improvement, and if this could be achieved throughout the OFE and for all mission tasks, then there should never be complaints concerning the piloting task. In practice, there has probably never been an aircraft this good, and Level 2 or even, on occasions, Level 3 characteristics have been features of operational aircraft. With a Level 2 aircraft, the pilot can still achieve adequate performance, but has to use moderate to extensive compensation and, therefore, workload. At the extreme of Level 2 (HQR 6) the mission is still flyable, but the pilot has little spare capacity for other duties and will not be able to sustain the flying for extended periods without the dangers that come from fatigue, i.e., the attendant safety hazards that follow from the increased risk of pilot error. These are the penalties of poor flying qualities. Beyond Level 2, the unacceptable should never be allowed in normal operational states, but this category is needed to describe the behaviour in emergency conditions associated with flight in severely degraded atmospheric conditions or following the loss of critical flight systems.

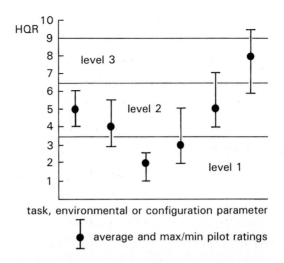

Fig. 2.36 Presentation of pilot handling qualities ratings showing variation with task, environmental or configuration parameter

The dilemma is that while performance targets can be defined on a mission requirement basis, the workload, and hence rating, can vary from pilot to pilot. The need for several opinions, to overcome the problem of pilot variability, increases the duration of a test programme and brings with it the need to resolve any strong differences of opinion. Pilot ratings will then typically be displayed as a mean and range as in Fig. 2.36. The range display is vital, for it shows not only the variability, but also whether the opinions cross the levels. Half ratings are allowed, except the 3.5 and 6.5 points; these points are not available when the pilot follows the HQR decision sequence properly (Fig. 2.35).

Quantifying quality objectively
While pilot-subjective opinion will always be the deciding factor, quantitative criteria are needed as design targets and to enable compliance demonstration throughout the design and certification phases. The most comprehensive set of requirements in existence is provided by the US Army's Aeronautical Design Standard for handling qualities – ADS-33 (Ref. 2.1), which will be referred to regularly throughout this text, particularly in Chapters 6, 7 and 8. During the initiation of these requirements, it was recognized that new criteria were urgently needed but could only ever be as valid as the underlying database from which they were developed. Hoh (Ref. 2.4), the principal author of ADS-33, commented that key questions needed to be asked of any existing test data.

(1) Were the data generated with similar manoeuvre precision and aggressiveness required in current and future operational missions?
(2) Were the data generated with outside visual cues and atmospheric disturbances relevant to and consistent with current operations?

Most of the existing data at that time (early 1980s) were eliminated when exposed to the scrutiny of these questions, and the facilities of several NATO countries were harnessed to support the development of a new and more appropriate database, notably

Canada (NAE, Ottawa), Germany (DLR, Braunschweig), UK (DRA Bedford, then RAE) and, of course, the United States itself, with the activity orchestrated by the US Army Aeroflightdynamics Directorate at the Ames Research Center.

The criteria in ADS-33 have been validated in development and any gaps represent areas where data are sparse or non-existent. To quote from ADS-33:

> *The requirements of this specification shall be applied in order to assure that no limitations on flight safety or on the capability to perform intended missions will result from deficiencies in flying qualities.*

For flight within the OFE, Level 1 handling qualities are required. Three innovations in ADS-33 requiring specification to ensure Level 1 handling are the *mission task element* (MTE), the *usable cue environment* (UCE) and the *response type* (e.g., rate command, attitude hold – RCAH). These can be seen to relate directly to three of the reference points discussed earlier in this chapter. Referring to Fig. 2.13 we see how, for slalom and sidestep MTEs, rate command response types are deemed adequate to provide Level 1 pitch or roll handling qualities for flight in conditions of a UCE 1. For low-speed operations however, the response type will need upgrading to attitude command, attitude hold (ACAH) for flight in the degraded visual environment of a UCE 2, while a translational rate command with position hold (TRCPH) is needed for flight in the IMC – like UCE 3. The task, the environment and the aircraft dynamics therefore interact to determine the flying qualities.

Frequency and amplitude – exposing the natural dimensions

At a deeper level the response types themselves can be classified further in terms of their frequency and amplitude characteristics, a perspective that we found useful in the modelling discipline described in the previous stop on this Tour. Figure 2.37 illustrates the structure, with the response classified into three levels of amplitude (small, moderate and large) and three frequency ranges, corresponding to long-, mid- and short-term behaviour. The zero frequency motion is identified as the trim line. In recognition of the multitude of cross-couplings inherent to helicopters, we have added the third dimension on the figure; to date, the criteria for cross-coupling requirements are considerably more immature than for the direct response. The boundary curve in the figure indicates the limits to practical flying, with higher frequency attitude and flight path motions restricted to small amplitude, and large amplitude motions restricted to

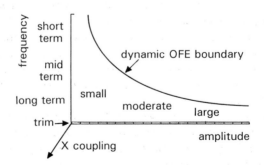

Fig. 2.37 Frequency and amplitude characterization of aircraft response

the lower frequency range. This representation will provide a convenient structure for developing quantitative response criteria later in Chapter 6.

Typical helicopter characteristics can now be discussed within the framework of this response-type classification. An unstable, low-frequency oscillation involving changes in speed and height characterizes the mid–long term, small amplitude response and stability of helicopter pitch motion. This mode can take the form of a mildly unstable pendulum-type motion in the hover, to a rapidly divergent 'phugoid' oscillation at high speed. Aircraft design and configuration parameters, e.g., cg location, rotor type and tailplane design, can have a marked effect on the stability of this mode in forward flight. At forward cg extremes, the oscillation can stabilize at moderate speeds, whereas with aft cg loadings for some configurations, particularly hingeless rotors or helicopters with small horizontal tailplanes, the oscillation can 'split' into two aperiodic divergences at high speed, with time to double amplitude less than 1 s in severe cases. The mode differs from the fixed-wing 'phugoid' in that speed changes during the climbs and dives induce pitching moments, which cause significant variations in fuselage and rotor incidence and thrust.

Stability – early surprises compared with aeroplanes

In the early days of helicopter testing, these differences were often a surprise to the fixed-wing test pilots. Research into helicopter flying qualities at the Royal Aircraft Establishment goes back to the 1940s when engineers and pilots were getting to grips with the theory and practice respectively. In these early days of helicopter research, one of the key concerns was stability, or rather the lack of it. Stewart and Zbrozek (Ref. 2.31) describe a loss-of-control incident on an S-51 helicopter at RAE in 1948. Quoting from the pilot's report in the reference:

> *When the observer said he was ready with his auto-observer, I pushed the stick forward about six inches and returned it quickly to its original position. The aircraft continued in straight and level flight for approximately three to five seconds before it slowly started a phugoid motion, with the nose dropping away slightly in the first instance. Each oscillation became greater, i.e. the dive and climb becoming steeper with every oscillation; it was accompanied by roll, at the bottom of the dive during the 'pull-out', it had maximum bank to the right.*
>
> *The observer intimated recovery action to be taken at the end of the third oscillation; as the aircraft came over the top from the climb to go into the dive, I eased the stick forward to help it over the top. The stick felt light and there appeared to be no additional response from the aircraft; as the aircraft commenced diving again, I eased the stick back to where I considered I had pushed it from, thinking that I would let the speed build up somewhat before easing the stick further back to pull out of the dive. Quite a steep dive developed and just as I was about to ease the stick back, probably three seconds after the previous stick movement, there was suddenly severe vibration throughout what seemed to be the whole machine. From then on until recovery was effected (I estimate five to ten seconds later), I have no clear recollection of what took place. I think that immediately after the vibration, the aircraft flicked sharply to the left and nearly on to its back; it then fell more or less the right way up but the fuselage was spinning, I think, to the right. It fell into a steep dive and repeated the performance again; I selected autorotation quite early during the proceedings. Once I saw the rotor rpm at 140, and later at 250. There*

Fig. 2.38 Fuselage failure on Sikorsky S-51 (Ref. 2.31)

were moments when the stick was very light and others when it was extremely heavy.
The machine, I think, did three of these manoeuvres; it seemed to want to recover
during the second dive but it actually responded to the controls during the third dive.
Height when straight and level was 400 feet above sea level (height loss 800 feet).
The aircraft responded normally to the controls when under control again; I flew
back to the airfield and landed.

The pilot had excited the phugoid mode with a longitudinal cyclic pulse; recovery
action was initiated at the end of the third oscillation, the aircraft increased speed
in a dive and during the pull-out the blades hit the droop stop, and eventually the
fuselage, causing a rapid uncontrollable rolling motion. The resulting erratic motions,
during which the pilot became disoriented, eventually settled down and the aircraft
was flown back to RAE and landed safely. The 'auto-observer' recorded a peak normal
acceleration of more than 4 g during the manoeuvre, causing severe buckling to occur
in the rear fuselage (Fig. 2.38). Two of the conclusions of the analysis of this incident
were

(1) '... large rapid movements of the controls are to be avoided, particularly at high
    speed'.
(2) 'some form of flight testing technique should be devised whereby the susceptibility of
    a helicopter to this trouble should be ascertained in the prototype stage'.

These conclusions are as relevant today as they were in the early days of helicopter
flight testing; the 'trouble' noted above is still a feature of unaugmented helicopters.
Today, however, there exist flying qualities criteria that define the boundaries of

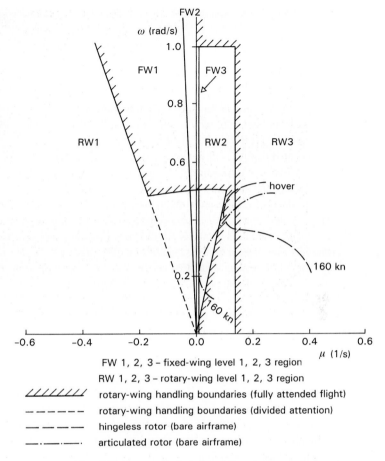

FW 1, 2, 3 – fixed-wing level 1, 2, 3 region

RW 1, 2, 3 – rotary-wing level 1, 2, 3 region

/////// rotary-wing handling boundaries (fully attended flight)

— — — — — rotary-wing handling boundaries (divided attention)

— —— —— — hingeless rotor (bare airframe)

—·——·——· articulated rotor (bare airframe)

Fig. 2.39 Long period pitch stability characteristics

acceptable mid-term pitch characteristics. Figure 2.39 illustrates the frequency/ damping requirements set down in ADS-33; the Level 1/2 and 2/3 boundaries are shown for both helicopters and fixed-wing aircraft (Ref. 2.3) for 'fully attended'[*] flight. Also included are the loci of characteristics for the two baseline simulation configurations, Helisim Lynx and Helisim Puma, in bare airframe or unstabilized configurations. Several points can be drawn out of this figure. First, there is a range where characteristics that are acceptable as Level 1 for helicopters are classified as Level 3 for fixed-wing aircraft. Secondly, for most of their flight envelopes, our two Helisim aircraft will not even meet the Level 3 requirements of the fixed-wing criteria. The fact is that it is impossible to build helicopters that, without augmentation, meet the fixed-wing standards; earlier in this chapter we discussed one of the reasons for this concerning the positive stability derivative M_w. But this is not a good reason for degrading the boundaries for safe flight. On the contrary, the boundaries in Fig. 2.39 are defined by flight results, which implies that rotary-wing pilots are willing to accept much less

[*] Pilot can devote full attention to attitude and flight path control.

than their fixed-wing counterparts. Hoh, in Ref. 2.4, has suggested two reasons for this:

(1) 'helicopter pilots are trained to cope with, and expect as normal, severe instabilities and cross-axis coupling;' and
(2) 'the tasks used in the evaluations were not particularly demanding'.

The two reasons go together and helicopters could not be used safely for anything but gentle tasks in benign conditions until feedback autostabilization could be designed and built to suppress the naturally divergent tendencies.

Included in Fig. 2.39 is the helicopter boundary for 'divided attention'[*] operations; this eliminates all unstable machines by requiring a damping ratio of 0.35. Thus, helicopters that need to operate in poor weather or where the pilot has to release the controls, or divert his attention to carry out a secondary task, need some form of artificial stabilization. This conclusion applies to both military and civil operations, the increased emphasis on safety for the latter providing an interesting counterpoint; criteria for civil helicopter flying qualities will be discussed further in Chapter 6.

The S-51 incident described above illustrates two important consequences of flying qualities deficiencies – that pilot disorientation and aircraft strength are the limiting factors, i.e., the things that eventually 'give', and can therefore terminate the situation following loss of control. This is the key to understanding that good flying qualities are 'mission critical'.

Pilot-in-the-loop control; attacking a manoeuvre

A pilot's most immediate impressions of a helicopter's flying qualities are likely to be formed as he or she attempts to maintain attitude and position in the hover, and later as the pilot manoeuvres and accelerates into forward flight. Here, the qualities of most interest are not the mid–long-term stability characteristics, but more the small–moderate–large amplitude, short-term response to control inputs (see Fig. 2.37). Consider the kinematics of a manoeuvre to change aircraft attitude. This may correspond, for example, to the initial phase of an acceleration from the hover (pitch) or a bank manoeuvre to turn in forward flight (roll). The so-called task portrait sketches in Fig. 2.40 illustrate the variations in pilot's control inputs (a), the attitudes (b) and rates (c) and include the manoeuvre (phase plane) portrait (d) and task signature diagrams ((e) and (f)), corresponding to three different pilot control strategies. The example assumes a simple rate response type. Case 1 corresponds to the pilot applying maximum control input as rapidly as possible and stabilizing out with an attitude change. Case 2 corresponds to the pilot manoeuvring more gently to achieve the same attitude. Case 3 corresponds to the pilot applying a much sharper maximum-pulse input, achieving much the same rate as in case 2 but settling to a smaller final attitude. For the third case, the input is so sharp that the aircraft does not have time to reach its steady-state rate response. The three cases are distinguished by the degree of aggressiveness and the size of the pilot input, i.e., by different frequency and amplitude content. The task signature diagrams (e) and (f) are constructed by computing the peak rate, p_{pk}, and associated attitude change $\Delta\phi$ for the different manoeuvres; each represents a point on the diagram. The ratio of peak rate to attitude change, shown in Fig 2.40(f), is a key parameter. Designated the 'quickness' parameter in ADS-33, this ratio has a maximum achievable value for a given attitude change. For large manoeuvres, the limit is

---

[*] Pilot required to perform non-control-related sidetasks for a moderate period of time.

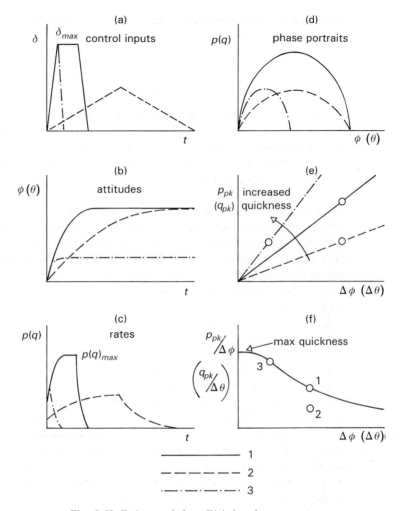

Fig. 2.40 Task portrait for roll/pitch and stop manoeuvre

naturally set by the maximum achievable rate or the attitude control power, $p(q)_{max}$; case 1 represents an example of such a situation. The quickness is a frequency measure and, for small amplitudes, represents the maximum 'closed-loop' frequency achievable from the aircraft. It is therefore, on the one hand, a measure of the inherent manoeuvre performance or agility of the aircraft and, on the other, a handling qualities parameter. If the maximum achievable quickness is too small, then the pilot may complain that the aircraft is too sluggish for tracking-type tasks; if the quickness is too high, then the pilot may complain of jerkiness or oversensitivity.

Bandwidth – a parameter for all seasons?
For the small amplitude, higher frequency end of the response spectrum, two classic measures of quality – the step response character and low-order-equivalent system (LOES) response – have proved deficient for capturing the important features that relate to tracking and pursuit-type tasks in helicopters. The equivalent systems approach adopted in the fixed-wing community has many attractions, but the rotorcraft's

non-classical response types really make the LOES a non-starter in most cases. Also, the detailed shape of the step response function appears to be sensitive to small imperfections in the control input shape and measurement inaccuracies. Strictly, of course, the small amplitude tracking behaviour should have little to do with the step response and much more to do with amplitude and phase at high frequency. Nevertheless, the direction taken by ADS-33, in this area, was clarified only after considerable debate and effort, and it is probably fair to say that there is still some controversy associated with the adoption of the so-called bandwidth criteria.

For simple response types, maximum quickness is actually a close approximation to this more fundamental handling qualities parameter – *bandwidth* (Ref. 2.32). This parameter will be discussed in more detail in Chapter 6, but some elaboration at this point is worthwhile. The bandwidth is that frequency beyond which closed-loop stability is threatened. That may seem a long step from the preceding discussion, and some additional exposure is necessary. For any closed-loop tracking task, the natural delays in the pilot's perceptual pathways, neuro-muscular and psycho-motor systems (Ref. 2.33), give rise to increasing control problems as the disturbance frequency increases. Without the application of pilot control lead, the closed-loop pilot/aircraft system will gradually lose stability as the pilot gain or disturbing frequency increases. The point of instability is commonly referred to as the crossover frequency and the bandwidth frequency corresponds to some lower value that provides an adequate stability margin. In practice, this is defined as the highest frequency at which the pilot can double his gain or allow a 135° phase lag between control input and aircraft attitude response without causing instability. The higher the bandwidth, the larger will be the aircraft's safety margin in high gain tracking tasks, but just as we have implied a possible upper limit on the quickness, so bandwidth may be limited by similar overresponsiveness.

We have introduced some new flying qualities parlance above, e.g., crossover, perceptual pathways, gain and lead/lag, and the reader will need to carry these concepts forward to later chapters for elaboration. The whole question of short-term attitude control has been discussed at some length because of its critical importance to any flying task; changing attitude tilts the rotor thrust vector and reorients the aircraft and hence the flight path vector. It is not surprising that handling criteria are most substantially developed on this topic. For the case of small amplitude, high-frequency pitch motions (see Fig. 2.37), bandwidth criteria have been developed for both fixed- and rotary-wing aircraft. For both types of aircraft the criteria are displayed in two-parameter form with the phase delay parameter, τ_p. Phase delay relates to the rate of change of phase with frequency above the crossover frequency and is also a measure of the equivalent time delay between attitude response and pilot control input. Fig. 2.41 tells a similar story to the comparison of fixed- and rotary-wing criteria for mid-term stability (cf. Fig. 2.39); there is a range in the τ_p, ω_{bw} plane where Level 1 helicopter characteristics correspond to Level 3 aeroplane characteristics. The boundaries in Fig. 2.41 are typical of the mission-oriented criteria found in modern specifications; they apply to air-combat tasks for helicopters and, more generally, to Category A flight phases for aeroplanes (see Ref. 2.3). They have been developed from the best available test data relating to current operational requirements. To a large extent, the striking differences between the fixed- and rotary wing relate directly to different task requirements; as yet, rotary-wing aircraft have not been required to deliver the performance of their fixed-wing counterparts. On the other hand, it would be very difficult to confer such bandwidth performance on a conventional helicopter from an engineering point of view, so a large

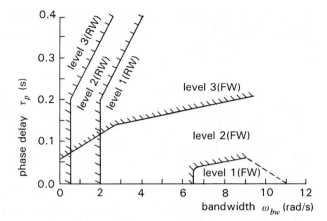

Fig. 2.41 Comparison of rotary- and fixed-wing aircraft pitch bandwidth requirements

degree of capability tailoring is inevitable. In later chapters, some of the configuration constraints and design limitations will be discussed in more detail.

Earlier, we dismissed equivalent low-order systems as being inadequate at characterizing helicopter attitude characteristics. While this is true for conventional helicopters without, or having limited authority, stability and control augmentation, future aircraft with task-tailored control laws can more usefully be described in this way. Later, in Chapters 6 and 7, we shall introduce the conceptual simulation model (CSM, Ref. 2.34), which is, in effect, a greatly simplified helicopter model in LOES form. The assumption underlying this model structure is that with active control, the flying qualities can be tailored in a wide range of different forms described now by simple equivalent systems. Flying qualities research at DRA using the CSM has been ongoing since the early 1980s and has enabled many of the desirable characteristics of future helicopters with active control technology (ACT) to be identified. This theme will be pursued in the later flying qualities chapters.

There is one helicopter flying characteristic that can, at least for the limited frequency range associated with pilot control, be described in terms of a simple first-order response – the vertical or heave axis in the hover. While it is recognized that the vertical axis dynamics are dominated by air mass and flapping motion at higher frequency, below about 5 rad/s the vertical velocity response ($\dot{h}$) to collective (δ_c) can be described by the LOES:

$$\frac{\dot{h}}{\delta_c} = \frac{K e^{-\tau_{h_{eq}} s}}{T_{h_{eq}} s + 1} \tag{2.83}$$

This formulation characterizes the first-order velocity response as a transfer function, with gain or control power K and time delay $T_{h_{eq}}$. The pure time delay $T_{h_{eq}}$ is an artifact included to capture any initial delay in achieving maximum vertical acceleration, e.g., due to rotor or air-mass dynamics. The acceptable flying qualities can then be defined in terms of the LOES parameters. Vertical axis flying qualities and flight path control in forward flight are also profoundly affected by the dynamic characteristics of the engine and rotorspeed governor system. Agile behaviour can be sustained only with rapid and sustained thrust and torque response, both of which are dependent on fast powerplant dynamics. As usual there is a trade-off, and too much agility can be unusable and wasteful. This will be a recurring theme of Chapter 7.

Flying a mission task element

Flying qualities parameters need to be physically meaningful and measurable. Assembled together as a requirement specification, they need to embrace the CACTUS rules (Ref. 2.35) outlined later in Chapter 6. Also in Chapter 6, the range of different criteria and the measurement of associated parameters in flight and simulation will be critically reviewed in the light of these underlying requirements. It needs to be re-emphasized that in most functional roles today, both military and civil helicopters need some form of artificial stability and control augmentation to achieve Level 1 flying qualities, which therefore become important drivers for both bare-airframe and automatic flight control system (AFCS) design. Before exploring the scope for artificial augmentation on this Tour, it is worth illustrating just how, in a demanding and fully attentive flying task, an MTE, flying qualities deficiencies can lead to reduced task performance and increased pilot workload.

Test techniques for the demonstration of flying qualities compliance need to exercise the aircraft to the limits of its performance. Figure 2.42 illustrates two NoE, hover to hover, repositioning manoeuvres: the quickhop and the sidestep. Tests conducted at the DRA Bedford in the mid-1980s demonstrated the importance of the task urgency or aggression factor (Ref. 2.36) on pilot workload and task performance. The manoeuvres were flown at increasing levels of aggression until the shortest possible task time was achieved. Start and finish position constraints, together with a height/track corridor, defined the acceptable flight path. Performance was increased by increasing the initial pitch or roll angle, to develop the maximum translational acceleration; both test aircraft, Lynx and Puma, were operated at relatively low weights allowing for accelerations over the ground of greater than 0.8 g ($\sim$40$°$ roll/pitch) corresponding to a hover thrust margin of about 30%. Figure 2.43 shows the recorded pilot HQRs as a function of task time for a Puma flying a 200-ft (60 m) sidestep. Above 11 s the pilot returned marginal Level 1 ratings; any reduction in task time below this resulted

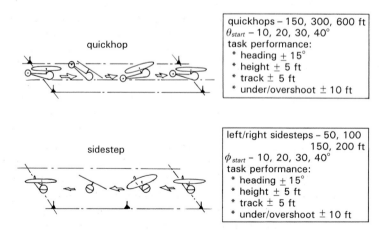

quickhops – 150, 300, 600 ft
θ_{start} – 10, 20, 30, 40$°$
task performance:
 * heading $\pm$ 15$°$
 * height $\pm$ 5 ft
 * track $\pm$ 5 ft
 * under/overshoot $\pm$ 10 ft

left/right sidesteps – 50, 100
150, 200 ft
ϕ_{start} – 10, 20, 30, 40$°$
task performance:
 * heading $\pm$ 15$°$
 * height $\pm$ 5 ft
 * track $\pm$ 5 ft
 * under/overshoot $\pm$ 10 ft

Fig. 2.42 Examples of low-speed mission task elements with performance requirements

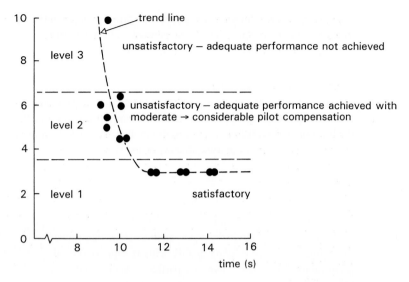

Fig. 2.43 Variations of pilot HQRs with task time for Puma 200-ft sidestep

in increasing workload. In fact, the pilot was unable to reduce the task time below 9 s and still achieve the flight path performance requirements. On one occasion the wheels hit the ground during the final recover to the hover; the pilot was applying full lateral cyclic, collective and pedal control, but, as a result of the roll and engine/rotorspeed governor response characteristics, the manoeuvre was not arrested in time. The low kinetic energy of the aircraft meant that no structural damage was incurred, but the pilot judged that he was 'out of control' and returned an HQR of 10 (Fig. 2.43).

The cliff edge and carefree handling

A combination of deficiencies in vehicle dynamics, the need for the pilot to monitor carefully critical parameters for proximity to flight limits, the poor outside visual references at high aircraft attitude angles and the overall pilot stress induced by the need to fly a tightly constrained flight path very close to the ground result in a Level 2/3 'situation'. Of course, the Puma, as a medium support helicopter, was not designed to fly 200-ft sidesteps in 8 s – the approximate limit for the test configuration. Nevertheless, pilots were inhibited from using the full performance (bank angles of 30° were the maximum measured) and many of the pilots' concerns are common to other types. A similar pattern emerged for the Lynx in the DRA tests and on aircraft used in trials conducted by the US Army (Ref. 2.37) during the same period. Also, the same trend appears for other MTEs, and is considered to represent a fundamental challenge to designers. Close to, say, within 20% of vehicle limits, it appears that the 'edge' is reached in several ways at the same time; flying qualities deficiencies are emerging strongly, just when the pilot has the greatest need for safe and predictable, or carefree, handling. The concept of carefree handling has been a familiar reality in aeroplane designs for some years, protecting against spin departure (e.g., Tornado) or deep stall (e.g., F-16) for example, but is yet to be implemented, at least in an active form, in helicopters. At the time of writing, another form of carefree handling, providing structural load alleviation, is being built into the computers of the fly-by-wire control system in the V22 Osprey. This topic is returned to in Chapter 7.

Agility factor

The DRA tests described above were part of a larger research programme aimed at providing a better understanding of the flying qualities deficiencies of current military types and quantifying future requirements. Of special interest was the impact of flying qualities on agility; the concept of agility will be developed further in Chapter 7 but, for this introductory Tour, a suitable definition is (Ref. 2.38)

> *the ability to adapt and respond rapidly and precisely with safety and with poise, to*
> *maximize mission effectiveness.*

A key question that the results of the above research raised concerned how the agility might be related to the flying qualities. One interpretation the author favours is that agility is indeed a flying quality. This is supported by the concept of the *agility factor:* if the performance used in a particular MTE could be normalized by the performance inherently available in the aircraft, then in the limit, this ratio would reveal the extent of usable performance. A more convenient way of computing this factor is to take the ratio of the theoretically ideal task time with the achieved task time. The ideal time is computed based on the assumption that the time to maximum acceleration is zero. So in the sidestep, or any similar lateral translational manoeuvre for example, the bank angle changes are achieved instantaneously. In a pure bank and stop manoeuvre, the roll rate would be assumed to develop instantaneously. The agility factor is useful for comparing the inherent agility of configurations with the same performance or competing to meet the same performance requirements. The calculation procedure and some of the factor's nuances will be elaborated on in Chapter 7. The Puma sidestep and quickhop MTE data converted to agility factors are shown with the HQRs in Fig 2.44. The trends shown previously in the time plots now appear even more dramatic; maximum agility factors for the Puma of 0.5–0.6 are achievable with borderline Level 2/3 HQRs only. The pilot can barely attain the adequate performance level, even with considerable workload. These tests were conducted in a clinical environment, with well-defined ground features and flown by skilled test pilots with opportunity to

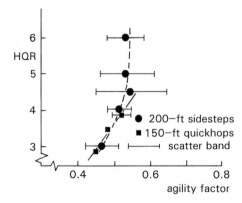

Fig. 2.44 HQRs versus agility factor for the Puma flying sidestep and quickhop MTEs

practice. In a real-world situation, the increased workload from other duties and the uncertainties of rapidly changing circumstances would inevitably lead to a further loss of agility or the increased risks of operation in the Level 3 regime; the pilot must choose in favour of safety or performance.

In agility factor experiments, the definition of the level of manoeuvre attack needs to be related to the key manoeuvre parameter, e.g., aircraft speed, attitude, turn rate or target motion. By increasing attack in an experiment, we are trying to reduce the time constant of the task, or reduce the task bandwidth. It is sufficient to define three levels – low, moderate and high, the lower corresponding to normal manoeuvring, and the upper to emergency manoeuvres.

Pilot's workload

The chief attributes of agility are speed, precision and safety and all can be eroded by the increased difficulties of the operational situation. Not only the time pressures, but also the atmospheric conditions (e.g., gustiness) and UCE (see Section 2.2) will affect the agility factor and achieved HQRs significantly. In many of these cases there is a close correlation between pilot control activity, task difficulty and pilot rating, and in such cases the level of control activity can be related to pilot workload. Figure 2.45 shows the pilot's lateral cyclic control for two different levels of aggression when flying a slalom MTE on the DRA's advanced flight simulator (AFS); the details of this and other experiments will be provided in Chapter 7, but for now the varying frequency and amplitude levels are highlighted. The HQR levels are also noted on the legend,

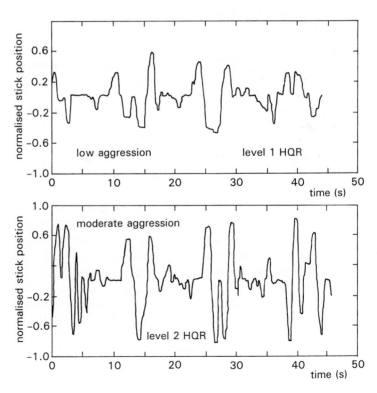

Fig. 2.45 Time histories of lateral cyclic in a lateral slalom MTE

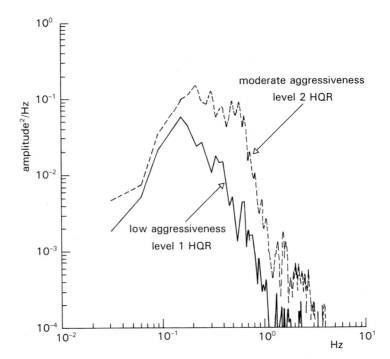

Fig. 2.46 Power spectrum of lateral cyclic in a lateral slalom MTE

indicating the degradation from Level 1 to 2 as the aggression is increased. In the case shown, the degradation corresponds to a task bandwidth increase.

Another way of representing the pilot control activity is in the frequency domain, and Fig. 2.46 illustrates the power spectral density function for the lateral cyclic, showing the amount of control 'energy' applied by the pilot at the different frequencies. The marked increase in effort for the higher aggression case is evident, particularly above 1.5 Hz. There is evidence that one of the critical parameters as far as the pilot workload is concerned is the ratio of aircraft bandwidth to task bandwidth. The latter is easy to comprehend for an aircraft flying, for example, a sinusoidal slalom, when the task bandwidth is related to the ground track geometry and the aircraft ground speed. Bandwidths for more angular MTE tracks are less obvious, but usually some ratio of speed, or mean speed, to distance will suffice. Figure 2.47 illustrates conceptually the expected trend. A workload metric, e.g., rms of control activity or frequency at which some proportion of the activity is accounted for, is plotted against the bandwidth ratio. As the ratio increases one expects the pilot's task to become easier, as shown. Conversely, as the ratio reduces, through either reduced aircraft bandwidth or increased task bandwidth, workload increases. There is a point at which the workload increases significantly, corresponding perhaps to pilot-induced oscillation onset, when the metric may no longer correlate with workload and where the control strategy is dominated by the so-called remnant, often reflecting confusion and a breakdown of the pilot acting as a quasi-linear element responding to task cue errors. Being able to detect incipient breakdown is important for establishing flying qualities boundaries and also for giving a pilot some advance warning of a potential high workload situation. Research in

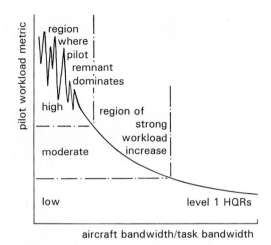

Fig. 2.47 Conceptual relationship between pilot workload and the bandwidth ratio

this field is still relatively immature, and most experiments rely heavily on subjective pilot opinion. To the author's knowledge, there are no reliable workload meters, the human equivalent of a mechanical health and usage monitoring system (HUMS), used in operational service.

Inceptors and displays

This tour of flying qualities would not be complete without some discussion on the other key characteristics associated with the air vehicle that have a primary effect on flight path control – the pilot's inceptors and displays. To dispel any myths that these are secondary issues it must be said that poor characteristics in either of these two areas can ruin otherwise excellent flying qualities. Of course, pilots can and will compensate for poor mechanical characteristics in cockpit controls, but the tactile and visual cues provided through these elements are essential for many flight phases. Sidestick controls and helmet-mounted displays are components of ACT and are likely to feature large in the cockpits of future helicopters (Figs 2.48, 2.49). Examples of recent research with these devices will be outlined in Chapter 7.

Operational benefits of flying qualities

So, what are the operational benefits of good flying qualities? Are they really significant or merely 'nice to have'? We have seen that one of the potential consequences of flying qualities deficiencies is loss of control, leading to structural damage, pilot disorientation and a crash. We have also seen that an aircraft that exhibits Level 1 characteristics in one situation can be Level 2 or even 3 in degraded or more demanding conditions. A question then arises as to the likelihood of a aircraft running into these situations in practice. This topic has recently received attention in the fixed-wing civil transport community in an attempt to quantify the probability of human error leading to a crash (Ref. 2.38). The same approach was taken to quantify the benefits of having baseline Level 1, as opposed to Level 2, flying qualities for military rotorcraft (Ref. 2.39). This research, which will be described in more detail in Chapter 7, derived a result that is summarized in Fig. 2.50. This shows the probability of achieving MTE success,

Fig. 2.48 CAE four-axis sidestick onboard the Canadian NRC variable stability Bell 205

Fig. 2.49 GEC biocular helmet-mounted display

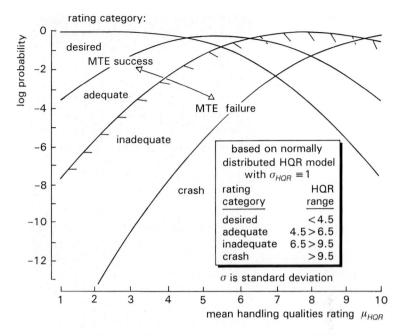

Fig. 2.50 Probability of rating category as a function of HQR

failure or loss of control (leading to a crash) as a function of mean HQR (derived, for example, from an ADS-33 objective assessment). The results are somewhat intuitive and fall out from fairly simple statistical analysis. There are a number of assumptions that need careful examination before the kind of results depicted in Fig. 2.50 can be substantiated, however, and these will be pursued further in Chapter 7. The approach, while somewhat controversial, has considerable appeal and opens up opportunities for providing a direct effectiveness measure for flying qualities.

Flying qualities review

A key emphasis on this stage of the Tour has been to highlight the importance of the relationship between flying qualities and the task or mission. Outside the context of a role and related tasks, the meaning of quality becomes vague and academic. Flying or handling qualities are not just stability and response properties of the air vehicle, but the synergy between what we have called the internal attributes of the aircraft and external influences. Flying qualities can be assessed objectively through analysis and clinical measurements, and subjectively through pilot opinion of the ability to fly MTEs within defined performance and workload constraints. The 1980s and 1990s saw considerable development in helicopter flying qualities, relevant to both design criteria and compliance demonstration, and Chapters 6 and 7 will present and discuss many of the new concepts in depth. There still exist gaps in the knowledge base however, largely due to an inadequate flight test database, and these areas will be highlighted. One of the important underdeveloped areas relates to the requirement for upper flying qualities limits. These are important for military roles requiring agility, where the assumption that more performance is always better is strongly countered by experience with oversensitive control response and unusable control powers. Agility will be covered in the section on special flying qualities in Chapter 7. The quantification of handling qualities degradation due to a variety of internal and external effects also represents a significant gap, and in Chapter 8 we discuss a number of the more significant issues.

It is recognized that without some form of stability and control augmentation system (SCAS), helicopters stand little chance of achieving Level 1 flying qualities for anything but the simplest of tasks. However, we need to be interested in the so-called bare-airframe flight dynamics for several reasons. First, the unaugmented characteristics form the baseline for SCAS design; the better they are known, the more likely that the SCAS design will work properly first time. Second, the case of failed augmentation systems has to be considered; the level of bare-airframe characteristics determine whether the SCAS is flight-safety or mission critical, i.e., whether the mission or even safe flight can be continued. Third, the better the flying qualities conferred by bare-airframe design, the less authority the SCAS requires, or the lower the gains in the feedback loops, and hence the more robust will the aircraft be to SCAS failures. And fourth, with a limited authority SCAS, any saturation in manoeuvring flight will expose the pilot to the bare-airframe characteristics; any problems associated with these conditions need to be well understood.

Clearly, SCAS performance is closely linked with the flight dynamics of the bare airframe and they both together form one of the drivers in the overall helicopter design, a subject that we now briefly visit on the last stop of this Tour.

2.5   DESIGN FOR FLYING QUALITIES; STABILITY AND CONTROL AUGMENTATION

In the helicopter design trade-off, flying qualities have often had to take a low priority. In the early days of rotorcraft, just as with fixed-wing aircraft, solving the basic control problem was the breakthrough required for the development to progress with pace, driven largely by performance considerations. The basic layout of the single rotor helicopter has remained the same since the early Sikorsky machines. What characterizes a modern helicopter is its higher performance (speed, payload), much improved reliability – hence greater safety, smoother ride and a suite of mission avionic systems that enable civil operations in poorer weather and military operations as an autonomous weapon system. Performance, reliability, comfort and functionality have been the drivers in helicopter development, and for many years flying quality was a by-product of the design, with deficiencies compensated for by highly trained pilots with a can-do attitude. As we have seen from our discussions earlier on this tour, helicopter flying characteristics are typically much poorer than fixed-wing aircraft in the same 'class'. In some cases, helicopters fall in the Level 3 quality area when built. The principal flying qualities deficiencies in the helicopter can be summarized as follows:

(1) *impurity of the primary response* in all axes, i.e., typically a mix of attitude or rate and varying significantly from hover to high speed;
(2) *strong cross-couplings* in all axes;
(3) the *degradation of response quality at flight envelope limits* and the lack of any natural carefree handling functions, e.g., the aerodynamic capability of the rotor typically exceeds the structural capability;

(4) the *stability* of a helicopter is characterized by a number of modes with low damping and frequency at low speed; as forward speed is increased, both longitudinal and lateral modes increase in frequency, as the tail surfaces contribute aerodynamic stiffness, but the modal damping can reduce and stability can often worsen, particularly with highly responsive hingeless rotors;

(5) the rotor presents a significant *filter to high bandwidth* control.

The combination of the above has always demanded great skill from helicopter pilots and coupled with today's requirements for extended operations in poor weather and visibility, and the need to relieve the piloting task in threat-intensive operations, led to the essential requirement for stability and control augmentation. Before discussing artificial stability, one first needs to look more closely at the key natural design features that contribute to flying qualities. The discussion will map directly onto the five headings in the above list and an attempt is made to illustrate how, even within the flying qualities discipline, compromises need to be made usually to satisfy both high- and low-speed requirements simultaneously.

Impurity of primary response

The helicopter rotor is sensitive to velocity perturbations in all directions and there is very little the rotor designer can do about this that doesn't compromise control response. Early attempts to build in natural dynamic couplings that neutralized the rotor from external disturbances (Refs 2.40, 2.41) resulted in complex rotor mechanisms that only partially succeeded in performing well, but, for better or worse, were never pursued to fruition and production. In reality, such endeavours were soon overtaken by the advances in 'electronic' stabilization. All motion axes of a helicopter have natural damping that resists the motion, providing a basic rate command control response in the very short term. However, soon after a control input is applied, the changes in incidence and sideslip give rise to velocity variations that alter the natural rate response characteristics in all axes. This can occur within a very short time ($O(1$ s$)$) as for the pitch axis response in high-speed flight, or longer (O(several seconds)) as for the yaw response in hover. The impurities require the pilot to stay in the loop to apply compensatory control inputs, as any manoeuvre develops. Apart from the main rotor sensitivity, the tail rotor and empennage sensitivities to main rotor wake effects can also introduce strong impurities into the control response. The size, location and incidence of the horizontal stabilizer can have a profound effect on the pilot's ability to establish trims in low-speed flight. Likewise, the tail rotor position, direction of rotation and proximity to the vertical stabilizer can significantly affect the pilot's ability to maintain heading in low-speed flight (Ref. 2.42). Both horizontal and vertical tail surfaces are practically redundant in hover and low-speed flight but provide natural stiffness and damping in high-speed flight to compensate for the unstable rotor and fuselage. The modelling of the interactional flowfields is clearly important for predicting response impurities and will be discussed further in Chapter 3.

Strong cross-couplings

Perhaps the greatest distinguishing feature of helicopters, and a bane in the designer's life, cross-couplings come in all shapes and sizes. On hingeless rotor helicopters in hover, the off-axis roll response from a pitch input can be as large as the on-axis response. At high-speed, the pitch response from collective can be as strong as from longitudinal cyclic. The yaw response from collective, due to the torque reaction, can require an equivalent tail rotor collective input to compensate. At high speed the pitch response from yaw can lead to dissimilar control strategies being required in right and left turns. These high levels of impurity again stem principally from the main rotor and its powerful wake and are inherent features of helicopters. During the 1970s and 1980s, several new designs underwent extensive flight test development to minimize the flying qualities deficiencies caused by cross-couplings and response impurities. The residual forces and moments and associated aircraft accelerations induced by these couplings can lead to serious shortcomings if high performance is being sought. For example, the saga of the empennage development for the AH64 (Ref. 2.43) and the AS 360 series helicopters (Ref. 2.44) indicate, on the one hand, how extensive the redesign to fix handling qualities can be, and, on the other, how much improvement can be obtained by careful attention to detail, e.g., in the aerodynamic characteristics of the horizontal and vertical stabilizers.

Response degradation at flight envelope limits
The boundary of the operational flight envelope should not be characterized by loss of control or performance; there should always be a safe control and performance margin for operation at the OFE limit. Most of these limit boundaries are not sign-posted however, and inadvertent excursions into the region between the OFE and the SFE boundary can and do happen, particularly when the crew's attention is diverted to other matters. Helicopters with low power margins can get caught in large-scale downdrafts behind buildings and other obstacles or terrain culture, making it very difficult for a pilot to arrest a rate of descent. Turning downwind can cause a helicopter to fly close to the vortex-ring region if the pilot judges his speed relative to the ground rather than the air. Both these examples can lead to a sharp reduction in lift and height and represent conditions most like wing stall for a fixed-wing aircraft. Hovering or manoeuvring at low speed close to obstacles in strong winds can also lead to loss of tail rotor control authority, or even, in exceptional cases, to a loss of cyclic control margins. Being 'out-of- (moment) control' close to obstacles can be as dangerous as losing lift. At high speed, or while manoeuvring in the mid-speed range, the rotor can experience local blade stall. While this is unlikely to have much effect on the overall lift, if the retreating blade stalls first, the aircraft will experience a nose-up pitching moment, further exacerbating the stall. Forward motion on the cyclic to correct the motion applies a further pitch increase on the retreating side of the disc, worsening the stall. There are very little data available on the handling qualities effects when the rotor is partially stalled in high-speed flight, but clearly flying qualities will degrade. Once again, the designer is forced to make a compromise. A low disc loading, highly twisted rotor serves hover and low-speed performance and handling, while a high disc loading, untwisted rotor gives better manoeuvrability and ride at high speed. From the designer's perspective, the alternate yaw control devices like the fenestron and Notar (Refs 2.5, 2.6) are attractive options to the open tail rotor if vulnerability is a major concern, even though handling and performance may be compromised.

Poor stability
The instabilities of the helicopter fall into two categories – those at low speed due to the rotor and those at high speed due to the rotor; the designer can do very little about the first with airframe design, but he can make flight at high speed almost as stable as a

fixed-wing aircraft. Unfortunately, if he chooses the latter option, he will almost certainly compromise control and agility. Building large enough fixed empennage stabilizers will always work but will, in turn, increase the demands on the rotor for manoeuvres. Selecting a rotor with zero or low equivalent hinge offset (e.g., most articulated rotors) will probably result in the pitch axis being marginally stable in high speed, but will again reduce the agility of the aircraft. On the other hand, a hingeless rotor, providing a roll time constant equivalent to a fixed-wing aircraft ($O(0.1$ s)) will also result in an unstable pitch mode with time to double of less than 1 s at high speed.

At low speed, without mechanical feedback, the single rotor helicopter is naturally unstable. The coupled pitch/roll, so-called pendulum instability is a product of the flapping rotor's response to velocity perturbations. The mode is actually a fairly docile one, and is easily controlled once the required strategy is learnt, but requires considerable attention by the pilot. However, if the outside world visual cues degrade, so that the pilot has difficulty perceiving attitude and velocity relative to the ground, then the hover task becomes increasingly difficult and Level 3 qualities are soon experienced.

The rotor as a control filter

The main rotor is the motivator for all but yaw control on the conventional helicopter, and before the fuselage can respond, the rotor must respond. The faster the rotorspeed, the faster the rotor flap response to control application and hence the faster the fuselage response. In many respects, the rotor acts like an actuator in the control circuit but there is one important difference. The rotor DoFs, the flap, lag and torsional motions, are considerably more complex than a simple servo system and can have low enough damping to threaten stability for high gain control tasks. Such potential problems are usually cured in the design of the SCAS, but often at the expense of introducing even further lags into the control loops. With typical actuator and rotor time constants, the overall effective time delay between pilot control input and rotor control demand can be greater than 100 ms. Such a delay can halve the response bandwidth capability of an 'instantaneous' rotor.

The five issues discussed above are compounded by the special problem associated with manoeuvring close to the ground and surrounding relief – providing an adequate field of view (FoV); the issue was expressed succinctly by Prouty (Ref. 2.45):

> The most important flying cue a pilot can have is a good view of the ground and everything around.

Field of view is a significant design compromise, most helicopters suffering from an inadequate FoV from a flying qualities perspective. Overhead panels in side-by-side cockpits obscure the view into turns and tandem seaters can be deficient in forward and downward views.

Fixing flying qualities deficiencies during flight test development can be very expensive and emphasizes the importance of accurate simulation, model testing and analytical tools in the design process. It also emphasizes the critical importance of validated design criteria – what constitutes good flying qualities for helicopters – and this book addresses this question directly in Chapter 6.

Artificial stability

It should be clear to the reader from the various discussions on this Tour that it is difficult to design and build helicopters that naturally exhibit Level 1 flying qualities. Pilots need help to fly and perform missions effectively in helicopters, and modern

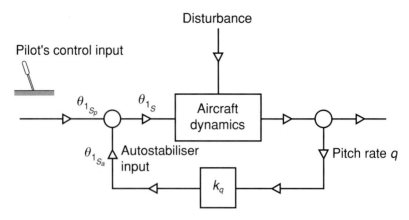

Fig. 2.51 Simple feedback augmenting pitch rate damping

SCAS and integrated displays go some way to providing this. Autostabilizers were first developed to increase the helicopter's operational envelope to include flight under instrument conditions. The first priority was to provide artificial stability to ensure that the aircraft would not wander off when the pilot's attention was divided with other tasks. If we consider the addition of rate damping in the pitch axis, we can write the feedback law in proportional form:

$$\theta_{1s_a} = k_q q \tag{2.84}$$

Figure 2.51 shows a block diagram of this feedback loop. With this proportional feedback working, as the helicopter flies through turbulence, every 1°/s of pitch rate change is counteracted by k_q° of longitudinal cyclic pitch θ_{1s}. The higher the value of gain k_q, the greater the 'artificial' stability conferred on the helicopter. The root loci in Fig. 2.52 illustrates how the high-speed unstable pitch mode can be stabilized through pitch rate feedback for Helisim Lynx. We can see that even with quite high values of gain magnitudes ($\sim$0.25), the aircraft is still marginally unstable. Gain magnitudes much higher than about 0.2 would not be acceptable because the limited authority of today's SCAS designs (typically about ±10% of actuator range) would result in the augmentation quickly saturating in manoeuvres or moderate turbulence. We can conclude from this discussion that rate feedback is insufficient to provide the levels of stability required for meeting Level 1 flying qualities in divided-attention operations. If we include attitude stabilization in the feedback loop, the control law can be written in proportional plus integral form

$$\theta_{1s_a} = k_q q + k_\theta \int q \, dt \tag{2.85}$$

Attitude feedback provides an effective stiffness in the pitch axis, and increasing k_θ serves to increase the frequency of the unstable pitch mode as shown in Fig 2.52. An appropriate combination of rate and attitude feedback can now be found to ensure Level 1 flying qualities and most modern SCAS designs incorporate both components. Rate and attitude feedbacks provide stability augmentation; but do nothing positive for control augmentation; in fact, the control response is reduced as the stability augmentation fights the pilot's actions as well as disturbances. Control augmentation is

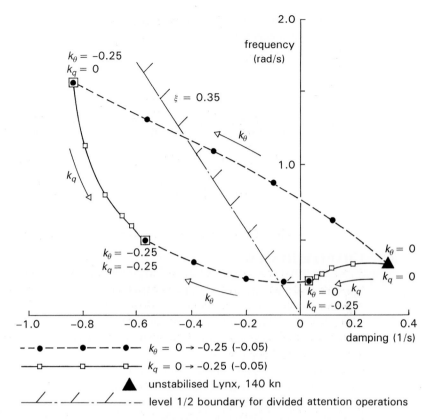

Fig. 2.52 Variation of long period pitch mode frequency and damping with autostabilizer gains for Lynx at 140 knots

accomplished by feeding forward the pilot's control signals into the SCAS, applying shaping functions or effectively disabling elements of the stability augmentation during manoeuvres. Different SCAS designs accomplish this in different ways; the Lynx system augments the initial response with a feedforward signal from the pilot's control, while the Puma system disables the attitude stabilization whenever the pilot moves the controls. More modern systems accomplish the same task with greater sophistication, but modern SCAS designs that interface with mechanical control systems will always be limited in their potential by the authority constraints designed to protect against failures. In the limit, increasing the authority of the augmentation system takes us towards ACT where the pilot's control (inceptor) inputs are combined with multiple sensor data in a digital computer to provide tailored response characteristics. ACT is still a developing technology for rotorcraft at the time of writing, but the potential benefits to both military performance and civil safety are considerable and can be classed under three general headings:

(1) task-tailored Level 1 flying qualities throughout the OFE, e.g., tailored for shipboard recoveries, underslung load positioning or air combat;
(2) carefree handling, ensuring safe operations at the edges of the OFE;
(3) integration with mission functions, e.g., navigation, HUMS.

The introduction of ACT into helicopters also offers the designer the opportunity to explore control-configured designs, in much the same way that fixed-wing military aircraft have developed over the last two decades. Even with the conventional single-rotor helicopter, ACT can free the designer to remove the empennage stabilizers altogether or alternatively to make them moving and under computer control. The main rotor could be made smaller or lighter if a carefree handling system were able to ensure a particular loading pattern in manoeuvres and at the OFE boundary. Of course it is with the more advanced rotorcraft concepts, with multiple control motivators, e.g., tilt rotors and thrust/lift compounding, that ACT will offer the greatest design freedoms and flying qualities enhancement. While this book has little to say about the flight dynamics of advanced rotorcraft, the author is conscious that the greatest strides in the future will be made with such configurations, if the 'market' can afford them or if the military requirements are strong enough.

2.6 CHAPTER REVIEW

The subject of flight dynamics is characterized by an interplay between theory and experiment. This Tour has attempted to highlight this interplay in a number of ways. Marking the four reference points early on the Tour – the environment, the vehicle, the task and the pilot – an attempt has been made to reveal the considerable scope of the subject and the skills required of the flight dynamics engineer. The importance of strong analytical tools, fundamental to the understanding of the behaviour of the helicopter's interacting subsystems, was emphasized in the modelling section. The powerful effect of aerodynamics on the flapping rotor was examined in some detail, with the resonant response highlighted as perhaps the single most important characteristic of rotor dynamics, enabling easy control of the rotor disc tilt. The modelling of flight dynamics was discussed within the framework of the frequency and amplitude of motion with three fundamental problems – trim, stability and response. The second major topic on the landscape of this Tour was flying quality, characterized by three key properties. Flying qualities are pilot-centred attributes; they are mission- and even task oriented, and they are ultimately the synergy between the internal attributes of the aircraft and the external environment in which it operates. Flying qualities are safety attributes but good flying qualities also allow the performance of the helicopter to be fully exploited. The remaining chapters of this book cover modelling and flying qualities in detail.

*The instrumented rotorhead of the DRA (RAE) research Puma
(Photograph courtesy of DRA Bedford)*

*The DRA (RAE) research Puma in trimmed flight over the
Bedfordshire countryside (Photograph courtesy of DRA Bedford)*

3 Modelling helicopter flight dynamics: building a simulation model

> *It is beyond dispute that the observed behaviour of aircraft is so complex
> and puzzling that, without a well developed theory, the subject could not
> be treated intelligently. Theory has at least three useful functions;*
> *a) it provides a background for the analysis of actual occurrences,*
> *b) it provides a rational basis for the planning of experiments and tests,
> thus securing economy of effort,*
> *c) it helps the designer to design intelligently.*
> *Theory, however, is never complete, final or exact. Like design and con-
> struction it is continually developing and adapting itself to circumstances.*
> *(Duncan 1952)*

3.1 INTRODUCTION AND SCOPE

The attributes of theory described by Duncan in this chapter's guiding quote have a
ring of eternal validity to them. With today's perspective we could add a little more.
Theory helps the flying qualities engineer to gain a deep understanding of the behaviour
of flight and the limiting conditions imposed by the demands of flying tasks, hence
providing insight and stimulating inspiration. The classic text by Duncan (Ref. 3.1)
was directed at fixed-wing aircraft, of course. Describing the flight behaviour of the
helicopter presents an even more difficult challenge to mathematical modelling. The
vehicle can be viewed as a complex arrangement of interacting subsystems, shown
in component form in Fig. 3.1. The problem is dominated by the rotor and this will
be reflected by the attention given to this component in the present chapter. The rotor
blades bend and twist under the influence of unsteady and nonlinear aerodynamic
loads, which are themselves a function of the blade motion. Figure 3.2 illustrates this
aeroelastic problem as a feedback system. The two feedback loops provide incidence
perturbations due to blade (and fuselage) motion and downwash, which are added to
those due to atmospheric disturbances and blade pitch control inputs. The calculation of
these two incidence perturbations dominates rotor modelling and hence features large
in this chapter. For the calculation of aerodynamic loads, we shall be concerned with
the blade motion relative to the air and hence the motion of the hub and fuselage as well
as the motion of the blades relative to the hub. Relative motion will be a recurring theme
in this chapter, which brings into focus the need for common reference frames. This
subject is given separate treatment in the appendix to this chapter, where the various
axes transformations required to derive the relative motion are set down. Expressions
for the accelerations of the fuselage centre of gravity and a rotor blade element are

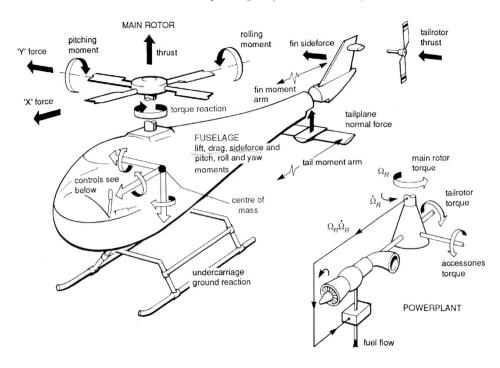

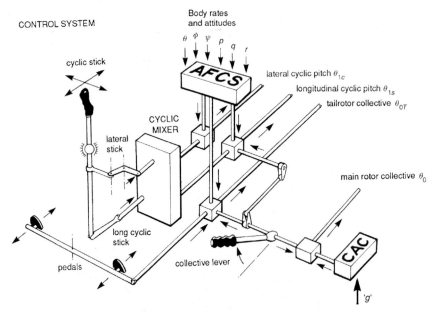

Fig. 3.1 The helicopter as an arrangement of interacting subsystems

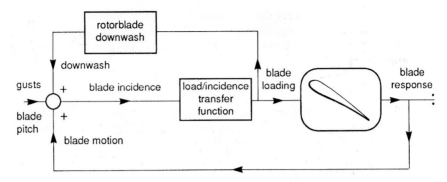

Fig. 3.2 Rotor blade aeroelasticity as a feedback problem

derived in the Appendix, Sections 3A.1 and 3A.4, respectively. Rotor blades operate in their own wake and that of their neighbour blades. Modelling these effects has probably consumed more research effort in the rotary-wing field than any other topic, ranging from simple momentum theory (Ref. 3.2) to three-dimensional flowfield solutions of the viscous fluid equations (Ref. 3.3).

The modelling requirements of blade motion and rotor downwash or inflow need to be related to the application. The terms *downwash* and *inflow* are used synonymously in this book; in some references the inflow includes components of free stream flow relative to the rotor, and not just the self-induced downwash. The rule of thumb, highlighted in Chapter 2, that models should be as simple as possible needs to be borne in mind. We refer back to Fig. 2.14 in the Introductory Tour, reproduced here in modified form (Fig. 3.3), to highlight the key dimensions – frequency and amplitude. In flight dynamics, a heuristic rule of thumb, which we shall work with, states that the

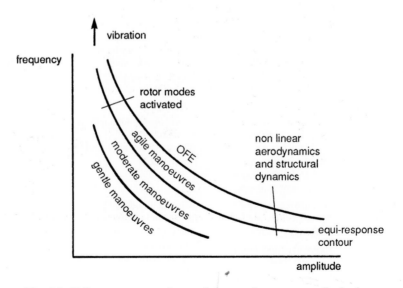

Fig. 3.3 Helicopter response characteristics on a frequency–amplitude plane

modelled frequency range in terms of forces and moments needs to extend to two or three times the range at which normal pilot and control system activity occurs. If we are solely concerned with the response to pilot control inputs for normal (correspond- ing to gentle to moderate actions) frequencies up to about 4 rad/s, then achievement of accuracy out to about 10 rad/s is probably good enough. With high gain feedback control systems that will be operating up to this latter frequency, modelling out to 25–30 rad/s may be required. The principal reason for this extended range stems from the fact that not only the controlled modes are of interest, but also the uncontrolled modes, associated with the rotors, actuators and transmission system, that could poten- tially lose stability in the striving to achieve high performance in the primary control loops. The required range will depend on a number of detailed factors, and some of these will emerge as we examine model fidelity in this and the later chapters. With respect to amplitude, the need to model gross manoeuvres defines the problem; in other words, the horizontal axis in Fig. 3.3 extends outwards to the boundary of the operational flight envelope (OFE).

It is convenient to describe the different degrees of rotor complexity in three levels, differentiated by the different application areas, as shown in Table 3.1.

Appended to the fuselage and drive system, basic Level 1 modelling defines the conventional six degree of freedom (DoF) flight mechanics formulation for the fuse- lage, with the quasi-steady rotor taking up its new position relative to the fuselage instantaneously. We have also included in this level the rotor DoFs in so-called multi- blade coordinate form (see Section 3.2.1), whereby the rotor dynamics are consolidated as a disc with coning and tilting freedoms. Perhaps the strongest distinguishing feature of Level 1 models is the analytically integrated aerodynamic loads giving closed form

Table 3.1 Levels of rotor mathematical modelling

	Level 1	Level 2	Level 3
Aerodynamics	linear 2-D dynamic inflow/local momentum theory analytically integrated loads	nonlinear (limited 3-D) dynamic inflow/local momentum theory local effects of blade vortex interaction unsteady 2-D compressibility numerically integrated loads	nonlinear 3-D full wake analysis (free or prescribed) unsteady 2-D compressibility numerically integrated loads
Dynamics	rigid blades (1) quasi-steady motion (2) 3 DoF flap (3) 6 DoF flap + lag (4) 6 DoF flap + lag + quasi-steady torsion	(1) rigid blades with options as in Level 1 (2) limited number of blade elastic modes	detailed structural representation as elastic modes or finite elements
Applications	parametric trends for flying qualities and performance studies well within operational flight envelope low bandwidth control	parametric trends for flying qualities and performance studies up to operational flight envelope medium bandwidth appropriate to high gain active flight control	rotor design rotor limit loads prediction vibration analysis rotor stability analysis up to safe flight envelope

expressions for the hub forces and moments. The aerodynamic downwash representation in our Level 1 models extends from simple momentum to dynamic inflow.

The analysis of flight dynamics problems through modelling is deferred until Chapters 4 and 5. Chapter 3 deals with model building. For the most part, the model elements are derived from approximations that allow analytic formulations. In this sense, the modelling is far from state of the art, compared with current standards of simulation modelling. This is particularly true regarding the rotor aerodynamics, but the so-called Level 1 modelling described in this chapter is aimed at describing the key features of helicopter flight and the important trends in behaviour with varying design parameters. In many cases, the simplified analytic modelling approximates reality to within 20%, and while this would be clearly inadequate for design purposes, it is ideal for establishing fundamental principles and trends.

There are three sections following. In Section 3.2, expressions for the forces and moments acting on the various components of the helicopter are derived; the main rotor, tail rotor, fuselage and empennage, powerplant and flight control system are initially considered in isolation, as far as this is possible. In Section 3.3, the combined forces and moments on these elements are assembled with the inertial and gravitational forces to form the overall helicopter equations of motion.

Section 3.4 of this chapter takes the reader briefly beyond the realms of Level 1 modelling to the more detailed and higher fidelity blade element and aeroelastic rotor formulations and the complexities of interactional aerodynamic modelling. The flight regimes where this, Level 2, modelling is required are discussed, and results of the kind of influence that aeroelasticity and detailed wake modelling have on flight dynamics are presented.

This chapter has an appendix concerned with defining the motion of the aircraft and rotor in terms of different axes systems as frames of reference. Section 3A.1 describes the inertial motion of an aircraft as a rigid body free to move in three translational and three rotational DoFs. Sections 3A.2 and 3A.3 develop supporting results for the orientation of the aircraft and the components of the gravitational force. Sections 3A.4 and 3A.5 focus on the rotor dynamics, deriving expressions for the acceleration and velocity of a blade element and discussing different axes systems used in the literature for describing the blade motion.

3.2 THE FORMULATION OF HELICOPTER FORCES AND MOMENTS IN LEVEL 1 MODELLING

In the following four subsections, analytic expressions for the forces and moments on the various helicopter components are derived. The forces and moments are referred to a system of body-fixed axes centred at the aircraft's centre of gravity/mass, as illustrated in Fig. 3.1. In general, the axes will be oriented at an angle relative to the principal axes of inertia, with the x direction pointing forward along some convenient fuselage reference line. The equations of motion for the six fuselage DoFs are assembled by applying Newton's laws of motion relating the applied forces and moments to the resulting translational and rotational accelerations. Expressions for the inertial velocities and accelerations in the fuselage-fixed axes system are derived in Appendix Section 3A.1, with the resulting equations of motion taking the classic form as given below.

Force equations

$$\dot{u} = -(wq - vr) + \frac{X}{M_a} - g \sin\theta \qquad (3.1)$$

$$\dot{v} = -(ur - wp) + \frac{Y}{M_a} + g \cos\theta \sin\phi \qquad (3.2)$$

$$\dot{w} = -(vp - uq) + \frac{Z}{M_a} + g \cos\theta \cos\phi \qquad (3.3)$$

Moment equations

$$I_{xx}\dot{p} = (I_{yy} - I_{zz})qr + I_{xz}(\dot{r} + pq) + L \qquad (3.4)$$

$$I_{yy}\dot{q} = (I_{zz} - I_{xx})rp + I_{xz}(r^2 - p^2) + M \qquad (3.5)$$

$$I_{zz}\dot{r} = (I_{xx} - I_{yy})pq + I_{xz}(\dot{p} - qr) + N \qquad (3.6)$$

where u, v and w and p, q and r are the inertial velocities in the moving axes system; ϕ, θ and ψ are the Euler rotations defining the orientation of the fuselage axes with respect to earth and hence the components of the gravitational force. I_{xx}, I_{yy}, etc., are the fuselage moments of inertia about the reference axes and M_a is the aircraft mass. The external forces and moments can be written as the sum of the contributions from the different aircraft components; thus, for the rolling moment

$$L = L_R + L_{TR} + L_f + L_{tp} + L_{fn} \qquad (3.7)$$

where the subscripts stand for: rotor, *R;* tail rotor, *TR;* fuselage, *f;* horizontal tailplane, *tp*; and vertical fin, *fn*.

In Chapters 4 and 5, we shall be concerned with the trim, stability and response solutions to eqns 3.1–3.6. Before we can address these issues we need to derive the expressions for the component forces and moments. The following four sections contain some fairly intense mathematical analyses for the reader who requires a deeper understanding of the aeromechanics of helicopters. The modelling is based essentially on the DRA's first generation, Level 1 simulation model Helisim (Ref. 3.4).

A few words on notation may be useful before we begin. First, the main rotor analysis is conducted in shaft axes, compared with the rotor-aligned, no-flapping or no-feathering systems. Appendix Section 3A.5 gives a comparison of some expressions in the three systems. Second, the reader will find the same variable name used for different states or parameters throughout the chapter. While the author accepts that there is some risk of confusion here, this is balanced against the need to maintain a degree of conformity with traditional practice. It is also expected that the serious reader of Chapters 3, 4 and 5 will easily cope with any potential ambiguities. Hence, for example, the variable r will be used for rotor radial position and aircraft yaw rate; the variable β will be used for flap angle and fuselage sideslip angle; the variable w will be used for blade displacement and aircraft inertial velocity along the z direction. A third point, and this applies more to the analysis of Chapters 4 and 5, relates to the use of capitals or lowercase for trim and perturbation quantities. For the work in the later modelling chapters we reserve capitals, with subscripts e (equilibrium), for trim states and lowercase letters for perturbation variables in the linear analysis. In

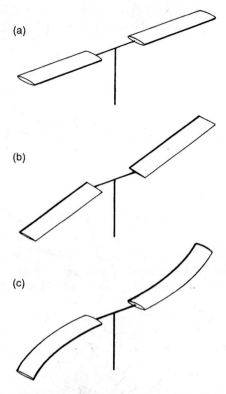

(a)

(b)

(c)

Fig. 3.4 Three flap arrangements: (a) teetering; (b) articulated; (c) hingeless

Chapter 3, where, in general, we will be dealing with variables from a zero reference, the conventional lowercase nomenclature is adopted. Possible ambiguities arise when comparing analysis from Chapters 3, 4 and 5, although the author believes that the scope for confusion is fairly minimal.

3.2.1 Main rotor

The mechanism of cyclic blade flapping provides indirect control of the direction of the rotor thrust and the rotor hub moments (i.e., the pilot has direct control only of blade pitch), hence it is the primary source of manoeuvre capability. Blade flap retention arrangements are generally of three kinds – teetering, articulated and hingeless, or more generally, bearingless (Fig. 3.4). The three different arrangements can appear very contrasting, but the amplitude of the flapping motion itself, in response to gusts and control inputs, is very similar. The primary difference lies in the hub moment capability. One of the key features of the Helisim model family is the use of a common analogue model for all three types – the so-called centre-spring equivalent rotor (CSER). We need to examine the elastic motion of blade flapping to establish the fidelity of this general approximation. The effects of blade lag and torsion dynamics are considered later in this section.

Blade flapping dynamics – introduction

We begin with a closer examination of the hingeless rotor. Figure 3.5 illustrates the out-of-plane bending, or flapping, of a typical rotating blade cantilevered to the rotor

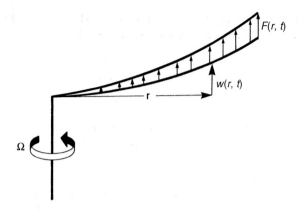

Fig. 3.5 Out-of-plane bending of a rotor blade

hub. Using the commonly accepted engineer's bending theory, the linearized equation of motion for the out-of-plane deflexion $w(r, t)$ takes the form of a partial differential equation in space radius r and time t, and can be written (Ref. 3.5) as

$$\frac{\partial^2}{\partial r^2}\left(EI\frac{\partial^2 w}{\partial r^2}\right) + m\frac{\partial^2 w}{\partial t^2} + \Omega^2\left[mr\frac{\partial w}{\partial r} - \frac{\partial^2 w}{\partial r^2}\int_r^R mr\,dr\right] = F(r, t) \quad (3.8)$$

where $EI(r)$ and $m(r)$ are the blade radial stiffness and mass distribution functions and Ω is the rotorspeed. The function $F(r, t)$ represents the radial distribution of the time-varying aerodynamic load, assumed here to act normal to the plane of rotation. As in the case of a non-rotating beam, the solution to eqn 3.8 can be written in separated variable form, as the summed product of mode shapes $S_n(r)$ and generalized coordinates $P_n(t)$, i.e.,

$$w(r, t) = \sum_{n=1}^{\infty} S_n(r)P_n(t) \quad (3.9)$$

with the time and spatial functions satisfying eqns 3.10 and 3.11 respectively, i.e.,

$$\frac{d^2 P_n(t)}{dt^2} + \Omega^2\lambda_n^2 P_n(t) = \frac{1}{I_n}\int_0^R F(r, t)S_n\,dr \quad (3.10)$$

$$I_n = \int_0^R m S_n^2\,dr$$

$$\frac{d^2}{dr^2}\left(EI\frac{d^2 S_n}{dr^2}\right) + \Omega^2\left[mr\frac{dS_n}{dr} - \frac{d^2 S_n}{dr^2}\int_r^R mr\,dr\right] - m\lambda_n^2\Omega^2 S_n = 0 \quad (3.11)$$

$$n = 1, 2, \ldots, \infty$$

I_n and λ_n are the modal inertias and natural frequencies. The mode shapes are linearly independent and have been orthogonalized, i.e.,

$$\int_0^R m S_p(r) S_n(r)\, dr = 0, \quad \int_0^R EI \frac{\partial^2 S_p}{\partial r^2} \frac{\partial^2 S_n}{\partial r^2}\, dr = 0; \qquad p \neq n \qquad (3.12)$$

The hub (denoted by subscript h) bending moment and shear force in the rotating system, denoted by the superscript (r), can then be written as

$$M_h^{(r)}(0, t) = \int_0^R \left[F(r, t) - m \left(\frac{\partial^2 w}{\partial t^2} + \Omega^2 w \right) \right] r\, dr \qquad (3.13)$$

$$V_h^{(r)}(0, t) = \int_0^R \left[F(r, t) - m \frac{\partial^2 w}{\partial t^2} \right] dr \qquad (3.14)$$

Substituting the expression for the aerodynamic loading $F(r, t)$ from eqn 3.8 into eqn 3.13, and after some reduction, the hub moment can be written as the sum of contributions from the different modes, i.e.,

$$M_h^{(r)}(0, t) = \Omega^2 \sum_{n=1}^{\infty} \left(\lambda_n^2 - 1 \right) P_n(t) \int_0^R mr S_n\, dr \qquad (3.15)$$

Retaining only the first mode gives

$$M_h^{(r)}(0, t) \approx \Omega^2 \left(\lambda_1^2 - 1 \right) P_1(t) \int_0^R mr S_1\, dr \qquad (3.16)$$

where P_1 is given by eqn 3.10 with $n = 1$. Equations 3.10 and 3.16 provide the solution for the first mode of flapping response of a rotor blade. How well this will approximate the complete solution for the blade response depends on the form of the aerodynamic load $F(r, t)$. From eqns 3.10 and 3.12, if the loading can be approximated by a distribution with the same shape as S_1, then the first mode response would suffice. Clearly this is not generally the case, but the higher mode responses can be expected to be less and less significant. It will be shown that the first mode frequency is always close to one-per-rev, and combined with the predominant forcing at one-per-rev the first flap mode generally does approximate the zero and one-per-rev blade dynamics and hub moments reasonably well, for the frequency range of interest in flight dynamics. The approximate model used in the Helisim formulation simplifies the first mode formulation even further to accommodate teetering and articulated rotors as well. The articulation and elasticity is assumed to be concentrated in a hinged spring at the centre of rotation (Fig. 3.6), otherwise the blade is straight and rigid; thus

$$S_1 = \frac{r}{R} \qquad (3.17)$$

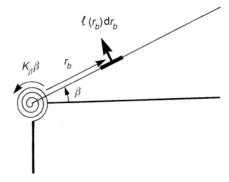

Fig. 3.6 The centre-spring equivalent rotor analogue

Such a shape, although not orthogonal to the elastic modes, does satisfy eqn 3.11 in a distributional sense. The centre-spring model is used below to represent all classes of retention system and contrasts with the offset-hinge and spring model used in a number of other studies. In the offset-hinge model, the hinge offset is largely determined from the natural frequency whereas in the centre-spring model, the stiffness is provided by the hinge spring. In many ways the models are equivalent, but they differ in some important features. It will be helpful to derive some of the characteristics of blade flapping before we compare the effectiveness of the different formulations. Further discussion is therefore deferred until later in this section.

The centre-spring equivalent rotor

Reference to Fig. 3.6 shows that the equation of motion for the blade flap angle $\beta_i(t)$ of the ith blade can be obtained by taking moments about the centre hinge with spring strength K_β; thus

$$\int_0^R r_b\{f_z(r_b) - ma_{zb}\}\, dr_b + K_\beta \beta_i = 0 \qquad (3.18)$$

The blade radial distance has now been written with a subscript b to distinguish it from similar variables. We have neglected the blade weight force in eqn 3.18; the mean lift and acceleration forces are typically one or two orders of magnitude higher. We follow the normal convention of setting the blade azimuth angle, ψ, to zero at the rear of the disc, with a positive direction following the rotor. The analysis in this book applies to a rotor rotating anticlockwise when viewed from above. From Fig. 3.7, the aerodynamic load $f_z(r_b, t)$ can be written in terms of the lift and drag forces as

$$f_z = -\ell\,\cos\phi - d\,\sin\phi \approx -\ell - d\phi \qquad (3.19)$$

where ϕ is the incidence angle between the rotor inflow and the plane normal to the rotor shaft. We are now working in the blade axes system, of course, as defined in Section 3A.4, where the z direction lies normal to the plane of no-pitch. The acceleration normal to the blade element, a_{zb}, includes the component of the gyroscopic effect due to the

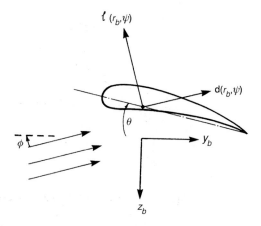

Fig. 3.7 Aerodynamic loads on a typical aerofoil section

rotation of the fuselage and hub, and is given approximately by (see the Appendix, Section 3A.4)

$$a_{zb} \approx r_b \left(2\Omega(p_{hw} \cos \psi_i - q_{hw} \sin \psi_i) + (\dot{q}_{hw} \cos \psi_i + \dot{p}_{hw} \sin \psi_i) - \Omega^2 \beta_i - \ddot{\beta_i} \right)$$

(3.20)

The angular velocities and accelerations have been referred to hub–wind axes in this formulation, and hence the subscript *hw*. Before expanding and reducing the hub moment in eqn 3.18 further, we need to review the range of approximations to be made for the aerodynamic lift force. The aerodynamic loads are in general unsteady, nonlinear and three-dimensional in character; our first approximation neglects these effects, and, in a wide range of flight cases, the approximations lead to a reasonable prediction of the overall behaviour of the rotor. So, our starting aerodynamic assumptions are as follows:

(1) *The rotor lift force is a linear function of local blade incidence and the drag force is a simple quadratic function of lift – both with constant coefficients.* Neglecting blade stall and compressibility can have a significant effect on the prediction of performance and dynamic behaviour at high forward speeds. Figure 2.10 illustrated the proximity of the local blade incidence to stall particularly at rotor azimuth angles 90° and 180°. Without these effects modelled, the rotor will be able to continue developing lift at low drag beyond the stall and drag divergence boundary, which is clearly unrealistic. The assumption of constant lift curve slope neglects the linear spanwise and one-per-rev timewise variations due to compressibility effects. The former can be accounted for to some extent by an effective rotor lift curve slope, particularly at low speed, but the azimuthal variations give rise to changes in cyclic and collective trim control angles in forward flight, which the constant linear model cannot simulate.

(2) *Unsteady (i.e., frequency dependent) aerodynamic effects are ignored.* Rotor unsteady aerodynamic effects can be conveniently divided into two problems – one involves the calculation of the response of the rotor blade lift and pitching moment to changes in local incidence, while the other involves the calculation of the unsteady local incidence due to the time variations of the rotor wake velocities. Both require additional DoFs to be taken into account. While the unsteady wake effects are

accounted for in a relatively crude but effective manner through the local/dynamic inflow theory described in this section, the time-dependent developments of blade lift and pitching moment are ignored, resulting in a small phase shift of rotor response to disturbances.

(3) *Tip losses and root cut-out effects are ignored.* The lift on a rotor blade reduces to zero at the blade tip and at the root end of the lifting portion of the blade. These effects can be accounted for when the fall-off is properly modelled at the root and tip, but an alternative is to carry out the load integrations between an effective root and tip. A tip loss factor of about 3% R is commonly used, while integrating from the start of the lifting blade accounts for most of the root loss. Both effects are small and accounts for only a few per cent of performance and response. Including them in the analysis increases the length of the equations significantly however, and can obscure some of the more significant effects. In the analysis that follows, we therefore omit these loss terms, recognizing that to achieve accurate predictions of power, for example, they need to be included.

(4) *Non-uniform spanwise inflow distribution is neglected.* The assumption of uniform inflow is a gross simplification, even in the hover, of the complex effects of the rotor wake, but provides a very effective approximation for predicting power and thrust. The use of uniform inflow stems from the assumption that the rotor is designed to develop minimum induced drag, and hence has ideal blade twist. In such an ideal case, the circulation would be constant along the blade span, with the only induced losses emanating from the tip and root vortices. Ideal twist, for a constant chord blade, is actually inversely proportional to radius, and the linear twist angles of $O(10°)$, found on most helicopters, give a reasonable, if not good, approximation to the effects of ideal twist over the outer lifting portion of the blades. The actual non-uniformity of the inflow has a similar shape to the bound circulation, increasing outboard and giving rise to an increase in drag compared with the uniform inflow theory. The blade pitch at the outer stations of a real blade will need to be increased relative to the uniform inflow blade to produce the same lift. This increase produces more lift inboard as well, and the resulting comparison of trim control angles may not be significantly different.

(5) *Reversed flow effects are ignored.* The reversed flow region occupies the small disc inboard on the retreating side of the disc, where the air flows over the blades from trailing to leading edge. Up to moderate forward speeds, the extent of this region is small and the associated dynamic pressures low, justifying its omission from the analysis of rotor forces. At higher speeds, the importance of the reversed flow region increases, resulting in an increment to the collective pitch required to provide the rotor thrust, but decreasing the profile drag and hence rotor torque.

These approximations make it possible to derive manageable analytic expressions for the flapping and rotor loads. Referring to Fig. 3.7, the aerodynamic loads can be written in the form

Lift:
$$\ell(\psi, r_b) = \frac{1}{2}\rho \left(U_T^2 + U_P^2\right) ca_0 \left(\theta + \frac{U_P}{U_T}\right) \qquad (3.21)$$

Drag:
$$d(\psi, r_b) = \frac{1}{2}\rho \left(U_T^2 + U_P^2\right) c\delta \qquad (3.22)$$

where

$$\delta = \delta_0 + \delta_2 C_T^2 \qquad (3.23)$$

We have made the assumption that the blade profile drag coefficient δ can be written in terms of a mean value plus a thrust-dependent term to account for blade incidence changes (Refs 3.6, 3.7). The non-dimensional in-plane and normal velocity components can be written as

$$\overline{U}_T = \overline{r}_b(1 + \overline{\omega}_x \beta) + \mu \sin \psi \tag{3.24}$$

$$\overline{U}_P = (\mu_z - \lambda_0 - \beta \mu \cos \psi) + \overline{r}_b(\overline{\omega}_y - \beta' - \lambda_1) \tag{3.25}$$

We have introduced into these expressions a number of new symbols that need definition:

$$\overline{r}_b = \frac{r_b}{R} \tag{3.26}$$

$$\mu = \frac{u_{hw}}{\Omega R} = \left(\frac{u_h^2 + v_h^2}{(\Omega R)^2}\right)^{1/2} \tag{3.27}$$

$$\mu_z = \frac{w_{hw}}{\Omega R} \tag{3.28}$$

The velocities u_{hw}, v_{hw} and w_{hw} are the hub velocities in the hub–wind system, oriented relative to the aircraft x-axis by the relative airspeed or wind direction in the x–y plane. β is the blade flap angle and θ is the blade pitch angle. The fuselage angular velocity components in the blade system, normalized by ΩR, are given by

$$\overline{\omega}_x = \overline{p}_{hw} \cos \psi_i - \overline{q}_{hw} \sin \psi_i$$

$$\overline{\omega}_y = \overline{p}_{hw} \sin \psi_i + \overline{q}_{hw} \cos \psi_i \tag{3.29}$$

The downwash, λ, normal to the plane of the rotor disc, is written in the form of a uniform and linearly varying distribution

$$\lambda = \frac{v_i}{\Omega R} = \lambda_0 + \lambda_1(\psi)\overline{r}_b \tag{3.30}$$

This simple formulation will be discussed in more detail later in this chapter.

We can now develop and expand eqn 3.18 to give the second-order differential equation of flapping motion for a single blade, with the prime indicating differentiation with respect to azimuth angle ψ:

$$\beta_i'' + \lambda_\beta^2 \beta_i = 2\left(\left(\overline{p}_{hw} + \frac{\overline{q}_{hw}'}{2}\right)\cos \psi_i - \left(\overline{q}_{hw} + \frac{\overline{p}_{hw}'}{2}\right)\sin \psi_i\right)$$

$$+ \frac{\gamma}{2}\int_0^1 (\overline{U}_T^2 \theta + \overline{U}_T \overline{U}_P)_i \overline{r}_b \, d\overline{r}_b \tag{3.31}$$

The blade Lock number γ is a fundamental parameter that expresses the ratio of aerodynamic to inertia forces acting on the blade; the flap frequency ratio, λ_β, is derived directly from the hub stiffness and the flap moment of inertia I_β:

$$\gamma = \frac{\rho c a_0 R^4}{I_\beta}, \quad \lambda_\beta^2 = 1 + \frac{K_\beta}{I_\beta \Omega^2}, \quad I_\beta = \int_0^R m r^2 \, dr \qquad (3.32)$$

where a_0 is the constant blade lift curve slope, ρ is the air density and c the blade chord.

Writing the blade pitch angle θ as a combination of applied pitch and linear twist, in the form

$$\theta = \theta_p + \bar{r}_b \theta_{tw} \qquad (3.33)$$

we may expand eqn 3.31 into the form

$$\beta_i'' + f_{\beta'} \gamma \beta_i' + \left(\lambda_\beta^2 + \gamma \mu \cos \psi_i f_\beta\right) \beta_i =$$

$$2\left(\left(\bar{p}_w + \frac{\bar{q}_w'}{2}\right) \cos \psi_i - \left(\bar{q}_w - \frac{\bar{p}_w'}{2}\right) \sin \psi_i\right)$$

$$+ \gamma \left[f_{\theta p} \theta_p + f_{\theta tw} \theta_{tw} + f_\lambda (\mu_z - \lambda_0) + f_\omega (\bar{\omega}_y - \lambda_1) \right] \qquad (3.34)$$

where the aerodynamic coefficients, f, are given by the expressions

$$f_{\beta'} = \frac{1 + \frac{4}{3}\mu \sin \psi_i}{8} \qquad (3.35)$$

$$f_\beta = f_\lambda = \frac{\frac{4}{3} + 2\mu \sin \psi_i}{8} \qquad (3.36)$$

$$f_{\theta p} = \frac{1 + \frac{8}{3}\mu \sin \psi_i + 2\mu^2 \sin^2 \psi_i}{8} \qquad (3.37)$$

$$f_{\theta tw} = \frac{\frac{4}{5} + 2\mu \sin \psi_i + \frac{4}{3}\mu^2 \sin^2 \psi_i}{8} \qquad (3.38)$$

$$f_\omega = \frac{1 + \frac{4}{3}\mu \sin \psi_i}{8} \qquad (3.39)$$

These aerodynamic coefficients have been expanded up to $O(\mu^2)$; neglecting higher order terms incurs errors of less than 10% in the flap response below μ of about 0.35. In Chapter 2, the Introductory Tour of this subject, we examined the solution of eqn 3.34 at the hover condition. The behaviour was discussed in some depth there, and to avoid duplication of the associated analysis we shall restrict ourselves to a short résumé of the key points from the material in Chapter 2.

(1) The blade flap response is dominated by the centrifugal stiffness, so that the natural frequency is always close to one-per-rev; even on hingeless rotors like the Lynx and Bo105, the flap frequency ratio, λ_β, is only about 10% higher than for a teetering rotor.

(2) The flap response to cyclic pitch is close to phase resonance, and hence is about 90° out of phase with the pitch control input; the stiffer the rotor, the smaller the phase

lag, but even the Lynx, with its moderately stiff rotor, has about 80° of lag between cyclic pitch and flap. The phase lag is proportional to the stiffness number (effectively the ratio of stiffness to blade aerodynamic moment), given by

$$S_\beta = 8 \left(\frac{\lambda_\beta^2 - 1}{\gamma} \right) \tag{3.40}$$

(3) There is a fundamental rotor resistance to fuselage rotations, due to the aerodynamic damping and gyroscopic forces. Rotating the fuselage with a pitch (q) or roll (p) rate leads to a disc rotation lagged behind the fuselage motion by a time given approximately by (see eqn 2.43)

$$\tau_\beta = \frac{16}{\gamma \Omega} \tag{3.41}$$

Hence the faster the rotorspeed, or the lighter the blades, for example, the higher is the rotor damping and the faster is the disc response to control inputs or fuselage motion.

(4) The rotor hub stiffness moment is proportional to the product of the spring strength and the flap angle; teetering rotors cannot therefore produce a hub moment, and hingeless rotors, as on the Bo105 and the Lynx, can develop hub moments about four times those for typical articulated rotors.

(5) The increased hub moment capability of hingeless rotors transforms into increased control sensitivity and damping and hence greater responsiveness at the expense of greater sensitivity to extraneous inputs such as gusts. The control power, or final steady-state rate per degree of cyclic, is independent of rotor stiffness to first order, since it is derived from the ratio of control sensitivity to damping, both of which increase in the same proportion with rotor stiffness.

(6) The flapping of rotors with Stiffness numbers up to about 0.3 is very similar – e.g., approximately 1° flap for 1° cyclic pitch.

The behaviour of a rotor with N_b blades will be described by the solution of a set of uncoupled differential equations of the form eqn 3.34, phased relative to each other. However, the wake and swash plate dynamics will couple implicitly the blade dynamics. We return to this aspect later, but for the moment, assume a decoupled system. Each equation has periodic coefficients in the forward flight case, but is linear in the flap DoFs (once again ignoring the effects of wake inflow). In Chapter 2, we examined the hover case and assumed that the blade dynamics were fast relative to the fuselage motion, hence enabling the approximation that the blade motion was essentially periodic with slowly varying coefficients. The rotor blades were effectively operating in two timescales, one associated with the rotor rotational speed and the other associated with the slower fuselage motion. Through this approximation, we were able to deduce many fundamental facets of rotor behaviour as noted above. It was also highlighted that the approximation breaks down when the frequencies of the rotor and fuselage modes approached one another, as can happen, for example, with hingeless rotors. This quasi-steady approximation can be approached from a more general perspective in the forward flight case by employing the so-called multi-blade coordinates (Refs 3.4, 3.8).

Multi-blade coordinates

We can introduce a transformation from the individual blade coordinates (IBCs) to the disc coordinates, or multi-blade coordinates (MBCs), as follows:

$$\beta_0 = \frac{1}{N_b} \sum_{i=1}^{N_b} \beta_i \tag{3.42}$$

$$\beta_{0d} = \frac{1}{N_b} \sum_{i=1}^{N_b} \beta_i (-1)^i \tag{3.43}$$

$$\beta_{jc} = \frac{2}{N_b} \sum_{i=1}^{N_b} \beta_i \ \cos j\psi_i \tag{3.44}$$

$$\beta_{js} = \frac{2}{N_b} \sum_{i=1}^{N_b} \beta_i \ \sin j\psi_i \tag{3.45}$$

or, in vector form, as

$$\boldsymbol{\beta}_I = \boldsymbol{L}_\beta \boldsymbol{\beta}_M$$

where, for a four-bladed rotor

$$\boldsymbol{\beta}_I = \{\beta_1, \ \beta_2, \ \beta_3, \ \beta_4\}, \qquad \boldsymbol{\beta}_M = \{\beta_0, \ \beta_{0d}, \ \beta_{1c}, \ \beta_{1s}\} \tag{3.46}$$

and

$$\boldsymbol{L}_\beta = \begin{bmatrix} 1 & -1 & \cos\psi & \sin\psi \\ 1 & 1 & \sin\psi & -\cos\psi \\ 1 & -1 & -\cos\psi & -\sin\psi \\ 1 & 1 & -\sin\psi & \cos\psi \end{bmatrix} \tag{3.47}$$

giving

$$\boldsymbol{L}_\beta^{-1} = \frac{1}{4} \begin{bmatrix} 1 & 1 & 1 & 1 \\ -1 & 1 & -1 & 1 \\ 2\cos\psi & 2\sin\psi & -2\cos\psi & -2\sin\psi \\ 2\sin\psi & -2\cos\psi & -2\sin\psi & 2\cos\psi \end{bmatrix} \tag{3.48}$$

In forming the matrix $\boldsymbol{L}_\beta$ we use the relationship between the individual blade azimuth angles, namely

$$\psi_i = \left[\psi - \frac{\pi}{2}(i-1)\right] \tag{3.49}$$

Once again, the reference zero angle for blade 1 is at the rear of the disc. The MBCs can be viewed as disc mode shapes (Fig. 3.8). The first, β_0, is referred to as coning – all the blades flap together in a cone. The first two cyclic modes β_{1c} and β_{1s} represent first harmonic longitudinal and lateral disc tilts, while the higher harmonics appear as a disc warping. For $N_b = 4$, the oddest mode of all is the differential coning, β_{0d}, which can be visualized as a mode with opposite pairs of blades moving in unison, but

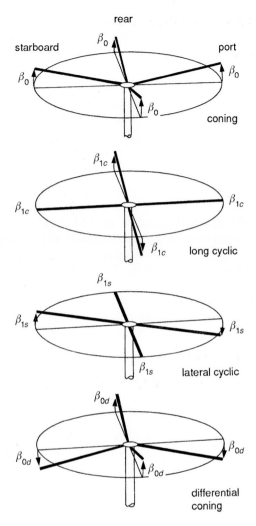

Fig. 3.8 The rotor disc in multi-blade coordinates

in opposition to neighbour pairs, as shown in Fig. 3.8. The transformation to MBCs has not involved any approximation; there are the same number of MBCs as there are IBCs, and the individual blade motions can be completely reconstituted from the MBC motions. There is one other important aspect that is worth highlighting. MBCs are not strictly equivalent to the harmonic coefficients in a Fourier expansion of the blade angle. In general, each blade will be forced and will respond with higher than a one-per-rev component (e.g., two-, three- and four-per-rev), yet with $N_b = 3$, only first harmonic MBCs will exist; the higher harmonics are then folded into the first harmonics. The real benefit of MBCs emerges when we conduct the coordinate transformation $\mathbf{L}_\beta$ on the uncoupled individual blade eqns 3.34, written in matrix form as

$$\boldsymbol{\beta}_I'' + \mathbf{C}_I(\psi)\boldsymbol{\beta}_I' + \mathbf{D}_I(\psi)\boldsymbol{\beta}_I = \mathbf{H}_I(\psi) \tag{3.50}$$

hence forming the MBC equations

$$\boldsymbol{\beta}''_M + \mathbf{C}_M(\psi)\boldsymbol{\beta}'_M + \mathbf{D}_M(\psi)\boldsymbol{\beta}_M = \mathbf{H}_M(\psi) \tag{3.51}$$

where the coefficient matrices are derived from the following expressions:

$$\mathbf{C}_M = \mathbf{L}_\beta^{-1}\left\{2\mathbf{L}'_\beta + \mathbf{C}_I\mathbf{L}_\beta\right\} \tag{3.52}$$

$$\mathbf{D}_M = \mathbf{L}_\beta^{-1}\left\{\mathbf{L}''_\beta + \mathbf{C}_I\mathbf{L}'_\beta + \mathbf{D}_I\mathbf{L}_\beta\right\} \tag{3.53}$$

$$\mathbf{H}_M = \mathbf{L}_\beta^{-1}\mathbf{H}_I \tag{3.54}$$

The MBC system described by eqn 3.51 can be distinguished from the IBC system in two important ways. First, the equations are now coupled, and second, the periodic terms in the coefficient matrices no longer contain first harmonic terms but have the lowest frequency content at $N_b/2$ per-rev (i.e., two for a four-bladed rotor). A common approximation is to neglect these terms, hence reducing eqn 3.51 to a set of ordinary differential equations with constant coefficients that can then be appended to the fuselage equations of motion allowing the wide range of linear stationary stability analysis tools to be brought to bear. In the absence of periodic terms, MBC equations take the form

$$\boldsymbol{\beta}''_M + \mathbf{C}_{M0}\boldsymbol{\beta}'_M + \mathbf{D}_{M0}\boldsymbol{\beta}_M = \mathbf{H}_{M0}(\psi) \tag{3.55}$$

where the constant coefficient matrices can be expanded, for a four-bladed rotor, as shown below:

$$\mathbf{C}_{M0} = \frac{\gamma}{8}\begin{bmatrix} 1 & 0 & 0 & \frac{2}{3}\mu \\ 0 & 1 & 0 & 0 \\ 0 & 0 & 1 & \frac{16}{\gamma} \\ \frac{4}{3}\mu & 0 & -\frac{16}{\gamma} & 1 \end{bmatrix} \tag{3.56}$$

$$\mathbf{D}_{M0} = \frac{\gamma}{8}\begin{bmatrix} \dfrac{8\lambda_\beta^2}{\gamma} & 0 & 0 & 0 \\ 0 & \dfrac{8\lambda_\beta^2}{\gamma} & 0 & 0 \\ \dfrac{4}{3}\mu & 0 & \dfrac{8(\lambda_\beta^2 - 1)}{\gamma} & 1 + \dfrac{\mu^2}{2} \\ 0 & 0 & -\left(1 - \dfrac{\mu^2}{2}\right) & \dfrac{8(\lambda_\beta^2 - 1)}{\gamma} \end{bmatrix} \tag{3.57}$$

$$\mathbf{H}_{MO} = \frac{\gamma}{8}
\begin{bmatrix}
\theta_0(1+\mu^2) + 4\theta_{tw}\left(\frac{1}{5} + \frac{\mu^2}{6}\right) + \frac{4}{3}\mu\theta_{1sw} + \frac{4}{3}(\mu_z - \lambda_0) + \frac{2}{3}\mu(\overline{p}_{hw} - \lambda_{1sw}) \\
\hline
0 \\
\hline
\frac{16}{\gamma}\left(\overline{p}_{hw} + \frac{\overline{q}'_{hw}}{2}\right) + \theta_{1cw}\left(1 + \frac{\mu^2}{2}\right) + (\overline{q}_{hw} - \lambda_{1cw}) \\
\hline
-\frac{16}{\gamma}\left(\overline{q}_{hw} - \frac{\overline{p}'_{hw}}{2}\right) + \frac{8}{3}\mu\theta_0 + 2\mu\theta_{tw} + \theta_{1sw}\left(1 + \frac{3}{2}\mu^2\right) + 2\mu(\mu_z - \lambda_0) + (\overline{p}_{hw} - \lambda_{1sw})
\end{bmatrix}$$

$$\tag{3.58}$$

The blade pitch angle and downwash functions have been written in the form

$$\theta_p = \theta_0 + \theta_{1c}\cos\psi + \theta_{1s}\sin\psi \tag{3.59}$$

$$\lambda = \lambda_0 + \overline{r}_b(\lambda_{1c}\cos\psi + \lambda_{1s}\sin\psi) \tag{3.60}$$

Some physical understanding of the MBC dynamics can be gleaned from a closer examination of the hover condition. The free response of the MBC then reveals the coning and differential coning as independent, uncoupled, DoFs with damping $\gamma/8$ and natural frequency λ_β, or approximately one-per-rev. The cyclic mode equations are coupled and can be expanded as

$$\beta''_{1c} + \frac{\gamma}{8}\beta'_{1c} + \left(\lambda_\beta^2 - 1\right)\beta_{1c} + 2\beta'_{1s} + \frac{\gamma}{8}\beta_{1s} = 0 \tag{3.61}$$

$$\beta''_{1s} + \frac{\gamma}{8}\beta'_{1s} + (\lambda_\beta^2 - 1)\beta_{1s} - 2\beta'_{1c} + \frac{\gamma}{8}\beta_{1c} = 0 \tag{3.62}$$

The eigenvalues of this cyclic flapping system are given by the roots of the following equation, and are shown sketched in Fig. 3.9:

$$\left(\lambda^2 + \frac{\gamma}{8}\lambda + \lambda_\beta^2 - 1\right)^2 + \left(2\lambda + \frac{\gamma}{8}\right)^2 = 0 \tag{3.63}$$

The two modes have been described as the flap precession (or regressing flap mode) and nutation (or advancing flap mode) to highlight the analogy with a gyroscope; both have the same damping factor as the coning mode but their frequencies are widely separated, the precession lying approximately at $(\lambda_\beta - 1)$ and the nutation well beyond this at $(\lambda_\beta + 1)$. While the nutation flap mode is unlikely to couple with the fuselage motions, the regressing flap mode frequency can be of the same order as the highest frequency fuselage modes. An often used approximation to this mode assumes that the inertia terms are zero and that the simpler, first-order formulation is adequate for describing the rotor flap as described in Chapter 2 (eqn 2.40). The motion tends to be more strongly coupled with the roll axis because of the lower time constant associated with roll than with pitch motion. The roll to pitch time constants are scaled by the ratio of the roll to pitch moment of inertia, a parameter with a typical value of about 0.25. We shall return to this approximation later in this chapter and in Chapter 5.

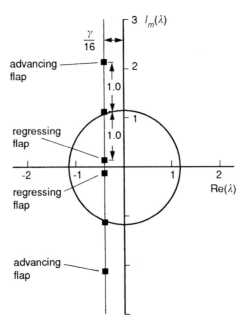

Fig. 3.9 Eigenvalues of a multi-blade coordinate rotor system

The differential coning is of little interest to us, except in the reconstruction of the individual blade motions; each pair of blade exerts the same effective load on the rotor hub, making this motion reactionless. Ignoring this mode, we see that the quasi-steady motion of the coning and cyclic flapping modes can be derived from eqn 3.55 and written in vector–matrix form as

$$\boldsymbol{\beta}_M = \mathbf{D}_{M0}^{-1}\mathbf{H}_{M0} \tag{3.64}$$

or expanded as

$$\boldsymbol{\beta}_M = \mathbf{A}_{\beta\theta}\boldsymbol{\theta} + \mathbf{A}_{\beta\lambda}\boldsymbol{\lambda} + \mathbf{A}_{\beta\omega}\boldsymbol{\omega} \tag{3.65}$$

where the subvectors are defined by

$$\boldsymbol{\beta}_M = \{\beta_0,\ \beta_{1c},\ \beta_{1s}\} \tag{3.66}$$

$$\boldsymbol{\theta} = \{\theta_0,\ \theta_{tw},\ \theta_{1sw},\ \theta_{1cw}\} \tag{3.67}$$

$$\boldsymbol{\lambda} = \{(\mu_z - \lambda_0),\ \lambda_{1sw},\ \lambda_{1cw}\} \tag{3.68}$$

$$\boldsymbol{\omega} = \{\bar{p}'_{hw},\ \bar{q}'_{hw},\ \bar{p}_{hw},\ \bar{q}_{hw}\} \tag{3.69}$$

and the coefficient matrices can be written as shown opposite in eqns 3.70, 3.71 and 3.72. Here

$$\eta_\beta = -\frac{1}{1 + S_\beta^2}$$

These quasi-steady flap equations can be used to calculate rotor trim conditions to $O(\mu^2)$ and also to approximate the rotor dynamics associated with low-frequency fuselage

$$A_{\beta\theta} = \frac{\gamma}{8\lambda_\beta^2}\left[\begin{array}{c:c:c:c}
1+\mu^2 & \dfrac{4}{5}+\dfrac{2}{3}\mu^2 & \dfrac{4}{3}\mu & 0 \\[2ex]
\hdashline
\eta_\beta\dfrac{4}{3}\mu\left(S_\beta(1+\mu^2)+\dfrac{16\lambda_\beta^2}{\gamma}\left(1+\dfrac{\mu^2}{2}\right)\right) & \eta_\beta 2\mu\left(\dfrac{8\lambda_\beta^2}{\gamma}\left(1+\dfrac{\mu^2}{2}\right)+\dfrac{8}{15}S_\beta\left(1+\dfrac{5}{2}\mu^2\right)\right) & \eta_\beta\left(\dfrac{8\lambda_\beta^2}{\gamma}(1+2\mu^2)+\left(\dfrac{4}{3}\mu\right)^2 S_\beta\right) & -\eta_\beta S_\beta\dfrac{8\lambda_\beta^2}{\gamma}\left(1+\dfrac{\mu^2}{2}\right) \\[2ex]
\hdashline
\eta_\beta\dfrac{4}{3}\mu\left(1+\dfrac{\mu^2}{2}-2S_\beta\dfrac{8\lambda_\beta^2}{\gamma}\right) & \eta_\beta 2\mu\left(\dfrac{8}{15}\left(1+\dfrac{\mu^2}{2}\right)-S_\beta\dfrac{8\lambda_\beta^2}{\gamma}\right) & \eta_\beta\left(\left(\dfrac{4}{3}\mu\right)^2 - S_\beta\dfrac{8\lambda_\beta^2}{\gamma}\left(1+\dfrac{3}{2}\mu^2\right)\right) & -\eta_\beta\dfrac{8\lambda_\beta^2}{\gamma}\left(1-\dfrac{\mu^4}{2}\right)
\end{array}\right] \quad (3.70)$$

$$A_{\beta\lambda} = \frac{\gamma}{8\lambda_\beta^2}\left[\begin{array}{c:c:c}
\dfrac{4}{3} & -\dfrac{2}{3}\mu & 0 \\[2ex]
\hdashline
\eta_\beta\mu\left(\left(\dfrac{4}{3}\right)^2 S_\beta+\dfrac{16\lambda_\beta^2}{\gamma}\left(1+\dfrac{\mu^2}{2}\right)\right) & -\eta_\beta\left(\dfrac{8\lambda_\beta^2}{\gamma}\left(1+\dfrac{\mu^2}{2}\right)+\dfrac{S_\beta}{2}\left(\dfrac{4}{3}\mu\right)^2\right) & \eta_\beta\dfrac{8\lambda_\beta^2}{\gamma}S_\beta \\[2ex]
\hdashline
\eta_\beta\mu\left(\left(\dfrac{4}{3}\right)^2\left(1-\dfrac{\mu^2}{2}\right)-S_\beta\dfrac{16\lambda_\beta^2}{\gamma}\right) & \eta_\beta\left(\dfrac{8\lambda_\beta^2}{\gamma}S_\beta - \dfrac{1}{2}\left(\dfrac{4}{3}\mu\right)^2\right) & \eta_\beta\dfrac{8\lambda_\beta^2}{\gamma}\left(1-\dfrac{\mu^2}{2}\right)
\end{array}\right] \quad (3.71)$$

$$A_{\beta\omega} = \frac{\gamma}{8\lambda_\beta^2}\left[\begin{array}{c:c:c:c}
0 & 0 & \dfrac{2}{3}\mu & 0 \\[2ex]
\hdashline
\eta_\beta\left(\dfrac{8\lambda_\beta}{\gamma}\right)\left(1+\dfrac{\mu^2}{2}\right) & -\eta_\beta\left(\dfrac{8\lambda_\beta}{\gamma}\right)^2 S_\beta & \eta_\beta\left(\dfrac{8\lambda_\beta^2}{\gamma}\left(1+\dfrac{\mu^2}{2}-\dfrac{16S_\beta}{\gamma}\right)+\dfrac{S_\beta}{2}\left(\dfrac{4}{3}\mu\right)^2\right) & -\eta_\beta\dfrac{8\lambda_\beta^2}{\gamma}\left(S_\beta+\dfrac{16}{\gamma}\left(1+\dfrac{\mu^2}{2}\right)\right) \\[2ex]
\hdashline
-\eta_\beta S_\beta\left(\dfrac{8\lambda_\beta}{\gamma}\right) & \eta_\beta\left(\dfrac{8\lambda_\beta}{\gamma}\right)^2\left(\dfrac{\mu^2}{2}-1\right) & \eta_\beta\dfrac{8\lambda_\beta^2}{\gamma}\left(\dfrac{16}{\gamma}\left(\dfrac{\mu^2}{2}-1\right)-S_\beta\right)+\dfrac{1}{2}\left(\dfrac{4}{3}\mu\right)^2 & \eta_\beta\dfrac{8\lambda_\beta^2}{\gamma}\left(\dfrac{16S_\beta}{\gamma}+\dfrac{\mu^2}{2}-1\right)
\end{array}\right] \quad (3.72)$$

motions. In this way the concept of flapping derivatives comes into play. These were introduced in Chapter 2 and examples were given in eqns 2.29–2.32; the primary flap control response and damping in the hover were derived as

$$\frac{\partial \beta_{1c}}{\partial \theta_{1s}} = -\frac{\partial \beta_{1s}}{\partial \theta_{1c}} = -\frac{1}{1 + S_\beta^2}$$

$$\frac{\partial \beta_{1c}}{\partial \overline{q}} = \frac{\partial \beta_{1s}}{\partial \overline{p}} = \frac{1}{1 + S_\beta^2} \left(S_\beta + \frac{16}{\gamma} \right)$$

showing the strong dependence of rotor flap damping on Lock number, compared with the weak dependence of flap response due to both control and shaft angular motion on rotor stiffness. To emphasize the point, we can conclude that conventional rotor types from teetering to hingeless, all flap in much the same way. Of course, a hingeless rotor will not need to flap nearly as much and the pilot might be expected to make smaller control inputs than with an articulated rotor, to produce the same hub moment and hence to fly the same manoeuvre.

The coupled rotor–body motions, whether quasi-steady or with first- or second-order flapping dynamics, are formed from coupling the hub motions with the rotor and driving the hub, and hence the fuselage, with the rotor forces. The expressions for the hub forces and moments in MBC form will now be derived.

Rotor forces and moments

Returning to the fundamental frames of reference given in Appendix 3A, in association with Fig. 3.10, we note that the hub forces in the hub–wind frame can be written as

$$X_{hw} = \sum_{i=1}^{N_b} \int_0^R \left\{ -(f_z - ma_{zb})_i \beta_i \cos \psi_i - (f_y - ma_{yb})_i \sin \psi_i + ma_{xb} \cos \psi_i \right\} dr_b$$

(3.73)

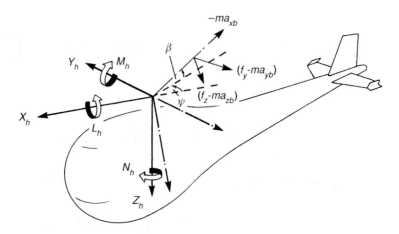

Fig. 3.10 The forces and moments acting on a rotor hub

$$Y_{hw} = \sum_{i=1}^{N_b} \int_0^R \left\{ (f_z - ma_{zb})_i \beta_i \, \sin \psi_i - (f_y - ma_{yb})_i \, \cos \psi_i - ma_{xb} \, \sin \psi_i \right\} dr_b$$

$$(3.74)$$

$$Z_{hw} = \sum_{i=1}^{N_b} \int_0^R (f_z - ma_{zb} + m_{xb}\beta_i)_i \, dr_b \qquad (3.75)$$

The expressions for the inertial accelerations of the blade element are derived in Appendix 3A. The aerodynamic loading is approximated by a simple lift and drag pair, with overall inflow angle assumed small, so that

$$f_z = -\ell \, \cos\phi - d \, \sin\phi \approx -\ell - d\phi \qquad (3.76)$$

$$f_y = d \, \cos\phi - \ell \, \sin\phi \approx d - \ell\phi \qquad (3.77)$$

Performing the integrations analytically using the approximations derived in eqns 3.21 and 3.22, we may write the forces in coefficient form as

$$\left(\frac{2C_{xw}}{a_0 s} \right) = \frac{X_{hw}}{\frac{1}{2}\rho(\Omega R)^2 \pi R^2 s a_0}$$

$$= \frac{1}{N_b} \sum_{i=1}^{N_b} F^{(1)}(\psi_i)\beta_i \, \cos \psi_i + F^{(2)}(\psi_i) \, \sin \psi_i \qquad (3.78)$$

$$\left(\frac{2C_{yw}}{a_0 s} \right) = \frac{Y_{hw}}{\frac{1}{2}\rho(\Omega R)^2 \pi R^2 s a_0}$$

$$= \frac{1}{N_b} \sum_{i=1}^{N_b} -F^{(1)}(\psi_i)\beta_i \, \sin \psi_i + F^{(2)}(\psi_i) \, \cos \psi_i \qquad (3.79)$$

$$\left(\frac{2C_{zw}}{a_0 s} \right) = \frac{Z_{hw}}{\frac{1}{2}\rho(\Omega R)^2 \, \pi R^2 s a_0} = \frac{1}{N_b} \sum_{i=1}^{N_b} -F^{(1)}(\psi_i) = -\left(\frac{2C_T}{a_0 s} \right) \qquad (3.80)$$

where

$$F^{(1)}(\psi_i) = -\int_0^1 \left\{ \overline{U}_T^2 \theta_i + \overline{U}_P \overline{U}_T \right\} d\bar{r}_b \qquad (3.81)$$

is the lift or normal force, and

$$F^{(2)}(\psi_i) = -\int_0^1 \left\{ \overline{U}_P \overline{U}_T \theta_i + \overline{U}_P^2 - \frac{\delta_i \overline{U}_T^2}{a_0} \right\} d\bar{r}_b \qquad (3.82)$$

is the in-plane force, comprising an induced and profile drag component. The rotor solidity s is defined as

$$s = \frac{N_b c}{\pi R} \tag{3.83}$$

The F functions can be expanded to give the expressions

$$
F^{(1)}(\psi) = \left(\frac{1}{3} + \mu \, \sin \psi + \mu^2 \, \sin^2 \psi \right) \theta_p + \left(\frac{1}{4} + \frac{2}{3} \mu \, \sin \psi + \frac{1}{2} \mu^2 \, \sin^2 \psi \right) \theta_{tw}
$$
$$
+ \left(\frac{1}{3} + \frac{\mu \, \sin \psi}{2} \right) (\overline{\omega}_y - \lambda_1 - \beta')
$$
$$
+ \left(\frac{1}{2} + \mu \, \sin \psi \right) (\mu_z - \lambda_0 - \beta\mu \, \cos \psi) \tag{3.84}
$$

$$
F^{(2)}(\psi) = \left\{ \left(\frac{1}{3} + \frac{1}{2} \mu \, \sin \psi \right) (\overline{\omega}_y - \lambda_1 - \beta') \right.
$$
$$
+ \left(\frac{1}{2} + \mu \, \sin \psi \right) (\mu_z - \lambda_0 - \beta\mu \, \cos \psi) \right\} \theta_p
$$
$$
+ \left\{ \left(\frac{1}{4} + \frac{\mu \, \sin \psi}{3} \right) (\overline{\omega}_y - \lambda_1 - \beta') \right.
$$
$$
+ \left(\frac{1}{3} + \frac{\mu \, \sin \psi}{2} \right) (\mu_z - \lambda_0 - \beta\mu \, \cos \psi) \right\} \theta_{tw}
$$
$$
+ (\mu_z - \lambda_0 - \beta\mu \, \cos \psi)^2 + (\mu_z - \lambda_0 - \beta\mu \, \cos \psi)(\overline{\omega}_y - \lambda_1 - \beta')
$$
$$
+ \frac{(\overline{\omega}_y - \lambda_1 - \beta')^2}{3} - \frac{\delta}{a_0} \left(\frac{1}{3} + \mu \, \sin \psi + \mu^2 \, \sin^2 \psi \right) \tag{3.85}
$$

This pair of normal and in-plane forces will produce vibratory (i.e., harmonics of rotorspeed) and quasi-steady loads at the hub. The quasi-steady components in the hub–wind axes system are of chief interest in flight dynamics and can be derived by expanding the loads in the rotating system, given by eqns 3.84 and 3.85 up to second harmonic; thus

$$
F^{(1)}(\psi) = F_0^{(1)} + F_{1c}^{(1)} \, \cos \psi + F_{1s}^{(1)} \, \sin \psi + F_{2c}^{(1)} \, \cos 2\psi + F_{2s}^{(1)} \, \sin 2\psi \tag{3.86}
$$

$$
F^{(2)}(\psi) = F_0^{(2)} + F_{1c}^{(2)} \, \cos \psi + F_{1s}^{(2)} \, \sin \psi + F_{2c}^{(2)} \, \cos 2\psi + F_{2s}^{(2)} \, \sin 2\psi \tag{3.87}
$$

Using eqns 3.78–3.80, we may write the hub force coefficients as

$$
\left(\frac{2C_{xw}}{a_0 s} \right) = \left(\frac{F_0^{(1)}}{2} + \frac{F_{2c}^{(1)}}{4} \right) \beta_{1cw} + \frac{F_{1c}^{(1)}}{2} \beta_0 + \frac{F_{2s}^{(1)}}{4} \beta_{1sw} + \frac{F_{1s}^{(2)}}{2} \tag{3.88}
$$

$$
\left(\frac{2C_{yw}}{a_0 s} \right) = \left(-\frac{F_0^{(1)}}{2} + \frac{F_{2c}^{(1)}}{4} \right) \beta_{1sw} - \frac{F_{1s}^{(1)}}{2} \beta_0 - \frac{F_{2s}^{(1)}}{4} \beta_{1cw} + \frac{F_{1c}^{(2)}}{2} \tag{3.89}
$$

$$\left(\frac{2C_{zw}}{a_0 s}\right) = -\left(\frac{2C_T}{a_0 s}\right) = -F_0^{(1)} \tag{3.90}$$

where the harmonic coefficients are given by the expressions

$$F_0^{(1)} = \theta_0 \left(\frac{1}{3} + \frac{\mu^2}{2}\right) + \frac{\mu}{2}\left(\theta_{1sw} + \frac{\overline{p}_{hw}}{2}\right) + \left(\frac{\mu_z - \lambda_0}{2}\right) + \frac{1}{4}(1 + \mu^2)\theta_{tw} \tag{3.91}$$

$$F_{1s}^{(1)} = \left(\frac{\alpha_{1sw}}{3} + \mu\left(\theta_0 + \mu_z - \lambda_0 + \frac{2}{3}\theta_{tw}\right)\right) \tag{3.92}$$

$$F_{1c}^{(1)} = \left(\frac{\alpha_{1cw}}{3} - \mu\frac{\beta_0}{2}\right) \tag{3.93}$$

$$F_{2s}^{(1)} = \frac{\mu}{2}\left(\frac{\alpha_{1cw}}{2} + \frac{\theta_{1cw} - \beta_{1sw}}{2} - \mu\beta_0\right) \tag{3.94}$$

$$F_{2c}^{(1)} = -\frac{\mu}{2}\left(\frac{\alpha_{1sw}}{2} + \frac{\theta_{1sw} + \beta_{1cw}}{2} + \mu\left(\theta_0 + \frac{\theta_{tw}}{2}\right)\right) \tag{3.95}$$

$$F_{1s}^{(2)} = \frac{\mu^2}{2}\beta_0\beta_{1sw} + \left(\mu_z - \lambda_0 - \frac{\mu}{4}\beta_{1cw}\right)(\alpha_{1sw} - \theta_{1sw}) - \frac{\mu}{4}\beta_{1sw}(\alpha_{1cw} - \theta_{1cw})$$

$$+ \theta_0\left(\frac{\alpha_{1sw} - \theta_{1sw}}{3} + \mu(\mu_z - \lambda_0) - \frac{\mu^2}{4}\beta_{1cw}\right)$$

$$+ \theta_{tw}\left(\frac{\alpha_{1sw} - \theta_{1sw}}{4} + \frac{\mu}{2}\left(\mu_z - \lambda_0 - \frac{\beta_{1cw}\mu}{4}\right)\right)$$

$$+ \theta_{1sw}\left(\frac{\mu_z - \lambda_0}{2} + \mu\left(\frac{3}{8}(\overline{p}_{hw} - \lambda_{1sw}) + \frac{\beta_{1cw}}{4}\right)\right)$$

$$+ \frac{\mu}{4}\theta_{1cw}\left(\frac{\overline{q}_{hw} - \lambda_{1cw}}{2} - \beta_{1sw} - \mu\beta_0\right) - \frac{\delta\mu}{a_0} \tag{3.96}$$

$$F_{1c}^{(2)} = (\alpha_{1cw} - \theta_{1cw} - 2\beta_0\mu)\left(\mu_z - \lambda_0 - \frac{3}{4}\mu\beta_{1cw}\right) - \frac{\mu}{4}\beta_{1sw}(\alpha_{1sw} - \theta_{1sw})$$

$$+ \theta_0\left(\frac{\alpha_{1cw} - \theta_{1cw}}{3} - \frac{\mu}{2}\left(\beta_0 + \frac{\mu}{2}\beta_{1sw}\right)\right)$$

$$+ \theta_{tw}\left(\frac{\alpha_{1cw} - \theta_{1cw}}{4} - \mu\left(\frac{\beta_0}{3} + \frac{\beta_{1sw}\mu}{8}\right)\right)$$

$$+ \theta_{1cw}\left(\frac{\mu_z - \lambda_0}{2} - \frac{\mu}{4}\left(\frac{\overline{p}_{hw} - \lambda_{1sw}}{2} - \beta_{1cw}\right)\right)$$

$$+ \frac{\mu}{4}\theta_{1sw}\left(\frac{\overline{q}_{hw} - \lambda_{1cw}}{2} - \beta_{1sw} - \mu\beta_0\right) \tag{3.97}$$

The effective blade incidence angles are given by

$$\alpha_{1sw} = \overline{p}_{hw} - \lambda_{1sw} + \beta_{1cw} + \theta_{1sw} \tag{3.98}$$

$$\alpha_{1cw} = \overline{q}_{hw} - \lambda_{1cw} - \beta_{1sw} + \theta_{1cw} \tag{3.99}$$

The foregoing expressions for the rotor forces highlight that in the non-rotating hub–wind–shaft axes system, a multitude of physical effects combine to produce the resultants. While the normal force, the rotor thrust, is given by a relatively simple equation, the in-plane forces are very complex indeed. However, some physical interpretation can be made. The $F_0^{(1)}\beta_{1cw}$ and $F_{1c}^{(1)}\beta_0$ components are the first harmonics of the product of the lift and flapping in the direction of motion and represent the contribution to X and Y from blades in the fore and aft positions. The terms $F_{1s}^{(2)}$ and $F_{1c}^{(2)}$ represent the contributions to X and Y from the induced and profile drag acting on the advancing and retreating blades. In hover, the combination of these effects reduces to the simple result that the in-plane contributions from the blade lift forces cancel, and the hub forces are given entirely by the tilt of the rotor thrust vector, i.e.,

$$C_{xw} = C_T\beta_{1cw} \tag{3.100}$$

$$C_{yw} = -C_T\beta_{1sw} \tag{3.101}$$

The assumption that the rotor thrust is normal to the disc throughout the flight envelope provides a common approximation in helicopter flight dynamics, effectively ignoring the many small contributions of the blade lift to the rotor in-plane forces given in the above equations. The approximation fails to model many effects however, particularly in lateral trims and dynamics. As an illustration, Fig. 3.11(a) shows a comparison of the rotor Y force in trim as a function of flight speed for the Helisim Bo105; the disc tilt approximation is grossly in error. The corresponding lateral cyclic comparison is shown in Fig. 3.11(b), indicating that the effect of the approximation on lateral trim is actually less significant. The disc tilt approximation is weakest in manoeuvres, particularly for teetering rotors or articulated rotors with small flapping hinge offsets, when the damping moment is dominated by the rotor lateral force rather than the hub moments. The most significant of the 3.90 series of equations is the first, the zeroth harmonic rotor thrust that appears in normalized form in eqn 3.90 itself. This simple equation is one of the most important in helicopter flight dynamics and we will return to it for more discussion when we explore the rotor downwash in the next section. To complete this rather lengthy derivation of the rotor forces and moments, we need to orient the hub–wind force components into shaft axes and derive the hub moments.

Using the transformation matrix derived in the Appendix, Section 3A.4, namely

$$\Delta = \begin{bmatrix} \cos\psi_w & -\sin\psi_w \\ \sin\psi_w & \cos\psi_w \end{bmatrix} \tag{3.102}$$

we can write the X, Y forces in the shaft axes system aligned along the fuselage nominal plane of symmetry,

$$\begin{bmatrix} C_x \\ C_y \end{bmatrix} = \Delta \begin{bmatrix} C_{xw} \\ C_{yw} \end{bmatrix} \tag{3.103}$$

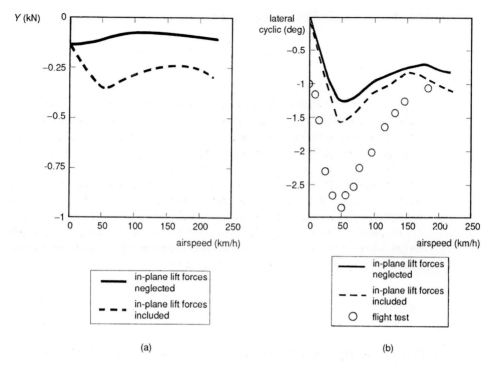

Fig. 3.11 Rotor side force and lateral cyclic variations in trimmed flight: (a) rotor side force (Bo105); (b) lateral cyclic pitch (Bo105)

The rotor hub roll (L) and pitch (M) moments in shaft axes, due to the rotor stiffness effect, are simple linear functions of the flapping angles in MBCs and can be written in the form

$$L_h = -\frac{N_b}{2} K_\beta \beta_{1s} \tag{3.104}$$

$$M_h = -\frac{N_b}{2} K_\beta \beta_{1c} \tag{3.105}$$

The disc flap angles can be obtained from the corresponding hub–wind values by applying the transformation

$$\begin{bmatrix} \beta_{1c} \\ \beta_{1s} \end{bmatrix} = \Delta \begin{bmatrix} \beta_{1cw} \\ \beta_{1sw} \end{bmatrix} \tag{3.106}$$

The hub stiffness can be written in terms of the flap frequency ratio, i.e.,

$$K_\beta = \left(\lambda_\beta^2 - 1 \right) I_\beta \Omega^2$$

showing the relationship between hub moment and flap frequency (cf. eqn 3.32). The equivalent K_β for a hingeless rotor can be three to four times that for an articulated rotor, and it is this amplification, rather than any significant difference in the magnitude of the flapping for the different rotor types, that produces the greater hub moments with hingeless rotors.

Rotor torque

The remaining moment produced by the rotor is the rotor torque and this produces a dominant component about the shaft axis, plus smaller components in pitch and roll due to the inclination of the disc to the plane normal to the shaft. Referring to Fig. 3.10, the torque moment, approximated by the yawing moment in the hub–wind axes, can be obtained by integrating the moments of the in-plane loads about the shaft axis

$$N_h = \sum_{i=1}^{N_b} \int_0^R r_b (f_y - ma_{yb})_i \, dr_b \tag{3.107}$$

We can neglect all the inertia terms except the accelerating torque caused by the rotor angular acceleration, hence reducing eqn 3.107 to the form,

$$N_h = \sum_{i=1}^{N_b} \int_0^R \{r_b(d - \ell\phi)\} \, dr_b + I_R \dot{\Omega} \tag{3.108}$$

where I_R is the moment of inertia of the rotor blades and hub about the shaft axis, plus any additional rotating components in the transmission system. Normalizing the torque equation gives

$$\frac{N_h}{\frac{1}{2}\rho(\Omega R)^2 \pi R^3 s a_0} = \frac{2C_Q}{a_0 s} + \frac{2}{\gamma}\left(\frac{I_R}{N_b I_\beta}\right)\overline{\Omega}' \tag{3.109}$$

where

$$\overline{\Omega}' = \frac{\dot{\Omega}}{\Omega^2} \tag{3.110}$$

and the aerodynamic torque coefficient can be written as

$$\frac{2C_Q}{a_0 s} = -\int_0^1 \overline{r}_b \left(\overline{U}_P \overline{U}_T \theta + \overline{U}_P^2 - \frac{\delta}{a_0}\overline{U}_T^2\right) d\overline{r}_b \equiv \left(\frac{2}{a_0 s}\right)\left(\frac{Q_R}{\rho(\Omega R)^2 \pi R^3}\right) \tag{3.111}$$

where Q_R is the rotor torque.

The above expression for torque can be expanded in a similar manner to the rotor forces earlier in this chapter. The resulting analysis and formulation is extensive and unwieldy, and a considerably simpler, but very effective, approximation can be derived by rearranging the terms in eqn 3.111 as follows.

Writing eqn 3.24 in the alternative approximate form

$$\overline{r}_b \approx \overline{U}_T - \mu \sin \psi \tag{3.112}$$

we may express the rotor torque in the form

$$\frac{2C_Q}{a_0 s} = -\int_0^1 (\overline{U}_T - \mu \sin \psi) \frac{\overline{U}_P}{\overline{U}_T} \overline{\ell} \, \overline{r}_b + \int_0^1 \overline{r}_b \overline{d} \, d\overline{r}_b \qquad (3.113)$$

where the normalized aerodynamic loads are given by the expressions

$$\overline{\ell} = \overline{U}_T^2 \theta + \overline{U}_P \overline{U}_T, \qquad \overline{d} = \frac{\delta}{a_0} \overline{U}_T^2 \qquad (3.114)$$

The three components of torque can then be written as

$$\frac{2C_Q}{a_0 s} = -\left(\int_0^1 \overline{U}_P \overline{\ell} \, d\overline{r}_b \right) + \left(\mu \sin \psi \int_0^1 \frac{\overline{U}_P}{\overline{U}_T} \overline{\ell} \, d\overline{r}_b \right) + \left(\int_0^1 \overline{r}_b \overline{d} \, d\overline{r}_b \right) \qquad (3.115)$$

Expanding eqn 3.115 and making further approximations to neglect small terms leads to the final equation for rotor aerodynamic torque, comprising the induced terms formed from the product of force and velocity and the profile torque, namely

$$\frac{2C_Q}{a_0 s} \approx -(\mu_Z - \lambda_0) \left(\frac{2C_T}{a_0 s} \right) + \mu \left(\frac{2C_{xw}}{a_0 s} \right) + \frac{\delta}{4a_0} (1 + 3\mu^2) \qquad (3.116)$$

The rotor disc tilt relative to the shaft results in components of the torque in the roll and pitch directions. Once again, only one-per-rev roll and pitch moments in the rotating frame of reference will transform through as steady moments in the hub–wind axes. Neglecting the harmonics of rotor torque, we see that the hub moments can therefore be approximated by the orientation of the steady torque through the one-per-rev disc title

$$L_{HQ} = -\frac{Q_R}{2} \beta_{1c} \qquad (3.117)$$

$$M_{HQ} = \frac{Q_R}{2} \beta_{1s} \qquad (3.118)$$

We shall return to the discussion of hub forces and moments later in Section 3.4 and Chapter 4. We still have considerable modelling ground to cover however, not only for the different helicopter components, but also with the main rotor to cover the details of the 'inner' dynamic elements. First, we take a closer look at rotor inflow.

Rotor inflow

The rotor inflow is the name given to the flowfield induced by the rotor at the rotor disc, thus contributing to the local blade incidence and dynamic pressure. In general, the induced flow at the rotor consists of components due to the shed vorticity from all the blades, extending into the far wake of the aircraft. To take account of these effects fully, a complex vortex wake, distorted by itself and the aircraft motion would need to be modelled. We shall assume that for flight dynamics analysis it is sufficient to consider the normal component of inflow, i.e., the rotor-induced downwash. We shall

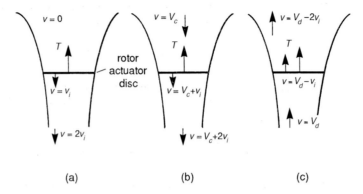

Fig. 3.12 Rotor flow states in axial motion: (a) hover; (b) climb; (c) descent

also make a number of fairly gross assumptions about the rotor and the character of the fluid motion in the wake in order to derive relatively simple formula for the downwash. The use of approximations to the rotor wake for flight dynamics applications has been the subject of two fairly comprehensive reviews of rotor inflow (Refs 3.9, 3.10), which deal with both quasi-static and dynamic effects; the reader is directed towards these works to gain a deeper understanding of the historical development of inflow modelling within the broader context of wake analysis. The simplest representation of the rotor wake is based on actuator disc theory, a mathematical artifact effectively representing a rotor with infinite number of blades, able to accelerate the air through the disc and to support a pressure jump across it. We begin by considering the rotor in axial flight.

Momentum theory for axial flight

Figures 3.12 (a)–(c) illustrate the flow states for the rotor in axial motion, i.e., when the resultant flow is always normal to the rotor disc, corresponding to hover, climbing or descending flight. The flow is assumed to be steady, inviscid and incompressible with a well-defined slipstream between the flowfield generated by the actuator disc (i.e., streamtube extending to infinity) and the external flow. Physically, this last condition is violated in descending flight when the flow is required to turn back on itself; we shall return to this point later. A further assumption we will make is that the pressure in the far wake returns to atmospheric. These assumptions are discussed in detail by Bramwell (Ref. 3.6) and Johnson (Ref. 3.7), and will not be laboured here. The simplest theory that allows us to derive the relationship between rotor thrust and torque and the rotor inflow is commonly known as momentum theory, utilizing the conservation laws of mass, momentum and energy. Our initial theoretical development will be based on the so-called global momentum theory, which assumes that the inflow is uniformly distributed over the rotor disc. Referring to Fig. 3.12, we note that T is the rotor thrust, v the velocity at various stations in the streamtube, v_i the inflow at the disc, V_c the climb velocity and V_d the rotor descent velocity.

First, we shall consider the hover and climb states (Figs 3.12(a), (b)). If $\dot{m}$ is the mass flow rate (constant at each station) and A_d the rotor disc area, then we can write the mass flow through the rotor as

$$\dot{m} = \rho A_d (V_c + v_i) \tag{3.119}$$

The rate of change of momentum between the undisturbed upstream conditions and the far wake can be equated to the rotor loading to give

$$T = \dot{m}(V_c + v_{i\infty}) - \dot{m}V_c = \dot{m}v_{i\infty} \tag{3.120}$$

where $V_{i\infty}$ is the induced flow in the fully developed wake.

The change in kinetic energy of the flow can be related to the work done by the rotor (actuator disc); thus

$$T\left(V_c + v_{i}\right) = \frac{1}{2}\dot{m}\left(V_c + v_{i\infty}\right)^2 - \frac{1}{2}\dot{m}V_c^2 = \frac{1}{2}\dot{m}\left(2V_c v_{i\infty} + v_{i\infty}^2\right) \tag{3.121}$$

From these relationships we can deduce that the induced velocity in the far wake is accelerated to twice the rotor inflow, i.e.,

$$v_{i\infty} = 2v_i \tag{3.122}$$

The expression for the rotor thrust can now be written directly in terms of the conditions at the rotor disc; hence

$$T = 2pA_d(V_c + v_i)v_i \tag{3.123}$$

Writing the inflow in normalized form

$$\lambda_i = \frac{v_i}{\Omega R} \tag{3.124}$$

we may express the hover-induced velocity (with $V_c = 0$) in terms of the rotor thrust coefficient, C_T, i.e.,

$$v_{i_{hover}} = \sqrt{\left(\frac{T}{2\rho A_d}\right)} \quad \text{or} \quad \lambda_{ih} = \sqrt{\left(\frac{C_T}{2}\right)} \tag{3.125}$$

The inflow in the climb situation can be written as

$$\lambda_i = \frac{C_T}{2(\mu_c + \lambda_i)} \tag{3.126}$$

or, derived from the positive solution of the quadratic

$$\lambda_{ih}^2 = (\mu_c + \lambda_i)\lambda_i \tag{3.127}$$

as

$$\lambda_i = -\frac{\mu_c}{2} + \sqrt{\left[\left(\frac{\mu_c}{2}\right)^2 + \lambda_{ih}^2\right]} \tag{3.128}$$

where

$$\mu_c = \frac{V_c}{\Omega R} \tag{3.129}$$

The case of vertical descent is more complicated. Strictly, the flow state satisfies the requirements for the application of momentum theory only in conditions where the wake is fully established above the rotor and the flow is upwards throughout the streamtube. This rotor condition is called the windmill brake state, in recognition of the similarity to a windmill, which extracts energy from the air (Fig. 3.12(c)). The work done by the rotor on the air is now negative and, following a similar analysis to that for the climb, the rotor thrust can be written as

$$T = 2\rho A_d (V_d - v_i) v_i \tag{3.130}$$

The inflow at the disc in the windmill brake state can therefore be written as

$$\lambda_i = \frac{\mu_d}{2} - \sqrt{\left[\left(\frac{\mu_d}{2}\right)^2 - \lambda_{ih}^2\right]} \tag{3.131}$$

where

$$\mu_d = \frac{V_d}{\Omega R} \tag{3.132}$$

The 'physical' solutions of eqns 3.128 and 3.131 are shown plotted as the full lines in Fig. 3.13. The dashed lines correspond to the 'unrealistic' solutions. These solutions include descent rates from hover through to the windmill brake condition, thus encompassing the so-called ideal autorotation condition when the inflow equals the descent rate. This region includes the vortex-ring condition where the wake beneath the rotor becomes entrained in the air moving upwards relative to the rotor outside the wake and, in turn, becoming part of the inflow above the rotor again. This circulating flow forms a toroidal-shaped vortex which has a very non-uniform and unsteady character, leading to large areas of high inflow in the centre of the disc and stall outboard. The vortex-ring condition is not amenable to modelling via momentum considerations alone. However, there is evidence that the mean inflow at the rotor can be approximated by a semi-empirical shaping function linking the helicopter and windmill rotor states shown in Fig. 3.13. The linear approximations suggested by Young (Ref. 3.11) are

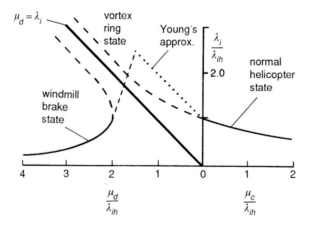

Fig. 3.13 Momentum theory solutions for rotor inflow in axial flight

shown in the figure as the chain dotted lines, and these match the test data gathered by Castles and Gray in the early 1950s (Ref. 3.12) reasonably well. Young's empirical relationships take the form

$$\lambda_i = \lambda_{i_h}\left(1 + \frac{\mu_d}{\lambda_{i_h}}\right) \qquad 0 \le -\mu_d \le -1.5\lambda_{i_h} \tag{3.133}$$

$$\lambda_i = \lambda_{i_h}\left(7 - 3\frac{\mu_d}{\lambda_{i_h}}\right) \qquad -1.5\lambda_{i_h} < -\mu_d \le -2\lambda_{i_h} \tag{3.134}$$

One of the important features of approximations like Young's is that they enable an estimate of the induced velocity in ideal autorotation to be derived. It should be noted that the dashed curve obtained from the momentum solution in Fig. 3.13 never actually crosses the autorotation line. Young's approximation estimates that the autorotation line is crossed at

$$\frac{\mu_d}{\lambda_{i_h}} = 1.8 \tag{3.135}$$

As pointed out by Bramwell (Ref. 3.6), the rotor thrust in this condition equates to the drag of a circular plate of the same diameter as the rotor, i.e., the rotor is descending with a rate of descent similar to that of a parachute.

Momentum theory in forward flight
In high-speed flight the downwash field of a rotor is similar to that of a fixed-wing aircraft with circular planform and the momentum approximations for deriving the induced flow at the wing apply (Ref. 3.13). Figure 3.14 illustrates the flow streamtube, with freestream velocity V at angle of incidence α to the disc, and the actuator disc inducing a velocity v_i at the rotor. The induced flow in the far wake is again twice the flow at the rotor (wing) and the conservation laws give the mass flux as

$$\dot{m} = \rho A_d V_{res} \tag{3.136}$$

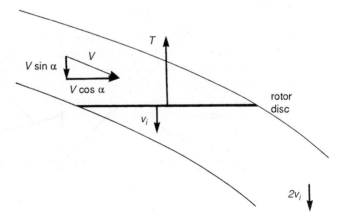

Fig. 3.14 Flow through a rotor in forward flight

and hence the rotor thrust (or wing lift) as

$$T = \dot{m}2v_i = 2\rho A_d V_{res} v_i \tag{3.137}$$

where the resultant velocity at the rotor is given by

$$V_{res}^2 = (V\cos\alpha_d)^2 + (V\sin\alpha_d + v_i)^2 \tag{3.138}$$

Normalizing velocities and rotor thrust in the usual way gives the general expression

$$\lambda_i = \frac{C_T}{2\sqrt{[\mu^2 + (\lambda_i - \mu_z)^2]}} \tag{3.139}$$

where

$$\mu = \frac{V\cos\alpha_d}{\Omega R}, \qquad \mu_z = -\frac{V\sin\alpha_d}{\Omega R} \tag{3.140}$$

and where α_d is the disc incidence, shown in Fig. 3.14. Strictly, eqn 3.139 applies to high-speed flight, where the downwash velocities are much smaller than in hover, but it can be seen that the solution also reduces to the cases of hover and axial motion in the limit when $\mu = 0$. In fact, this general equation is a reasonable approximation to the mean value of rotor inflow across a wide range of flight conditions, including steep descent, and also provides an estimate of the induced power required.

Summarizing, we see that the rotor inflow can be approximated in hover and high-speed flight by the formulae

$$V = 0, \qquad v_i = \sqrt{\left(\frac{T}{2A_d\rho}\right)} \tag{3.141}$$

$$V \gg v_i \qquad v_i = \frac{T}{2VA_d\rho} \tag{3.142}$$

showing the dependence on the square root of disc loading in hover, and proportional to disc loading in forward flight.

Between hover and μ values of about 0.1 (about 40 knots for Lynx), the mean normal component of the rotor wake velocities is still high, but now gives rise to fairly strong non-uniformities along the longitudinal, or, more generally, the flight axis of the disc. Several approximations to this non-uniformity were derived in the early developments of rotor aerodynamic theory using the vortex form of actuator disc theory (Refs 3.14–3.16). It was shown that a good approximation to the inflow could be achieved with a first harmonic with a linear variation along the disc determined by the wake angle relative to the disc, given by

$$\lambda_i = \lambda_0 + \frac{r_b}{R}\lambda_{1cw}\cos\psi_w \tag{3.143}$$

where

$$\lambda_{1cw} = \lambda_0\tan\left(\frac{\chi}{2}\right), \qquad \chi < \frac{\pi}{2}$$

$$\lambda_{1cw} = \lambda_0 \cot\left(\frac{\chi}{2}\right), \qquad \chi > \frac{\pi}{2} \tag{3.144}$$

and the wake angle, χ, is given by

$$\chi = \tan^{-1}\left(\frac{\mu}{\lambda_0 - \mu_z}\right) \tag{3.145}$$

where λ_0 is the uniform component of inflow as given by eqn 3.139.

The solution of eqn 3.144 can be combined with that of eqn 3.139 to give the results shown in Fig. 3.15 where, again, α_d is the disc incidence and V is the resultant velocity of the free stream relative to the rotor. The solution curves for the (non-physical) vertical descent cases are included. It can be seen that the non-uniform component is approximately equal to the uniform component in high-speed straight and level flight, i.e., the inflow is zero at the front of the disc. In low-speed steep descent, the non-uniform component varies strongly with speed and is also of similar magnitude to the uniform component. Longitudinal variations in blade incidence lead to first harmonic lateral flapping and hence rolling moments. Flight in steep descent is often characterized by high vibration, strong and erratic rolling moments and, as the vortex-ring region is entered, loss of vertical control power and high rates of descent (Ref. 3.17). The

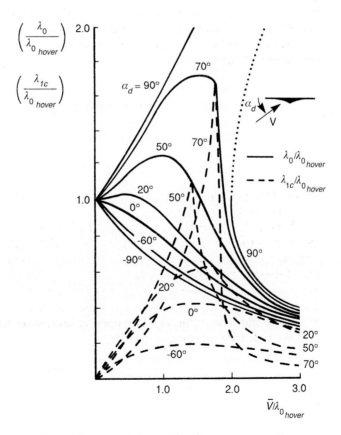

Fig. 3.15 General inflow solution from momentum theory

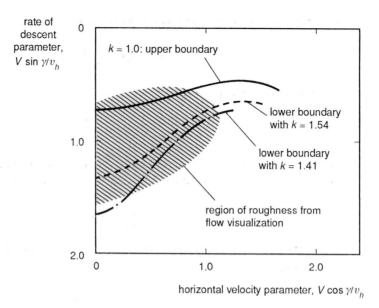

Fig. 3.16 Vortex-ring boundaries (Ref. 3.19)

simple uniform/non-uniform inflow model given above begins to account for some of these effects (e.g., power settling, Ref. 3.18) but cannot be regarded as a proper representation of either the causal physics or flight dynamics effects; in particular, the dramatic loss of control power caused by the build up of the toroidal vortex ring is not captured by the simple model, and recourse to empiricism is required to model this effect. An effective analysis to predict the boundaries of the vortex-ring state, using momentum theory, was conducted in the early 1970s (Ref. 3.19) and extended in the 1990s using classical vortex theory (Ref. 3.20). Wolkovitch's results are summarized in Fig. 3.16, showing the predicted upper and lower boundaries as a function of normalized horizontal velocity; the so-called region of roughness measured previously by Drees (Ref. 3.21) is also shown. The parameter k shown on Fig. 3.16 is an empirical constant scaling the downward velocity of the wake vorticity. The lower boundary is set at a value of $k < 2$, i.e., before the wake is fully contracted, indicating breakdown of the protective tube of vorticity a finite distance below the rotor. Knowledge of the boundary locations is valuable for including appropriate flags in simulation models (e.g., Helisim). Once again though, the simple momentum and vortex theories are inadequate at modelling the flow and predicting flight dynamics within the vortex-ring region. We shall return to this topic in Chapters 4 and 5 when discussing trim and control response.

The momentum theory used to formulate the expressions for the rotor inflow is strictly applicable only in steady flight when the rotor is trimmed and in slowly varying conditions. We can, however, gain an appreciation of the effects of inflow on rotor thrust during manoeuvres through the concept of the lift deficiency function (Ref. 3.7). When the rotor thrust changes, the inflow changes in sympathy, increasing for increasing thrust and decreasing for decreasing thrust. Considering the thrust changes as perturbations on the mean component, we can write

$$\delta C_T = \delta C_{T_{QS}} + \left(\frac{\partial C_T}{\partial \lambda_i}\right)_{QS} \delta \lambda_i \tag{3.146}$$

where, from the thrust equation (eqn 3.91)

$$\left(\frac{\partial C_T}{\partial \lambda_i}\right)_{QS} = -\frac{a_0 s}{4} \tag{3.147}$$

and where the quasi-steady thrust coefficient changes without change in the inflow. Assuming that the inflow changes are due entirely to thrust changes, we can write

$$\delta \lambda_i = \frac{\partial \lambda_i}{\partial C_T} \delta C_T \tag{3.148}$$

The derivatives of inflow with thrust have simple approximate forms at hover and in forward flight

$$\frac{\partial \lambda_i}{\partial C_T} = \frac{1}{4\lambda}, \quad \mu = 0 \tag{3.149}$$

$$\frac{\partial \lambda_i}{\partial C_T} \approx \frac{1}{2\mu}, \quad \mu > 0.2 \tag{3.150}$$

Combining these relationships, we can write the thrust changes as the product of a deficiency function and the quasi-steady thrust change, i.e.,

$$\delta C_T = C' \delta C_{T_{QS}} \tag{3.151}$$

where

$$C' = \frac{1}{1 + \frac{a_0 s}{16 \lambda_i}}, \quad \mu = 0 \tag{3.152}$$

and

$$C' = \frac{1}{1 + \frac{a_0 s}{8\mu}} \quad \mu > 0.2 \tag{3.153}$$

Rotor thrust changes are therefore reduced to about 60–70% in hover and 80% in the mid-speed range due to the effects of inflow. This would apply, for example, to the thrust changes due to control inputs. It is important to note that these deficiency functions do not apply to the thrust changes from changes in rotor velocities. In particular, when the vertical velocity component changes, there are additional inflow perturbations that lead to even further lift reductions. In hover, the deficiency function for vertical velocity changes is half that due to collective pitch changes, i.e.,

$$C'_{\mu_z} = \frac{C'}{2}, \quad \mu = 0 \tag{3.154}$$

In forward flight the lift loss is recovered and eqn 3.151 also applies to the vertical velocity perturbations. This simple analysis demonstrates how the gust sensitivity of rotors increases strongly from hover to mid speed, but levels out to the constant quasi-steady value at high speed (see discussion on vertical gust response in Chapter 2).

Because the inflow depends on the thrust and the thrust depends on the inflow, an iterative solution is required. Defining the zero function g_0 as

$$g_0 = \lambda_0 - \left(\frac{C_T}{2\Lambda^{1/2}} \right) \tag{3.155}$$

where

$$\Lambda = \mu^2 + (\lambda_0 - \mu_z)^2 \tag{3.156}$$

and recalling that the thrust coefficient can be written as (eqn 3.91)

$$C_T = \frac{a_0 s}{2} \left(\theta_0 \left(\frac{1}{3} + \frac{\mu^2}{2} \right) + \frac{\mu}{2} \left(\theta_{1sw} + \frac{\overline{p}_w}{2} \right) + \left(\frac{\mu_z - \lambda_0}{2} \right) + \frac{1}{4} \left(1 + \mu^2 \right) \theta_{tw} \right) \tag{3.157}$$

Newton's iterative scheme gives

$$\lambda_{0_{j+1}} = \lambda_{0_j} + f_j h_j(\lambda_{0_j}) \tag{3.158}$$

where

$$h_j = - \left(\frac{g_0}{\mathrm{d}g_0/\mathrm{d}\lambda_0} \right)_{\lambda=\lambda_{0_j}} \tag{3.159}$$

i.e.,

$$h_j = - \frac{\left(2\lambda_{0_j}\Lambda^{1/2} - C_T \right)\Lambda}{2\Lambda^{3/2} + \frac{a_0 s}{4}\Lambda - C_T(\mu_z - \lambda_{0_j})} \tag{3.160}$$

For most flight conditions the above scheme should provide rapid estimates of the inflow at time t_{j+1} from a knowledge of conditions at time t_j. The stability of the algorithm is determined by the variation of the function g_0 and the initial value of λ_0. However, in certain flight conditions near the hover, the iteration can diverge, and the damping constant f is included to stabilize the calculation; a value of 0.6 for f appears to be a reasonable compromise between achieving stability and rapid convergence (Ref. 3.4).

A further approximation involved in the above inflow formulation is the assumption that the freestream velocity component normal to the disc (i.e., $V \sin \alpha_d$) is the same as μ_z. This is a reasonable approximation for small flapping angles, and even for the larger angles typical of low-speed manoeuvres the errors are small because of the insensitivity of the inflow to disc incidence (see Fig. 3.15). The approximation is convenient because there is no requirement to know the disc tilt or rotor flapping relative to the shaft to compute the inflow, hence leading to a further simplification in the iteration procedure.

The simple momentum inflow derived above is effective in predicting the gross and slowly varying uniform and rectangular, wake-induced, inflow components. In practice, the inflow distribution varies with flight condition and unsteady rotor loading (e.g., in manoeuvres) in a much more complex manner. Intuitively, we can imagine the inflow varying around the disc and along the blades, continuously satisfying local flow balance conditions and conservation principles. Locally, the flow must respond to local changes in blade loading, so if, for example, there are one-per-rev rotor forces and moments, we might expect the inflow to be related to these. We can also expect the inflow to take a finite time to develop as the air mass is accelerated to its new velocity. Also, the rotor wake is far more complex and discrete than the uniform flow in a streamtube assumption of momentum theory. It is known that local blade–vortex interactions can cause very large local perturbations in blade inflow and hence incidence. These can be sufficient to stall the blade in certain conditions and are important for predicting rotor stall boundaries and the resulting flight dynamics at the flight envelope limits. We shall return to this last topic later in the discussion on advanced, high-fidelity modelling. Before leaving inflow however, we shall examine the theoretical developments needed to improve the prediction of the non-uniform and unsteady components.

Local-differential momentum theory and dynamic inflow
We begin by considering the simple momentum theory applied to the rotor disc element shown in Fig. 3.17. We make the gross assumption that the relationship between the change in momentum and the work done by the load across the element applies locally as well as globally, giving the equations for the mass flow through the element and the thrust differential as shown in eqns 3.161 and 3.162.

$$\mathrm{d}\dot{m} = \rho V_{r_b} \, \mathrm{d}r_b \, \mathrm{d}\psi \qquad (3.161)$$

$$\mathrm{d}T = \mathrm{d}\dot{m} \, 2v_i \qquad (3.162)$$

Using the two-dimensional blade element theory, these can be combined into the form

$$\frac{N_b}{2\pi} \left(\frac{1}{2}\rho a_0 c \left(\theta \overline{U}_T^2 + \overline{U}_T \overline{U}_p \right) \mathrm{d}r_b \, \mathrm{d}\psi \right) = 2\rho r_b \left(\mu^2 + (\lambda_i - \mu_z)^2 \right)^{1/2} \lambda_i \, \mathrm{d}r_b \, \mathrm{d}\psi$$

$$(3.163)$$

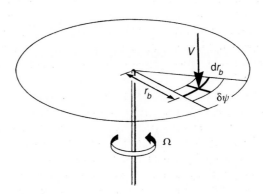

Fig. 3.17 Local momentum theory applied to a rotor disc

Integrating around the disc and along the blades leads to the solution for the mean uniform component of inflow derived earlier. If, instead of averaging the load around the disc, we apply the momentum balance to the one-per-rev components of the load and inflow, then expressions for the non-uniform inflow can be derived. Writing the first harmonic inflow in the form

$$\lambda_i = \lambda_0 + \bar{r}_b(\lambda_{1c}\cos\psi + \lambda_{1s}\sin\psi) \tag{3.164}$$

eqn 3.163 can be expanded to give a first harmonic balance which, in hover, results in the expressions

$$\lambda_{1c} = \frac{3a_0 s}{16}\frac{1}{\lambda_0}F_{1c}^{(1)} \tag{3.165}$$

and

$$\lambda_{1s} = \frac{3a_0 s}{16}\frac{1}{\lambda_0}F_{1s}^{(1)} \tag{3.166}$$

where the F loadings are given by eqns 3.92 and 3.93. These one-per-rev lift forces are closely related to the aerodynamic moments at the hub in the non-rotating fuselage frame – the pitching moment C_{Ma} and the rolling moment C_{La}, i.e.,

$$\frac{2C_{La}}{a_0 s} = -\frac{3}{8}F_{1s}^{(1)} \tag{3.167}$$

$$\frac{2C_{Ma}}{a_0 S} = -\frac{3}{8}F_{1c}^{(1)} \tag{3.168}$$

These hub moments are already functions of the non-uniform inflow distributions; hence, just as with the rotor thrust and the uniform inflow, we find that the moments are reduced by a similar moment deficiency factor

$$C_{La} = C_1' C_{LaQS} \tag{3.169}$$

$$C_{Ma} = C_1' C_{MaQS} \tag{3.170}$$

where, as before, the deficiency factors are given by

$$C_1' = \frac{1}{1 + \frac{a_0 s}{16\lambda_0}} \tag{3.171}$$

in hover, with typical value 0.6, and

$$C_1' = \frac{1}{1 + \frac{a_0 s}{8\mu}} \tag{3.172}$$

in forward flight, with typical value of 0.8 when $\mu = 0.3$. In hover, the first harmonic inflow components given by eqns 3.165 and 3.166 can be expanded as

$$\lambda_{1c} = C'_1 \frac{a_0 s}{16\lambda_0}(\theta_{1c} - \beta_{1s} + \bar{q}) \tag{3.173}$$

$$\lambda_{1s} = C'_1 \frac{a_0 s}{16\lambda_0}(\theta_{1s} + \beta_{1c} + \bar{p}) \tag{3.174}$$

As the rotor blade develops an aerodynamic moment, the flowfield responds with the linear, harmonic distributions derived above. The associated deficiency factors have often been cited as the cause of mismatches between theory and test (Refs 3.9,3.22–3.29), and there is no doubt that the resulting overall effects on flight dynamics can be significant. The assumptions are fragile however, and the theory can, at best, be regarded as providing a very approximate solution to a complex problem. More recent developments, with more detailed spatial and temporal inflow distributions, are likely to offer even higher fidelity in rotor modelling (see Pitt and Peters, Ref 3.26, and the series of Peters' papers from 1983, Refs 3.27–3.29).

The inflow analysis outlined above has ignored any time dependency other than the quasi-steady effects and harmonic variations. In reality, there will always be a transient lag in the build-up or decay of the inflow field; in effect, the flow is a dynamic element in its own right. An extension of momentum theory has also been made to include the dynamics of an 'apparent' mass of fluid, first by Carpenter and Fridovitch in 1953 (Ref. 3.30). To introduce this theory, we return to axial flight; Carpenter and Fridovitch suggested that the transient inflow could be taken into account by including an accelerated mass of air occupying 63.7% of the air mass of the circumscribed sphere of the rotor. Thus, we write the thrust balancing the mass flow through the rotor to include an apparent mass term

$$T = 0.637\rho \frac{4}{3}\pi R^3 \dot{v}_i + 2A_d \rho v_i (V_c + v_i) \tag{3.175}$$

To understand how this additional effect contributes to the motion, we can linearize eqn 3.175 about a steady hover trim; writing

$$\lambda_i = \lambda_{i_{trim}} + \delta\lambda_i \tag{3.176}$$

and

$$C_T = C_{T_{trim}} + \delta C_T \tag{3.177}$$

the perturbation equation takes the form

$$\tau_\lambda \dot{\lambda}_i + \delta\lambda_i = \lambda_{C_T} \delta C_T \tag{3.178}$$

where the time constant and the steady-state inflow gain are given by

$$\tau_\lambda = \frac{0.849}{4\lambda_{i_{trim}}\Omega}, \qquad \lambda_{C_T} = \frac{1}{4\lambda_{i_{trim}}} \tag{3.179}$$

For typical rotors, moderately loaded in the hover, the time constant for the uniform inflow works out at about 0.1 s. The time taken for small adjustments in uniform inflow is therefore very rapid, according to simple momentum considerations, but this estimate is clearly a linear function of the 'apparent mass'. Since this early work, the concept of dynamic inflow has been developed by a number of researchers, but it is the work of Peters, stemming from the early Ref. 3.23 and continuing through to Ref. 3.29, that has provided the most coherent perspective on the subject from a fluid mechanics standpoint. The general formulation of a 3-DoF dynamic inflow model can be written in the form

$$
\begin{bmatrix} \mathbf{M} \end{bmatrix} \begin{Bmatrix} \dot{\lambda}_0 \\ \dot{\lambda}_{1s} \\ \dot{\lambda}_{1c} \end{Bmatrix} + \begin{bmatrix} \mathbf{L} \end{bmatrix}^{-1} \begin{Bmatrix} \lambda_0 \\ \lambda_{1s} \\ \lambda_{1c} \end{Bmatrix} = \begin{Bmatrix} C_T \\ C_L \\ C_M \end{Bmatrix}
\tag{3.180}
$$

The matrices $\mathbf{M}$ and $\mathbf{L}$ are the apparent mass and gain functions respectively; C_T, C_L and C_M are the thrust, rolling and pitching aerodynamic moment perturbations inducing the uniform and first harmonic inflow changes. The mass and gain matrices can be derived from a number of different theories (e.g., actuator disc, vortex theory). In the most recent work, Peters has extended the modelling to an unsteady three-dimensional finite-state wake (Ref. 3.29) which embraces the traditional theories of Theordorsen and Lowey (Ref. 3.31). Dynamic inflow will be discussed again in the context of stability and control derivatives in Chapter 4, and the reader is referred to Refs 3.28 and 3.29 for full details of the aerodynamic theory.

Before discussing additional rotor dynamic DoFs and progressing on to other helicopter components, we return to the centre-spring model for a further examination of its merits as a general approximation.

Rotor flapping–further considerations of the centre-spring approximation
The centre-spring equivalent rotor, a rigid blade analogue for modelling all types of blade flap retention systems, was originally proposed by Sissingh (Ref. 3.32) and has considerable appeal because of the relatively simple expressions, particularly for hub moments, that result. However, even for moderately stiff hingeless rotors like those on the Lynx and Bo105, the blade shape is rather a gross approximation to the elastic deformation, and a more common approximation used to model such blades is the offset-hinge and spring analogue originally introduced by Young (Ref. 3.33). Figure 3.18 illustrates the comparison between the centre-spring, offset-hinge and spring and a typical first elastic mode shape. Young proposed a method for determining the values of offset-hinge and spring strength, the latter from the non-rotating natural flap frequency, which is then made up with the offset to match the rotating frequency. The ratio of offset to spring strength is not unique and other methods for establishing the mix have been proposed; for example, Bramwell (Ref. 3.34) derives an expression for the offset e in terms of the first elastic mode frequency ratio λ_1 in the form

$$
e = \frac{\lambda_1^2 - 1}{\lambda_1}
\tag{3.181}
$$

with the spring strength in this case being zero. In Reichert's method (Ref. 3.35), the offset hinge is located by extending the first mode tip tangent to meet the undeformed reference line. The first elastic mode frequency is then made up with the addition of a spring, which can have a negative stiffness. Approximate modelling options therefore range from the centre spring out to Bramwell's limit with no spring. The questions

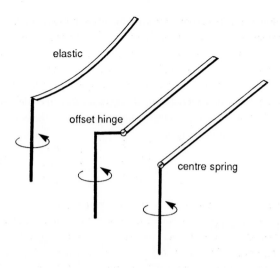

Fig. 3.18 Different approximate models for a hingeless rotor blade

that naturally arise are, first, whether these different options are equivalent or what are the important differences in the modelling of flapping motion and hub moments, and second, which is the most appropriate model for flight dynamics applications? We will try to address these questions in the following discussion.

We refer to the analysis of elastic blade flapping at the beginning of Chapter 3 and the series of equations from 3.8 to 3.16, developing the approximate expression for the hub flap moment due to rotor stiffness in the form

$$M_h^{(r)}(0, t) \approx \Omega^2 \left(\lambda_1^2 - 1 \right) P_1(t) \int_0^R mr\, S_1\, dr \tag{3.182}$$

where S_1 is the first elastic mode shape and P_1 is the time-dependent blade tip deflection. The 'mode shape' of the offset-hinge model, with flap hinge at eR, can be written in the form

$$\begin{aligned} S_1(r) &= 0 & 0 \le r \le eR \\ S_1(r) &= \frac{r - eR}{R(1 - e)} & eR \le r \le R \end{aligned} \tag{3.183}$$

If we substitute eqn 3.183 into eqn 3.182, we obtain the hub flap moment

$$M_h^{(r)}(0, t) = \Omega^2 I_\beta \left(\lambda_\beta^2 - 1 \right) \beta(t) \left(1 + \frac{eRM_\beta}{I_\beta} \right) \tag{3.184}$$

where

$$I_\beta = \int_{eR}^R m(r - eR)^2\, dr, \quad M_\beta = \int_{eR}^R m(r - eR)\, dr \tag{3.185}$$

and the tip deflection is approximately related to the flapping angles by the linear expression

$$P_1(t) \approx R\beta_1(t) \approx R(1 - e)\beta(t) \tag{3.186}$$

The expression for the flap frequency ratio λ_β can be derived from the same method of analysis used for the centre-spring model. Thus, the equation for the flapping motion can be written in the form

$$\beta'' + \lambda_\beta^2 \beta = \left(1 + \frac{eRM_\beta}{I_\beta}\right)\sigma_x + \frac{\gamma}{2} \int_{eR}^{R} (\overline{U}_T^2 \theta + \overline{U}_T \overline{U}_P)(\overline{r}_b - e)\, \mathrm{d}\overline{r}_b \tag{3.187}$$

where, as before

$$\beta' = \frac{\mathrm{d}\beta}{\mathrm{d}\psi}$$

and the Lock number is given by

$$\gamma = \frac{\rho ca_0 R^4}{I_\beta} \tag{3.188}$$

The in-plane and normal velocity components at the disc are given by (cf. eqns 3.24 and 3.25)

$$\overline{U}_T = \overline{r}_b(1 + \overline{\omega}_x \beta) + \mu \sin\psi$$
$$\overline{U}_P = \mu_z - \lambda_0 - \mu\beta \cos\psi + \overline{r}_b(\overline{\omega}_y - \lambda_1) - (\overline{r}_b - e)\beta' \tag{3.189}$$

and the combined inertial acceleration function is given by the expression

$$\sigma_x = (\overline{p}' - 2\overline{q}) \sin\psi + (\overline{q}' + 2\overline{p}) \cos\psi \tag{3.190}$$

Finally, the flap frequency ratio is made up of a contribution from the spring stiffness and another from the offset hinge, given by

$$\lambda_\beta^2 = 1 + \frac{K_\beta}{I_\beta \Omega^2} + \frac{eRM_\beta}{I_\beta} \tag{3.191}$$

The hub moment given by eqn 3.184 is clearly in phase with the blade tip deflection. However, a more detailed analysis of the dynamics of the offset-hinge model developed by Bramwell (Ref. 3.34) reveals that this simple phase relationship is not strictly true for the offset-hinge model. Referring to Fig. 3.19, the hub flap moment can be written as the sum of three components, i.e.,

$$M^{(r)}(0, t) = K_\beta \beta - eRS_z + \int_0^{eR} F(r, t)r\, \mathrm{d}r \tag{3.192}$$

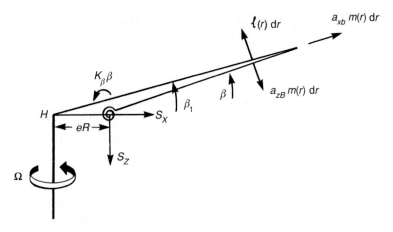

Fig. 3.19 The offset-hinge model of rotorblade flapping

The shear force at the flap hinge is given by the balance of integrated aerodynamic ($F(r, t)$) and inertial loads on the blade; thus

$$S_z = -\int_{eR}^{R} [F(r, t) - m(r - eR)\ddot{\beta}]\, dr \qquad (3.193)$$

If we assume a first harmonic flap response so that

$$\ddot{\beta} = -\Omega^2 \beta \qquad (3.194)$$

then the flap moment takes the form

$$M^{(r)}(0, t) = \Omega^2 I_\beta \left(\lambda_\beta^2 - 1\right) \beta(t) + eR \int_{eR}^{R} F(r, t)\, dr + \int_{0}^{eR} F(r, t)\, dr \qquad (3.195)$$

The third component due to the lift on the flap arm is $O(e^3)$ in the hover and will be neglected. The result given by eqn 3.195 indicates that the hub moment will be out of phase with blade flapping to the extent that any first harmonic aerodynamic load is out of phase with flap. Before examining this phase relationship in a little more detail, we need to explain the inconsistency between Young's result above in eqn 3.184 and the correct expression given by eqn 3.195. To uncover the anomaly it is necessary to return to the primitive expression for the hub flap moment derived from bending theory (cf. eqn. 3.13):

$$M_h^{(r)}(0, t) = \int_{0}^{R} \left[F(r, t) - m \left(\frac{\partial^2 w}{\partial t^2} + \Omega^2 w \right) \right] r\, dr \qquad (3.196)$$

Using eqns 3.9 and 3.10, the hub moment can then be written in the form

$$M^{(r)}(0,\ t) = \int_0^R F(r,\ t)r\ dr - \sum_{n=1}^{\infty} \frac{\int_0^R mr\,S_n\ dr}{\int_0^R m S_n^2\ dr} \int_0^R F(r,\ t)S_n\ dr$$

$$+ \Omega^2 \sum_{n=1}^{\infty} \left(\lambda_n^2 - 1\right) P_n \int_0^R mr\,S_n\ dr \tag{3.197}$$

If an infinite set of modes is included in the hub moment expression, then the first two terms in eqn 3.197 cancel, leaving each modal moment in phase with its corresponding blade tip deflection. With only a finite number of modes included, this is no longer the case (Bramwell, Ref. 3.34). In particular, if only the first elastic mode is retained, then the hub flap moment has a residual

$$M^{(r)}(0,\ t) = \int_0^R F(r,\ t)\left(r - \frac{\int_0^R mr\,S_1\ dr}{\int_0^R m S_1^2\ dr} S_1\right) dr + \Omega^2 \left(\lambda_1^2 - 1\right) P_1 \int_0^R mr\,S_1\ dr \tag{3.198}$$

When the aerodynamic loading has the same shape as the first mode, i.e.,

$$F(r,\ t) \propto m S_1 \tag{3.199}$$

then the first term in eqn 3.198 vanishes and the hub moment expression reduces to that given by Young (Ref. 3.33). These conditions will not, in general, be satisfied since, even in hover, there are r^2 terms in the aerodynamic loading. Substituting the mode shape for the offset hinge, given by eqn 3.183, into eqn 3.198, leads to the correct hub moment with the out-of-phase aerodynamic component as given by eqn 3.195. Neglecting the effect of the in-plane loads, we see that the roll and pitch hub flap moments applied to the fuselage from a single blade in non-rotating coordinates, are given by the transformation

$$L_h = -M^{(r)} \sin \psi \tag{3.200}$$

$$M_h = -M^{(r)} \cos \psi \tag{3.201}$$

Substituting for the aerodynamic loads in eqns 3.200 and 3.201 and expanding to give the quasi-steady (zeroth harmonic) components, leads to the hover result

$$\frac{2L_h}{I_\beta \Omega^2} = -\left(\lambda_\beta^2 - 1\right)\beta_{1s} - \frac{eRM_\beta}{I_\beta}\left(1 + \frac{eRM_{\beta 0}}{M_\beta}\right)(\overline{p}' - 2\overline{q})$$

$$- e\frac{\gamma}{2}\left(\frac{\overline{p} + \beta_{1c}\left(1 - \frac{3}{2}e\right) + \theta_{1s}}{3}\right) \tag{3.202}$$

$$\frac{2M_h}{I_\beta \Omega^2} = -\left(\lambda_\beta^2 - 1\right)\beta_{1c} - \frac{eRM_\beta}{I_\beta}\left(1 + \frac{eRM_{\beta_0}}{M_\beta}\right)(\bar{q}' + 2\bar{p})$$

$$- e\frac{\gamma}{2}\left(\frac{\bar{q} + \beta_{1s}\left(1 - \frac{3}{2}e\right) + \theta_{1c}}{3}\right) \tag{3.203}$$

The blade mass coefficient is given by

$$M_{\beta_0} = \int_{eR}^{R} m\, d\bar{r}_b \tag{3.204}$$

The inertial and aerodynamic components proportional to the offset e in the above are clearly absent in the centre-spring model when the hub moment is always in phase with the flapping. The extent to which the additional terms are out of phase with the flapping can be estimated by examining the hub moment derivatives. By far the most significant variations with offset appear in the control coupling derivatives. Expressions for the flapping derivatives can be derived from the harmonic solutions to the flapping equations; hence

$$\frac{\partial \beta_{1c}}{\partial \theta_{1c}} = \beta_{1c\theta_{1c}} = \beta_{1s\theta_{1s}} = \frac{S_\beta}{d_\beta}\left(1 - \frac{4}{3}e\right) \tag{3.205}$$

$$\frac{\partial \beta_{1c}}{\partial \theta_{1s}} = \beta_{1c\theta_{1s}} = -\beta_{1s\theta_{1c}} = -\frac{1}{d_\beta}\left(1 - \frac{8}{3}e\right)\left(1 - \frac{4}{3}e\right) \tag{3.206}$$

where

$$d_\beta = S_\beta^2 + \left(1 - \frac{8}{3}e\right)^2 \tag{3.207}$$

The hub roll moment control derivatives can therefore be written to an accuracy of $O(e^2)$ in the form

$$\frac{2L_{h\theta_{1c}}}{I_\beta \Omega^2}\left(\frac{8}{\gamma}\right) = -S_\beta \beta_{1s\theta_{1c}} - eR\frac{4}{3}\beta_{1c\theta_{1c}}\left(1 - \frac{3}{2}e\right) \tag{3.208}$$

$$\frac{2L_{h\theta_{1s}}}{I_\beta \Omega^2}\left(\frac{8}{\gamma}\right) = -S_\beta \beta_{1s\theta_{1s}} - eR\frac{4}{3}\left[1 + \beta_{1c\theta_{1s}}\left(1 - \frac{3}{2}e\right)\right] \tag{3.209}$$

To compare numerical values for the roll control derivatives with various combinations of offset and spring stiffness, it is assumed that the flap frequency ratio λ_β and the blade Lock number remain constant throughout. These would normally be set using the corresponding values for the first elastic flap mode frequency and the modal inertia given by eqn 3.11. The values selected are otherwise arbitrary and uses of the offset-spring model in the literature are not consistent in this regard. We chose to draw our comparison for a moderately stiff rotor, with $\gamma_\beta^2 = 1.2$ and $S_\beta = 0.2$. Figure 3.20 shows a cross-plot of the flap control derivatives for values of offset e extending out to

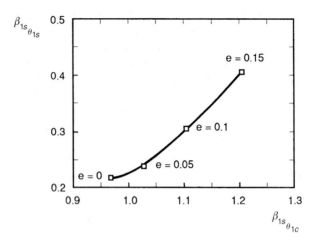

Fig. 3.20 Cross-plot of rotor flap control derivatives

0.15. With $e = 0$, the flap frequency ratio is augmented entirely with the centre spring; at $e = 0.15$, the offset alone determines the augmented frequency ratio. The result shows that the rotor flapping changes in character as hinge offset is increased, with the flap/control phase angle decreasing from about 80° for the centre-spring configuration to about 70° with 15% offset. The corresponding roll and pitch hub moment derivatives are illustrated in Fig. 3.21 for the same case. Figure 3.21 shows that over the range of offset-hinge values considered, the primary control derivative increases by 50% while the cross-coupling derivative increases by over 100%. The second curve in Fig. 3.21 shows the variation of the hub moment in phase with the flapping. It can be seen that more than 50% of the change in the primary roll moment derivative is due to the aerodynamic moment from disc flapping in the longitudinal direction. These moments

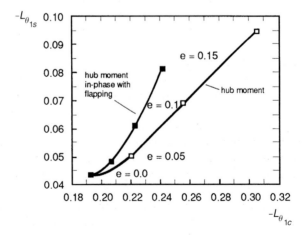

Fig. 3.21 Cross-plot of roll control derivatives as a function of flap hinge offset

could not be developed from just the first mode of an elastic blade and are a special feature of large offset-hinge rotors.

The results indicate that there is no simple equivalence between the centre-spring model and the offset-hinge model. Even with Young's approximation, where the aerodynamic shear force at the hinge is neglected, the flapping is amplified as shown above. A degree of equivalence, at least for control moments, can be achieved by varying the blade inertia as the offset hinge is increased, hence increasing the effective Lock number, but the relationship is not obvious. Even so, the noticeable decrease in control phasing, coupled with the out-of-phase moments, gives rise to a dynamic behaviour which is not representative of the first elastic flap mode. On the other hand, the appeal of the centre-spring model is its simplicity, coupled with the preservation of the correct phasing between control and flapping and between flapping and hub moment. The major weakness of the centre-spring model is the crude approximation to the blade shape and corresponding tip deflection and velocity, aspects where the offset-hinge model is more representative.

The selection of parameters for the centre-spring model is relatively straightforward. In the case of hingeless or bearingless rotors, the spring strength and blade inertia are chosen to match the first elastic mode frequency ratio and modal inertia respectively. For articulated rotors, the spring strength is again selected to give the correct flap frequency ratio, but now the inertia is changed to match the rotor blade Lock number about the real offset flap hinge.

It needs to be remembered that the rigid blade models discussed above are only approximations to the motion of an elastic blade and specifically to the first cantilever flap mode. In reality, the blade responds by deforming in all of its modes, although the contribution of higher bending modes to the quasi-steady hub moments is usually assumed to be small enough to be neglected. As part of a study of hingeless rotors, Shupe (Refs 3.36 and 3.37) examined the effects of the second flap bending mode on flight dynamics. Because this mode often has a frequency close to three-per-rev, it can have a significant forced response, even at one-per-rev, and Shupe has argued that the inclusion of this effect is important at high speed. This brings us to the domain of aeroelasticity and we defer further discussion until Section 3.4, where we shall explore higher fidelity modelling issues in more detail.

Rotor blades need to lag and twist in addition to flap, and here we discuss briefly the potential contributions of these DoFs to helicopter flight dynamics.

Rotor in-plane motion – lead–lag

Rigid or elastic lead–lag blade motion attenuates the in-plane forces on the rotor. On articulated rotors, the rigid-blade lead–lag motion revolves about an offset hinge, necessary to enable the applied torque to rotate the rotor. On hingeless rotors, lead–lag takes the form of in-plane bending. Because the in-plane aerodynamic damping forces are low, it is usual to find mechanical dampers attached to the lead–lag hinge. Additional mechanical in-plane damping is even found on some hingeless rotors. A comprehensive discussion on the significance of lead–lag on blade stability and loads is provided by Johnson in Ref. 3.7. For most flight mechanics analysis, the presence of lead–lag motion contributes little to the overall response and stability of the helicopter. There is one aspect that is relevant and needs to be referred to, however. To aid the discussion, the coupled equations of flap/lead–lag motion are required; for the present purposes, we assume that the flap and lag blade inertias are equal and describe the

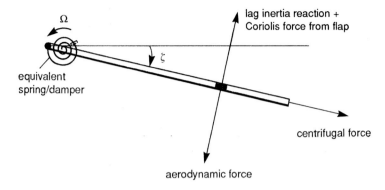

Fig. 3.22 Rotor blade lag motion

coupled motion in the simplified form:

$$\beta'' + \lambda_\beta^2 \beta - 2\beta\dot{\zeta} = M_F \tag{3.210}$$

$$\zeta'' + C_\zeta\dot{\zeta} + \lambda_\zeta^2\zeta + 2\beta\dot{\beta} = M_L \tag{3.211}$$

We assume that both the flap (β) and lead–lag (ζ) motion can be approximated by the centre-spring equivalent model as illustrated in Figs 3.6 and 3.22. The direct inertial forces are balanced by restoring moments; in the case of the lag motion, the centrifugal stiffening works only with an offset lag hinge (or centre-spring emulation of centrifugal stiffness). If the lag hinge offset is e_ζ, then the frequency is given by

$$\lambda_\zeta^2 = \frac{3}{2}\left(\frac{e_\zeta}{1 - e_\zeta}\right) \tag{3.212}$$

The natural lag frequency λ_ζ is typically about 0.25Ω for articulated rotors; hingeless rotors can have subcritical ($< \Omega$, e.g., Lynx, Bo105) or supercritical ($> \Omega$, e.g., propellers) lag frequencies, but λ_ζ should be far removed from Ω to reduce the amount of in-plane lag response to excitation. The flap and lag equations above have a similar form. We have included a mechanical viscous lag damper C_ζ for completeness. M_F and M_L are the aerodynamic flap and lag moments. Flap and lag motions are coupled, dynamically through the Coriolis forces in eqns 3.210 and 3.211, and aerodynamically from the variations in rotor blade lift and drag forces. The Coriolis effects are caused by blade elements moving radially as the rotor flaps and lags. Because of the lower inherent damping in lag, the Coriolis moment tends to be more significant in the lag equation due to flap motion. In addition, the lag aerodynamic moment M_L will be strongly influenced by in-plane lift forces caused by application of blade pitch and variations in induced inflow. The impact of these effects will be felt in the frequency range associated with the coupled rotor/fuselage motions. In terms of MBCs, the regressing and advancing lag modes will be located at ($\Omega - \lambda_\zeta$) and ($\Omega + \lambda_\zeta$) respectively. A typical layout of the uncoupled flap and lag modes is shown on the complex eigenvalue plane in Fig. 3.23. The flap modes are well damped and located far into the left plane. In contrast, the lag modes are often weakly damped, even with mechanical

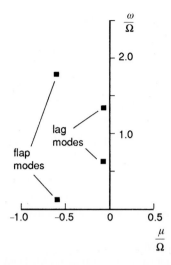

Fig. 3.23 Flap and lag mode eigenvalues

dampers, and are more susceptible to being driven unstable. The most common form of stability problem associated with the lag DoF is ground resonance, whereby the coupled rotor/fuselage/undercarriage system develops a form of 'flutter'; the in-plane rotation of the rotor centre of mass resonates with the fuselage/undercarriage system.

Another potential problem, seemingly less well understood, arises through the coupling of rotor and fuselage motions in flight. Several references examined this topic in the early days of hingeless rotor development (Refs 3.38, 3.39), when the emphasis was on avoiding any hinges or bearings at the rotor hub to simplify the design and maintenance procedures. Control of rotor in-plane motion and loads through feedback of roll motion to cyclic pitch was postulated. This design feature has never been exploited, but the sensitivity of lag motion to attitude feedback control has emerged as a major consideration in the design of autostabilization systems. The problem is discussed in Ref. 3.6 and can be attributed to the combination of aerodynamic effects due to cyclic pitch and the powerful Coriolis moment in eqn 3.211. Both the regressing and advancing lag modes are at risk here. In Ref. 3.40, Curtiss discusses the physical origin of the couplings and shows an example where the advancing lag mode actually goes unstable at a relatively low value of gain in a roll rate to lateral cyclic feedback control system ($-0.2°/°/s$). In contrast, the roll regressing mode can be driven unstable at higher values of roll attitude feedback gain. The results of Ref. 3.40 and the later Bo105 study by Tischler (Ref. 3.41) give clear messages to the designers of autostabilizers and, particularly, high gain active control systems for helicopters. Designs will need to be evaluated with models that include the lead–lag dynamics before implementation on an aircraft. However, the modelling requirements for specific applications are likely to be considerably more complex than is implied by the simple analysis outlined above. Pitch–flap–lag couplings, nonlinear mechanical lag damping and pre-cone are examples of features of relatively small importance in themselves, but which can have a powerful effect on the form of the coupled rotor/fuselage modes.

Of course, one of the key driving mechanisms in the coupling process is the development of in-plane aerodynamic loads caused by blade pitch; any additional

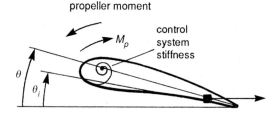

Fig. 3.24 Rotor blade pitch motion

dynamic blade twist and pitch effects will also contribute to the overall coupled motion, but blade pitch effects have such a profound first-order effect on flapping itself that it is in this context that they are now discussed.

Rotor blade pitch

In previous analysis in this chapter the blade pitch angle was assumed to be prescribed at the pitch bearing in terms of the cyclic and collective applied through the swash plate. Later, in Section 3.4, the effects of blade elastic torsion are referred to, but there are aspects of rigid blade pitch motion that can be addressed prior to this. Consider a centrally hinged blade with a torsional spring to simulate control system stiffness, K_θ, as shown in Fig. 3.24. For simplicity, we assume coincident hinges and centre of mass and elastic axis so that pitch–flap coupling is absent. The equation of motion for rigid blade pitch takes the form

$$\theta'' + \lambda_\theta^2 \theta = M_p + \omega_\theta^2 \theta_i \qquad (3.213)$$

where the pitch natural frequency is given by

$$\lambda_\theta^2 = 1 + \omega_\theta^2 \qquad (3.214)$$

where M_P is the normalized applied moment and θ_i is the applied blade pitch. The natural frequency for free pitch motion (i.e., with zero control system stiffness) is one-per-rev; on account of the so-called propeller moment contribution to the restoring moment. This effect is illustrated in Fig. 3.25 where mass elements along the chord line experience in-plane inertial moments due to small components of the large centrifugal force field. For rigid control systems, $\theta = \theta_i$. The control system stiffness is usually relatively high, giving values of ω_θ between 2 and 6 Ω. In this range we usually find the first elastic torsion mode frequency, the response of which can dominate that of

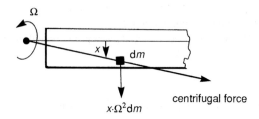

Fig. 3.25 Coriolis forces acting to twist a rotor blade

the rigid blade component. A similar form to eqn 3.213 will apply to the first elastic mode, which will have a nearly linear variation along the blade radius. This aspect will be considered later in Section 3.4, but there are two aspects that are relevant to both rigid and elastic blade torsion which will be addressed here.

First, we consider the gyroscopic contribution to the applied moment M_P. Just as we found with flap motion earlier in this chapter, as the rotor shaft rotates under the action of pitch and roll moments, so the rotor blade will experience nose-up gyroscopic pitching moments of magnitude given by the expression

$$M_{P(gyro)} = -2(\overline{p} \sin \psi + \overline{q} \cos \psi) \tag{3.215}$$

The induced cyclic pitch response can then be written as

$$\delta\theta_{1s} = \frac{-2\overline{p}}{\lambda_\theta^2 - 1}, \qquad \delta\theta_{1c} = \frac{-2\overline{q}}{\lambda_\theta^2 - 1} \tag{3.216}$$

where p and q are the helicopter roll and pitch rates, with the bar signifying normalization by Ω. For low blade torsional or swash plate stiffness, the magnitude of the gyroscopic pitch effects can therefore be significant. More than a degree of induced cyclic can occur with a soft torsional rotor rolling rapidly (Ref. 3.42).

The second aspect concerns the location of the pitch bearing relative to the flap and lag hinges. If the pitch application takes place outboard of the flap and lag hinges, then there is no kinematic coupling from pitch into the other rotor DoFs. However, with an inboard pitch bearing, the application of pitch causes in-plane motion with a flapped blade and out-of-plane motion for a lagged blade. The additional motion also results in an increased effective pitch inertia and hence reduced torsional frequency. These effects are most significant with hingeless rotors that have large effective hinge offsets. On the Lynx, the sequence of rotations is essentially flap/lag followed by pitch, while the reverse is the case for the Bo105 helicopter (Figs 3.26(a) and (b)). The arrangement of the flap and lag real or virtual hinges is also important for coupling of these motions into pitch. Reference 3.7 describes the various structural mechanisms that contribute to these couplings, noting that the case of matched flap and lag stiffness close to the blade root minimizes the induced torsional moments (e.g., Westland Lynx).

As already noted, any discussion of blade torsion would be deficient without consideration of blade elastic effects and we shall return to these briefly later. However, the number of parameters governing the dynamics is large and includes the location of the elastic axis relative to the mass axis and aerodynamic centre, the stiffness distribution and any pre-cone and twist. Introducing this degree of complexity into the structural dynamics also calls for a consistent approach to the blade section aerodynamics, including chordwise pitching moments and unsteady aerodynamics. These are all topics for further discussion in Section 3.4.

Before we proceed with detailing the modelling of the other rotorcraft components, there is one final rotor-related aerodynamic effect to be considered – ground effect.

Ground effect on inflow and induced power

Operating helicopters close to the ground introduces a range of special characteristics in the flight dynamic behaviour. The most significant is the effect on the induced velocity

(a)

(b)

Fig. 3.26 Lynx (a) and Bo105 (b) rotor hubs

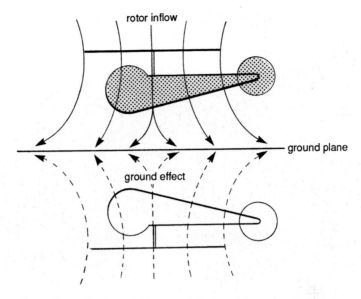

Fig. 3.27 Ground effect on a helicopter in hovering flight

at the rotor and hence the rotor thrust and power required. A succinct analysis of the principal effects from momentum considerations was reported in Ref. 3.43, where, in addition, comparison with test data provided useful validation for a relatively simple theory. Close to the ground, the rotor downwash field is strongly influenced by the surface as shown in Fig. 3.27. In Ref. 3.43, Cheeseman and Bennett modelled the ground plane influence with a rotor of equal and opposite strength, in momentum terms, at an equidistance below the ground (Fig. 3.27). This mirror image was achieved with a simple fluid source that, according to potential flow theory, served to reduce the inflow v_i at the rotor disc in hover by an amount given by

$$\delta v_i = \frac{A_d v_i}{16\pi z_g^2} \tag{3.217}$$

where z_g is the distance of the ground below the rotor disc and A_d is the rotor disc area. The rotor thrust, at constant power, can be written as the ratio of the induced velocity out-of-ground effect (oge) to the induced velocity in-ground effect (ige). Reference 3.43 goes on to derive an approximation for the equivalent thrust change in forward flight with velocity V, the approximation reducing to the correct expression in hover, given by eqn 3.218.

$$\frac{T_{ige}}{T_{oge}} = \frac{1}{\left[1 - \frac{1}{16}\left(\frac{R}{z_g}\right)^2 / \left(1 + \left(\frac{V}{v_i}\right)^2\right)\right]} \tag{3.218}$$

Figure 3.28 illustrates the variation in normalized thrust as a function of rotor height above ground and forward velocity. Ground effect is most significant in hover, and, below heights of the order of a rotor radius, thrust increments of between 5 and 15% are predicted. In forward flight, ground effect becomes insignificant above normalized

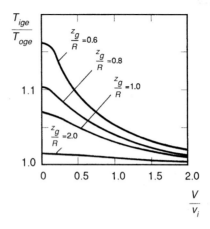

Fig. 3.28 Influence of ground effect on rotor thrust (Ref. 3.43)

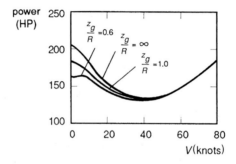

Fig. 3.29 Influence of ground effect on power (Ref. 3.43)

speeds of 2. Simple momentum considerations are unable to predict any influence of blade loading on ground effect. By combining momentum theory with blade element theory, it can be shown that (Ref. 3.43) increasing blade loading typically reduces ground effect such that a 10% increase in blade loading reduces the ige thrust increment by about 10%. Another interesting result from these predictions is that the increase in power required as a helicopter transitions oge is greater than the decrease in power due to the reduction in induced velocity. Figure 3.29, from Ref. 3.43, illustrates the point, showing the variation in power required as a function of forward speed, and reflects practical observations that a power increase is required as a helicopter flies off the ground cushion (Ref. 3.44). Further discussion of ground effect, particularly the effects on non-uniform inflow and hub moments, can be found in Ref. 3.45.

3.2.2   The tail rotor

The tail rotor operates in a complex flowfield, particularly in low-speed flight, inground effect, sideways flight and in the transition to forward flight. The wake of the main rotor, together with the disturbed air shed from the main rotor hub, rear fuselage and vertical stabilizer, interacts with the tail rotor to create a strongly non-uniform flowfield that can dominate the tail rotor loading and control requirements. The basic

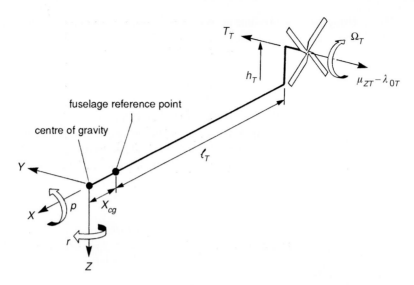

Fig. 3.30 Sketch of tail rotor subsystem

equations for tail rotor forces and moments are similar to those for the main rotor, but a high-fidelity tail rotor model will require a sophisticated formulation for the normal and in-plane components of local induced inflow. Initially, we shall ignore the non-uniform effects described above and derive the tail rotor forces and moments from simple considerations. The interactional effects will be discussed in more detail in Section 3.4.2. The relatively small thrust developed by the tail rotor, compared with the main rotor (between 500 and 1000 lb (2220 and 4440 N) for a Lynx-class helicopter), means that the X and Z components of force are also relatively small and, as a first approximation, we shall ignore these.

Referring to the tail rotor subsystem in Fig. 3.30, we note that the tail rotor sideforce can be written in the form

$$Y_T = \rho(\Omega_T R_T)^2 s_T a_{0_T} (\pi R_T^2) \left(\frac{C_{T_T}}{a_{0_T} s_T} \right) F_T \qquad (3.219)$$

where Ω_T and R_T are the tail rotor speed and radius, s_T and a_{0_T} the solidity and mean lift curve slope, and C_{T_T} the thrust coefficient given by eqn 3.220:

$$C_{T_T} = \frac{T_T}{\rho(\Omega_T R_T)^2(\pi R_T^2)} \qquad (3.220)$$

The scaling factor F_T is introduced here as an empirical fin blockage factor, related to the ratio of fin area S_{fn} to tail rotor area (Ref. 3.46):

$$F_T = 1 - \frac{3}{4} \frac{S_{fn}}{\pi R_T^2} \qquad (3.221)$$

Using the same two-dimensional blade element theory applied to the main rotor thrust derivation, we can write the tail rotor thrust coefficient as

$$\left(\frac{2C_{T_T}}{a_{0_T}s_T}\right) = \frac{\theta_{0_T}^*}{3}\left(1 + \frac{3}{2}\mu_T^2\right) + \frac{(\mu_{Z_T} - \lambda_{0_T})}{2} + \frac{\mu_T}{2}\theta_{1s_T}^* \tag{3.222}$$

where $\theta_{0_T}^*$ and $\theta_{1s_T}^*$ are the effective tail rotor collective and cyclic pitch respectively. Tail rotors are usually designed with a built-in coupling between flap and pitch, the δ_3 angle, defined by the parameter $k_3 = \tan\delta_3$, hence producing additional pitch inputs when the rotor disc cones and tilts (in MBC parlance). This coupling is designed to reduce transient flapping angles and blade stresses. However, it also results in reduced control sensitivity; the relationship can be written in the form

$$\theta_{0_T}^* = \theta_{0_T} + k_3\beta_{0_T}$$

$$\theta_{1s_T}^* = \theta_{1s_T} + k_3\beta_{1s_T} \tag{3.223}$$

where θ_{0_T} and θ_{1s_T} are the control-system- and pilot-applied control inputs; the cyclic inputs are usually zero as tail rotors are not normally fitted with a tilting swash plate. Note that the cyclic change is applied at the same azimuth as the flapping, rather than with the 90° phase shift as with swash-plate-applied cyclic on the main rotor; the δ_3-applied cyclic is therefore fairly ineffective at reducing disc tilt and is actually likely to give rise to more first harmonic cyclic flapping than would otherwise occur. Again, using the main rotor derivations, particularly the coning relationship in eqn 3.64, we note that the effective collective pitch may be written as

$$\theta_{0_T}^* = \frac{\theta_{0_T} + k_3\left(\frac{\gamma}{8\lambda_\beta^2}\right)_T \frac{4}{3}(\mu_z - \lambda_{0_T})}{1 - k_3\left(\frac{\gamma}{8\lambda_\beta^2}\right)_T (1 + \mu_T^2)} \tag{3.224}$$

The δ_3 angle is typically set to $-45°$, which reduces the tail rotor control effectiveness significantly. The cyclic flap angles can be written in the hub–wind axes form (using the cyclic relationships in eqn 3.64)

$$\beta_{1sw_T} = -\frac{\frac{8}{3}\mu_T\left[k_3 + \left(\frac{\gamma}{16\lambda_\beta^2}\right)_T\right]\theta_{0_T} + 2\mu_T\left[k_3 + \left(\frac{\gamma}{16\lambda_\beta^2}\right)_T\left(\frac{4}{3}\right)^2\right](\mu_{Z_T} - \lambda_{0_T})}{\left[1 + k_3\left(\frac{\gamma}{8\lambda_\beta^2}\right)_T\left(\frac{4}{3}\mu_T\right)^2 + k_3^2\left(1 + 2\mu_T^2\right)\right]} \tag{3.225}$$

$$\beta_{1cw_T} = -\frac{8}{3}\mu_T\theta_{0_T} - 2\mu_T(\mu_{Z_T} - \lambda_{0_T}) - k_3\left(1 + 2\mu_T^2\right)\beta_{1sw_T} \tag{3.226}$$

The tail rotor hub aerodynamic velocities are given by

$$\mu_T = \frac{\left[u^2 + \left(w - k_{\lambda_T}\lambda_0 + q(l_T + x_{cg})\right)^2\right]^{1/2}}{\Omega_T R_T} \tag{3.227}$$

$$\mu_{z_T} = \frac{(-v + (l_T + x_{cg})r - h_T p)}{\Omega_T R_T} \tag{3.228}$$

where the velocities of the tail rotor hub relative to the aircraft centre of gravity have been taken into account, and the factor k_{λ_T} scales the normal component of main rotor inflow at the tail rotor (at this point no time lag is included, but see later in Chapter 4). The tail rotor uniform inflow is given by the expression

$$\lambda_{0_T} = \frac{C_{T_T}}{2\left[\mu_T^2 + \left(\mu_{z_T} - \lambda_{0_T}\right)^2\right]^{1/2}} \tag{3.229}$$

The inflow is determined iteratively in conjunction with the tail rotor thrust coefficient. In the above equations we have assumed that the tail rotor has zero hinge moment, a valid approximation for the rotor forces. For teetering hub blade retention (e.g., in the Bo105), the coning angle at the hub centre can be assumed to be zero and no collective pitch reductions occur.

The tail rotor torque can be derived using the same assumptions as for the main rotor, i.e.,

$$Q_T = \frac{1}{2}\rho(\Omega_T R_T)^2 \pi R_T^3\, a_{0_T}\, s_T \left(\frac{2C_{Q_T}}{a_{0_T} s_T}\right) \tag{3.230}$$

with induced and profile torque components as defined by

$$\left(\frac{2C_{Q_T}}{a_{0_T} s_T}\right) = (\mu_{z_T} - \lambda_{0_T})\left(\frac{2C_{T_T}}{a_{0_T} s_T}\right) + \frac{\delta_T}{4a_{0_T}}\left(1 + 3\mu_T^2\right) \tag{3.231}$$

The mean blade drag coefficient is written as

$$\delta_T = \delta_{0_T} + \delta_{2_T} C_{T_T}^2 \tag{3.232}$$

While the tail rotor torque is quite small, the high rotorspeed results in a significant power consumption, which can be as much as 30% of the main rotor power and is given by the expression

$$P_T = Q_T \Omega_T \tag{3.233}$$

The tail rotor forces and moments referred to the aircraft centre of gravity are given approximately by the expressions

$$X_T \approx T_T \beta_{1c_T} \tag{3.234}$$

$$Y_T = T_T \tag{3.235}$$

$$Z_T \approx -T_T \beta_{1s_T} \tag{3.236}$$

$$L_T \approx h_T Y_T \tag{3.237}$$

$$M_T \approx (l_T + x_{cg})Z_T - Q_T \tag{3.238}$$

$$N_T = -(l_T + x_{cg})Y_T \tag{3.239}$$

The above expressions undoubtedly reflect a crude approximation to the complex aerodynamic environment in which the tail rotor normally operates, both in low- and high-speed flight. We revisit the complexities of interactional aerodynamics briefly in Section 3.4.2.

3.2.3   Fuselage and empennage

The fuselage aerodynamic forces and moments

The flow around the fuselage and empennage is characterized by strong nonlinearities and distorted by the influence of the main rotor wake. The associated forces and moments due to the surface pressures and skin friction are therefore complex functions of flight speed and direction. While computer modelling of the integrated flowfield is no longer an impossible task, most of the flight mechanics modelling to date has been based on empirical fitting of wind tunnel test data, gathered at a limited range of dynamic pressure and fuselage angles of incidence, at model (Ref. 3.47) or full scale (Ref. 3.48). Assuming 'similar' fluid dynamics at the test and full-scale flight conditions, we note that the forces at a general flight speed, or dynamic pressure, can be estimated from the data at the measured conditions through the relationship

$$F(V_f, \rho_f) = F(V_{test}, \rho_{test}) \left(\frac{\rho_f V_f^2 S}{\rho_{test} V_{test}^2 S_{test}} \right) \tag{3.240}$$

where the subscript *test* refers to the tunnel test conditions and S is a reference area. Most of the published test data have been measured on isolated fuselage shapes, although the findings of Ref. 3.47 have shed light on the principal effects of rotor wake/fuselage interaction and the approximate formulation outlined below is based on this work.

The three most significant components in forward flight are the fuselage drag, which dominates the power requirement at high speed, and the pitching and yawing moment changes with incidence and sideslip respectively. The fuselage rolling moment is usually small except for configurations with deep hulls where the fuselage aerodynamic centre can be significantly below the aircraft centre of gravity. At lower speeds, the fuselage aerodynamic loads are correspondingly smaller, although significant effects will be the sideforce in sideways flight and the vertical load and yawing moment due to the main rotor wake. The fuselage moments are generally destabilizing, resulting from the greater planform and side area ahead of the aircraft centre of gravity. These two points will not, in general, be coincident. In addition, wind tunnel test data are relative to a third point, generally referred to as the 'fuselage aerodynamic reference point', to be distinguished from our 'fuselage reference point' below the main rotor hub on the aircraft x-axis. Fuselage aerodynamic data measured in a wind tunnel are usually presented in wind tunnel axes as lift, drag, sideforce and corresponding moments about the tunnel-fixed reference system. We assume that the transformation from wind tunnel to fuselage axes has been applied so that we work with forces in the moving fuselage axes system. The effect of rotor downwash can be approximately taken into account by assuming the fuselage is immersed in the uniform component, through the assumption of superposition; hence, the fuselage incidence and velocity can be written as

$$\alpha_f = \tan^{-1}\left(\frac{w}{u}\right), \quad V_f = (u^2 + v^2 + w^2)^{1/2}, \quad \lambda_0 < 0 \quad (3.241)$$

$$\alpha_f = \tan^{-1}\left(\frac{w_\lambda}{u}\right), \quad V_f = (u^2 + v^2 + w_\lambda^2)^{1/2}, \quad \lambda_0 > 0 \quad (3.242)$$

where

$$w_\lambda = w - k_{\lambda f}\Omega R\lambda_0 \quad (3.243)$$

and $k_{\lambda f}$ is a constant taking into account the increase in downwash at the fuselage relative to the rotor disc. The fuselage sideslip angle is defined as

$$\beta_f = \sin^{-1}\left(\frac{v}{V_f}\right) \quad (3.244)$$

The forces and moments may now be written in the generalized form:

$$X_f = \frac{1}{2}\rho V_f^2 S_p C_{xf}(\alpha_f, \beta_f) \quad (3.245)$$

$$Z_f = \frac{1}{2}\rho V_f^2 S_p C_{zf}(\alpha_f, \beta_f) \quad (3.246)$$

$$M_f = \frac{1}{2}\rho V_f^2 S_p l_f C_{mf}(\alpha_f, \beta_f) \quad (3.247)$$

$$Y_f = \frac{1}{2}\rho V_f^2 S_s C_{yf}(\alpha_f, \beta_f) \quad (3.248)$$

$$L_f = \frac{1}{2}\rho V_f^2 S_s l_f C_{lf}(\alpha_f, \beta_f) \quad (3.249)$$

$$N_f = \frac{1}{2}\rho V_f^2 S_s l_f C_{nf}(\alpha_f, \beta_f) \quad (3.250)$$

where S_p and S_s are the plan and side areas of the helicopter fuselage, respectively. Typically, the force and moment coefficients will be derived from table look-up functions of incidence and sideslip. Piecewise linear variations illustrating the main characteristics over the incidence and sideslip range, $-180°$ to $+180°$, are shown in Fig. 3.31. The yaw moment is sometimes defined as two functions corresponding to forward and rearward flight, i.e.,

$$C_{nf} = C_{nf_a}, \quad u > 0$$

$$C_{nf} = C_{nf_b}, \quad u < 0 \quad (3.251)$$

The fuselage X force has a minimum value at small angles of incidence, and is practically zero in vertical flight ($\alpha = 90°$). The pitching moment increases linearly with incidence up to some moderate value when flow separation at the leeward fuselage hull causes a loss in circulatory lift and moment and a corresponding loading reversal. The Y and Z forces have similar shapes, rising to maximum values at $90°$ of incidence

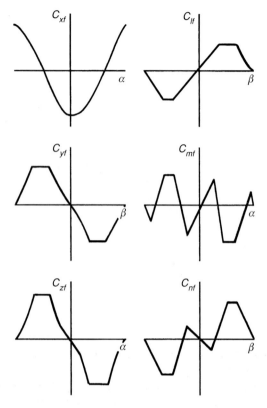

Fig. 3.31 Typical variation of fuselage aerodynamic force coefficients with incidence angles

and sideslip respectively. The breakpoints shown in Fig. 3.31 are very much dependent on the particular fuselage configuration. To account for local, more strongly nonlinear effects, smaller incidence and sideslip intervals would certainly be required. Numerical values for Lynx, Bo105 and Puma fuselage aerodynamic coefficients are given in Appendix 4B. Here the data are taken from wind tunnel tests conducted so that X, Z and M varied only with angle of incidence and Y and N varied only with angle of sideslip. Using these data in conjuction with the equations 3.241–3.250, one should be careful to acknowledge the absence of the cross effects, e.g., the variation of X force with sideslip. The simplest expedient is to delete the v^2 term in the expression for V_f, eqns 3.241 and 3.242. A more general approach could be to assume a simple $\cos \beta$ shape so that the X force becomes zero at $\beta = 90°$.

   The above discussion has been restricted to essentially steady effects whereas, in practice, the relatively bulbous shapes of typical helicopter fuselages, with irregular contours (e.g., engine and rotor shaft cowlings), give rise to important unsteady separation effects that are difficult to simulate accurately at model scale; unsteady effects in manoeuvring flight are also difficult to account for. The problem is exacerbated by the immersion in the rotor downwash at low speed. Sophisticated wind-tunnel and computer modelling techniques are available nowadays but are often very expensive, and confidence in such techniques is still reduced by lack of full-scale validation data.

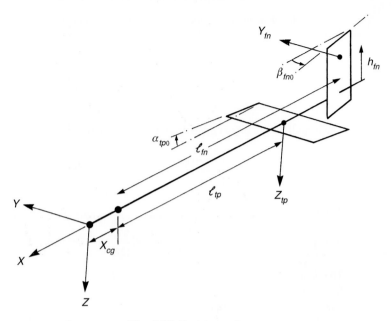

Fig. 3.32 Empennage layout

The empennage aerodynamic forces and moments

The horizontal tailplane and vertical fin, together forming the empennage of a helicopter, perform two principal functions. In steady forward flight, the horizontal tailplane generates a trim load that reduces the main rotor fore–aft flapping; similarly, the vertical fin generates a sideforce and yawing moment serving to reduce the tail rotor thrust requirement. In manoeuvres, the tail surfaces provide pitch and yaw damping and stiffness and enhance the pitch and directional stability. As with the fuselage, the force and moments can be expressed in terms of coefficients that are functions of incidence and sideslip angles. Referring to the physical layout in Fig. 3.32, we note that the principal components are the tailplane normal force, denoted Z_{tp}, and given by

$$Z_{tp} = \frac{1}{2}\rho V_{tp}^2 S_{tp} C_{z_{tp}}(\alpha_{tp}, \beta_{tp}) \tag{3.252}$$

which gives rise to a pitching moment at the centre of gravity, i.e.,

$$M_{tp} = (l_{tp} + x_{cg})Z_{tp} \tag{3.253}$$

and the fin sideforce, denoted by Y_{fn}, i.e.,

$$Y_{fn} = \frac{1}{2}\rho V_{fn}^2 S_{fn} C_{y_{fn}}(\alpha_{fn}, \beta_{fn}) \tag{3.254}$$

which gives rise to a yawing moment at the centre of gravity, i.e.,

$$N_{fn} = -(l_{fn} + x_{cg})Y_{fn} \tag{3.255}$$

where S_{tp} and S_{fn} are the tailplane and fin areas respectively.

The local incidence at the tailplane, assumed constant across its span, may be written as

$$\alpha_{tp} = \alpha_{tp0} + \tan^{-1}\left[\frac{w + q(l_{tp} + x_{cg}) - k_{\lambda_{tp}}\Omega R\lambda_0}{u}\right], \qquad u \geq 0 \quad (3.256)$$

$$(\alpha_{tp})_{reverse} = (\alpha_{tp})_{forward} + \pi, \qquad u < 0 \quad (3.257)$$

The local flow velocity at the tail can be written in the form

$$\mu_{tp}^2 = \left[\frac{u^2 + (w + q(l_{tp} + x_{cg}) - k_{\lambda_{tp}}\Omega R\lambda_0)^2}{(\Omega R)^2}\right] \quad (3.258)$$

where

$$\mu_{tp} = \frac{V_{tp}}{\Omega R} \quad (3.259)$$

The parameter $k_{\lambda tp}$ defines the amplification of the main rotor wake uniform velocity from the rotor disc to the tail. The tailplane incidence setting is denoted by α_{tp0}. The main rotor wake will impinge on the horizontal tail surface only when the wake angle falls between χ_1 and χ_2, given by (see Fig. 3.33)

$$\chi_1 = \tan^{-1}\left(\frac{l_{tp} - R}{h_r - h_{tp}}\right) \quad \text{and} \quad \chi_2 = \tan^{-1}\left(\frac{l_{tp}}{h_r - h_{tp}}\right) \quad (3.260)$$

otherwise, $k_{\lambda tp}$ can be set to zero.

In Ref. 3.49, Loftin gives wind tunnel measurements for a NACA 0012 aerofoil section for the complete range of incidence, $-180° < \alpha < 180°$. From these data, an approximation to the normal force coefficient can be derived in the form

$$|C_{z_{tp}}| \leq C_{z_{tpl}} \qquad C_{z_{tp}}(\alpha_{tp}) = -a_{0_{tp}} \sin\alpha_{tp} \quad (3.261)$$

$$|C_{z_{tp}}| > C_{z_{tpl}} \qquad C_{z_{tp}}(\alpha_{tp}) = -C_{z_{tpl}}\frac{\sin\alpha_{tp}}{|\sin\alpha_{tp}|} \quad (3.262)$$

where a_{0tp} is, effectively, the slope of the tailplane lift coefficient curve for small angles of incidence. The value of this parameter is assumed to be a mean value for the whole surface. Typically, helicopter tailplanes are fairly low aspect ratio surfaces, sometimes having endplates to increase the effective angle of attack. Values of a_{0tp} between 3.5

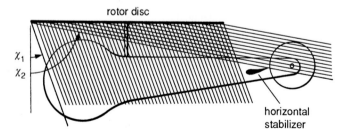

Fig. 3.33 Influence of rotor downwash on tail surfaces

and 4.5 are typical. The constant limit value C_{ztpl} is approximately 2 for the NACA 0012 aerofoil, corresponding to the drag coefficient in two-dimensional vertical flow.

The above formulation, leading to constant rotor downwash over the tailplane, can be improved in two relatively simple respects. First, the spanwise variation of downwash at the tail can be derived from the lateral distribution of downwash from the main rotor, λ_{1s}. Second, the downwash at time t at the tail can be estimated from the loading conditions at time $t - \delta t$ on the main rotor, where δt is the time taken for the flow to reach the tail. This effect manifests itself in an acceleration derivative, or force and moment due to rate of change of incidence, and is discussed in more detail in Chapter 4. The lateral variation in downwash over the horizontal tail generates a roll moment and can also lead to a strong variation of pitching moment with sideslip, as discussed by Cooper (Ref. 3.50) and Curtiss and McKillip (Ref. 3.51).

The local angle of sideslip and velocity (in x–y plane) at the vertical fin may be written in the form

$$\beta_{fn} = \beta_{fn_0} + \sin^{-1}\left[\frac{v - r(l_{fn} + x_{cg}) + h_{fn}p}{\mu_{fn}(\Omega R)}\right] \tag{3.263}$$

$$\mu_{fn}^2 = \left[\frac{(v - r(l_{fn} + x_{cg}))^2 + u^2}{(\Omega R)^2}\right] \tag{3.264}$$

$$\mu_{fn} = \frac{V_{fn}}{\Omega R} \tag{3.265}$$

The loading on the vertical surface can be derived in much the same way as the tailplane, either as a simple analytic function or via a table look-up. One additional complexity, characteristic of helicopter fins, is that they are sometimes quite thick aerofoil sections, carrying within them the tail rotor torque tube. The lift generated at small values of incidence on aerofoils with thickness ratios greater than about 20% can be negated by the lower surface suction near the trailing edge, as discussed by Hoerner and Borst (Ref. 3.52). Figure 3.34, approximated from wind tunnel measurements on the SA 330 Puma (Ref. 3.53), shows how the fin sideforce varies with sideslip angle; over the first 5° of incidence, no lift (sideforce) is produced. This effect partly explains the loss of directional stability and attendant weak Dutch roll damping on the Puma, an aspect that will be the subject of further discussion in Chapters 4 and 5.

The forces generated by the empennage at small values of incidence and sideslip can be represented either by look-up tables or by high-order polynomials, e.g., the Puma

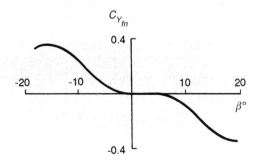

Fig. 3.34 Variation of vertical stabilizer sideforce with sideslip–Puma

fin sideforce requires at least a fifth-order function to match the strongly nonlinear feature illustrated in Fig. 3.34 (see the Appendix, Section 4B.1).

3.2.4 Powerplant and rotor governor

In this section we derive a simplified model for a helicopter's rotorspeed and associated engine and rotor governor dynamics based on the Helisim powerplant model (Ref. 3.4). The rotorspeed of a turbine-powered helicopter is normally automatically governed to operate over a fairly narrow range with the steady-state relationship given by the equation

$$Q_e = -K_3(\Omega - \Omega_i) \qquad (3.266)$$

where Q_e is the turbine engine torque output at the rotor gearbox, Ω is the rotorspeed and Ω_i is the so-called idling rotorspeed, corresponding to approximately zero engine torque. Equation 3.266 is sometimes described as the droop law of the rotor, the droop constant K_3 indicating the reduction in steady-state rotorspeed between autorotation and full power (e.g., in climb or high-speed flight). The rotor control system enforces this droop to prevent any 'hunting' that might be experienced should the control law attempt to maintain constant rotorspeed. Rotorspeed control systems typically have two components, one relating the change (or error) in rotorspeed with the fuel flow, ω_f, to the engine, i.e., in transfer function form

$$\xrightarrow{\Omega} [G_e(s)] \xrightarrow{\quad} \atop \omega_f \qquad (3.267)$$

the second relating the fuel input to the required engine torque output

$$\xrightarrow{\omega_f} [H_e(s)] \xrightarrow{\quad} \atop Q_e \qquad (3.268)$$

The most simple representative form for the fuel control system transfer functions is a first-order lag

$$\frac{\overline{\omega}_f(s)}{\overline{\Omega}(s)} = G_e(s) = \frac{K_{e_1}}{1 + \tau_{e_1} s} \qquad (3.269)$$

where a bar above a quantity signifies the Laplace transform.

The gain K_{e1} can be selected to give a prescribed rotorspeed droop (e.g., between 5 and 10%) from flight idle fuel flow to maximum contingency fuel flow; we write the ratio of these two values in the form

$$\frac{\omega_{f_{max}}}{\omega_{f_{idle}}} \equiv \omega_{f_{mi}} \qquad (3.270)$$

The time constant τ_{e1} will determine how quickly the fuel is pumped to the turbine and, for a fast engine response, needs to be $O(0.1\text{ s})$.

The engine torque response to the fuel injection can be written as a lead–lag element

$$\frac{\overline{Q}_e(s)}{\overline{\omega}_f(s)} = H_e(s) = K_{e2} \left(\frac{1 + \tau_{e_2} s}{1 + \tau_{e_3} s} \right) \qquad (3.271)$$

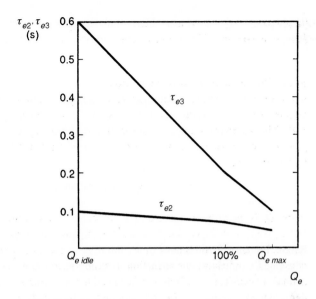

Fig. 3.35 Variation of engine time constants with torque

The gain K_{e2} can be set to give, say, 100% Q_e at some value of fuel flow ω_f (e.g., 75% ω_{fmax}), thus allowing a margin for maximum contingency torque. In the engine model used in Ref 3.4, the time constants in this dynamic element are a function of engine torque. Figure 3.35 illustrates the piecewise relationship showing tighter control at the engine power limit. Linear approximations for the lag and lead constants can be written in the form

$$\tau_{e_2} = \tau_{e_2}(Q_e) \approx \tau_{2_0} + \tau_{2_1} Q_e$$

$$\tau_{e_3} = \tau_{e_3}(Q_e) \approx \tau_{3_0} + \tau_{3_1} Q_e \tag{3.272}$$

where the time constant coefficients change values at $Q_e = 100\%$.

Coupling the two-engine/rotor subsystems gives the transfer function equation

$$\frac{\overline{Q_e}}{\overline{\Omega}} = G_e(s)H_e(s) \tag{3.273}$$

or, in time-domain, differential form

$$\ddot{Q}_e = -\frac{1}{\tau_{e_1}\tau_{e_3}} \left\{ (\tau_{e_1} + \tau_{e_3}) \dot{Q}_e + Q_e + K_3 \left(\Omega - \Omega_i + \tau_{e_2}\dot{\Omega} \right) \right\} \tag{3.274}$$

where

$$K_3 = K_{e_1} K_{e_2} = -\frac{Q_{emax}}{\Omega_i(1 - \Omega_{mi})} \tag{3.275}$$

and

$$\Omega_{mi} = \frac{\Omega_m}{\Omega_i} \tag{3.276}$$

This second-order, nonlinear differential equation is activated by a change in rotor speed and acceleration. These changes initially come through the dynamics of the rotor/transmission system, assumed here to be represented by a simple equation relating the rotor acceleration (relative to the fuselage, $\dot{\Omega} - \dot{r}$) to the applied torque, i.e., the difference between the applied engine torque and the combination of main rotor Q_R and tail rotor torque Q_T, referred to the main rotor through the gearing g_T, i.e.,

$$\dot{\Omega} = \dot{r} + \frac{1}{I_R}(Q_e - Q_R - g_T Q_T) \qquad (3.277)$$

where I_R is the combined moment of inertia of the rotor hub and blades and rotating transmission through to the free turbine, or clutch if the rotor is disconnected as in autorotation.

3.2.5 Flight control system

The flight control system model includes the pilot's controls, mechanical linkages, actuation system and control rods; it also includes any augmentation through feedback control and hence will, in general, encompass the sensors, computing element and any additional actuation in parallel and/or in series with those driven by the mechanical inputs from the pilot. This description corresponds to the classical layout found in most contemporary helicopters. Discussion on the modelling requirements for full authority, digital, active control systems are not covered in this book. We refer to Fig. 3.36 as we develop the model of the flight control system, from the rotors through to the cockpit controls – the cyclic, collective and pedals. In the following analysis, the cockpit controls are represented by the variable η, with appropriate subscripts; in all cases,

$$0 \leq \eta \leq 1 \qquad (3.278)$$

with the positive sense defined by a positive increase in the corresponding rotor blade angle (see Fig. 3.36). The automatic flight control system (AFCS) is usually made up of stability and control augmentation system (SCAS) functions, applied through series actuators, and autopilot functions applied through parallel actuators. In this section we consider only the modelling of the SCAS.

Pitch and roll control

The swash plate concept was introduced in Chapter 2 (Fig. 2.5) as one of the key innovations in helicopter development, allowing one-per-rev variations in rotor blade pitch to be input in a quasi-steady manner from the actuators. The approximately 90° phase shift between cyclic pitch and the cyclic flapping response comes as a result of forcing the rotor with lift changes at resonance. In practice, cyclic pitch can be applied through a variety of mechanisms; the conventional swash plate is by far the most common, but Kaman helicopters incorporate aerodynamic surfaces in the form of trailing edge flaps and cyclic control in the Westland Lynx is effected through the so-called 'dangleberry', with the blade control rods running inside the rotor shaft. Whatever the physical mechanism, cyclic pitch requires very little energy to apply at one-per-rev, and, for our purposes, a generalized swash plate is considered, with a minimum of three actuators to provide the capability of tilting the swash plate at an arbitrary angle relative to the rotor shaft.

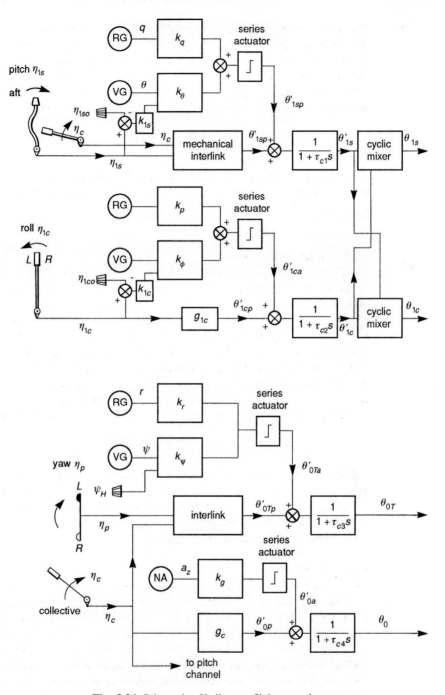

Fig. 3.36 Schematic of helicopter flight control system

Progressing downwards along the control rods (assumed rigid) from the blades, and through the rotating swash plate, we come to the so-called mixing unit. This combines the actuation outputs from the two cyclic controls with a phase angle. For articulated and hingeless rotor configurations, even in the hover, the phase lag between cyclic pitch and flap is less than 90° and, to achieve a pure pitch or roll control, the pilot needs to apply a coupled input. As the forward speed increases, the response coupling changes due to the increased aerodynamic damping effects. A single mixing is usually selected as a compromise between these different conditions and can be written in the form

$$\begin{bmatrix} \theta_{1s} \\ \theta_{1c} \end{bmatrix} = \begin{bmatrix} \cos\psi_f & \sin\psi_f \\ -\sin\psi_f & \cos\psi_f \end{bmatrix} \begin{bmatrix} \theta'_{1s} \\ \theta'_{1c} \end{bmatrix} \tag{3.279}$$

where ψ_f is the mixing angle, usually between 8° and 12°, and a prime simply denotes the cyclic angle before mixing.

The next stage in the reverse sequence is the actuation itself. Most modern helicopters incorporate powered flying controls through hydraulic actuation. The actuation system is quite a complicated mechanism with its own feedback control designed to ensure that the response and stability to control inputs has good performance and stability characteristics. The actuation system has inherent nonlinearities at both small and large amplitudes, including rate limiting when the pilot demands more than the hydraulic system can supply. Typical rate limits are of the order 100% of full actuator range per second. Helicopters fitted with an AFCS usually incorporate a limited authority series actuation system driven by the voltage outputs of the SCAS element. As shown in Fig. 3.36, these augmentation inputs to the actuators are limited to amplitudes of the order ±10% of the full actuator throw. For our purposes we assume that each actuation element can be represented by a first-order lag, although it has to be recognized that this is a crude approximation to the complex behaviour of a complcated servo-elastic system; hence, we write the cyclic actuator outputs as the sum of pilot (subscript p) and AFCS (subscript a) inputs in the transfer function form

$$\overline{\theta}'_{1s} = \frac{\overline{\theta}'_{1s_p} + \overline{\theta}'_{1s_a}}{1 + \tau_{c1}s} \tag{3.280}$$

$$\overline{\theta}'_{1c} = \frac{\overline{\theta}'_{1c_p} + \overline{\theta}'_{1c_a}}{1 + \tau_{c2}s} \tag{3.281}$$

The time constants τ_{c1} and τ_{c2} are typically between 25 and 100 ms, giving actuation bandwidths between 40 and 10 rad/s. For systems operating at the lower end of this bandwidth range, we can expect the actuation to inhibit rapid control action by the pilot.

The mechanical control runs connect the actuators to the pilot's cyclic stick through a series of levers and pulleys. At the stick itself, an artificial feel system is usually incorporated to provide the pilot with stick centring tactile cues. A simple spring with a breakout force is the most common form of feel system found in helicopters, with a constant spring gradient, independent of flight condition or manoeuvre state. If we neglect the dynamics of these elements, then the relationship for roll and pitch cyclic can be written in the simple algebraic form as

$$\theta'_{1s_p} = g_{1s_0} + g_{1s_1}\eta_{1s} + (g_{sc_0} + g_{sc_1}\eta_{1s})\eta_c \tag{3.282}$$

$$\theta'_{1c_p} = g_{1c0} + g_{1c_1}\eta_{1c} + (g_{cc0} + g_{cc_1}\eta_{1c})\eta_c \tag{3.283}$$

where the g coefficients are the gains and offsets, and η_{1c} and η_{1s} are the pilot's cyclic stick inputs. Included in the above equations are simple interlinks between the collective and cyclic, so that a collective input from the pilot also drives the cyclic control runs. In this way, collective to roll and pitch couplings can be minimized. The coefficients in eqns 3.282–3.283 can conveniently be expressed in terms of four parameters:

θ_{1s_0} – the pitch at zero cyclic stick and zero collective lever
θ_{1s_1} – the pitch at maximum cyclic stick and zero collective lever
θ_{1s_2} – he pitch at zero cyclic stick and maximum collective lever
θ_{1s_3} – the pitch at maximum cyclic stick and maximum collective lever

The coefficients can therefore be written as

$$g_{1s0} = \theta_{1s0}$$
$$g_{1s_1} = \theta_{1s_1} - \theta_{1s0}$$
$$g_{1c0} = \theta_{1s_2} - \theta_{1s_1}$$
$$g_{1c_1} = (\theta_{1s_3} - \theta_{1s_2}) - (\theta_{1s_1} - \theta_{1s0}) \tag{3.284}$$

This analysis assumes a linear relationship between control movement and actuator input. In practice, the mechanical system will exhibit some nonlinearities particularly at the extremities of control throw due to the geometry of the linkage, and look-up tables will be a more appropriate representation. For example, Fig. 3.37 illustrates the cyclic/collective interlink functionality for the Lynx helicopter (Ref. 3.54).

With regard to the autostabilizer inputs, these will, in general, be complex functions of sensor and control inputs with various filters arranged to stabilize the feedback

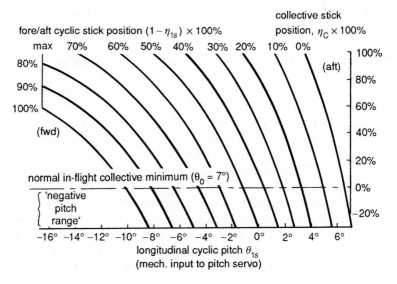

Fig. 3.37 Geometry of mechanical interlink between collective and cyclic for Lynx (Ref. 3.54)

dynamics and protect against sensor noise. For the present purposes we shall assume that the autostabilizer adds feedback control signals proportional to attitude and angular rate, together with a feedforward signal proportional to the pilot's control input, referred to some adjustable datum (Fig. 3.36). This allows the zero or mid-range of the autostabilizer to be reset by the pilot during flight. This would be necessary, for example, if the attitude gains were high enough to cause saturation as the speed increases from hover to high speed. Other systems automatically disengage the attitude stabilization when the pilot moves his control, thus obviating the need for a pilot-adjustable zero (e.g., Puma). The simple proportional autostabilizer can be described by the equations

$$\theta_{1s_a} = k_\theta \theta + k_q q + k_{1s} \left(\eta_{1s} - \eta_{1s_0} \right) \tag{3.285}$$

$$\theta_{1c_a} = k_\phi \phi + k_p p + k_{1c} \left(\eta_{1c} - \eta_{1c_0} \right) \tag{3.286}$$

In Chapter 4, we shall demonstrate how rate stabilization alone is typically inadequate for stabilizing a helicopter's unstable pitch motion. However, with a combination of fairly modest values of rate and attitude gains, $k \, (O(0.1))$, a helicopter can be stabilized throughout its OFE, and a pilot can fly 'hands off' or at least with some divided attention, hence allowing certification in IFR conditions. However, a low-authority AFCS will quickly saturate in aggressive manoeuvres, or during flight in moderate to severe turbulence, and can be regarded only as an aid to steady flight.

Yaw control

In a similar way, the pilot and autostabilizer commands are input to the yaw actuator servo in the simplified first-order transfer function form

$$\bar{\theta}_{0T} = \frac{\bar{\theta}_{0T_p} + \bar{\theta}_{0T_a}}{1 + \tau_{c_3} s} \tag{3.287}$$

The gearing between the actuator input and the yaw control run variable, η_{cT}, can be written as

$$\theta_{0T_p} = g_{T_0} + g_{T_1} \eta_{cT} \tag{3.288}$$

where the control run is generally proportional to both pedal, η_p, and collective lever, η_c, inputs, in the form

$$\eta_{cT} = g_{cT_0} (1 - \eta_p) + \left(1 - 2 g_{cT_0} \right) \eta_c \tag{3.289}$$

In eqn 3.289, the collective lever accounts for the normal mechanical interlink between collective and yaw to reduce yaw excursions following power inputs. Equation 3.289 is a linear approximation to a relationship that can become strongly nonlinear at the extremes of the control range, when the interlink geometry reduces the sensitivity. Figure 3.38 illustrates the nonlinear variation for the Lynx helicopter (Ref. 3.54).

Heave control

Finally, the main rotor collective pitch output from the main rotor servos, achieved through raising and lowering the swash plate, can be written in terms of the mechanical

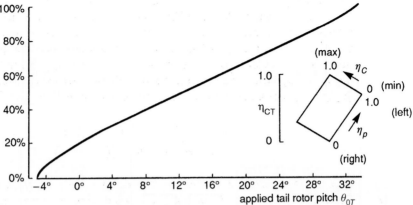

Fig. 3.38 Geometry of mechanical link between tail rotor control run and cockpit controls for Lynx (Ref. 3.54)

and electrical inputs from the pilot and autostabilizer respectively, namely

$$\bar{\theta}_0 = \frac{\bar{\theta}_{0_p} + \bar{\theta}_{0_a}}{1 + \tau_{c_4} s} \tag{3.290}$$

The gearing with the collective lever is written as

$$\theta_{0_p} = g_{c_0} + g_{c_1} \eta_c \tag{3.291}$$

For most modern helicopters, there is no autostabilizer component in the collective channel, but for completeness we include here a simple model of the so-called collective acceleration control (Ref. 3.55) found in the Lynx. An error signal proportional to the normal acceleration is fed back to the collective, i.e.,

$$\theta_{0_a} = k_g a_z \tag{3.292}$$

For the Lynx, this system was implemented to provide dissimilar redundancy in the SCAS. At high speeds, the collective is a very effective pitch control on hingeless rotor helicopters, and this additional loop supplements the cyclic stabilization of aircraft pitch attitude and rate.

3.3 INTEGRATED EQUATIONS OF MOTION OF THE HELICOPTER

In the preceding sections the equations for the individual helicopter subsystems have been derived. A working simulation model requires the integration of the subsystems in sequential or concurrent form, depending on the processing architecture. Figure 3.39 illustrates a typical arrangement showing how the component forces and moments depend on the aircraft motion, controls and atmospheric disturbances. The general

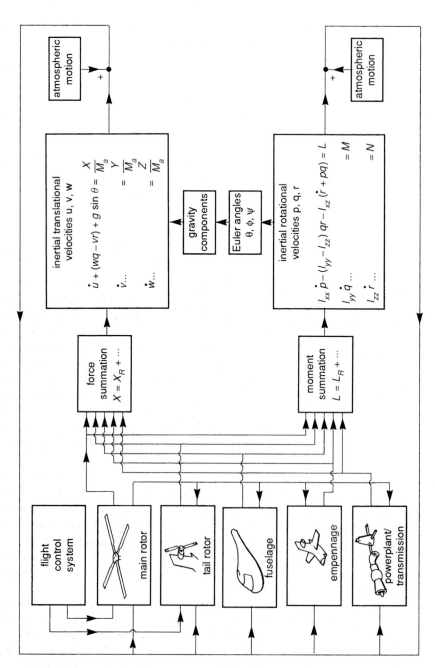

Fig. 3.39 The integrated helicopter simulation model

nonlinear equations of motion take the form

$$\dot{\mathbf{x}} = \mathbf{F}(\mathbf{x}, \mathbf{u}, t) \qquad (3.293)$$

where the state vector $\mathbf{x}$ has components from the fuselage $\mathbf{x}_f$, rotors $\mathbf{x}_r$, engine/rotorspeed $\mathbf{x}_p$ and control actuation $\mathbf{x}_c$ subsystems, i.e.,

$$\mathbf{x} = \{\mathbf{x}_f, \mathbf{x}_r, \mathbf{x}_p, \mathbf{x}_c\} \qquad (3.294)$$

$$\mathbf{x}_f = \{u, w, q, \theta, v, p, \phi, r\} \qquad (3.295)$$

$$\mathbf{x}_r = \{\beta_0, \beta_{1c}, \beta_{1s}, \lambda_0, \lambda_{1c}, \lambda_{1s}\} \qquad (3.296)$$

$$\mathbf{x}_p = \{\Omega, Q_e, \dot{Q}_e\} \qquad (3.297)$$

$$\mathbf{x}_c = (\theta_0, \theta_{1s}, \theta_{1c}, \theta_{0T}) \qquad (3.298)$$

where we have assumed only first-order flapping dynamics.

SCAS inputs apart, the control vector is made up of main and tail rotor cockpit controls,

$$\mathbf{u} = (\eta_0, \eta_{1s}, \eta_{1c}, \eta_{0T}) \qquad (3.299)$$

Written in the explicit form of eqn 3.293, the helicopter dynamic system is described as instantaneous and non-stationary. The instantaneous property of the system refers to the fact that there are no hysteretic or more general hereditary effects in the formulation as derived in this chapter. In practice, of course, the rotor wake can contain strong hereditary effects, resulting in loads on the various components that are functions of past motion. These effects are usually ignored in Level 1 model formulations, but we shall return to this discussion later in Section 3.4. The non-stationary dynamic property refers to the condition that the solution depends on the instant at which the motion is intitiated through the explicit dependence on time t. One effect, included in this category, would be the dependence on the variation of the atmospheric velocities–wind gusts and turbulence. Another arises from the appearance of aerodynamic terms in eqn 3.293 which vary with rotor azimuth.

The solution of eqn 3.293 depends upon the initial conditions – usually the helicopter trim state – and the time histories of controls and atmospheric disturbances. The trim conditions can be calculated by setting the rates of change of the state vector to zero and solving the resultant algebraic equations. However, with only four controls, only four of the flight states can be defined; the values of the remaining 17 variables from eqn 3.293 are typically computed numerically. Generally, the trim states are unique, i.e., for a given set of control positions there is only one steady-state solution of the equations of motion.

The conventional method of solving for the time variations of the simulation equations is through forward numerical integration. At each time step, the forces and moments on the various components are computed and consolidated to produce the total force and moment at the aircraft centre of mass (see Fig. 3.39). The simplest integration scheme will then derive the motion of the aircraft at the end of the next time step by assuming some particular form for the accelerations. Some integration methods smooth the response over several time steps, while others step backwards and forwards through the equations to achieve the smoothest response. These various elaborations are required to ensure efficient convergence and sufficient accuracy and will be required when particular dynamic properties are present in the system (Ref. 3.56). In recent years, the use of inverse simulation has been gaining favour, particularly for model validation research and for comparing different aircraft flying the same manoeuvre. With inverse simulation, instead of the control positions being prescribed as functions of time, some subset of the aircraft dynamic response is defined and the controls required to fly the manoeuvre computed. The whole area of trim and response will be discussed in more detail in Chapters 4 and 5, along with the third 'problem' of flight dynamics – stability. In these chapters, the analysis will largely be confined to what we have described as Level 1 modelling, as set down in detail in Chapter 3. However, we have made the point on several occasions that a higher level of modelling fidelity is required for predicting flight dynamics in some areas of the flight envelope. Before we proceed to discuss modelling applications, we need to review and discuss some of the missing aeromechanics effects, beyond Level 1 modelling.

3.4   BEYOND LEVEL 1 MODELLING

'Theory is never complete, final or exact. Like design and construction it is continually developing and adapting itself to circumstances'. We consider again Duncan's introductory words and reflect that the topic of this final section in this model-building chapter could well form the subject of a book in its own right. In fact, higher levels of modelling are strictly outside the intended scope of the present book, but we shall attempt to discuss briefly some of the important factors and issues that need to be considered as the modelling domain expands to encompass 'higher' DoFs, nonlinearities and unsteady effects. The motivation for improving a simulation model comes from a requirement for greater accuracy or a wider range of application, or perhaps both. We have already stated that the so-called Level 1 modelling of this chapter, augmented with 'correction' factors for particular types, should be quite adequate for defining trends and preliminary design work and should certainly be adequate for gaining a first-order understanding of helicopter flight dynamics. In Chapters 4 and 5 comparison with test data will confirm this, but the features that make the Level 1 rotor modelling so amenable to analysis – rigid blades, linear aerodynamics and trapezoidal wake structure – are also the source of its limitations. Figure 3.40, for example, taken from Ref. 3.57, compares the rotor incidence distribution for the Puma helicopter (viewed from below) derived from flight measurements of rotor blade leading edge pressure, with the Level 1 Helisim prediction. The flight condition is a straight and level trim at 100 knots. While there are similarities in the two contour plots, theory fails to capture many of the details in the flight measurements. The region of high incidence on the retreating side is more extensive and further outboard in the flight results, and there is a clearly defined ridge in the flight measurement caused by the blade vortex interaction, which is, of course, completely missed by Helisim. At this 100-knot trim condition, Helisim may well predict the controls to trim reasonably accurately, simply because the integrated forces and moments tend to smooth out the effects of the detailed differences apparent in Fig. 3.40. However, there are a wide range of problems where the details become significant in the predictive capability of modelling. Examples include the pitch-up effect of blade stall in 'high *g*' manoeuvres, the transient rotor torque excursions in rapid rolls, the effects of blade icing or battle damage on power and control

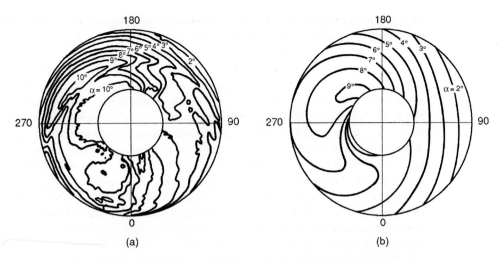

Fig. 3.40 Comparison of rotor incidence distribution measured on the DRA research Puma with theory: (a) flight; (b) Helisim (Ref. 3.57)

margins. If we consider the effects of the rotor wake on the tail rotor and empennage, then the simple trapezoidal downwash model fails to predict important effects, for example tail rotor control margins in quartering flight or the strong couplings induced by the wake effects on the rear fuselage and empennage, particularly in manoeuvres. High-fidelity simulation requires that these effects can be predicted, and to achieve this we need to consider the modelling elements at Levels 2 and 3, as described in Table 3.1. The following qualitative discussions will draw heavily from the published works of selected contributors to the field of enhanced rotorcraft modelling. The author is all too aware of the enormous amount of published work and achievements by a great number of researchers in Europe and the United States in recent years, particularly to rotor aeroelastic modelling, and a complete review is surely the topic for another text. The aim here is to draw the readers' attention to selected advances that lay emphasis on physical understanding.

3.4.1 Rotor aerodynamics and dynamics

Rotor aerodynamics
The linear aerodynamic theory used in Level 1 rotor modelling is a crude approximation to reality and, while quite effective at predicting trends and gross effects, has an air of sterility when compared with the rich and varied content of the fluid dynamics of the real flow through rotors. Compressibility, unsteadiness, three-dimensional and viscous effects have captured the attention of several generations of helicopter engineers; they are vital ingredients for rotor design, but the extent of the more 'academic' interest in real aerodynamic effects is a measure of the scientific challenge intrinsic to rotor modelling. It is convenient to frame the following discussion into two parts – the prediction of the local rotor blade angle of incidence and the prediction of the local rotor blade lift, drag and pitching moment. While the two problems are part of the same feedback system, e.g., the incidence depends on the lift and the lift depends

on the incidence, separating the discussion provides the opportunity to distinguish between some of the critical issues in both topics.

Modelling section lift, drag and pitching moment

The rotor blade section loading actions of interest are the lift, drag and pitching moment. All three are important and all three can signal limiting effects in terms of blade flap, lag and torsion response. A common approximation to real flow effects assumes two-dimensional, quasi-steady variations with local incidence and Mach number uniquely determining the blade loading. In Ref. 3.58, Prouty gives an account of empirical findings based on analysis of a wide range of two-dimensional aerofoil test data. Key parameters defining the performance and behaviour of an aerofoil section are the maximum achievable lift coefficient C_{Lmax} and the drag divergence Mach number M_d. Both depend critically on the geometry of the aerofoil, as expected, and hence on the type of rotor stall. Prouty identifies three types of stall to which rotor blades are prone – thin aerofoil, leading edge and trailing edge stall.

Prouty's findings suggest that aerofoil sections with thickness-to-chord ratios greater than about 8% will normally experience trailing edge stall and, at their best, achieve values of C_{Lmax} up to about 1.6. For thinner aerofoils, leading edge stall is more likely, with a C_{Lmax} that increases with thickness/chord up to about 1.8. The general effects of trailing and leading edge stall on lift, drag and moment coefficients are sketched in Fig. 3.41, where these are shown as functions of incidence.

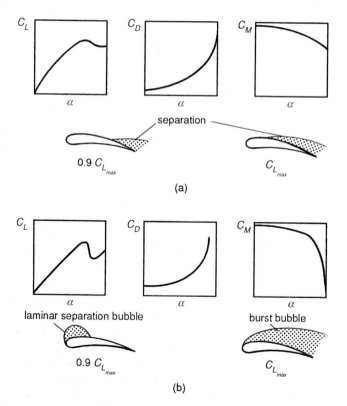

Fig. 3.41 Types of aerofoil stall: (a) trailing edge stall; (b) leading edge stall

Trailing edge stall is characterized by a gradual increase in the region of separated flow moving forward from the trailing edge. Leading edge stall is triggered by the bursting of a laminar separation bubble over the nose of the aerofoil giving rise to sharp changes in lift, drag and pitching moment. Generally, thin aerofoils are favoured for their performance (high M_d) at high Mach number on the advancing side of the disc, and thicker aerofoils are favoured for their performance (high C_{Lmax}) at low Mach number and high incidence on the retreating side. Most helicopter blades are therefore designed as a compromise between these two conflicting requirements and may experience both types of stall within the operating envelope. Reference 3.59 describes the evolution of the cambered aerofoil sections adopted for the Lynx helicopter, showing a favourable all-round comparison with the thicker, symmetrical NACA 0012 section. The latter was typical of aerofoil sections used on helicopter rotors before the 1970s.

In blade element rotor simulation models, the lift, drag and pitching moment coefficients are usually stored in table look-up form as nonlinear functions of incidence and Mach number, with the data tables derived from either wind tunnel tests or theoretical predictions. In Ref. 3.6, Bramwell reports on the effects of swirl and other three-dimensional, in-plane effects on section characteristics with significant changes in C_{Lmax} particularly at the higher Mach numbers. Also, in Ref. 3.60, Leishman draws attention to the powerful effects of sweep angle on C_{Lmax}. Generally, however, for a large extent of the rotor radius, the two-dimensional approximation is relatively accurate. An exception is close to the tip, where three-dimensional effects due to the interaction of the upper and lower surface flows result in marked changes in the chordwise pressure distribution for a given incidence and Mach number. Accurate modelling of the tip aerodynamics is still the subject of intense research and renewed impetus with the advent of novel tip sections and planforms.

In forward flight and manoeuvres, the section incidence and Mach number are changing continuously and we need to consider the effects of aerodynamic unsteadiness on the section characteristics. In a series of papers (e.g., Refs 3.61 – 3.64) Beddoes and Leishman have reported the development of an indicial theory for unsteady compressible aerodynamics applicable to both attached and separated flow, for the computation of section lift, drag and pitching moment. In attached flow, the shed wake in the vicinity of the aerofoil induces a time-dependent circulatory force on the section, with a transient growth corresponding to about five chord lengths. A non-circulatory lift also develops (due to the airfoil virtual mass) and decays to zero in approximately the same spatial scale. Both effects are approximated in the Beddoes model by combinations of exponential functions (Ref. 3.63) responding to arbitrary motions of an aerofoil in pitch and heave. To account for the response of the aerofoil to its passage through the wake and individual vortices of other blades, the method also models the loading actions due to arbitrary variations in the incidence of the airflow (Ref. 3.64). A special feature of these developments has been the extension to the modelling of separated flow and the prediction of dynamic stall (see earlier paper by Johnson and Ham for discussion of the physics of dynamic stall, Ref. 3.65). In unsteady motion, the passage of shed vorticity over the aerofoil upper surface following leading edge stall gives rise to a delay in both lift and moment 'break', resulting in an overshoot of lift to well beyond the normal quasi-steady value of C_{Lmax}. Beddoes has encapsulated this effect in a semi-empirical model, summarized in Fig. 3.42, taken from Ref. 3.62. To quote from Ref. 3.62

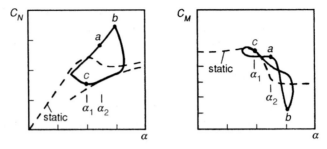

Static data idealised:

C_M break at α_1, separation stabilised at α_2

Dynamic stall progression:

Exceed α_1 without separation, start time delay.

(a)   Time delay (τ_1) exceeded. Vortex shed from L.E.
        lift maintained, moment diverges ~ C.P. = $f(\alpha_1, t)$

(b)   Time delay (τ_2) exceeded. Vortex passes T.E.
        lift decays ~ reflected in moment variation.

(c)   Flow reattaches for $\alpha < \alpha_1$

Fig. 3.42 Time delay model for dynamic stall (Ref. 3.62) (T.E., trailing edge; L.E., leading edge; C.P., centre of pressure)

For each Mach number the angle of attack (α_1) which delimits attached flow is determined by the break in pitching moment and a further angle (α_2) is used to represent the condition where flow separation and hence centre of pressure is stabilised. In application, when the local value of angle of attack exceeds α_1 the onset of separation is assumed to be delayed for a finite period of time (τ_1) during which the lift and moment behave as appropriate for attached flow. When this time delay is exceeded, flow separation is assumed to be initiated by the shedding of a vortex from the surface of the aerofoil and after a period of time (τ_2), during which the vortex traverses the chord, it passes free of the surface. In this interval, lift is generated by the vortex and the overall level maintained equivalent to that for fully attached flow but the centre of pressure moves aft as a function of both angle of attack and time. When the vortex passes free of the surface, the lift decays rapidly to a value appropriate to fully separated flow assuming that the angle of attack is still sufficiently high. If and when the angle of attack reduces below the value α_1 re-attachment of the flow is represented by the attached flow model, re-initialised to account for the current lift deficiency.

Beddoes goes on to suggest ways that the method can be extended to account for trailing edge stall and also in compressible conditions, when stall is more often triggered by shock wave–boundary layer interaction (Ref. 3.62). Unsteady aerodynamic effects are essential ingredients to understanding many rotor characteristics at high speed and in manoeuvres, and have found practical application in current-generation loads, vibration and aeroelastic stability prediction models. The impact on flight dynamics is less well explored, but two important considerations provide evidence that for some problems, unsteady aerodynamic effects may need to be simulated in real-time applications. First, we consider the azimuthal extent of the development of unsteady lift and moment. The linear potential theory discussed above predicts a time to reach steady-state lift following a step change in incidence of about 5–10 chord lengths, equating to between $10°$ and $20°$ azimuth. Even the lower frequency one-per-rev incidence changes associated with cyclic pitch will lead to a not-insignificant phase lag, depending on the rotorspeed. Phase lags as low as $5°$ between control inputs and lift change can have a significant effect on pitch to roll cross-coupling. Second, modelling the trigger to blade stall correctly is important for simulating flight in gross manoeuvres, when the azimuthal/radial location of initial stall can determine the evolution of the separated flow and hence the effect on pitch and roll hub moments particularly. Dynamic, rather than quasi-steady stall, is, of course, the norm in forward flight and manoeuvring conditions.

With two-dimensional test data tables, three-dimensional and low-frequency unsteady corrections and an empirical stall model, deriving the section forces and pitching moment is a relatively straightforward computational task. A much more significant task is involved in estimating the local incidence.

Modelling local incidence

The local incidence at azimuth station ψ and radial station r can be expanded as a linear combination of contributions from a number of sources, as indicated by eqn 3.300:

$$\alpha(\psi, r; t) = \alpha_{pitch} + \alpha_{twist} + \alpha_{flap} + \alpha_{w_h} + \alpha_{inflow} \qquad (3.300)$$

The component α_{pitch} is the contribution from the physical pitch of the blade applied through the swash plate and pitch control system. The α_{twist} component includes contributions from both static and dynamic twist; the latter will be discussed below in the Rotor dynamics subsection. The α_{flap} component due to rigid blade motion has been fully modelled within the Level 1 framework; again we shall return to the elastic flap contribution below. The α_{w_h} component corresponds to the inclination of incident flow at the hub. Within the body of this chapter, the modelling of α_{inflow} has been limited to momentum theory, which although very effective, is a gross simplification of the real helical vortex wake of a rotor. Downwash, in the form of vorticity, is shed from a rotor blade in two ways, one associated with the shedding of a (spanwise) vortex wake due to the time-varying lift on the blade, the other associated with the trailing vorticity due to the spanwise variation in blade lift. We have already discussed the inflow component associated with the near (shed) vortex wake due to unsteady motion; it was implicit in the indicial theory of Beddoes and Leishman. Modelling the trailing vortex system and its effect on the inflow at the rotor disc has been the subject of research since the early days of rotor development. Bramwell (Ref. 3.6) presents a comprehensive review of activities up to the early 1970s, when the emphasis was on what can be described as 'prescribed' wakes, i.e., the position of the vortex lines or sheets are prescribed in space and the induced velocity at the disc derived using the Biot–Savart law. The strength of the vorticity is a function of the lift when the vorticity was shed from the rotor, which is itself a function of the inflow.

Solving the prescribed wake problem thus requires an iterative procedure. Free wake analysis allows the wake vorticity to interact with itself and hence the position of the wake becomes a third unknown in the problem; a free wake will tend to

roll-up with time and hence gives a more realistic picture of the flowfield downstream
of the rotor. Whether prescribed or free, vortex wakes are computationally intensive
to model and have not, to date, found application in flight simulation. As distributed
flowfield singularities, they also represent only approximate solutions to the underlying
equations of fluid dynamics. In recent years, comprehensive rotor analysis models are
beginning to adopt more extensive solutions to the three-dimensional flowfield, using
so-called computational fluid dynamics techniques (Ref. 3.3). The complexity of such
tools and the potential of the achievable accuracy may be somewhat bewildering to the
flight dynamicist, and a real need remains for simpler approximations that have more
tractable forms with the facility for deriving linearized perturbations for stability analysis. Earlier, in Section 3.2.1, we referred to the recent development of wake models that
exhibit these features (Refs 3.28, 3.29), the so-called finite-state wake structures. Here,
the inflow at the rotor is modelled as a series of modal functions in space-time, each
satisfying the rotor boundary conditions and the underlying continuity and momentum
equations, through the relationship with the blade lift distribution. The theory results
in a series of ordinary differential equations for the coupled inflow/lift which can be
appended to the rotor dynamic model. Comparison with test results for rotor inflow in
trimmed flight (Ref. 3.29) shows good agreement and encourage further development
and application with this class of rotor aerodynamic model.

Rotor dynamics

Several of the important components of local blade incidence stem from the motion
and shape of the blade relative to the hub. A characteristic of Level 1 (flight dynamics)
modelling is the approximation of rigid blade motion for flap, lag and torsion. We
have seen how the CSER can be used to represent the different types of flap retention
system – teetering, articulated or bearingless. In MBC form, the dynamics of one-
per-rev disc tilting are apparently well represented. Since the hub moments of interest
are produced by the one-per-rev flapping, this level of approximation would appear to
be adequate for problems in the frequency range of interest to the flight dynamicist.
However, a significant simplification in the centre-spring approximation involves the
relationship between the disc tilt and the hub moment. We have suggested earlier that
the linear relationship is a powerful attribute of the centre-spring model; if we look
more closely at the potential effects of elastic blade motion, we see that what appears
to be a strength of the approximation in many cases is actually a weakness in others.
With the centre-spring model, it can easily be shown that the moment computed from
the disc tilt and the hub moment computed from the integrated aerodynamic loads are
always in balance, and hence always in phase. More generally, for both articulated
and hingeless rotor approximate models, this is not the case. Consider the blade flap
moment (in rotating axes) at the hub, given by eqn 3.15, but expanded to show how
the time-dependent generalized (modal) coordinates can be written as a summation of
harmonics with coefficients a_m and b_m, as in eqn 3.301:

$$M^{(r)}(0, t) = \Omega^2 \sum_{n=1}^{\infty} \left(\lambda_n^2 - 1 \right) \int_0^R mrS_n(r)\, dr \left(\sum_{m=1}^{\infty} a_m \cos m\psi + b_m \sin m\psi \right)$$

$$(3.301)$$

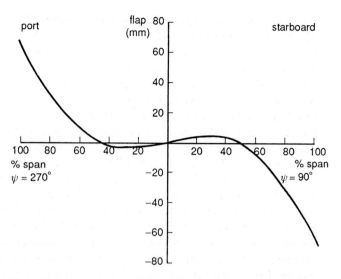

Fig. 3.43 Rotor blade shape at the advancing (90°) and retreating (270°) azimuth angles for Lynx at 150 knots

Each mode will contribute to the rotating hub moment through the different harmonics, but only the first harmonic contribution to each will be transmitted through to result in quasi-steady fuselage motions. The extent of the contribution of higher modes to the hub moment depends entirely on the character of the aerodynamic forcing; the stronger the radial nonlinearity in the one-per-rev aerodynamic forces, the greater will be the excitation of the higher modes. Of course, the higher the frequency of the mode, the more attentuated will be its one-per-rev tip response, but equally, the hub moment for a given tip deflection will be greater for the higher modes. An illustration of the potential magnitude of contributions to the hub moment from higher elastic modes is provided by Fig. 3.43. The blade bending at azimuth stations 90° and 270° are shown for the Lynx rotor in trim at 150 knots, derived from the DRA aeroelastic rotor model. The rotor model used to compute the results shown includes first and second flap modes, first torsion and first lead–lag. The shape of the blade highlights the strong contribution from the second flap mode in the trim condition, with the 'node' (zero displacement) at about 50% radius. In fact, the hub moment, defined by the curvature of the blade at the hub, has an opposite sign to the tilt of the disc. The total hub rolling moment (in fuselage axes), computed from either the modal curvature or integrated aerodynamics, is about -1000 N m (to port), clearly in opposition to the disc tilt to starboard. For the case with only the fundamental flap mode retained, the hub moment derived from the first mode curvature is about $+2000$ N m (to starboard), while the aerodynamic moment integrates to about -600 N m (cf. Fig. 3.43). This result argues strongly for a harmony in the model between aerodynamic and dynamic formulations, particularly for high-speed flight ($\mu > 0.3$) when nonlinear aerodynamics and hence the effects of higher modes are likely to become more pronounced.

Shupe, in Ref. 3.36, presents results on the effect of the second flap mode over a wide range of conditions, supporting the above conclusion that the influence of the loading on the shape of the hingeless blade at high speed is significant, and higher order modes need to be included in simulation modelling for flight dynamics. Shupe also noted the powerful effects of blade twist on the distribution of out-of-plane bending between the first and second flap modes; twist tends to pull the blade loading inboard, hence leading to a radial aerodynamic distribution with a shape more like the second flap mode. A subtle effect that should be noted here is that the response of the second flap mode to one-per-rev aerodynamic loads will not feature the 90° phase shift associated with the first flap mode. The natural frequency of mode 'flap 2' is an order of magnitude higher than that for mode 'flap 1' and the phase lag at one-per-rev will be very small. Hence, lateral cyclic (θ_{1c}) will lead primarily to longitudinal disc tilt (β_{1c}) in mode 'flap 2', thus having a stronger effect on cross-coupling than the direct response. The influence of the second flap mode in flight dynamics is yet to be fully explored and remains a research topic worthy of further investigation.

Blade dynamic twist will clearly have a major effect on local blade incidence, flapping and hub moments and can arise from a number of sources. Any offset of the blade chordwise centre of mass or elastic axis from the quarter chord will result in a coupling of the flap and torsion DoFs in the elastic modes. The shift of the chordwise aerodynamic centre due to compressibility, stall or by design through swept tip planforms will also be a source of torsional moments from the section aerodynamic pitching moment. References 3.66–3.68 report results of flight dynamics simulation models that incorporate elastic modes, paying particular attention to the effect of elastic torsion. For the cases studied in both Ref. 3.67, using the FLIGHTLAB simulation model, and Ref. 3.68, using the UM-GENHEL simulation model, elastic torsion was shown to have a negligible effect on aircraft trim, stability and dynamic response; comparisons were made with test data for the articulated rotor UH-60 helicopter in hover and forward flight. Articulated rotor helicopters are normally designed so that the blade pitch control is positioned outboard of the flap and lag hinges, thereby reducing kinematic couplings. On hingeless rotors, combined flap and lag bending outboard of the pitch control will produce torsional moments leading to elastic twist of the whole blade or flexing of the control system. This feature was described in the context of the design of the Westland Lynx helicopter in Ref. 3.69. The combination of an inboard flapping element with high lag stiffness and a circular section element with matched flap and lag stiffnesses outboard of the feathering hinge resulted in a minimization of torsion–flap–lag coupling on Lynx. For both articulated and hingeless rotors, it should be clear that the potential for elastic couplings and/or forced torsional response is quite high, and even with designs that have emphasized the reduction of the sources of coupled torsional moments, we can expect the combination of many small elastic and particularly unsteady aerodynamic effects to lead to both transient and steady-state elastic twist.

Aeroelastic effects clearly complicate rotor dynamics but are likely to be an important ingredient and a common feature of future high-fidelity rotorcraft simulations. It will be clear to the serious student of the subject that most of the approximations lie in the formulation of the aerodynamic theory, particularly the dynamic inflow, but the degree of 'aeroelastic' modelling required to complete the feedback loop correctly is not well researched. As new rotor designs with tailored elastic properties and 'flexible' surfaces become mature enough for application, we should expect an associated increase in the motivation for understanding and developing more general and definitive rules for the effects of aeroelastics on flight stability and control.

3.4.2 Interactional aerodynamics

The helicopter is characterized by an abundance of interactional aerodynamic effects, often unseen in design but powerful in their, usually adverse, effects in flight. A principal source of interactions is the main rotor wake as it descends over the fuselage, empennage and through the tail rotor disc. The main rotor wake also interacts with the ground and with itself, in vortex-ring conditions. The modelling problem is therefore largely an extension of the problem of predicting the wake effects at the rotor disc; for interactional aerodynamics we are interested in the development of the wake within approximately one rotor diameter of the rotor. In this space-time frame, the wake is in unsteady transition between its early form as identifiable vorticity and fully developed rolled-up form, and presents a formidable modelling problem.

In recent times a number of factors have combined to increase the significance of interactional aerodynamics – higher disc loadings resulting in stronger downwash, more compact configurations often with relatively large fuselage and empennage areas and the increased use of helicopters in low level, nap-of-the-earth operations. From a design perspective, the most useful information relating to interactional aerodynamics is located in the reports of full and model scale testing. In Ref. 3.44, Prouty discusses a number of datasets showing the effects of rotor downwash on the empennage. A review of test results from a period of activity at Boeing helicopters is reported by Sheridan in Ref. 3.70. In this reference, interactions are classified into downstream (e.g., rotor/empennage upset loads, tail rotor/loss of effectiveness), localized (e.g., rotor/fuselage download, tail rotor/fin blockage), ground proximity (e.g., trim power, unsteady loads from ground vortex) and external interaction (e.g., helicopter/helicopter upset loads, ground winds) categories. One problem that has received considerable attention through testing is the interaction of the rotor downwash with the rear fuselage (tailboom) at low speed. In Ref. 3.71, Brocklehurst describes the successful implementation of fuselage strakes to control the separation of the circulatory flow caused by the downwash flowing over the tailboom in sideways flight. Reference 3.72 discusses a number of similar test programmes on US helicopters. In all these cases the use of the strakes reduced the tail rotor control and power requirements, hence recovering the flight envelope from the restrictions caused by the high tailboom sideforces.

The interaction of the main rotor wake with the tail rotor has been the subject of an extensive test programme at the DRA Bedford (Refs 3.73, 3.74), aimed at providing data for interactional modelling developments. In Ref. 3.73, from an analysis of Lynx flight test data with an instrumented tail rotor, Ellin identified a number of regions of the flight envelope where the interactional aerodynamics could be categorized. Particular attention was paid to the so-called quartering-flight problem, where the tail rotor control requirements for trim can be considerably different from calculations based on an essentially isolated tail rotor. Figure 3.44 shows a plan view of the helicopter in quartering flight – hovering with a wind from about 45° to starboard. There exists a fairly narrow range of wind directions when the tail rotor is exposed to the powerful effect of the advancing blade tip vortices as they are swept downstream. A similar situation will arise in quartering flight from the left, although the tail rotor control margins are considerably greater for this lower (tail rotor) power condition. From a detailed study of tail rotor pressure data, Ellin was able to identify the passage of individual main rotor tip vortices through the tail rotor disc. Based on this evidence, Ellin constructed a Beddoes main rotor wake (Ref. 3.75) and was able to model, in a semi-empirical manner, the effect of the main rotor vortices on the tail rotor control

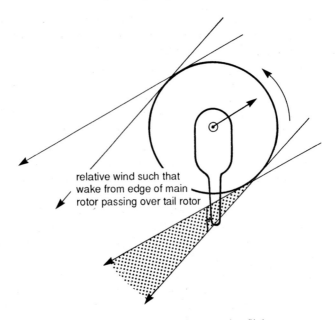

Fig. 3.44 The tail rotor in quartering flight

margin. Effectively the advancing blade tip vortices introduce a powerful in-plane velocity component at the tail rotor disc. For the case of the Lynx Mk 5, with its 'top-forward' tail rotor rotation direction, this leads to a reduction in dynamic pressure and an increase in control angle and power to achieve the same rotor thrust. Tail rotors with 'top-aft' rotations (e.g., Lynx Mk 7) do not suffer from this problem, and the control requirements, at least in right quartering flight, can actually be improved in some circumstances, although interactions with the aerodynamics of the vertical fin are also an important ingredient of this complex problem. Figure 3.45 shows the pedal control margin for Lynx Mk 5 hovering in a wind from all directions 'around the clock' out to 30 knots. Figure 3.45(a) presents Ellin's flight measurements. The limiting condition corresponding to right quartering flight is shown as the 10% margin contour. The

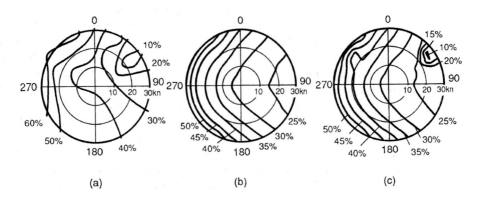

Fig. 3.45 Comparison of the tail rotor pedal margin measured on the DRA research Lynx with theory: (a) flight; (b) Helisim; (c) Helisim corrected (Ref. 3.73)

situation in left quartering flight manifests itself in a drawing out of the 60% contour as shown, although the situation is further complicated in left flight by the tail rotor experiencing vortex-ring flow states. Figure 3.45(b) shows the same result predicted by Helisim with an isolated tail rotor; clearly none of the non-uniformities caused by the interactions with the main rotor wake and fin is present. In comparison, Fig. 3.45(c) shows the Helisim pedal margin results after correction of the dynamic pressure experienced by the tail rotor, using the Beddoes main rotor wake. The non-uniformities in quartering flight are now well predicted, although in flight to the left, the predicted margin is still 10–15% greater than in flight. The results of Ellin's research point towards the direction of improved modelling for main rotor wake/tail rotor interactions, although achieving real-time operation with the kind of prescribed wake used remains a significant task.

A similar investigation into the effects of main rotor wake/tail rotor interaction on yaw control effectiveness is reported in Ref. 3.76, using the University of Maryland Advanced Rotor Code (UMARC). For predicting the distribution of main rotor wake velocity perturbations behind the rotor, a free wake model was used and correlated against wind-tunnel test data. In general, a good comparison was found, except for the critical positions close to the main rotor tip vortices where peak velocities some 100% greater than predicted were measured. Correlation of predicted tail rotor control margin at the critical quartering flight azimuths was reasonable, although theory typically underestimated the control margins by about 10–15%. The UMARC analysis was conducted on an SH-2 helicopter with top-forward tail rotor rotation and the positive effects of main rotor wake/tail rotor interaction were predicted to be much stronger in theory than measured in flight. The Maryland research in this area represents one of the first applications of comprehensive rotor modelling to wake/tail interactions and their effects on flying qualities.

The series of papers by Curtiss and his co-workers at Princeton University report another important set of findings in the area of interactional aerodynamics; in this case, special attention was paid to the effect of the main rotor wake on the empennage (Refs 3.55, 3.77, 3.78). Reference 3.78 compares results using a 'flat' prescribed wake (Ref. 3.79) with a free wake (Ref. 3.80) for predicting the induced velocity distribution at the location of the horizontal stabilizer for a UH-60 helicopter. Comparison of the non-dimensional downwash (normalized by momentum value of uniform downwash at the disc) predicted by the two methods, as a function of lateral displacement at the tail surface, is shown in Fig. 3.46. The UH-60 tailplane has a full span of about $0.5R$. The simpler flat wake captures most of the features in the considerably more complex free wake model, although the peak velocities from the rolled-up wake on the advancing and retreating sides are overestimated by about 30% with the flat wake. The much stronger induced flow on the advancing side of the disc is clearly predicted by both models. The upwash outside the rotor disc ($y/R > 1.0$) is also predicted by both models. One of the applications studied in Ref. 3.78 involved the prediction of cross-coupling from sideslip into pitch, a characteristic known to feature quite large on the UH-60. From Figure 3.46, we can deduce that sideslip will give rise to significant variations in the levels of downwash at the horizontal stabilizer – a sideslip of 15°, for example, will cause a shift in the downwash pattern by about $0.25R$, to left or right. Figure 3.47 compares the pitch rate response to a pedal doublet input at 100 knots; the flight test results are also plotted for comparison (Ref. 3.78). It can be seen that the powerful pitching moment, developing during the first second of the manoeuvre, is reasonably

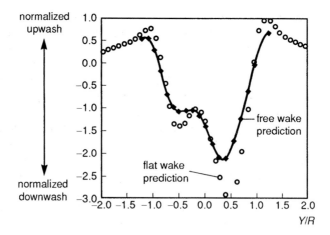

Fig. 3.46 Comparison of flat and free wake predictions for normalized downwash at the horizontal stabilizer location; UH-60, $\mu = 0.2$ (Ref. 3.78)

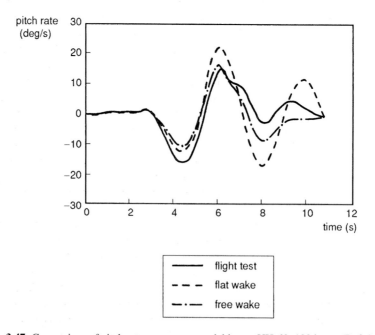

Fig. 3.47 Comparison of pitch rate response to pedal input; UH-60, 100 knots (Ref. 3.78)

well predicted by both interactional aerodynamic models. As an aside, we would not expect to see any pitch response from the Helisim model until the yawing and rolling motions had developed. The free wake model appears to match flight test fairly well until the motion has decayed after about 10 s, while the flat wake underpredicts the oscillatory damping.

Ultimately, the value of interactional aerodynamic modelling will be measured by its effectiveness at predicting the degrading or enhancing effects on operational performance. To reiterate, the motivation for developing an increased modelling capability for use in design and requirements capture, in terms of the potential pay-off, is very high. Much of the redesign effort on helicopters over the last 30 years has been driven by the unexpected negative impact of interactional problems (Ref. 3.81), and there is a real need for renewed efforts to improve the predictive capability of modelling. This must, of course, be matched by the gathering of appropriate validation test data.

At the time of writing, 'operational' simulation models with a comprehensive treatment of nonlinear, unsteady rotor and interactional aerodynamics are becoming commonplace in industry, government research laboratories and in academia. Some of these have been referred to above. The computational power to run blade-element rotor models, with elastic modes and quite sophisticated aerodynamic effects, in real time, is now available and affordable. The domain of flight dynamics is rapidly overlapping with the prediction of loads, vibration, rotor aeroelastic stability and aeroacoustics. Yet the overall effects on our understanding of helicopter flight dynamics, stemming from the vigorous developments in recent years, does not appear to have been cumulative. This is partly because of the human factor – the reservoirs of knowledge are people rather than reports and journal papers – but there is another important issue. In the author's view, the pace associated with our ability to computer-model detailed fluid and structural dynamics has far outpaced our ability to understand the underlying causal physics. Even if the 'perfect' simulation model existed, its effective use in requirements capture, design and development would need to be underpinned by our ability to interpret the outputs meaningfully. While the perfect model does not yet exist, it is the vision of many rotorcraft engineers, but the achievement of this goal will need to be accompanied by two companion activities in the author's view, or not realized at all. First, recalling how important the interplay between theory and experiment has been in the development of rotorcraft, confidence in simulation modelling will increase only through validation against test data. High-quality measurements of surface and flowfield aerodynamics and component loads are difficult and expensive to make and are often available only for commercially sensitive programmes. The focus needs to be on 'generic' test data, with an emphasis on manoeuvring flight and into areas at flight envelope boundaries where nonlinearities govern dynamic behaviour. Second, there needs to be renewed emphasis on the development of narrow range approximations that truly expose cause and effect and, just like the critical missing jigsaw piece, provide significant insight and understanding. However, the skills required to build a simulation model and those required to derive analytic approximations, while complementary, are quite different, and it is a mistake to assume that the former begets the latter. The importance of these integrated modelling skills needs to be recognized in university courses and industrial training programmes or there is a real danger that the analytical skills will be lost in favour of computational skills. Chapters 4 and 5 are concerned with working with simulation models, where validation and analytic approximation feature strongly.

APPENDIX 3A FRAMES OF REFERENCE AND COORDINATE TRANSFORMATIONS

3A.1 The inertial motion of the aircraft

In this section we shall derive the equations of translational and rotational motion of a helicopter assumed to be a rigid body, referred to an axes system fixed at the centre of

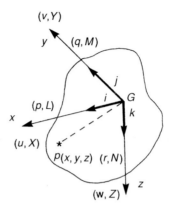

Fig. 3A.1 The fuselage-fixed reference axes system

mass of the aircraft (assumed to be fixed in the aircraft). The axes, illustrated in Fig. 3A.1, move with time-varying velocity components u, v, w and p, q, r, under the action of applied forces X, Y, Z and L, M, N.

The evolutionary equations of motion can be derived by equating the rates of change of the linear and angular momentum to the applied forces and moments. Assuming constant mass, the equations are conveniently constructed by selecting an arbitrary material point, P, inside the fuselage and by deriving the expression for the absolute acceleration of this point. The acceleration can then be integrated over the fuselage volume to derive the effective change in angular momentum and hence the total inertia force. A similar process leads to the angular acceleration and corresponding inertial moment. The centre of the moving axes is located at the helicopter's centre of mass, G. As the helicopter translates and rotates, the axes therefore remain fixed to material points in the fuselage. This is an approximation since the flapping and lagging motion of the rotor cause its centre of mass to shift and wobble about some mean position, but we shall neglect this effect, the mass of the blades being typically $<5\%$ of the total mass of the helicopter. In Fig 3A.1, $\mathbf{i}$, $\mathbf{j}$, $\mathbf{k}$ are unit vectors along the x, y and z axes respectively.

We can derive the expression for the absolute acceleration of the material point P by summing together the acceleration of P relative to G and the acceleration of G relative to fixed earth. The process is initiated by considering the position vector of the point P relative to G, namely

$$\mathbf{r}_{p/g} = x\mathbf{i} + y\mathbf{j} + z\mathbf{k} \tag{3A.1}$$

The velocity can then be written as

$$\mathbf{v}_{p/g} = \dot{\mathbf{r}}_{p/g} = (\dot{x}\mathbf{i} + \dot{y}\mathbf{j} + \dot{z}\mathbf{k}) + (x\dot{\mathbf{i}} + y\dot{\mathbf{j}} + z\dot{\mathbf{k}}) \tag{3A.2}$$

Since the reference axes system is moving, the unit vectors change direction and therefore have time derivatives; these can be derived by considering small changes in the angles $\delta\theta$, about each axis. Hence

$$\delta\mathbf{i} = \mathbf{j}\delta\theta_z - \mathbf{k}\delta\theta_y \tag{3A.3}$$

and

$$\frac{di}{dt} = \dot{i} = \mathbf{j}\frac{d\theta_z}{dt} - \mathbf{k}\frac{d\theta_y}{dt} = r\mathbf{j} - q\mathbf{k} \tag{3A.4}$$

Defining the angular velocity vector as

$$\boldsymbol{\omega}_g = p\mathbf{i} + q\mathbf{j} + r\mathbf{k} \tag{3A.5}$$

we note from eqn 3A.4 that the unit vector derivatives can be written as the vector product

$$\dot{\mathbf{i}} = \boldsymbol{\omega}_g \wedge \mathbf{i} \tag{3A.6}$$

with similar forms about the $\mathbf{j}$ and $\mathbf{k}$ axes.

Since the fuselage is assumed to be rigid, the distance of the material point P from the centre of mass is fixed and the velocity of P relative to G can be written as

$$\mathbf{v}_{P/g} = \boldsymbol{\omega}_g \wedge \mathbf{r}_P \tag{3A.7}$$

or in expanded form as

$$\mathbf{v}_{P/g} = (qz - ry)\mathbf{i} + (rx - pz)\mathbf{j} + (py - qx)\mathbf{k} = u_{P/g}\mathbf{i} + v_{P/g}\mathbf{j} + w_{P/g}\mathbf{k} \tag{3A.8}$$

Similarly, the acceleration of P relative to G can be written as

$$\mathbf{a}_{P/g} = \dot{\mathbf{v}}_{P/g} = (\dot{u}_{P/g}\mathbf{i} + \dot{v}_{P/g}\mathbf{j} + \dot{w}_{P/g}\mathbf{k}) + (u_{P/g}\dot{\mathbf{i}} + v_{P/g}\dot{\mathbf{j}} + w_{P/g}\dot{\mathbf{k}})$$
$$= \mathbf{a}_{P/g_{rel}} + \boldsymbol{\omega}_g \wedge \mathbf{v}_P \tag{3A.9}$$

or, in expanded form, as

$$\mathbf{a}_{P/g} = (\dot{u}_{P/g} - rv_{P/g} + qw_{P/g})\mathbf{i} + (\dot{v}_{P/g} - pw_{P/g} + ru_{P/g})\mathbf{j}$$
$$+ (\dot{w}_{P/g} - qu_{P/g} + pv_{P/g})\mathbf{k} \tag{3A.10}$$

Writing the inertial velocity (relative to fixed earth) of the aircraft centre of mass, G, in component form as

$$\mathbf{v}_g = u\mathbf{i} + v\mathbf{j} + w\mathbf{k} \tag{3A.11}$$

we can write the velocity of P relative to the earth reference as

$$\mathbf{v}_P = (u - ry + qz)\mathbf{i} + (v - pz + rx)\mathbf{j} + (w - qx + py)\mathbf{k} \tag{3A.12}$$

Similarly, the acceleration of P takes the form

$$\mathbf{a}_P = \mathbf{a}_{P_{rel}} + \boldsymbol{\omega}_g \wedge \mathbf{v}_P \tag{3A.13}$$

or

$$\mathbf{a}_P = a_x\mathbf{i} + a_y\mathbf{j} + a_z\mathbf{k} \tag{3A.14}$$

with components

$$a_x = \dot{u} - rv + qw - x(q^2 + r^2) + y(pq - \dot{r}) + z(pr + \dot{q}) \qquad (3\text{A}.15)$$

$$a_y = \dot{v} - pw + ru - y(p^2 + r^2) + z(qr - \dot{p}) + x(pq + \dot{r}) \qquad (3\text{A}.16)$$

$$a_z = \dot{w} - qu + pv - z(p^2 + r^2) + x(pr - \dot{q}) + y(qr + \dot{p}) \qquad (3\text{A}.17)$$

These are the components of acceleration of a point distance x, y, z from the centre of mass when the velocity components of the axes are given by $u(t)$, $v(t)$, $w(t)$ and $p(t)$, $q(t)$, $r(t)$.

We now assume that the sum of the external forces acting on the aircraft can be written in component form acting at the centre of mass, i.e.,

$$\mathbf{F}_g = X\mathbf{i} + Y\mathbf{j} + Z\mathbf{k} \qquad (3\text{A}.18)$$

If the material point, P, consists of an element of mass dm, then the total inertia force acting on the fuselage is the sum of all elemental forces; the equations of motion thus take the component forms

$$X = \int_{body} a_x \, dm \qquad (3\text{A}.19)$$

$$Y = \int_{body} a_y \, dm \qquad (3\text{A}.20)$$

$$Z = \int_{body} a_z \, dm \qquad (3\text{A}.21)$$

Since G is the centre of mass, then by definition

$$\int_{body} x \, dm = \int_{body} y \, dm = \int_{body} z \, dm = 0 \qquad (3\text{A}.22)$$

and the mass of the aircraft is given by

$$M_a = \int_{body} dm \qquad (3\text{A}.23)$$

The translational equations of motion of the aircraft are therefore given by the relatively simple equations

$$X = M_a(\dot{u} - rv + qw)$$
$$Y = M_a(\dot{v} - pw + ru)$$
$$Z = M_a(\dot{w} - qu + pv) \qquad (3\text{A}.24)$$

Thus, in addition to the linear acceleration of the centre of mass, the inertial loading is composed of the centrifugal terms when the aircraft is manoeuvring with rotational motion. For the rotational motion itself, the external moment vector about the centre of mass can be written in the form

$$\mathbf{M}_g = L\mathbf{i} + M\mathbf{j} + N\mathbf{k} \tag{3A.25}$$

The integrated inertial moment can be written as

$$\int_{body} \mathbf{r}_p \wedge \mathbf{a}_p \, dm = \left[\int_{body} (ya_z - za_y) \, dm \right] \mathbf{i} + \left[\int_{body} (za_x - xa_z) \, dm \right] \mathbf{j}$$

$$+ \left[\int_{body} (xa_y - ya_x) \, dm \right] \mathbf{k} \tag{3A.26}$$

Considering the component of rolling motion about the fuselage x-axis, we have

$$L = \int_{body} (ya_z - za_y) \, dm \tag{3A.27}$$

and substituting for a_y and a_z we obtain

$$L = \dot{p} \int_{body} (y^2 + z^2) \, dm - qr \int_{body} (z^2 - y^2) \, dm + (r^2 - q^2) \int_{body} yz \, dm$$

$$- (pq + \dot{r}) \int_{body} xz \, dm + (pr - \dot{q}) \int_{body} xy \, dm \tag{3A.28}$$

Defining the moments and product (I_{xz}) of inertia as

$$x\text{-axis:} \quad I_{xx} = \int_{body} (y^2 + z^2) \, dm \tag{3A.29}$$

$$y\text{-axis:} \quad I_{yy} = \int_{body} (x^2 + z^2) \, dm \tag{3A.30}$$

$$z\text{-axis:} \quad I_{zz} = \int_{body} (x^2 + y^2) \, dm \tag{3A.31}$$

$$xz\text{-axes:} \quad I_{xz} = \int_{body} xz \, dm \tag{3A.32}$$

we note that the external moments can finally be equated to the inertial moments in the form

$$L = I_{xx}\dot{p} - (I_{yy} - I_{zz})qr - I_{xz}(pq + \dot{r})$$
$$M = I_{yy}\dot{q} - (I_{zz} - I_{xx})pr + I_{xz}(p^2 - r^2)$$
$$N = I_{zz}\dot{r} - (I_{xx} - I_{yy})pq - I_{xz}(\dot{p} - rq) \tag{3A.33}$$

which are the rotational equations of motion of the aircraft.

The product of inertia, I_{xz}, is retained because of the characteristic asymmetry of the fuselage shape in the xz plane, giving typical values of I_{xz} comparable to I_{xx}.

3A.2   The orientation problem – angular coordinates of the aircraft

The helicopter fuselage can take up a new position by rotations about three independent directions. The new position is not unique, since the finite orientations are not vector quantities, and the rotation sequence is not permutable. The standard sequence used in flight dynamics is yaw, ψ, pitch, θ, and roll, ϕ, as illustrated in Fig. 3A.2. We can consider the initial position as a quite general one and the fuselage is first rotated about the z-axis (unit vector $\mathbf{k_0}$) through the angle ψ (yaw). The unit vectors in the rotated frame can be related to those in the original frame by the transformation Ψ, i.e.,

$$
\begin{bmatrix} \mathbf{i_1} \\ \mathbf{j_1} \\ \mathbf{k_0} \end{bmatrix} = \begin{bmatrix} \cos\psi & \sin\psi & 0 \\ -\sin\psi & \cos\psi & 0 \\ 0 & 0 & 1 \end{bmatrix} \begin{bmatrix} \mathbf{i_0} \\ \mathbf{j_0} \\ \mathbf{k_0} \end{bmatrix} \quad \text{or} \quad \{b\} = \Psi\{a\} \tag{3A.34}
$$

Next, the fuselage is rotated about the new y-axis (unit vector $\mathbf{j_1}$) through the (pitch) angle θ, i.e.,

$$
\begin{bmatrix} \mathbf{i_2} \\ \mathbf{j_1} \\ \mathbf{k_1} \end{bmatrix} = \begin{bmatrix} \cos\theta & 0 & -\sin\theta \\ 0 & 1 & 0 \\ \sin\theta & 0 & \cos\theta \end{bmatrix} \begin{bmatrix} \mathbf{i_1} \\ \mathbf{j_1} \\ \mathbf{k_0} \end{bmatrix} \quad \text{or} \quad \{c\} = \Theta\{b\} \tag{3A.35}
$$

Finally, the rotation is about the x-axis (roll), through the angle ϕ, i.e.,

$$
\begin{bmatrix} \mathbf{i_2} \\ \mathbf{j_2} \\ \mathbf{k_2} \end{bmatrix} = \begin{bmatrix} 1 & 0 & 0 \\ 0 & \cos\phi & \sin\phi \\ 0 & -\sin\phi & \cos\phi \end{bmatrix} \begin{bmatrix} \mathbf{i_2} \\ \mathbf{j_1} \\ \mathbf{k_1} \end{bmatrix} \quad \text{or} \quad \{d\} = \Phi\{c\} \tag{3A.36}
$$

Any vector, $\mathbf{d}$, in the new axes system can therefore be related to the components in the original system by the relationship

$$
\{d\} = \Phi\,\Theta\,\Psi\,\{a\} = \Gamma\,\{a\} \tag{3A.37}
$$

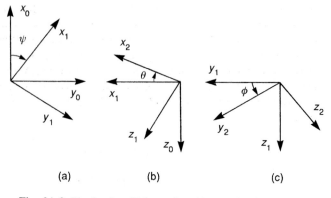

(a)   (b)   (c)

Fig. 3A.2 The fuselage Euler angles: (a) yaw; (b) pitch; (c) roll

Since all the transformation matrices are themselves orthogonal, i.e.,

$$\Psi^T = \Psi^{-1}, \text{ etc.} \tag{3A.38}$$

the product is also orthogonal, hence

$$\Gamma^T = \Gamma^{-1} \tag{3A.39}$$

where

$$\Gamma = \begin{bmatrix} \cos\theta \cos\psi & \cos\theta \sin\psi & -\sin\theta \\ \sin\phi \sin\theta \cos\psi - & \sin\phi \sin\theta \sin\psi + & \sin\phi \cos\theta \\ \cos\phi \sin\psi & \cos\phi \cos\psi & \\ \cos\phi \sin\theta \cos\psi + & \cos\phi \sin\theta \sin\psi - & \cos\phi \cos\theta \\ \sin\phi \sin\psi & \sin\phi \cos\psi & \end{bmatrix} \tag{3A.40}$$

Of particular interest is the relationship between the time rate of change of the orientation angles and the fuselage angular velocities in the body axes system, i.e.,

$$\omega_g = p\mathbf{i}_2 + q\mathbf{j}_2 + r\mathbf{k}_2$$
$$= \dot{\psi}\mathbf{k}_0 + \dot{\theta}\mathbf{j}_1 + \dot{\phi}\mathbf{i}_2 \tag{3A.41}$$

Using eqns 3A.34–3A.36, we can derive

$$p = \dot{\phi} - \dot{\psi} \sin\theta$$
$$q = \dot{\theta} \cos\phi + \dot{\psi} \sin\phi \cos\theta$$
$$r = -\dot{\theta} \sin\phi + \dot{\psi} \cos\phi \cos\theta \tag{3A.42}$$

3A.3 Components of gravitational acceleration along the aircraft axes

The relationships derived in Appendix Section 3A.2 are particularly important in flight dynamics as the gravitational components appear in the equations of motion in terms of the so-called Euler angles, θ, ϕ, ψ, while the aerodynamic forces are referenced directly to the fuselage angular motion. We assume for helicopter flight dynamics that the gravitational force always acts in the vertical sense and the components in the fuselage-fixed axes are therefore easily obtained with reference to the transformation matrix given in eqn 3A.40. The gravitational acceleration components along the fuselage x, y and z axes can therefore be written in terms of the Euler roll and pitch angles as

$$a_{x_g} = -g \sin\theta$$
$$a_{y_g} = g \cos\theta \sin\phi$$
$$a_{z_g} = g \cos\theta \cos\phi \tag{3A.43}$$

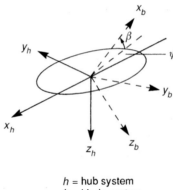

h = hub system
b = blade system

Fig. 3A.3 The hub and blade reference axes systems

3A.4   The rotor system – kinematics of a blade element

The components of velocity and acceleration of a blade element relative to the air through which it is travelling, and the inertial axes system, are important for calculating the blade dynamics and loads. When the hub is fixed, the only accelerations experienced by a flapping blade are due to the centrifugal force and out-of-plane motion. When the hub is free to translate and rotate, then the velocities and accelerations of the hub contribute to the accelerations at a blade element. We begin with an analysis of the transformation between vectors in the non-rotating hub reference system and vectors in the blade axes system.

Figure 3A.3 illustrates the hub reference axes, with the x and y directions oriented parallel to the fuselage axes centred at the centre of mass. The z direction is directed downwards along the rotor shaft, which, in turn, is tilted forward relative to the fuselage z-axis by an angle γ_s. The blade referenced axes system has the positive x direction along the blade quarter chord line. The zero azimuth position is conventionally at the rear of the disc as shown in the figure, with the positive rotation anticlockwise when viewed from above, i.e., in the negative sense about the z-axis. Positive flapping is upwards. The positive y and z directions are such that the blade and hub systems align when the flapping is zero and the azimuth angle is $180°$.

We shall derive the relationship between components in the rotating and non-rotating systems by considering the unit vectors. The orientation sequence is first azimuth, then flap. Translational and angular velocities and accelerations in the hub system can be related to the blade system by the transformation

$$\begin{bmatrix} \mathbf{i}_h \\ \mathbf{j}_h \\ \mathbf{k}_h \end{bmatrix} = \begin{bmatrix} -\cos\psi & -\sin\psi & 0 \\ \sin\psi & -\cos\psi & 0 \\ 0 & 0 & 1 \end{bmatrix} \begin{bmatrix} \cos\beta & 0 & \sin\beta \\ 0 & 1 & 0 \\ -\sin\beta & 0 & \cos\beta \end{bmatrix} \begin{bmatrix} \mathbf{i}_b \\ \mathbf{j}_b \\ \mathbf{k}_b \end{bmatrix} \qquad (3A.44)$$

or, in expanded form

$$\begin{bmatrix} \mathbf{i}_h \\ \mathbf{j}_h \\ \mathbf{k}_h \end{bmatrix} = \begin{bmatrix} -\cos\psi\cos\beta & -\sin\psi & -\cos\psi\sin\beta \\ \sin\psi\cos\beta & -\cos\psi & \sin\psi\sin\beta \\ -\sin\beta & 0 & \cos\beta \end{bmatrix} \begin{bmatrix} \mathbf{i}_b \\ \mathbf{j}_b \\ \mathbf{k}_b \end{bmatrix} \qquad (3A.45)$$

The hub velocity components in the hub reference system are related to the velocities of the centre of mass, u, v and w through the transformation

$$\begin{bmatrix} u_h \\ v_h \\ w_h \end{bmatrix} = \begin{bmatrix} \cos \gamma_s & 0 & \sin \gamma_s \\ 0 & 1 & 0 \\ -\sin \gamma_s & 0 & \cos \gamma_s \end{bmatrix} \begin{bmatrix} u - qh_R \\ v + ph_R + rx_{cg} \\ w - qx_{cg} \end{bmatrix} \tag{3A.46}$$

where γ_s is the forward tilt of the rotor shaft, and h_R and x_{cg} are the distances of the rotor hub relative to the aircraft centre of mass, along the negative z direction and forward x direction (fuselage reference axes) respectively.

It is more convenient, in the derivation of rotor kinematics and loads, to refer to a non-rotating hub axes system which is aligned with the resultant velocity in the plane of the rotor disc; we refer to this system as the hub–wind system, with subscript hw. The translational velocity vector of the hub can therefore be written with just two components, i.e.,

$$\mathbf{v}_{hw} = u_{hw} \mathbf{i}_{hw} + w_{hw} \mathbf{k}_{hw} \tag{3A.47}$$

The angular velocity of the hub takes the form

$$\boldsymbol{\omega}_{hw} = p_{hw} \mathbf{i}_{hw} + q_{hw} \mathbf{j}_{hw} + r_{hw} \mathbf{k}_{hw} \tag{3A.48}$$

The hub–wind velocities are given by the relationships

$$u_{hw} = u_h \cos \psi_w + v_h \sin \psi_w = \left(u_h^2 + v_h^2 \right)^{1/2}$$

$$v_{hw} = 0 \tag{3A.49}$$

$$w_{hw} = w_h$$

$$\begin{bmatrix} p_{hw} \\ q_{hw} \end{bmatrix} = \begin{bmatrix} \cos \psi_w & \sin \psi_w \\ -\sin \psi_w & \cos \psi_w \end{bmatrix} \begin{bmatrix} p \\ q \end{bmatrix} \tag{3A.50}$$

$$r_{hw} = r + \dot{\psi}_w \tag{3A.51}$$

where the rotor 'sideslip' angle ψ_w is defined by the expressions

$$\cos \psi_w = \frac{u_h}{\sqrt{u_h^2 + v_h^2}}, \qquad \sin \psi_w = \frac{v_h}{\sqrt{u_h^2 + v_h^2}} \tag{3A.52}$$

We now write the angular velocity components transformed to the rotating system as

$$\begin{bmatrix} \omega_x \\ \omega_y \end{bmatrix} = \begin{bmatrix} \cos \psi & -\sin \psi \\ \sin \psi & \cos \psi \end{bmatrix} \begin{bmatrix} p_{hw} \\ q_{hw} \end{bmatrix} \tag{3A.53}$$

Using the transformation matrix in eqn 3A.45, and assuming that the flapping angle β remains small so that $\cos \beta \approx 1$ and $\sin \beta \approx \beta$, we note that the velocities at blade station r_b, in the blade axes system, may be written as

$$u_b = -u_{hw} \cos \psi - w_{hw} \beta$$

$$v_b = -u_{hw} \sin \psi - r_b (\Omega - r_{hw} + \beta \omega_x)$$

$$w_b = -u_{hw} \beta \cos \psi + w_{hw} + r_b (\omega_y - \dot{\beta}) \qquad (3A.54)$$

Similarly, the blade accelerations can be derived, but in this case the number of terms increases considerably. The dominant effects are due to the acceleration of the blade element relative to the hub, with the centrifugal and Coriolis inertia forces giving values typically greater than 500 g at the blade tip. Blade normal accelerations are an order of magnitude smaller than this, but are still an order of magnitude greater than the mean accelerations of the aircraft centre of mass and rotor hub. We shall therefore neglect the translational accelerations of the hub and many of the smaller nonlinear terms due to products of hub and blade velocities. The approximate acceleration components at the blade station are then given by the expressions

$$a_{xb} = r_b \left(-(\Omega - r_{hw})^2 + 2\dot{\beta}\omega_y - 2(\Omega - r_{hw})\beta\omega_x \right)$$

$$a_{yb} = r_b \left(-(\dot{\Omega} - \dot{r}_{hw}) - \beta(\dot{q}_{hw} \sin \psi - \dot{p}_{hw} \cos \psi) + r_{hw}\beta\omega_y \right)$$

$$a_{zb} = r_b \left(2\Omega\omega_x + \left(\dot{q}_{hw} \cos \psi + \dot{p}_{hw} \sin \psi \right) - r_{hw}\omega_x - \underline{(\Omega - r_{hw})^2 \beta} - \ddot{\beta} \right)$$

$$(3A.55)$$

The underscored components are the primary effects due to centrifugal and Coriolis forces and the angular accelerations of the hub, although the latter are also quite small in most cases. For practical purposes, we can usually make the additional assumption that the rotorspeed is much higher than the fuselage yaw rate, so that

$$(\Omega - r_{hw}) \approx \Omega \qquad (3A.56)$$

3A.5 Rotor reference planes – hub, tip path and no-feathering

In rotor dynamics analysis, three natural reference axes systems have found application in various texts and reports – the hub (or shaft) system, the tip-path plane (or no-flapping) system and the no-feathering system. These are illustrated in Fig. 3A.4, where the hub plane has been drawn horizontal for convenience. In this book we consistently use the hub system but it is useful to compare expressions for key rotor quantities in the three systems. The motivation for adopting the rotor-oriented no-flapping or no-feathering systems is that they greatly simplify the expressions for the rotor X and Y forces, as shown in Ref. 3A.1 The no-feathering axes are equivalent to the, so-called, control axes when the rotor pitch/flap coupling is zero. The control axis is aligned along the swash plate.

Assuming small angles, the normalized velocities in the rotor systems are related to those in the hub system by the approximate relationships

$$\mu_{tp} = \mu_h + \mu_{zh} \beta_{1c}$$

$$\mu_{ztp} = \mu_{zh} - \mu_h \beta_{1c} \qquad (3A.57)$$

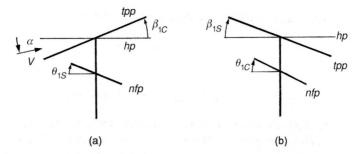

Fig. 3A.4 Reference planes for rotor dynamics: (a) longitudinal plane; (b) lateral plane

and

$$\mu_{nf} = \mu_h - \mu_{zh}\,\theta_{1s}$$

$$\mu_{znf} = \mu_{zh} + \mu_h\,\theta_{1s} \tag{3A.58}$$

Similarly, the disc incidences are given by the expressions

$$\alpha_{tp} = \alpha_h - \beta_{1c}$$

$$\alpha_{nf} = \alpha_h + \theta_{1s} \tag{3A.59}$$

and the non-rotating rotor forces are given as

$$X_{nf} = X_h - T\,\theta_{1s}$$

$$X_{tp} = X_h + T\beta_{1c} \tag{3A.60}$$

$$Y_{tp} = Y_h - T\beta_{1s}$$

$$Y_{nf} = Y_h - T\theta_{1c} \tag{3A.61}$$

where it is assumed that the rotor thrust T and Z forces in the three systems have the same magnitude and opposite directions.

In hover, the alignment of the tip-path plane and the no-feathering plane highlights the equivalence of flapping and feathering. These expressions are valid only for rotors with flap articulation at the centre of rotation. Elastic motion of hingeless rotors and flapping of articulated rotors with offset flap hinges cannot be described with these rotor axes systems. It should also be noted that the induced inflow discussed earlier in this chapter, λ, is strictly referred to the tip-path plane, giving the inflow normal to the hub plane as

$$\lambda_h = \lambda_{tp} - \mu\beta_{1c} \tag{3A.62}$$

This effect is taken into account in the derivation of the rotor torque given in this chapter (eqn 3.116), but not in the iterative calculation of λ. The small flap angle approximation will give negligible errors for trim flight, but could be more significant during manoeuvres when the flapping angles are large.

*The Empire Test Pilot School's Lynx in an agile pitch manoeuvre –
complementary to the cover picture (Photograph courtesy of
DTEO Boscombe Down and the Controller HMSO)*

4 Modelling helicopter flight dynamics: trim and stability analysis

> *The challenge and responsibility of modern engineering practice demand a high level of creative activity which, in turn, requires the support of strong analytical capability. The primary focus should be on the engineering significance of physical quantities with the mathematical structure acting in a supporting role.*
>
> *(Meriam 1966)*

4.1 INTRODUCTION AND SCOPE

Meriam's words of advice at the head of this chapter should act as a guiding light for engineers wishing to strengthen their skills in flight dynamics (Ref. 4.1). In Chapter 3 we sought to describe the physics and mathematics required for building a simulation model of helicopter flight behaviour. This chapter takes the products of this work and develops various forms of analysis to gain insight into how helicopters behave the way they do, hence establishing the engineering significance of the physics. Within the framework illustrated in our reference Fig. 4.1, the mechanics of helicopter flight can be described in terms of three aspects – trim, stability and response – as shown by the regions highlighted in the figure. The trim problem concerns the control positions required to hold the helicopter in equilibrium. The aircraft may be climbing, turning and may be at large angles of incidence and sideslip, but if the three translational velocity components are constant with the controls fixed, then the aircraft is in trim. Strictly, climbing and diving flight cannot be described as trim conditions, because the changing air density will require continual corrections to the controls. Provided the rates of climb or descent are relatively small, however, the helicopter will be, practically speaking, in trim. Stability is concerned with the behaviour of the aircraft following a disturbance from trim. Classically, static stability is determined by the initial tendency (i.e., will the aircraft tend to return to, or depart from, the initial trim?), while dynamic stability concerns longer term effects. These are useful physical concepts and will be embraced within the more general theory of the stability of the natural modes of motion, developed from the linear theory of flight mechanics. Response to pilot control inputs and to atmospheric disturbances are essentially nonlinear problems, but some insight can be gained from extending the linear approximations to finite amplitude motion. We shall return to response in Chapter 5. Trim, stability and response make up the flying characteristics. Later in Chapters 6 and 7, the reader will find that these flying characteristics are part of the domain of flying qualities. These later chapters will be concerned with how to quantify and measure flying quality, while here in Chapters 4 and 5 we are more interested in the physical mechanisms that generate the response.

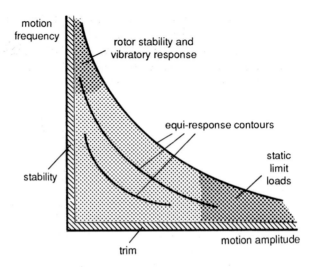

Fig. 4.1 The territory of helicopter flight mechanics

Typical problems tackled by the flight dynamicist through mathematical modelling include

(1) determination of the control margins at the operational flight envelope (OFE) and safe flight envelope (SFE);
(2) design of flight control laws that confer Level 1 handling qualities throughout the OFE;
(3) simulation of the effects of tail rotor drive failure in forward flight – establish the pitch, roll and yaw excursions after 3 s;
(4) derivation of the sensitivity of roll attitude bandwidth to rotor flapping stiffness;
(5) establishing the tailplane size required to ensure natural pitch stability at high speed;
(6) determination of the effects of main rotor blade twist on power required for various missions;
(7) establishing the maximum take-off weight, hence payload, of a twin-engine helicopter while conforming to the civil certification requirements for fly-away capability following a single engine failure;
(8) assessment and comparison of various candidate aircraft's ability to meet the flying qualities standard – ADS-33.

Of course, we could continue adding more tasks, but the range of problems has, hopefully, been adequately demonstrated with the above list. Setting down this 'short list' of activities, some of which the author has been intimately involved with over the past 20 years, serves as a reminder of the importance of modelling in aircraft design and development – relying on experiment to tackle these problems would be prohibitively expensive. This is, of course, not to devalue or diminish the importance of flight testing.

Before we engage the supporting mathematics for describing the trim and stability problems, it may be useful for the reader to explore how all three are encapsulated in the relatively simple problem of heave motion in vertical flight. The key equations taken from Chapter 3 relate to the thrust coefficient C_T and uniform component of inflow λ_0 through the rotor (eqns 3.91, 3.139):

$$\frac{2C_T}{a_0 s} = \frac{\theta_0}{3} + \frac{\mu_z - \lambda_0}{2} + \frac{\theta_{tw}}{4} \tag{4.1}$$

$$\lambda_0 = \frac{C_T}{2(\lambda_0 - \mu_z)} \tag{4.2}$$

This approximation of uniform rotor inflow is strictly applicable only when the blade twist has the ideal variation, inversely proportional to radius, giving constant circulation across the rotor and minimum induced drag. Linear blade washout of 10° or more generally gives a reasonably good approximation to the ideal loading.

In its simplest form, the trim problem amounts to determining the collective pitch θ_0 required to hold a hover, which is often written in terms of the equivalent pitch at the three-quarter radius, rather than at the rotor hub; i.e., from eqns 4.1 and 4.2, we can write

$$\theta_{\frac{3}{4}R} = \theta_0 + \frac{3}{4}\theta_{tw} = 3\left(\frac{2C_T}{a_0 s} + \frac{1}{2}\sqrt{\left(\frac{C_T}{2}\right)}\right) \tag{4.3}$$

For moderate values of thrust coefficient ($C_T -0.007$) and typical values of solidity ($s = N_b c / \pi R \sim 0.09$), the collective required to hover is approximately doubled by the presence of the induced velocity. The rotor torque required is then the sum of the induced and profile contributions (eqn 3.116)

$$C_Q = C_{Q_i} + C_{Q_p} = \frac{C_T^{1.5}}{\sqrt{2}} + C_{Q_p} \tag{4.4}$$

showing the nonlinear relationship between torque and thrust in hover.

The trim problem is generally formulated as a set of nonlinear algebraic equations. In the case examined, the unique solution was obtained by simple rearrangement. In a more general trim, when the relevant equations are coupled, this will not be as straightforward and recourse to numerical solutions will be necessary.

Analysis of the dynamic stability and response problems requires the formulation of the equation of motion relating the normal acceleration to the applied thrust

$$\dot{w} = \frac{Z}{M_a} = -\frac{T}{M_a} \tag{4.5}$$

Stability and response characteristics may be assessed (in the first approximation) by analysis of the linearized form of the nonlinear eqn 4.5. We write the normal velocity ($w = \Omega R \mu_z$ in hover) as the sum of a trim or equilibrium value (subscript e) and a perturbation value

$$w = W_e + \delta w \tag{4.6}$$

If we assume that the Z force acting on the helicopter in the hover is an analytic function of the control θ_0 and normal velocity w, together with their time rates of change, then the force can be expanded as a Taylor series about the trim value (Ref. 4.2), in the form

$$Z = Z_e + \frac{\partial Z}{\partial w}\delta w + \frac{\partial Z}{\partial \theta_0}\delta\theta_0 + \frac{1}{2}\frac{\partial^2 Z}{\partial w^2}w^2 + \cdots + \frac{\partial Z}{\partial \dot{w}}\delta\dot{w} + \cdots \qquad (4.7)$$

In the simple form of thrust equation given by eqn 4.5, there are no unsteady aerodynamic effects and hence there are no explicit acceleration derivatives. For small and slow changes in w (i.e., δw) and θ_0 (i.e., $\delta\theta_0$), the first two (linear) perturbation terms in eqn 4.7 will approximate the changes in the applied force, i.e.,

$$Z \approx Z_e + \frac{\partial Z}{\partial w}\delta w + \frac{\partial Z}{\partial \theta_0}\delta\theta_0 \qquad (4.8)$$

The stability problem concerns the nature of the solution of the homogeneous equation

$$\dot{w} - Z_w w = 0 \qquad (4.9)$$

where we have subsumed the aircraft mass M_a within the heave damping derivative Z_w without any dressing, which is normal practice in helicopter flight dynamics, i.e.,

$$Z_w \equiv \frac{Z_w}{M_a} \qquad (4.10)$$

In eqn 4.9, we have used lowercase w for the perturbation in heave velocity away from the trim condition (cf. eqn 4.8 $\delta w \rightarrow w$, assumed small). This will be the general practice throughout this book, lowercase u, v and w, p, q and r denoting either total or perturbation velocities, depending upon the context. It is clear that the solution of eqn 4.9 will be stable if and only if Z_w is strictly negative, as then the solution will be a simple exponential subsidence.

The heave damping derivative can be estimated from the derivative of thrust coefficient with rotor heave velocity

$$\frac{\partial C_T}{\partial \mu_z} = \frac{1}{2}\left(1 - \frac{\partial \lambda_0}{\partial \mu_z}\right) = \frac{2a_0 s \lambda_0}{16\lambda_0 + a_0 s} \qquad (4.11)$$

giving the result

$$Z_w = -\frac{2a_0 A_b \rho(\Omega R)\lambda_0}{(16\lambda_0 + a_0 s)M_a} \qquad (4.12)$$

which ensures stability. The damping derivative, or the heave eigenvalue (see Appendix 4B), typically has a value of between -0.25 and -0.4 (1/s) and, from eqn 4.12, is a linear function of lift curve slope, a_0, and is inversely proportional to blade loading (M_a/A_b). The natural time constant of helicopter vertical motion in hover is therefore relatively large, falling between 4 and 2.5 s.

The response to small collective control inputs is governed by the inhomogeneous linear differential equation

$$\dot{w} - Z_w w = Z_{\theta_0}\theta_0(t) \qquad (4.13)$$

where the thrust derivative

$$\frac{\partial C_T}{\partial \theta_0} = \frac{8}{3} \left(\frac{a_0 s \lambda_0}{16 \lambda_0 + a_0 s} \right) \qquad (4.14)$$

is used to determine the control derivative

$$Z_{\theta_0} = -\frac{8}{3} \frac{a_0 A_b \rho (\Omega R)^2 \lambda_0}{(16 \lambda_0 + a_0 s) M_a} \qquad (4.15)$$

The ratio of the control derivative to the heave damping gives the steady-state response in heave velocity to a step change in collective pitch as

$$w_{ss} = -\frac{4}{3} \Omega R \, \theta_0 \qquad (4.16)$$

The rate sensitivity, or the steady-state rate per degree of collective, is seen to be a function of tip speed only. The rate of climb following a step input in collective is therefore independent of disc loading, lift curve slope, air density and solidity according to the simplifying assumptions of momentum theory. These assumptions, of which uniform inflow and constant lift curve slope are probably the most significant, were discussed at the beginning of Chapter 3.

The nature of the response to a vertical gust was described in some detail in Chapter 2, the equation of motion taking the form

$$\dot{w} - Z_w w = Z_w w_g(t) \qquad (4.17)$$

The initial vertical acceleration is given by the product of the heave damping and the gust strength. A vertical gust of 5 m/s gives rise to a bump of about 0.2 g for the higher levels of vertical damping. Reducing the blade loading has a powerful effect on the sensitivity to vertical gusts according to eqn 4.12, although overall, the helicopter is relatively insensitive to vertical gusts in the hover.

Helicopter vertical motion in hover is probably the simplest to analyse, but even here our simplifying approximations break down at higher frequencies and amplitudes, as unsteady aerodynamics, blade stall and rotor dynamic effects alter the details of the motion considerably. We shall return to this example later in Chapter 5. More general helicopter motions, in both hover and forward flight, tend to be coupled, and adequate single degree of freedom (DoF) descriptions are a rarity. As we progress through Chapters 4 and 5 however, the approach outlined above will form the pattern – that is, taking the basic nonlinear equations from Chapter 3 for trim and then linearizing for stability, control and small perturbation response analysis.

Chapter 4 is structured as follows. The techniques for describing and analysing trim and stability are set down in Sections 4.2 and 4.3 respectively. The expressions for the general trim problem will be derived, i.e., a turning, climbing/descending, sideslipping manoeuvre. Stability analysis requires linearization about a trim point and an examination of the eigenvalues and eigenvectors of the system. The key, 6 DoF, stability and control derivatives will be highlighted and their physical significance described. The natural modes of motion predicted from 6 DoF theory are also described. One of the major aids to physical interpretation of helicopter dynamic behaviour comes from the various approximations to the full equations of motion. Section 4.3 deals with this topic, principally with linear, narrow range approximations that highlight how the various aerodynamic effects interact to shape the natural modes of helicopter motion. Working with modelling approximations is at the heart of a flight dynamics engineer's practice, and we aim to give this area ample attention in both Chapters 4 and 5 to help the serious reader develop the required skills. The underlying mathematical methods used draw heavily on the theory of finite dimensional vector spaces, and Appendix 4A presents a summary of the key results required to gain maximum value from this chapter.

The theory of the stability of helicopter motion will be continued in Chapter 5, with special emphasis on constrained motion. The response problem is inherently nonlinear and typical behaviour will also be described in Chapter 5, with solutions from forward and inverse simulation. Discussion on some of the important differences between results using quasi-steady and higher order rotor models is also deferred until Chapter 5.

In order that some of the fundamental physical concepts of helicopter flight mechanics can be discussed in terms of analytical expressions, it is necessary to make gross approximations regarding the rotor dynamic and aerodynamic behaviour. We include all the assumptions associated with Level 1 modelling as discussed at the beginning of Chapter 3, and then go further to assume a simple trapezoidal downwash field and ignore the in-plane lift loads in the calculation of rotor forces and moments. These latter effects can be important, but assuming that the lift forces are normal to the disc plane leads to a significant simplification in the trim and stability analyses. In most cases, this assumption leads to results that are 80% or more of the answer derived from considerably more complex rotor modelling and the resulting approximate theory can be used to gain the first-order insight into flight dynamics, which is particularly useful for the prediction of trends and in preliminary design.

We have already referred to Appendix 4A, containing the background theory of vector–matrix mechanics; two additional appendices complete the Chapter 4 series. Appendix Section 4B.1 presents the configuration datasets, including aerodynamic, structural, mass and geometric properties, for the three aircraft used in this book – the Lynx, Puma and Bo105. Appendix Section 4B.2 presents, in graphical form, the complete set of stability and control derivatives for the three aircraft predicted from two-sided numerical perturbations applied to the full Helisim nonlinear equations of motion. In the second edition of the book, a new Appendix Section 4B.3, presenting these derivatives and associated eigenvalues, is included. An analysis of the trim orientation problem is given in Appendix 4C.

4.2   TRIM ANALYSIS

The simplest trim concept is portrayed in Figs 4.2(a)–(c). The helicopter, flying forward in straight trimmed flight, is assumed to consist of a main and tail rotor with a fuselage experiencing only a drag force. The rotor is assumed to be teetering in flap, with no moments transmitted through the hub to the fuselage, and the centre of mass lies on the shaft, below the rotor. Assuming the fuselage pitch and roll attitudes are small, the following elementary model of trim can be constructed.

The balance of forces in the vertical direction gives the thrust approximately equal to the weight

$$T \approx W \tag{4.18}$$

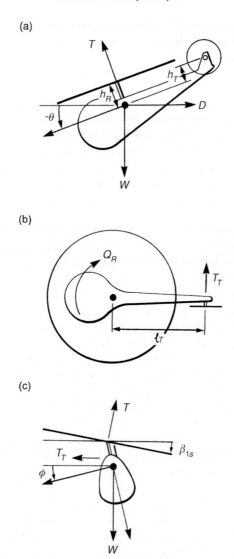

Fig. 4.2 Simple consideration of trim in hover: (a) longitudinal (view from port); (b) yaw (view from above); (c) roll (view from front)

This condition actually holds true up to moderate forward speeds for most helicopters. Balancing the forces along forward fuselage axis gives the approximate pitch angle as the ratio of drag to thrust

$$\theta \approx -\frac{D}{T} \tag{4.19}$$

Since the thrust remains essentially constant in trimmed straight flight, the pitch angle follows the drag and varies as the square of forward speed. In our simple model, the absence of any aerodynamic pitching moment from the fuselage or tail requires that the hub moment is zero, or that the disc has zero longitudinal flapping.

From Fig. 4.2(b), the tail rotor thrust can be written as the main rotor torque divided by the tail arm

$$T_T \approx \frac{Q_R}{l_T} \qquad (4.20)$$

The tail rotor thrust therefore has the same form as the main rotor torque, with the bucket at minimum main rotor power. In practice, the vertical fin is usually designed to produce a sideforce in forward flight, hence reducing the thrust required from the tail rotor. Figure 4.2(c) then shows the balance of rolling moment from the main and tail rotors, to give the lateral disc flapping

$$\beta_{1s} \approx \frac{h_T T_T}{h_R T} \qquad (4.21)$$

Thus, the disc tilts to port, for anticlockwise rotors, and the disc tilt varies as the tail rotor thrust.

The balance of sideforce gives the bank angle

$$\phi \approx \frac{T_T}{M_a g}\left(1 - \frac{h_T}{h_R}\right) \qquad (4.22)$$

If the tail rotor is located at the same height above the fuselage reference line as the main rotor, then the required bank angle is zero, for this simple helicopter design. In practice, the two terms in the numerator of eqn 4.22 are of the same order and the neglected in-plane lift forces have a significant influence on the resulting bank angle.

From the force and moment balance can be derived the required control angles – main/tail rotor collectives producing the required thrusts and the lateral cyclic from the lateral disc tilt.

4.2.1 The general trim problem

The elementary analysis outlined above illustrates the primary mechanisms of trim and provides some insight into the required pilot trim strategy, but is too crude to be of any real practical use. The most general trim condition resembles a spin mode illustrated in Fig. 4.3. The spin axis is always directed vertically in the trim, thus ensuring that the rates of change of the Euler angles θ and ϕ are both zero, and hence the gravitational force components are constant. The aircraft can be climbing or descending and flying out of lateral balance with sideslip. The general condition requires that the rate of change of magnitude of the velocity vector is identically zero. Considering eqns 3.1–3.6 from Chapter 3, we see that the trim forms reduce to

$$-(W_e Q_e - V_e R_e) + \frac{X_e}{M_a} - g\sin\Theta_e = 0 \qquad (4.23)$$

$$-(U_e R_e - W_e P_e) + \frac{Y_e}{M_a} + g\cos\Theta_e \sin\Phi_e = 0 \qquad (4.24)$$

$$-(V_e P_e - U_e Q_e) + \frac{Z_e}{M_a} + g\cos\Theta_e \cos\Phi_e = 0 \qquad (4.25)$$

$$(I_{yy} - I_{zz})Q_e R_e + I_{xz}P_e Q_e + L_e = 0 \qquad (4.26)$$

$$(I_{zz} - I_{xx})R_e P_e + I_{xz}\left(R_e^2 - P_e^2\right) + M_e = 0 \qquad (4.27)$$

$$(I_{xx} - I_{yy})P_e Q_e + I_{xz}Q_e R_e + N_e = 0 \qquad (4.28)$$

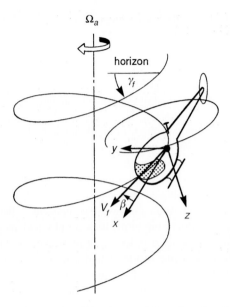

Fig. 4.3 The general trim condition of an aircraft

where the reader is reminded that the subscript e refers to the equilibrium condition. For the case where the turn rate is zero, the applied aerodynamic loads, X_e, Y_e and Z_e, balance the gravitational force components and the applied moments L_e, M_e and N_e are zero. For a non-zero turn rate, the non-zero inertial forces and moments (centrifugal, Coriolis, gyroscopic) are included in the trim balance. For our first-order approximation, we assume that the applied forces and moments are functions of the translational velocities (u, v, w), the angular velocities (p, q, r) and the rotor controls $(\theta_0, \theta_{1s}, \theta_{1c}, \theta_{0T})$. The Euler angles are given by the relationship between the body axis angular rates and the rate of change of Euler angle Ψ, the turn rate about the vertical axis, given in eqn 3A.42, i.e.,

$$P_e = -\dot{\Psi}_e \sin \Theta_e \tag{4.29}$$

$$Q_e = \dot{\Psi}_e \sin \Phi_e \cos \Theta_e \tag{4.30}$$

$$R_e = \dot{\Psi}_e \cos \Phi_e \cos \Theta_e \tag{4.31}$$

The combination of 13 unknowns and 9 equations means that to define a unique solution, four of the variables may be viewed as arbitrary and must be prescribed. The prescription is itself somewhat arbitrary, although particular groupings have become more popular and convenient than others. We shall concern ourselves with the classic case where the four prescribed trim states are defined as in Fig. 4.3, i.e.,

V_{fe}	flight speed
γ_{fe}	flight path angle
$\Omega_{a_e} = \dot{\Psi}_e$	turn rate
β_e	sideslip

In Appendix 4C, the relationships between the prescribed trim conditions and the body axis aerodynamic velocities are derived. In particular, an expression for the track angle between the projection of the fuselage x-axis and the projection of the flight velocity vector, both onto the horizontal plane, is given by the numerical solution of a nonlinear equation. Since the trim eqns 4.23–4.28 are nonlinear, and are usually solved iteratively, initial values of some of the unknown flight states need to be estimated before they are calculated. In the following sequence of calculations, initial values are estimated for the Euler pitch and roll angles Θ_e and Φ_e, the rotorspeed Ω_R, the main and tail rotor uniform downwash components λ_0 and λ_{0T} and the main rotor lateral flapping angle β_{1s}.

The solution of the trim problem can be found by using a number of different techniques, many of which are available as closed software packages, that find the minimum of a set of nonlinear equations within defined constraints. The sequential process outlined below and summarized in Fig. 4.4 is recognized as rather inefficient in view of the multiple iteration loops – one for pitch, one for roll, one for rotorspeed and one for each of the downwash components – but it does enable us to describe a sequence of partial trims, provides some physical insight into the trim process and can assist in identifying 'trim locks', or regions of the flight envelope where it becomes difficult or even impossible to find a trim solution. The process is expanded as a sequence in Fig. 4.5. The first stage is the computation of the aerodynamic velocities, enabling the fuselage forces and moments to be calculated, using the initial estimates of aerodynamic incidence angles. The three iteration loops can then be cycled.

4.2.2   Longitudinal partial trim

The main rotor thrust coefficient, longitudinal flapping and fuselage pitch attitude are calculated from the three longitudinal equations (A, B and C in Fig. 4.5). At this point, a comparison with the previous estimated value of pitch attitude is made; if the new estimate is close to the previous one, defined by the tolerance v_θ, then the partial longitudinal trim is held and the process moves on to the lateral/directional trim. If the iteration has not converged to within the tolerance, the process returns to the start and repeats until convergence is satisfied. Note that the new estimate of pitch attitude in Fig. 4.5 is given by

$$\Theta_{e_i} = \Theta_{e_{i-1}} + k_{\theta_i}\left(g_{\theta_{i-1}} - \Theta_{e_{i-1}}\right)$$ (4.32)

where

$$\Theta_e = g_\theta\left(\Theta_e, \Phi_e, V_{f_e}, \gamma_{f_e}, \Omega_{a_e}, \beta_e, \Omega_R\right)$$ (4.33)

In some cases the iteration can diverge away from, rather than converge towards, the true solution, and the value of the 'damping factor' k_θ can be selected to ensure convergence; the smaller (<1) the k factor, the slower, but more stable, is the iterative process. The key calculations in this longitudinal phase of the trim algorithm are the thrust coefficient, the longitudinal disc flapping and the pitch angle itself. For straight flight, the thrust remains relatively constant; in turning manoeuvres, the inertial term in the normal acceleration a_z, predominantly the $U_e Q_e$ term, will result in an increased thrust. The longitudinal flapping is derived from a more complicated expression, but essentially the rotor needs to flap to balance the resultant of the aerodynamic moments from the fuselage and empennage in straight flight. If the tail rotor is canted then an additional flap component will be required. Many helicopters are designed with a

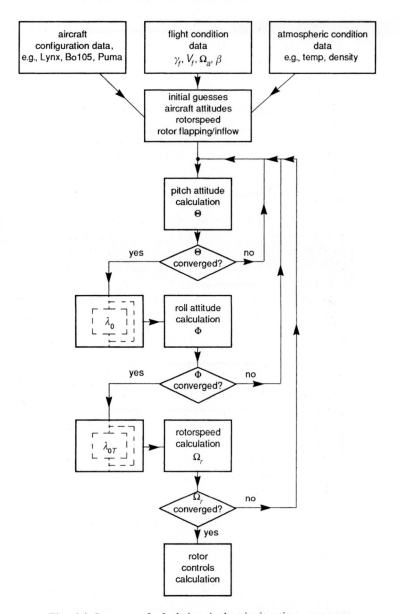

Fig. 4.4 Sequence of calculations in the trim iteration – summary

forward main rotor shaft tilt so that, at the cruise condition, the fuselage is level and the one-per-rev longitudinal flapping is zero or very small. We have already noted that the pitch angle is essentially derived from the ratio of drag to thrust, hence exhibiting a quadratic form with forward speed.

Figure 4.6(a) illustrates the variation of pitch angle with speed for Helisim Bo105 together with a comparison against the DLR flight measurements. Note the hover pitch attitude of about 3°, due to the forward shaft tilt. The transition region is typically characterized by an increase in pitch angle as the main rotor downwash impinges on

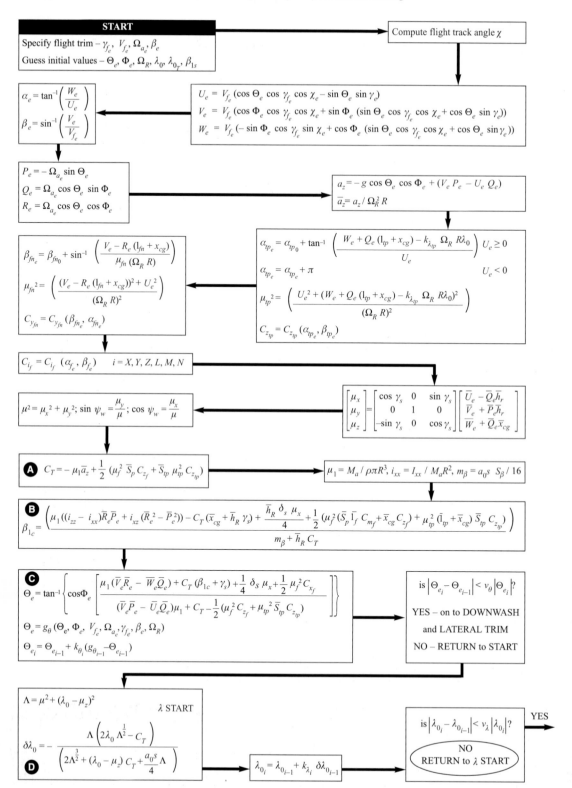

Fig. 4.5 Part I – Sequence of calculations in the trim iteration – expanded form

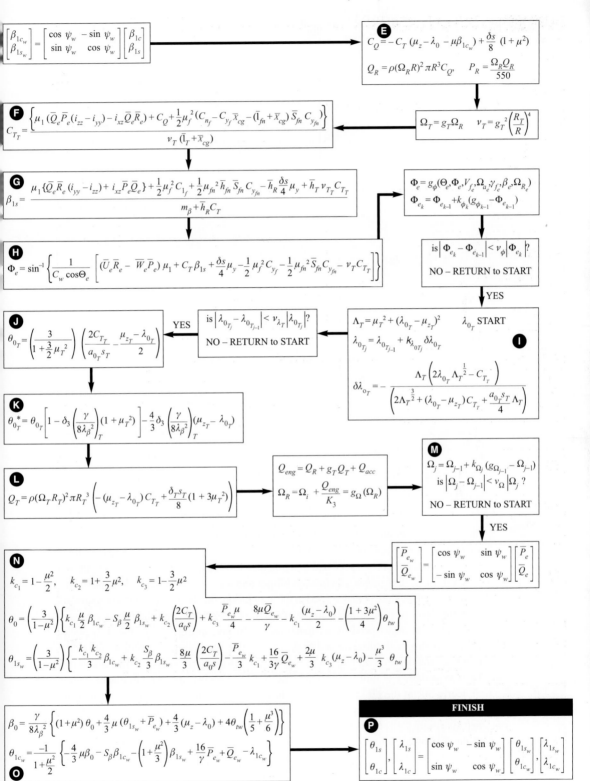

E
$$C_Q = -C_T (\mu_z - \lambda_0 - \mu\beta_{1c_w}) + \frac{\delta s}{8}(1+\mu^2)$$
$$Q_R = \rho(\Omega_R R)^2 \pi R^3 C_Q, \qquad P_R = \frac{\Omega_R Q_R}{550}$$

$$\Omega_T = g_T \Omega_R \qquad v_T = g_T^2 \left(\frac{R_T}{R}\right)^4$$

F
$$C_{T_T} = \frac{\left\{\mu_1(\bar{Q}_e \bar{P}_e(i_{zz}-i_{yy}) - i_{xz}\bar{Q}_e \bar{R}_e) + C_Q + \frac{1}{2}\mu_f^2(C_{n_f} - C_{y_f}\bar{x}_{cg} - (\bar{l}_{fn}+\bar{x}_{cg})\bar{S}_{fn}C_{y_{fn}})\right\}}{v_T(\bar{l}_T + \bar{x}_{cg})}$$

G
$$\beta_{1s} = \frac{\mu_1\{\bar{Q}_e\bar{R}_e(i_{yy}-i_{zz}) + i_{xz}\bar{P}_e\bar{Q}_e\} + \frac{1}{2}\mu_f^2 C_{l_f} + \frac{1}{2}\mu_{fn}^2\bar{h}_{fn}\bar{S}_{fn}C_{y_{fn}} - \bar{h}_R\frac{\delta s}{4}\mu_y + \bar{h}_T v_{T_T} C_{T_T}}{m_\beta + \bar{h}_R C_T}$$

$$\Phi_e = g_\phi(\Theta_e,\Phi_e,V_{f_e},\Omega_{a_e},\gamma_{f_e},\beta_e,\Omega_{R_e})$$
$$\Phi_{e_k} = \Phi_{e_{k-1}} + k_{\phi_k}(g_{\phi_{k-1}} - \Phi_{e_{k-1}})$$

is $\left|\Phi_{e_k} - \Phi_{e_{k-1}}\right| < v_\phi \left|\Phi_{e_k}\right|$?

NO – RETURN to START

H
$$\Phi_e = \sin^{-1}\left\{\frac{1}{C_w\cos\Theta_e}\left[(\bar{U}_e\bar{R}_e - \bar{W}_e\bar{P}_e)\mu_1 + C_T\beta_{1s} + \frac{\delta s}{4}\mu_y - \frac{1}{2}\mu_f^2 C_{y_f} - \frac{1}{2}\mu_{fn}^2\bar{S}_{fn}C_{y_{fn}} - v_T C_{T_T}\right]\right\}$$

↓ YES

I
$$\Lambda_T = \mu_T^2 + (\lambda_{0_T} - \mu_{z_T})^2 \qquad \lambda_{0_T} \text{ START}$$
$$\lambda_{0_{Tj}} = \lambda_{0_{Tj-1}} + k_{\lambda_{0Tj}}\delta\lambda_{0_T}$$
$$\delta\lambda_{0_T} = -\frac{\Lambda_T\left(2\lambda_{0_T}\Lambda_T^{\frac{1}{2}} - C_{T_T}\right)}{\left(2\Lambda_T^{\frac{3}{2}} + (\lambda_{0_T}-\mu_{z_T})C_{T_T} + \frac{a_{0_T}s_T}{4}\Lambda_T\right)}$$

is $\left|\lambda_{0_{Tj}} - \lambda_{0_{Tj-1}}\right| < v_{\lambda_T}\left|\lambda_{0_{Tj}}\right|$?

YES ←

NO – RETURN to START

J
$$\theta_{0_T} = \left(\frac{3}{1+\frac{3}{2}\mu_T^2}\right)\left(\frac{2C_{T_T}}{a_{0_T}s_T} - \frac{\mu_{z_T}-\lambda_{0_T}}{2}\right)$$

K
$$\theta_{0_T}^* = \theta_{0_T}\left[1 - \delta_3\left(\frac{\gamma}{8\lambda_\beta^2}\right)_T(1+\mu_T^2)\right] - \frac{4}{3}\delta_3\left(\frac{\gamma}{8\lambda_\beta^2}\right)_T(\mu_{z_T} - \lambda_{0_T})$$

L
$$Q_T = \rho(\Omega_T R_T)^2 \pi R_T^3\left(-(\mu_{z_T}-\lambda_{0_T})C_{T_T} + \frac{\delta_T s_T}{8}(1+3\mu_T^2)\right)$$

$$Q_{eng} = Q_R + g_T Q_T + Q_{acc}$$
$$\Omega_R = \Omega_i + \frac{Q_{eng}}{K_3} = g_\Omega(\Omega_R)$$

M
$$\Omega_j = \Omega_{j-1} + k_{\Omega_j}(g_{\Omega_{j-1}} - \Omega_{j-1})$$
is $\left|\Omega_j - \Omega_{j-1}\right| < v_\Omega\left|\Omega_j\right|$?

NO – RETURN to START

↓ YES

$$\begin{bmatrix}\bar{P}_{e_w} \\ \bar{Q}_{e_w}\end{bmatrix} = \begin{bmatrix}\cos\psi_w & \sin\psi_w \\ -\sin\psi_w & \cos\psi_w\end{bmatrix}\begin{bmatrix}\bar{P}_e \\ \bar{Q}_e\end{bmatrix}$$

N
$$k_{c_1} = 1 - \frac{\mu^2}{2}, \qquad k_{c_2} = 1 + \frac{3}{2}\mu^2, \qquad k_{c_3} = 1 - \frac{3}{2}\mu^2$$

$$\theta_0 = \left(\frac{3}{1-\mu^2}\right)\left\{k_{c_1}\frac{\mu}{2}\beta_{1c_w} - S_\beta\frac{\mu}{2}\beta_{1s_w} + k_{c_2}\left(\frac{2C_T}{a_0 s}\right) + k_{c_3}\frac{\bar{P}_{e_w}\mu}{4} - \frac{8\mu\bar{Q}_{e_w}}{\gamma} - k_{c_1}\frac{(\mu_z - \lambda_0)}{2} - \left(\frac{1+3\mu^2}{4}\right)\theta_{tw}\right\}$$

$$\theta_{1s_w} = \left(\frac{3}{1-\mu^2}\right)\left\{-\frac{k_{c_1}k_{c_2}}{3}\beta_{1c_w} + k_{c_2}\frac{S_\beta}{3}\beta_{1s_w} - \frac{8\mu}{3}\left(\frac{2C_T}{a_0 s}\right) - \frac{\bar{P}_{e_w}}{3}k_{c_1} + \frac{16}{3\gamma}\bar{Q}_{e_w} + \frac{2\mu}{3}k_{c_3}(\mu_z - \lambda_0) - \frac{\mu^3}{3}\theta_{tw}\right\}$$

O
$$\beta_0 = \frac{\gamma}{8\lambda_\beta^2}\left\{(1+\mu^2)\theta_0 + \frac{4}{3}\mu(\theta_{1s_w} + \bar{P}_{e_w}) + \frac{4}{3}(\mu_z - \lambda_0) + 4\theta_{tw}\left(\frac{1}{5} + \frac{\mu^3}{6}\right)\right\}$$

$$\theta_{1c_w} = \frac{-1}{1+\frac{\mu^2}{2}}\left\{-\frac{4}{3}\mu\beta_0 - S_\beta\beta_{1c_w} - \left(1+\frac{\mu^2}{3}\right)\beta_{1s_w} + \frac{16}{\gamma}\bar{P}_{e_w} + \bar{Q}_{e_w} - \lambda_{1c_w}\right\}$$

FINISH

P
$$\begin{bmatrix}\theta_{1s} \\ \theta_{1c}\end{bmatrix}, \begin{bmatrix}\lambda_{1s} \\ \lambda_{1c}\end{bmatrix} = \begin{bmatrix}\cos\psi_w & -\sin\psi_w \\ \sin\psi_w & \cos\psi_w\end{bmatrix}\begin{bmatrix}\theta_{1s_w} \\ \theta_{1c_w}\end{bmatrix}, \begin{bmatrix}\lambda_{1s_w} \\ \lambda_{1c_w}\end{bmatrix}$$

Fig. 4.5 Part II

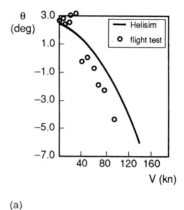

(a)

(b)

Fig. 4.6 Pitch angle in trim: (a) Trim pitch angle as a function of forward
speed – comparison of flight and theory; (b) Trim pitch angle as a function of turn rate

the horizontal stabilizer; this effect is evident in the flight, but not well predicted by
simulation. Some helicopters feature a movable horizontal stabilizer to reduce this
pitch-up tendency at low speed and to maintain a level fuselage in high-speed flight. In
forward flight, the comparison between flight and theory suggests a higher full-scale
value of fuselage drag than that used in the simulation; this is typical of the comparison
with this level of modelling, with the simulation underpredicting the fuselage nose-
down pitch at high speed by as much as 2°. In non-straight flight, the trim pitch
angle will vary with turn rate and flight path angle. The strongest variations occur in
climbing and descending flight, and Fig. 4.6(b) illustrates the kind of effect found on
Lynx in climbs. The turn rate extends out to 0.4 rad/s, corresponding to a bank angle
of nearly 60° at the 80-knot condition shown. As the climb rate increases, indicated
by the increasingly negative flight path angle, the pitch angle to trim rises markedly; a
negative flight path angle of −0.15 rad corresponds to a climb rate of about 1200 ft/min
at the 80 knots trim speed. The increased pitch attitude at this steep bank angle is
required to maintain zero sideslip. If the nose were set on the horizon in this condition

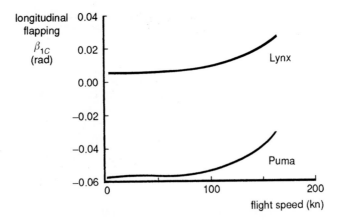

Fig. 4.7 Longitudinal flapping for Lynx and Puma as a function of forward speed

then the sideslip angle would correspond to the pitch angle shown in Fig. 4.6(b); the correct pitch attitude is achieved by balancing the turn with pedals, rather than pulling back on the cyclic stick.

One further point on longitudinal trim relates to the differences between helicopters with different rotor types. Since the pitch angle is determined primarily by the ratio of drag to lift on the whole vehicle, we should not expect to find any significant differences in pitch attitude to trim between hingeless and articulated rotor helicopters, but we might expect to see differences in longitudinal flapping angle. Note the previous observation that the longitudinal flapping will compensate for any residual moment on the other helicopter components. Figure 4.7 compares the longitudinal disc tilt for the Lynx and Puma across the speed range. The large difference in trim flap at hover is partly due to the different baseline centre of gravity (cg) locations for the two aircraft. The Puma cg lies practically under the hub at the fuselage reference point; the hover flap-back then almost equates to the forward shaft tilt. For the Lynx, with its aft cg lying practically on the shaft axis, the hover flap is close to zero. For both aircraft, as forward speed increases, the disc tilts further forward, implying that the residual pitch moment from both aircraft is nose up (i.e., from the horizontal stabilizer). The change in disc tilt for both aircraft is only about 1.5° across the speed range.

4.2.3 Lateral/directional partial trim

Satisfaction of the longitudinal trim at this stage in Fig. 4.5 does not guarantee a valid trim; estimates of the lateral trim have been used and the process now has to continue with the aim of correcting both of these. Having derived a new estimate for the lateral trim, the longitudinal cycle will then need to be repeated until all six force and moment equations balance properly. But the next stage in Fig. 4.5 involves the calculation of a new value for the main rotor downwash (D in Fig. 4.5), which is itself an iterative process (see Chapter 3), and the estimation of the main rotor torque and power required (E). With these calculations performed, the tail rotor thrust can be estimated from the yawing moment equation (F), the lateral flapping corrected from the rolling moment

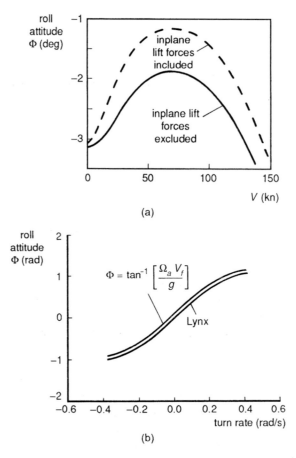

(a)

(b)

Fig. 4.8 Roll angle in trim: (a) Trim roll angle as a function of forward speed; (b) Trim roll angle as a function of turn rate

equation (G) and the new value of roll attitude derived from the sideforce equation (H). A check is now made on the convergence of the roll attitude in the same way as described for the pitch attitude, with defined convergence tolerance and damping factor. For both pitch and roll attitude, the number of iterations required, and hence the speed of convergence, depends critically on the initial guesses; clearly, the further away from the correct solution that the initial guess is, the longer will convergence take. For straight flight, setting the initial values to zero is usually adequate for fairly rapid convergence. Figure 4.8(a) shows the variation in roll attitude with forward speed for the Lynx, illustrating the powerful effect of adding the in-plane lift loads in the calculation of rotor sideforce (see Chapter 3). In turning flight, the bank angle will become large, and an initial guess based on the rules of simple circular motion is usually sufficient to ensure rapid convergence, i.e.,

$$\Phi_{e1} = \tan^{-1}\left[\frac{\Omega_{ae} V_{fe}}{g}\right] \qquad (4.34)$$

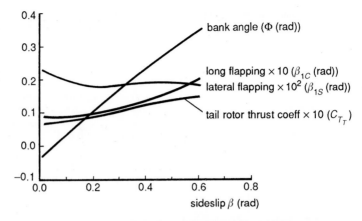

Fig. 4.9 Lynx trimmed in sideslipping flight at 100 knots

Figure 4.8(b) shows how the Lynx roll attitude varies with turn rate at the 80-knot trim point. The approximate result given by eqn 4.34 is plotted for comparison and shows how accurate by this simple kinematic relationship predicts the Lynx result.

At large turn rates, in forward flight, the roll attitude iteration can become sensitive to the sign of the error between the initial guess and the correct solution. If the initial Φ results in a lateral acceleration greater than the weight component, then this simple trim procedure will diverge, no matter how much damping is added. The trim iteration will converge only when a Φ estimate greater than the solution is introduced. These details will need to be considered when a simple trim algorithm is used, but they are usually catered for in the more sophisticated nonlinear numerical search algorithms. In sideslip flight, the bank angle also varies significantly as shown in Fig. 4.9, where Lynx trim results for bank angle, flapping angles and tail rotor thrust coefficient are plotted. The bank angle is approximately linear with sideslip up to about 30°, with both aerodynamic and gravitational sideforces on the aircraft varying approximately as $\sin \beta$. Longitudinal flapping increases at a greater rate than lateral flapping, as the rotor thrust is tilted further forward to compensate for the increased drag in sideslip flight.

4.2.4　Rotorspeed/torque partial trim

When the lateral and directional trim have converged, the tail rotor downwash is calculated (I), followed by the tail rotor collective (J), including the effect of the δ_3 pitch/flap coupling (K), and tail rotor torque (L). The total engine torque required can now be calculated, from which the rotorspeed can be updated using the droop law (M in Fig. 4.5). The rotorspeed calculation is the final stage in the iterative cycle and the whole sequential process described above must be repeated until convergence is achieved.

The remaining calculations in Fig. 4.5 determine the main rotor control angles, first in the hub/wind axes system (N, O) (see the Appendix Section 3A.4), followed by a transformation into hub axes, to give the swash plate control outputs (P). We shall return to discuss the controls to trim below in Section 4.2.6.

Table 4.1 Trim forces and moments – Lynx at 80 knots in climbing turn ($\gamma_{fe} = -0.15$ rad, $\Omega_{ae} = 0.4$ rad/s)

Component	X(N)	Y(N)	Z(N)	L(N m)	M(N m)	N(N m)
Gravity	−5647.92	35035.54	23087.88	0.00	0.00	0.00
Inertial	1735.41	−38456.29	58781.41	86.49	−18.87	49.80
Rotor	5921.18	−415.68	−82034.80	−4239.18	1045.06	28827.72
Fuselage	−2008.32	0.00	225.79	0.00	−571.94	0.00
Tailplane	0.00	0.00	−60.291	0.00	−454.238	0.00
Fin	0.00	374.801	0.00	201.013	0.00	−2830.976
Tail rotor	0.00	3457.164	0.00	3951.677	0.00	−26046.555
Total	**0.3556**	**−4.4629**	**−0.0136**	**0.0005**	**0.0002**	**−0.0098**

4.2.5 Balance of forces and moments

Trim is concerned with balancing the forces and moments acting on the aircraft. A typical trim is given by Table 4.1, where the various contributions to the forces and moments are given for a Lynx in a climbing turn (case $\gamma_{fe} = -0.15$ rad, $\Omega_{ae} = 0.4$ rad/s in Fig. 4.6(b)).

For our approximate model, many of the second-order effects have been neglected as can be seen in Table 4.1. (e.g., the X force from the empennage and X and Z force from the tail rotor, the fuselage rolling moment and tail rotor pitching moment). The inertial force components along the body axes are seen to be large, arising from the centrifugal force due to the angular motion of the aircraft. For the case shown, the trim tolerances were set at values that left the residual forces and moments as shown in the 'Totals' row. With zero initial value for pitch attitude and roll attitude set by eqn 4.34, convergence can usually be achieved to these levels of force within a few iterations.

4.2.6 Control angles to support the forces and moments

At this point in the trim algorithm, the various forces and moments on the components are, practically speaking, balanced, and now we have to look at the internal rotor equations to compute the controls required to hold these forces. The main rotor control angles are derived from the inverse of the flapping angle calculations given in the 3.60 series of equations (see Chapter 3), coupled with the thrust coefficient equation for the collective pitch. Figures 4.10(a)–(d) show a comparison between flight and theory of the main and tail rotor controls for the Bo105 as a function of forward speed. The errors give an indication of the level of fidelity achievable with the Level 1 modelling of Chapter 3. The nonlinear aerodynamic and blade twist effects increase the collective pitch required in flight relative to Helisim. As noted above in the discussion on longitudinal trim, the downwash over the tail causes a pitch up in the low-speed regime, giving rise to an increase in the required forward cyclic; the comparison for the Bo105 is actually very good in the mid-speed region (Fig. 4.10(a)). Also at low speed and into the transition region, the inflow roll increases the left cyclic required, revealing a failing in the simple trapezoidal model of longitudinal inflow predicted by the Glauert representation (Fig. 4.10(b)). The comparison of main rotor collective pitch is illustrated in Fig. 4.10(c). The underprediction by about 10% in hover, increasing to over 30% at high speed, is typical of linear aerodynamic theory. The tail rotor pitch is also usually underpredicted (Fig. 4.10(d)) as a combined result of missing tail rotor losses

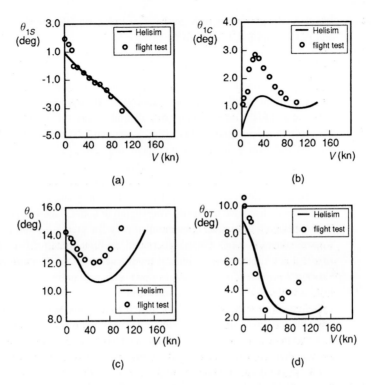

Fig. 4.10 Bo105 control angles in level trimmed flight: (a) longitudinal cyclic; (b) lateral cyclic; (c) main rotor collective; (d) tail rotor collective

and underpredicted main rotor torque, most noticeably at high speed. At moderate-to high-speed flight, the absence of tail rotor flapping and the powerful interactions with the aerodynamics of the rear fuselage and vertical fin increase the modelling discrepancies.

The power required, shown in Fig. 4.11 for the Bo105, has the characteristic bucket profile as a function of forward speed, reflecting the reduction in induced (rotor drag) power and increase in parasite (fuselage drag) power as speed increases. At high

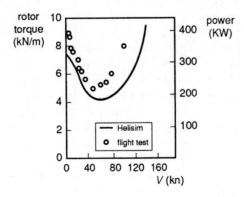

Fig. 4.11 Bo105 power required as a function of forward speed

speed, nonlinear rotor aerodynamic terms have a significant effect on collective pitch
and power required, leading to the gross errors with the simplified Level 1 modelling
as shown by the comparison with Bo105 flight test data in Figs 4.10(c) and 4.11. For
moderate rates of climb and descent, the Level 1 theory predicts the basic trends in
power required and control angles fairly well. The rotor is particularly efficient in
climbing flight. While the power required to climb a fixed-wing aircraft is approximately equal to the rate of change of potential energy, the increased mass flow through
the rotor of a helicopter reduces the power required to half this value. For similar
reasons, the rotor is inefficient in descent, the power reduction corresponding to only
half the rate of loss of potential energy. These simple results are explicable through the
momentum theory of Level 1 modelling. In steep descent, however, strongly nonlinear
aerodynamic effects dominate the trim (and stability and response) requirements. We
have already discussed the vortex-ring region in both Chapters 2 and 3 and highlighted
the inadequacy of simple momentum theory for predicting the power required and
response characteristics. On the other hand, for higher rates of descent, between vortex
ring and autorotation, the empirical modifications to momentum theory discussed in
Chapter 3 provide a reasonable interpolation between the helicopter and windmill solutions to the momentum equations for rotor inflow. An analysis of trim requirements
in helicopter descending flight is reported by Heyson in Ref. 4.3. At steep angles of
descent, and flight speeds of about 1.5 times the hover induced velocity, the power
required to increase the rate of descent actually increases. Figure 4.12, taken from Ref.
4.3, illustrates the power required as a function of glide slope angle for several different
values of flight speed along the glide slope. The reference velocity w_h and power P_h
correspond to the hover values, otherwise the results are quite general. At steep angles
of descent ($>60°$), the power required to increase the rate of descent at a constant speed
increases. Also shown, in Fig. 4.12(b), is a power contour for the rotor pitched up by

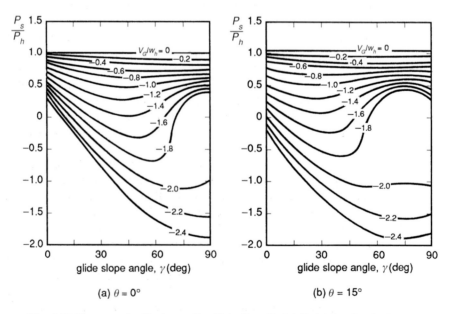

Fig. 4.12 Power required in descending flight (from Ref. 4.3): (a) $\theta = 0°$; (b) $\theta = 15°$

15°, showing the expanded region of increased power required as a function of flight path angle. Heyson refers to this, and the associated response characteristics, as power settling, and comments on the operational significance of this strong nonlinearity. To quote from Heyson's report,

> *A pilot flying a steep approach generally flies with reference to the ground either visually or through the use of some avionic system. Although he can sense sidewinds as a drift, his perception of a headwind or tailwind is poor. Even a light tail wind can produce a major difference between the glideslope with respect to the surrounding air mass and the geometric glideslope. If the flight is stabilised near one of the minimum power points, Fig. 12 shows many combinations of γ, θ, and V_G/w_h for which a tailwind induced change of only 10° or 15° in γ increases the required shaft power by 50 to 100% of the installed power. In the presence of such a major increase in required power, the helicopter settles, thus increasing the glide slope and still further increasing the required power.*

With the above discussion on steep descent, we have strayed into the response domain, showing the importance of predictable trim characteristics to the pilot's flying task. We shall return to this aspect in Chapters 5 and 7.

Predicting the trim control angles required, the power required and the steady loads on the various components forms a basis for calculating such static characteristics as the control margins at the flight envelope boundary, payload and range capabilities and limit loads on the tail boom in sideslipping flight. Achievement of accurate estimates (e.g., to within a few per cent of the true values) of such parameters will almost certainly require more detailed modelling than that described in the above analysis. The main rotor forces are a more complex function of rotor motion and the aerodynamics can be strongly nonlinear in high speed or at high thrust coefficient. The fuselage and empennage forces can be strongly influenced by the rotor wake and the tail rotor flapping can tilt the disc and thrust vector and change the power required. In some cases these will be first-order effects and cannot be ignored. Certainly, the component interactions will tend to spoil the simple sequential nature of the algorithm described above, giving rise to many more potential convergence problems and demanding more sophistication in the iterative solution. For trend predictions, however, the simple theory can be remarkably accurate; the characteristic shapes of the trim control curves is evidence of this. Examination of the effects of small changes from some baseline configuration can also provide useful insight into the sensitivities to design configuration or flight state parameters.

Trim solutions are generally unique, with a fixed set of control positions defining each equilibrium condition. A natural question that arises out of the study of trim is what happens if the aircraft is disturbed by a small amount from the trim? This could happen with a small gust or nudge of the controls. Will the aircraft immediately settle into a new trim, return to the original trim or depart away from the trim state in an unstable fashion? These questions cannot be answered from analysis of the trim equations; they require the full dynamic equations of motion from which the time evolution of the flight trajectory and fuselage attitudes can be determined. While a wide flight envelope simulator will usually require the full nonlinear equations, the answers to our questions regarding the effects of small perturbations can generally be found through analysis of the linearized equations using the concepts of the stability and control derivatives.

4.3 STABILITY ANALYSIS

The concept of stability of motion in a dynamic system is a very intuitive one that grasped the imagination of early pioneers of aviation. The supporting archetypal theory for flight stability was developed in the very early days of manned flight (Refs 4.2 and 4.4). The concept that stability and control were unlikely partners, the latter gaining from shortcomings in the former, was also recognized in these same early days, such that marginal stability, or even instability, was actually a useful property when considering the required piloting effort. Since that time much has been written on stability and control and much of the theoretical foundation for the stability of low-speed fixed-wing aircraft was already well developed by the time that early helicopters were in serious development. The first helicopters were unstable but the presence of mild instabilities at low speed was probably something of a blessing because control power was fairly marginal on these early craft. While fixed-wing aircraft have developed and can easily be conferred with high levels of natural stability and compatible levels of control, basically helicopters are still naturally unstable and require some level of artificial stabilization to ensure safe control in poor weather and when flying under instrument flight rules (IFR). The degree of stability required in helicopters to ensure safe operations is an important flying quality consideration, and will be discussed in some detail later in Chapter 6.

Understanding the flight behaviour of helicopters, why they are so difficult to build with natural stability, and developing rational explanations for the many curious dynamic characteristics, cannot be achieved simply through deriving the equations of Chapter 3, or even by building a simulation model. These are necessary but insufficient activities. The development of a deep understanding of flight behaviour comes from the intellectual interplay between theory and practice, with an emphasis on handson practice and analytical theory. Most of the understanding of stability and control has come from relatively simple theoretical approximations that permit expression of cause and effect in parametric form. Coincidentally, the publication of the earliest texts that provided a definitive treatment of both fixed- and rotary-wing stability and control occurred in 1952 (Refs 4.2, 4.5). Both of these texts deal with fundamental concepts in analytical terms that are still valid today. While our ability to model more and more complex representations of the aerodynamics and dynamics of aircraft seems to extend every year, our understanding of why things happen the way they do essentially comes from simple theory melded with a good physical understanding.

With these words of introduction we embark on this section on stability (and control) with the guiding light in search of simple approximations to complex behaviour. We shall draw heavily on the theory of linear dynamic systems but the underlying vector–matrix theoretical methods, including a discussion on eigenanalysis, are contained in Appendix 4A, to which the unfamiliar reader is referred. Features of the classical description of aircraft stability are the concepts of static and dynamic stability, the former relating to the immediate behaviour following a disturbance, the latter to the longer term behaviour. These are useful but elementary concepts, particularly for fixed-wing aircraft, drawing parallels with stiffness and damping in a simple mechanical analogue, but the distinction is blurred in the study of helicopter motion because of the stronger couplings between longitudinal and lateral motions. The perspective we shall take here is to draw the distinction between local and global stability – the former relating to the stability of motion following small disturbances from a trim condition, the latter relating to larger, potentially unbounded motion. Of course, unbounded motion is only a theoretical concept, and ultimately the issue is likely to be one of strength rather than stability in this case. Analysis tools for large nonlinear motions of aircraft are limited and tend either to be based on the assumption that the motion is nearly linear (i.e., nonlinearities are weak), so that approximating describing functions can be used, or to be applicable to very special forms of strong nonlinearity that can be described analytically.

Nonlinear analysis of fixed-wing aircraft has been stimulated by such phenomena as stall (including deep stall), spinning, inertial coupling and wing rocking. The need to understand the flight dynamics in these situations has led to extensive research into analytical methods that are able to predict the various kinds of departure, particularly during the 1970s and 1980s (Ref. 4.6). Helicopter flight dynamics also has its share of essentially nonlinear phenomena including vortex-ring state, main rotor wake–tail rotor interactions, rotor stall and rotor wake–empennage interactions. Much less constructive analytic work has been done on these nonlinear problems, and many potentially fruitful areas of research need attention. The methods developed for fixed-wing analysis will be equally applicable to helicopters and, just as the transfer of basic linear analysis techniques gave the helicopter engineers considerable leverage in early days, so too will the describing function and bifurcation techniques that have enabled so much insight into the dynamics of fixed-wing combat aircraft. Nonlinear problems are considerably more difficult than linear ones, one consolation being that they are usually considerably more interesting too, but little has been published to date on nonlinear helicopter flight dynamics.

In this chapter we restrict the discussion to linear analysis. We shall consider classical 6 DoF motion in detail. This level of approximation is generally good for low-moderate frequency, handling qualities analysis. The assumption underlying the 6 DoF theory is that the higher order rotor and inflow dynamics are much faster than the fuselage motions and have time to reach their steady state well within the typical time constants of the whole aircraft response modes. This topic has been discussed in the Tour of Chapter 2 and the conditions for validity are outlined in Appendix 4A.

4.3.1   Linearization

Consider the helicopter equations of motion described in nonlinear form, given by

$$\dot{\mathbf{x}} = \mathbf{F}(\mathbf{x}, \mathbf{u}, t) \tag{4.35}$$

In 6 DoF form, the motion states and controls are

$$\mathbf{x} = \{u, w, q, \theta, v, p, \phi, r, \psi\}$$

where u, v and w are the translational velocities along the three orthogonal directions of the fuselage fixed axes system described in Appendix 3A; p, q and r are the angular velocities about the x-, y- and z-axes and θ, ϕ and ψ are the Euler angles, defining the orientation of the body axes relative to the earth.

The control vector has four components: main rotor collective, longitudinal cyclic, lateral cyclic and tail rotor collective

$$\mathbf{u} = \{\theta_0, \theta_{1s}, \theta_{1c}, \theta_{0T}\}$$

The expanded form of eqn 4.35 can be written as eqn 4.36 combined with the Euler angles, eqn 4.37 (inverse of eqn 3A.42), as derived in Chapter 3 and Appendix 3A.

$$\dot{u} = -(wq - vr) + \frac{X}{M_a} - g \sin \theta$$

$$\dot{v} = -(ur - wp) + \frac{Y}{M_a} + g \cos \theta \sin \phi$$

$$\dot{w} = -(vp - uq) + \frac{Z}{M_a} + g \cos \theta \cos \phi$$

$$I_{xx} \dot{p} = (I_{yy} - I_{zz})qr + I_{xz}(\dot{r} + pq) + L$$

$$I_{yy} \dot{q} = (I_{zz} - I_{xx})rp + I_{xz}(r^2 - p^2) + M$$

$$I_{zz} \dot{r} = (I_{xx} - I_{yy})pq + I_{xz}(\dot{p} - qr) + N \qquad (4.36)$$

$$\dot{\phi} = p + q \sin \phi \tan \theta + r \cos \phi \tan \theta$$

$$\dot{\theta} = q \cos \phi - r \sin \phi$$

$$\dot{\psi} = q \sin \phi \sec \theta + r \cos \phi \sec \theta \qquad (4.37)$$

Using small perturbation theory, we assume that during disturbed motion, the helicopter behaviour can be described as a perturbation from the trim, written in the form

$$\mathbf{x} = \mathbf{x}_e + \delta \mathbf{x} \qquad (4.38)$$

A fundamental assumption of linearization is that the external forces X, Y and Z and moments L, M and N can be represented as analytic functions of the disturbed motion variables and their derivatives. Taylor's theorem for analytic functions then implies that if the force and moment functions (i.e., the aerodynamic loadings) and all its derivatives are known at any one point (the trim condition), then the behaviour of that function anywhere in its analytic range can be estimated from an expansion of the function in a series about the known point. The requirement that the aerodynamic and dynamic loads be analytic functions of the motion and control variables is generally valid, but features such as hysteresis and sharp discontinuities are examples of non-analytic behaviour where the process will break down. Linearization amounts to neglecting all except the linear terms in the expansion. The validity of linearization depends on the behaviour of the forces at small amplitude, i.e., as the motion and control disturbances become very small, the dominant effect should be a linear one. The forces can then be written in the approximate form

$$X = X_e + \frac{\partial X}{\partial u} \delta u + \frac{\partial X}{\partial w} \delta w + \cdots + \frac{\partial X}{\partial \theta_0} \delta \theta_0 + \cdots, \text{etc.} \qquad (4.39)$$

All six forces and moments can be expanded in this manner. The linear approximation also contains terms in the rates of change of motion and control variables with time, but we shall neglect these initially. The partial nature of the derivatives indicates that they are obtained with all the other DoFs held fixed – this is simply another manifestation of the linearity assumption. For further analysis we shall drop the perturbation notation, hence referring to the perturbed variables by their regular characters u, v, w, etc., and write the derivatives in the form, e.g.,

$$\frac{\partial X}{\partial u} = X_u, \qquad \frac{\partial L}{\partial \theta_{1c}} = L_{\theta_{1c}}, \qquad \text{etc.} \tag{4.40}$$

The linearized equations of motion for the full 6 DoFs, describing perturbed motion about a general trim condition, can then be written as

$$\dot{\mathbf{x}} - \mathbf{A}\mathbf{x} = \mathbf{B}\mathbf{u}(t) + \mathbf{f}(t) \tag{4.41}$$

where the additional function $\mathbf{f}(t)$ has been included to represent atmospheric and other disturbances. Following from eqn 4.40, the so-called system and control matrices are derived from the partial derivatives of the nonlinear function $\mathbf{F}$, i.e.,

$$\mathbf{A} = \left(\frac{\partial \mathbf{F}}{\partial \mathbf{x}} \right)_{\mathbf{x}=\mathbf{x}_e} \tag{4.42}$$

and

$$\mathbf{B} = \left(\frac{\partial \mathbf{F}}{\partial \mathbf{u}} \right)_{\mathbf{x}=\mathbf{x}_e} \tag{4.43}$$

In fully expanded form, the system and control matrices can be written as shown in eqns 4.44 and 4.45 on page 212. In eqn 4.44 the heading angle ψ has been omitted, the direction of flight in the horizontal plane having no effect on the aerodynamic or dynamic forces and moments. The derivatives are written in semi-normalized form, i.e.,

$$X_u \equiv \frac{X_u}{M_a} \tag{4.46}$$

where M_a is the aircraft mass, and

$$L'_p = \frac{I_{zz}}{I_{xx}I_{zz} - I_{xz}^2} L_p + \frac{I_{xz}}{I_{xx}I_{zz} - I_{xz}^2} N_p \tag{4.47}$$

$$N'_r = \frac{I_{xz}}{I_{xx}I_{zz} - I_{xz}^2} L_r + \frac{I_{xx}}{I_{xx}I_{zz} - I_{xz}^2} N_r \tag{4.48}$$

I_{xx} and I_{zz} are the roll and yaw moments of inertia and I_{xz} is the roll/yaw product of inertia. The k constants in the inertia terms in eqn 4.44 are given by the expressions

$$k_1 = \frac{I_{xz}(I_{zz} + I_{xx} - I_{yy})}{I_{xx}I_{zz} - I_{xz}^2} \tag{4.49}$$

$$k_2 = \frac{I_{zz}(I_{zz} - I_{yy}) + I_{xz}^2}{I_{xx}I_{zz} - I_{xz}^2} \tag{4.50}$$

$$k_3 = \frac{I_{xx}(I_{yy} - I_{xx}) - I_{xz}^2}{I_{xx}I_{zz} - I_{xz}^2} \tag{4.51}$$

$$\mathbf{A} = \begin{bmatrix}
X_u & X_w - Q_e & X_q - W_e & -g\cos\Theta_e & X_v + R_e & X_p & 0 & X_r + V_e \\[2pt]
Z_u + Q_e & Z_w & Z_q + U_e & -g\cos\Phi_e\sin\Theta_e & Z_v - P_e & Z_p - V_e & -g\sin\Phi_e\cos\Theta_e & Z_r \\[2pt]
M_u & M_w & M_q & 0 & M_v & \begin{subarray}{l} M_p - 2P_eI_{xz}I_{yy} \\ -R_e(I_{xx}-I_{zz})I_{yy} \end{subarray} & 0 & \begin{subarray}{l} M_r + 2R_eI_{xz}I_{yy} \\ -P_e(I_{xx}-I_{zz})I_{yy} \end{subarray} \\[2pt]
0 & 0 & \cos\Theta_e & 0 & 0 & 0 & -\Omega_a\cos\Theta_e & -\sin\Theta_e \\[2pt]
Y_u - R_e & Y_w + P_e & Y_q & -g\sin\Phi_e\sin\Theta_e & Y_v & Y_p + W_e & g\cos\Phi_e\cos\Theta_e & Y_r - U_e \\[2pt]
L'_u & L'_w & L'_q + k_1P_e - k_2R_e & 0 & L'_v & L'_p + k_1Q_e & 0 & L'_r - k_2Q_e \\[2pt]
0 & 0 & \sin\Phi_e\tan\Theta_e & \Omega_a\sec\Theta_e & 0 & 1 & 0 & \cos\Phi_e\tan\Theta_e \\[2pt]
N'_u & N'_w & N'_q - k_1R_e - k_3P_e & 0 & N'_v & N'_p - k_3Q_e & 0 & N'_r - k_1Q_e
\end{bmatrix}$$

$$\tag{4.44}$$

$$\mathbf{B} = \begin{bmatrix}
X_{\theta_0} & X_{\theta_{1s}} & X_{\theta_{1c}} & X_{\theta_{0T}} \\[2pt]
Z_{\theta_0} & Z_{\theta_{1s}} & Z_{\theta_{1c}} & Z_{\theta_{0T}} \\[2pt]
M_{\theta_0} & M_{\theta_{1s}} & M_{\theta_{1c}} & M_{\theta_{0T}} \\[2pt]
0 & 0 & 0 & 0 \\[2pt]
Y_{\theta_0} & Y_{\theta_{1s}} & Y_{\theta_{1c}} & Y_{\theta_{0T}} \\[2pt]
L'_{\theta_0} & L'_{\theta_{1s}} & L'_{\theta_{1c}} & L'_{\theta_{0T}} \\[2pt]
0 & 0 & 0 & 0 \\[2pt]
N'_{\theta_0} & N'_{\theta_{1s}} & N'_{\theta_{1c}} & N'_{\theta_{0T}}
\end{bmatrix}$$

$$\tag{4.45}$$

In addition to the linearized aerodynamic forces and moment, eqn 4.44 also contains perturbational inertial, gravitational and kinematic effects linearized about the trim condition defined by

$$\Phi_e, \Theta_e, U_e, V_e, W_e, P_e, Q_e, R_e$$

The trim angular velocities are given in terms of the aircraft turn rate in eqns 4.29–4.31.

Equation 4.41 is the fundamental linearized form for describing the stability and response of small motion about a trim condition. The coefficients in the **A** and **B** matrices represent the slope of the forces and moments at the trim point reflecting the strict definition of the stability and control derivatives. Analytic differentiation of the force and moment expressions is required to deliver the exact values of the derivatives. In practice, two other methods for derivative calculation are more commonly used, leading to equivalent linearizations for finite amplitude motion. The first method is simply the numerical differencing equivalent to analytic differentiation. The forces and moments are perturbed by each of the states in turn, either one-sided or two-sided, as illustrated conceptually in Fig. 4.13; the effect of increasing the perturbation size is illustrated in the hypothetical case shown in Fig. 4.13(b), where the strong nonlinearity gives rise to a significant difference with the small perturbation case in Fig. 4.13(a). The numerical derivatives will converge to the analytic, true, values as the perturbation size reduces to zero. If there is any significant nonlinearity at small amplitude, then the slope at the trim may not give the best 'fit' to the force over the amplitude range of interest. Often, larger perturbation values are used to ensure the best overall linearization over the range of motion amplitude of interest in a particular application, e.g., order 1 m/s for velocities and 0.1 rad for controls, attitudes and rates. In each case it is important to estimate the degree of nonlinearity over the range of interest, as the derivative value used can have a significant effect on stability and response characteristics.

Before we examine the derivatives themselves in more detail we should refer to the second 'numerical' method for deriving derivative estimates. This involves a fitting or model-matching process whereby a linear model structure is used to fit the response of the nonlinear simulation model. This method can also be applied to flight data and is described under the general heading – system identification. We discussed the approach briefly in Chapter 2 and we shall give more attention to applications in Chapter 5. The system identification approach seeks to find the best overall model fit and, as such, will embody the effects of any nonlinearities and couplings into the equivalent

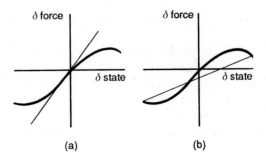

Fig. 4.13 Derivative calculation by backward–forward differencing: (a) small perturbation; (b) large perturbation

derivative estimates. The states are no longer perturbed independently; instead, the nonlinear model, or test aircraft, is excited by the controls so that the aircraft responds in some 'optimal' manner that leads to the maximum identifiability of the derivatives. The derivatives are varied as a group until the best fit is obtained. How these estimates relate to the pure analytic and numerical equivalents will depend on a number of factors, including the degree of nonlinearity, the correlation between states in the response and the extent of the measurement noise on the test data. In this chapter we shall discuss only the analytic and numeric methods of derivative estimation, returning briefly to the global system identification approach in the applications in Chapter 5.

4.3.2 The derivatives

There are 36 stability derivatives and 24 control derivatives in the standard 6 DoF set. In this section we shall discuss a limited number of the more important derivatives and their variation with configuration and flight condition parameters. The complete set of numerical derivatives for all three reference aircraft are contained as charts in the Appendix, Section 4B.2, and the reader may find it useful to refer to these as the discussion unfolds. It should be noted that the derivatives plotted in Appendix 4B include the inertial and gravitational effects from eqn 4.44. For example, the elements Z_q and Y_r tend to be dominated by the forward velocity term U_e. Each derivative is made up of a contribution from the different aircraft components – the main rotor, fuselage, etc. In view of the dominant nature of the rotor in helicopter flight dynamics, we shall give particular, but certainly not exclusive, attention to main rotor derivatives in the following discussion. The three most significant rotor disc variables are the rotor thrust T and the two multi-blade coordinate disc tilts β_{1c} and β_{1s}. During disturbed motion these rotor states will vary according to the algebraic relationships derived in Chapter 3 (eqns 3.90, 3.65). Considering the simple approximation that the rotor thrust is normal to the disc, for small flapping angles, the rotor X and Y forces take the form

$$X_R = T\beta_{1c}, \qquad Y_R = -T\beta_{1s} \tag{4.52}$$

The derivatives with respect to any motion or control variable can then be written as, for example,

$$\frac{\partial X_R}{\partial u} = \frac{\partial T}{\partial u}\beta_{1c} + T\frac{\partial \beta_{1c}}{\partial u} \tag{4.53}$$

Rotor force and moment derivatives are therefore closely related to individual thrust and flapping derivatives. Many of the derivatives are strongly nonlinear functions of velocity, particularly the velocity derivatives themselves. The derivatives are also nonlinear functions of the changes in downwash during perturbed motion, and can be written as a linear combination of the individual effects, as in the thrust coefficient change with advance ratio, given by

$$\frac{\partial C_T}{\partial \mu} = \left(\frac{\partial C_T}{\partial \mu}\right)_{\lambda=const} + \frac{\partial C_T}{\partial \lambda_0}\frac{\partial \lambda_0}{\partial \mu} + \frac{\partial C_T}{\partial \lambda_{1s}}\frac{\partial \lambda_{1s}}{\partial \mu} + \frac{\partial C_T}{\partial \lambda_{1c}}\frac{\partial \lambda_{1c}}{\partial \mu} \tag{4.54}$$

where C_T is the thrust coefficient and μ the advance ratio defined by

$$C_T = \frac{T}{\rho(\Omega R)^2 \pi R^2}, \qquad \mu = \frac{V}{\Omega R} \tag{4.55}$$

and the λs are the components of the rotor induced inflow in the harmonic, trapezoidal form

$$\lambda_i = \frac{w_i}{\Omega R} = \lambda_0 + \frac{r}{R}(\lambda_{1s} \sin \psi + \lambda_{1c} \cos \psi) \tag{4.56}$$

The thrust coefficient partial derivative with respect to μ can be written as

$$\left(\frac{\partial C_T}{\partial \mu}\right)_{\lambda=const} = \frac{a_0 s}{2}\left[\mu\left(\theta_0 + \frac{\theta_{tw}}{2}\right) + \frac{\theta_{1s_w}}{2}\right] \tag{4.57}$$

The rotor force, moment and flapping equations as derived in Chapter 3 are expressed in terms of the advance ratio in hub/wind axes. The relationships between the velocity components at the aircraft centre of mass and the rotor in-plane and out-of-plane velocities are given in Chapter 3, Section 3A.4. It is not the intention here to derive general analytic expressions for the derivatives; hence, we shall not be concerned with the full details of the transformation from rotor to fuselage axes except where this is important for enhancing our understanding.

The translational velocity derivatives

Velocity perturbations give rise to rotor flapping, changes in rotor lift and drag and the incidence and sideslip angles of the flow around the fuselage and empennage. Although we can see from the equations in the 3.70 series of Chapter 3 that the flapping appears to be a strongly nonlinear function of forward velocity, the longitudinal cyclic required to trim, as shown in Fig. 4.10(a), is actually fairly linear up to moderate forward speeds. This gives evidence that the moment required to trim the flapping at various speeds is fairly constant and hence the primary longitudinal flapping derivative with forward speed is also relatively constant. The orientation between the fuselage axes and rotor hub/wind axes depends on the shaft tilt, rotor flapping and sideslip angle; hence a u velocity perturbation in the fuselage system, say, will transform to give μ_x, μ_y and μ_z disturbances in the rotor axes. This complicates interpretation. For example, the rotor force response to μ_z perturbations is much stronger than the response to the in-plane velocities, and the resolution of this force through only small angles can be the same order of magnitude as the in-plane loads. This is demonstrated in the derivatives X_u and Z_u at low speed where the initial tendency is to vary in the opposite direction to the general trend in forward flight.

The derivatives X_u, Y_v, X_v and Y_u (M_v and L_u)

The four derivatives X_u, Y_v, X_v and Y_u are closely associated with each other at low speed. They are shown as a function of speed for the Lynx in straight and level flight in Fig. 4.14. In high-speed flight the coupling derivatives are fairly insignificant and the direct force dampings X_u and Y_v are practically linear with speed and reflect the drag and sideforce on the rotor–fuselage combination respectively. At hover and at low speed, all four derivatives are the same order of magnitude. The direct derivatives are principally due to the disc tilts to aft and port following perturbations in u and v

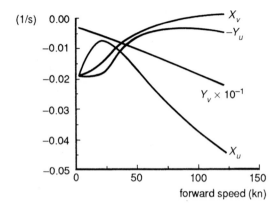

Fig. 4.14 Variation of force/velocity derivatives with forward speed

(see eqn 4.53). The coupling derivatives are less obvious, and we have to look into the theory of non-uniform inflow, described in Chapter 3, for an explanation to the surprisingly large values of X_v and Y_u extending to about 40 knots forward flight. At the hover condition, a perturbation in forward velocity u leads to a strong variation in wake angle χ (eqn 3.145) and hence non-uniform inflow λ_{1c}. An approximation to the increase in inflow at the rear of the disc can be derived from eqn 3.144:

$$\frac{\partial \lambda_{1c}}{\partial \mu} \approx \frac{1}{2} \tag{4.58}$$

For every 1 m/s increase in forward velocity, the downwash increases by 0.5 m/s at the rear of the disc. The linear variation of inflow along the blade radius results in a uniform incidence change; hence the effect is identical to cyclic pitch in the hover. The direct rotor response to a longitudinal incidence distribution is therefore a lateral disc tilt β_{1s}. The derivative of lateral flapping with inflow can be derived from eqn 3.71, as

$$\frac{\partial \beta_{1s}}{\partial \lambda_{1c}} = -\frac{1}{1 + S_\beta^2}, \quad S_\beta = \frac{8\left(\lambda_\beta^2 - 1\right)}{\gamma} \tag{4.59}$$

where the Stiffness number is given in terms of the flap frequency ratio and Lock number.

The Stiffness number ranges up to values of about 0.3 for current helicopters; hence the lateral flapping derivative in eqn 4.58 is close to unity and a perturbation of 1 m/s in forward velocity leads to about 0.2°–0.3° lateral disc tilt to starboard, depending on the rotorspeed. Similar arguments can be made to explain the low speed variation of X_v and the same effect will be reflected in the moment derivatives M_v and L_u. These variations in non-uniform inflow can be expected to impact the coupling of lateral and longitudinal motions at low speed. We shall return to this topic later when discussing the natural modes of motion.

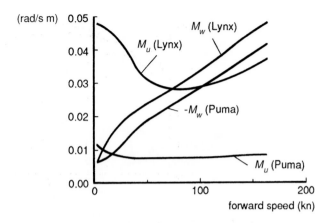

Fig. 4.15 Variation of longitudinal static stability derivatives with forward speed

The derivatives M_u and M_w

The derivatives M_u and M_w, the so-called speed and incidence static stability derivatives, have a major effect on longitudinal stability and hence handling qualities. For fixed-wing aircraft flying at low subsonic speeds, the speed stability derivative is practically zero – all the aerodynamic moments are proportional to dynamic pressure and the derivative works out to be proportional to the trim value of aerodynamic pitching moment, i.e., zero. With a helicopter, the main rotor moments due to speed changes are roughly constant across the speed range, but the aerodynamic loads on the fuselage and empennage are strong functions of forward velocity. In particular, the normal load on the horizontal stabilizer gives a strong pitching moment at the centre of mass and this component provides a contribution to M_u proportional to the trim load on the tail. Figure 4.15 compares the variation of the two static stability derivatives with speed for Lynx and Puma. The fourfold increase in magnitude of M_u for the Lynx relative to the Puma is a result of the much higher rotor moments generated by the hingeless rotor for the same velocity perturbation. Both aircraft exhibit static speed stability; an increase in forward speed causes the disc to flap back, together with an increase in the download on the tailplane, resulting in a nose-up pitching moment and a tendency to reduce speed. This positive (apparent) speed stability is important for good handling qualities in forward flight (see Chapter 6), but can degrade dynamic stability in both hover and forward flight (see the later section on stability of the natural modes). Comparing the incidence stability derivative M_w for the two aircraft, we can see similar orders of magnitude, but the Lynx exhibits instability while the Puma is stable. This derivative was discussed at some length in Chapter 2 (see Figs 2.25 and 2.26). In forward flight, a positive perturbation in normal velocity, w, causes a greater increase in lift on the advancing than on the retreating side of the disc. The disc flaps back giving rise to a positive, nose-up, destabilizing, pitching moment. This effect does not change in character between an articulated rotor (Puma) and a hingeless rotor (Lynx), but the magnitude is scaled by the hub stiffness. The pitching moments arise from three major sources – the main rotor, the tailplane and the fuselage (Fig. 4.16), written as shown in eqn 4.60.

$$M = M_R + M_{tp} + M_f \tag{4.60}$$

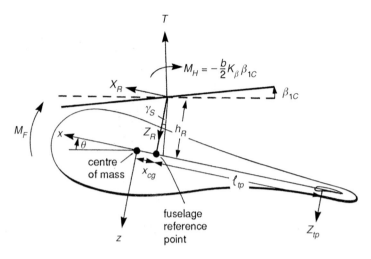

Fig. 4.16 Sketch showing pitching moments at the aircraft centre of mass

In very approximate form, the rotor moment can be written as a combination of a moment proportional to the disc tilt and one proportional to the rotor thrust, i.e.,

$$M \approx -\left(\frac{N_b}{2}K_\beta + h_R T\right)\beta_{1c} - (x_{cg} + h_R\gamma_s)T + (l_{tp} + x_{cg})Z_{tp} + M_f \qquad (4.61)$$

The pitching moments from the rotor, tailplane and fuselage are shown in Fig. 4.16. The contribution of the tail to M_w is always stabilizing – with a positive incidence change, the tail lift increases (Z_{tp} reduces) resulting in a nose-down pitch moment. The importance of the horizontal tail to the derivative M_w and helicopter pitch stability is outlined in Ref. 4.7, where the sizing of the tail for the YUH-61A is discussed. The contribution from the fuselage is nearly always destabilizing; typically the aerodynamic centre of the fuselage is forward of the centre of mass. The overall contribution from the main rotor depends on the balance between the first two terms in eqn 4.61. We have already stated that the disc always flaps back with a positive (upward) perturbation in w, but the thrust also increases; hence the second term, due to the offset of the thrust from the centre of mass, is actually stabilizing for configurations with forward centre of mass and shaft tilt. This is the major effect for fixed-wing aircraft, where the distance between the centre of mass and the aerodynamic centre of the whole aircraft is referred to as the static margin. For small offset articulated rotors, with a centre of mass well forward of the shaft, the thrust offset effect can be as large as the hub moment term in eqn 4.61, resulting in a fairly small overall rotor moment. This is the case for the Puma, with our baseline configuration having a forward centre of mass location; also, the flap hinge offset is only 3.8% of the rotor radius. For hingeless rotors with aft centre of mass, both thrust offset and hub moment effects are destabilizing, with the hub moment due to flapping dominating. All three contributions to the incidence stability vary approximately linearly with speed above about 40 knots. Figure 4.17 illustrates the contributions from the different components to M_w at the 120-knot high-speed condition. The overall magnitude of all three components is greater for the Lynx, reflecting the much smaller pitch moment of inertia (which normalizes the derivative) for that aircraft compared with the Puma.

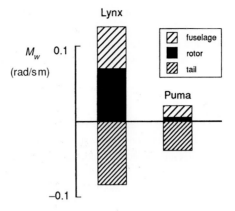

Fig. 4.17 Contributions to the static stability derivative M_w at 120 knots for Lynx and Puma

The derivatives $M_{\dot{w}}$, M_v and $M_{\dot{v}}$

Before leaving the pitching moment derivatives with speed, it should be noted that the incidence at the empennage is a combination of the fuselage incidence and the effect of the rotor downwash at the tail. This effect will normally be taken into account when perturbing the forces and moments with the w perturbation; the rotor thrust and the downwash will change, resulting in an incidence perturbation at the tail. The magnitude of the downwash at the tail depends on the distance between the tail and rotor. Let α_f be the fuselage incidence, μ the advance ratio and λ_{tp} the downwash at the tail. We can write the incidence at the tail in the form

$$\alpha_{tp} = \alpha_f - \frac{\lambda_{tp}}{\mu} \tag{4.62}$$

The downwash at the tail at time t was generated at the rotor disc at time $t - \ell_{tp}/U_e$ earlier. If we assume that this time increment is small compared with the response time, we can write

$$\lambda_{tp}(t) \approx k_{tp}\lambda\left(t - \frac{\ell_{tp}}{U_e}\right) \approx k_{tp}\left(\lambda(t) - \frac{d\lambda}{dt}\frac{\ell_{tp}}{U_e}\right) \tag{4.63}$$

where k_{tp} is the amplification factor on the downwash. The incidence at the tail therefore depends explicitly on the rate of change of rotor inflow with time. Applying the theory of small perturbations, we can write this downwash acceleration as a linear combination of the rates of change of aircraft states and controls, i.e.,

$$\frac{d\lambda}{dt} \approx \frac{\partial\lambda}{\partial\theta_0}\dot{\theta}_0 + \frac{\partial\lambda}{\partial w}\dot{w} + \cdots \tag{4.64}$$

Thus, we find the appearance of acceleration derivatives like $M_{\dot{w}}$ in the longitudinal motion, for which analytic expressions are relatively straightforward to derive from the thrust coefficient and uniform inflow equations. The presence of non-uniform inflow and wake contraction complicates the overall effect, reducing the validity of the above simple approximation. Nevertheless, the physical mechanism is very similar to that found on fixed-wing aircraft where the downwash lag at the tail, attributed entirely to incidence changes on the main wing, leads to an effective acceleration derivative.

Any lateral variation in rotor downwash at the tail will also lead to changes in pitching moments during yaw manoeuvres. This effect is discussed in Refs 4.8, 4.9 and 4.10 where relatively simple flat wake models are shown to be effective in modelling the pitching moment due to sideslip (see Chapter 3), leading to the derivatives M_v and $M_{\dot{v}}$, in a similar fashion to the effect from w perturbations.

The derivative Z_w

The heave damping derivative Z_w has already been discussed in some depth in the Introductory Tour to this book (Chapter 2) and earlier on in the present chapter. While the fuselage and empennage will undoubtedly contribute in high-speed flight, the main rotor tends to dominate Z_w throughout the flight envelope and can be approximated by the thrust coefficient derivative in eqn 4.65

$$Z_w = -\frac{\rho(\Omega R)\pi R^2}{M_a}\frac{\partial C_T}{\partial \mu_z} \tag{4.65}$$

Expressions for the thrust coefficient and uniform inflow component λ_0 were derived in Chapter 3 (eqn 3.90 series) in the form given below in eqns 4.66 and 4.67:

$$C_T = \frac{a_0 s}{2}\left[\theta_0\left(\frac{1}{3}+\frac{\mu^2}{2}\right)+\frac{\mu}{2}\left(\theta_{1sw}+\frac{\overline{p}_w}{2}\right)+\left(\frac{\mu_z-\lambda_0}{2}\right)+\frac{1}{4}(1+\mu^2)\theta_{tw}\right] \tag{4.66}$$

$$\lambda_0 = \frac{C_T}{2\sqrt{[\mu^2+(\lambda_0-\mu_z)^2]}} \tag{4.67}$$

The thrust coefficient is therefore proportional to the normal velocity component, μ_z, as expected from the assumed linear aerodynamics, but the induced inflow will also vary during vertical perturbations such that

$$\frac{\partial C_T}{\partial \mu_z} = \left(\frac{\partial C_T}{\partial \mu_z}\right)_{\lambda=0}+\frac{\partial C_T}{\partial \lambda}\frac{\partial \lambda}{\partial \mu_z} = \frac{a_0 s}{4}\left(1-\frac{\partial \lambda}{\partial \mu_z}\right) \tag{4.68}$$

Good approximations for heave damping in hover and forward flight ($\mu > 0.15$) can be obtained and are summarized in Table 4.2.

It can be seen that rotor blade loading, defined as the aircraft mass divided by the blade area (M_a/A_b), is a very important parameter defining the heave damping at hover

Table 4.2 Approximations for heave damping derivative Z_w

Hover	Forward flight
$v_{i_0}^2 \approx \dfrac{T}{2\rho A_d}$	$v_{i_\mu} \approx \dfrac{T}{2\rho A_d V'}$
$\dfrac{\partial C_T}{\partial \mu_z} \approx \dfrac{2a_0 s \lambda_0}{16\lambda_0+a_0 s}$	$\dfrac{\partial C_T}{\partial \mu_z} \approx \dfrac{2a_0 s \mu}{8\mu+a_0 s}$
$Z_w \approx -\dfrac{2a_0 A_b \rho(\Omega R)\lambda_0}{(16\lambda_0+a_0 s)M_a}$	$Z_w \approx -\dfrac{\rho a_0 \mu(\Omega R)A_b}{2M_a}\left(\dfrac{4}{8\mu+a_0 s}\right)$

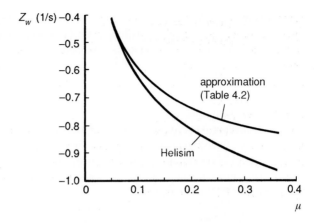

Fig. 4.18 Comparison of Z_w approximate and 'exact' results for Lynx

and in forward flight. The derivative Z_w represents the initial acceleration following an abrupt vertical gust and is inversely proportional to blade loading (see Chapter 2, eqn 2.63). The much higher typical blade loadings on rotorcraft, compared with fixed-wing aircraft of similar weight, partly account for the smaller values of Z_w, and hence lower gust sensitivity, for helicopters in forward flight. A second major factor is disguised in the variation of the heave damping with speed. The forward flight approximation in Table 4.2 is shown plotted against the Lynx value from the Helisim simulation model in Fig. 4.18; the 10% difference at the higher values of advance ratio is attributed to the fuselage and tail. The variation is seen to level off at higher speeds, while the gust sensitivity of fixed-wing aircraft continues to increase linearly with speed (see Fig. 2.28). As discussed in Chapter 2, the reason for the asymptotic behaviour of the helicopter damping stems from the increased harmonic distribution of the airloads as the speed increases. The thrust coefficient can be written as

$$\frac{2C_T}{a_0 s} = \int_0^1 \left(\overline{U}_T^2 \theta + \overline{U}_p \overline{U}_T \right) \, d\overline{r} \qquad (4.69)$$

and the in-plane and normal velocity components can be approximated by the expressions

$$\overline{U}_T \approx \overline{r} + \mu \, \sin \psi, \qquad \overline{U}_p = \mu_z - \lambda_i - \mu \beta \, \cos \psi - \overline{r} \beta' \qquad (4.70)$$

The harmonic components of thrust in the expanded form of eqn 4.69 therefore define the level of quasi-steady and vibratory loads that reach the fuselage. Perturbations in w show up in the second term in parenthesis in eqn 4.69. The component that increases linearly with forward speed is also a one-per-rev loading. Hence, while the zero-per-rev or quasi-steady term levels off at higher speeds, the vibratory response to a gust at N_b-per-rev, where N_b is the number of blades, continues to increase. While these loads do not result in significant flight path or attitude changes, and therefore are unlikely to cause handling problems, they do affect the overall ride quality. Further discussion on the general topic of ride quality is contained in Chapter 5.

The derivatives L_v, N_v

The remaining velocity derivatives belong to the lateral/directional DoFs and the most significant are the sideslip derivatives – the dihedral effect L_v and the weathercock stability N_v. The magnitude of these two moments as sideslip increases determines the lateral/directional static stability characteristics. A positive value of N_v is stabilizing, while a negative value of L_v is stabilizing. Both have the same kind of effects on rotary-wing as on fixed-wing aircraft but with rotary-wing aircraft the new component is the tail rotor which can contribute strongly to both. The magnitude of the tail rotor contribution to the dihedral effect depends on the height of the tail rotor above the aircraft centre of mass. The fuselage can also contribute to L_v if the aerodynamic centre is offset vertically from the centre of mass, as in the case of deep fuselage hulls, which typically leads to a negative L_v component. But once again, the main rotor is usually the dominant effect, especially for helicopters with hingeless rotors, where all the main rotor moments are magnified roughly proportionally with the rotor stiffness. In hover, the derivative L_v is generated by similar aerodynamics to those of the pitch derivative M_u, and as forward flight increases some of the basic similarities remain. As the blades are exposed to the velocity perturbation, the advancing blade experiences an increase in lift, the retreating blade a decrease, and the one-per-rev flapping response occurs approximately $90°$ around the azimuth, giving a rolling moment to port (starboard for clockwise rotors) for a lateral velocity perturbation and a pitch-up moment in response to a longitudinal velocity perturbation. The extent of the flap response depends on the rotor stiffness, the Lock number and also the trim lift on the rotor blades. To examine the flap derivatives we can refer back to eqns 3.70, 3.71 and 3.72 from Chapter 3. At hover, we can write

$$\left(\frac{\partial \beta_{1s}}{\partial \mu_y}\right)_{\mu=0} = -\left(\frac{\partial \beta_{1c}}{\partial \mu_x}\right)_{\mu=0} = \frac{\gamma}{8\lambda_\beta^2} \eta_\beta \left\{ \frac{4}{3}\left(S_\beta + \frac{16\lambda_\beta^2}{\gamma}\right)\theta_0 \right.$$

$$\left. + \left(\left(\frac{4}{3}\right)^2 S_\beta + \frac{16\lambda_\beta^2}{\gamma}\right)(\mu_z - \lambda_0)\right\}$$

(4.71)

or, for the special case of a rotor hinged at the hub centre, the flap response depends only on the trim lift on the rotor blades, i.e.,

$$\left(\frac{\partial \beta_{1s}}{\partial \mu_y}\right)_{\mu=0} = \frac{8}{3}\theta_0 - 2\lambda_0$$

(4.72)

The dihedral effect can therefore potentially change sign for teetering rotors at low and negative rotor thrust conditions that are outside the operational flight envelope for such aircraft, because of such reversals of flap response and the associated hub moments.

The directional stability derivative N_v is critically important for both static and dynamic stability of helicopters. The main contributors are the tail rotor, the vertical fin and the fuselage. The latter is usually destabilizing with the fuselage centre of pressure behind the centre of mass; both the tail rotor and vertical fin are stabilizing (i.e., positive). All are approximately linear with speed up to moderate forward speeds. However, the contribution from the tail rotor is similar to the heave damping on the main rotor, arising from a change in tail rotor thrust due to a change in velocity normal to the disc, and levels off at high speed; the contributions from the fin and fuselage continue to increase in the positive and negative senses respectively. The weathercock

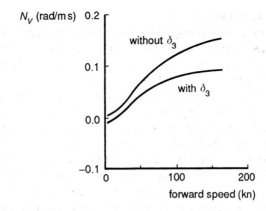

N_v (rad/m s)

without δ_3

with δ_3

forward speed (kn)

Fig. 4.19 Effect of tail rotor δ_3 angle on weathercock stability derivative N_v

stability is strongly influenced by the tail rotor δ_3 angle (see Chapter 3, Section 3.2.2) which reduces blade pitch as a function of blade flapping. On the four-bladed Lynx tail rotor, changes in tail rotor thrust lead to changes in rotor coning and hence changes in tail rotor collective. Figure 4.19 illustrates the comparison of N_v for the Lynx, with and without δ_3, showing that δ_3 produces a reduction of about 40% at high speed.

For the Lynx, the low-speed values of N_v tend to be dominated by the inertia coupling with roll (and hence the much stronger dihedral effect L_v) through the product of inertia I_{xz} (see eqn 4.48). The reduced effectiveness of the tail rotor to directional stability makes the contribution of the vertical fin all the more important. For helicopters with high set tail rotors, these vertical surfaces also carry the tail rotor drive shaft and can have high ratios of thickness to chord. Aerofoil sections having this property can exhibit a flattening or even reversal of the lift curve slope at small incidence values (Ref. 4.11). In such cases it can be expected that the yawing moment due to sideslip will exhibit a strong nonlinearity with sideslip velocity. The Puma features this characteristic and the associated effects on stability have been discussed in Ref. 4.12; the fin aerodynamics are summarized in the Appendix, Section 4B.1. Figure 4.20 shows how the value of

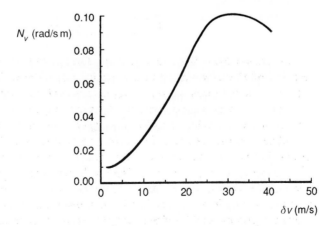

N_v (rad/s m)

δv (m/s)

Fig. 4.20 Variation of derivative N_v with 'v' velocity perturbation for Puma

N_v for the Puma varies with the magnitude of the velocity perturbation at a trim speed of 120 knots. At small amplitude, up to about 5 m/s perturbation (corresponding to about 5° of sideslip), the fin contributes nothing to the weathercock stability. As the perturbation is increased up to 30 m/s (corresponding to nearly 30° of sideslip) there is a tenfold increase in N_v. The effects of this nonlinearity on stability and response will be discussed later, but are obviously significant and need to be taken into account in the linearization process.

The derivatives N_u, N_w, L_u, L_w

These four derivatives play an important role in coupling the low-frequency longitudinal and lateral motions of the helicopter. The yawing moment derivatives stem largely from the changes in main rotor torque with velocity perturbations, although there is also an effect from the fin (N_u), similar to the contribution of the horizontal stabilizer to M_u. The N_w effect can be quite significant since torque changes to vertical velocity are similar to the direct control coupling–torque changes from collective inputs. The L_u effect reflects the changes in lateral cyclic to trim with forward speed, being dominated by the main rotor effect at low speed. Forward velocity perturbations increase the incidence on the forward part of the (coned) disc and reduce the incidence at the rear. The disc will therefore tilt to starboard for anticlockwise rotors (port for clockwise rotors). As forward speed increases, the four derivatives show similar trends and the dominating main rotor components are closely related through the shaft tilt and product of inertia.

The angular velocity derivatives

Our discussion on derivatives with respect to roll, pitch and yaw rate covers three distinct groups – the force derivatives, the roll/pitch moment derivatives due to roll and pitch and the roll/yaw derivatives due to yaw and pitch. Derivatives in the first group largely share their positions in the system matrix (eqn 4.44) with the trim inertial velocity components. In some cases the inertial velocities are so dominant that the aerodynamic effects are negligible (e.g., Z_q, Y_r). In other cases the aerodynamic effects are important to primary response characteristics. Two such examples are X_q and Y_p.

The derivatives X_q, Y_p

These derivatives are dominated by the main rotor contributions. For teetering rotors and low flap hinge-offset rotors, the changes in rotor hub X and Y forces are the primary contribution to the pitch and roll moments about the aircraft centre of mass. Hence the derivatives X_q and Y_p can contribute significantly to aircraft pitch and roll damping. The basic physical effects for the two derivatives are the same and can be understood from an analysis of a teetering rotor in hovering flight. If we assume that the thrust acts normal to the disc in manoeuvres, and ignoring the small drag forces, then the rotor X force can be written as the tilt of the thrust vector:

$$X = T\beta_{1c} \qquad (4.73)$$

The pitch rate derivative is then simply related to the derivative of flapping with respect to pitch rate. As the aircraft pitches, the rotor disc lags behind the shaft by an amount proportional to the pitch rate. This effect was modelled in Chapter 3 and the relationships were set down in eqns 3.71 and 3.72. For a centrally hinged rotor with zero spring stiffness, the disc lags behind the shaft by an amount given by the expression

$$\frac{\partial \beta_{1c}}{\partial q} = \frac{16}{\gamma \Omega} \tag{4.74}$$

The Lock number γ is the ratio of aerodynamic to inertia forces acting on the rotor blade; hence the disc will flap more with heavy blades of low aspect ratio. Equation 4.74 implies that the force during pitching produces a pitch damping moment about the centre of mass that opposes the pitch rate. However, the assumption that the thrust remains normal to the disc has ignored the effect of the in-plane lift forces due to the inclination of the lift vectors on individual blade sections. To examine this effect in more detail we need to recall the expressions for the rotor hub forces from Chapter 3, eqns 3.88–3.99. Considering longitudinal motion only, thus dropping the hub/wind dressings, the normalized X rotor force can be written as

$$\left(\frac{2C_x}{a_0 s} \right) = \frac{F_0^{(1)}}{2} \beta_{1c} + \frac{F_{1s}^{(2)}}{2} \tag{4.75}$$

The first term in eqn. 4.75 represents the contribution from the fore and aft blades to the X force when the disc is tilted and is related to the rotor thrust coefficient by the expression

$$F_0^{(1)} = - \left(\frac{2C_T}{a_0 s} \right) \tag{4.76}$$

This effect accounts for only half of the approximation given by eqn 4.73. Additional effects come from the rotor blades in the lateral positions and here the contributions are from the in-plane tilt of the lift force, i.e.,

$$F_{1s}^{(2)} = \left(\frac{\theta_0}{3} - \lambda_0 \right) \beta_{1c} - \frac{\lambda_0}{2} \theta_{1s} \tag{4.77}$$

During a pitch manoeuvre from the hover, the cyclic pitch can be written as (see eqn 3.72)

$$\theta_{1s} = -\beta_{1c} + \frac{16}{\gamma \Omega} q \tag{4.78}$$

Hence, substituting eqn 4.78 into eqn 4.77 and then into eqn 4.75, the force derivative can be written in the form

$$\frac{\partial C_x}{\partial q} = C_T \left(1.5 - \frac{\theta_0}{12 C_T / a_0 s} \right) \frac{\partial \beta_{1c}}{\partial q} = C_T \left(1.5 - \frac{\theta_0}{12 C_T / a_0 s} \right) \frac{16}{\gamma \Omega} \tag{4.79}$$

We can see that the thrust is inclined relative to the disc during pitching manoeuvres due to the in-plane loads when the blades are in the lateral position. The scaling coefficient given in eqn 4.79 reduces in the hover to

$$\left(1.5 - \frac{\theta_0}{12 C_T / a_0 s} \right)_{hover} = \left(1 - \frac{a_0 s}{8 \sqrt{(2 C_T)}} \right) \tag{4.80}$$

and has been described as the 'Amer effect' (Ref. 4.13). Further discussion can be found in Bramwell (Ref. 4.14) and in the early paper by Sissingh (Ref. 4.15). Although our analysis has been confined to hover, the approximation in eqn 4.79 is reasonably good up to moderate forward speeds. The effect is most pronounced in conditions of high collective pitch setting and low thrust, e.g., high-power climb, where the rotor damping can reduce by as much as 50%. In autorotation, the Amer effect almost disappears.

The derivatives M_q, L_p, M_p, L_q

The direct and coupled damping derivatives are collectively one of the most important groupings in the system matrix. Primary damping derivatives reflect short-term, small and moderate amplitude, handling characteristics, while the cross-dampings play a dominant role in the level of pitch–roll and roll–pitch couplings. They are the most potent derivatives in handling qualities terms, yet because of their close association with short-term rotor stability and response, they can also be unreliable as handling parameters. We shall discuss this issue later in Chapter 5 and in more detail in Chapter 7, but first we need to explore the many physical mechanisms that make up these derivatives. There has already been some discussion on the roll damping derivative in Chapter 2, when some of the fundamental concepts of rotor dynamics were introduced. The reader is referred back to the Tour (Section 2.3) for a refresher.

Taking the pitch moment as our example for the following elucidation, we write the rotor moment about the centre of mass in the approximate form

$$M_R = -\left(N_b\frac{K_\beta}{2} + Th_R\right)\beta_{1c} \tag{4.81}$$

where K_β is the flapping stiffness, T the rotor thrust and h_R the rotor vertical displacement from the centre of mass. In this simple analysis we have ignored the moment due to the in-plane rotor loads, but we shall discuss the effects of these later in this section. The rotor moment therefore has two components – one due to the moment of the thrust vector tilt from the centre of mass, the other from the hub moment arising from real or effective rotor stiffness. Effective stiffness arises from any flap hinge offset, where the hub moment is generated by the offset of the blade lift shear force at the flap hinge. According to eqn 4.81, the rotor moment is proportional to, and hence in phase with, the rotor disc tilt (for the centre-spring rotor). The relative contributions of the two components depend on the rotor stiffness. The hub pitch moment can be expanded in the form (see eqns 3.104, 3.105)

$$M_h = -N_b\frac{K_\beta}{2}\beta_{1c} = -\frac{N_b}{2}\Omega^2 I_\beta\left(\lambda_\beta^2 - 1\right)\beta_{1c} \tag{4.82}$$

and the corresponding roll moment as

$$L_h = -N_b\frac{K_\beta}{2}\beta_{1s} = -\frac{N_b}{2}\Omega^2 I_\beta\left(\lambda_\beta^2 - 1\right)\beta_{1s} \tag{4.83}$$

The hub moment derivatives can therefore be derived directly from the flapping derivatives. Since the quasi-steady assumption indicates that the disc tilt reaches its steady-state value before the fuselage begins to move, the flap derivatives can be obtained from the matrix in eqn 3.72; thus, in hover, where the flap effects are symmetrical, we can write

$$\frac{\partial \beta_{1c}}{\partial \overline{q}} = \frac{\partial \beta_{1s}}{\partial \overline{p}} = \frac{1}{1 + S_\beta^2} \left(S_\beta + \frac{16}{\gamma} \right) \tag{4.84}$$

$$\frac{\partial \beta_{1c}}{\partial \overline{p}} = -\frac{\partial \beta_{1s}}{\partial \overline{q}} = \frac{1}{1 + S_\beta^2} \left(S_\beta \frac{16}{\gamma} - 1 \right) \tag{4.85}$$

The Stiffness number S_β is given by eqn 4.59.

The variation of the flap damping derivatives in eqns 4.84 and 4.85 with the fundamental stiffness and Lock parameters has been discussed in Chapter 2 (see Figs 2.21 (b) and (c)). For values of Stiffness number up to about 0.4, corresponding to the practical limits employed in most current helicopters, the direct flap derivative is fairly constant, so that helicopters with hingeless rotors flap in very much the same way as helicopters with teetering rotors. The Lock number has a much more dramatic effect on the direct flap motion. Looking at the coupling derivatives, we can see a linear variation over the same range of Stiffness number, with rotors having low Lock number experiencing a reversal of sign. This effect is manifested in the Bo105 helicopter, as illustrated in the derivative charts of the Appendix, Section 4B.2, where the Lock number of 5 and Stiffness number of about 0.4 result in a practical cancelling of the rolling moment due to pitch rate L_q. The rotor Lock number is critically important to the degree of pitch–roll coupling.

From the theory of flap dynamics derived in Chapter 3, we can explain the presence of the two terms in the flap derivatives. The primary mechanism for flap and rotor damping derives from the second term in parenthesis in eqn 4.84 and is caused by the aerodynamic moment generated by the flapping rate (at azimuth positions 90° and 270°) that occurs when the rotor is pitching. The disc precesses as a result of the aerodynamic action at these azimuth stations, and lags behind the rotor shaft by the angle $(16/\gamma\Omega) \times$ pitch rate. The primary mechanism for coupling is the change in one-per-rev aerodynamic lift generated when the rotor pitches or rolls (second term in eqn 4.85), adding an effective cyclic pitch. Both effects are relatively insensitive to changes in rotor stiffness. The additional terms in S_β in eqns 4.84 and 4.85 arise from the fact that the flap response is less than 90° out of phase with the applied aerodynamic load. The direct aerodynamic effects, giving rise to the longitudinal and lateral flapping, therefore couple into the lateral and longitudinal flapping respectively. The effect on the coupling is especially strong since the direct flap derivative provides a component in the coupling sense through the sign of the phase angle between aerodynamic load and flap response.

Combining the flap derivatives with the hub moments in eqns 4.82 and 4.83, we can derive approximate expressions for the rotor hub moment derivatives, in seminormalized form, for small values of S_β:

$$(M_q)_h \approx -\frac{N_b S_\beta I_\beta \Omega}{I_{yy}} \left(1 + S_\beta \frac{\gamma}{16} \right) \tag{4.86}$$

$$(L_p)_h \approx -\frac{N_b S_\beta I_\beta \Omega}{I_{xx}} \left(1 + S_\beta \frac{\gamma}{16} \right) \tag{4.87}$$

$$(L_q)_h \approx \frac{N_b S_\beta I_\beta \Omega}{I_{xx}} \left(S_\beta - \frac{\gamma}{16} \right) \qquad (4.88)$$

$$(M_p)_h \approx -\frac{N_b S_\beta I_\beta \Omega}{I_{yy}} \left(S_\beta - \frac{\gamma}{16} \right) \qquad (4.89)$$

The hub moment derivatives are therefore scaled by the Stiffness number, but otherwise follow the same behaviour as the flap derivatives. They also increase with blade number and rotorspeed. It is interesting to compare the magnitude of the hub moment with the thrust tilt contribution to the rotor derivatives. For the Lynx, the hub moment represents about 80% of the total pitch and roll damping. For the Puma, the fraction is nearer 30% and the overall magnitude is about 25% of that for the Lynx. Such is the powerful effect of rotor flap stiffness on all the hub moment derivatives reflected in the values of S_β for the Lynx and Puma of 0.22 and 0.044 respectively. As can be seen from the derivative charts in the Appendix, Section 4B.2, except for the pitch damping, most of the rate derivatives discussed above are fairly constant over the speed range, reflecting the insensitivity with forward speed of rotor response to equivalent cyclic pitch change. The pitch damping derivative M_q also has a significant stabilizing contribution from the horizontal tailplane, amounting to about 40% of the total at high speed.

Before leaving the roll and pitch moment derivatives, it is important that we consider the influence of the in-plane rotor loads on the moments transmitted to the fuselage. In our previous discussion of the force derivatives X_q and Y_p, we have seen how the 'Amer effect' reduces the effective rotor damping, most significantly on teetering rotors in low-thrust flight conditions. An additional effect stems from the orientation of the in-plane loads relative to the shaft when the rotor disc is tilted with one-per-rev flapping. The effect is illustrated in Fig. 4.21, showing the component of rotor torque oriented as a pitching moment with lateral flapping (the same effect gives a rolling moment with longitudinal flapping). The incremental hub moments can be written in terms of the product of the steady torque component and the disc tilt; hence, for four-bladed rotors

$$(\delta L)_{torque} = -\frac{Q_R}{2} \beta_{1c} \qquad (4.90)$$

$$(\delta M)_{torque} = \frac{Q_R}{2} \beta_{1s} \qquad (4.91)$$

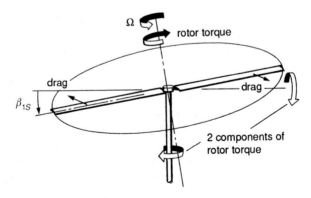

Fig. 4.21 Source of rotor hub couple due to inclination of rotor torque to the shaft

These moments will then combine with the thrust vector tilt and hub moment to give the total rotor moment. To examine the contribution of all three effects to the derivatives, we compare the breakdown for the Puma and Lynx. The Helisim predicted hover torque for the Puma and Lynx work out at about 31 000 N m and 18 000 N m respectively. The corresponding rotor thrusts are 57 000 N and 42 000 N and the effective spring stiffnesses 48 000 N m/rad and 166 000 N m/rad. The resultant derivative breakdown can then be written in the form

Puma   Lynx

$$M_q = -6.62\frac{\partial \beta_{1c}}{\partial q} + 0.46\frac{\partial \beta_{1s}}{\partial q} \qquad M_q = -27.82\frac{\partial \beta_{1c}}{\partial q} + 0.66\frac{\partial \beta_{1s}}{\partial q} \qquad (4.92)$$

$$M_p = -6.62\frac{\partial \beta_{1c}}{\partial p} + 0.46\frac{\partial \beta_{1s}}{\partial p} \qquad M_p = -27.82\frac{\partial \beta_{1c}}{\partial p} + 0.66\frac{\partial \beta_{1s}}{\partial p} \qquad (4.93)$$

The effect of the torque moment on the direct damping derivative is therefore negligible. In the case of the coupling derivative, the effect appears to be of concern only for articulated rotors, and then only for rotors with very light blades (see the low Lock number cases in Figs 2.21(b) and (c)).

The derivatives N_r, L_r, N_p

The final set of rate derivatives have little in common in terms of their physical makeup but share, along with their 'big brother' L_p, the property of having a primary influence on the character of the lateral/directional stability and control characteristics of the helicopter. We begin with a discussion of the yaw damping derivative N_r. In our previous discussion of the force derivatives, we rather dismissed the sideforce due to yaw rate Y_r, since the inertial effect due to forward speed $(U_e r)$ was so dominating. The aerodynamic contribution to Y_r, however, is directly linked to the yaw damping and is dominated by the loads on the tail rotor and vertical fin. Assuming that these components are at approximately the same location, we can write the yaw damping as

$$N_r \approx -\ell_t \frac{M_a}{I_{zz}} Y_r \qquad (4.94)$$

In the hover, our theory predicts that N_r is almost entirely due to the tail rotor, with a numerical value of between -0.25 and -0.4, depending on the tail rotor design parameters, akin to the effect of main rotor design parameters on Z_w (see Table 4.2). The low value of yaw damping is reduced even further (by about 30%) by the effect of the mechanical δ_3 coupling built into tail rotors to reduce transient flapping. The fin 'blockage' effects on the tail rotor can reduce N_r by another 10–30% depending on the separation and relative cover of the tail rotor from the fin. Yaw motion in the hover is therefore very lightly damped with a time constant of several seconds.

    In low-speed manoeuvres the effect of the main rotor downwash over the tail boom can have a strong effect on the yaw damping. The flow inclination over the tail boom can give rise to strong circulatory loads for deep, slender tail booms. This effect has been explored in terms of tail rotor control margins in sideways flight (Refs 4.9, 4.16 and 4.17), and the associated tail boom loads in steady conditions, from which we can deduce the kind of effects that might be expected in manoeuvres. The magnitude of the moment from the sideloads on the tail boom in a yaw manoeuvre depends on a number of factors, including the strength and distribution of main rotor downwash, the tail boom 'thickness ratio' and the location of any strakes to control the flow separation points (Ref. 4.16). A fixed strake, located to one side of the tail boom (for example, to reduce the tail rotor power requirement in right sideways flight), is likely to cause significant asymmetry in yaw manoeuvres. Main rotors with low values of static twist will have downwash distributions that increase significantly towards the rotor tips, leading to a tail boom centre of pressure in manoeuvres that is well aft of the aircraft centre of mass. The overall effect is quite complex and will depend on the direction of flight, but increments to the yaw damping derivative could be quite high, perhaps even as much as 100%.

As forward speed increases, so does the yaw damping in an approximately proportional way up to moderate speeds, before levelling off at high speed, again akin to the heave damping on the main rotor. The reduced value of N_r for the Puma, compared with Lynx and Bo105, shown in Appendix 4B, stems from the low fin effectiveness at small sideslip angles discussed earlier in the context of the weathercock stability derivative N_v. For larger sideslip excursions the derivative increases to the same level as the other aircraft.

One small additional modifying effect to the yaw damping is related to the rotor-speed governor sensing a yaw rate as an effective change in rotorspeed. At low forward speeds, the yaw rate can be as high as 1 rad/s, or between 2 and 4% of the rotorspeed. This will translate into a power change, hence a torque change and a yaw reaction on the fuselage serving to increase the yaw damping, with a magnitude depending on the gain and droop in the rotorspeed governor control loop.

N_p and L_r couple the yaw and roll DoFs together. The rolling moment due to yaw rate has its physical origin in the vertical offset of the tail rotor thrust and vertical fin sideforce from the aircraft centre of mass. L_r should therefore be positive, with the tail rotor thrust increasing to starboard as the aircraft nose yaws to starboard. However, if the offsets are small and the product of inertia I_{xz} relatively high, so that the contribution of N_r to L_r increases, L_r can change sign, a situation occurring in the Lynx, as shown in the derivative charts of the Appendix, Section 4B.2. The derivative N_p is more significant and although the aerodynamic effects from the main and tail rotor are relatively small, any product of inertia I_{xz} will couple the roll into yaw with powerful consequences. This effect can be seen most clearly for the Lynx and Bo105 helicopters. The large negative values of N_p cause an adverse yaw effect, turning the aircraft away from the direction of the roll (hence turn). In the next section we shall see how this effect influences the stability characteristics of the lateral/directional motion. Before leaving N_p and the stability derivatives however, it is worth discussing the observed effect of large torque changes during rapid roll manoeuvres (Refs 4.18, 4.19). On some helicopters this effect can be so severe that overtorquing can occur and the issue is given attention in the cautionary notes in aircrew manuals. The effect can be represented as an effective N_p. During low- to moderate-amplitude manoeuvres, the changes in rotor torque caused by the drag increments on the blades are relatively benign. However, as the roll rate is increased, the rotor blades can stall, particularly when rolling to the retreating side of the disc (e.g., a roll rate of 90°/s will generate a local incidence change of about 3° at the blade tip). The resulting transient rotor torque change can now be significant and lead to large demands on the engine. The situation is exacerbated by the changes in longitudinal flapping and hence pitching moment with roll rate as the blades stall. Within the structure of the Level 1 Helisim model, this effect cannot be modelled explicitly; a blade element rotor model with nonlinear aerodynamics is required. The nonlinear nature of the phenomenon, to an extent, also makes it inappropriate to use an equivalent linearization, particularly to model the onset of the effect as roll rate increases.

The control derivatives

Of the 24 control derivatives, we have selected the 11 most significant to discuss in detail and have arranged these into four groups: collective force, collective moment, cyclic moment and tail rotor collective force and moment.

The derivatives Z_{θ_0}, $Z_{\theta_{1s}}$

The derivative of thrust with main rotor collective and longitudinal cyclic can be obtained from the thrust and uniform inflow equations already introduced earlier in this chapter as eqns 4.66, 4.67. Approximations for hover and forward flight can be written in the form

$$\mu = 0: \quad \frac{\partial C_T}{\partial \theta_0} \approx \frac{8}{3}\left(\frac{a_0 s \lambda_0}{16\lambda_0 + a_0 s}\right), \quad Z_{\theta_0} \approx -\frac{8}{3}\frac{a_0 A_b \rho (\Omega R)^2 \lambda_0}{(16\lambda_0 + a_0 s) M_a} \tag{4.95}$$

$$\mu > 0.1: \quad \frac{\partial C_T}{\partial \theta_0} \approx \frac{4}{3}\left(\frac{a_0 s \mu (1 + 1.5\mu^2)}{8\mu + a_0 s}\right), \quad Z_{\theta_0} \approx -\frac{4}{3}\frac{a_0 A_b \rho (\Omega R)^2 \mu}{(8\mu + a_0 s) M_a}(1 + 1.5\mu^2) \tag{4.96}$$

From the derivative charts in the Appendix, Section 4B.2, it can be seen that this Z-force control derivative doubles in magnitude from hover to high-speed flight. This is the heave control sensitivity, and as with the heave damping derivative Z_w, it is primarily influenced by the blade loading and tip speed. The reader is reminded that the force derivatives are in semi-normalized form, i.e., divided by the aircraft mass. The thrust sensitivity for all three case aircraft is about 0.15 $g/°$ collective. Unlike the heave damping, the control sensitivity continues to increase with forward speed, reflecting the fact that the blade lift due to collective pitch changes divides into constant and two-per-rev components, while the lift due to vertical gusts is dominated by the one-per-rev incidence changes.

The thrust change with longitudinal cyclic is zero in the hover, and the approximation for forward flight can be written as

$$\mu > 0.1: \quad \frac{\partial C_T}{\partial \theta_{1s}} \approx \frac{2 a_0 s \mu^2}{8\mu + a_0 s}, \quad Z_{\theta_{1s}} \approx -\frac{2 a_0 A_b \rho (\Omega R)^2 \mu^2}{(8\mu + a_0 s) M_a} \tag{4.97}$$

As forward speed increases, the change in lift from aft cyclic on the advancing blade is greater than the corresponding decrease on the retreating side, due to the differential dynamic pressure. As with the collective derivative at higher speeds, $Z_{\theta_{1s}}$ increases almost linearly with speed, reaching levels at high speed very similar to the collective sensitivity in hover.

The derivatives M_{θ_0}, L_{θ_0}

Pitch and roll generated by the application of collective pitch arise from two physical sources. First, the change in rotor thrust (already discussed above) will give rise to a moment when the thrust line is offset from the aircraft centre of mass. Second, any change in flapping caused by collective will generate a hub moment proportional to the flap angle. It is the second of these effects that we shall focus on here. Referring to the flap response matrices from Chapter 3 (eqn 3.70), we can derive the main effect of collective pitch on flap by considering the behaviour of a teetering rotor at moderate forward speed. Hence we assume that

$$\lambda_\beta^2 = 1, \qquad \mu^2 \ll 1 \tag{4.98}$$

so that the expressions for the longitudinal and lateral flapping derivatives simplify to

$$\frac{\partial \beta_{1c}}{\partial \theta_0} \approx -\frac{8}{3}\mu \tag{4.99}$$

and

$$\frac{\partial \beta_{1s}}{\partial \theta_0} \approx -\frac{\gamma}{6}\mu \tag{4.100}$$

The aft flapping from increased collective develops from the greater increase in lift on the advancing blade, than on the retreating blade in forward flight. The effect grows in strength as forward speed increases, hence the approximate proportionality with speed. From the charts in the Appendix, Section 4B.2, we can observe that the effect is considerably stronger for the hingeless rotor configurations, as expected. In high-speed flight, the pitching moment from collective is of the same magnitude as the cyclic moment, illustrating the powerful effect of the differential loading from collective. Increasing collective also causes the disc to tilt to starboard (to port on the Puma). The physical mechanism is less obvious than for the pitching moment and according to the approximation in eqn 4.100, the degree of lateral flap for a change in collective pitch is actually a function of the rotor Lock number. The disc tilt arises from the rotor coning, which results in an increase in lift on the front of the disc and a decrease at the rear when the blades cone up (e.g., following an increase in collective pitch). The amount of lateral flapping depends on the coning, which itself is a function of the rotor Lock number. Once again, the resultant rolling moment will depend on the balance between thrust changes and disc tilt effects, which will vary from aircraft to aircraft (see control derivative charts in the Appendix, Section 4B.2).

The derivatives $M_{\theta_{1s}}$, $M_{\theta_{1c}}$, $L_{\theta_{1s}}$, $L_{\theta_{1c}}$

The dominant rotor moments are proportional to the disc tilt for the centre-spring equivalent rotor and can be written in the form

$$M_R \approx -\left(\frac{N_b}{2}K_\beta + h_R T\right)\beta_{1c}, \qquad L_R \approx -\left(\frac{N_b}{2}K_\beta + h_R T\right)\beta_{1s} \tag{4.101}$$

The cyclic control derivatives can therefore be approximated by the moment coefficient in parenthesis multiplied by the flap derivatives. We gave some attention to these functions in Chapter 2 of this book. The direct and coupled flap responses to cyclic

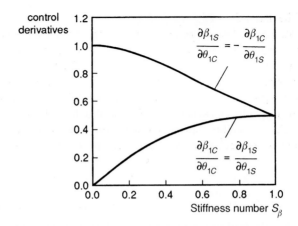

Fig. 4.22 Variation of rotor flap derivatives with Stiffness number

control inputs are practically independent of forward speed and can be written in the form

$$\frac{\partial \beta_{1c}}{\partial \theta_{1s}} \approx -\frac{\partial \beta_{1s}}{\partial \theta_{1c}} \approx -\frac{1}{1+S_\beta^2} \tag{4.102}$$

$$\frac{\partial \beta_{1c}}{\partial \theta_{1c}} \approx \frac{\partial \beta_{1s}}{\partial \theta_{1s}} \approx \frac{S_\beta}{1+S_\beta^2} \tag{4.103}$$

where the Stiffness number is given by eqn 4.59. The variations of direct and cross-coupled flap derivatives with stiffness were illustrated in Fig. 2.21 (a), and are repeated here as Fig. 4.22 for the reader's convenience. Up to Stiffness numbers of about 0.3 the direct flap derivative remains within a few percent of unity. The implication is that current so-called hingeless or 'semi-rigid' rotors (e.g., Lynx and Bo105) flap in much the same way as a teetering rotor following a cyclic control input $-1°$ direct flap for $1°$ cyclic. The cross-flap derivative arises with non-zero stiffness, because the natural frequency of flap motion is then less than one-per-rev, resulting in a flap response phase of less than $90°$. The phase angle is given by $\tan^{-1} S_\beta$, hence varying up to about $17°$ for $S_\beta = 0.3$. Although the cross-control derivative can be, at most, about 30% of the direct derivative, when considering pitch to roll coupling, this can result in a much greater coupled aircraft response to control inputs, because of the high ratio of pitch to roll inertias. This is evidenced by the coupling derivative $L_{\theta_{1s}}$ for the Lynx and Bo105 in the charts of the Appendix, Section 4B.2, which is actually of higher magnitude than the direct control moment $M_{\theta_{1s}}$. The aircraft will therefore experience a greater initial roll than pitch acceleration following a step longitudinal cyclic input. The cyclic controls are usually 'mixed' at the swash plate partly to cancel out this initial coupling. We have already discussed the cross-coupling effects from pitch and roll rates and, according to the simple rotor theory discussed here, the total short-term coupling will be a combination of the two effects.

The derivatives $Y_{\theta_{0T}}$, $L_{\theta_{0T}}$, $N_{\theta_{0T}}$

According to the simple actuator disc model of the tail rotor, the control derivatives all derive from the change in tail rotor thrust due to collective pitch. As with the derivatives N_v and N_r, the control derivative also decreases by as much as 30% as a result of the action of a mechanical δ_3 hinge set to take off 1° pitch for every 1° flap. The exact value depends upon the tail rotor Lock number. In the derivative charts in the Appendix, Section 4B.2, we can see that the force derivative $Y_{\theta_{0T}}$ for the Bo105 is about 20% higher than the corresponding values for the Lynx and Puma. The Bo105 sports a teetering tail rotor so that the δ_3 effect works only to counteract cyclic flapping. The control derivatives increase with speed in much the same way as the main rotor collective Z-force derivative, roughly doubling the hover value at high speed as the V^2 aerodynamics take effect.

The effects of non-uniform rotor inflow on damping and control derivatives

In Chapter 3 we introduced the concept of non-uniform inflow derived from local momentum theory applied to the rotor disc (see eqns 3.161, 3.181). Just as the uniform inflow balances the rotor thrust, so the flow needs to react to any hub moments generated by the rotor and a first approximation is given by a one-per-rev variation. The non-uniform inflow components can be written in the form given by

$$\lambda_{1c} = \left(1 - C_1'\right)\left(\theta_{1c} - \beta_{1s} + q\right) \tag{4.104}$$

$$\lambda_{1s} = \left(1 - C_1'\right)\left(\theta_{1s} - \beta_{1c} + p\right) \tag{4.105}$$

where the lift deficiency factor in the hover takes the form

$$C_1' = \frac{1}{1 + a_0 s / 16\lambda_0} \tag{4.106}$$

The non-uniform inflow has a direct effect on the flapping motion and hence on the moment derivatives. The effect was investigated in Refs 4.20 and 4.21 where a simple scaling of the rotor Lock number was shown to reflect the main features of the hub moment modification. We can write the flap derivatives as a linear combination of partial effects, as shown for the flap damping below:

$$\frac{\partial \beta_{1c}}{\partial q} = \left(\frac{\partial \beta_{1c}}{\partial q}\right)_{ui} + \frac{\partial \beta_{1c}}{\partial \lambda_{1s}}\frac{\partial \lambda_{1s}}{\partial q} + \frac{\partial \beta_{1c}}{\partial \lambda_{1c}}\frac{\partial \lambda_{1c}}{\partial q} \tag{4.107}$$

The subscript ui indicates that the derivative is calculated with uniform inflow only. Using the expressions for the flap derivatives set down earlier in this section, we can write the corrected flap derivatives in the form

$$\beta_{1c_q} = \beta_{1s_p} = \frac{S_\beta}{\Omega} + \frac{16}{\Omega\gamma^*} + S_\beta C_2'\beta_{1s_q} \tag{4.108}$$

$$\beta_{1s_q} = -\beta_{1c_p} = \frac{1}{\Omega} - S_\beta\frac{16}{\Omega\gamma^*} - S_\beta C_2'\beta_{1c_q} \tag{4.109}$$

where the equivalent Lock number has been reduced by the lift deficiency factor, i.e.,

$$\gamma^* = C_1'\gamma \tag{4.110}$$

and the new C coefficient is given by the expression

$$C'_2 = \frac{1 - C'_1}{C'_1} = \frac{a_0 s}{16\lambda_0} \tag{4.111}$$

Equation 4.108 shows the first important effect of non-uniform inflow that manifests itself even on rotors with zero hub stiffness. When the helicopter is pitching, the rotor lags behind the shaft by the amount given in eqn 4.108. This flapping motion causes an imbalance of moments which has a maximum and minimum on the advancing and retreating blades. This aerodynamic moment, caused by the flapping rate, gives rise to a wake reaction and the development of a non-uniform, laterally distributed component of downwash, λ_{1s}, serving to reduce the incidence and lift on the advancing and retreating blades. In turn, the blades flap further in the front and aft of the disc, giving an increased pitch damping M_q. The same arguments follow for rolling motion. The effect is quite significant in the hover, where the lift deficiency factor can be as low as 0.6.

By rearranging eqns 4.108 and 4.109, we may write the flap derivatives in the form

$$\beta_{1c_q} = \beta_{1s_p} = \frac{C'_3}{\Omega}\left[S_\beta + \frac{16}{\gamma^*} + S_\beta C'_2\left(1 - S_\beta\frac{16}{\gamma^*}\right)\right] \tag{4.112}$$

$$\beta_{1s_q} = -\beta_{1c_p} = \frac{C'_3}{\Omega}\left[1 - S_\beta\frac{16}{\gamma^*} - S_\beta C'_2\left(S_\beta + \frac{16}{\gamma^*}\right)\right] \tag{4.113}$$

where the third C coefficient takes the form

$$C'_3 = \frac{1}{1 + (C'_2 S_\beta)^2} \tag{4.114}$$

and can be approximated by unity.

The new terms in parentheses in eqns 4.112 and 4.113 represent the coupling components of flapping due to the non-uniform inflow and can make a significant contribution to the lateral (longitudinal) flapping due to pitch (roll) rate, and hence to the coupled rate derivatives L_q and M_p.

A similar analysis leads to the control derivatives, which can be written in the forms

$$\beta_{1c_{\theta 1s}} = -\beta_{1s_{\theta 1c}} = -C'_3\left(1 - C'_2 S_\beta^2\right) \tag{4.115}$$

$$\beta_{1c_{\theta 1c}} = \beta_{1s_{\theta 1s}} = C'_3\frac{S_\beta}{C'_1} \tag{4.116}$$

In the hover, for a rotor with zero flap stiffness, the aerodynamic moment due to flapping is exactly equal to that from the applied cyclic pitch; hence there is no non-uniform inflow in this case. The coupled flap/control response, given by eqn 4.116, is the only significant effect for the control moments indicating an increase of the coupled flapping of about 60% when the lift deficiency factor is 0.6.

Some reflections on derivatives

Stability and control derivatives aid the understanding of helicopter flight dynamics and the preceding qualitative discussion, supported by elementary analysis, has been

aimed at helping the reader to grasp some of the basic physical concepts and mechanisms at work in rotorcraft dynamics. Earlier in this chapter, the point was made that
there are three quite distinct approaches to estimating stability and control derivatives:
analytic, numerical, backward–forward differencing scheme and system identication
techniques. We have discussed some analytic properties of derivatives in the preceding
sections, and the derivative charts in the Appendix, Section 4B.3, illustrate numerical
estimates from the Helisim nonlinear simulation model. A discussion on flight estimated values of the Bo105 and Puma stability and control derivatives is contained in
the reported work of AGARD WG18 – Rotorcraft System Identification (Refs 4.22,
4.23). Responses to multistep control inputs were matched by 6 DoF model structures by a number of different system identification approaches. Broadly speaking,
estimates of primary damping and control derivatives compared favourably with the
Level 1 modelling described in this book. For cross-coupling derivatives and, to some
extent, the lower frequency velocity derivatives, the comparisons are much poorer,
however. In some cases this can clearly be attributed to missing features in the modelling, but in other cases, the combination of a lack of information in the test data
and an inappropriate model structure (e.g., overparameterized model) suggests that the
flight estimates are more in error. The work of AGARD WG18 represents a landmark
in the application of system identification techniques to helicopters, and the reported
results and continuing analysis of the unique, high-quality flight test data have the
potential for contributing to significant increased understanding of helicopter flight
dynamics. Selected results from this work will be discussed in Chapter 5, and the comparisons of estimated and predicted stability characteristics are included in the next
section.

Derivatives are, by definition, one-dimensional views of helicopter behaviour,
which appeal to the principle of superposition, and we need to combine the various constituent motions in order to understand how the unconstrained flight trajectory develops and to analyse the stability of helicopter motion. We should note,
however, that superposition no longer applies in the presence of nonlinear effects,
however small, and in this respect, we are necessarily in the realms of approximate
science.

4.3.3   The natural modes of motion

For small-amplitude stability analysis, helicopter motion can be considered to comprise a linear combination of natural modes, each having its own unique frequency,
damping and distribution of the response variables. The linear approximation that allows this interpretation is extremely powerful in enhancing physical understanding of
the complex motions in disturbed flight. The mathematical analysis of linear dynamic
systems is summarized in Appendix 4A, but we need to review some of the key results
to set the scene for the following discussion. Free motion of the helicopter is described
by the homogeneous form of eqn 4.41

$$\dot{\mathbf{x}} - \mathbf{A}\mathbf{x} = \mathbf{0} \qquad\qquad (4.117)$$

subject to initial conditions

$$\mathbf{x}(0) = \mathbf{x}_0 \qquad\qquad (4.118)$$

The solution of the initial value problem can be written as

$$\mathbf{x}(t) = \mathbf{W} \operatorname{diag} \left[\exp(\lambda_i t) \right] \mathbf{W}^{-1} \mathbf{x}_0 = \mathbf{Y}(t) \mathbf{x}_0 \qquad (4.119)$$

The eigenvalues, λ, of the matrix $\mathbf{A}$ satisfy the equation

$$\det [\lambda \mathbf{I} - \mathbf{A}] = 0 \qquad (4.120)$$

and the eigenvectors $\mathbf{w}$, arranged in columns to form the square matrix $\mathbf{W}$, are the special vectors of the matrix $\mathbf{A}$ that satisfy the relation

$$\mathbf{A} \mathbf{w}_i = \lambda_i \mathbf{w}_i \qquad (4.121)$$

The solution can be written in the alternative form

$$\mathbf{x}(t) = \sum_{i=0}^{n} (\mathbf{v}_i^T \mathbf{x}_0) \, \exp(\lambda_i t) \mathbf{w}_i \qquad (4.122)$$

where the $\mathbf{v}$ vectors are the eigenvectors of the transpose of $\mathbf{A}$ (columns of $\mathbf{W}^{-1}$), i.e.,

$$\mathbf{A}^T \mathbf{v}_j = \lambda_j \mathbf{v}_j \qquad (4.123)$$

The free motion is therefore shown in eqn 4.122 to be a linear combination of natural modes, each with an exponential character in time defined by the eigenvalue, and a distribution among the states, defined by the eigenvector.

The full 6 DoF helicopter equations are ninth order, usually arranged as $[u, w, q, \theta, v, p, \phi, r, \psi]$, but since the heading angle ψ appears only in the kinematic equation relating the rate of change of Euler angle ψ to the fuselage rates p, q, r, this equation is usually omitted for the purpose of stability analysis. Note that, for the ninth-order system including the yaw angle, the additional eigenvalue is zero (there is no aerodynamic or gravitational reaction to a change in heading) and the associated eigenvector is $\{0, 0, 0, 0, 0, 0, 0, 0, 1\}$.

The eight natural modes are described as linearly independent so that no single mode can be made up of a linear combination of the others and, if a single mode is excited precisely the motion will remain in that mode only. The eigenvalues and eigenvectors can be complex numbers, so that a mode has an oscillatory character, and such a mode will then be described by two of the eigenvalues appearing as conjugate complex pairs. If all the modes were oscillatory, then there would only be four in total. The stability of the helicopter can now be discussed in terms of the stability of the individual modes, which is determined entirely by the signs of the real parts of the eigenvalues. A positive real part indicates instability, a negative real part stability. As one might expect, helicopter handling qualities, or the pilot's perception of how well a helicopter can be flown in a task, are strongly influenced by the stability of the natural modes. In some cases (for some tasks), a small amount of instability may be acceptable; in others it may be necessary to require a defined level of stability. Eigenvalues can be illustrated as points in the complex plane, and the variation of an eigenvalue with some flight condition or aircraft configuration parameter portrayed as a root locus. The eigenvalues are given as the solutions of the determinantal eqn 4.120, which can also be written in the alternate polynomial form as the characteristic equation

$$\lambda^n + a_{n-1}\lambda^{n-1} + \cdots + a_1\lambda_1 + a_0 = 0 \qquad (4.124)$$

or as the product of individual factors

$$(\lambda - \lambda_n)(\lambda - \lambda_{n-1})(\cdots\cdots)(\lambda - \lambda_1) = 0 \qquad (4.125)$$

The coefficients of the characteristic equation are nonlinear functions of the stability derivatives discussed in the previous section. Before we discuss helicopter eigenvalues and vectors, we need one further analysis tool that will prove indispensable for relating the stability characteristics to the derivatives.

Although eigenanalysis is a simple computational task, the eighth-order system is far too complex to deal with analytically, and we need to work analytically to glean any meaningful understanding. We have seen from the discussion in the previous section that many of the coupled longitudinal/lateral derivatives are quite strong and are likely to have a major influence on the response characteristics. As far as stability is concerned however, we shall make a first approximation that the eigenvalues fall into two sets – longitudinal and lateral, and append the analysis with a discussion of the effects of coupling. Even grouping into two fourth-order sets presents a formidable analysis problem, and to gain maximum physical understanding we shall strive to reduce the approximations for the modes even further to the lowest order possible. The conditions of validity of these reduced order modelling approximations are described in Appendix 4A where the method of weakly coupled systems is discussed (Ref. 4.24). In the present context the method is used to isolate, where possible, the different natural modes according to the dominant constituent motions. The partitioning works only when there exists a natural separation of the modes in the complex plane. Effectively, approximations to the eigenvalues of slow modes can be estimated by assuming that the faster modes behave in a quasi-steady manner. Likewise, approximations to the fast modes can be derived by assuming that the slower modes do not react in the fast time scale. A second condition requires that the coupling effects between the contributing motions are small. The theory is covered in Appendix 4A and the reader is encouraged to assimilate this before tackling the examples described later in this section.

Figures 4.23, 4.24 and 4.25 illustrate the eigenvalues for the Lynx, Bo105 and Puma, respectively, as predicted by the Helisim theory. The pair of figures for each aircraft shows both the coupled longitudinal/lateral eigenvalues and the corresponding uncoupled values. The predicted stability characteristics of Lynx and Bo105 are very similar. Looking first at the coupled results for these two aircraft, we see that an unstable phugoid-type oscillation persists throughout the speed range, with time to double amplitude varying from about 2.5 s in the hover to just under 2 s at high speed. At the hover condition, this phugoid mode is actually a coupled longitudinal/lateral oscillation and is partnered by a similar lateral/longitudinal oscillation which develops into the classical Dutch roll oscillation in forward flight, with the frequency increasing strongly with speed. Apart from a weakly oscillatory heave/yaw oscillation in hover, the other modes are all subsidences having distinct characters at hover and low speed – roll, pitch, heave and yaw, but developing into more-coupled modes in forward flight, e.g., the roll/yaw spiral mode. The principal distinction between the coupled (Figs 4.23(a) and 4.24(a)) and the uncoupled (Figs 4.23(b) and 4.24(b)) cases lies in the

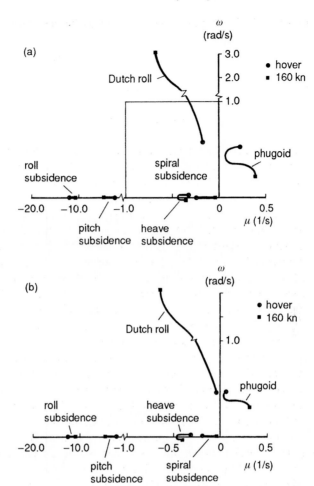

Fig. 4.23 Loci of Lynx eigenvalues as a function of forward speed: (a) coupled; (b) uncoupled

stability of the oscillatory modes at low speed where the coupled case shows a much higher level of instability. This effect can be shown to be almost exclusively due to the coupling effects of the non-uniform inflow caused by the change in wake angle induced by speed perturbations; the important derivatives are M_v and L_u, caused by the coupled rotor flapping response to lateral and longitudinal distributions of first harmonic inflow respectively. The unstable mode is a coupled pitch/roll oscillation with similar ratios of p to q and v to u, in the eigenvector.

The Puma comparison is shown in Figs 4.25(a) and (b). The greater instability for the coupled case at low speed is again present, and now we also see the short-term roll and pitch subsidences combined into a weak oscillation at low speed and hover. As speed increases the coupling effects again reduce, at least as far as stability is concerned. Here we are not discussing response and we should expect the coupled response characteristics to be strong at high speed, we shall return to this topic in the next chapter. At higher speeds the modes of the Puma, with its articulated rotor,

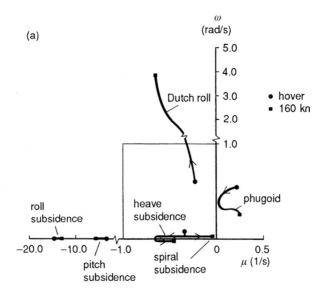

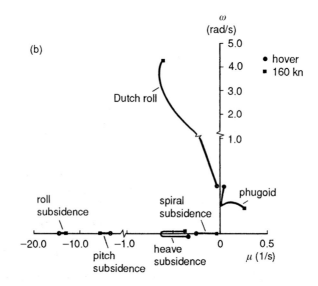

Fig. 4.24 Loci of Bo105 eigenvalues as a function of forward speed: (a) coupled; (b) uncoupled

resemble the classical fixed-wing set: pitch short period, phugoid, Dutch roll, spiral and roll subsidence. An interesting feature of the Puma stability characteristics is the dramatic change in stability of the Dutch roll from mid- to high speed. We shall discuss this later in the section.

Apart from relatively local, although important, effects, the significance of coupling for stability is sufficiently low to allow a meaningful investigation based on the uncoupled results, and hence we shall concentrate on approximating the characteristics

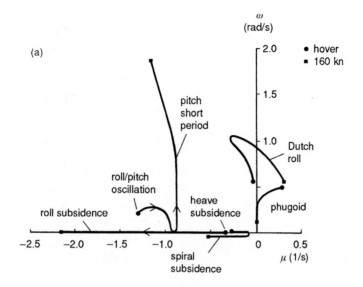

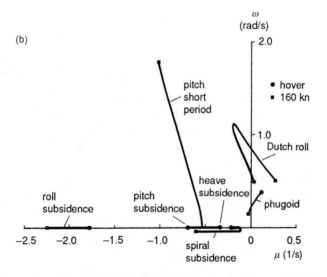

Fig. 4.25 Loci of Puma eigenvalues as a function of forward speed: (a) coupled; (b) uncoupled

illustrated in Figs 4.23(b)–4.25(b) and begin with the stability of longitudinal flight dynamics.

The longitudinal modes

Hovering dynamics have long presented a challenge to reduced order modelling. The eigenvectors for the unstable hover 'phugoid' for the three aircraft are given in Table 4.3 and highlight that the contribution of the normal velocity w to this oscillation is less than 10% of the forward velocity u.

Table 4.3 Eigenvectors for the hover phugoid oscillation

	Lynx magnitude/phase (degrees)	Bo105 magnitude/phase (degrees)	Puma magnitude/phase (degrees)
u	1.0	1.0	1.0
w	0.08/−1.24	0.036/9.35	0.021/75.8
q	0.024/−13.7	0.027/−7.8	0.017/−35
θ	0.049/−97.0	0.053/−94	0.042/−107.7

This suggests that a valid approximation to this mode could be achieved by neglecting the vertical motion and analysing stability with the simple system given by the surge and pitch equations

$$\dot{u} - X_u u + g\theta = 0 \tag{4.126}$$

$$\dot{q} - M_u u - M_q q = 0 \tag{4.127}$$

The small X_q derivative has also been neglected in this first approximation. In vector–matrix form, this equation can be written as

$$\frac{\mathrm{d}}{\mathrm{d}t}\begin{bmatrix} u \\ \theta \\ \hline q \end{bmatrix} - \begin{bmatrix} X_u & -g & 0 \\ 0 & 0 & 1 \\ \hline M_u & 0 & M_q \end{bmatrix}\begin{bmatrix} u \\ \theta \\ \hline q \end{bmatrix} = \mathbf{0} \tag{4.128}$$

The partitioning has been added to indicate the approximating subsystems – the relatively high-frequency pitch subsidence and the low-frequency phugoid. Unfortunately, the first weakly coupled approximation indicates that the mode damping is given entirely by the derivative X_u, hence predicting a stable oscillation. We can achieve much better accuracy in this case by extending the analysis to the second approximation (see Appendix 4A), so that the approximating characteristic equation for the low-frequency oscillation becomes

$$\begin{vmatrix} \lambda - X_u & g \\ \dfrac{M_u}{M_q}\left(1 + \lambda/M_q\right) & \lambda \end{vmatrix} = 0 \tag{4.129}$$

or in expanded form as the quadratic equation

$$\lambda^2 - \left(X_u + g\frac{M_u}{M_q^2}\right)\lambda - g\frac{M_u}{M_q} = 0 \tag{4.130}$$

The approximate phugoid frequency and damping are therefore given by the simple expressions

$$\omega_p^2 \approx -g\frac{M_u}{M_q} \tag{4.131}$$

$$2\zeta_p\omega_p = -\left(X_u + g\frac{M_u}{M_q^2}\right) \tag{4.132}$$

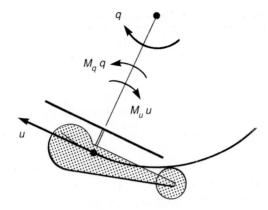

Fig. 4.26 Simple representation of unstable pitch phugoid in hover

The ratio of the pitching moments due to speed (speed stability) and pitch rate (damping) play an important role in both the frequency and damping of the oscillation. This mode can be visualized in the form of a helicopter rotating like a pendulum about a virtual hinge (Fig. 4.26). The frequency of the pendulum is given by

$$\omega^2 = \frac{g}{\ell} \tag{4.133}$$

where ℓ is the length of the pendulum (i.e., distance of helicopter centre of mass below the virtual hinge). This length determines the ratio of u to q in the eigenvector for this mode (cf. eqn 4.131). Comparison of the approximations given above with the 'exact' uncoupled phugoid roots is given in Table 4.4.

There is good agreement, particularly for the Lynx and Bo105. The smaller pitch damping for the Puma results in a more unstable motion, characteristic of articulated rotor helicopters at low speed. The speed stability derivative M_u is approximately proportional to the flapping derivative

$$\frac{\partial \beta_{1c}}{\partial \mu} = \frac{8}{3}\theta_0 - 2\lambda_0 \tag{4.134}$$

and scaled by the hub moment. The amount of flapback for a speed increment therefore depends only on the rotor loading, defined by the collective and inflow components. This derivative is the source of the instability and dominates the damping in eqn 4.132.

Table 4.4 Comparison of 'exact' and approximate hover phugoid eigenvalues

	Lynx	Bo105	Puma
X_u	−0.02	−0.021	−0.0176
M_u	0.047	0.105	0.0113
M_q	−1.9	−3.75	−0.451
Im(λ) exact	0.474	0.515	0.382
Im(λ) approx.	0.489	0.524	0.42
Re(λ) exact	0.056	0.034	0.116
Re(λ) approx.	0.054	0.026	0.264

As the helicopter passes through the 'trough' of the oscillation, the velocity u and pitch rate q are both at a maximum, the former leading to an increased pitch-up, the latter leading to a pitch-down, which, in turn, leads to a further increase in the u velocity component. The strong coupling of these effects results in the conventional helicopter configuration always being naturally unstable in hover.

The concept of the helicopter oscillating like a pendulum is discussed by Prouty in Ref. 4.9, where the approximate expression for frequency of the motion is given by

$$\omega_p \approx \frac{\sqrt{\left(g \frac{C_T}{a_0 s} \gamma\right)}}{\sqrt{R}} \tag{4.135}$$

From eqn 4.133 we can approximate the location of the virtual point of rotation above the helicopter

$$\ell \approx \frac{R}{(C_T/a_0 s)\gamma} \approx 10R \tag{4.136}$$

The expression for the approximate damping given in eqn 4.132 provides a clue as to the likely effects of feedback control designed to stabilize this mode. The addition of pitch rate feedback (i.e., increase of M_q) would seem to be fairly ineffective and would never be able to add more damping to this mode than the natural source from the small X_u. Feeding back velocity, hence augmenting M_u, would appear to have a much more powerful effect; a similar result could be obtained through attitude feedback, hence adding effective derivatives X_θ and M_θ.

The longitudinal pitch and heave subsidences hold no secrets at low speed and the eigenvalues are directly related to the damping derivatives in those two axes. The comparisons are shown in Table 4.5.

The hover approximations hold good for predicting stability at low speed, but in the moderate- to high-speed range, pitch and heave become coupled through the 'other' static stability derivative M_w, rendering the hover approximations invalid. From Figs 4.23(b), 4.24(b) and 4.25(b) we can see that the stability characteristics for the Lynx and Bo105 are quite different from that of the Puma. As might be expected, this is due to the different rotor types with the hingeless rotors exhibiting a much more unstable phugoid mode at high speed, while the articulated rotor Puma features a classical short-period pitch/heave oscillation and neutrally stable phugoid. At high speed, the normal velocity w features in both the long and short period modes and this makes it difficult to partition the longitudinal system matrix into subsystems based on the conventional aircraft states $\{u, w, q, \theta\}$. In Ref.4.25 it is shown that a more suitable partitioning can be found by recognizing that the motion in the long period mode is associated more

Table 4.5 Comparison of 'exact' and approximate longitudinal subsidences

	Lynx	Bo105	Puma
λ_{pitch}	−2.025	−3.836	−0.691
M_q	−1.896	−3.747	−0.451
λ_{heave}	−0.313	−0.323	−0.328
Z_w	−0.311	−0.322	−0.32

with the vertical velocity component

$$w_0 = w - U_e\theta \tag{4.137}$$

Transforming the longitudinal equations into the new variables then enables a partitioning as shown in eqn 4.138:

$$\frac{d}{dt}\begin{bmatrix} u \\ w_0 \\ \hline w \\ q \end{bmatrix} - \begin{bmatrix} X_u & g\cos\Theta_e/U_e & X_w - g\cos\Theta_e/U_e & X_q - W_e \\ Z_u & g\sin\Theta_e/U_e & Z_w - g\sin\Theta_e/U_e & Z_q \\ \hline Z_u & g\sin\Theta_e/U_e & Z_w - g\sin\Theta_e/U_e & Z_q + U_e \\ M_u & 0 & M_w & M_q \end{bmatrix}\begin{bmatrix} u \\ w_0 \\ \hline w \\ q \end{bmatrix} = 0 \tag{4.138}$$

Following the weakly coupled system theory in Appendix 4A, we note that the approximating characteristic equation for the low-frequency oscillation can be written as

$$\lambda^2 + 2\zeta_p\omega_p\lambda + \omega_p^2 = 0 \tag{4.139}$$

where the frequency and damping are given by the expressions (assuming Z_q small and neglecting $g\sin\Theta_e$)

$$2\zeta_p\omega_p = \left\{ \begin{array}{l} \left(X_w - \dfrac{8}{U_e}\cos\Theta_e\right)(Z_uM_q - M_u(Z_q + U_e)) \\[2mm] -X_u + \dfrac{+(X_q - W_e)(Z_wM_u - M_wZ_u)}{M_qZ_w - M_w(Z_q + U_e)} \end{array} \right\} \tag{4.140}$$

$$\omega_p^2 = -\frac{g}{U_e}\cos\Theta_e\left\{Z_u - \frac{Z_w(Z_uM_q - M_u(Z_q + U_e))}{M_qZ_w - M_w(Z_q + U_e)}\right\} \tag{4.141}$$

Similarly, the approximate characteristic equation for predicting the stability of the short period mode is given by

$$\lambda^2 + 2\zeta_{sp}\omega_{sp}\lambda + \omega_{sp}^2 = 0 \tag{4.142}$$

where the frequency and damping are given by the expressions

$$2\zeta_{sp}\omega_{sp} = -(Z_w + M_q) \tag{4.143}$$

$$\omega_{sp}^2 = Z_wM_q - (Z_q + U_e)M_w \tag{4.144}$$

The strong coupling of the translational velocities with the angular velocities in both short and long period modes actually results in the conditions for weak coupling being invalid for our hingeless rotor helicopters. The powerful M_u and M_w effects result in strong coupling between all the DoFs and the phugoid instability cannot be predicted using eqn 4.140. For the Puma helicopter, on the other hand, the natural modes are more classical, and very similar to a fixed-wing aircraft with two oscillatory modes becoming more widely separated as speed increases. Table 4.6 shows a comparison of the approximations for the phugoid

Table 4.6 Comparison of 'exact' and approximate longitudinal eigenvalues for Puma (exact results shown in parenthesis)

	120 knots	140 knots	160 knots
Re (λ_p) (1/s)	−0.025 (−0.0176)	−0.023 (−0.019)	−0.021 (−0.019)
Im (λ_p) (rad/s)	0.151 (0.159)	0.139 (0.147)	0.12 (0.13)
Re (λ_{sp}) (1/s)	−0.91 (−0.906)	− 0.96 (−0.96)	−1.01 (−1.01)
Im (λ_{sp}) (rad/s)	1.39 (1.4)	1.6 (1.58)	1.82 (1.8)

and short period stability characteristics at high speed with the exact longitudinal subset results. The agreement is very good, particularly for the short period mode.

The short period mode involves a rapid incidence adjustment with little change in forward speed, and has a frequency of about 2 rad/s at high speed for the Puma. Increasing the pitch stiffness M_w increases this frequency. Key configuration parameters that affect the magnitude of this derivative are the tailplane effectiveness (moment arm × tail area × tail lift slope) and the aircraft centre of mass location. As noted in the earlier section on derivatives, the hub moment contribution to M_w is always positive (destabilizing), which accounts for the strong positive values for both Lynx and Bo105 and the associated major change in character of the longitudinal modes.

The significant influence of the aircraft centre of mass (centre of gravity (cg)) location on longitudinal stability is illustrated in Fig. 4.27, which shows the eigenvalue of the Lynx phugoid mode for forward ($0.035R$), mid (0.0) and aft ($-0.035R$) centre of mass locations as a function of forward speed. With the aft centre of mass, the mode has become severely unstable with a time to double amplitude of less than 1 s. At this condition the short period approximation, eqn 4.142, becomes useful for predicting this change in the stability

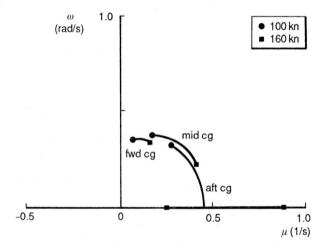

Fig. 4.27 Effect of centre of mass location on the stability of the longitudinal phugoid for Lynx

Table 4.7 Lynx stability characteristics with aft centre of mass

	120 knots	140 knots	160 knots
$Z_w + M_q$	−3.49	−3.66	−3.84
$Z_w M_q - M_w(Z_q + U_e)$	−1.49	−2.33	−3.214
λ approx. (1/s)	−3.87 +0.384	−4.213 +0.553	−4.45 +0.7
λ exact (1/s)	−3.89 +0.339 ± 0.311i	−4.238 +0.421 ± 0.217i	−4.586 +0.647

characteristics. The stiffness part of the short period approximation, given by eqn 4.144, is sometimes referred to as the manoeuvre margin of the aircraft (or the position of the aerodynamic centre relative to the centre of mass during a manoeuvre), and we can see from Table 4.7 that this parameter has become negative at high speed for the aft centre of mass case, due entirely to the strongly positive M_w. The divergence is actually well predicted by the short period approximation, along with the strong pitch subsidence dominated by the derivative M_q.

The lateral/directional modes

The lateral/directional motion of a helicopter in forward flight is classically composed of a roll/yaw/sideslip (Dutch roll) oscillation and two aperiodic subsidences commonly referred to as the roll and spiral modes. In hover, the modes have a broadly similar character, but different modal content. The roll subsidence mode is well characterized by the roll damping L_p at hover and, with some exceptions, throughout the speed envelope. The spiral mode in hover is largely made up of yaw motion (stability determined by yaw damping N_r) and the oscillatory mode could better be described as the lateral phugoid, in recognition of the similarity with the longitudinal phugoid mode already discussed. While the frequencies of the two hover oscillations are very similar, one big difference with the lateral phugoid is that the mode is predicted to be stable (for Bo105 and Lynx) or almost stable (Puma), on account of the strong contribution of yaw motion to the mode. The ratio of yaw to roll in this mode is typically about 2 for all three aircraft, rendering approximations based on a similar analysis to that conducted for the pitch phugoid unsuitable. We have to move into forward flight to find the Dutch roll mode more amenable to reduced order stability analysis, but even then there are complications that arise due to the roll/yaw ratio. For our case aircraft, the Lynx and Bo105 again exhibit similar characteristics to one another, while the Puma exhibits more individual behaviour, although not principally because of its articulated rotor. We shall return to the Puma later in this section, but first we examine the more conventional behaviour as typified by the Lynx.

Finding a suitable partitioning for approximating the lateral/directional modes requires the introduction of a new state variable into the lateral DoFs. With longitudinal motion we found that a partitioning into phugoid/short period subsets required the introduction of the vertical velocity, in place of the Euler pitch angle θ. The basic problem is the same; where both the short period and phugoid involved excursions in w, both the spiral and Dutch roll mode typically involve excursions in the lateral velocity v as well as roll and yaw motion. However, it can be shown that the spiral mode is characterized by excursions in the sway velocity component (Refs 4.12, 4.26)

$$v_0 \approx v + U_e \psi \tag{4.145}$$

Transforming the lateral equations to replace the roll angle ϕ with the sway velocity leads to the new lateral/directional subset in the form

$$\frac{d}{dt}\begin{bmatrix} \dot{v}_0 \\ \hline v \\ \hline \dot{v} \\ \hline p \end{bmatrix} - \begin{bmatrix} 0 & 0 & Y_v & g \\ \hline 0 & 0 & 1 & 0 \\ \hline -N_r & -U_e N_v & N_r + Y_v & g - N_p U_e \\ \hline L_r/U_e & L_v & -L_r/U_e & L_p \end{bmatrix}\begin{bmatrix} \dot{v}_0 \\ \hline v \\ \hline \dot{v} \\ \hline p \end{bmatrix} = \mathbf{0} \tag{4.146}$$

where we have neglected the small derivatives Y_p and Y_r.

The partitioning shown leads to three levels of approximation with regions in the complex plane approximately bounded by the radii – $O(0.1 \text{ rad/s})$ for the spiral mode, $O(1 \text{ rad/s})$ for the Dutch roll mode and $O(10 \text{ rad/s})$ for the roll subsidence. The analysis for three-level systems described by the matrix

$$\mathbf{A} = \begin{bmatrix} \mathbf{A}_{11} & \mathbf{A}_{12} & \mathbf{A}_{13} \\ \mathbf{A}_{21} & \mathbf{A}_{22} & \mathbf{A}_{23} \\ \mathbf{A}_{31} & \mathbf{A}_{32} & \mathbf{A}_{33} \end{bmatrix} \tag{4.147}$$

is given in Appendix 4A, where the approximation for the lowest order eigenvalue is given by the modified subsystem

$$\mathbf{A}_{11}^* = \mathbf{A}_{11} - [\mathbf{A}_{12} \quad \mathbf{A}_{13}]\begin{bmatrix} \mathbf{A}_{22} & \mathbf{A}_{23} \\ \mathbf{A}_{32} & \mathbf{A}_{33} \end{bmatrix}^{-1}\begin{bmatrix} \mathbf{A}_{21} \\ \mathbf{A}_{31} \end{bmatrix} \tag{4.148}$$

The stability of the spiral mode is therefore approximated by the expression

$$\lambda_s \approx \frac{g}{L_p}\frac{(L_v N_r - N_v L_r)}{(U_e N_v + \sigma_s L_v)} \tag{4.149}$$

where

$$\sigma_s = \frac{g - N_p U_e}{L_p} \tag{4.150}$$

This simple approximation gives the same result as the Bairstow approximation (Ref. 4.2), obtained from the lowest order terms of the characteristic equation.

The middle-level approximation for the Dutch roll mode takes the form

$$\lambda^2 + 2\zeta_d \omega_d \lambda + \omega_d^2 = 0 \tag{4.151}$$

where the damping is given by

$$2\zeta_d \omega_d \approx -\left(N_r + Y_v + \sigma_d\left\{\frac{L_r}{U_e} - \frac{L_v}{L_p}\right\}\right)\Bigg/\left(1 - \frac{\sigma_d L_r}{L_p U_e}\right) \tag{4.152}$$

and the frequency by the expression

$$\omega_d^2 \approx (U_e N_v + \sigma_d L_v)/\left(1 - \frac{\sigma_d L_r}{L_p U_e}\right) \tag{4.153}$$

with

$$\sigma_d = \sigma_s \tag{4.154}$$

In the derivation of this approximation we have extended the analysis to second-order terms (see Appendix 4A) to model the destabilizing effects of the dihedral effect shown in eqn 4.152.

Finally, the roll mode at the third level is given by

$$\lambda_r \approx L_p \tag{4.155}$$

The accuracy of this set of approximations can be illustrated for the case of the Lynx at a forward flight speed of 120 knots, as shown below:

$$\lambda_{s_{approx}} = -0.039/s \qquad \lambda_{s_{exact}} = -0.042/s$$
$$2\zeta_d\omega_{d\,approx} = 1.32/s \qquad 2\zeta_d\omega_{d\,exact} = 1.23/s$$
$$\omega_{d_{approx}} = 2.66\ \text{rad/s} \qquad \omega_{d_{exact}} = 2.57\ \text{rad/s}$$
$$\lambda_{r_{approx}} = -10.3/s \qquad \lambda_{r_{exact}} = -10.63/s$$

The approximate eigenvalues are mostly well within 10% of the full subset predictions, which provides confidence in their worth, which actually holds good from moderate to high speeds for both Lynx and Bo105. The validity of this approximation for the Dutch roll oscillation depends upon the coupling between roll and yaw. The key coupling derivatives are N_p and L_v, both of which are large and negative for our two hingeless rotor helicopters. The yaw due to roll derivative is augmented by the inertia coupling effects in eqn 4.48 (for the Lynx, $I_{xz} = 2767\ \text{kg/m}^2$; for the Bo105, $I_{xz} = 660\ \text{kg/m}^2$). The simplest approximation to the Dutch roll mode results when the coupling is zero so that the motion is essentially a yaw/sideslip oscillation. The yaw rate then exhibits a 90° phase lag relative to the sideslip and the damping is given by the first two terms in the numerator of eqn 4.152 (i.e., $N_r + Y_v$). A negative value of N_p tends to destabilize the oscillation by superimposing a roll motion into the mode such that the term $N_p p$ effectively adds negative damping. The eigenvector for the Dutch roll mode of the Lynx at 120 knots, shown below, illustrates that, while the yaw rate is still close to 90° out of phase with sideslip, the roll/yaw ratio is 0.5 with the yawing moment due to roll rate being almost in anti-phase with the sideslip.

$$v \quad 1.0\ \text{m/s}; \quad p \quad 0.02\ \text{rad/s}\ (160°); \quad r \quad 0.04\ \text{rad/s}\ (-80°)$$

The approximations described above break down when the roll/yaw ratio in the Dutch roll oscillation is high. Such a situation occurs for the Puma, and we close Chapter 4 with a discussion of this case.

We refer back to Fig. 4.25(b) where the loci of the Puma eigenvalues are plotted with speed. Above 100 knots, the Dutch roll mode becomes less and less stable until at high speed the damping changes sign. At 120 knots, the Puma Dutch roll eigenvector is

$$v \quad 1.0\ \text{m/s}; \quad p \quad 0.04\ \text{rad/s}\ (150°); \quad r \quad 0.01\ \text{rad/s}\ (-70°)$$

which, compared with the Lynx mode shape, contains considerably more roll motion with a roll/yaw ratio of about 4, eight times that for the Lynx. The reason for the 'unusual'

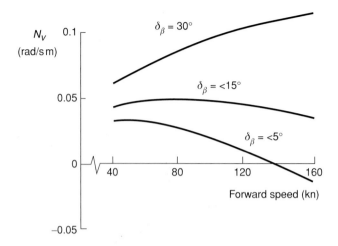

Fig. 4.28 Variation of weathercock stability derivative N_v with speed for different sideslip perturbations for Puma

behaviour of the Dutch roll mode for the Puma can be attributed to the derivative N_v. In the previous discussion on this weathercock stability derivative, we observed that the Puma value was influenced by the strong nonlinearity in the force characteristics of the vertical fin with sideslip. At small angles of sideslip the fin sideforce is practically zero, due to the strong suction on the 'undersurface' of the thick aerofoil section (Refs 4.11, 4.12). For larger angles of sideslip, the circulatory lift force builds up in the normal way. The value of the fin contribution to N_v therefore depends upon the amplitude of the perturbation used to generate the derivative (as with the yaw damping N_r, to a lesser extent). In Fig. 4.28, the variation of N_v with speed is shown for three different perturbation levels corresponding to $<5°$, $15°$ and $30°$ of sideslip. For the small amplitude case, the directional stability changes sign at about 140 knots and is the reason for the loss of Dutch roll stability illustrated in Fig. 4.25(b). For the large amplitude perturbations, the derivative increases with speed, indicating that the vertical fin is fully effective for this level of sideslip. Figure 4.29 presents the loci of Dutch roll eigenvalues for the three perturbation sizes as a function of speed, revealing the dramatic effect of the weathercock stability parameter. The mode remains stable for the case of the high sideslip perturbation level. It appears that the Puma is predicted to be unstable for small amplitudes and stable for large amplitude motion. These are classic conditions for so-called limit cycle oscillations, where we would expect the oscillation to limit in amplitude at some finite value with the mode initially dominated by roll and later, as the amplitude grows, to settle into a more conventional yaw/roll motion. We shall return to the nature of this motion when discussing response, in Chapter 5.

Comparison with flight

Table 4.8 presents a comparison of the stability characteristics of Puma and Bo105 Helisim with flight estimates derived from the work of AGARD WG18 (Refs 4.22, 4.23). Generally speaking, the comparison of modal frequencies is very good, while dampings are less well predicted, particularly for the weakly damped or unstable phugoid and spiral modes. The pitch/heave subsidences for the Bo105 show remarkable agreement while the roll subsidence appears to be overpredicted by theory, although this is largely attributed to the compensating effect of an added time delay in the adopted model structure used to derive the flight estimates (Ref. 4.22). This aspect is returned to in Chapter 5 when results are

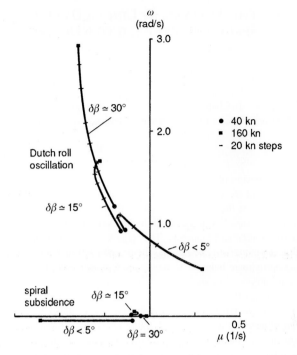

Fig. 4.29 Loci of Dutch roll and spiral mode eigenvalues with speed for different sideslip perturbations for Puma

Table 4.8 Comparison of flight estimates and theoretical predictions of Puma and Bo105 stability characteristics

	Puma		Bo105	
Mode of motion	flight estimate[+]	Helisim prediction	flight estimate[++]	Helisim prediction
Phugoid	[0.001, 0.27]	[0.047, 0.22]	[−0.15, 0.33]	[−0.058, 0.3]
Short period	[0.934, 1.4]	[0.622, 1.28]	(4.36), (0.6)	(4.25), (0.653)
Spiral	(0.0055)	(0.12)	(0.02)	(0.024)
Dutch roll	[0.147, 1.35]	[0.162, 1.004]	[0.14, 2.5]	[0.214, 2.64]
Roll subsidence	(2.07)	(1.683)	(8.49)	(13.72)

Shorthand notation:
λ Complex variation $\mu \pm i\omega$;
$[\zeta, \omega_n]$ damping ratio and natural frequency associated with roots of $\lambda^2 + 2\zeta\omega_n\lambda + \omega_n^2$;
$(1/\tau)$ inverse of time constant τ in root $(\lambda - 1/\tau)$;
[+]Puma flight estimates from Glasgow/DRA analysis in Ref. 4.22;
[++]Bo105 flight estimates from DLR analysis in Ref. 4.22.

presented from the different model structures used for modelling roll response to lateral cyclic.

The above discussions have concentrated on 6 DoF motion analysis. There are several areas in helicopter flight dynamics where important effects are missed by folding the rotor dynamics and other higher order effects into the fuselage motions in quasi-steady form. These will be addressed in the context of constrained stability and aircraft response analysis in Chapter 5.

APPENDIX 4A   THE ANALYSIS OF LINEAR DYNAMIC SYSTEMS (WITH SPECIAL REFERENCE TO 6 DoF HELICOPTER FLIGHT)

The application of Newton's laws of motion to a helicopter in flight leads to the assembly of a set of nonlinear differential equations for the evolution of the aircraft response trajectory and attitude with time. The motion is referred to an orthogonal axes system fixed at the aircraft's cg. In Chapter 3 we have discussed how these equations can be combined together in first-order vector form, with state vector $\mathbf{x}(t)$ of dimension n, and written as

$$\dot{\mathbf{x}} = \mathbf{F}(\mathbf{x}, \mathbf{u}, t) \tag{4A.1}$$

The dimension of the dynamic system depends upon the number of DoFs included. For the moment we will consider the general case of dimension n. The solution of eqn 4A.1 depends upon the initial conditions of the motion state vector and the time variation of the vector function $\mathbf{F}(\mathbf{x}, \mathbf{u}, t)$, which includes the aerodynamic loads, gravitational forces and inertial forces and moments. The trajectory can be computed using any of a number of different numerical integration schemes which time march through a simulation, achieving an approximate balance of the component accelerations with the applied forces and moments at every time step. While this is an efficient process for solving eqn 4A.1, numerical integration offers little insight into the physics of the aircraft flight behaviour. We need to turn to analytic solutions to deliver a deeper understanding between cause and effect. Unfortunately, the scope for deriving analytic solutions of general nonlinear differential equations as in eqn 4A.1 is extremely limited; only in special cases can functional forms be found and, even then, the range of validity is likely to be very small. Fortunately, the same is not true for linearized versions of eqn 4A.1, and much of the understanding of complex dynamic aircraft motions gleaned over the past 80 years has been obtained from studying linear approximations to the general nonlinear motion. Texts that provide suitable background reading and deeper understanding of the underlying theory of dynamic systems are Refs 4A.1–4A.3. The essence of linearization is the assumption that the motion can be considered as a perturbation about a trim or equilibrium condition; provided that the perturbations are small, the function $\mathbf{F}$ can usually be expanded in terms of the motion and control variables (as discussed earlier in this chapter) and the response written in the form

$$\mathbf{x} = \mathbf{X}_e + \delta\mathbf{x} \tag{4A.2}$$

where $\mathbf{X}_e$ is the equilibrium value of the state vector and $\delta\mathbf{x}$ is the perturbation. For convenience, we will drop the δ and write the perturbation equations in the linearized form

$$\dot{\mathbf{x}} - \mathbf{A}\mathbf{x} = \mathbf{B}\mathbf{u}(t) + \mathbf{f}(t) \tag{4A.3}$$

where the $(n \times n)$ state matrix $\mathbf{A}$ is given by

$$\mathbf{A} = \left(\frac{\partial \mathbf{F}}{\partial \mathbf{x}}\right)_{\mathbf{x}=\mathbf{x}_e} \tag{4A.4}$$

and the $(n \times m)$ control matrix $\mathbf{B}$ is given by

$$\mathbf{B} = \left(\frac{\partial \mathbf{F}}{\partial \mathbf{u}}\right)_{\mathbf{x}=\mathbf{x}_e} \tag{4A.5}$$

and where we have assumed without much loss of generality that the function $\mathbf{F}$ is differentiable with all first derivatives bounded for bounded values of flight trajectory $\mathbf{x}$ and time t.

We write the initial value at time $t = 0$, as

$$\mathbf{x}(0) = \mathbf{x}_0 \tag{4A.6}$$

The flight state vector $\mathbf{x}$ is a vector in n-dimensional space, where n is the number of independent components. As an example of eqn 4A.3, consider the longitudinal motion of the helicopter, uncoupled from lateral/directional dynamics, and with rotor and other 'higher' DoFs subsumed into the fourth-order rigid body equations. The linearized equations of motion for perturbations from straight flight can be written in the form

$$\frac{d}{dt}\begin{bmatrix} u \\ w \\ q \\ \theta \end{bmatrix} - \begin{bmatrix} X_u & X_w & X_q - W_e & -g\cos\Theta_e \\ Z_u & Z_w & Z_q + U_e & -g\sin\Theta_e \\ M_u & M_w & M_q & 0 \\ 0 & 0 & 1 & 0 \end{bmatrix}\begin{bmatrix} u \\ w \\ q \\ \theta \end{bmatrix} = \begin{bmatrix} X_{\theta_0} & X_{\theta_{1s}} \\ Z_{\theta_0} & Z_{\theta_{1s}} \\ M_{\theta_0} & M_{\theta_{1s}} \\ 0 & 0 \end{bmatrix}\begin{bmatrix} \theta_0(t) \\ \theta_{1s}(t) \end{bmatrix}$$

$$\tag{4A.7}$$

$$+ \begin{bmatrix} X_w \\ Z_w \\ M_w \\ 0 \end{bmatrix}\begin{bmatrix} w_g(t) \end{bmatrix}$$

Longitudinal motion is here described by the four-vector with elements u (forward velocity perturbation from trim U_e), w (normal velocity perturbation from trim W_e), q (pitch rate) and θ (pitch attitude perturbation from trim Θ_e). As an illustration of the forcing function, we have included both collective and cyclic rotor controls and a normal gust field w_g. In this tutorial-style appendix, we will use the example given by eqn 4A.7 to illustrate the physical significance of theoretical results as they are derived.

Equation 4A.3 is valid for calculating the perturbed responses from a trim point, but in the homogeneous form, with no forcing function, it can be used to quantify the stability characteristics for small motions of the nonlinear dynamic system described by eqn 4A.1. This is a most important application and underpins most of the understanding of flight dynamics. The free motion solutions of eqn 4A.3 take the form of exponential functions; the signs of the real parts determine the stability with positive values indicating instability. The theory of the stability of motion for linear dynamic systems can be most succinctly expressed using linear algebra and the concepts of eigenvalues and eigenvectors.

Consider the free motion given by

$$\dot{\mathbf{x}} - \mathbf{A}\mathbf{x} = \mathbf{0} \tag{4A.8}$$

With the intention of simplifying the equations, we introduce the transformation

$$\mathbf{x} = \mathbf{W}\mathbf{y} \tag{4A.9}$$

so that eqn 4A.8 can be written as

$$\dot{\mathbf{y}} - \boldsymbol{\Lambda}\mathbf{y} = \mathbf{0} \tag{4A.10}$$

with

$$\boldsymbol{\Lambda} = \mathbf{W}^{-1}\mathbf{A}\mathbf{W} \tag{4A.11}$$

For a given matrix $\mathbf{A}$, there is a unique transformation matrix $\mathbf{W}$ that reduces $\mathbf{A}$ to a canonical form, $\boldsymbol{\Lambda}$, most often diagonal, so that eqn 4A.10 can usually be written as a series of uncoupled equations

$$\dot{y}_i - \lambda_i y_i = 0, \quad i = 1, 2, \ldots, n \tag{4A.12}$$

with solutions

$$y_i = y_{i_0} e^{\lambda_i t} \tag{4A.13}$$

Collected together in vector form, the solution can be written as

$$\mathbf{y} = \text{diag}\left[\exp(\lambda_i t)\right] \mathbf{y}_0 \tag{4A.14}$$

Transforming back to the flight state vector $\mathbf{x}$, we obtain

$$\mathbf{x}(t) = \mathbf{W} \, \text{diag}\left[\exp(\lambda_i t)\right] \mathbf{W}^{-1} \mathbf{x}_0 = \mathbf{Y}(t)\mathbf{x}_0 \equiv \exp(\mathbf{A}t)\mathbf{x}_0 \tag{4A.15}$$

where the principal matrix solution $\mathbf{Y}(t)$ is defined as

$$\mathbf{Y}(t) = \mathbf{0}, \quad t < 0, \qquad \mathbf{Y}(t) = \mathbf{W} \, \text{diag}\left[\exp(\lambda_i t)\right] \mathbf{W}^{-1}, \quad t \geq 0 \tag{4A.16}$$

We need to stop here and take stock. The transformation matrix $\mathbf{W}$ and the set of numbers λ have a special meaning in linear algebra; if $\mathbf{w}_i$ is a column of $\mathbf{W}$ then the pairs $[\mathbf{w}_i, \lambda_i]$ are the eigenvectors and eigenvalues of the matrix $\mathbf{A}$. The eigenvectors are special in that when they are transformed by the matrix $\mathbf{A}$, all that happens is that they change in length, as given by the equation

$$\mathbf{A}\mathbf{w}_i = \lambda_i \mathbf{w}_i \tag{4A.17}$$

No other vectors in the space on which $\mathbf{A}$ operates are quite like the eigenvectors. Their special property makes them suitable as basis vectors for describing more general motion. The associated eigenvalues are the real or complex scalars given by the n solutions of the polynomial

$$\det[\lambda \mathbf{I} - \mathbf{A}] = 0 \tag{4A.18}$$

The free motion of a helicopter is therefore described by a linear combination of simple exponential motions, each with a mode shape given by the eigenvector and a trajectory envelope defined by the eigenvalue. Each mode is linearly independent of the others, i.e., the motion in a mode is unique and cannot be made up from a combination of other modes. Earlier in this chapter, a full discussion on the character of the modes of motion and how they vary with flight state and aircraft configuration was given. Below, in Figs 4A.1 and 4A.2, we illustrate the eigenvalue and eigenvector associated with the longitudinal short period mode of the Puma flying straight and level at 100 knots; we have included the modal content of all eight state vector components $u, w, q, \theta, v, p, \phi, r$. The eigenvalue, illustrated in Fig. 4A.1, is given by

$$\lambda_{sp} = -1.0 \pm 1.3\mathrm{i} \tag{4A.19}$$

The negative unit real part gives a time to half amplitude of

$$t_{1/2} = \frac{\ln(2)}{\text{Re}(\lambda_{sp})} = 0.69 \text{ s} \tag{4A.20}$$

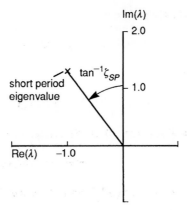

Fig. 4A.1 Longitudinal short period eigenvalue – Puma at 100 knots

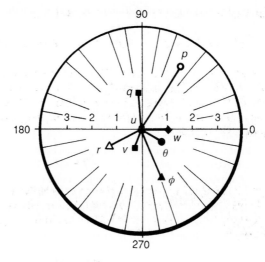

Fig. 4A.2 Longitudinal short period eigenvector – Puma at 100 knots

The short period frequency is given by

$$\omega_{sp} = \text{Im}(\lambda_{sp}) = 1.3 \text{ rad/s} \tag{4A.21}$$

and, finally, the damping ratio is given by

$$\zeta_{sp} = -\frac{\text{Re}(\lambda_{sp})}{\omega_{sp}} = 0.769 \tag{4A.22}$$

We choose to present the angular and translational rates in the eigenvectors shown in Fig. 4A.2 in deg/s and in m/s respectively. Because the mode is oscillatory, each component has a magnitude and phase and Fig. 4A.2 is shown in polar form. During the short period oscillation, the ratio of the magnitudes of the exponential envelope of the state variables remains constant. Although the mode is described as a pitch short period, it can be seen in Fig. 4A.2 that the roll and sideslip coupling content is significant, with roll about twice the magnitude of pitch. Pitch rate is roughly in quadrature with heave velocity.

The eigenvectors are particularly useful for interpreting the behaviour of the free response of the aircraft to initial condition disturbances, but they can also provide key information on the response to controls and atmospheric disturbances. The complete solution to the homogeneous eqn 4A.3 can be written in the form

$$
\mathbf{x}(t) = \mathbf{Y}(t)\mathbf{x}_0 + \int_0^t \mathbf{Y}(t-\tau)(\mathbf{Bu}(\tau) + \mathbf{f}(\tau))\, d\tau \tag{4A.23}
$$

or expanded as

$$
\mathbf{x}(t) = \sum_{i=1}^n \left[(\mathbf{v}_i^{\mathrm{T}}\mathbf{x}_0) \exp(\lambda_i t) + \int_0^t \left(\mathbf{v}_i^{\mathrm{T}} (\mathbf{Bu}(\tau) + \mathbf{f}(\tau)) \exp[\lambda_i(t-\tau)] \right) d\tau \right] \mathbf{w}_i \tag{4A.24}
$$

where $\mathbf{v}$ is the eigenvector of the matrix $\mathbf{A}^{\mathrm{T}}$, i.e., $\mathbf{v}_j^{\mathrm{T}}$ are the rows of $\mathbf{W}^{-1}$ ($\mathbf{V}^{\mathrm{T}} = \mathbf{W}^{-1}$) so that

$$
\mathbf{A}^{\mathrm{T}}\mathbf{v}_j = \lambda_j\mathbf{v}_j \tag{4A.25}
$$

The dual vectors $\mathbf{w}$ and $\mathbf{v}$ satisfy the bi-orthogonality relationship

$$
\mathbf{v}_j^{\mathrm{T}}\mathbf{w}_k = 0, \quad j \neq k \tag{4A.26}
$$

Equations 4A.24 and 4A.26 give us useful information about the system response. For example, if the initial conditions or forcing functions are distributed throughout the states with the same ratios as an eigenvector, the response will remain in that eigenvector. The mode participation factors, in the particular integral component of the solution, given by

$$
\mathbf{v}_i^{\mathrm{T}} (\mathbf{Bu}(\tau) + \mathbf{f}(\tau)) \tag{4A.27}
$$

determine the contribution of the response in each mode $\mathbf{w}_i$.

A special case is the solution for the case of a periodic forcing function of the form

$$
\mathbf{f}(t) = \mathbf{F}e^{i\omega t} = \mathbf{F}(\cos \omega t + i \sin \omega t) \tag{4A.28}
$$

The steady-state response at the input frequency ω is given by

$$
\mathbf{x}(t) = \mathbf{X}e^{i\omega t}
$$

$$
\mathbf{X} = [i\omega\mathbf{I} - \mathbf{A}]^{-1}\mathbf{F} = \mathbf{W}(i\omega\mathbf{I} - \mathbf{\Lambda})^{-1}\mathbf{V}^{\mathrm{T}}\mathbf{F} \equiv \sum_{j=1}^n \frac{\mathbf{w}_j(\mathbf{v}_j^{\mathrm{T}}\mathbf{F})}{i\omega - \lambda_j} \tag{4A.29}
$$

The frequency response function $\mathbf{X}$ is derived from the (Laplace) transfer function (of the complex variable s) evaluated on the imaginary axis. The transfer function for a given input (i)/output (o) pair can be written in the general form

$$
\frac{x_o}{x_i}(s) = \frac{N(s)}{D(s)} \tag{4A.30}
$$

The response vector $\mathbf{X}$ is generally complex with a magnitude and phase relative to the input function $\mathbf{F}$. Figure 4A.3 illustrates how the magnitude of the frequency response as a function of frequency can be represented as the value of the so-called transfer function,

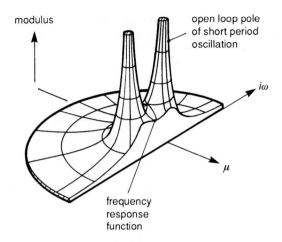

modulus

open loop pole
of short period
oscillation

$i\omega$

μ

frequency
response
function

Fig. 4A.3 Frequency response as the transfer function evaluated on imaginary axis

when $s = i\omega$. In Fig. 4A.3 a single oscillatory mode is shown, designated the pitch short period. In practice, for the 6 DoF helicopter model, all eight poles would be present, but the superposition principle also applies to the transfer function. The peaks in the frequency response correspond to the modes of the system (i.e., the roots of the denominator $D(s) = 0$ in eqn 4A.30) that are set back in either the right-hand side or the left-hand side of the complex plane, depending on whether the eigenvalue real parts are positive (unstable) or negative (stable). The troughs in the frequency response function correspond to the zeros of the transfer function, or the eigenvalues for the case of infinite gain when a feedback loop between input and output is closed (i.e., the roots of the numerator $N(s) = 0$ in eqn 4A.30). Ultimately, at a high enough frequency, the gain will typically roll-off to zero as the order of $D(s)$ is higher than $N(s)$. The phase between input and output varies across the frequency range, with a series of ramp-like 180° changes as each mode is traversed; for modes in close proximity the picture is more complicated.

For the case when the system modes are widely separated, a useful approximation can sometimes be applied that effectively partitions the system into a series of weakly coupled subsystems (Ref. 4A.4). We illustrate the technique by considering the n-dimensional homogeneous system, partitioned into two levels of subsystem, with states $\mathbf{x}_1$ and $\mathbf{x}_2$, with dimensions ℓ and m, such that $n = \ell + m$

$$\left[\begin{array}{c} \dot{\mathbf{x}}_1 \\ \hline \dot{\mathbf{x}}_2 \end{array} \right] - \left[\begin{array}{c|c} \mathbf{A}_{11} & \mathbf{A}_{12} \\ \hline \mathbf{A}_{21} & \mathbf{A}_{22} \end{array} \right] \left[\begin{array}{c} \mathbf{x}_1 \\ \hline \mathbf{x}_2 \end{array} \right] = \mathbf{0} \tag{4A.31}$$

Equation 4A.31 can be expanded into two first-order equations and the eigenvalues can be determined from either of the alternative forms of characteristic equation

$$f_1(\lambda) = \left| \lambda \mathbf{I} - \mathbf{A}_{11} - \mathbf{A}_{12}(\lambda \mathbf{I} - \mathbf{A}_{22})^{-1} \mathbf{A}_{21} \right| = 0 \tag{4A.32}$$

$$f_2(\lambda) = \left| \lambda \mathbf{I} - \mathbf{A}_{22} - \mathbf{A}_{21}(\lambda \mathbf{I} - \mathbf{A}_{11})^{-1} \mathbf{A}_{12} \right| = 0 \tag{4A.33}$$

Using the expansion of a matrix inverse (Ref. 4A.4), we can write

$$(\lambda \mathbf{I} - \mathbf{A}_{22})^{-1} = -\mathbf{A}_{22}^{-1}(\mathbf{I} + \lambda \mathbf{A}_{22}^{-1} + \lambda^2 \mathbf{A}_{22}^{-2} + \cdots) \tag{4A.34}$$

We assume that the eigenvalues of the subsystems $\mathbf{A}_{11}\{\lambda_1^{(1)}, \lambda_1^{(2)}, \ldots, \lambda_1^{(\ell)}\}$ and $\mathbf{A}_{22}\{\lambda_2^{(1)}, \lambda_2^{(2)}, \ldots, \lambda_2^{(m)}\}$ are widely separated in modulus. Specifically, the eigenvalues of $\mathbf{A}_{11}$ lie within the circle of radius r ($r = \max|\lambda_1^{(i)}|$), and the eigenvalues of $\mathbf{A}_{22}$ lie without the circle of radius R ($R = \min|\lambda_2^{(j)}|$). We have assumed that the eigenvalues of smaller modulus belong to the matrix $\mathbf{A}_{11}$. The method of weakly coupled systems is based on the hypothesis that the solutions to the characteristic eqn 4A.32 can be approximated by the roots of the first $m + 1$ terms and the solution to the characteristic eqn 4A.33 can be approximated by the roots of the last $\ell + 1$ terms, solved separately. It is shown in Ref. 4A.4 that this hypothesis is valid when two conditions are satisfied:

(1) the eigenvalues form two disjoint sets separated as described above, i.e.,

$$\left[\frac{r}{R}\right] \ll 1 \tag{4A.35}$$

(2) the coupling terms are small, such that if γ and δ are the maximum elements of the coupling matrices $\mathbf{A}_{12}$ and $\mathbf{A}_{21}$, then

$$\left[\frac{\ell\gamma\delta}{R^2}\right] \ll 1 \tag{4A.36}$$

When these weak coupling conditions are satisfied, the eigenvalues of the complete system can be approximated by the two polynomials

$$f_1(\lambda) = |\lambda\mathbf{I} - \mathbf{A}_{11} + \mathbf{A}_{12}\mathbf{A}_{22}^{-1}\mathbf{A}_{21}| = 0 \tag{4A.37}$$

$$f_2(\lambda) = |\lambda\mathbf{I} - \mathbf{A}_{22}| = 0 \tag{4A.38}$$

According to eqns 4A.37 and 4A.38, the larger eigenvalue set is approximated by the roots of $\mathbf{A}_{22}$ and is unaffected by the slower dynamic subsystem $\mathbf{A}_{11}$. Conversely, the smaller roots, characterizing the slower dynamic subsystem $\mathbf{A}_{11}$, are strongly affected by the behaviour of the faster subsystem $\mathbf{A}_{22}$. In the short term, motion in the slow modes does not develop enough to affect the overall motion, while in the longer term, the faster modes have reached their steady-state values and can be represented by quasisteady effects.

The method has been used extensively earlier in this chapter, but here we provide an illustration by looking more closely at the longitudinal motion of the Puma helicopter in forward flight, given by the homogeneous form of eqn 4A.7, i.e.,

$$\frac{d}{dt}\begin{bmatrix} u \\ w \\ q \\ \theta \end{bmatrix} - \begin{bmatrix} X_u & X_w & X_q - W_e & -g\cos\Theta_e \\ Z_u & Z_w & Z_q + U_e & -g\sin\Theta_e \\ M_u & M_w & M_q & 0 \\ 0 & 0 & 1 & 0 \end{bmatrix}\begin{bmatrix} u \\ w \\ q \\ \theta \end{bmatrix} = \mathbf{0} \tag{4A.39}$$

The eigenvalues of the longitudinal subsystem are the classical short period and phugoid modes with numerical values for the 100-knot flight condition given by

Phugoid: $\lambda_{1,2} = -0.0103 \pm 0.176i$ (4A.40)

Short period: $\lambda_{3,4} = -1.0 \pm 1.30i$ (4A.41)

While these modes are clearly well separated in magnitude ($r/R = O(0.2)$), the form of the dynamic system given by eqn 4A.39 does not lend itself to partitioning as it stands. The phugoid mode is essentially an exchange of potential and kinetic energy, with excursions in forward velocity and vertical velocity, while the short period mode is a rapid incidence adjustment with only small changes in speed. This classical form of the two longitudinal modes does not always characterize helicopter motion however; earlier in this chapter, it was shown that the approximation breaks down for helicopters with stiff rotors. For articulated rotors, the equations can be recast into more appropriate coordinates to enable an effective partitioning to be achieved. The phugoid mode can be better represented in terms of the forward velocity u and vertical velocity

$$w_0 = w - U_e \theta \tag{4A.42}$$

Equation 4A.39 can then be recast in the partitioned form

$$
\frac{d}{dt}
\begin{bmatrix} u \\ w_0 \\ \hline w \\ q \end{bmatrix}
-
\left[
\begin{array}{cc|cc}
X_u & g\cos\Theta_e/U_e & X_w - g\cos\Theta_e/U_e & X_q - W_e \\
Z_u & g\sin\Theta_e/U_e & Z_w - g\sin\Theta_e/U_e & Z_q \\
\hline
Z_u & g\sin\Theta_e/U_e & Z_w - g\sin\Theta_e/U_e & Z_q + U_e \\
M_u & 0 & M_w & M_q
\end{array}
\right]
\begin{bmatrix} u \\ w_0 \\ \hline w \\ q \end{bmatrix}
= 0
\tag{4A.43}
$$

The approximating polynomials for the phugoid and short period modes can then be derived using eqns 4A.37 and 4A.38, namely

Low-modulus phugoid (assuming Z_q small):

$$
f_1(\lambda) = \lambda^2 +
\left\{
-X_u +
\frac{
\left(X_w - \dfrac{g}{U_e}\cos\Theta_e\right)\left(Z_u M_q - M_u\left(Z_q + U_e\right)\right)
+ (X_q - W_e)(Z_w M_u - M_w Z_u)
}{
M_q Z_w - M_w(Z_q + U_e)
}
\right\} \lambda
$$

$$
- \frac{g}{U_e}\cos\Theta_e
\left\{
Z_u - \frac{Z_w(Z_u M_q - M_u(Z_q + U_e))}{M_q Z_w - M_w(Z_q + U_e)}
\right\} = 0
\tag{4A.44}
$$

High-modulus short period:

$$
f_2(\lambda) = \lambda^2 - (Z_w + M_q)\lambda + Z_w M_q - M_w(Z_q + U_e) = 0
\tag{4A.45}
$$

A comparison of the exact and approximate eigenvalues is shown in Table 4A.1, using the derivatives shown in the charts of Appendix 4B. The two different 'exact' results are given for the fully coupled longitudinal and lateral equations and the uncoupled longitudinal set. Comparisons are shown for two flight speeds – 100 knots and 140 knots. A first point to note is that the coupling with lateral motion significantly reduces the phugoid damping,

Table 4A.1 Comparison of exact and approximate eigenvalues for longitudinal modes of motion

	Forward flight speed 100 knots		Forward flight speed 140 knots	
	Phugoid	Short period	Phugoid	Short period
Exact coupled	$-0.0103 \pm 0.176i$	$-1.0 \pm 1.30i$	$-0.0006 \pm 0.14i$	$-1.124 \pm 1.693i$
Exact uncoupled	$-0.0153 \pm 0.177i$	$-0.849 \pm 1.17i$	$-0.0187 \pm 0.147i$	$-0.96 \pm 1.583i$
Weakly coupled approximation	$-0.025 \pm 0.175i$	$-0.85 \pm 1.47i$	$-0.022 \pm 0.147i$	$-0.963 \pm 1.873i$

particularly at 140 knots where the oscillation is almost neutrally stable. The converse is true for the short period mode. The weakly coupled approximation fares much better at the higher speed and appears to converge towards the exact, uncoupled results. The approximations do not predict the growing loss of phugoid stability as a result of coupling with lateral dynamics, however. The higher the forward speed, the more the helicopter phugoid resembles the fixed-wing phugoid where the approximation works very well for aircraft with strongly positive manoeuvre margins (the constant term in eqn 4A.45 with negative M_w).

The approximations given by eqns 4A.44 and 4A.45 are examples of many that are discussed in Chapter 4 and that serve to provide additional physical insight into complex behaviour at a variety of flight conditions. The importance of the speed stability derivative M_u in the damping and frequency of the phugoid is highlighted by the expressions. For a low-speed fixed-wing aircraft, M_u is typically zero, while the effect of pitching moments due to speed effects dominates the helicopter phugoid. The last term in eqn 4A.45 represents the manoeuvre margin and the approximation breaks down long before instability occurs at positive values of the static stability derivative M_w (Ref. 4A.5).

To complete this appendix we present two additional results from the theory of weakly coupled systems. For cases where the system partitions naturally into three levels

$$
\begin{bmatrix}
\mathbf{A}_{11} & \mathbf{A}_{12} & \mathbf{A}_{13} \\
\mathbf{A}_{21} & \mathbf{A}_{22} & \mathbf{A}_{23} \\
\mathbf{A}_{31} & \mathbf{A}_{32} & \mathbf{A}_{33}
\end{bmatrix}
$$

then the approximating polynomials take the form (see Ref. 4A.6)

$$
\mathbf{A}_{11}^{*} = \mathbf{A}_{11} - [\mathbf{A}_{12}\mathbf{A}_{13}]
\begin{bmatrix}
\mathbf{A}_{22} & \mathbf{A}_{23} \\
\mathbf{A}_{32} & \mathbf{A}_{33}
\end{bmatrix}^{-1}
\begin{bmatrix}
\mathbf{A}_{21} \\
\mathbf{A}_{31}
\end{bmatrix}
\tag{4A.46}
$$

$$
\mathbf{A}_{22}^{*} = \mathbf{A}_{22} - \mathbf{A}_{23}\mathbf{A}_{33}^{-1}\mathbf{A}_{32}
\tag{4A.47}
$$

$$
\mathbf{A}_{33}^{*} = \mathbf{A}_{33}
\tag{4A.48}
$$

Similar conditions for weak coupling apply to the three levels of subsystem $\mathbf{A}_{11}$, $\mathbf{A}_{22}$ and $\mathbf{A}_{33}$.

The second result concerns cases where a second-order approximation is required to determine an accurate estimate of the low-order eigenvalue. Writing the expanded inverse, eqn 4A.34, in the approximate form

$$
(\lambda \mathbf{I} - \mathbf{A}_{22})^{-1} \approx -\mathbf{A}_{22}^{-1}\left(\mathbf{I} + \lambda \mathbf{A}_{22}^{-1}\right)
\tag{4A.49}
$$

leads to the low-order approximation

$$
f_1(\lambda) = \det\left[\lambda \mathbf{I} - \mathbf{A}_{11} + \mathbf{A}_{12}\mathbf{A}_{22}^{-1}(\mathbf{I} + \lambda \mathbf{A}_{22}^{-1})\mathbf{A}_{21}\right]
\tag{4A.50}
$$

Both these extensions to the more basic technique are employed in the analysis of Chapter 4.

APPENDIX 4B   THE THREE CASE HELICOPTERS: LYNX, Bo105 AND PUMA

4B.1   Aircraft configuration parameters

The DRA (RAE) research Lynx, ZD559

The Westland Lynx Mk 7 is a twin engine, utility/battlefield helicopter in the 4.5-ton category currently in service with the British Army Air Corps. The DRA Research Lynx (Fig. 4B.1) was delivered off the production line to RAE as an Mk 5 in 1985 and modified to Mk 7 standard in 1992. The aircraft is fitted with a comprehensive instrumentation suite and digital recording system. Special features include a strain-gauge fatigue usage monitoring fit, and pressure- and strain-instrumented rotor blades for fitment on both the main and tail rotor. The aircraft has been used extensively in a research programme to calibrate agility standards of future helicopter types. The four-bladed hingeless rotor is capable of producing

Fig. 4B.1 DRA research Lynx ZD559 in flight

Table 4B.1 Configuration data – Lynx

a_0	6.0/rad	I_{zz}	12 208.8 kg m^2	x_{cg}	-0.0198
a_{0T}	6.0/rad	K_β	166 352 N m/rad	δ_0	0.009
α_{tp0}	-0.0175 rad	l_{fn}	7.48 m	δ_2	37.983
β_{fn0}	-0.0524 rad	l_{tp}	7.66 m	δ_3	$-45°$
c	0.391 m	l_T	7.66 m	δ_{T0}	0.008
g_T	5.8	M_a	4313.7 kg	δ_{T2}	5.334
h_R	1.274 m	N_b	4		
h_T	1.146 m	R	6.4 m	γ	7.12
I_β	678.14 kg m^2	R_T	1.106 m	γ_s	0.0698 rad
I_{xx}	2767.1 kg m^2	S_{fn}	1.107 m^2	λ_β^2	1.193
I_{xz}	2034.8 kg m^2	S_{tp}	1.197 m^2	θ_{tw}	-0.14 rad
I_{yy}	13 904.5 kg m^2	s_T	0.208	Ω_{idle}	35.63 rad/s

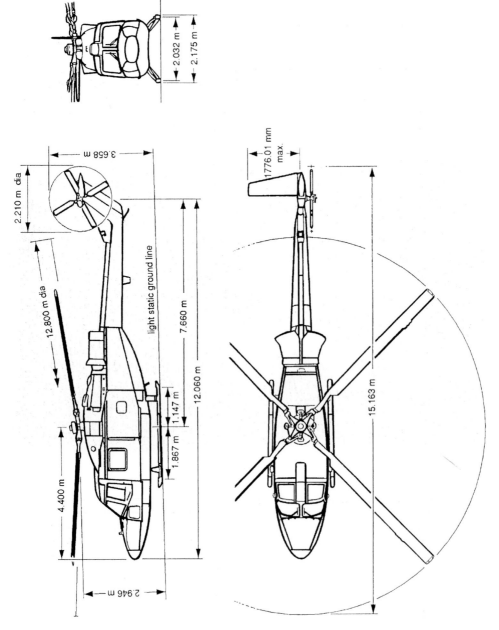

Fig. 4B.2 DRA research Lynx ZD559 three-view drawing

large control moments and hence angular accelerations. A 1960's design, the Lynx embodies many features with significant innovation for its age – hingeless rotor with cambered aerofoil sections (RAE 9615, 9617), titanium monoblock rotor head and conformal gears.

A three-view drawing of the aircraft is shown in Fig. 4B.2. The physical characteristics of the aircraft used to construct the Helisim simulation model are provided in Table 4B.1.

The DLR research Bo105, S123

The Eurocopter Deutschland (formerly MBB) Bo105 is a twin engine helicopter in the 2.5-ton class, fulfilling a number of roles in transport, offshore, police and battlefield operations. The DLR Braunschweig operate two Bo105 aircraft. The first is a standard serial type (Bo105-S123), shown in Fig. 4B.3. The second aircraft is a specially modified fly-by-wire/light in-flight simulator – the ATTHEs (advanced technology testing helicopter system), Bo105-S3. The Bo105 features a four-bladed hingeless rotor with a key innovation for a 1960's design – fibre-reinforced composite rotor blades.

Fig. 4B.3 DLR research Bo105 S123 in flight

Table 4B.2 Configuration data – Bo105

a_0	6.113/rad	I_{zz}	4099 kg m^2	x_{cg}	0.0163
a_{0T}	5.7/rad	K_β	113 330 N m/rad	δ_0	0.0074
α_{tp0}	0.0698 rad	l_{fn}	5.416 m	δ_2	38.66
β_{fn0}	−0.08116 rad	l_{tp}	4.56 m	δ_3	−45°
c	0.27 m	l_T	6 m	δ_{T0}	0.008
g_T	5.25	M_a	2200 kg	δ_{T2}	9.5
h_R	1.48 m	N_b	4		
h_T	1.72 m	R	4.91 m	γ	5.087
I_β	231.7 kg m^2	R_T	0.95 m	γ_s	0.0524 rad
I_{xx}	1433 kg m^2	S_{fn}	0.805 m^2	λ_β^2	1.248
I_{xz}	660 kg m^2	S_{tp}	0.803 m^2	θ_{tw}	−0.14 rad
I_{yy}	4973 kg m^2	s_T	0.12	Ω_{idle}	44.4 rad/s

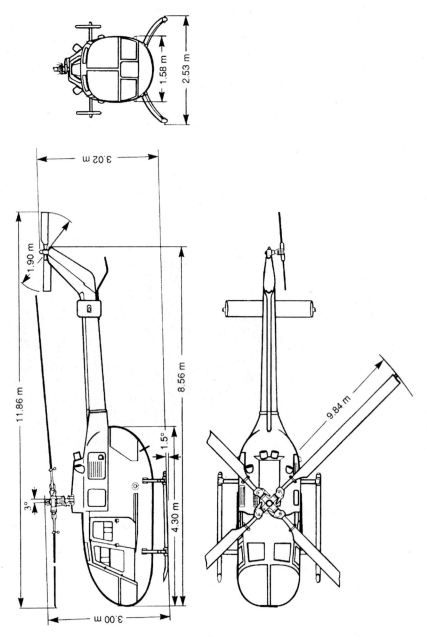

Fig. 4B.4 DLR research Bo105 S123 three-view drawing

A three-view drawing of the aircraft is shown in a Fig. 4B.4, and the physical characteristics of the aircraft used to construct the Helisim simulation model are provided in Table 4B.2.

The DRA (RAE) research Puma, SA330

The SA 330 Puma is a twin-engine, medium-support helicopter in the 6-ton category, manufactured by Eurocopter France (ECF) (formerly Aerospatiale, formerly Sud Aviation), and in service with a number of civil operators and armed forces, including the Royal Air Force, to support battlefield operations. The DRA (RAE) research Puma XW241 (Fig. 4B.5) was one of the early development aircraft acquired by RAE in 1974 and extensively instrumented for flight dynamics and rotor aerodynamics research. With its original analogue data acquisition system, the Puma provided direct support during the 1970s to the development of new rotor aerofoils through the measurement of surface pressures on modified blade profiles. In

Fig. 4B.5 DRA research Puma XW241 in flight

Table 4B.3 Configuration data – Puma

a_0	5.73/rad	I_{zz}	25 889 kg m^2	x_{cg}	0.005
a_{0T}	5.73/rad	K_β	48 149 N m/rad	δ_0	0.008
α_{tp0}	−0.0262 rad	l_{fn}	9 m	δ_2	9.5
β_{fn0}	0.0175 rad	l_{tp}	9 m	δ_3	−45°
c	0.5401 m	l_T	9 m	δ_{T0}	0.008
g_T	4.82	M_a	5805 kg	δ_{T2}	9.5
h_R	2.157 m	N_b	4		
h_T	1.587 m	R	7.5 m	γ	9.374
I_β	1280 kg m^2	R_T	1.56 m	γ_s	0.0873 rad
I_{xx}	9638 kg m^2	S_{fn}	1.395 m^2	λ_β^2	1.052
I_{xz}	2226 kg m^2	S_{tp}	1.34 m^2	θ_{tw}	−0.14 rad
I_{yy}	33 240 kg m^2	s_T	0.19	Ω_{idle}	27 rad/s

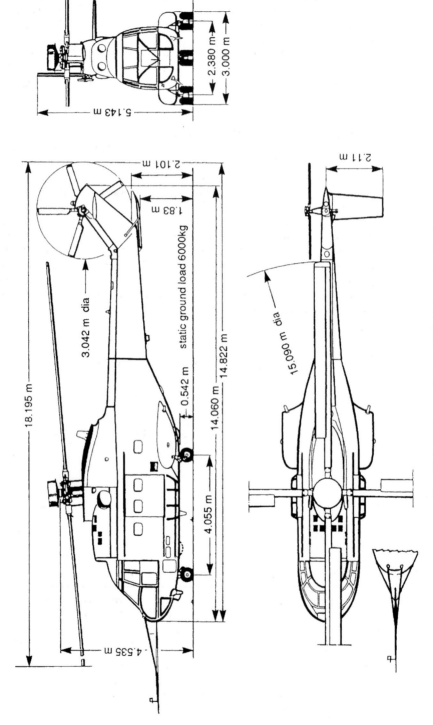

Fig. 4B.6 DRA research Puma XW241 three-view drawing

the early 1980s a digital PCM system was installed in the aircraft and a research programme to support simulation model validation and handling qualities was initiated. Over the period between 1981 and 1988, more than 150 h of flight testing was carried out to gather basic flight mechanics data throughout the flight envelope of the aircraft (Ref. 4B.1). The aircraft was retired from RAE service in 1989.

A three-view drawing of the aircraft in its experimental configuration is shown in Fig. 4B.6. The aircraft has a four-bladed articulated man rotor (modified NACA 0012 section, 3.8% flapping hinge offset). The physical characteristics of the aircraft used to construct the Helisim simulation model are provided in Table 4B.3.

Fuselage aerodynamic characteristics

Chapter 3 developed a generalized form for the aerodynamic forces and moments acting on the fuselage; Table 4B.4 presents a set of values of force and moment coefficients, giving one-dimensional, piecewise linear variations with incidence and sideslip. These values have been found to reflect the characteristics of a wide range of fuselage shapes; they are used in Helisim to represent the large angle approximations.

Small angle approximations $(-20° < (\alpha_f, \beta_f) < 20°)$ for the fuselage aerodynamics of the Lynx, Bo105 and Puma helicopters, based on wind tunnel measurements, are given in eqns 4B.1 – 4B.15. The forces and moments (in N, N/rad, N m, etc.) are given as functions of incidence and sideslip at a speed of 30.48 m/s (100 ft/s). The increased order of the polynomial approximations for the Bo105 and Puma is based on more extensive curve fitting applied to the wind tunnel test data. The small angle approximations should be fared into the large angle piecewise forms.

Lynx

$$X_{f100} = -1112.06 + 3113.75\,\alpha_f^2 \tag{4B.1}$$

$$X_{f100} = -8896.44\,\beta_f \tag{4B.2}$$

$$Z_{f100} = -4225.81\,\alpha_f \tag{4B.3}$$

$$M_{f100} = 10\,168.65\,\alpha_f \tag{4B.4}$$

$$N_{f100} = -10\,168.65\,\beta_f \tag{4B.5}$$

Table 4B.4 Generalized fuselage aerodynamic coefficients

α_f	−180	−160	−90	−30	0	20	90	160	180		
C_{xf}	0.1	0.08	0.0	−0.07	−0.08	−0.07	0.0	0.08	0.1		
α_f	−180	−160	−120	−60	−20	0	20	60	120	160	180
C_{zf}	0.0	0.15	1.3	1.3	0.15	0.0	−0.15	−1.3	−1.3	−0.15	0.0
α_f	−205	−160	−130	−60	−25	25	60	130	155	200	
C_{mf}	0.02	−0.03	0.1	0.1	−0.04	0.02	−0.1	−0.1	0.02	−0.03	
β_f	−90	−70	−25	0	25	70	90				
C_{nfa}	−0.1	−0.1	0.005	0.0	−0.005	0.1	0.1				
β_f	−90	−60	0	60	90						
C_{nfb}	−0.1	−0.1	0.0	0.1	0.1						

Bo105

$$X_{f100} = -580.6 - 454.0\,\alpha_f + 6.2\,\alpha_f^2 + 4648.9\,\alpha_f^3 \tag{4B.6}$$

$$Y_{f100} = -6.9 - 2399.0\,\beta_f - 1.7\,\beta_f^2 + 12.7\,\beta_f^3 \tag{4B.7}$$

$$Z_{f100} = -51.1 - 1202.0\,\alpha_f + 1515.7\,\alpha_f^2 - 604.2\,\alpha_f^3 \tag{4B.8}$$

$$M_{f100} = -1191.8 + 12\,752.0\,\alpha_f + 8201.3\,\alpha_f^2 - 5796.7\,\alpha_f^3 \tag{4B.9}$$

$$N_{f100} = -10\,028.0\,\beta_f \tag{4B.10}$$

Puma

$$X_{f100} = -822.9 + 44.5\,\alpha_f + 911.9\,\alpha_f^2 + 1663.6\,\alpha_f^3 \tag{4B.11}$$

$$Y_{f100} = -11\,672.0\,\beta_f \tag{4B.12}$$

$$Z_{f100} = -458.2 - 5693.7\,\alpha_f + 2077.3\,\alpha_f^2 - 3958.9\,\alpha_f^3 \tag{4B.13}$$

$$M_{f100} = -1065.7 + 8745.0\,\alpha_f + 12\,473.5\,\alpha_f^2 - 10\,033.0\,\alpha_f^3 \tag{4B.14}$$

$$N_{f100} = -24\,269.2\,\beta_f + 97\,619.0\,\beta_f^3 \tag{4B.15}$$

Empennage aerodynamic characteristics

Following the convention and notation used in Chapter 3, small angle approximations ($-20° < (\alpha_{tp}, \beta_{fn}) < 20°$) for the vertical tailplane and horizontal fin (normal) aerodynamic force coefficients are given by the following equations. As for the fuselage forces, the Puma approximations have been curve fitted to greater fidelity over the range of small incidence and sideslip angles.

Lynx

$$C_{ztp} = -3.5\,\alpha_{tp} \tag{4B.16}$$

$$C_{yfn} = -3.5\,\beta_{fn} \tag{4B.17}$$

Bo105

$$C_{ztp} = -3.262\,\alpha_{tp} \tag{4B.18}$$

$$C_{yfn} = -2.704\,\beta_{fn} \tag{4B.19}$$

Puma

$$C_{ztp} = -3.7\left(\alpha_{tp} - 3.92\,\alpha_{tp}^3\right) \tag{4B.20}$$

$$C_{yfn} = -3.5\left(11.143\,\beta_{fn}^3 - 85.714\,\beta_{fn}^5\right) \tag{4B.21}$$

4B.2   Stability and control derivatives

The stability and control derivatives predicted by Helisim for the three subject helicopters are shown in Figs 4B.7–4B.13 as functions of forward speed. The flight conditions correspond to sea level ($\rho = 1.227$ kg/m^3) with zero sideslip and turn rate, from hover to 140 knots. Figures 4B.7 and 4B.8 show the direct longitudinal and lateral derivatives respectively. Figures 4B.9 and 4B.10 show the lateral to longitudinal, and longitudinal to lateral, coupling derivatives respectively. Figures 4B.11 and 4B.12 illustrate the longitudinal and lateral main rotor control derivatives, and Fig. 4B.13 shows the tail rotor control derivatives. As an aid to interpreting the derivative charts, the following points should be noted:

(1) The force derivatives are normalized by aircraft mass, and the moment derivatives are normalized by the moments of inertia.

(2) For the moment derivatives, pre-multiplication by the inertia matrix has been carried out so that the derivatives shown include the effects of the product of inertia I_{xz} (see eqns 4.47–4.51).

(3) The derivative units are as follows:

force/translational velocity	*e.g.* X_u	1/s
force/angular velocity	*e.g.* X_q	m/s. rad
moment/translational velocity	*e.g.* M_u	rad/s. m
moment/angular velocity	*e.g.* M_q	1/s
force/control	*e.g.* $X_{\theta 0}$	m/s^2 rad
moment/control	*e.g.* $M_{\theta 0}$	1/s^2

(4) the force/angular velocity derivatives, as shown in the figures, include the trim velocities, e.g.,

$$Z_q \equiv Z_q + U_e$$

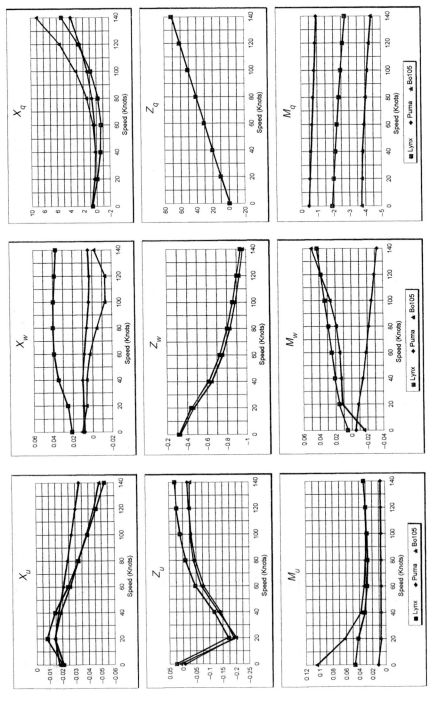

Fig. 4B.7 Stability derivatives – longitudinal

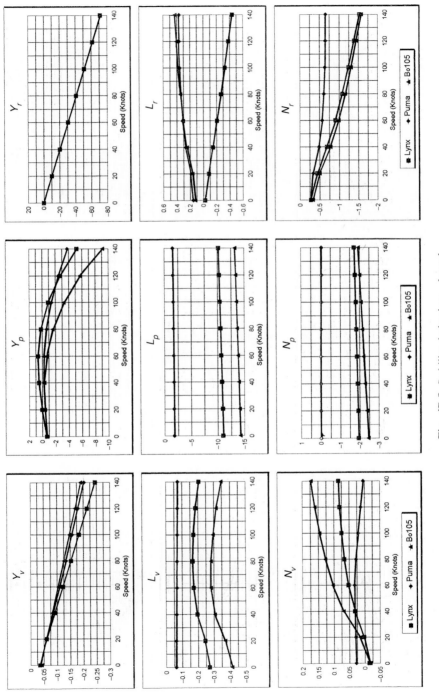

Fig. 4B.8 Stability derivatives – lateral

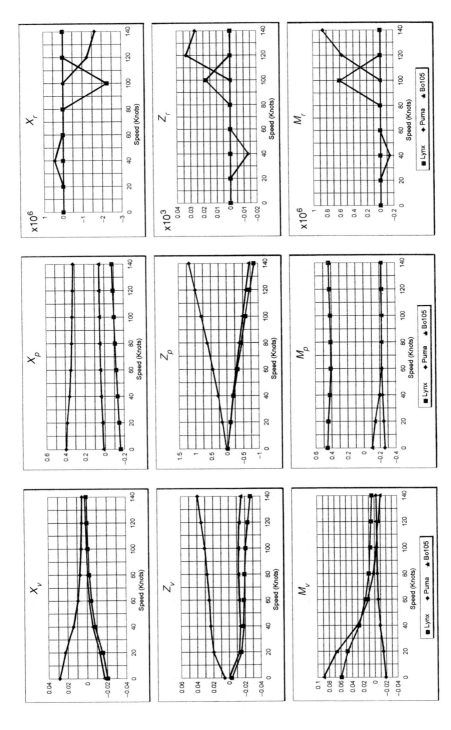

Fig. 4B.9 Stability derivatives – lateral into longitudinal

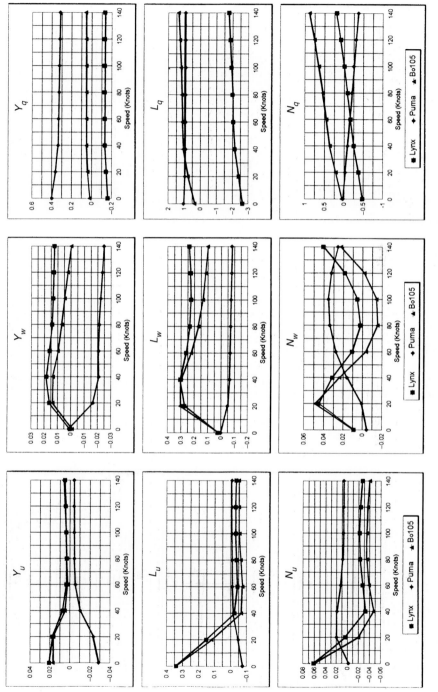

Fig. 4B.10 Stability derivatives – longitudinal into lateral

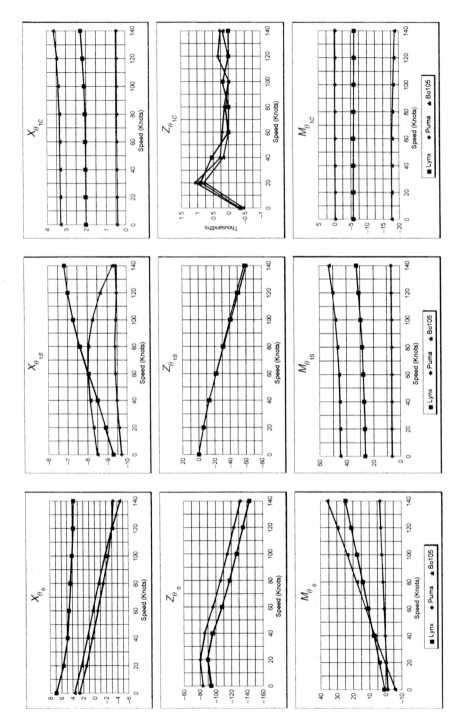

Fig. 4B.11 Control derivatives – main rotor longitudinal

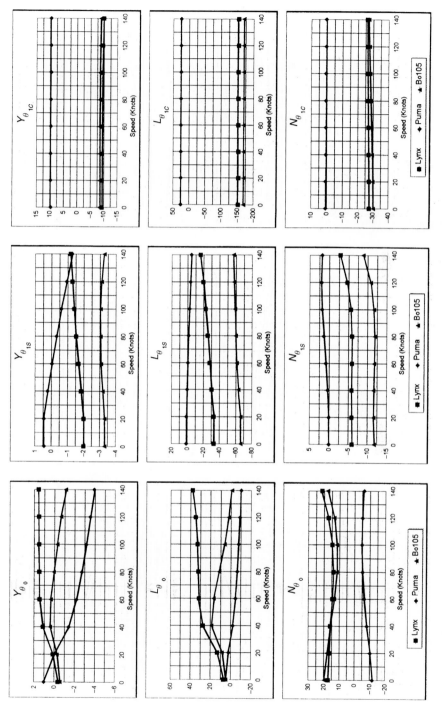

Fig. 4B.12 Control derivatives – main rotor lateral

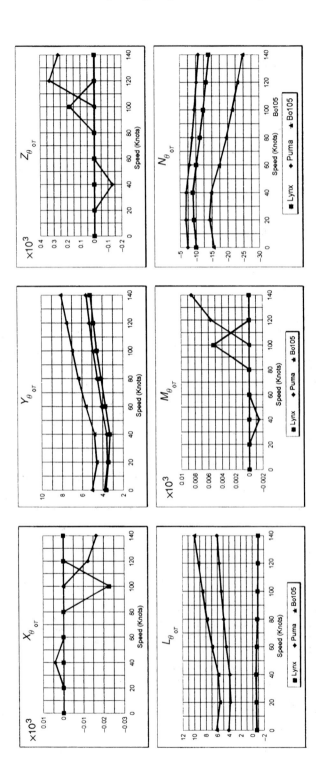

Fig. 4B.13 Control derivatives – tail rotor

4B.3 Tables of stability and control derivatives and system eigenvalues

The stability and control matrices appear in the linearized equations $\frac{dx}{dt} - Ax = Bu$, where, from eqns 4.44 and 4.45, the system matrix A and control matrix B can be written in general form as

$$A = \begin{bmatrix}
X_u & X_w - Q_e & X_q - W_e & -g\cos\Theta_e & X_v + R_e & X_p & 0 & X_r + V_e \\
Z_u + Q_e & Z_w & Z_q + U_e & -g\cos\Phi_e\sin\Theta_e & Z_v - P_e & Z_p - V_e & -g\sin\Phi_e\cos\Theta_e & Z_r \\
M_u & M_w & M_q & 0 & M_v & \begin{matrix}M_p - 2P_e I_{xz}/I_{yy}\\ -R_e(I_{xx}-I_{zz})/I_{yy}\end{matrix} & 0 & \begin{matrix}M_r + 2R_e I_{xz}/I_{yy}\\ -P_e(I_{xx}-I_{zz})/I_{yy}\end{matrix} \\
0 & 0 & \cos\Phi_e & 0 & 0 & 0 & -\sin\Phi_e & 0 \\
Y_u - R_e & Y_w + P_e & Y_q & -g\sin\Phi_e\sin\Theta_e & Y_v & Y_p + W_e & g\cos\Phi_e\cos\Theta_e & Y_r - U_e \\
L'_u & L'_w & L'_q + k_1 P_e - k_2 R_e & 0 & L'_v & L'_p + k_1 Q_e & 0 & L'_r - k_2 Q_e \\
0 & 0 & \sin\Phi_e\tan\Theta_e & \Omega_a\sec\Theta_e & 0 & 1 & 0 & \cos\Phi_e\tan\Theta_e \\
N'_u & N'_w & N'_q - k_1 R_e - k_3 P_e & 0 & N'_v & N'_p + k_3 Q_e & 0 & N'_r - k_1 Q_e
\end{bmatrix}$$

$$B = \begin{bmatrix}
X_{\theta_0} & X_{\theta_{1s}} & X_{\theta_{1c}} & X_{\theta_{0T}} \\
Z_{\theta_0} & Z_{\theta_{1s}} & Z_{\theta_{1c}} & Z_{\theta_{0T}} \\
M_{\theta_0} & M_{\theta_{1s}} & M_{\theta_{1c}} & M_{\theta_{0T}} \\
0 & 0 & 0 & 0 \\
Y_{\theta_0} & Y_{\theta_{1s}} & Y_{\theta_{1c}} & Y_{\theta_{0T}} \\
L'_{\theta_0} & L'_{\theta_{1s}} & L'_{\theta_{1c}} & L'_{\theta_{0T}} \\
0 & 0 & 0 & 0 \\
N'_{\theta_0} & N'_{\theta_{1s}} & N'_{\theta_{1c}} & N'_{\theta_{0T}}
\end{bmatrix}$$

In the following tables, the A and B matrices and the associated eigenvalues of A are listed for the Lynx, Puma and Bo105 at flight speeds from hover to 140 knots in straight and level flight, i.e., $\Omega_a = P_e = Q_e = R_e = 0$. Eigenvalues are given for both with and without longitudinal-lateral/directional couplings.

Note – the Lynx derivatives in the first edition were computed with the tail rotor δ_3 inadvertently set to zero; the new results have $\delta_3 = 45°$, affecting primarily the lateral/directional stability and control derivatives.

LYNX

Lynx V = 0 kts

A matrix

```
-0.0199    0.0215    0.6674 -9.7837 -0.0205  -0.1600  0.0000    0.0000
 0.0237   -0.3108    0.0134 -0.7215 -0.0028  -0.0054  0.5208    0.0000
 0.0468    0.0055   -1.8954  0.0000  0.0588   0.4562  0.0000    0.0000
 0.0000    0.0000    0.9985  0.0000  0.0000   0.0000  0.0000    0.0532
 0.0207    0.0002   -0.1609  0.0380 -0.0351  -0.6840  9.7697    0.0995
 0.3397    0.0236   -2.6449  0.0000 -0.2715 -10.9759  0.0000   -0.0203
 0.0000    0.0000   -0.0039  0.0000  0.0000   1.0000  0.0000    0.0737
 0.0609    0.0089   -0.4766  0.0000 -0.0137  -1.9367  0.0000   -0.2743
```

B matrix

```
  6.9417  -9.2860    2.0164  0.0000
-93.9179  -0.0020   -0.0003  0.0000
  0.9554  26.4011   -5.7326  0.0000
  0.0000   0.0000    0.0000  0.0000
 -0.3563  -2.0164   -9.2862  3.6770
  7.0476 -33.2120 -152.9537 -0.7358
  0.0000   0.0000    0.0000  0.0000
 17.3054  -5.9909  -27.5911 -9.9111
```

eigenvalues (coupled)	eigenvalues (decoupled)
0.2394 ± 0.5337i	0.0556 ± 0.4743i
−0.1703 ± 0.6027i	−0.0414 ± 0.4715i
−0.2451	−0.1843
−0.3110	−0.3127
−2.2194	−2.0247
−10.8741	−11.0182

Lynx V = 20 kts

A matrix

```
-0.0082    0.0254   -0.0685 -9.7868 -0.0158  -0.1480 0.00000    0.0000
-0.1723   -0.4346   10.4965 -0.6792 -0.0150  -0.1044 0.45450    0.0000
 0.0417    0.0157   -2.0012  0.0000  0.0482   0.4441 0.00000    0.0000
 0.0000    0.0000    0.9989  0.0000  0.0000   0.0000 0.00000    0.0464
 0.0173    0.0161   -0.1435  0.0311 -0.0604   0.0308 9.77607  -10.1108
 0.1531    0.2739   -2.4044  0.0000 -0.2439 -10.9208 0.00000   -0.0793
 0.0000    0.0000   -0.0032  0.0000  0.0000   1.0000 0.00000    0.0694
 0.0037    0.0455   -0.3753  0.0000  0.0025  -1.9201 0.00000   -0.4404
```

B matrix

```
  5.6326  -8.9083    2.0273  0.0000
-89.9908  -6.0809    0.0010  0.0000
  3.8558  26.6794   -5.7663  0.0000
  0.0000   0.0000    0.0000  0.0000
  0.1249  -2.0098   -9.3275  3.4515
 13.2029 -32.8252 -153.5913 -0.6907
  0.0000   0.0000    0.0000  0.0000
 16.5240  -5.9080  -27.5007 -9.3029
```

eigenvalues (coupled)	eigenvalues (decoupled)
0.1273 ± 0.5157i	0.0471 ± 0.4396i
−0.0526	−0.0986
−0.2213 ± 0.8272i	−0.1637 ± 0.7956i
−0.3554	−0.3556
−2.4185	−2.1826
−10.8511	−10.9956

Lynx V = 40 kts

A matrix

```
-0.0146    0.0347   -0.5681 -9.7934 -0.0083  -0.1321 0.0000    0.0000
-0.1186   -0.6156   20.6855 -0.5779 -0.0180  -0.2022 0.3519    0.0000
 0.0319    0.0212   -2.1033  0.0000  0.0277   0.4210 0.0000    0.0000
 0.0000    0.0000    0.9994  0.0000  0.0000   0.0000 0.0000    0.0359
 0.0070    0.0184   -0.1303  0.0205 -0.0915   0.5342 9.7869 -20.3077
-0.0255    0.3040   -2.1361  0.0000 -0.1949 -10.7839 0.0000  -0.1441
 0.0000    0.0000   -0.0021  0.0000  0.0000   1.0000 0.0000   0.0590
-0.0325    0.0314   -0.2522  0.0000  0.0316  -1.8857 0.0000  -0.68597
```

B matrix

```
  4.8686  -8.5123     2.0305   0.0000
-95.5241 -12.7586     0.0003   0.0000
  7.2883  27.0667    -5.7827   0.0000
  0.0000   0.0000     0.0000   0.0000
  1.1239  -1.8435    -9.3132   3.3289
 27.3295 -30.1532  -153.4552  -0.6662
  0.0000   0.0000     0.0000   0.0000
 15.9423  -5.8252   -27.2699  -8.9726
```

eigenvalues (coupled)	eigenvalues (decoupled)
0.0878 ± 0.4135i	0.0542 ± 0.3933i
−0.0053	−0.0571
−0.3321 ± 1.2240i	−0.3098 ± 1.1926i
−0.3896	−0.4457
−2.6712	−2.3962
−10.7402	−10.8845

Lynx V = 60 kts

A matrix

```
-0.0243    0.0392   -0.6705 -9.8014 -0.0041  -0.1190 0.0000    0.0000
-0.0467   -0.7285   30.8640 -0.4200 -0.0186  -0.3216 0.3117    0.0000
 0.0280    0.0248   -2.2156  0.0000  0.0159   0.4108 0.0000    0.0000
 0.0000    0.0000    0.9995  0.0000  0.0000   0.0000 0.0000    0.0318
 0.0035    0.0159   -0.1293  0.0133 -0.1228   0.6465 9.7964 -30.5334
-0.0437    0.2611   -2.0532  0.0000 -0.1713 -10.6565 0.0000  -0.2069
 0.0000    0.0000   -0.0014  0.0000  0.0000   1.0000 0.0000   0.0429
-0.0273    0.0109   -0.1661  0.0000  0.0529  -1.8568 0.0000  -0.9039
```

B matrix

```
    4.6289  -8.0560     2.0386    0.0000
 -107.3896 -21.2288     0.0000    0.0000
   10.7004  27.6889    -5.8115    0.0000
    0.0000   0.0000     0.0000    0.0000
    1.4472  -1.6712    -9.3018    3.7509
   31.4636 -27.4424  -153.3177   -0.7505
    0.0000   0.0000     0.0000    0.0000
   14.5826  -5.9178   -27.0369  -10.1087
```

eigenvalues (coupled)	eigenvalues (decoupled)
0.1058 ± 0.3816i	0.0736 ± 0.3823i
−0.0262	−0.0428
−0.4055	−0.4253 ± 1.5923i
−0.4355 ± 1.6130i	−0.4723
−2.9217	−2.6433
−10.6387	−10.7897

Lynx V = 80 kts

A Matrix

```
 -0.0322   0.0403   -0.2262  -9.8081  -0.0021  -0.1086  0.0000    0.0000
 -0.0010  -0.8018   41.0936  -0.2113  -0.0194  -0.4512  0.3223    0.0000
  0.0271   0.0288   -2.3350   0.0000   0.0104   0.4102  0.0000    0.0000
  0.0000   0.0000    0.9995   0.0000   0.0000   0.0000  0.0000    0.0329
  0.0032   0.0143   -0.1287   0.0069  -0.1535   0.2134  9.8028  -40.7844
 -0.0371   0.2344   -1.9959   0.0000  -0.1659 -10.5388  0.0000   -0.2668
  0.0000   0.0000   -0.0007   0.0000   0.0000   1.0000  0.0000    0.0215
 -0.0227   0.0025   -0.0877   0.0000   0.0662  -1.8331  0.0000   -1.0840
```

B Matrix

```
    4.3447  -7.6327     2.0578    0.0000
 -117.7857 -30.3913     0.0000    0.0000
   14.0778  28.5401    -5.8552    0.0000
    0.0000   0.0000     0.0000    0.0000
    1.4988  -1.5282    -9.3201    4.1854
   32.0714 -25.0312  -153.2298   -0.8376
    0.0000   0.0000     0.0000    0.0000
   13.9462  -5.9565   -26.8073  -11.2811
```

eigenvalues (coupled)	eigenvalues (decoupled)
0.1357 ± 0.3772i	0.1037 ± 0.3832i
−0.0330	−0.0396
−0.4035	−0.4582
−0.5151 ± 1.9608i	−0.5095 ± 1.9513i
−3.1945	−2.9182
−10.5556	−10.7176

Lynx V = 100 kts

A Matrix

```
-0.0393    0.0398    0.8831  -9.8103 -0.0010  -0.0997 0.0000    0.0000
 0.0104   -0.8564   51.3352   0.0397 -0.0210  -0.5854 0.3744    0.0000
 0.0279    0.0334   -2.4604   0.0000  0.0075   0.4148 0.0000    0.0000
 0.0000    0.0000    0.9993   0.0000  0.0000   0.0000 0.0000    0.0382
 0.0037    0.0134   -0.1282  -0.0015 -0.1838  -0.8825 9.8032 -51.0333
-0.0327    0.2252   -1.9302   0.0000 -0.1713 -10.4201 0.0000   -0.3253
 0.0000    0.0000    0.0002   0.0000  0.0000   1.0000 0.0000   -0.0040
-0.0219    0.0056   -0.0044   0.0000  0.0751  -1.8067 0.0000   -1.2436
```

B Matrix

```
    4.0394  -7.2845     2.0955    0.0000
 -126.8300 -39.8088     0.0000    0.0000
   17.4865  29.6369    -5.9169    0.0000
    0.0000   0.0000     0.0000    0.0000
    1.5127  -1.4002    -9.4000    4.5569
   32.9346 -22.4516  -153.2494   -0.9119
    0.0000   0.0000     0.0000    0.0000
   14.7283  -5.6161   -26.5849  -12.2824
```

eigenvalues (coupled)	eigenvalues (decoupled)
0.1799 ± 0.3731i	0.1466 ± 0.3847i
−0.0365	−0.0404
−0.3912	−0.4356
−0.5773 ± 2.2781i	−0.5726 ± 2.2763i
−3.4965	−3.2136
−10.4845	−10.6621

Lynx V = 120 kts

A Matrix

```
-0.0460    0.0385    2.7192  -9.8052 -0.0001  -0.0916 0.0000    0.0000
 0.0221   -0.9008   61.5464   0.3205 -0.0236  -0.7219 0.4681    0.0000
 0.0299    0.0380   -2.5919   0.0000  0.0058   0.4225 0.0000    0.0000
 0.0000    0.0000    0.9989   0.0000  0.0000   0.0000 0.0000    0.0477
 0.0043    0.0129   -0.1283  -0.0152 -0.2142  -2.7024 9.7940 -61.2455
-0.0320    0.2281   -1.8534   0.0000 -0.1847 -10.2992 0.0000   -0.3827
 0.0000    0.0000    0.0016   0.0000  0.0000   1.0000 0.0000   -0.0327
-0.0237    0.0187    0.0877   0.0000  0.0811  -1.7721 0.0000   -1.3896
```

B Matrix

```
    3.8024  -7.0223     2.1602    0.0000
 -135.2500 -49.3051     0.0001    0.0000
   20.9344  30.9867    -6.0002    0.0000
    0.0000   0.0000     0.0000    0.0000
    1.5360  -1.2845    -9.5747    4.8851
   34.9038 -19.4471  -153.4332   -0.9776
    0.0000   0.0000     0.0000    0.0000
   17.1838  -4.6280   -26.3702  -13.1671
```

eigenvalues (coupled)	eigenvalues (decoupled)
0.2391 ± 0.3576i	0.2010 ± 0.3787i
−0.0397	−0.0429
−0.3780	−0.4184
−0.6235 ± 2.5705i	−0.6164 ± 2.5713i
−3.8269	−3.5222
−10.4281	−10.6271

Lynx V = 140 kts

A Matrix

```
-0.0525   0.0370    5.2710  -9.7910  0.0000  -0.0838  0.0000   0.0000
 0.0286  -0.9392   71.6880   0.6160 -0.0272  -0.8596  0.6083   0.0000
 0.0328   0.0426   -2.7297   0.0000  0.0047   0.4327  0.0000   0.0000
 0.0000   0.0000    0.9981   0.0000  0.0000   0.0000  0.0000   0.0621
 0.0052   0.0127   -0.1297  -0.0383 -0.2446  -5.2343  9.7720 -71.3836
-0.0338   0.2396   -1.7657   0.0000 -0.2050 -10.1775  0.0000  -0.4394
 0.0000   0.0000    0.0039   0.0000  0.0000   1.0000  0.0000  -0.0629
-0.0269   0.0406    0.1928   0.0000  0.0851  -1.7264  0.0000  -1.5264
```

B Matrix

```
   3.6956  -6.8427     2.2599    0.0000
-143.5034 -58.7853     0.0001    0.0000
  24.4192  32.5904    -6.1083    0.0000
   0.0000   0.0000     0.0000    0.0000
   1.5764  -1.1831    -9.8730    5.1875
  38.1461 -15.8917  -153.8247   -1.0381
   0.0000   0.0000     0.0000    0.0000
  21.4497  -2.7783   -26.1582  -13.9821
```

eigenvalues (coupled)	eigenvalues (decoupled)
0.3123 ± 0.3175i	0.2637 ± 0.3580i
−0.0430	−0.0464
−0.3675	−0.4099
−0.6538 ± 2.8411i	−0.6414 ± 2.8379i
−4.1847	−3.8388
−10.3916	−10.6193

PUMA

Puma V = 0 kts

A matrix

```
-0.0176   0.0076    0.6717  -9.8063  0.0287   0.3966  0.0000   0.0000
-0.0092  -0.3195    0.0126  -0.2803  0.0059   0.0210 -0.4532   0.0000
 0.0113  -0.0057   -0.4506   0.0000 -0.0193  -0.2667  0.0000   0.0000
 0.0000   0.0000    0.9989   0.0000  0.0000   0.0000  0.0000  -0.0467
-0.0287   0.0012    0.3973  -0.0129 -0.0374  -0.6983  9.7953   0.1415
-0.0684  -0.0009    0.9462   0.0000 -0.0491  -1.6119  0.0000   0.0713
 0.0000   0.0000    0.0013   0.0000  0.0000   1.0000  0.0000   0.0285
-0.0082  -0.0050    0.1107   0.0000  0.0249  -0.1361  0.0000  -0.2850
```

B matrix

```
   2.5041  -9.7041   0.4273   0.0000
 -84.7599  -0.0019  -0.0004   0.0000
  -1.4979   6.5240  -0.2873   0.0000
   0.0000   0.0000   0.0000   0.0000
   1.0110   0.4273   9.7042   3.8463
   1.3939   1.0185  23.1286   1.9123
   0.0000   0.0000   0.0000   0.0000
 -12.1328   0.1196   2.7198  -7.6343
```

eigenvalues (coupled)	eigenvalues (decoupled)
0.2772 ± 0.5008i	0.1154 ± 0.3814i
−0.0410 ± 0.5691i	0.0424 ± 0.5019i
−0.2697	−0.2355
−0.3262	−0.3279
−1.2990 ± 0.2020i	−0.6908
	−1.7838

Puma V = 20 kts

A matrix

```
 -0.0143   0.0083   0.3923  -9.8071   0.0226   0.3803   0.0000   0.0000
 -0.1960  -0.4515  10.6478  -0.2534   0.0196   0.1577  -0.3887   0.0000
  0.0070  -0.0096  -0.5195   0.0000  -0.0152  -0.2561   0.0000   0.0000
  0.0000   0.0000   0.9992   0.0000   0.0000   0.0000   0.0000  -0.0396
 -0.0234  -0.0167   0.3609  -0.0097  -0.0592  -0.4329   9.7991 -10.1149
 -0.0391  -0.0440   0.8680   0.0000  -0.0492  -1.6035   0.0000   0.0878
  0.0000   0.0000   0.0010   0.0000   0.0000   1.0000   0.0000   0.0258
  0.0173  -0.0032   0.0482   0.0000   0.0265  -0.1383   0.0000  -0.3526
```

B matrix

```
   1.2644  -9.6244   0.4308   0.0000
 -81.7293  -6.2038   0.0008   0.0000
  -0.6852   6.4851  -0.2897   0.0000
   0.0000   0.00000  0.0000   0.0000
   0.0166   0.4350   9.7707   3.5755
  -0.7952   0.9708  23.2954   1.7777
   0.0000   0.0000   0.0000   0.0000
 -10.7813   0.1326   2.7028  -7.0969
```

eigenvalues (coupled)	eigenvalues (decoupled)
0.1206 ± 0.4429i	0.0809 ± 0.3364i
−0.0685 ± 0.7055i	−0.0433 ± 0.7116i
−0.1089	−0.1491
−0.4635	−0.5737 ± 0.2943i
−1.2663 ± 0.2583i	−0.6908

Puma V = 40 kts

A matrix

−0.0173	0.0091	0.2635	−9.8085	0.0138	0.3575	0.0000	0.0000
−0.1434	−0.6300	20.7701	−0.1884	0.0232	0.2926	−0.2924	0.0000
0.0065	−0.0156	−0.5974	0.0000	−0.0093	−0.2413	0.0000	0.0000
0.0000	0.0000	0.9995	0.0000	0.0000	0.0000	0.0000	−0.0298
−0.0101	−0.0207	0.3368	−0.0054	−0.0867	−0.3036	9.8040	−20.3363
−0.0117	−0.0541	0.8079	0.0000	−0.0503	−1.5802	0.0000	0.1246
0.0000	0.0000	0.0005	0.0000	0.0000	1.0000	0.0000	0.0192
0.0188	0.0118	−0.0232	0.0000	0.0315	−0.1300	0.0000	−0.4998

B matrix

0.2349	−9.5134	0.4313	0.0000
−87.3055	−13.0706	0.0004	0.0000
0.0124	6.4310	−0.2904	0.0000
0.0000	0.0000	0.0000	0.0000
−1.4329	0.2270	9.7444	3.6518
−4.0991	0.4344	23.2602	1.8156
0.0000	0.0000	0.0000	0.0000
−8.7652	0.4391	2.6618	−7.2483

eigenvalues (coupled)	eigenvalues (decoupled)
0.0275 ± 0.3185i	0.0301 ± 0.2944i
−0.0976	−0.1299
−0.1543 ± 0.9181i	−0.1372 ± 0.9193i
−0.9817	−0.6525 ± 0.5363i
−1.0394 ± 0.2798i	−1.7625

Puma V = 60 kts

A matrix

−0.0210	0.0073	0.3765	−9.8099	0.0091	0.3432	0.0000	0.0000
−0.0795	−0.7421	30.8776	−0.0907	0.0250	0.4551	−0.2475	0.0000
0.0066	−0.0200	−0.6761	0.0000	−0.0062	−0.2323	0.0000	0.0000
0.0000	0.0000	0.9996	0.0000	0.0000	0.0000	0.0000	−0.0252
−0.0056	−0.0206	0.3314	−0.0022	−0.1124	−0.4077	9.8067	−30.5902
−0.0068	−0.0541	0.7943	0.0000	−0.0525	−1.5530	0.0000	0.1467
0.0000	0.0000	0.0002	0.0000	0.0000	1.0000	0.0000	0.0092
0.0120	0.0242	−0.0788	0.0000	0.0304	−0.1347	0.0000	−0.5884

B matrix

−0.7331	−9.4443	0.4325	0.0000
−98.2248	−21.6573	0.0000	0.0000
0.6804	6.4103	−0.2917	0.0000
0.0000	0.0000	0.0000	0.0000
−2.2031	−0.0321	9.7074	4.1752
−5.9234	−0.2301	23.2099	2.0758
0.0000	0.0000	0.0000	0.0000
−7.1877	0.9154	2.6188	−8.2872

eigenvalues (coupled)	eigenvalues (decoupled)
0.0016 ± 0.2511i	0.0004 ± 0.2416i
−0.1189	−0.1244
−0.2057 ± 1.0360i	−0.1867 ± 1.0628i
−0.8944 ± 0.7315i	−0.7200 ± 0.7579i
−1.3772	−1.7559

Puma V = 80 kts

A Matrix

```
−0.0242    0.0047    0.7972  −9.8103    0.0069    0.3347    0.0000    0.0000
−0.0477   −0.8162   41.0540    0.0312    0.0274    0.6321   −0.2402    0.0000
 0.0066   −0.0238   −0.7534    0.0000   −0.0047   −0.2273    0.0000    0.0000
 0.0000    0.0000    0.9997    0.0000    0.0000    0.0000    0.0000   −0.0244
−0.0046   −0.0211    0.3279    0.0007   −0.1367   −0.8151    9.8073  −40.8630
−0.0063   −0.0558    0.7861    0.0000   −0.0552   −1.5234    0.0000    0.1586
 0.0000    0.0000    0.0000    0.0000    0.0000    1.0000    0.0000   −0.0031
 0.0088    0.0304   −0.1337    0.0000    0.0251   −0.1448    0.0000   −0.6366
```

B Matrix

```
  −1.7044    −9.4462    0.4351    0.0000
−108.0689   −30.9385    0.0002    0.0000
   1.3741     6.4432   −0.2938    0.0000
   0.0000     0.0000    0.0000    0.0000
  −2.6987    −0.3062    9.6749    4.6260
  −7.1694    −0.9469   23.1637    2.2999
   0.0000     0.0000    0.0000    0.0000
  −6.2441     1.3939    2.5758   −9.1820
```

eigenvalues (coupled)	eigenvalues (decoupled)
−0.0085 ± 0.2074i	−0.0106 ± 0.2030i
−0.1358	−0.1348
−0.1854 ± 1.0546i	−0.1955 ± 1.1176i
−0.9252 ± 1.0503i	−0.7863 ± 0.9666i
−1.5163	−1.7709

Puma V = 100 kts

A Matrix

```
−0.0273    0.0027    1.5563  −9.8089    0.0058    0.3287    0.0000    0.0000
−0.0316   −0.8714   51.2609    0.1683    0.0306    0.8161   −0.2598    0.0000
 0.0067   −0.0275   −0.8289    0.0000   −0.0040   −0.2243    0.0000    0.0000
 0.0000    0.0000    0.9996    0.0000    0.0000    0.0000    0.0000   −0.0264
−0.0042   −0.0221    0.3240    0.0044   −0.1602   −1.5564    9.8054  −51.1379
−0.0061   −0.0590    0.7766    0.0000   −0.0584   −1.4915    0.0000    0.1653
 0.0000    0.0000   −0.0004    0.0000    0.0000    1.0000    0.0000   −0.0171
 0.0076    0.0313   −0.1912    0.0000    0.0176   −0.1551    0.0000   −0.6639
```

B Matrix

```
 -2.5532  -9.4968   0.4400    0.0000
-116.9950 -40.5003  0.0000    0.0000
  2.0420    6.5254  -0.2972    0.0000
  0.0000    0.0000   0.0000    0.0000
 -3.1281   -0.6053   9.6625    5.0042
 -8.3224   -1.7510  23.1411    2.4880
  0.0000    0.0000   0.0000    0.0000
 -5.9403    1.7385   2.5342   -9.9326
```

eigenvalues (coupled)	eigenvalues (decoupled)
−0.0103 ± 0.1760i	−0.0152 ± 0.1772i
−0.1072 ± 1.0231i	−0.1630
−0.1667	−0.1687 ± 1.0864i
−0.9990 ± 1.3006i	−0.8485 ± 1.1691i
−1.6435	−1.8153

Puma V = 120 kts

A Matrix

```
-0.0303   0.0021   2.6514  -9.8054   0.0051   0.3230   0.0000   0.0000
-0.0231  -0.9159  61.4724   0.3121   0.0346   1.0041  -0.3036   0.0000
 0.0069  -0.0314  -0.9017   0.0000  -0.0036  -0.2220   0.0000   0.0000
 0.0000   0.0000   0.9995   0.0000   0.0000   0.0000   0.0000  -0.0309
-0.0041  -0.0234   0.3202   0.0096  -0.1834  -2.6293   9.8007  -61.4031
-0.0057  -0.0630   0.7660   0.0000  -0.0619  -1.4583   0.0000   0.1686
 0.0000   0.0000  -0.0009   0.0000   0.0000   1.0000   0.0000  -0.0318
 0.0072   0.0277  -0.2524   0.0000   0.0087  -0.1614   0.0000  -0.6779
```

B Matrix

```
 -3.1457   -9.5395   0.4479    0.0000
-125.6242 -50.1868   0.0001    0.0000
  2.6325    6.6388  -0.3021    0.0000
  0.0000    0.0000   0.0000    0.0000
 -3.5316   -0.9271   9.6842    5.3451
 -9.4816   -2.6468  23.1589    2.6574
  0.0000    0.0000   0.0000    0.0000
 -6.3276    1.8255   2.4952  -10.6091
```

eigenvalues (coupled)	eigenvalues (decoupled)
−0.0031 ± 0.9409i	−0.0176 ± 0.1596i
−0.0067 ± 0.1538i	−0.0965 ± 0.9659i
−0.2256	−0.2255
−1.0699 ± 1.5057i	−0.9064 ± 1.3729i
−1.7824	−1.9009

Puma V = 140 kts

A Matrix

```
-0.0331    0.0035    4.0531  -9.7998   0.0045    0.3157   0.0000    0.0000
-0.0191   -0.9537   71.6757   0.4557   0.0394    1.1947  -0.3726    0.0000
 0.0072   -0.0357   -0.9712   0.0000  -0.0033   -0.2196   0.0000    0.0000
 0.0000    0.0000    0.9992   0.0000   0.0000    0.0000   0.0000   -0.0380
-0.0040   -0.0245    0.3172   0.0173  -0.2064   -4.0045   9.7927  -71.6501
-0.0050   -0.0672    0.7554   0.0000  -0.0657   -1.4244   0.0000    0.1696
 0.0000    0.0000   -0.0017   0.0000   0.0000    1.0000   0.0000   -0.0465
 0.0071    0.0196   -0.3186   0.0000  -0.0010   -0.1610   0.0000   -0.6825
```

B Matrix

```
  -3.3699   -9.5029    0.4596    0.0000
-134.3683  -59.9275    0.0000    0.0000
   3.1013    6.7591   -0.3088    0.0000
   0.0000    0.0000    0.0000    0.0000
  -3.9131   -1.2629    9.7500    5.6673
 -10.6581   -3.6219   23.2288    2.8177
   0.0000    0.0000    0.0000    0.0000
  -7.5059    1.5310    2.4590  -11.2488
```

eigenvalues (coupled)	eigenvalues (decoupled)
0.1297 ± 0.7846i	0.0516 ± 0.7605i
−0.0005 ± 0.1399i	−0.0187 ± 0.1466i
−0.3385	−0.3756
−1.1238 ± 1.6931i	−0.9604 ± 1.5827i
−1.9435	−2.0409

Bo105

Bo105 V = 0 kts

A matrix

```
-0.0211    0.0113    0.7086  -9.8029  -0.0170    0.0183   0.0000    0.0000
 0.0091   -0.3220   -0.0311  -0.3838  -0.0008   -0.0006   0.4445    0.0000
 0.1045   -0.0151   -3.7472   0.0000   0.0900   -0.0972   0.0000    0.0000
 0.0000    0.0000    0.9990   0.0000   0.0000    0.0000   0.0000    0.0453
 0.0170   -0.0010    0.0182   0.0168  -0.0405   -0.7365   9.7927    0.1017
 0.3402    0.0155    0.3688   0.0000  -0.4114  -14.1949   0.0000    0.1277
 0.0000    0.0000   -0.0017   0.0000   0.0000    1.0000   0.0000    0.0392
 0.0607    0.0088    0.0656   0.0000  -0.0173   -2.4296   0.0000   -0.3185
```

B matrix

```
  3.6533   -8.4769    3.3079    0.0000
-92.9573   -0.0020   -0.0004    0.0000
 -5.6815   44.9965  -17.5587    0.0000
  0.0000    0.0000    0.0000    0.0000
 -0.5527   -3.3079   -8.4770    5.0433
  4.9933  -66.3704 -170.0832    6.1969
  0.0000    0.0000    0.0000    0.0000
 19.7319  -11.8019  -30.2442  -15.4596
```

eigenvalues (coupled)	eigenvalues (decoupled)
−14.2112	−3.8362
−3.8365	0.0343 ± 0.5141i
0.2361 ± 0.5248i	−0.3227
−0.2098 ± 0.5993i	−14.2136
−0.3246 ± 0.0053i	−0.0338 ± 0.5107i
	−0.2728

Bo105 V = 20 kts

A matrix
```
-0.0144    0.0066    0.3366  -9.8046 -0.0124    0.0277 0.0000    0.0000
-0.1988   -0.4579   10.5118  -0.3362 -0.0123   -0.0891 0.3763    0.0000
 0.0625    0.0124   -3.7648   0.0000  0.0673   -0.1342 0.0000    0.0000
 0.0000    0.0000    0.9993   0.0000  0.0000    0.0000 0.0000    0.0384
 0.0162    0.0137    0.0405   0.0129 -0.0579   -0.3853 9.7974  -10.1266
 0.1194    0.3051    0.7778   0.0000 -0.3693  -14.0962 0.0000    0.1694
 0.0000    0.0000   -0.0013   0.0000  0.0000    1.0000 0.0000    0.0343
-0.0200    0.0488    0.2180   0.0000  0.0150   -2.3684 0.0000   -0.4951
```

B matrix
```
  2.3711  -8.2871    3.3178    0.0000
-89.2133  -6.2995    0.0009    0.0000
  0.3168  44.9332  -17.6247    0.0000
  0.0000   0.0000    0.0000    0.0000
 -0.2429  -3.3151   -8.4930    4.6002
  8.5867 -65.9671 -170.3937    5.6525
  0.0000   0.0000    0.0000    0.0000
 17.9170 -11.6194  -29.8891  -14.1015
```

eigenvalues (coupled)	eigenvalues (decoupled)
−14.1327	−3.8506
−3.8553	0.0155 ± 0.3991i
−0.2762 ± 0.9791i	−0.4174
0.0663 ± 0.4906i	−14.1412
−0.4486	−0.2050 ± 0.9845i
−0.0298	−0.0980

Bo105 V = 40 kts

A matrix
```
-0.0185    0.0063    0.2281  -9.8080 -0.0054    0.0399 0.0000    0.0000
-0.1329   -0.6443   20.6582  -0.2187 -0.0136   -0.1716 0.2856    0.0000
 0.0384    0.0140   -3.8153   0.0000  0.0307   -0.1885 0.0000    0.0000
 0.0000    0.0000    0.9996   0.0000  0.0000    0.0000 0.0000   -0.0291
 0.0047    0.0136    0.0523   0.0064 -0.0815   -0.2839 9.8038  -20.3319
-0.0662    0.3002    1.0841   0.0000 -0.3050  -13.9073 0.0000    0.2507
 0.0000    0.0000   -0.0007   0.0000  0.0000    1.0000 0.0000    0.0223
-0.0470    0.0240    0.3649   0.0000  0.0685   -2.2679 0.0000   -0.7706
```

B matrix

```
    1.2496  −8.1445     3.3196    0.0000
  −96.3119 −13.4400     0.0003    0.0002
    6.7362  45.2389   −17.6777    0.0000
    0.0000   0.0000     0.0000    0.0000
    0.4085  −3.1839    −8.4690    4.8124
   18.8799 −63.2958  −170.1924    5.9132
    0.0000   0.0000     0.0000    0.0000
   16.0420 −11.5423   −29.4425  −14.7520
```

eigenvalues (coupled)	eigenvalues (decoupled)
−13.9468	−3.9300
−3.9368	0.0091 ± 0.3172i
−0.4000 ± 1.5964i	−0.5664
−0.5727	−13.9624
0.0100 ± 0.3548i	−0.3743 ± 1.5762i
−0.0013	−0.0485

Bo105 V = 60 kts

A matrix

```
 −0.0259    0.0031     0.5799  −9.8104  −0.0019    0.0469  0.0000    0.0000
 −0.0681   −0.7526    30.8197  −0.0352  −0.0124   −0.2691  0.2558    0.0000
  0.0331    0.0155    −3.8998   0.0000   0.0118   −0.2152  0.0000    0.0000
  0.0000    0.0000     0.9997   0.0000   0.0000    0.0000  0.0000    0.0261
  0.0022    0.0099     0.0520   0.0009  −0.1041   −0.6331  9.8070  −30.5612
 −0.0752    0.2249     1.1333   0.0000  −0.2813  −13.7516  0.0000    0.3075
  0.0000    0.0000    −0.0001   0.0000   0.0000    1.0000  0.0000    0.0036
 −0.0387   −0.0035     0.4570   0.0000   0.1041   −2.1940  0.0000   −0.9847
```

B matrix

```
    0.3538   −8.0368     3.3259    0.0000
 −108.0952  −22.2694     0.0000    0.0000
   12.2033   45.8401   −17.7685    0.0000
    0.0000    0.0000     0.0000    0.0000
    0.3505   −3.0808    −8.4369    5.6547
   16.7547  −61.2387  −169.9602    6.9482
    0.0000    0.0000     0.0000    0.0000
   13.4132  −11.7472   −28.9891  −17.3339
```

eigenvalues (coupled)	eigenvalues (decoupled)
−13.8140	−4.0680
−4.0705	0.0107 ± 0.3015i
−0.4979 ± 2.1473i	−0.6317
−0.6414	−13.8323
0.0107 ± 0.3154i	−0.4864 ± 2.1301i
−0.0185	−0.0353

Bol05 V = 80 kts

A Matrix

```
 -0.0325  -0.0041    1.5364  -9.8084 -0.0003   0.0505 0.0000    0.0000
 -0.0394  -0.8260   41.0254   0.2001 -0.0116  -0.3738 0.2705    0.0000
  0.0309   0.0195   -3.9954   0.0000  0.0029  -0.2254 0.0000    0.0000
  0.0000   0.0000    0.9996   0.0000  0.0000   0.0000 0.0000    0.0276
  0.0026   0.0070    0.0519  -0.0054 -0.1259  -1.5838 9.8046  -40.7909
 -0.0639   0.1723    1.1687   0.0000 -0.2812 -13.6124 0.0000    0.3474
  0.0000   0.0000    0.0006   0.0000  0.0000   1.0000 0.0000   -0.0204
 -0.0347  -0.0153    0.5476   0.0000  0.1283  -2.1328 0.0000   -1.1568
```

B Matrix

```
  -0.6669  -8.0493    3.3486    0.0000
-118.0582 -31.6696    0.0000    0.0000
  17.7191  46.8384  -17.9056    0.0000
   0.0000   0.0000    0.0000    0.0000
   0.0658  -3.0325   -8.4263    6.3678
  11.3877 -59.9090 -169.7559    7.8245
   0.0000   0.0000    0.0000    0.0000
  11.8374 -11.9066  -28.5328  -19.5199
```

eigenvalues (coupled)	eigenvalues (decoupled)
−13.7201	−4.2569
−4.2542	0.0174 ± 0.2997i
−0.5648 ± 2.6403i	−0.6319
−0.6533	−13.7404
−0.0162 ± 0.3051i	−0.5611 ± 2.6299i
−0.0243	−0.0324

Bol05 V = 100 kts

A Matrix

```
 -0.0386  -0.0126    3.1987  -9.7989  0.0007   0.0526 0.0000    0.0000
 -0.0275  -0.8818   51.2163   0.4745 -0.0116  -0.4818 0.3198    0.0000
  0.0300   0.0270   -4.1011   0.0000 -0.0023  -0.2278 0.0000    0.0000
  0.0000   0.0000    0.9995   0.0000  0.0000   0.0000 0.0000    0.0326
  0.0036   0.0047    0.0530  -0.0155 -0.1472  -3.2372 9.7937  -50.9843
 -0.0561   0.1382    1.2229   0.0000 -0.2937 -13.4749 0.0000    0.3785
  0.0000   0.0000    0.0016   0.0000  0.0000   1.0000 0.0000   -0.0484
 -0.0346  -0.0140    0.6470   0.0000  0.1465  -2.0724 0.0000   -1.3068
```

B Matrix

```
  -1.8276  -8.2366    3.3998    0.0000
-126.7841 -41.2529    0.0000    0.0000
  23.5122  48.3289  -18.0972    0.0000
   0.0000   0.0000    0.0000    0.0000
  -0.2703  -3.0357   -8.4657    6.9809
   6.3316 -58.8399 -169.6293    8.5778
   0.0000   0.0000    0.0000    0.0000
  11.8610 -11.6747  -28.0705  -21.3991
```

eigenvalues (coupled)	eigenvalues (decoupled)
−13.6522	−4.5177
−4.5113	0.0363 ± 0.3087i
−0.6103 ± 3.0909i	−0.5764
−0.6048	−13.675
0.0329 ± 0.3096i	−0.6106 ± 3.0871i
−0.0271	−0.0328

Bo105 V = 120 kts

A Matrix

```
-0.0443  -0.0128    5.6197  -9.7796  0.0014   0.0540 0.0000    0.0000
-0.0243  -0.9240   61.3411   0.7760 -0.0126  -0.5914 0.4027    0.0000
 0.0305   0.0386   -4.2184   0.0000 -0.0062  -0.2259 0.0000    0.0000
 0.0000   0.0000    0.9992   0.0000  0.0000   0.0000 0.0000    0.0412
 0.0047   0.0025    0.0553  -0.0320 -0.1684  -5.6461 9.7713  -61.1001
-0.0523   0.1160    1.2962   0.0000 -0.3148 -13.3381 0.0000    0.4042
 0.0000   0.0000    0.0033   0.0000  0.0000   1.0000 0.0000   -0.0794
-0.0367  -0.0019    0.7586   0.0000  0.1612  -2.0062 0.0000   -1.4436
```

B Matrix

```
  -3.1064   -8.6332     3.4922    0.0000
-135.0215  -50.8604     0.0000    0.0000
  29.5811   50.3616   -18.3518    0.0000
   0.0000    0.0000     0.0000    0.0000
  -0.6407   -3.1012    -8.5830    7.5485
   2.2099  -57.8314  -169.6238    9.2753
   0.0000    0.0000     0.0000    0.0000
  13.6405  -10.7871   -27.5920  -23.1392
```

eigenvalues (coupled)	eigenvalues (decoupled)
−13.6137	−4.8670
−4.8596	0.0832 ± 0.3240i
−0.6376 ± 3.5057i	−0.4862
−0.5127	−13.6394
0.0768 ± 0.3230i	−0.6381 ± 3.5072i
−0.0293	−0.0344

Bo105 V = 140 kts

A Matrix

```
-0.0483  -0.0021    9.0876  -9.7448  0.0022   0.0558 0.0000    0.0000
-0.0268  -0.9610   71.3231   1.1306 -0.0149  -0.7019 0.5336    0.0000
 0.0326   0.0498   -4.3502   0.0000 -0.0098  -0.2218 0.0000    0.0000
 0.0000   0.0000    0.9985   0.0000  0.0000   0.0000 0.0000    0.0548
 0.0059  -0.0001    0.0590  -0.0620 -0.1898  -9.0982 9.7302  -71.0636
-0.0511   0.1015    1.3886   0.0000 -0.3455 -13.2145 0.0000    0.4264
 0.0000   0.0000    0.0064   0.0000  0.0000   1.0000 0.0000   -0.1160
-0.0398   0.0226    0.8882   0.0000  0.1732  -1.9288 0.0000   -1.5715
```

B Matrix

```
  -4.4500    -9.2674       3.6495     0.0000
-143.1715   -60.3623       0.0002     0.0000
  35.8480    52.9534     -18.6836     0.0000
   0.0000     0.0000       0.0000     0.0000
  -1.0897    -3.2669      -8.8310     8.1049
  -0.8726   -56.8271    -169.8386     9.9589
   0.0000     0.0000       0.0000     0.0000
  17.8367    -8.8322     -27.0839   -24.8447
```

eigenvalues (coupled)	eigenvalues (decoupled)
−13.6301	−5.2455
−5.2399	0.1573 ± 0.3225i
−0.6430 ± 3.8845i	−0.4285
−0.4482	−13.6595
0.1503 ± 0.3215i	−0.6396 ± 3.8893i
−0.0318	−0.0372

APPENDIX 4C   THE TRIM ORIENTATION PROBLEM

In this section we derive the relationship between the flight trim parameters and the velocities in the fuselage axes system, for use earlier in the chapter. Figure 4C.1 shows the trim velocity vector $\mathbf{V}_{fe}$ of the aircraft with positive components along the fuselage axes directions x, y and z given by U_e, V_e and W_e respectively, where subscript e denotes equilibrium.

The trim condition is defined in terms of the trim velocity V_{fe}, the flight path angle γ_{fe}, the sideslip angle β_e and the angular velocity about the vertical axis Ω_{ae}. The latter plays no part in the translational velocity derivations. The incidence and sideslip angles are defined as

$$\alpha_e = \tan^{-1}\left(\frac{W_e}{U_e}\right) \tag{4C.1}$$

$$\beta_e = \sin^{-1}\left(\frac{V_e}{V_{fe}}\right) \tag{4C.2}$$

The sequence of discrete orientations required to derive the fuselage velocities in terms of the trim variables is shown below in Fig. 4C.2.

The axes are first rotated about the horizontal y-axis through the flight path angle. Next, the axes are rotated through the track angle, positive to port (corresponding to a positive sideslip angle) giving the orientation of the horizontal velocity component relative to the projected aircraft axes. The final two rotations about the Euler pitch and roll angles bring the axes into alignment with the aircraft axes as defined in Chapter 3, Section 3A.1.

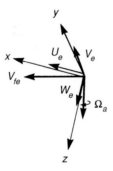

Fig. 4C.1 Flight velocity vector relative to the fuselage axes in trim

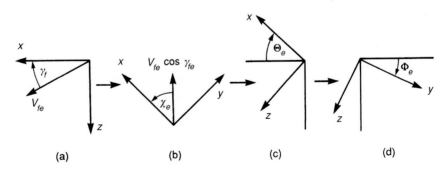

Fig. 4C.2 Sequence of orientations from velocity vector to fuselage axes in trim:
(a) rotation to horizontal through flight path angle γ_f; (b) rotation through track angle χ;
(c) rotation through Euler pitch angle Θ; (d) rotation through Euler roll angle Φ

The trim velocity components in the fuselage-fixed axis system may then be written as

$$U_e = V_{fe} \left(\cos \Theta_e \cos \gamma_{fe} \cos \chi_e - \sin \Theta_e \sin \gamma_e \right)$$

(4C.3)

$$V_e = V_{fe} \left(\cos \Phi_e \cos \gamma_{fe} \sin \chi_e + \sin \Phi_e \left(\sin \Theta_e \cos \gamma_{fe} \cos \chi_e + \cos \Theta_e \sin \gamma_{fe} \right) \right)$$

(4C.4)

$$W_e = V_{fe} \left(-\sin \Phi_e \cos \gamma_{fe} \sin \chi_e + \cos \Phi_e \left(\sin \Theta_e \cos \gamma_{fe} \cos \chi_e + \cos \Theta_e \sin \gamma_{fe} \right) \right)$$

(4C.5)

The track angle is related to the sideslip angle through eqn 4C.4 above. The track angle is then given by the solution of a quadratic, i.e.,

$$\sin \chi_e = -k_{\chi 4} \pm \frac{1}{2} \sqrt{\left(k_{\chi 4}^2 - k_{\chi 5} \right)}$$

(4C.6)

where

$$k_{\chi 4} = \frac{-k_{\chi 2} k_{\chi 3}}{k_{\chi 1}^2 + k_{\chi 2}^2}$$

(4C.7)

$$k_{\chi 5} = \frac{k_{\chi 3}^2 - k_{\chi 1}^2}{k_{\chi 1}^2 + k_{\chi 2}^2}$$

(4C.8)

and the k coefficients are

$$k_{\chi 1} = \sin \Phi_e \sin \Theta_e \cos \gamma_{fe}$$

(4C.9)

$$k_{\chi 2} = \cos \Phi_e \cos \gamma_{fe}$$

(4C.10)

$$k_{\chi 3} = \sin \beta_e - \sin \Phi_e \cos \Theta_e \sin \gamma_{fe}$$

(4C.11)

Only one of the solutions of eqn 4C.6 will be physically valid in a particular case.

Finally, the relationship between the fuselage angular velocities and the trim vertical rotation rate can be easily derived using the same transformation as for the gravitational forces. Hence,

$$P_e = -\Omega_{ae} \sin \Theta_e$$

(4C.12)

$$Q_e = \Omega_{ae} \cos \Theta_e \sin \Phi_e$$

(4C.13)

$$R_e = \Omega_{ae} \cos \Theta_e \cos \Phi_e$$

(4C.14)

Note that for conventional level turns, the roll rate p is small and the pitch and yaw rates are the dominant components, with the ratio between the two dependent upon the trim bank angle. In trim at high climb or descent rates the pitch angle can be significantly different from zero, increasing the roll rate in the turn manoeuvre.

*The German DLR variable stability (fly-by-wire/light) Bo105 S3
(Photograph courtesy of DLR Braunschweig)*

5 Modelling helicopter flight dynamics: stability under constraint and response analysis

Everybody's simulation model is guilty until proved innocent. (Thomas H. Lawrence at the 50th Annual Forum of the AHS, Washington, 1994)

5.1 INTRODUCTION AND SCOPE

Continuing the theme of 'working with models', this chapter deals with two related topics – stability under constraint and response. The response to controls and atmospheric disturbances is the third in the trilogy of helicopter flight mechanics topics; where Chapter 4 focused on trim and natural stability, response analysis is given prime attention in Section 5.3. Understandably, a helicopter's response characteristics can dominate a pilot's impression of flying qualities in applied flying tasks or mission task elements. A pilot may, for example, be able to compensate for reduced stability provided the response to controls is immediate and sufficiently large. He may also be quite oblivious to the 'trim-ability' of the aircraft when active on the controls. What he will be concerned with is the helicopter's ability to be flown smoothly, and with agility if required, from place to place, and also the associated flying workload to compensate for cross-couplings, atmospheric disturbances and poor stability. Quantifying the quality of these response characteristics has been the subject of an extensive international research programme, initiated in the early 1980s. Chapter 6 deals with these in detail, but in the present chapter we shall examine the principal aerodynamic and dynamic effects, mostly unique to the helicopter, which lead to the various response characteristics. Response by its very nature is a nonlinear problem, but insight can be gained from investigating small amplitude response through the linearized equations of perturbed motion. This is particularly true for situations where the pilot, human or automatic, is attempting to constrain the motion – to apply strong control – to achieve a task. We discuss this class of problems in Section 5.2, with an emphasis on the kind of changes in the pilot/vehicle stability that can come about, therefore maintaining some continuity with the material in Chapter 4. Section 5.3 follows with an examination of the characteristics of the helicopter's response to clinical control inputs, and the chapter is concluded with a brief discussion on helicopter response to atmospheric disturbances.

Most of the theory in both Chapters 4 and 5 is concerned with six degree of freedom (DoF) motion, from which considerable insight into helicopter flight dynamics can be gained. However, when the domain of interest on the frequency/amplitude plane includes 'higher order' DoFs associated with the rotors, engines, transmission and flight control system, the theory can become severely limited, and recourse to more complexity is essential. Selected topics that require this greater complexity will be featured in this chapter.

5.2 STABILITY UNDER CONSTRAINT

Both civil and military helicopters are required to operate in confined spaces, often in conditions of poor visibility and in the presence of disturbed atmospheric conditions. To assist the pilot with flight path (guidance) and attitude (stabilization) control, some helicopters are fitted with automatic stability and control augmentation systems (SCAS) that, through a control law, feed back a combination of errors in aircraft states to the rotor controls. The same effect can be achieved by the pilot, and depending on the level of SCAS sophistication and the task the share of the workload falling on the pilot can vary from low to very high. The combination of aircraft, SCAS and pilot, coupled together into a single dynamic system, can exhibit stability characteristics profoundly different than the natural behaviour discussed in detail in Chapter 4. An obvious aim of the SCAS and the pilot is to improve stability and task performance, and in most situations the control strategy to achieve this is conceptually straightforward – proportional control to cancel primary errors, rate control to quicken the response and integral action to cancel steady-state errors. In some situations, however, the natural control option does not always lead to improved stability and response, e.g., tight control of one constituent motion can drive another unstable. It is of interest to be able to predict such behaviour and to understand the physical mechanisms at work. A potential barrier to physical insight in such situations, however, is the increased dimension of the problem. The sketch of the Lynx SCAS in Chapter 3 (see Fig. 3.36) highlights the complexity of a relatively simple automatic system. Integrating the SCAS with the aircraft will lead to a dynamic system of much higher order than that of the aircraft itself; the pilot behaviour will be even more complex and the scope for deriving further understanding of the dynamic behaviour diminishes, as the complexity and order of the integrated system model increases.

One solution to this dilemma was first discussed by Neumark in Ref. 5.1, who identified that it was possible to imagine control so strong that one or more of the motion variables could actually be held at equilibrium or some other prescribed values. The behaviour of the remaining, unconstrained, variables would then be described by a reduced order dynamical system with dimension even less than the order of the natural aircraft. This concept of constrained flight has considerable appeal for the analysis of motion under strong pilot or SCAS control because of the potential for deriving physical understanding from tractable, low-order analytic solutions. Neumark's attention was drawn to solving the problem of speed stability for fixed-wing aircraft operating below minimum drag speed; by applying strong control of flight path with elevator, the pilot effectively drives the aircraft into speed instability. In a later report, Pinsker (Ref. 5.2) demonstrated how, through strong control of roll attitude on fixed-wing aircraft with relatively high values of aileron-yaw, a pilot could drive the effective directional stiffness negative, leading to nose-slice departure characteristics. Clearly, a helicopter pilot has only four controls to cope with 6 DoFs. If the operational situation demands that the pilot constrains some motions more tightly than others, then there is always a question over the stability of the unconstrained motion. If strong control is required to maintain a level attitude for example, then flight path accuracy may suffer and vice versa. Should strong control of some variables lead to a destabilizing of others, then the pilot should soon recognize this and subsequently share his workload between constraining the primary DoFs and compensating for residual motions of the weakly constrained DoFs. This apparent loss of stability can be described as a pilot-induced oscillation (PIO). However, this form of PIO can be insidious in two respects: first, where the unconstrained motion departs slowly, making it difficult for the pilot to identify the departure until well developed, and second, where a rapid loss of stability occurs with only a small increase in pilot gain. With helicopters, the relatively loose coupling of the rotor and fuselage (compared with the wing and fuselage of a fixed-wing aircraft) can exacerbate the problem of constrained flight to the extent that coupled rotor/fuselage motions can occur which have a limited effect on the aircraft flight path, yet cause significant attitude excursions.

In the following analyses, we shall deal with strong attitude control and strong flight path control separately. We shall make extensive use of the theory of weakly coupled systems (Ref. 5.3), which was used to investigate motion under constraint in Ref. 5.4. The theory is described in Appendix 4A and has already been utilized in Chapter 4 in the derivation of approximations for a helicopter's natural stability characteristics. The reader is referred to these sections of the book and the references for further elucidation. The method is ideally suited to the analysis of strongly controlled aircraft, when the dynamic motions tend to split into two types – those under control and those not – with the latter tending to form into new modes with stability characteristics quite different from those of the uncontrolled aircraft. In control theory terms, these modes become the zeros of the closed-loop system in the limit of infinite gain.

5.2.1 Attitude constraint

To illustrate the principal effects of strong attitude control, we first examine pitch control and simplify the analysis by considering the longitudinal subset only. The essential features are preserved under this decoupled approximation. Strong control is assumed to be applied by the pilot or SCAS through simple proportional and rate feedback of pitch attitude to the longitudinal cyclic pitch

$$\theta_{1s} = k_\theta \theta + k_q q, \quad k_\theta, k_q < 0 \tag{5.1}$$

where the gains k are measured in deg/deg (deg/deg s) or rad/rad (rad/rad s). Typical values used in limited authority SCAS systems are $O(0.1)$, whereas pilots can adopt gains an order of magnitude greater than this in tight tracking tasks. We make the assumption that, for high values of gain, the pitch attitude θ and rate q, motions separate off from the flight path translational velocities u and w, leaving these latter variables to dominate the unconstrained modes. This line of argument leads to a partitioning of the longitudinal system matrix (subset of eqn 4.45) in the form

$$\begin{bmatrix} X_u & X_w & X_q - W_e + k_q X_{\theta_{1s}} & -g\cos\Theta_e + k_\theta X_{\theta_{1s}} \\ Z_u & Z_w & Z_q - U_e + k_q Z_{\theta_{1s}} & -g\sin\Theta_e + k_\theta Z_{\theta_{1s}} \\ \hline M_u & M_w & M_q + k_q M_{\theta_{1s}} & k_\theta M_{\theta_{1s}} \\ 0 & 0 & 1 & 0 \end{bmatrix} \tag{5.2}$$

The derivatives have now been augmented by the control terms as shown. Before deriving the approximating polynomials for the low modulus (u, w) and high modulus (θ, q) subsystems, we note that the transfer function of the attitude response to longitudinal cyclic can be written in the form

$$\frac{\theta(s)}{\theta_{1s}(s)} = \frac{(\lambda k_q + k_\theta)R(s)}{D(s)} \tag{5.3}$$

where the polynomial $D(s)$ is the characteristic equation for the longitudinal open-loop eigenvalues. The eigenvalues of the closed-loop system are given by the expression

$$D(\lambda) - (\lambda k_q + k_\theta)R(\lambda) = 0 \tag{5.4}$$

where the polynomial $R(\lambda)$ gives the closed-loop zeros, or the eigenvalues for infinite control gains, and can be written in the expanded form

$$R(\lambda) = \lambda^2 - \left(X_u + Z_w - \frac{1}{M_{\theta_{1s}}}\left(M_u X_{\theta_{1s}} + M_w Z_{\theta_{1s}}\right)\right)\lambda$$

$$+ \left(X_u Z_w - Z_u X_w + \frac{1}{M_{\theta_{1s}}}\left(X_{\theta_{1s}}(Z_u M_w - M_u Z_w) + Z_{\theta_{1s}}(M_u X_w - M_w X_u)\right)\right) \tag{5.5}$$

Equation 5.5 signifies that there are two finite zeros for strong attitude control and, referring to eqn 5.4, we can see a further zero at the origin for strong control of pitch rate. Figure 5.1 shows a sketch of the loci of longitudinal eigenvalues for variations in k_θ and k_q. The forms of the loci are applicable to hingeless rotor configurations, which exhibit two damped aperiodic modes throughout the speed range. Articulated rotor helicopters, whose short-term dynamics are characterized by a short period oscillation, would exhibit a similar pattern of zeros. The two finite zeros shared by both the attitude and rate control loops are given by the roots of eqn 5.5 and both remain stable over the forward flight envelope. For strong control, we can derive approximations for the closed-loop poles of the augmented system matrix in eqn 5.2 from the weakly coupled

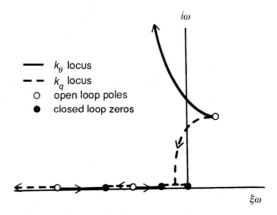

Fig. 5.1 Root loci for longitudinal stability characteristics with varying attitude and rate feedback gains

approximations to the low- (unconstrained motion) and high (constrained motion) order subsystems, as defined by the partitioning shown in eqn 5.2. The general form of the approximating quadratic is written as

$$\lambda^2 + 2\zeta\omega\lambda + \omega^2 = 0 \tag{5.6}$$

For the low-order subsystem, we can write

$$2\zeta\omega = -\left(X_u + Z_w - \frac{1}{M_{\theta_{1s}}}\left(M_u X_{\theta_{1s}} + M_w Z_{\theta_{1s}} + \frac{M_u g}{k_\theta}\right)\right) \tag{5.7}$$

$$\omega^2 = \left(X_u Z_w - Z_u X_w + \frac{1}{M_{\theta_{1s}}}\left(\left(X_{\theta_{1s}} - \frac{g}{k_\theta}\right)\right.\right.$$

$$\left.\left. \times(Z_u M_w - M_u Z_w) + Z_{\theta_{1s}}(M_u X_w - M_w X_u)\right)\right) \tag{5.8}$$

and for the high-order subsystem

$$\lambda^2 - \left(M_q + k_q M_{\theta_{1s}}\right)\lambda - k_\theta M_{\theta_{1s}} = 0 \tag{5.9}$$

The approximations work well for moderate to high levels of feedback gain (k $O(1)$). For weak control (k $O(0.1)$) however, the approximations given above will not produce accurate results. To progress here we should have to derive an analytic extension to the approximations for the open-loop poles derived in Chapter 4. The terms in eqns 5.7 and 5.8 that are independent of the control derivatives reflect the crude, but effective, approximation found by perfectly constraining the pitch attitude. The two subsidences of the low-order approximation, which emerge from strong attitude control, are essentially a speed mode (dominated by u), with almost neutral stability, and a heave mode (dominated by w), with time to half amplitude given approximately by the heave damping Z_w. The strongly controlled mode, with stability given by eqn 5.9, exhibits an increase in frequency proportional to the square root of attitude feedback gain, and in damping proportional to the rate feedback gain. The shift of the (open-loop) pitch and heave modes to (closed-loop) heave and speed modes at high gain is accompanied by a reduction in the stability of these flight trajectory motions, but the overall coupled aircraft/controller system remains stable.

A concern with strong attitude control is actually not so much with the unconstrained motion, but rather with the behaviour of the constrained motion when the presence of higher order modes with eigenvalues further out into the complex plane is taken into account. The problem is best illustrated with reference to strong control of roll attitude and we restrict the discussion to the hover, although the principles again extend to forward flight. Figure 5.2 illustrates the varying stability characteristics of Helisim Lynx in hover with the simple proportional feedback loop defined by

$$\theta_{1c} = k_\phi\phi, \quad k_\phi > 0 \tag{5.10}$$

where, once again, the control may be effected by an automatic SCAS and/or by the pilot. The scale on Fig. 5.2 has been deliberately chosen for comparison with later results. The cluster of pole-zeros around the origin is of little interest in the present discussion; all these eigenvalues lie within a circle of radius <1 rad/s and represent

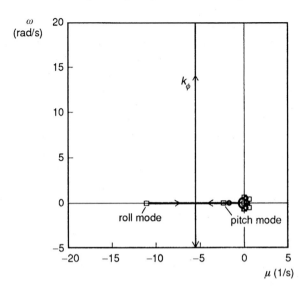

Fig. 5.2 Root loci for varying roll attitude feedback gain for 6 DoF Lynx in hover

the unconstrained, coupled lateral and longitudinal modes, none of which is threatened with instability by the effects of high gain. However, any system modes that lie in the path of the strongly controlled mode (shown as the locus increasing with frequency and offset by approximately $L_p/2$ from the imaginary axis) can have a significant effect on overall stability. Table 5.1 shows a comparison of the coupled system eigenvalues for the 6 DoF 'rigid body dynamics' and 9 DoF coupled dynamics cases, the latter including the flapping Dofs as multi-blade coordinates (see Chapter 3, eqns 3.55–3.63).

Table 5.1 Eigenvalues for 6 DoF and 9 DoF motions – Lynx in hover

Mode type	6 DoF uncoupled lat/long	6 DoF coupled lat/long	9 DoF first-order flap	9 DoF second-order flap
Yaw subsidence	−0.184	−0.245	−0.245	−0.245
Heave subsidence	−0.313	−0.311	−0.311	−0.311
Long	0.056	0.239	0.24	0.24
phugoid	±0.474i	±0.534i	±0.532i	±0.532i
Roll/yaw	−0.041	−0.17	−0.171	−0.17
oscillation	±0.472i	±0.603i	±0.606i	±0.606i
Pitch subsidence	−2.025	−2.219	−2.602	−2.59
Roll subsidence	−11.02	−10.87	−11.744	−13.473
Roll/regressing	–	–	−7.927	−8.272
flap			±10.446i	±10.235i
Coning	–	–	−47.78	−15.854
				±35.545i
Advancing	–	–	–	−15.5
flap				±71.06i

Both first-order and second-order flapping dynamics have been included for comparison, illustrating that the regressing flap mode is reasonably well predicted by neglecting the acceleration effects in the multi-blade coordinates β_{1c} and β_{1s}.

The similar modulus of the roll subsidence and regressing flap modes (both lie roughly on the complex plane circle with radius 10 rad/s), together with the presence of appreciable roll motion in the latter, signals that the use of the 6 DoF weakly coupled approximation for analysing the effects of strong roll control on the stability of lateral motion is unlikely to be valid. Observing the location of the regressing flap mode eigenvalue from Table 5.1, and referring to Fig. 5.2, we can also see that the regressing flap mode lies in the path of the root locus of the strongly controlled roll mode – another clear indication that the situation is bound to change with the addition of higher order flapping dynamics. The root loci of roll attitude feedback to lateral cyclic for the second-order 9 DoF model is shown in Fig. 5.3, revealing that the stability is, as anticipated, changed markedly by the addition of the regressing flap mode. The fuselage eigenvalues no longer coalesce and stiffen the roll axis response into a high-frequency oscillation, but instead the high-gain response energy becomes entrained in a coupled roll/flap mode which becomes unstable at relatively low values of gain, depending on the aircraft configuration. For the Lynx in hover illustrated in Fig. 5.3, the critical value of gain is just below unity. In practice, the response amplitude will be limited by nonlinearities in the actuation system for SCAS operation, or by the pilot reducing his gain and backing out of the control loop. Even for lower values of attitude gain, typical of those found in a SCAS ($O(0.2)$), the stability of the coupled mode will reduce to levels that could cause concern for flight through turbulence. To alleviate this problem, it is a fairly common practice to introduce notch filters into the SCAS that reduce the feedback signals and response around the frequency of the coupled fuselage/rotor modes. A similar problem arises through the coupling of the roll mode with the regressing lag mode when the roll rate feedback gain is increased, developing into a mode that typically has lower damping than the regressing flap mode;

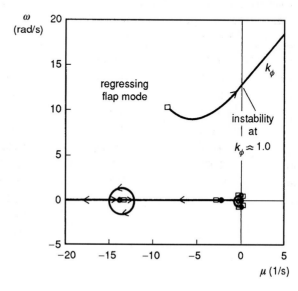

Fig. 5.3 Root loci for varying roll attitude feedback gain for 9 DoF Lynx in hover

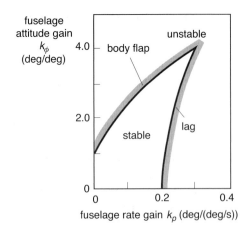

Fig. 5.4 Stability limits as a function of roll feedback gain (Ref. 5.5)

this problem has already been discussed in Chapter 3 (see Section 3.2.1 on lead–lag dynamics) where the work of Curtiss was highlighted (Ref. 5.5). Figure 5.4, taken from Ref. 5.5, shows Curtiss's estimate of the stability boundary for rate and attitude feedback control. The example helicopter in Ref. 5.5 has an articulated rotor, but the level of attitude gain that drives the fuselage–flap mode unstable is very similar to Lynx, i.e., about $1°/°$ for zero rate gain. The level of stabilization through rate damping before the fuselage–lag mode is driven unstable is even lower, according to Curtiss, at $0.2°/°$ s.

High-gain attitude control is therefore seen to present a problem for pilots and SCAS designers. We can obtain some insight into the loss of damping in the roll/flap regressing mode through an approximate stability analysis of the coupled system at the point of instability. We make the assumption that the first-order representation of multi-blade flapping dynamics is adequate for predicting the behaviour of the regressing flap mode. We also neglect the low-modulus fuselage dynamics. From Chapter 3, the equations of motion for the coupled rotor–fuselage angular motion in hover are given by

$$
\begin{bmatrix} -2 & \gamma/8 \\ \gamma/8 & 2 \end{bmatrix} \begin{bmatrix} \beta'_{1c} \\ \beta'_{1s} \end{bmatrix} - \begin{bmatrix} \gamma/8 & -\left(\lambda_\beta^2 - 1\right) \\ -\left(\lambda_\beta^2 - 1\right) & -\gamma/8 \end{bmatrix} \begin{bmatrix} \beta_{1c} \\ \beta_{1s} \end{bmatrix} = \begin{bmatrix} -2\overline{q} + (\theta_{1s} + \overline{p})\gamma/8 \\ 2\overline{p} + (\theta_{1c} + \overline{q})\gamma/8 \end{bmatrix}
$$

$$(5.11)$$

$$
I_{yy}\,\overline{q}' = -\frac{M_\beta}{\Omega^2}\beta_{1c} \tag{5.12}
$$

$$
I_{xx}\,\overline{p}' = -\frac{M_\beta}{\Omega^2}\beta_{1s} \tag{5.13}
$$

where the fuselage rates are normalized by the rotorspeed, i.e.,

$$
\overline{p} = p/\Omega \tag{5.14}
$$

$$
\overline{q} = q/\Omega \tag{5.15}
$$

and the prime denotes differentiation with respect to azimuth angle ψ, e.g.,

$$\beta'_{1c} = \frac{d\beta_{1c}}{d\psi} \tag{5.16}$$

The hub moment about the aircraft centre of mass is approximated by the expression

$$M_\beta = \left(\frac{N_b}{2} K_\beta + Th_R\right) \tag{5.17}$$

which includes the moments due to thrust vector tilt and hub stiffness.

Equations 5.11–5.13 represent a fourth-order coupled system and when the attitude feedback law, given by eqn 5.10, is included, the order of the system increases to 5. For high values of feedback gain we make the assumption that the coupled roll regressing flap mode and the roll subsidence mode in Fig. 5.3 define the high-modulus system so that no further reduction is possible due to the similar modulus of the eigenvalues of these modes. This third-order system has a characteristic equation of the form

$$\lambda^3 + \alpha_2 \lambda^2 + \alpha_1 \lambda + \alpha_0 = 0 \tag{5.18}$$

The condition for stability of the coupled fuselage–flap mode can be written in the form of a determinant inequality (Ref. 5.6)

$$\begin{vmatrix} a_2 & 1 \\ a_0 & a_1 \end{vmatrix} > 0, \quad a_1 a_2 - a_0 > 0 \tag{5.19}$$

After rearrangement of terms and application of reasonably general further approximations, the condition for stability in terms of the roll attitude feedback gain can be written in the form

$$k_\phi < \frac{2\Omega^2}{M_\beta/I_{xx}} \frac{1}{\left(4 + (\gamma/8)^2\right)^2} \tag{5.20}$$

For the Lynx, the value of attitude gain at the stability boundary is estimated to be approximately $0.8°/°$ from eqn 5.20. The relatively high value of hub stiffness reduces the allowable level of feedback gain, but conversely, the relatively high rotorspeed on the Lynx serves to increase the usable range of feedback gain. On the Puma, with its slower turning, articulated rotor with higher Lock number, the allowable gain range increases to about $2°/°$. Slow, stiff rotors would clearly be the most susceptible to the destabilizing effect of roll attitude to lateral cyclic feedback gain.

Later, in Chapter 6, we shall discuss some of the handling qualities considerations for attitude control. The potential for stability problems in high-gain tracking tasks will be seen to be closely related to the shape of the attitude frequency response at frequencies around the upper end of the range which pilots normally operate in closed-loop tasks. The presence of the rotor and other higher order dynamic elements introduces a lag between the pilot and the aircraft's response to controls. The pilot introduces an even further delay through neuro-muscular effects and the combination of the two effects reduces the amplitude, and increases the slope of the phase, of the attitude response at high frequency, both of which can lead to a deterioration in the pilot's perception of aircraft handling. Further discussion on the influence of SCAS gains on rotor–fuselage stability can be found in Refs 5.7 and 5.8; Ref. 5.9 discusses the same problem through the influence of the pilot modelled as a simple dynamic system.

Now we turn to the second area of application of strong control and stability under constraint, where the pilot or automatic controller is attempting to constrain the flight path, to fly along virtual rails in the sky. We shall see that this is possible only at considerable expense to the stability of the unconstrained modes, dominated in this case by the aircraft attitudes.

5.2.2   Flight path constraint

Longitudinal motion
Consider helicopter flight where the pilot is using the cyclic or collective to maintain constant vertical velocity $w_0 = w - U_e\theta$. In the case of zero perturbation from trim, so that the pilot holds a constant flight path, we can write the constraint in the form

$$w = U_e\theta \tag{5.21}$$

When the aircraft pitches, the flight path therefore remains straight. We can imagine control so strong that the dynamics in the heave axis is described by the simple algebraic relation

$$Z_u u + Z_w w \approx 0 \tag{5.22}$$

and the dynamics of surge motion is described by the differential equation

$$\frac{du}{dt} - X_u u + \frac{g}{U_e}w = 0 \tag{5.23}$$

In this constrained flight, heave and surge velocity perturbations are related through the ratio of heave derivatives; thus

$$w \approx -\frac{Z_u}{Z_w}u \tag{5.24}$$

and the single unconstrained degree of surge freedom is described by the first-order system

$$\frac{du}{dt} - \left(X_u + \frac{g}{U_e}\frac{Z_u}{Z_w}\right)u = 0 \tag{5.25}$$

The condition for speed stability can therefore be written in the form

$$X_u + \frac{g}{U_e}\frac{Z_u}{Z_w} < 0 \tag{5.26}$$

For the Lynx, and most other helicopters, this condition is violated below about 60 knots as the changes in rotor thrust with forward speed perturbations become more and more influenced by the strong changes in rotor inflow. Hence, whether the pilot applies cyclic or collective to maintain a constant flight path below approximately minimum power speed, there is a risk of the speed diverging unless controlled. In practice, the pilot would normally use both cyclic and collective to maintain speed and flight path angle at low speed. At higher forward speeds, while the speed mode is never well damped, it becomes stable and is dominated by the drag of the aircraft (derivative X_u). At steep descent angles the control problems become more acute, as the vertical response to both collective and cyclic reverses, i.e., the resulting change in flight path angle following an aft cyclic or up collective step is downward (Refs 5.10, 5.11).

Strong control of the aircraft flight path has an even more powerful influence on the pitch attitude modes of the aircraft, which also change character under vertical motion constraint. Consider the feedback control between vertical velocity and longitudinal cyclic pitch, given by the simple proportional control law

$$\theta_{1s} = k_{w_0} w_0 \tag{5.27}$$

Rearranging the longitudinal system matrix in eqn 4.138 to shift the heave (w_0) variable to the lowest level (i.e., highest modulus) leads to the modified system matrix, with partitioning as shown, namely

$$
\begin{bmatrix}
X_u & X_w - g\cos\Theta_e/U_e & X_q - W_e & g\cos\Theta_e/U_e \\
\hline
Z_u & Z_w & U_e & k_{w_0} Z_{\theta_{1s}} \\
M_u & M_w & M_q & k_{w_0} M_{\theta_{1s}} \\
\hline
Z_u & Z_w & 0 & k_{w_0} Z_{\theta_{1s}}
\end{bmatrix}
\tag{5.28}
$$

The damping in the low-modulus speed mode is approximated by the expression

$$\lambda_1 = \left(X_u + \frac{g}{U_e} \frac{1}{Z_w\, k_{w_0}\, M_{\theta_{1s}}} \left(M_u Z_w - Z_u \left(M_w + k_{w_0} M_{\theta_{1s}} \right) \right) \right) \tag{5.29}$$

and we note that as the feedback gain increases, the stability becomes asymptotic to the value given by the simpler approximation in eqn 5.26. If we make the reasonable assumptions

$$\left| M_u Z_w \right| \gg \left| Z_u M_w \right|, \qquad \left| \frac{g}{U_e} Z_u \right| \gg \left| X_u Z_w \right| \tag{5.30}$$

then the condition for stability is given by the expression

$$k_{w_0} < \frac{M_u Z_w}{Z_u M_{\theta_{1s}}} \tag{5.31}$$

For the Lynx at 40 knots, a gain of 0.35°/(m/s) would be sufficient to destabilize the speed mode.

The approximation for the mid-modulus pitch 'attitude' mode is given by the expression

$$\lambda_2^2 - M_q \lambda_2 + M_{\theta_{1s}} = 0 \tag{5.32}$$

where we have used the additional approximations

$$M_{\theta_{1s}} \gg U_e M_w \qquad (5.33)$$

$$Z_{\theta_{1s}} \approx U_e Z_w \qquad (5.34)$$

In the form of eqn 5.32, the approximation is therefore able to estimate only the location of the pitch mode at infinite gain. The mode is predicted to be stable, with a damping given by the pitch damping derivative and frequency by the pitch control sensitivity derivative. The mode has the appearance of a pendulum mode, with the fuselage rocking beneath the rotor, the latter remaining fixed relative to the flight path. An obvious question that arises from the above analysis relates to how influential this mode is likely to be on handling qualities and, hence, what the character of the mode is likely to be at lower values of gain. Figure 5.5 shows the root loci for the 3 DoF longitudinal modes of the Lynx at a 60 knots level flight condition. The locus near the origin is the eigenvalue of the speed mode already discussed, with the closed-loop zero at the origin (neutral stability). The root moving out to the left on the real axis corresponds to the strongly controlled vertical mode. The finite zero, characterizing the pitch mode,

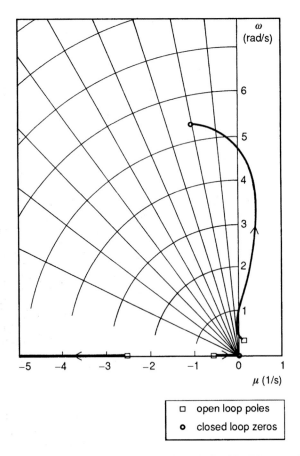

Fig. 5.5 Root loci for varying vertical velocity gain for 6 DoF Lynx at 60 knots

is shown located where approximate theory predicts, with a damping ratio of about 0.2. Of particular interest is the locus of this root as the feedback gain is increased, showing how, for a substantial range of the locus (up to a gain of approximately unity), the mode is driven increasingly unstable. This characteristic is likely to inhibit strong control of the flight path in the vertical plane far more than the loss of stability in the speed mode.

The moderately high frequency of the pitch mode at high values of gain suggests that there is the potential for coupling with the regressing flap mode. Figure 5.6 illustrates the same root loci for the 9 DoF Lynx model; the lower level oscillatory root on Fig. 5.6 represents the Dutch roll oscillation. The roll regressing mode is not shown on the axes range, but is actually hardly affected by the control. What does happen is that the pitch mode now becomes neutrally stable in the limit, with correspondingly more significant excursions into the unstable range at lower values of gain. Even when the open-loop pole has been stabilized through pitch attitude and rate feedback, we can expect the same general trend, with instability occurring at relatively low values of gain. The physical source of the instability is the reduction in incidence stability (M_w) resulting from the control of vertical velocity with cyclic pitch; a positive change

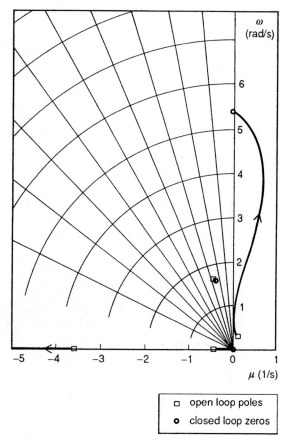

Fig. 5.6 Root loci for varying vertical velocity gain for 9 DoF Lynx at 60 knots

in incidence will indicate an increased rate of descent and will be counteracted by a positive (aft) cyclic, hence reducing static stability. A more natural piloting strategy is to use collective for flight path control and cyclic for speed and attitude control. At low to moderate speeds this strategy will always be preferred, and sufficient collective margin should be available to negate any pilot concerns not to over-torque the rotor/engine/transmission. At higher speeds however, when the power margins are much smaller, and the flight path response to cyclic is stronger, direct control of flight path with cyclic is more instinctive. The results shown in Figs 5.5 and 5.6 suggest a potential conflict between attitude control and flight path control in these conditions. If the pilot tightens up the flight path control (e.g., air-to-air refuelling or target tracking), then a PIO might develop. We have already seen evidence of PIOs from the analysis of strong attitude control in the last section. Both effects are ultimately caused by the loose coupling between rotor and fuselage and are inherently more significant with helicopters than with fixed-wing aircraft. PIOs represent a limit to safe flight for both types of aircraft and criteria are needed to ensure that designs are not PIO prone. This issue will be addressed further in Chapter 6.

The examples discussed above have highlighted a conflict between attitude control (or stability) and flight path control (or guidance) for the helicopter pilot. This conflict is most vividly demonstrated by an analysis of constrained flight in the horizontal plane (Ref. 5.12).

Lateral motion

We consider a simple model of a helicopter being flown along a prescribed flight path in two-dimensional, horizontal flight (Fig. 5.7). The key points can be made with the most elementary simulation of the helicopter flight dynamics (i.e., quasi-steady rotor dynamics). It is assumed that the pilot is maintaining height and balance with collective and pedals, respectively. The equation for the rolling motion is given in terms of the lateral flapping β_{1s}:

$$I_{xx}\ddot{\phi} = -M_\beta\,\beta_{1s} \tag{5.35}$$

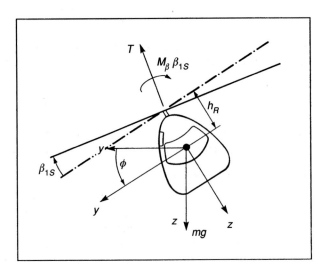

Fig. 5.7 Helicopter force balance in simple lateral manoeuvre

where M_β is the rolling moment per unit flapping given by

$$M_\beta = \left(\frac{N_b K_\beta}{2} + h_R T \right) \tag{5.36}$$

The rotor thrust T varies during the manoeuvre to maintain horizontal flight. As described in Chapter 3, the hub stiffness K_β can be written in terms of the flap frequency ratio λ_β^2, flap moment of inertia I_β and rotorspeed Ω, in the form

$$K_\beta = \left(\lambda_\beta^2 - 1 \right) I_\beta \Omega^2 \tag{5.37}$$

The equations of force balance in earth axes can be written as

$$T \cos (\phi - \beta_{1s}) = mg \tag{5.38}$$

$$T \sin (\phi - \beta_{1s}) = m\ddot{y} \tag{5.39}$$

where $y(t)$ is the lateral flight path displacement. Combining eqns 5.35, 5.38 and 5.39, and assuming small roll and lateral flapping angles (including constant rotor thrust), we obtain the second-order equation for the roll angle ϕ

$$\frac{d^2\phi}{dt^2} + \omega_\phi^2 \phi = \omega_\phi^2 v \tag{5.40}$$

Where v is the normalized sideforce (i.e., lateral acceleration) given in linearized form by

$$v = \ddot{y}/g \tag{5.41}$$

and the 'natural' frequency ω_ϕ is related to the rotor moment coefficient by the expression

$$\omega_\phi = \sqrt{\frac{M_\beta}{I_{xx}}} = \frac{1}{\sqrt{\tau_\beta \tau_p}} \tag{5.42}$$

The rotor and fuselage roll time constants are given in terms of more fundamental rotor parameters, the rotor Lock number (γ) and the roll damping derivative (L_p)

$$\tau_\beta = \frac{16}{\gamma\Omega}, \qquad \tau_p = -\frac{1}{L_p} \tag{5.43}$$

The reader should note that the assumption of constant frequency ω_ϕ implies that the thrust changes are small compared with the hub component of the rotor moment. The frequency ω_ϕ is equal to the natural frequency of the roll-regressive flap mode discussed in Chapters 2 and 4.

Equation 5.40 holds for general small amplitude lateral manoeuvres and can be used to estimate the rotor forces and moments, hence control activity, required to fly a manoeuvre characterized by the lateral flight path $y(t)$. This represents a simple case of so-called inverse simulation (Ref. 5.13), whereby the flight path is prescribed and the equations of motion solved for the loads and controls. A significant difference between rotary and fixed-wing aircraft modelled in this way is that the inertia term in eqn 5.40 vanishes for fixed-wing aircraft, with the sideforce then being simply proportional to the roll angle. For helicopters, the loose coupling between rotor and fuselage leads to the presence of a mode, with dynamics described by eqn 5.40, with frequency ω_ϕ, representing an oscillation of the aircraft relative to the rotor, while the rotor maintains the prescribed orientation in space. We have already seen evidence of a similar mode in the previous analysis of constrained vertical motion. Oscillations of the fuselage in this mode, therefore, have no effect on the flight path of the aircraft. It should be noted that this 'mode' is not a feature of unconstrained flight, where the two natural modes are a roll subsidence (magnitude L_p) and a neutral mode (magnitude 0), representing the indifference of the aircraft dynamics to heading or lateral position. The degree of excitation of the 'new' free mode depends upon the frequency content of the flight path excursions and hence the sideforce v. For example, when the prescribed flight paths are genuinely orthogonal to the free oscillation (i.e., combinations of sine waves), then the response will be uncontaminated by the free oscillation. In practice, slalom-type manoeuvres, while similar in character to sine waves, can have significantly different load requirements at the turning points, and the scope for excitation of the free oscillation is potentially high. A further important point to note about the character of the solution to eqn 5.40 is that as the frequency of the flight path approaches the natural frequency ω_ϕ, the roll angle approaches a 'resonance' condition. To understand what happens in practice, we must look at the equation for 'forward' rather than 'inverse' simulation. This can be written in terms of the lateral cyclic control input θ_{1c} forcing the flight path sideforce v, in the approximate form

$$\frac{d^2v}{dt^2} - L_p \frac{dv}{dt} = \left\{ \frac{d^2\theta_{1c}}{dt^2} + \omega_\phi^2 \theta_{1c} \right\} \tag{5.44}$$

The derivation of this equation depends on the assumption that the rotor responds to control action and fuselage angular rate in a quasi-steady manner, taking up a new disc tilt instantaneously. In reality the rotor responds with a time constant equal to $(16/\gamma\Omega)$, but for the purposes of the present argument this delay will be neglected. The presence of the control acceleration term on the right-hand side of eqn 5.44 is crucial to what happens close to the natural frequency. In the limit, when the input frequency is at the natural frequency, the flight path response is zero due to the cancelling of the control terms, hence the implication of the theoretical artefact, in eqn 5.40, that the roll angle would grow unbounded at that frequency. As the pilot moves his stick at the critical frequency, the rotor disc remains horizontal and the fuselage wobbles beneath. We found a similar effect for longitudinal motion. For cyclic control inputs at slightly lower frequencies, the sideforces are still very small and large control displacements are required to generate the turning moments. Stick movements at frequencies slightly higher than ω_ϕ produce small forces of the opposite sign, acting in the wrong direction. Hence, despite intense stick activity the pilot may not be able to fly the desired track. The difference between what the pilot can do and what he is trying to do increases sharply with the severity of the desired manoeuvre, and the upper limit of what can usefully or safely be accomplished, in terms of task 'bandwidth', is determined by the frequency ω_ϕ. For current helicopters, the natural frequency ω_ϕ varies from 6 rad/s for low hinge offset, slowly rotating rotors, to 12 rad/s for hingeless rotors with higher rotor speeds.

Two questions arise out of the above simple analysis. First, as the severity of the task increases, what influences the cut-off frequency beyond which control activity becomes unreasonably high and hence can this cut-off frequency be predicted? Second, how does the pilot cope with the unconstrained oscillations, if indeed they manifest themselves in practice. A useful parameter in the context of the first question is the ratio of aircraft to task natural frequencies. The task natural frequency ω_t can, in general, be derived from a frequency analysis of the flight path variation, but for simple slalom manoeuvres the value is approximately related to the inverse of the task time. It is suggested in Ref. 5.12 that a meaningful upper limit to task frequency can be written in the form

$$\frac{\omega_\varphi}{\omega_t} > 2n_v \tag{5.45}$$

where n_v is the number of flight path changes required in a given task. A two-sided slalom, for example, as illustrated in Fig. 5.8, contains five such distinct changes; hence at a minimum, for slalom manoeuvres

$$\frac{\omega_\varphi}{\omega_t} > 10 \tag{5.46}$$

This suggests that a pilot flying a reasonably agile aircraft, with $\omega_\phi = 10$ rad/s, could be expected to experience control problems when trying to fly a two-sided slalom in less than about 6 s. A pilot flying a less agile aircraft ($\omega_\phi = 7$ rad/s) might experience similar control problems in a 9-s slalom. This 50% increase in usable performance for an agile helicopter clearly has very important implications for military and some civil operations.

Figures 5.9(a)–(c) show results from an inverse simulation of Helisim Lynx fitted with different rotor types flying a slalom mission task element (Ref. 5.12). Three rotor configurations are shown on the figures corresponding to $\omega_\phi = 11.8$ (standard Lynx), $\omega_\phi = 7.5$ (articulated) and $\omega_\phi = 4.5$ (teetering). The aspect ratio of the slalom, defined as the overall width to length, is 0.077, the maximum value achievable by the teetering rotor configuration before the lateral cyclic reaches the control limits; the flight speed is 60 knots. Figures 5.9(a) and (b) show comparisons of the roll attitude and rate responses respectively. The attitude changes, not surprisingly, are very similar for the three cases,

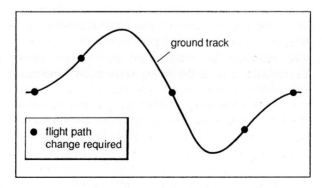

Fig. 5.8 Flight path changes in a slalom manoeuvre (Ref. 5.12)

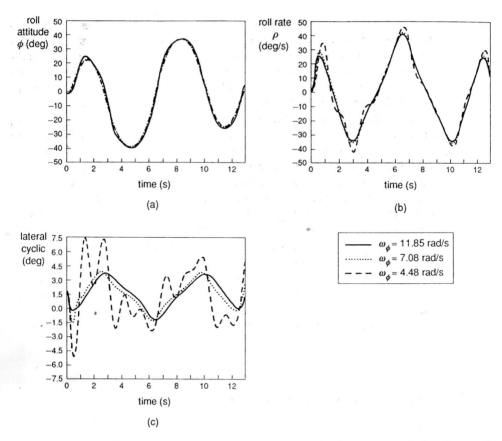

Fig. 5.9 Results from Helisim Lynx constrained to fly a lateral slalom (Ref. 5.12): (a) roll attitude; (b) roll rate; (c) lateral cyclic

as are the rates, although we can now perceive the presence of higher frequency motion in the signal for the teetering and articulated rotors. For the teetering rotor, roll rate peaks some 10 – 20% higher than required with the hingeless rotor can be observed – entirely a result of the component of the free mode in the aircraft response. Figure 5.9(c) illustrates the lateral cyclic required to fly the 0.077 slalom, the limiting aspect ratio for the teetering configuration. The extent of the excitation of the roll oscillation for the different cases, and hence the higher frequency 'stabilization' control inputs, is shown clearly in the time histories of lateral cyclic. The difference between the three rotor configurations is now very striking. The Lynx, with its standard hingeless rotor, requires 30% of maximum control throw, while the articulated rotor requires slightly more at about 35%.

    In Ref. 5.12, the above analysis is extended to examine pilot workload metrics based on control activity. The premise is that the conflict between guidance and stabilization is a primary source of workload for pilots as they attempt to fly manoeuvres beyond the critical aircraft/task bandwidth ratio. Both time and frequency domain workload metrics are discussed in Ref. 5.12, and correlation between inverse simulation and Lynx flight test results is shown to be good; the limiting slalom for both cases was about 0.11. This line of research to determine reliable workload metrics for predicting critical flying qualities boundaries is, in many ways, still quite formative, and detailed coverage, in this book, of the current status of the various approaches is therefore considered to be inappropriate.

The topic of stability under constraint is an important one in flight dynamics, and the examples described in this section have illustrated how relatively simple analysis, made possible by certain sensible approximations, can sometimes expose the physical nature of a stability boundary. This appears fortuitous but there is actually a deeper underlying reason why these simple models work so well in predicting problems. Generally, the pilot will fly a helicopter using a broad range of control inputs, in terms of both frequency and amplitude, to accomplish a task, often adopting a control strategy that appears unnecessarily complex. There is some evidence that this strategy is important for the pilot to maintain a required level of attention to the flying task, and hence, paradoxically, plentiful spare capacity for coping with emergencies. Continuous exercise of a wide repertoire of control strategies is therefore a sign of a healthy situation with the pilot adopting low to moderate levels of workload to stay in command. A pilot experiencing stability problems, particularly those that are self-induced, will typically concentrate more and more of his or her effort in a narrow frequency range as he or she becomes locked into what is effectively a man–machine limit cycle. Again paradoxically, these very structured patterns of activity in pilot control activity are usually a sign that handling qualities are deteriorating and workload is increasing.

5.3   ANALYSIS OF RESPONSE TO CONTROLS

5.3.1   General

Previous sections have focused on trim and stability analysis. The analysis of flight behaviour following control or disturbance inputs is characterized under the general heading 'Response', and is the last topic of this series of modelling sections. Along with the trim and stability analysis, response forms a bridge between the model building activities in Chapter 3 and the flying qualities analysis of Chapters 6 and 7. In the following sections, results will be presented from so-called system identification techniques, and readers unfamiliar with these methods are strongly encouraged to devote some time to familiarizing themselves with the different tools (Ref. 5.14). Making sense of helicopter dynamic flight test data in the validation context requires a combination of experience (e.g., knowing what to expect) and analysis tools that help to isolate cause and effect, and hence provide understanding. System identification methods provide a rational and systematic approach to this process of gaining better understanding.

Before proceeding with a study of four different response topics, we need to recall the basic equations for helicopter response, given in earlier chapters and Appendix 4A. The nonlinear equations for the motion of the fuselage, rotor and other dynamic elements, combined into the state vector $\mathbf{x}(t)$, in terms of the applied controls $\mathbf{u}(t)$ and disturbances $\mathbf{f}(t)$, can be written as

$$\frac{d\mathbf{x}}{dt} = \mathbf{F}\big(\mathbf{x}(t), \mathbf{u}(t), \mathbf{f}(t); t\big) \tag{5.47}$$

with solution as a function of time given by

$$\mathbf{x}(t) = \mathbf{x}(0) + \int_0^t \mathbf{F}(\mathbf{x}(\tau), \mathbf{u}(\tau), \mathbf{f}(\tau); \tau) \, d\tau \qquad (5.48)$$

In a forward simulation, eqn 5.48 is solved numerically by prescribing the form of the evolution of $\mathbf{F}(t)$ over each time interval and integrating. For small enough time intervals, a linear or low-order polynomial form for $\mathbf{F}(t)$ generally gives rapid convergence. Alternatively, a process of prediction and correction can be devised by iterating on the solution at each time step. The selection of which technique to use will usually not be critical. Exceptions occur for systems with particular characteristics (Ref. 5.15), leading to premature numerical instabilities, largely determined by the distribution of eigenvalues; inclusion of rotor and other higher order dynamic modes in eqn 5.47 can sometimes lead to such problems and care needs to be taken to establish a sufficiently robust integration method, particularly for real-time simulation, when a constraint will be to achieve the maximum integration cycle time. We will not dwell on these clearly important issues here, but refer the reader to any one of the numerous texts on numerical analysis.

The solution of the linearized form of eqn 5.47 can be written in either of two forms (see Appendix 4A)

$$\mathbf{x}(t) = \mathbf{Y}(t)\mathbf{x}_0 + \int_0^t \mathbf{Y}(t - \tau)\big(\mathbf{B}\mathbf{u}(\tau) + \mathbf{f}(\tau)\big) \, d\tau \qquad (5.49)$$

$$\mathbf{x}(t) = \sum_{i=1}^n \left[(\mathbf{v}_i^\mathrm{T}\mathbf{x}_0) \exp(\lambda_i t) + \int_0^t \left(\mathbf{v}_i^\mathrm{T}\big(\mathbf{B}\mathbf{u}(\tau) + \mathbf{f}(\tau)\big) \exp[\lambda_i(t - \tau)] \right) d\tau \right] \mathbf{w}_i$$

$$(5.50)$$

where the principal matrix solution $\mathbf{Y}(t)$ is given by

$$\mathbf{Y}(t) = \mathbf{0}, \quad t < 0, \qquad \mathbf{Y}(t) = \mathbf{W} \, \text{diag}[\exp(\lambda_i^f)]\mathbf{V}^\mathrm{T}, \quad t \geq 0 \qquad (5.51)$$

$\mathbf{W}$ is the matrix of right-hand eigenvectors of the system matrix $\mathbf{A}$, $\mathbf{V}^\mathrm{T} = \mathbf{W}^{-1}$ is the matrix of eigenvectors of $\mathbf{A}^\mathrm{T}$, and λ_i are the corresponding eigenvalues; $\mathbf{B}$ is the control matrix. The utility of the linearized response solutions depends on the degree of nonlinearity and the input and response amplitude. In general terms, the linear formulation is considerably more amenable to analysis, and we shall regularly use linear approximations in the following sections to gain improved understanding. In particular, the ability to estimate trends through closed-form analytic solutions, exploited fully in the analysis of stability, highlights the power of linear analysis and, unless a nonlinearity is obviously playing a significant role, equivalent linear systems analysis is always preferred in the first instance.

It is inevitable that the following treatment has to be selective; we shall examine response characteristics in different axes individually, concerning ourselves chiefly with direct response to controls. In several cases, comparisons between flight and Helisim simulation are shown and reference is made to the AGARD Working Group 18

(Rotorcraft System Identification) flight test databases for the DRA Puma and DLR Bo105 helicopters (Ref. 5.14).

5.3.2   Heave response to collective control inputs

Response to collective in hover

In this first example, we examine in some detail the apparently straightforward case of a helicopter's vertical response to collective in hover. In both Chapters 2 and 4 we have already discussed the quasi-steady approximation for helicopter vertical motion given by the first-order equation in vertical velocity

$$\dot{w} - Z_w w = Z_{\theta_0} \theta_0 \tag{5.52}$$

where the heave damping and control sensitivity derivatives are given from momentum theory in terms of the blade loading M_a/A_b, tip speed Ω and hover-induced velocity λ_0 (or thrust coefficient C_T), by the expressions (see Chapter 4)

$$Z_w = -\frac{2a_0 A_b \rho(\Omega R)\lambda_0}{(16\lambda_0 + a_0 s)M_a} \tag{5.53}$$

$$Z_{\theta_0} = -\frac{8}{3}\frac{a_0 A_b \rho(\Omega R)^2 \lambda_0}{(16\lambda_0 + a_0 s)M_a} \tag{5.54}$$

A comparison of the vertical acceleration response to a collective step input derived from eqn 5.52 with flight measurements on the DRA research Puma is shown in Fig. 5.10. It can be seen that the quasi-steady model fails to capture some of the detail in the response shape in the short term, although the longer term decay is reasonably well predicted. For low-frequency collective inputs, the quasi-steady model is expected to give fairly high fidelity for handling qualities evaluations, but at moderate to high frequencies, the fidelity will be degraded. In particular, the transient thrust peaks observed in response to sharp collective inputs will be smoothed over. The significance of this effect was first examined in detail by Carpenter and Fridovitch in the

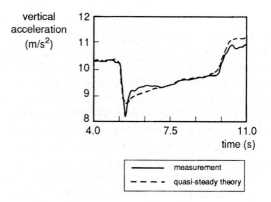

Fig. 5.10 Comparison of quasi-steady theory and flight measurement of vertical acceleration response to a step collective pitch input for Puma in hover (Ref. 5.16)

early 1950s, in the context of the performance characteristics of rotorcraft during jump take-offs (Ref. 5.17). Measurements were made of rotor thrust following the application of sharp and large collective inputs and compared with results predicted by a dynamic rotor coning/inflow nonlinear simulation model. The thrust changes T were modelled by momentum theory, extended to include the unsteady effects on an apparent mass of air m_{am}, defined by the circumscribed sphere of the rotor.

$$T = m_{am} \frac{dv_i}{dt} + 2\pi R^2 \rho v_i \left(v_i - w + \frac{2}{3} R \frac{d\beta}{dt} \right) \tag{5.55}$$

where

$$m_{am} = 0.637 \rho \frac{4}{3} \pi R^3 \tag{5.56}$$

In eqn 5.55 we have added the effect of aircraft vertical motion w, not included in the test stand constraints in Ref. 5.17; v_i is the induced velocity and β the blade flapping. Thrust overshoots of nearly 100% were measured and fairly well predicted by this relatively simple theory, with the inflow build-up, as simulated by eqn 5.55, accounting for a significant proportion of this effect. A rational basis for this form of inflow modelling and the associated azimuthal non-uniformities first appeared in the literature in the early 1970s, largely in the context of the prediction of hub moments (Ref. 5.18), and later with the seminal work of Pitt and Peters in Ref. 5.19. The research work on dynamic inflow by Pitt and Peters, and the further developments by Peters and his co-workers, has already received attention in Chapters 3 and 4 of this book. Here we observe that Ref. 5.19 simulated the rotor loading with a linear combination of polynomial functions that satisfied the rotor blade tip boundary conditions and also satisfied the underlying unsteady potential flow equation. The Carpenter–Fridovitch apparent mass approximation was validated by Ref. 5.19, but a 'corrected' and reduced value was proposed as an alternative that better matched the loading conditions inboard on the rotor.

Based on the work of Refs 5.17 and 5.19, an extensive analysis of the flight dynamics of helicopter vertical motion in hover, including the effects of aircraft motion, rotor flapping and inflow, was conducted by Chen and Hindson and reported in Ref. 5.20. Using a linearized form of eqn 5.55 in the form

$$\delta T = m_{am} \dot{v}_i + 2\pi R^2 \rho \left\{ 2\lambda_0 (2v_i - w) + \frac{2}{3} R \dot{\beta} \right\} \tag{5.57}$$

Chen and Hindson predicted the behaviour of the integrated 3 DoF system and presented comparisons with flight test data measured on a CH-47 helicopter. The large transient thrust overshoots were confirmed and shown to be strong functions of rotor trim conditions and Lock number. The theory in Ref. 5.20 was developed in the context of evaluating the effects of rotor dynamics on the performance of high-gain digital flight control, where dynamic behaviour over a relatively wide frequency range would potentially affect the performance of the control system. One of the observations in Ref. 5.20 was related to the very short-term effect of blade flapping on the fuselage response. Physically, as the lift develops on a rotor blade, the inertial reaction at the hub depends critically on the mass distribution relative to the aerodynamic centre. For the case where the inertial reaction at the hub is downward for an increased lift, the aircraft normal acceleration to collective pitch transfer function will exhibit a so-called non-minimum phase characteristic; although the eventual response is in the same direction as the input, the initial response is in the opposite direction. One of the consequences of very-high-gain feedback control with such systems is instability. In practice, the gain level necessary to cause concern from this effect is likely to be well outside the range required for control of helicopter vertical motion.

An extensive comparison of the behaviour of the Chen–Hindson model with flight test data was conducted by Houston on the RAE research Puma in the late 1980s and reported in Refs 5.16, 5.21–5.23, and we continue this case study with an exposition of Houston's research findings. They demonstrate the utility of applied system identification, and also highlight some of the ever-present pitfalls. The linearized derivative equations of motion for the 3 DoF – vertical velocity w, uniform inflow v_i and rotor coning β_0 – can be written in the form (Ref. 5.20)

$$\frac{d}{dt}\begin{bmatrix} v_i \\ \beta_0 \\ \dot{\beta}_0 \\ w \end{bmatrix} - \begin{bmatrix} I_{v_i} & 0 & I_{\dot{\beta}_0} & I_w \\ 0 & 0 & 1 & 0 \\ F_{v_i} & F_{\beta_0} & F_{\dot{\beta}_0} & F_w \\ Z_{v_i} & Z_{\beta_0} & Z_{\dot{\beta}_0} & Z_w \end{bmatrix}\begin{bmatrix} v_i \\ \beta_0 \\ \dot{\beta}_0 \\ w \end{bmatrix} = \begin{bmatrix} I_{\theta_0} \\ 0 \\ F_{\theta_0} \\ Z_{\theta_0} \end{bmatrix}[\theta_0] \tag{5.58}$$

where the inflow I, coning F and heave Z derivatives are given by the expressions

$$I_{v_i} = -4k_1\left(\lambda_0 + \frac{a_0 s}{16}\right)C_0, \qquad I_{\dot{\beta}_0} = -\frac{4}{3}Rk_1\left(\lambda_0 + \frac{a_0 s}{8}\right)C_0 = -\frac{2R}{3}I_w \tag{5.59}$$

$$F_w = -\frac{\Omega\gamma}{k_2 R}\left(\frac{1}{6} - \frac{N_b M_\beta}{4M_a R}\right), \qquad F_{\beta_0} = -\frac{\Omega^2}{k_2}, \qquad F_{\dot{\beta}_0} = -\frac{\Omega\gamma}{k_2}\left(\frac{1}{8} - \frac{N_b M_\beta}{6M_a R}\right) \tag{5.60}$$

$$Z_{v_i} = -Z_w = \frac{N_b \Omega\gamma}{k_2 R M_a}\left(\frac{I_\beta}{4R} - \frac{M_\beta}{6}\right), \qquad Z_{\beta_0} = -\frac{N_b M_\beta \Omega^2}{k_2 M_a},$$

$$Z_{\dot{\beta}_0} = \frac{N_b \Omega\gamma}{k_2 M_a}\left(\frac{I_\beta}{6R} - \frac{M_\beta}{8}\right) \tag{5.61}$$

$$I_{\theta_0} = \frac{25\pi\Omega^2 Ra_0 s}{256}C_0, \qquad F_{\theta_0} = \frac{\Omega^2\gamma}{k_2}\left(\frac{1}{8} - \frac{N_b M_\beta}{6M_a R}\right),$$

$$Z_{\theta_0} = -\frac{N_b \Omega^2\gamma}{k_2 M_a}\left(\frac{I_\beta}{6R} - \frac{M_\beta}{8}\right) \tag{5.62}$$

The coefficient C_0 is equal to 0.64 for the Carpenter–Fridovitch apparent mass and unity for the Pitt–Peters value. M_β and I_β are the first and second mass moments of a rotor blade about the centre hinge and the k coefficients are given by the expressions

$$k_1 = \frac{75\pi\Omega}{128}, \qquad k_2 = 1 - \frac{N_b M_\beta^2}{M_a I_\beta} \tag{5.63}$$

where M_a is the mass of the aircraft.

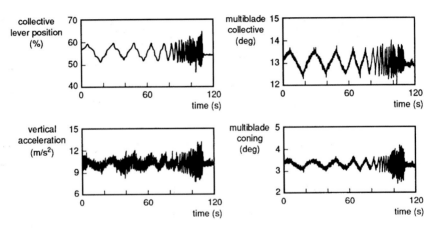

Fig. 5.11 Puma collective frequency response in hover (Ref. 5.21)

The blade Lock number and hover value of inflow are given by the expressions

$$\gamma = \frac{\rho c a_0 R^4}{I_\beta}, \qquad \lambda_0 = \sqrt{\left(\frac{C_T}{2}\right)} \qquad (5.64)$$

The expressions for the Z derivatives indicate that the resultant values are determined by the difference between two inertial effects of similar magnitude. Accurate estimates of the mass moments are therefore required to obtain fidelity in the heave DoF. In Houston's analysis the rotorspeed was assumed to be constant, a valid approximation for short-term response modelling. In Ref. 5.24, results are reported that include the effects of rotorspeed indicating that this variable can be prescribed in the above 3 DoF model with little loss of accuracy.

Flight tests were conducted on the Puma to measure the vertical motion and rotor coning in response to a collective frequency sweep. Figure 5.11 shows a sample of the test data with collective lever, rotor coning angle (derived from multi-blade coordinate analysis) and normal acceleration at the aircraft centre of mass (Ref. 5.21). The test input was applied over a wide frequency range from less than 0.1 Hz out to 3.5 Hz. The test data were converted to the frequency domain using fast Fourier transform techniques for transfer function modelling. Figure 5.12 shows an example of the magnitude, phase and coherence of the acceleration and coning response along with the fitted 3 DoF model, derived from a least-squares fit of magnitude and phase. The coherence function indicates strong linearity up to about 2 Hz with some degradation up to about 3 Hz, above which the coherence collapses. The increasing response magnitude in both heave and coning DoFs is a characteristic of the effects of inflow dynamics. The identified model parameters and modal stability characteristics are shown in Table 5.2, compared with the theoretical predictions for the Puma using the derivative expressions in eqns 5.59–5.62. The percentage spread gives the range of values estimated from six different sets of test data.

Key observations from the comparisons in Table 5.2 are that theory predicts the inflow derivatives reasonably accurately but overestimates the coning derivatives by over 30%; the heave derivatives are typically of opposite sign and the predicted stability

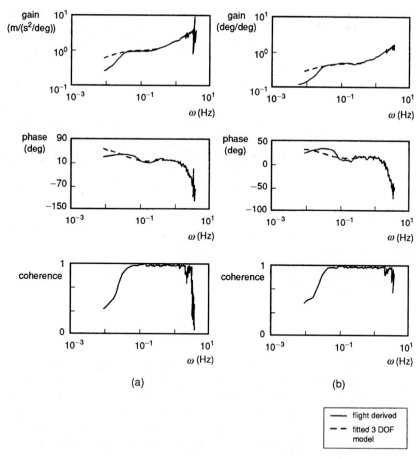

Fig. 5.12 Comparison of equivalent system fit and flight measurements of Puma frequency response to collective in hover (Ref. 5.16): (a) vertical acceleration; (b) multiblade coning

is considerably greater than that estimated from flight. In an attempt to reconcile these differences, Houston examined the effects of a range of gross model 'corrections' based on physical reasoning of both aerodynamic and structural effects, but effectively distorting the derivatives in eqns 5.59–5.62. Close examination of Houston's analysis shows that two of the correction factors effectively compensated for each other, hence resulting in somewhat arbitrary final values. We can proceed along similar lines by noting that each of the derivative groups – inflow, coning and heave – have similar errors in the theoretical predictions, suggesting that improved theoretical predictions may be obtained for each group separately. Considering the inflow derivatives, we can see that the inflow/collective derivative is predicted to within 2% of the flight estimate. This gives confidence in the Carpenter–Fridovitch value of the apparent mass coefficient C_0. The other key parameter in eqn 5.59 is the hover value of rotor inflow. An empirical correction factor of 0.7 applied to this value, together with a 2% reduction in C_0, leads to the modified theoretical estimates given in Table 5.2, now all within 10% of the flight values. The coning derivatives are a strong function of blade Lock number as shown in eqn 5.60; a 30% reduction in blade Lock number from 9.37 to 6.56, brings

Table 5.2 Comparison of theoretical predictions and flight estimates of Puma derivatives and stability characteristics

Derivative	Theoretical value	Flight estimate	Spread (%)	Modified theory
I_{vi}	−11.44	−8.55	−16.4, 17.3	−10.0
I_{β_0}	−39.27	−35.34	−10.9, 7.9	−35.67
I_w	7.86	7.07	−10.9, 7.9	7.13
I_{θ_0}	589.37	578.83	−4.1, 2.8	580.2
F_{vi}	−5.69	−4.11	−24.3, 16.9	−4.2
F_{β_0}	−794.0	−803.72	−10.0, 10.6	−794.0
$F_{\dot{\beta}_0}$	−32.16	−22.52	−21.2, 12.9	−23.6
F_w	5.69	4.11	−24.3, 16.9	4.2
F_{θ_0}	887.68	638.58	−13.8, 10.4	650.6
Z_{vi}	−0.168	0.449	−38.1, 42.1	0.383
Z_{β_0}	−177.8	−109.41	−30.0, 23.8	−109.41
$Z_{\dot{\beta}_0}$	−1.618	2.619	−69.0, 58.1	1.56
Z_w	0.168	−0.449	−38.1, 42.1	−0.383
Z_{θ_0}	44.66	−44.39	−39.7, 48.2	−42.9
	Eigenvalue	Eigenvalue		Modified theory
Inflow mode	−19.74	−12.35		−11.51
Coning mode	−11.7 ± 19.61i	−9.51 ± 22.83i		−9.04 ± 23.62i
Heave mode	−0.303	−0.159		−0.151

the coning derivatives all within 5% of the flight values, again shown in Table 5.2. The heave derivatives are strongly dependent on the inertia distribution of the rotor blade as already discussed. We can see from eqn 5.61 that the heave due to coning is proportional to the first mass moment M_β. Using this simple relation to estimate a corrected value for M_β, a 30% reduction from 300 to 200 kg m^2 is obtained. The revised values for the heave derivatives are now much closer to the flight estimates as shown in Table 5.2, with the heave damping within 15% and control sensitivity within 4%. The application of model parameter distortion techniques in validation studies gives an indication of the extent of the model deficiencies. The modified rotor parameters can be understood qualitatively in terms of several missing effects – non-uniform inflow, tip and root losses, blade elasticity and unsteady aerodynamics, and also inaccuracies in the estimates of blade structural parameters. For the present example, the correction consistency across the full set of parameters is a good indication that the modifications are physically meaningful.

The example does highlight potential problems with parameter identification when measurements are deficient; in the present case, no inflow measurements were available and the coherence of the frequency response functions was seen to decrease sharply above about 3 Hz, which is where the coning mode natural frequency occurs (see Table 5.2). The test data are barely adequate to cover the frequency range of interest in the defined model structure and it is remarkable how well the coning mode characteristics are estimated. In the time domain, the estimated 3 DoF model is now able to reflect much of the detail missed by the quasi-steady model. Figure 5.13 illustrates a comparison of time responses of coning and normal acceleration following a 1° step in collective pitch. The longer term mismatch, appearing after about 5 s, is possibly due to the effects of unmodelled rotorspeed changes. In the short term, the transient flapping and acceleration overshoots are perfectly captured. The significance of these

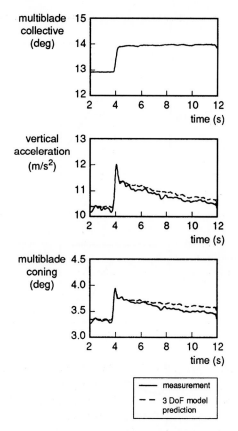

Fig. 5.13 Comparison of 3 DoF estimated model and flight measurements of response to collective for Puma in hover (Ref. 5.16)

higher order modelling effects has been confirmed more recently in a validation study of the Ames Genhel simulation model with flight test data from a UH-60 helicopter in hovering flight (Ref. 5.25).

The RAE research reported by Houston was motivated by the need for robust criteria for vertical axis handling qualities. At that time, the international effort to develop new handling qualities standards included several contending options. In the event, a simple model structure was adopted for the low-frequency control strategies required in gross bob-up type manoeuvres. This topic is discussed in more detail in Section 6.5. For high-gain feedback control studies however, a 3 DoF model should be used to address the design constraints associated, for example, with very precise height keeping.

Response to collective in forward flight

The response to collective in forward flight is considerably more complicated than in hover. While collective pitch remains the principal control for vertical velocity and flight path angle up to moderate forward speed, pilots normally use a combination of collective and cyclic to achieve transient flight path changes in high-speed flight. Also, collective pitch induces powerful pitch and roll moments in forward flight. Figure 5.14

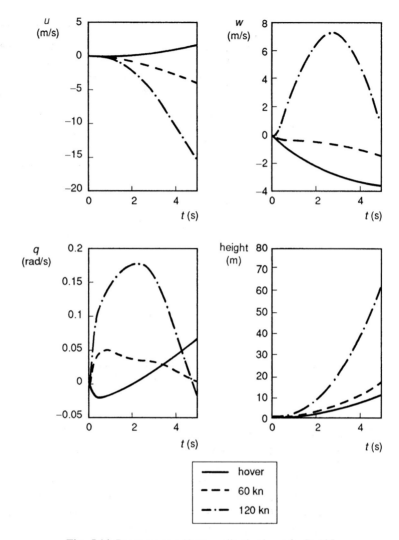

Fig. 5.14 Response to a 1° step collective input for Bo105

compares the longitudinal response of the Helisim Bo105 to collective steps in hover, at 60 and 120 knots. Comparisons are made between perturbations in forward velocity u, heave velocity w, pitch rate q and height h. We draw particular attention to the comparison of the w-velocity component and the relationship with climb rate. In hover, the aircraft reaches its steady rate of climb of about 4 m/s (750 ft/min) in about 5 s. At 120 knots the heave velocity component is initially negative, but almost immediately reverses and increases to about 7 m/s in only 3 s. The height response shows that the aircraft has achieved a climb rate of 20 m/s (approx. 3800 ft/min) after about 4 s. The powerful pitching moment generated by the collective input (M_{θ_0}), together with the pitch instability (M_w), has caused the aircraft to zoom climb, achieving a pitch rate of about 10°/s after only 2 s. Thus, the aircraft climbs while the heave velocity (climb rate, $V\theta$) increases positively. The pitching response due to collective is well predicted by our Level 1 Helisim model as shown in Fig. 5.15, which compares flight and simulation for

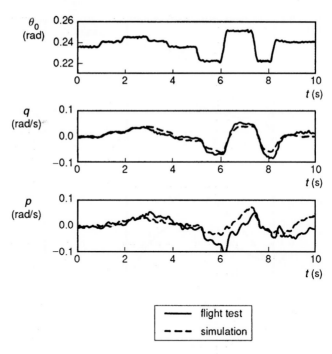

θ_0 (rad)

q (rad/s)

p (rad/s)

flight test
simulation

Fig. 5.15 Response to a 3211 collective input for Bo105 at 80 knots: comparison of flight and simulation

the Bo105 excited by a collective (modified) 3211 input at 80 knots. The 3211 test input, in standard and modified forms, was developed by the DLR (Ref. 5.14) as a general-purpose test input with a wide frequency range and good return to trim properties. Figure 5.15 compares the pitch and roll response to the collective disturbance. Roll response is less well predicted than pitch, with the simulated amplitude only 50% of the flight measurement, although the trends are correct. The roll response will be largely affected by the change in main rotor incidence in the longitudinal plane, caused by elastic coning, pitch rate and non-uniform inflow effects; the Helisim rigid blade approximation will not, of course, simulate the curvature of the blade.

5.3.3   Pitch and roll response to cyclic pitch control inputs
In Chapter 2 we discussed the mechanism of cyclic pitch, cyclic flapping and the resulting hub moments generated by the tilt of the rotor disc. In Chapters 3 and 4, the aeromechanics associated with cyclic flapping was modelled in detail. In this section, we build on this extensive groundwork and examine features of the attitude response to cyclic pitch, chiefly with the Helisim Bo105 as a reference configuration. Simulated responses have been computed using the full nonlinear version of Helisim with the control inputs at the rotor (θ_{1s} and θ_{1c}), and are presented in this form, unless otherwise stated.

Response to step inputs in hover – general features
Figure 5.16 shows the pitch and roll response to 1° cyclic step inputs at the hover. It can be seen that the direct response after 1s, i.e., q to θ_{1s} and p to θ_{1c}, has a similar magnitude

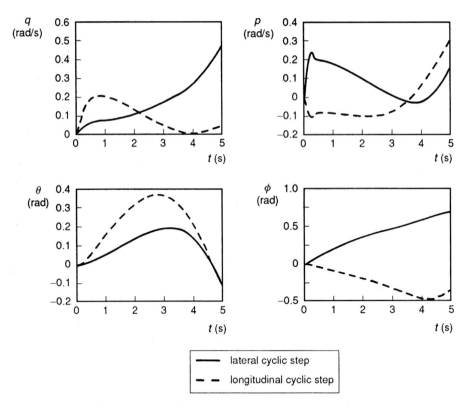

Fig. 5.16 Pitch and roll response to 1° cyclic pitch steps in hover

for both pitch and roll – in the present case about 12°/s per deg of cyclic. However, while the hub moments and rate sensitivities (i.e., the steady-state rate response per degree of cyclic) are similar, the control sensitivities and dampings are scaled by the respective inertias (pitch moment of inertia is more than three times the roll moment of inertia). Thus the maximum roll rate response is achieved in about one-third the time it takes for the maximum pitch rate to be reached. The short-term cross-coupled responses (q and p) exhibit similar features, with about 40% of the corresponding direct rates (p and q) reached less than 1 s into the manoeuvre. The accompanying sketches in Fig. 5.17 illustrate the various influences on the helicopter in the first few seconds of response from the hover condition. The initial snatch acceleration is followed by a rapid growth to maximum rate when the control moment and damping moment effectively balance. The cross-coupled control moments are reinforced by the coupled damping moments in the same time frame. As the aircraft accelerates translationally, the restoring moments due to surge and sway velocities come into play. However, these effects are counteracted by cross-coupled moments due to the development of non-uniform induced velocities normal to the rotor disc (λ_{1s} and λ_{1c}). After only 3–4 s, the aircraft has rolled to nearly 30° following either pitch or roll control inputs, albeit in opposite directions. Transient motion from the hover is dominated by the main rotor dynamics and aerodynamics. In the very short time scale (<1 s), the attitude response is strongly influenced by the rotor flapping dynamics and we need to examine this effect in more detail.

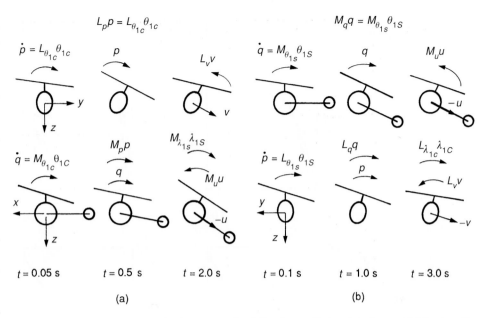

$$L_p p = L_{\theta_{1c}} \theta_{1c} \qquad \qquad M_q q = M_{\theta_{1s}} \theta_{1s}$$

$$\dot{p} = L_{\theta_{1c}} \theta_{1c} \qquad p \qquad L_v v \qquad \dot{q} = M_{\theta_{1s}} \theta_{1s} \qquad q \qquad M_u u$$

$$\dot{q} = M_{\theta_{1c}} \theta_{1c} \qquad M_p p \qquad M_{\lambda_{1s}} \lambda_{1s} \qquad \dot{p} = L_{\theta_{1s}} \theta_{1s} \qquad L_q q \qquad L_{\lambda_{1c}} \lambda_{1c}$$

$$q \qquad M_u u \qquad \qquad p \qquad L_v v$$

| $t = 0.05$ s | $t = 0.5$ s | $t = 2.0$ s | $t = 0.1$ s | $t = 1.0$ s | $t = 3.0$ s |

(a)   (b)

Fig. 5.17 Sketches of helicopter motion following cyclic inputs in hover: (a) lateral cyclic step; (b) longitudinal cyclic step

Effects of rotor dynamics

The rotor theory of Chapter 3 described three forms of the multi-blade coordinate representation of blade flap motion – quasi-steady, first-and second order. Figure 5.18 compares the short-term response with the three different rotor models to a step lateral cyclic input in hover. The responses of all three models become indistinguishable after about 1 s, but the presence of the rotor regressing flap mode (Bo105 frequency approximately 13 rad/s – see Chapter 4) is most noticeable in the roll rate response in the first quarter of a second, giving rise to a 20% overshoot compared with the deadbeat response of the quasi-steady model. The large-scale inset figure shows the comparison over the first 0.1 s, highlighting the higher order dynamic effects, including the very fast dynamics of the advancing flap mode in the second-order representation. With rotor dynamics included, the maximum angular acceleration (hence hub moment) occurs after about 50 ms, or after about 120° rotor revolution for the Bo105. The quasi-steady approximation, which predicts the maximum hub moment at $t = 0$, is therefore valid only for low-frequency dynamics (below about 10 rad/s). We have already seen earlier in this chapter how rotor dynamics have a profound effect on the stability of the rotor/fuselage modes under the influence of strong attitude control. This is a direct result of the effective time delay caused by the rotor transient shown in Fig. 5.18. Unless otherwise stated, the examples shown in this section have been derived using the first-order rotor flap approximation.

Step responses in hover – effect of key rotor parameters

Two of the fundamental rotor parameters affecting helicopter angular motion are the effective flap stiffness, reflected in the flap frequency ratio λ_β, and the rotor Lock

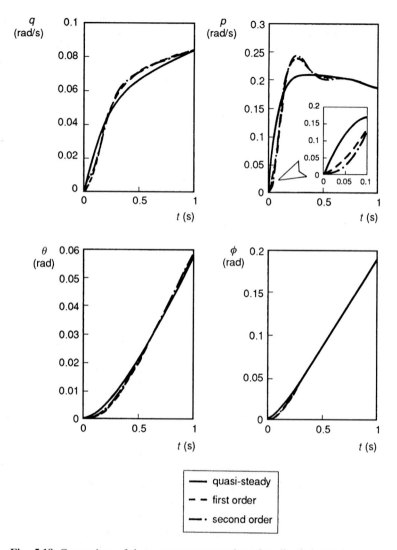

Fig. 5.18 Comparison of short-term response to lateral cyclic pitch step in hover

number γ, given by

$$\lambda_\beta^2 = 1 + \frac{K_\beta}{I_\beta \Omega^2}, \qquad \gamma = \frac{\rho c a_0 R^4}{I_\beta} \tag{5.65}$$

The effects of these parameters on helicopter stability have already been discussed in Chapter 4, and in Chapter 2 we briefly examined the effects on dynamic response. Figures 5.19 and 5.20 compare responses to step lateral cyclic inputs for the Bo105 in hover with varying λ_β and γ respectively, across ranges of values found in current operational helicopters. The effect of rotor flapping stiffness is felt primarily in the short term. The initial angular acceleration decreases and the time to reach maximum roll rate increases as the hub stiffness is reduced. On the other hand, the rate sensitivity

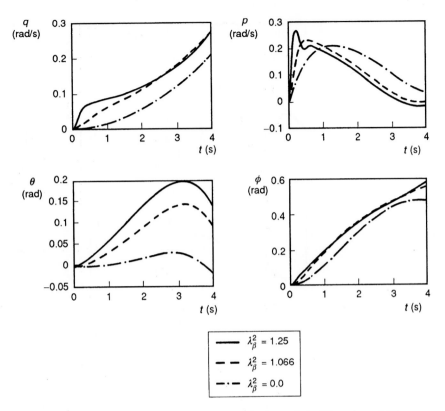

Fig. 5.19 Response characteristics with varying rotor flap stiffness ($\gamma = 5.09$)

is practically the same for all three rotors. It is also apparent that the influence of the regressing flap mode on the short-term response is reduced as the rotor stiffness decreases. To achieve an equivalent attitude response for the standard Bo105 in the first second with the two softer rotors would require larger inputs with more complex shapes. The stiffer rotor can be described as giving a more agile response requiring a simpler control strategy, a point already highlighted in Section 5.2.2. One of the negative aspects of rotor stiffness is the much stronger cross-coupling, also observed in Fig. 5.19. The strong effects of control and rate coupling (see derivatives $M_{\theta_{1c}}$ and M_p in Chapter 4) can be seen in the pitch response of the standard Bo105 rotor configuration. The ratio of pitch attitude excursions for the three rotors after only 2 s is 7:4:1. Cross-coupling must be compensated for by the pilot, a task that clearly adds to the workload. We shall examine another negative aspect of stiff rotors later, degraded pitch stability. As always, the optimum rotor stiffness will depend on the application, but with active flight control augmentation most of the negative effects of stiffer rotors can be virtually eliminated.

The selection of rotor Lock number is also application dependent, and the results in Fig. 5.20 illustrate the principal effects in the hover; in the cases shown, the Lock number was varied by changing the rotor blade inertia with compensating changes to rotor stiffness (K_β in eqn 5.65) to maintain constant λ_β. Roll control sensitivity (i.e., initial acceleration) is unaffected by Lock number, i.e., all three rotors flap by the same amount following the application of cyclic (actually the same happens for the three

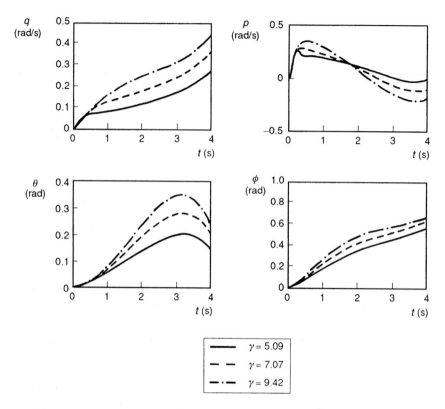

Fig. 5.20 Response characteristics with varying rotor Lock number ($\lambda_\beta^2 = 1.25$)

rotors of different stiffness, but in those cases the hub moments are also scaled by the stiffness). The major effect of Lock number is to change the rate sensitivity, the lighter blades associated with the higher values of γ resulting in lower values of gyroscopic damping retarding the hub control moment. The lower damping also increases the attitude response time constant as shown. Lock number also has a significant effect on cross-coupling as illustrated by the pitch response in Fig. 5.20.

Response variations with forward speed

One of the characteristics of helicopters is the widely varying response characteristics as a function of forward speed. Since aircraft attitude essentially defines the direction of the rotor thrust, flight path control effectiveness also varies with speed. This adds to pilot workload especially during manoeuvres involving large speed changes. We choose the attitude response to longitudinal cyclic to illustrate the various effects. Figure 5.21 compares the pitch and roll rate response to a step cyclic input from three different trimmed flight conditions – hover, 60 and 120 knots. We have already discussed the features of the step response from hover. At 60 knots, the pitch response is almost pure rate, sustained for more than 3 s. At 120 knots, the pitch rate continues to increase after the initial transient due to the control input. This continued pitch-up is essentially caused by the strongly positive pitching moment with incidence (M_w) on the Bo105. The pitch motion eventually subsides as the speed decreases under the influence of the very high nose-up pitch attitude in this zoom climb manoeuvre. The coupled roll

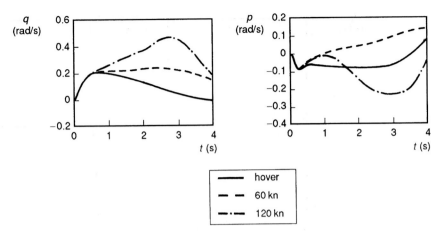

Fig. 5.21 Variation of response characteristics with forward speed

response has developed a new character at the high-speed condition, where the yaw response of the aircraft begins to have a stronger influence. The rotor torque decreases as the aircraft pitches up giving rise to a nose left sideslip, exposing the rotor to a powerful dihedral effect rolling the aircraft to port, and initiating the transient motion in the Dutch roll mode. Clearly, the pilot cannot fly by cyclic alone.

Stability versus agility – contribution of the horizontal tailplane

We have already seen in Figs 5.18 and 5.19 how effective the hingeless rotor can be in rolling the aircraft rapidly. This high level of controllability also applies to the pitch axis, although the accelerations are scaled down by the higher moment of inertia. Hingeless rotor helicopters are often described as agile because of their crisp attitude response characteristics. In Chapter 4 this point was emphasized by the relative magnitude of the control and damping derivatives of helicopters with different flap retention systems. We also examined stability in Chapter 4 and noted the significant decrease in longitudinal stability for hingeless rotor helicopters because of the positive pitch moment with incidence. For hingeless rotor helicopters, stability and agility clearly conflict. One of the natural ways of augmenting longitudinal stability is by increasing the tailplane effectiveness, normally achieved through an increase in tail area. The effects of tailplane size on agility and stability are illustrated by the results in Fig. 5.22 and Table 5.3. Three cases are compared, first with the tailplane removed altogether, second with the nominal Bo105 tail size of about 1% of the rotor disc area and third with the tail area increased threefold. The time responses shown in Fig. 5.22 are the fuselage pitch rate and vertical displacement following a step input of about 1° in longitudinal cyclic from straight and level trimmed flight at 120 knots. In Table 5.3 the principal aerodynamic pitching moment derivatives and eigenvalues of short-term pitch modes for the three configurations are compared. The pitch damping varies by about 10–20% from the standard Bo105 while the static stability changes by several hundred per cent.

If we measure agility in terms of the height of obstacle that can be cleared in a given time, the no-tailplane case is considerably more agile. This configuration is also very unstable, however, with a time-to-double amplitude of less than 1 s. Although not

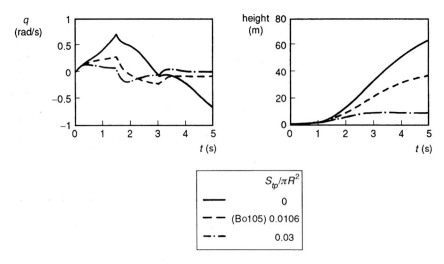

Fig. 5.22 Contribution of the horizontal stabilizer to stability and agility ($V = 120$ knots)

shown in Fig. 5.22, in clearing a 50-m obstacle, the tailless aircraft has decelerated almost to the hover and rolled over by about 60° only 4 s into the manoeuvre. Pilot control activity to compensate for these transients would be extensive. At the other extreme, the large tail aircraft is stable with a crisp pitch rate response, but only manages to pop up by about 10 m in the same time. To achieve the same flight path (height) change as that of the tailless aircraft would require a control input about four times as large; put simply, the price of increased stability is less agility, the tailplane introducing a powerful stabilizing stiffness (M_w) and damping (M_q). One way to circumvent this dichotomy is to use a moving tailplane, providing stability against atmospheric disturbances and agility in response to pilot control inputs.

Comparison with flight

In the preceding subsections, we have seen some of the characteristics of the predicted Helisim behaviour in response to cyclic control. We complete the section with a discussion on correlation with flight test data. Figures 5.23 and 5.24 show a comparison of simulation and flight test for the Bo105 aircraft disturbed from a 80-knot trim condition. The pilot inputs are 3211 multi-steps at the lateral stick (Fig. 5.23) and longitudinal stick (Fig. 5.24); in the figures these control inputs have been transformed to the swash plate cyclics derived from sensors on the blade pitch bearings. The cyclic control signals contain harmonic components characteristic of combining

Table 5.3 Variation of longitudinal stability characteristics with tailplane size (Bo105 – 120 knots)

$S_{tp}/\pi R^2$	M_q	M_w	λ_{sp}
0.0	−3.8	0.131	0.77, 0.284
0.0106	−4.48	0.039	$0.077 \pm 0.323i$
0.03	−5.0	−0.133	$-0.029 \pm 0.134i$

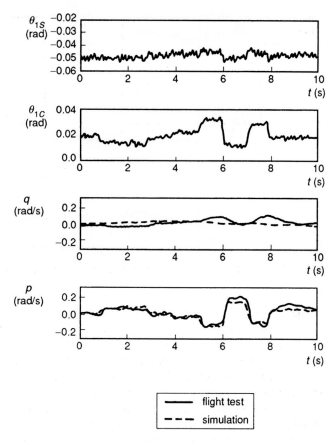

Fig. 5.23 Comparison of flight and simulation response to lateral cyclic 3211 inputs for Bo105 at 80 knots

together individual blade angles into multi-blade form, when each blade has a slightly different mean position. The magnitude of the control inputs at the rotor is about 1° in the direct axis and about 0.3° in the coupling axis, giving an indication of the swash plate phasing on the Bo105. The comparison of the direct, or on-axis, response is good, with amplitudes somewhat overestimated by simulation. The coupled, or off-axis, response comparisons are much poorer. The swash plate phasing appears to work almost perfectly in cancelling the coupled response in simulation, while the flight data show an appreciable coupling, particularly roll in the longitudinal manoeuvre, a consequence of the low ratio of roll to pitch aircraft moments of inertia. This deficiency of the Level 1 model to capture pitch–roll and roll–pitch cross-coupling appears to be a common feature of current modelling standards and has been attributed to the absence of various rotor modelling sources including a proper representation of dynamic inflow, unsteady aero-dynamics (changing the effective control phasing) and torsional dynamics. Whatever the explanation, and it may well be different in different cases, it does seem that cross-couplings are sensitive to a large number of small effects, e.g., 5° of swash plate phasing can lead to a roll acceleration as much as 40% of the pitch acceleration following a step longitudinal cyclic input.

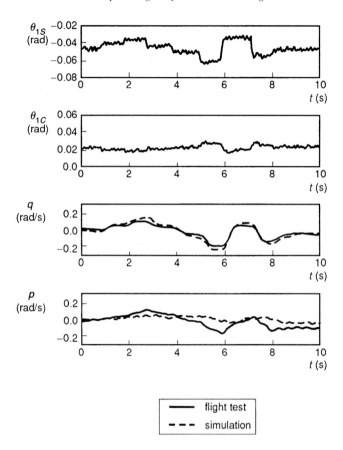

Fig. 5.24 Comparison of flight and simulation response to longitudinal cyclic 3211 inputs for Bo105 at 80 knots

Another topic that has received attention in recent years concerns the adequacy of low-order models in flight dynamics. In Ref. 5.7, Tischler is concerned chiefly with the modelling requirements for designing and analysing the behaviour of automatic flight control systems with high-bandwidth performance. Tischler used the DLR's AGARD Bo105 test database to explore the fidelity level of different model structures for the roll attitude response (ϕ) to lateral cyclic stick (η_{1c}). Parametric transfer function models were identified from flight data using the frequency sweep test data, covering the range 0.7–30 rad/s. Good fidelity over a modelling frequency range of 2–18 rad/s was judged to be required for performing control law design. The least-squares fits of the baseline seventh-order model and band-limited quasi-steady model are shown in Figs 5.25 and 5.26, respectively. The identified transfer function for the baseline model is given by eqn 5.66, capturing the coupled roll/regressive flap dynamics, the regressive lead–lag dynamics, the Dutch roll mode, roll angle integration (0) and actuator dynamics, the latter modelled as a simple time delay:

$$\frac{\overline{\phi}}{\overline{\eta}_{1c}}(s) = \frac{2.62[0.413, 3.07][0.0696, 16.2]e^{-0.0225s}}{(0)[0.277, 2.75][0.0421, 15.8][0.509, 13.7]} \qquad (5.66)$$

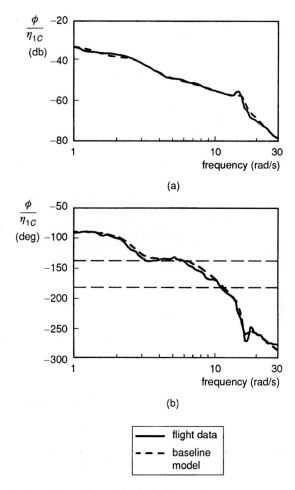

Fig. 5.25 Comparison of open-loop roll attitude frequency response; seventh-order baseline identified model versus flight test for Bo105 at 80 knots (a) magnitude; (b) phase (Ref. 5.7)

where the parentheses signify

$$[\zeta, \omega] \rightarrow s^2 + 2\zeta\omega s + \omega^2, \qquad (1/T) \rightarrow s + (1/T) \tag{5.67}$$

The frequencies of the Dutch roll (2.75 rad/s) and roll/flap regressing mode (13.7 rad/s) compare well with the corresponding Helisim predictions of 2.64 rad/s and 13.1 rad/s, respectively. A 23-ms actuator lag has also been identified from the flight test data. From Fig. 5.25 we can see that both the amplitude and phase of the frequency response are captured well by the seventh-order baseline model.

The band-limited quasi-steady model shown in Fig. 5.26 is modelled by the much simpler transfer function given in eqn 5.68:

$$\frac{\overline{\phi}}{\overline{\eta}_{1c}}(s) = \frac{0.3 \, e^{-0.0838s}}{(0)(14.6)} \tag{5.68}$$

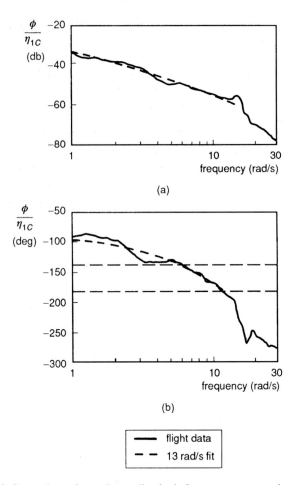

Fig. 5.26 Comparison of open-loop roll attitude frequency response; band-limited identified quasi-steady model versus flight test for Bo105 at 80 knots (Ref. 5.7): (a) magnitude; (b) phase

The response is now modelled by a simple exponential lag, characterized by the damping derivative L_p, supplemented by a pure time delay to account for the unmodelled lags. The frequency range for which this band-limited model is valid was established by Tischler in Ref. 5.7 by identifying the frequency at which the least-squares fit error began to diverge. This occurred at a frequency of about 14 rad/s, above which the estimated parameters became very sensitive to the selected frequency. This sensitivity is usually an indication that the model structure is inappropriate. The estimated roll damping in eqn 5.68 is 14.6/s, compared with the Helisim value of 13.7/s; a total delay of over 80 ms now accounts for both actuation and rotor response. Tischler concludes that the useful frequency range for such quasi-steady (roll and pitch response) models extends almost up to the regressing flap mode. In his work with the AGARD test database, Tischler has demonstrated the power and utility of frequency domain identification and transfer function modelling. Additional work supporting the conclusions of Ref. 5.7 can be found in Refs 5.26 and 5.27.

One of the recurring issues regarding the modelling of roll and pitch response to cyclic pitch concerns the need for inclusion of rotor DoFs. We have addressed this on many occasions throughout this book, and Tischler's work shows clearly that the need for rotor modelling depends critically on the application. This topic was the subject of a theoretical review by Hanson in Ref. 5.28, who also discussed the merits of various approximations to the higher order rotor effects and the importance of rotor dynamics in system identification. References 5.29 and 5.30 also report results of frequency domain fitting of flight test data, in this case the pitch response of the DRA Puma to longitudinal cyclic inputs. Once again, the inclusion of an effective time delay was required to obtain sensible estimates of the 6 DoF model parameters – the stability and control derivatives. Without any time delay in the model structure, Ref. 5.29 reports an estimated value of Puma pitch damping (M_q) of −0.353, compared with the Helisim prediction of −0.835. Including an effective time delay in the estimation process results in an identified M_q of −0.823. This result is typical of many reported studies where flight estimates of key physical parameters appear to be unrealistic, simply because the model structure is inappropriate.

An important modelling element omitted from the forward flight Bo105 and Puma results discussed above is the effect of non-uniform dynamic inflow. We have seen from Chapters 3 and 4 that 'unsteady' momentum theory predicts the presence of powerful non-uniform effects in response to the development of aerodynamic hub moments. References 5.25, 5.31 and 5.32 report comparisons of predictions from blade element rotor models with the NASA UH-60 hover test database (Ref. 5.25), where the inflow effects are predicted to be strongest. The results from all three references are in close agreement. Figure 5.27, from Ref. 5.32, illustrates a comparison between flight and the

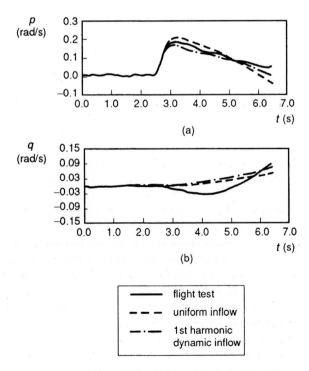

Fig. 5.27 Comparison of flight and simulation response to roll cyclic step, showing contribution of dynamic inflow for UH-60 in hover (Ref. 5.32): (a) roll response; (b) pitch response

FLIGHTLAB simulation model of the response to a 1-inch step input in lateral cyclic from hover. Dynamic inflow is seen to reduce the peak roll rate response in the first half second by about 25%. The inflow effects do not appear, however, to improve the cross-coupling predictions.

5.3.4 Yaw/roll response to pedal control inputs

In this section we examine the characteristics of the coupled yaw/roll response to pedal control inputs in forward flight; attention will be focused on comparison with test data from the DRA Puma helicopter. Yaw/roll motions are coupled through a variety of different physical mechanisms. Even at the hover, any vertical offset of the tail rotor from the aircraft centre of gravity will give rise to a rolling moment from tail rotor collective. As forward speed increases, the forces and moments reflected in the various coupled stability derivatives, e.g., dihedral L_v, adverse yaw N_p, combine to form the character of the Dutch roll mode discussed in Chapter 4. Figure 5.28, taken from Ref. 5.33, illustrates the comparison of yaw, roll and sideslip responses from flight and Helisim following a 3211 multi-step pedal control input. It can be seen that the simulation overpredicts the initial response in all 3 DoFs and also appears to overpredict the damping and period of the free oscillation in the longer term. In Chapter 4 we examined approximations to this mode, concluding that for both the Puma and Bo105, a 3 DoF yaw/roll/sideslip model was necessary but that, provided the sideways motion was small compared with sideslip, a second-order approximation was adequate; with this approximation, the stability is then characterized by the roots of the equation

$$\lambda^2 + 2\zeta_d\omega_d\lambda + \omega_d^2 = 0 \tag{5.69}$$

where the damping is given by

$$2\zeta_d\omega_d \approx -\left(N_r + Y_v + \sigma_d\left\{\frac{L_r}{U_e} - \frac{L_v}{L_p}\right\}\right) \bigg/ \left(1 + \frac{\sigma_d L_r}{L_p U_e}\right) \tag{5.70}$$

and the frequency by the expression

$$\omega_d^2 \approx (U_e N_v + \sigma_d L_v) \bigg/ \left(1 + \frac{\sigma_d L_r}{L_p U_e}\right) \tag{5.71}$$

with

$$\sigma_d = (g - N_p V)/L_p \tag{5.72}$$

If the 'true' values of the stability and control derivatives were known, then this kind of approximation may be able to help to explain where the modelling deficiencies lie. Estimates of the Puma derivatives derived by the DLR using the test data in Fig. 5.28 are shown in Table 5.4, along with Dutch roll eigenvalues for three different cases – the fully coupled 6 DoF motion, lateral subset and the approximation given by eqn 5.69. It can be seen that the latter accounts for about 80% of the damping and more than 90% of the frequency for the flight results (compare $\lambda^{(1)}$ with $\lambda^{(3)}$) and therefore serves as a representative model of Dutch roll motion; note that theory overpredicts the damping by more than 60% and underpredicts the frequency by 20%.

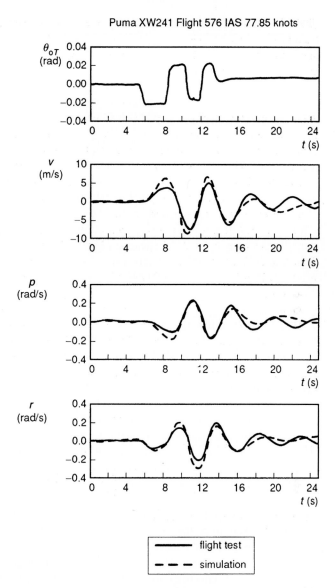

Fig. 5.28 Response to pedal 3211 input – comparison of flight and simulation for Puma at 80 knots (Ref. 5.33)

Figure 5.29 shows a comparison between flight measurements and the 3 DoF second-order approximation using the flight-estimated derivatives in Table 5.4. The comparison is noticeably better in the short term, but the damping appears now to be slightly underpredicted, consistent with the comparison already noted between $\lambda^{(1)}$ with $\lambda^{(3)}$. Looking more closely at the derivatives in Table 5.4, we see that the most striking mismatch between flight and theory is the overprediction of the yaw damping and control sensitivity by about 70% and the underprediction of the roll damping and dihedral effects by 30 and 20%, respectively. The Helisim prediction of adverse yaw

Table 5.4 Dutch roll oscillation characteristics

Derivative	Flight test – DLR	Helisim
Y_v	−0.135 (0.0019)	−0.125
L_v	−0.066 (0.0012)	−0.055
N_v	0.027 (0.0002)	0.0216
L_p	−2.527 (0.0534)	−1.677
N_p	−0.395 (0.0092)	−0.174
L_r	−0.259 (0.0343)	0.142
N_r	−0.362 (0.0065)	−0.57
L_{lat}	−0.051 (0.0012)	−0.043
N_{lat}	−0.008 (0.0002)	−0.0047
L_{ped}	0.011 (0.0007)	0.0109
N_{ped}	−0.022 (0.0001)	−0.0436
τ_{lat}	0.125	0.0
$\lambda^{(1)}$	−0.104 ± 1.37i	−0.163 ± 1.017i
$\lambda^{(2)}$	−0.089 ± 1.27i	−0.166 ± 1.08i
$2\zeta\omega$	0.1674	0.390
ω^2	1.842	1.417
$\lambda^{(3)}$	−0.081 ± 1.34i	−0.199 ± 1.199i

$\lambda^{(1)}$ Dutch roll (fully coupled).
$\lambda^{(2)}$ Dutch roll (lateral subset).
$\lambda^{(3)}$ Dutch roll (second-order roll/yaw/sideslip) approximation.
Numbers in parentheses give the standard deviation of the estimated derivatives.

N_p is less than half the value estimated from the flight data. A simple adjustment to the yaw and roll moments of inertia, albeit by a significant amount, would bring the theoretical predictions of damping and control sensitivity much closer to the flight estimates. Similarly, the product of inertia I_{xz} has a direct effect on the adverse yaw. Moments of inertia are notoriously difficult to estimate and even more difficult to measure (particularly roll and yaw), and errors in the values used in the simulation model of as much as 30% are possible. However, the larger discrepancies in the yaw axis are unlikely to be due solely to incorrect configuration data. The absence of any interactional aerodynamics between the main rotor wake/fuselage/empennage and tail rotor is likely to be the cause of some of the model deficiency. Typical effects unmodelled in the Level 1 standard described in Chapter 3 include reductions in the dynamic pressure in the rotor/fuselage wake at the empennage/tail rotor and sidewash effects giving rise to effective $\dot{v}$ acceleration derivatives (akin to $M_{\dot{w}}$ from the horizontal tailplane).

The approximation for the Dutch roll damping given by eqn 5.70 can be further reduced to expose effective damping derivatives in yaw and sideslip:

$$N_{r_{effective}} = N_r + N_p \frac{VL_v}{L_p^2} \qquad (5.73)$$

$$Y_{v_{effective}} = Y_v - g \frac{L_v}{L_p^2} \qquad (5.74)$$

In both cases the additional effect due to rolling motion is destabilizing, with the adverse yaw effect reducing the effective yaw damping by half. The adverse yaw is almost entirely a result of the high value of the product of inertia I_{xz}, coupling the

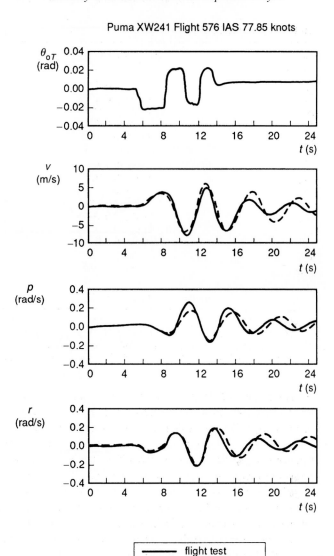

Fig. 5.29 Response to pedal 3211 input – comparison of flight and identified model – Puma at 80 knots (Ref. 5.33)

roll damping into the yaw motion. The damping decrements due to rolling manifest themselves as a moment (eqn 5.73) and a force (eqn 5.74) reinforcing the motion at the effective centre of the oscillation. This interpretation is possible because of the closely coupled nature of the motion. The yaw, roll and sideslip motions are locked in a tight phase relationship in the Puma Dutch roll – sideslip leading yaw rate by 90° and roll rate lagging behind yaw rate by 180°. Hence, as the aircraft nose swings to starboard with a positive yaw rate, the aircraft is also rolling to port (induced by the dihedral effect from the positive sideslip) thus generating an adverse yaw N_p in the same direction as the yaw rate.

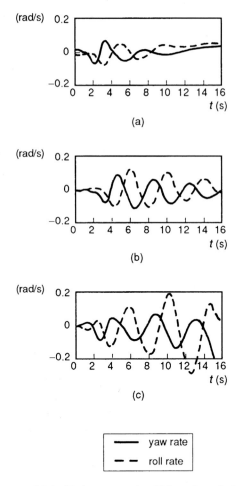

Fig. 5.30 Response to pedal double input – varying flight path angle for Puma at 100 knots (Ref. 5.35): (a) Run 463/09/13 $\gamma = -0.1$ (descent); (b) Run 467/10/11 $\gamma = 0$ (level); (c) Run 464/01/01 $\gamma = 0.1$ (climb)

The powerful effect of the damping decrement from adverse yaw can be even more vividly illustrated with an example taken from Refs 5.34 and 5.35. Figure 5.30 presents a selection of Puma flight results comparing the roll and yaw response in the Dutch roll mode at 100 knots in descending, level and climbing flight; the control input is a pedal doublet in all three cases. It can be seen that the stability of the oscillation is affected dramatically by the flight path angle. In descent, the motion has virtually decayed after about 10 s. In the same time frame in climbing flight, the pilot is about to intervene to inhibit an apparently violent departure. A noticeable feature of the response in the three conditions is the changing ratio of roll to yaw. Reference 5.35 discusses this issue and points out that when the roll and yaw motions are approximately 180° out of phase, the effective damping can be written in the form

$$N_{r_{effective}} = N_r + \left| \frac{p}{r} \right| N_p \qquad (5.75)$$

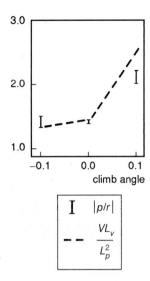

Fig. 5.31 Variation of Dutch roll oscillation roll/yaw ratio with flight path angle (Ref. 5.35)

Figure 5.31 compares the variation of the ratio p/r in the three conditions with the approximation in eqn 5.73 ($V L_v / L_p^2$), providing additional validation of this relatively simple approximation for a complex mode. The ratio of roll to yaw motion in the Dutch roll mode was discussed at the end of Chapter 4 in the context of the SA330 Puma. There it was shown how small perturbation linear analysis predicted Dutch roll instability at about 120 knots. When larger sideslip perturbations were used to calculate the derivatives however, the nonlinearity in the yawing moment with sideslip led to a much larger value for the weathercock stability and a stable Dutch roll (see Fig. 4.28). This strong nonlinearity leads to the development of a limit cycle in the Puma at high speed. Figure 5.32 compares the Puma Dutch roll response for the small perturbation linear Helisim with the full nonlinear Helisim, following a 5 m/s initial disturbance in sideslip from a trim condition of 140 knots. The linear model predicts a rapidly growing unstable motion with roll rates of more than 70°/s developing after only three oscillation cycles. The nonlinear response, which is representative of flight behaviour, indicates a limit cycle with the oscillation sustained at roll and yaw rate levels of about 5°/s and 10°/s, respectively; the sideslip excursions are about 10°, a result consistent with the stability change in Fig. 4.28 lying between sideslip perturbations of 5° and 15°.

The Dutch roll is often described as a 'nuisance' mode, in that its presence confers nothing useful to the response to pedal or lateral cyclic controls. The Dutch roll mode also tends to become rather easily excited by main rotor collective and longitudinal cyclic control inputs, on account of the rotor/engine torque reactions on the fuselage. In Chapter 6, criteria for the requirements on Dutch roll damping and frequency are presented and it is apparent that most helicopters naturally lie in the unsatisfactory area, largely due to the relatively high value for the ratio of dihedral to weathercock stability. Some form of yaw axis artificial stability is therefore quite a common feature of helicopters required to fly in poor weather or where the pilot is required to fly with divided attention.

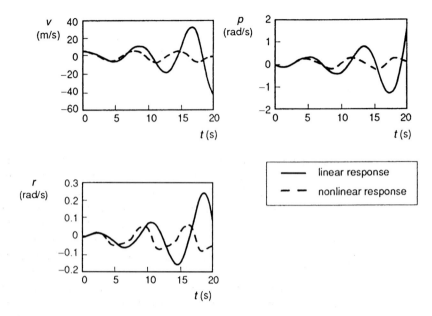

Fig. 5.32 Response to a pedal doublet input – comparison of linear and nonlinear solutions for Puma at 120 knots

5.4 RESPONSE TO ATMOSPHERIC DISTURBANCES

Helicopter flying qualities criteria try to take account of the influence of atmospheric disturbances on the response of the aircraft in terms of required control margins close to the edge of the operational flight envelope and the consequent pilot fatigue caused by the increased workload. We can obtain a coarse understanding of the effects of gusts on helicopter response through linear analysis in terms of the aerodynamic derivatives. In Chapter 2 we gave a brief discourse on response to vertical gusts, which we can recall here to introduce the subject. Assuming a first-order initial heave response to vertical gusts, we can write the equation of motion in the form of eqn 5.76:

$$\frac{dw}{dt} - Z_w w = Z_w w_g \tag{5.76}$$

The heave damping derivative Z_w now defines both the transient response and the gust input gain. The initial normal acceleration in response to a sharp edge gust is given by the expression

$$\left(\frac{dw}{dt}\right)_{t=0} = Z_w w_g \tag{5.77}$$

In Chapter 2, and later in Chapter 4, we derived approximate expressions for the magnitude of the derivative Z_w, and hence the initial heave bump, for hover and forward flight in the forms

Hover:

$$Z_w = -\frac{2a_0 A_b \rho (\Omega R)\lambda_i}{(16\lambda_i + a_0 s)M_a} \tag{5.78}$$

Forward flight:

$$Z_w = -\frac{\rho a_0 V A_b}{2M_a}\left(\frac{4}{8\mu + a_0 s}\right) \tag{5.79}$$

A key parameter in the above expressions is the blade loading (M_a/A_b), to which the gust response is inversely proportional. The much higher blade loadings on rotorcraft, compared with wing loadings on fixed-wing aircraft, are by far the single most significant reason why helicopters are less sensitive to gusts than are the corresponding fixed-wing aircraft of the same weight and size. An important feature of helicopter gust response in hover, according to eqn 5.78, is the alleviation due to the build-up of rotor inflow. However, as we have already seen in Section 5.3.2, rotor inflow has a time constant of about 0.1 s, hence the alleviation will not be as significant in practice. In forward flight the gust sensitivity is relatively constant above speeds of about 120 knots. This saturation effect is due to the cyclic blade loadings; the loadings proportional to forward speed are dominated by the one-per-rev lift. A similar analysis can be conducted for the response of the helicopter in surge and sway with velocity perturbations in three directions. This approach assumes that the whole helicopter is immersed in the gust field instantaneously, thus ignoring any penetration effects or the cyclic nature of the disturbance caused by the rotating blades. An approximation to the effects of spatial variations in the gust strength can also be included in the form of linear variations across the scale of the fuselage and rotor through effective rate derivatives (e.g., M_q, L_p, N_r). In adopting this approach care must be taken to include only the aerodynamic components of these derivatives to derive the gust gains.

While the 6 DoF derivatives provide a useful starting point for understanding helicopter gust response, the modelling problem is considerably more complex. Early work on the analysis of helicopter gust response in the 1960s and 1970s (Refs 5.36–5.41) examined the various alleviation factors due to rotor dynamics and penetration effects, drawing essentially on analysis tools developed for fixed-wing applications. More recently (late 1980s and 1990s) attention has been paid to understanding response with turbulence models more representative of helicopter operating environments, e.g., nap of the earth and recovery to ships (Refs 5.42, 5.43). These two periods of activity are not obviously linked and the underlying subject of ride qualities has received much less attention than handling qualities in recent years; as such there has not been a coherent development of the subject of helicopter (whole-body) response to gusts and turbulence. What can be said is that the subject is considerably more complex than the response to pilot's controls and requires a different analytical framework for describing and solving the problems. The approach we take in this section is to divide the response problem into three parts and to present an overview: first, the characterization and modelling of atmospheric disturbances for helicopter applications; second, the modelling of helicopter response; third, the derivation of suitable ride qualities. A flavour running through this overview will be taken from current UK research to develop a unified analytic framework for describing and solving the problems contained in all three elements (Ref. 5.44).

Modelling atmospheric disturbances

In the UK Airworthiness Defence Standard (Ref. 5.45), turbulence intensity is characterized by four bands: light (0–3 ft/s), moderate (3–6 ft/s), severe (6–12 ft/s) and extreme (12–24 ft/s). A statistical approach to turbulence refers to the probability of equalling or exceeding given intensities at defined heights above different kinds of terrain. A second important property of turbulence is the relationship between the intensity and the spatial or temporal scale or duration, which also varies with height above terrain. Such a classification has obvious attractions for design and certification purposes, and the extensions of fixed-wing methods to helicopters are generally applicable at operating heights above about 200 ft. Below this height, the dearth of measurements of three-dimensional atmospheric disturbances means that the characterization of turbulence is less well understood, with the distinct exception of airflow around man-made constructions (Ref. 5.46). For the purposes of this discussion, we circumvent this dearth of knowledge and concern ourselves only with the general modelling issues rather than specific cases, but it is important to note that extensions of results from fixed-wing studies may well not apply to helicopters. Another area where this read-across has to be reinterpreted is the treatment of turbulence scale length. In fixed-wing work a common approximation assumes a 'frozen-field' of disturbances, such that the scale length and duration are related directly through the forward speed of the aircraft. For helicopters in hover or flying through winds at low speed, this approach is clearly not valid and it is more appropriate to consider the aircraft flying through (or hovering in) a steady wind with the turbulence superimposed.

The most common form of turbulence model involves the decomposition of the velocity into frequency components, where the rms of the aircraft response can be related to the rms intensity of the turbulence. In Ref. 5.45, this power spectral density (PSD) method is recommended for investigations of general handling qualities in continuous turbulence. The PSD contains information about the excitation energy within the atmosphere as a function of frequency (or spatial wavelength), and several models exist based on measurements of real turbulence. For example, the von Karman PSD of the vertical component of turbulence takes the form (Refs 5.47, 5.48)

$$\Phi_{w_g}(\nu) = \sigma_{w_g}^2 \frac{L_w}{\pi} \frac{1 + \frac{8}{3}(1.339 L_w \nu)^2}{\left(1 + (1.339 L_w \nu)^2\right)^{11/6}} \tag{5.80}$$

where the wavenumber $\nu = $ frequency/airspeed, L is the turbulence scale and σ is the rms of the intensity. The von Karman method assumes that the disturbance has Gaussian properties and the extensive theory of stationary random processes can be brought to bear when considering the response of an aircraft as a linear system. This is clearly a strength of the approach but it also reveals a weakness. A significant shortcoming of the basic PSD approach is its inability to model any detailed structure in the disturbance. Large peaks and intermittent features are smoothed over as the amplitude and phase characteristics are assumed to be uniformly distributed across the spectrum. Any phase correlations in the turbulence record are lost in the PSD process, hence removing the capability of sinusoidal components reinforcing one another. Atmospheric disturbances with highly structured character, corresponding for example to shear layers exhibiting sharp velocity gradients, are clearly important for helicopter applications in the wake of hills and structures, and a different form of modelling is required in these cases.

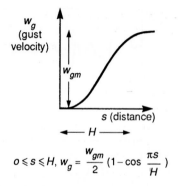

$$0 \leqslant s \leqslant H, \; w_g = \frac{w_{gm}}{2} \left(1 - \cos \frac{\pi s}{H}\right)$$

Fig. 5.33 Elemental ramp gust used in the statistical discrete gust approach

The statistical discrete gust (SDG) approach to turbulence modelling was developed by Jones at DRA (RAE) for fixed-wing applications (Refs 5.49–5.51), essentially to cater for more structured disturbances, and appears to be ideally suited for low-level helicopter applications. In Ref. 5.45, the SDG method is recommended for the assessment of helicopter response to, and recovery from, large disturbances. The basis of the SDG approach is an elemental ramp gust (Fig. 5.33) with gradient distance (scale) H and gust amplitude (intensity) w_g. A non-Gaussian turbulence record can be reconstituted as an aggregate of discrete gusts of different shapes and sizes; different elemental shapes, with self-similar characteristics (Ref. 5.49), can be used for different forms of turbulence. One of the properties of turbulence, correctly modelled by the PSD approach, and that the SDG method must preserve, is the shape of the PSD itself which appears to fit measured data well. This so-called energy constraint (Ref. 5.50) is satisfied by the self-similar relationship between the gust amplitude and length in the aggregate used to build the equivalent SDG model. For example, with the VK spectrum, the relationship takes the form $w_g \propto H^{1/3}$ for gusts with length small compared with the reference spectral scale L.

Where the PSD approach adopts frequency-domain, linear analysis, the SDG approach is essentially a time-domain, nonlinear technique. Since the early development of the SDG method, a theory of general transient signal analysis has been developed, providing a rational framework for its use. The basis of this analysis is the so-called 'Wavelet' transform, akin to the Fourier transform, but returning a new time-domain function of scale and intensity (Ref. 5.51). The SDG elements can now be interpreted as a particular class of wavelet, and a turbulence time history can be decomposed into a combination of wavelets adopting the so-called adaptive wavelet analysis (Ref. 5.48). These new techniques provide considerably more flexibility in the modelling of structured turbulence and should find regular use in helicopter response analysis.

To close this brief review of turbulence modelling, Fig. 5.34 illustrates the comparison between two forms of turbulence record reconstruction (Ref. 5.48). The measurements exhibit features common to real atmospheric disturbances – sharp velocity gradients associated with shear layers and periods of relative quiescence (Fig. 5.34(a)). Figure 5.34(b) shows the turbulence reconstructed using measured amplitude components with random phase components from the PSD model. From a PSD perspective, the information in Fig. 5.34(b) is identical to the measurements, but it can be seen that all structured features in the measurements have apparently disappeared in the

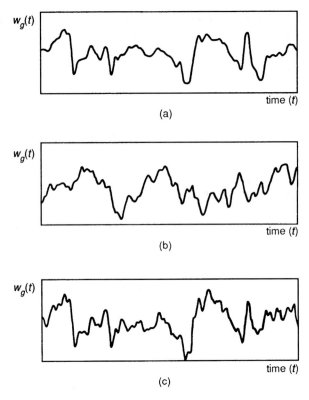

$w_g(t)$

time (*t*)

(a)

$w_g(t)$

time (*t*)

(b)

$w_g(t)$

time (*t*)

(c)

Fig. 5.34 Influence of Fourier amplitude and phase on the structure of atmospheric turbulence: (a) measured atmospheric turbulence; (b) reconstruction using measured amplitude components; (c) reconstruction using measured phase components

reconstruction process. Figure 5.34(c), on the other hand, has been reconstructed using the measured phase components with random amplitude components from the PSD model. The structure has been preserved in this process, clear evidence of the non-Gaussian characteristics of these turbulence measurements, where the real phase correlation has preserved the reinforcement of energy present in the concentrated events. These structured features of turbulence are important for helicopter work. They occur in the nap-of-the-earth and close to oil rigs and ships. Their scales can also be quite small and, at low helicopter speeds, scale lengths as small as the rotor radius can influence ride and handling.

Modelling helicopter response
Consideration of the response of helicopters to atmospheric disturbances needs to take account of a number of factors. We have seen from the simple Level 1 modelling described in Chapter 3 that the rotor response to in-plane and out-of-plane velocity perturbations is distributed over the frequencies associated with the harmonics of rotorspeed. In high-speed flight the force response at the rotor hub tends to be dominated by the n-per-rev and $2n$-per-rev components, and many studies have focused on important fatigue and hub vibratory loading problems (e.g., Refs 5.52, 5.53). Only

the zero harmonic forces and the first harmonic moments lead to zero-frequency hub and fuselage response and thus affect the piloting task directly. Several studies (e.g., Refs 5.38–5.40) have concentrated on investigating factors that alleviate the fuselage response relative to the sharp bump predicted by 'instantaneous' models, typified, for example, by eqn 5.77. Rotor–fuselage penetration effects coupled with any gust ramp characteristics tend to dominate the alleviation with secondary effects due to rotor dynamics and blade elasticity. Rotor unsteady aerodynamics can also have a significant influence on helicopter response, particularly the inflow/wake dynamics (see Chapter 3). An important aspect covered in several published works concerns the cyclo-stationary nature of the rotor blade response (Refs 5.42, 5.43, 5.54, 5.55). Essentially, the radial distribution of turbulence effects varies periodically and the gust velocity environments at the rotor hub and rotor blade tip are therefore substantially different. If a helicopter is flying through a sinusoidal vertical gust field with scale L and intensity w_{gm}, then the turbulence velocities experienced at the rotor hub and blade tip are given by the expressions

$$w_g^{(h)}(t) = w_{gm} \sin \left\{ \frac{2\pi V t}{L} \right\} \tag{5.81}$$

$$w_g^{(t)}(t) = w_{gm} \sin \left\{ \frac{2\pi V t}{L} - \frac{2\pi R}{L} \cos \Omega t \right\} \tag{5.82}$$

where V is the combined forward velocity of the aircraft and gust field. Response studies that include only hub-fixed turbulence models (eqn 5.81) and assume total immersion of the rotor at any instant clearly ignore much of the local detail in the way the individual blades experience the gust field. At low speed with scales $O(R)$, this approximation becomes invalid, and recourse to more detailed modelling is required. Assuming the gust field varies linearly across the rotor allows the disturbance to be incorporated as an effective pitch (roll) rate or non-uniform inflow component. This level of approximation can be regarded as providing an interim level of accuracy for gust scales that are larger than the rotor but that still vary significantly across the disc at any given time. In Ref. 5.56, a study is reported on the validity of various approximations to the way in which rotor blades respond to turbulence, suitable for incorporation into a real-time simulation model. The study concluded that the modelling of two-dimensional turbulence effects is likely to be required, and that approximating the turbulence intensity over a whole blade by the value at the 3/4 radius would provide adequate levels of accuracy.

In addition to characterizing the atmospheric disturbance, the SDG approach, augmented with the transient wavelet analysis, provides a useful insight into helicopter response. The concept of the tuned response is illustrated in Fig. 5.35. Associated with each ramp gust input (Fig. 5.35(a)) we assume the response variable of interest has a single dominant peak, of amplitude γ, as shown in Fig. 5.35(b). If the helicopter model is excited with each member of the family of equi-probable gusts, according to the von Karman PSD, then we find, in general, that the response peak function takes the form given in Fig. 5.35(c). There exists a tuned gust length $\overline{H}$ that produces a 'resonant' response from the helicopter. This transient response resonance is the equivalent of the resonance frequency in the frequency domain representation, and can be used to quantify a helicopter's ride qualities.

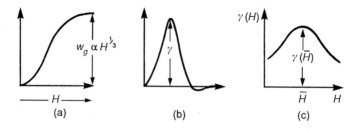

Fig. 5.35 Transient response analysis using the SDG method: (a) gust input; (b) aircraft response; (c) tuned response function

Ride qualities

The third aspect in this section is concerned with the sensitivity of aircraft, crew, weapon system, passengers or equipment to atmospheric disturbances, taken together under the general heading – ride qualities. Reference 5.57 discusses the parameters used to quantify ride bumpiness for military fixed-wing aircraft, in terms of the normal acceleration response. For helicopter applications, the meaning of ride qualities, in terms of which flight parameters are important, is a powerful function of the aircraft role and flight condition. For example, the design of a civil transport helicopter required to cruise at 160 knots may well consider the critical case as the number of, say, 1/2 g vertical bumps per minute in the passenger cabin when flying through severe turbulence. For an attack helicopter the critical case may be the attitude perturbations in the hover, while cargo helicopters operating at low speed with underslung loads may have flight path displacement as the design case. For the first example quoted, a direct parallel can be drawn from fixed-wing experience. In Ref. 5.57 Jones promotes the application of the SDG method to aircraft ride qualities in the following way. We have already introduced the concept of the tuned gust, producing the maximum or tuned transient response. Based on tuned gust analysis, the predicted rate of occurrence of vertical bumps can be written in the form

$$n_y = n_0 \exp\left(-\frac{y}{\beta\overline{\gamma}}\right), \qquad n_0 = \frac{\alpha}{\lambda\overline{H}} \qquad (5.83)$$

where

n_y is the average number of aircraft normal acceleration peaks with magnitude greater than y, per unit distance flown;

α and β reflect statistical properties of the patch of turbulence through which the aircraft is flying;

$\overline{H}$ is the tuned gust scale (length);

$\overline{\gamma}$ is the tuned response (Fig. 5.35(c));

λ is the gust length sensitivity.

In addition to its relative simplicity, this kind of formulation has the advantage that it caters for structured turbulence and hence structured aircraft response. The approach can be extended to cases where the gust field is better represented by gust pairs and other more complicated patterns, with associated complex tuning functions. The basis of the SDG method is the assumption that structured atmospheric disturbance is more correctly and more efficiently modelled by localized transient features. The wavelet

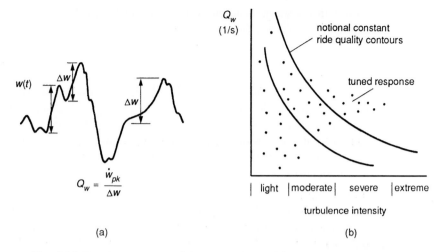

Fig. 5.36 Transient response quickness as a ride qualities parameter: (a) quickness extraction; (b) quickness chart

analysis has provided a sound theoretical framework for extending the forms of transient disturbance and response shape to more deterministic analyses. At the same time, new handling qualities criteria are being developed that characterize the response in moderate amplitude manoeuvres, also in terms of transient response. The so-called attitude 'quickness' (Ref. 5.58), to be discussed in more detail in Chapter 6, represents a transient property of an aircraft's response to pilot control inputs. The same concept can be extended to the analysis of the aircraft response to discrete gusts, as summarized in Fig. 5.36. The response quickness, shown in Fig. 5.36(a) applied to the normal velocity response, is extracted from a signal by identifying significant changes (Δw – strictly the integral of acceleration) and estimating the associated maximum or peak rate of change, in this case peak normal acceleration ($a_{z\,pk}$); the quickness associated with the event is then given by the discrete parameter

$$Q_W = \frac{a_{z\,pk}}{\Delta w} \tag{5.84}$$

Clearly, each discrete gust has an associated quickness, which actually approximates to $1/L$ in the limit of a linear ramp gust. Quickness values can then be plotted as points on charts as shown in Fig. 5.36(b). In this case we have plotted the values as a function of the gust input intensity, assuming a unique relationship between the input–output pair. Bradley *et al.* (Ref. 5.44) have shown that the quickness points group along the tuning lines, related to $\bar{\gamma}$, as shown in hypothetical form in Fig. 5.36(b). Also shown in the figure are contours of equi-responsiveness or equi-comfort which suggest a possible format for specifying ride quality.

The ongoing research on the topic of ride qualities is likely to produce alternative approaches to modelling and analysing disturbance and response, derived as ever from different perspectives and experiences. The key to more general acceptance will certainly be validation with real-world experience and test data, and it is in this area that the major gaps lie and much more work needs to be done. There are very few sets of test data available, and perhaps none that is fully documented, that characterize the disturbance and the helicopter response to turbulence in low-level nap-of-the-earth flight; it is a prime area for future research. Data are important to validate simulation modelling and also to establish new ride criteria that can be used with confidence in the design of new aircraft and the associated automatic flight control systems. The current specification standards for rotorcraft handling do not make any significant distinction between the performance associated with the response to controls and disturbances. Clearly, an aircraft which is naturally agile is also likely to be naturally bumpy, and an active control system will need to have design features that cope with both handling and ride quality requirements. Fortunately, the handling qualities standards, perhaps the more important of the two, are now, in general, better understood, having been the subject of intense investigations over the last 15 years. Handling qualities forms the subject of the remaining two chapters of this book.

*The Canadian NRC variable stability (fly-by-wire) Bell 205 during
a handling qualities evaluation near Ottawa
(Photograph from the author's collection)*

6 Flying qualities: objective assessment and criteria development

> *Experience has shown that a large percentage, perhaps as much as 65%, of the life-cycle cost of an aircraft is committed during the early design and definition phases of a new development program. It is clear, furthermore, that the handling qualities of military helicopters are also largely committed in these early definition phases and, with them, much of the mission capability of the vehicle. For these reasons, sound design standards are of paramount importance both in achieving desired performance and avoiding unnecessary program cost. ADS-33 provides this sound guidance in areas of flying qualities, and the authority of the new standards is anchored in a unique base of advanced simulation studies and in-flight validation studies, developed under the TTCP collaboration.*
>
> *(From the TTCP* Achievement Award, Handling Qualities Requirements for Military Rotorcraft, 1994)*

6.1 GENERAL INTRODUCTION TO FLYING QUALITIES

In Chapter 2, the Introductory Tour of this book, we described an incident that occurred in the early days of helicopter flying qualities testing at the Royal Aircraft Establishment (Ref. 6.1). An S-51 helicopter was being flown to determine the longitudinal stability and control characteristics, when the pilot lost control of the aircraft. The aircraft continued to fly in a series of gross manoeuvres before it self-righted and the pilot was able to regain control and land safely. The incident highlighted the potential consequences of poor handling qualities – pilot disorientation and structural damage. In the case described the crew were fortunate; in many other circumstances these consequences can lead to a crash and loss of life. Good flying qualities play a major role in contributing to flight safety. But flying qualities also need to enhance performance, and this tension between improving safety and performance in concert is ever present in the work of the flying qualities engineer. An arguable generalization is that military requirements lean towards an emphasis on performance while civil requirements are more safety oriented. We shall address this tension with some fresh insight later in this chapter, but first we need to bring out the scope of the topic.

The 'original' definition of handling qualities by Cooper and Harper (Ref. 6.2),

> *Those qualities or characteristics of an aircraft that govern the ease and precision with which a pilot is able to perform the tasks required in support of an aircraft role*

---

* The Technical Cooperation Programme (UK, United States, Canada, Australia, New Zealand).

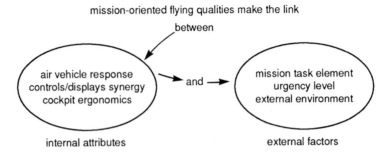

Fig. 6.1 Mission oriented flying qualities (Ref. 6.3)

still holds good today, but needs to be elaborated to reveal the scope of present day and future usage. Figure 6.1 attempts to do this by illustrating the range of influences – those external to and those internal to the aircraft and pilot. This allows us to highlight flying qualities as the synergy between these two groups of influencing factors. To emphasize this point, it could be argued that without a complete description of the influencing factors, it is ambiguous to talk about flying qualities. It could also be argued that there is no such thing as a Level 1 or a Level 2 aircraft, in the Cooper–Harper parlance. The quality can be referenced only to a particular mission (or even mission task element (MTE)) in particular visual cues, etc. In a pedantic sense this argument is hard to counter, but we take a more liberal approach in this book by examining each facet, each influencing factor, separately, and by discussing relevant quality criteria as a sequence of one-dimensional perspectives. We shall return to the discussion again in Chapter 7.

In the preceding paragraph, the terms handling qualities and flying qualities have deliberately been used interchangeably and this flexibility is generally adopted throughout the book. At the risk of distracting the reader, it is fair to say, however, that there is far from universal agreement on this issue. In an attempt to clarify and emphasize the task dependencies, Key, as discussed in Ref. 6.4, proposes a distinction whereby flying qualities are defined as the aircraft's stability and control characteristics (i.e., the internal attributes), while handling qualities are defined with the task and environment included (external influences). While it is tempting to align with this perspective, the author resists on the basis that provided the important influence of task and environment are recognized, there seems no good reason to relegate flying to be a subset of handling.

In the structure of current Civil and Military Requirements, good flying qualities are conferred to ensure that safe flight is guaranteed throughout the operational flight envelope (OFE). The concept of flying quality requires a measurement scale to judge an aircraft's suitability for role, and most of the efforts of flying qualities engineers over the years have been directed at the development of appropriate scales and metrics, underpinned and substantiated by flight test data. The most developed and widely recognized quality scale is that due to Cooper and Harper (Ref. 6.2). Goodness, or quality, according to Cooper–Harper, can be measured on a scale spanning three levels (Fig. 6.2). Aircraft are normally required to be Level 1 throughout the OFE (Refs 6.5, 6.6); Level 2 is acceptable in failed and emergency situations but Level 3 is considered unacceptable. The achievement of Level 1 quality signifies that a minimum required standard has been met or exceeded in design, and can be

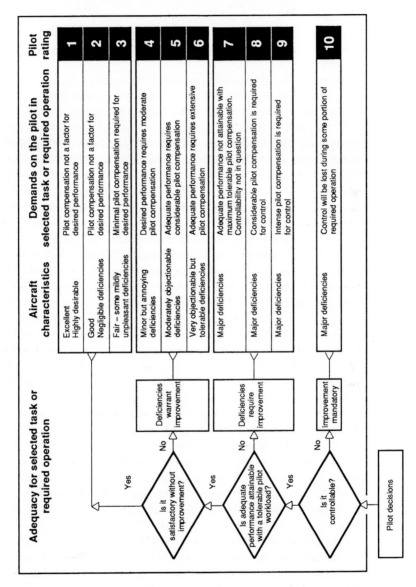

Fig. 6.2 The Cooper–Harper handling qualities rating scale (Ref. 6.2)

expected to be achieved regularly in operational use, measured in terms of task performance and pilot workload. Compliance flight testing is required to demonstrate that a helicopter meets the required standard and involves clinical measurements of flying qualities parameters for which good values are known from experience. It also involves the performance of pilot-in-the-loop MTEs, along with the acquisition of subjective comments and pilot ratings. The emphasis on minimum requirements is important and is made to ensure that manufacturers are not unduly constrained when conducting their design trade studies. Establishing the quality of flying, therefore, requires objective and subjective assessments which are the principal topics of Chapters 6 and 7 respectively.

We refer to the subjective pilot ratings given on the Cooper–Harper scale as handling qualities ratings (HQRs). HQRs are awarded by pilots for an aircraft flying an MTE, and are determined by the pilot following through the decision tree shown in Fig. 6.2, to arrive at his or her rating based on their judgement of task performance achieved and pilot workload expended. The task performance requirements will have been set at desired and adequate levels and the pilot should have sufficient task cues to support the judgement of how well he or she has done. In a well-defined experiment this will usually be the case, but poorly defined flight tests can lead to increased scatter between pilots on account of the variations in perceived performance (see Chapter 7 for more discussion on this topic). Task performance can be measured, be it flight path accuracy, tracking performance or landing scatter and the results plotted to give a picture of the relative values of flying quality. The other side of the coin, pilot workload, is much more difficult to quantify, but we ask test pilots to describe their workload in terms of the compensation they are required to apply, with qualifiers – minimal, moderate, considerable, extensive and maximum. The scale implies an attempt to determine how much spare capacity the pilot has to accomplish other mission duties or to think ahead and react quickly in emergencies. The dual concepts of 'attentional demand' and 'excess control capacity' have been introduced by McRuer (Ref. 6.7) to distinguish and measure the contributing factors to the pilot workload. However subjective, and hence flawed by the variability of pilot training and skill, the use of HQRs may seem, along with supporting pilot comment and task performance results, it has dominated flying qualities research since the 1960s. It is recognized that at least three pilots should participate in a flying qualities experiment (preferably more, ideally five or six) and that HQRs can be plotted with mean, max and min shown; a range of more than two or three pilot ratings should alert the flying qualities engineer to a fault in experimental design. These are detailed issues and will be addressed later in Chapter 7, but the reader should register the implied caution; misuse of HQRs and the Cooper–Harper scale is all too easy and too commonly found.

Task performance requirements drive HQRs and give modern military flying qualities standards, like the US Army's ADS-33C (Ref. 6.5), a mission orientation. The flying qualities are intended to support the task. In a hierarchical manner, ADS-33C defines the response types (i.e., the short-term character of response to control input) required to achieve Level 1 or 2 handling qualities for a wide variety of different MTEs, in different usable cue environments (UCE) for normal and failed states, with full and divided pilot attention. Criteria are defined for both hover/low speed and forward flight, in recognition of the different MTEs and related pilot control strategies in the two operating regimes. Within these flight phases, the criteria can be further related to the level of aggressiveness used by the pilot in attacking a manoeuvre or MTE. At a deeper level, the response characteristics are broken down in terms of amplitude and frequency range, from the small amplitude, higher frequency requirements set by criteria like equivalent low-order system response or bandwidth, to the large amplitude manoeuvre requirements set by control power. By comparison, the equivalent fixed-wing requirements, MIL-F-8785C (Ref. 6.6), take a somewhat different perspective, with flight phases and aircraft categories, but the basic message is the same – how to establish flying quality in mission-related tasks. The innovations of ADS-33 are many and varied and will be covered in this chapter; one of the many significant departures from its predecessor, MIL-H-8501A, is that there is no categorization according to aircraft size or, explicitly, according to role, but only by the required MTEs. This emphasizes the multi-role nature of helicopters and gives the new specification document generic value. Without doubt, ADS-33 has resulted in a significant increase in attention to flying qualities in the procurement process and manufacturing since its first publication in the mid-1980s, and will possibly be perceived in later years as marking a watershed in helicopter development. The new flying qualities methodology is best illustrated by Key's diagram (Ref. 6.8) shown in Fig. 6.3.

In this figure the role of the manufacturer is highlighted, and several of the ADS-33 innovations that will be discussed in more detail later in this chapter are brought out, e.g., response types and the UCE.

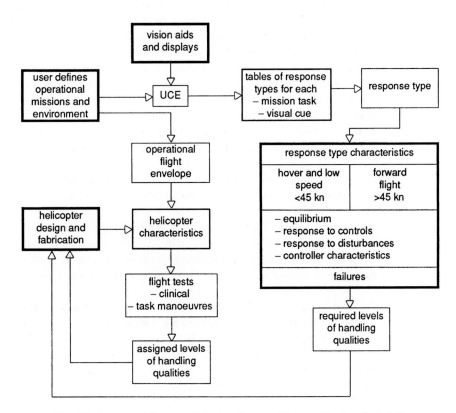

Fig. 6.3 Conceptual framework for handling qualities specification (Ref. 6.8)

In a similar timeframe to the development of ADS-33C, the revision to the UK's Def Stan 00970 was undertaken (Ref. 6.9). This document maintained the UK tradition of stating mandatory requirements in qualitative terms only, backed up with advisory leaflets to provide guidance on how the characteristics may be achieved. An example of the 970 requirements, relating to response to control inputs, reads

> *10.1.2 The flying controls should not be over-sensitive, to an extent that leads directly to difficulty in establishing or maintaining any desired flight condition, or that promotes pilot induced oscillations (leaflet 600/7, para 3).*

With this intentionally catch-all parlance, Def Stan 970 leaves it to the manufacturer to decide how this is to be achieved. In contrast, the criteria in ADS-33 quantify responsiveness and sensitivity and lay down mandatory quality boundaries on measurable parameters. A comparison of the philosophy in the two distinct approaches could occupy much of this chapter but the author is reluctant to embark on such a venture. In the author's view, however, one thing that does need to be stressed is that the resources applied to the development of ADS-33, and the harnessing of the best international facilities, have resulted in a breakthrough in the development of helicopter flying qualities – all based on the creation of a new flying qualities test database, the absence of which has hindered several previous initiatives over the last 25 years. Def Stan 970 complements the more substantiated US requirements, and those areas where 970 provides additional insight will be highlighted in this chapter.

If we turn to flying qualities requirements for civil helicopters, we find safety a much more significant driver and the requirements are once again more qualitative in nature (Refs 6.10–6.13). Of major concern are the safety of operations in the ever-decreasing weather minima and the ability of the pilot to recover to safe flight following major system failures. Handling qualities research efforts have therefore been focused on the development of requirements for artificial stability to support IFR flight and flight test procedures for recovering from failures. The increased emphasis on military flying qualities requirements in recent years has also prompted a closer examination of the potential of the new criteria formats for civil applications. One such review is reported in Ref. 6.12, and some of the ideas arising from this study will be sampled throughout Chapters 6 and 7.

This chapter is primarily about how Level 1 helicopters should behave and how to test for compliance, not how they are made. Design issues are touched on occasionally in the context of criteria development but will not be central to the discussion. The reader is referred to Chapters 4 and 5 for implicit design considerations through the analysis of trim, stability and response. However, the subject of design for helicopter flying qualities, including bare airframe and stability and control augmentation, is left for a future book and perhaps to an author closer to the manufacturing disciplines.

6.2 Introduction and Scope: The Objective Measurement of Quality

This chapter is concerned with those flying qualities characteristics that can be quantified in parametric, and hence, numerical, terms. A range of new concepts in quality discrimination were established during the 1980s and are now taking a firm hold in the development of new projects in both Europe and the United States. Some background interpretation and discussion of the development rationale are provided from the author's own perspective, particularly relating to the quantitative criteria in the US Army's ADS-33. Before ADS-33, the existing mandatory and even advisory design criteria were so ill-matched to the high performance helicopter that achieving compliance with these in simulation provided little insight into problems that might occur in flight. Furthermore, aircraft that demonstrated compliance during flight test could still be unfit for their intended role. These two paradoxical situations have prevailed since design criteria were first written down and their continued existence can be tolerated only on two counts. First, there is the argument that criteria should not constrain the design creativity unduly and, second, that handling qualities of new designs in new roles should not be prejudiced by a limited database derived from older types. These two points should serve to alert us to the need for living requirements criteria that are robust; the term robust is applicable in this context to requirements that meet the so-called CACTUS rules (Ref. 6.14), namely

(1) Complete – covering all missions, flight phases and response characteristics, i.e., all the internal and external influencing factors;
(2) Appropriate – the criteria formats should be robust enough to discern quality in the intended range of application (e.g., frequency domain rather than time domain criteria for pilot-induced oscillation (PIO) boundaries);
(3) Correct – all Level 1/2 and 2/3 quality boundaries should be positioned correctly;
(4) Testable – from design through to certification;
(5) Unambiguous – clear and simple, easy to interpret; perhaps the most challenging of the rules, and vital for widespread acceptance;
(6) Substantiated – drawn and configured from a database derived from similar types performing similar roles; perhaps the single most important rule that underpins the credibility of new criteria.

Striving to meet the CACTUS rules is recognized as a continuing challenge for the flying qualities engineer as roles develop and new data become available. The criteria discussed in this chapter conform to the rules to varying degrees, some strongly, some hardly at all; we shall attempt to reflect on these different levels of conformity as the chapter progresses.

Turning to the framework of ADS-33, Table. 6.1 provides an overview snapshot for selected MTEs. The figure links together the key innovations of the specification – the response types (RT), the mission task elements (MTE) and the usable cue environment (UCE). The UCE, derived from pilot subjective ratings of the quality of visual task cues, will be discussed in more detail later in Chapter 7. Its introduction into ADS-33 draws attention to the need for different flying qualities in different visual conditions, in particular in so-called degraded visual environments (DVE), when flying close to the ground. A UCE of 1 corresponds to conditions where the pilot has very good visual cues to support the control of attitude and velocity, while a UCE 3 corresponds to conditions where the pilot can make small and gentle corrections only because of deficient visual cues. Table 6.1 tells us that to achieve Level 1 flying qualities in the selected MTEs, rate command is adequate in a UCE 1 while the requirements become more demanding in poorer UCEs. Attitude command is required for pitch and roll in a UCE of 2 and translational rate command with position and height hold for a UCE 3.

The RT relates to the character of the response in the first few seconds following a pilot-applied step control input. Figure 6.4 shows how the attitude varies for the

Table 6.1 Response-type requirements in different usable cue environments for selected MTEs

UCE	Response types in hover/low-speed flight	Response types in forward flight
UCE 3	TRC + RCDH + RCHH + PH	
UCE 2	ACAH + RCDH + RCHH + PH	RC + TC
UCE 1	RC	

Response types		Selected MTEs
RC	– rate command	rapid bob up/down, hover turn,
TC	– turn coordination (applies to yaw and pitch response)	rapid transition to precision hover, sonar dunking, rapid
ACAH	– attitude command, attitude hold (roll and pitch)	sidestep, rapid accel-decel,
RCDH	– rate command, direction hold (yaw)	target aquisition and tracking,
RCHH	– rate command, height hold (heave)	divided attention tasks
PH	– position hold (horizontal plane)	
TRC	– translational rate command	

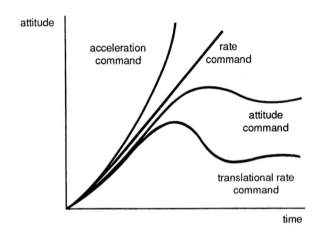

Fig. 6.4 Attitude response type following step cyclic control input

different types in pure form, including for completeness, the acceleration RT. Rate Command (RC) response is generally regarded as the simplest practical type found with conventional helicopters. The definition of rate command in ADS-33 actually allows for variations in the response away from the pure rate to include the variety of current helicopters that do not fall neatly into the pure categories but still exhibit satisfactory handling qualities. A basic requirement is that the initial and final cockpit controller force, following an attitude change, shall be the same sign. As the RTs become more directly related to translational response, two associated factors impact the pilotage. First, attitude command (AC) is easier to fly than RC, and translational rate command (TRC) is easier to fly than attitude command, attitude hold (ACAH). With TRC, not only is the attitude loop automatically closed, thus relieving the pilot of the higher gain-attitude stabilization, but also the velocity feedback loop is automatically closed, reducing piloting essentially to a steering task. Second, the additional stability is achieved at the expense of manoeuvrability and agility. The highest performance can, in principle, be achieved with an acceleration command RT through a direct force/moment inceptor, but the pilot would have to work so hard (performing three mental integrations) to achieve flight path accuracy that the additional performance 'available' would almost certainly be wasted. As the RT becomes progressively more stable, the available manoeuvre performance envelope reduces. This is completely in accord with the need for the higher levels of augmentation, of course. Pilots will not normally require high performance in DVE. The different requirements highlighted by Table 6.1 reinforce the importance of task in the quality of flying. But it is not sufficient to define the RT; the detailed character of the response in the short–long term and at small–large amplitude needs quantifying. We need a framework for this deeper study of response quality.

Figure 6.5 illustrates a convenient division of aircraft response characteristics on one of our reference diagrams, showing response frequency against amplitude. The discriminators – large, medium and small for amplitude and long, mid and short term for frequency – are intended to encompass, in a meaningful and systematic way, all of the task demands the pilot is likely to encounter. The framework includes the zero and very low frequency trim, and the zero and very low amplitude stability areas. A third dimension, cross-coupling, is added to highlight that direct response characteristics are insufficient to describe response quality fully. The hyperbolic-like boundary shows how the manoeuvre envelope of an aircraft is constrained – as the amplitude increases, then various physical mechanisms come into play that limit the speed at which the manoeuvre can be accomplished, e.g., control range, actuator rate, rotor thrust/moment capability, etc. The boundary represents the dynamic OFE and flying qualities criteria are required across the full range of frequency and amplitude.

This chapter is primarily concerned with the characteristics required to confer a helicopter with Level 1 flying qualities, although we shall give some attention to Level 2 characteristics since most operational helicopters spend a considerable time in Level 2.

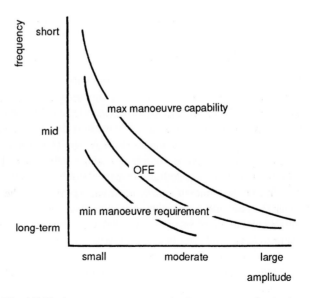

Fig. 6.5 Equi-response contours on the frequency–amplitude plane

Within the sections, each of the four primary response axes will be discussed – roll, pitch, yaw and heave, along with the variety of different cross-coupling mechanisms. The other important internal factors, inceptors and displays, will be discussed in Chapter 7; they both have strong influences in flying qualities, particularly for flight in degraded visual conditions and at flight envelope limits where tactile cueing through the pilot's controls is particularly important. Where appropriate, some comparison with fixed-wing aircraft criteria and quality boundaries will be made. A as the reader embarks on this chapter, it is worth noting that while considerable progress has been made with helicopter flying qualities criteria between the mid-1980s and early-1990s, the evolutionary process is continuing. At the time of writing the second edition of this book the same is true, and some of the developments, particularly the requirements criteria in ADS-33, are summarized in Section 6.9. Chapter 8 addresses degraded handling qualities. The author has had to be selective with the material covered in view of the considerable amount of relevant published work in the literature. This comment is particularly germane to the coverage given to ADS-33 criteria; readers are earnestly referred to these design guidelines and associated references in the open literature for a more precise and complete definition of handling boundaries. We begin with the roll axis and this will allow us to introduce and develop a range of concepts also applicable to the other axes of control.

6.3   ROLL AXIS RESPONSE CRITERIA

The ability to generate rolling moments about the aircraft's centre of gravity serves three purposes. First, to enable the pilot to trim out residual moments from the fuselage, empennage and tail rotor, e.g., in a pure hover, sideslipping flight, slope landings, hovering in side-winds. Second, so that the rotor thrust vector can be reoriented to manoeuvre in the lateral plane, e.g., repositioning sidestep, attitude regulation in tight flight path control. Third, so that the pilot can counteract the effects of atmospheric disturbances. All three can make different demands on the aircraft, and flying qualities criteria must try to embrace them in a complementary way.

6.3.1   Task margin and manoeuvre quickness

The roll axis has probably received more attention than any other over the years, possibly as a carry-over from the extensive research database in fixed-wing flying qualities, but also because roll control arguably exhibits the purest characteristics and is most amenable to analysis. A comprehensive review of roll flying qualities is contained in Ref. 6.15. In this work, Heffley and his co-authors introduced the concept of the 'task signature' or 'task portrait', discussed in Chapter 2 of this book (see Fig. 2.40), and also the 'task margin', or the control margin beyond that required for the task in hand. The basic ideas are summarized conceptually in Fig. 6.6, which shows how the roll rate requirements vary with manoeuvre amplitude (i.e., change in roll angle). The manoeuvre demand limit line is defined by the tasks required of the helicopter in the particular mission. The task margin is the additional vehicle capability required for emergency operations. The manoeuvre amplitude range can conveniently be divided into the three ranges discussed earlier – small, moderate and large – corresponding to precise tracking, discrete manoeuvring and maximum manoeuvre tasks, as shown.

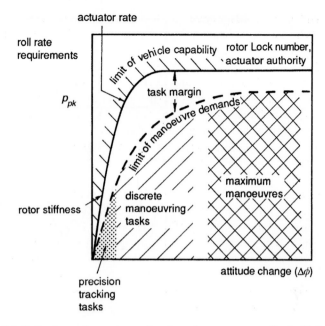

Fig. 6.6 Roll rate requirements as a function of manoeuvre amplitude (Ref. 6.15)

Highlighted in Fig. 6.6 are the principal design features that define the outer limit of vehicle capability – rotor stiffness in the small amplitude range and Lock number in the high amplitude range; the actuation rate and authority limits also define the shape of the capability boundary in the moderate to high amplitude range. To convert Fig. 6.6 into a form compatible with the frequency/amplitude diagram in Fig. 6.5 requires us to look back at the very simple task signature concept. Figure 6.7 shows the time histories of lateral cyclic, roll attitude and rate for a Lynx flying a slalom MTE (Ref. 6.16). The manoeuvre kinematics can be loosely interpreted as a sequence of attitude changes each associated with a particular roll rate peak, emphasized in the phase plane portraits in Fig. 6.8. For the case of the Lynx, roll control is essentially rate command, so that the attitude rate follows the control activity reasonably closely (see Fig. 6.9). The task signature portrait in Fig. 6.10 (a) shows selected rate peaks plotted against the corresponding attitude change during the slalom. Each point represents a discrete manoeuvre change accomplished with a certain level of aggression or attack. Points that lie on the same 'spoke' lines correspond to similar levels of attack by the pilot. We reserve the descriptors 'attack' and 'aggression' for the pilot behaviour, and use the expression 'quickness' to describe this temporal property of the manoeuvre. Manoeuvre quickness, or in the present case roll attitude quickness, is the ratio of peak rate to attitude change during a discrete manoeuvre and was first proposed in Ref. 6.15 as an alternate flying qualities or control effectiveness parameter:

$$\text{roll attitude quickness} = \frac{p_{pk}}{\Delta\phi} \tag{6.1}$$

The data in Fig. 6.10 (a) are transformed into quickness values in Fig. 6.10 (b). If we transform the generalized boundaries on Fig. 6.6 into quickness, by plotting the slope

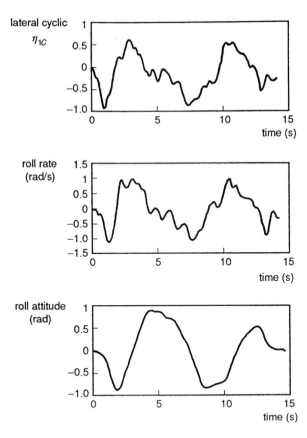

Fig. 6.7 Control and response time histories for Lynx flying slalom

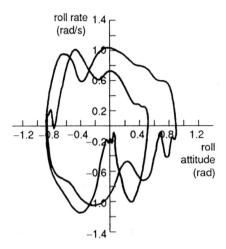

Fig. 6.8 Phase plane portrait for Lynx flying slalom manoeuvre

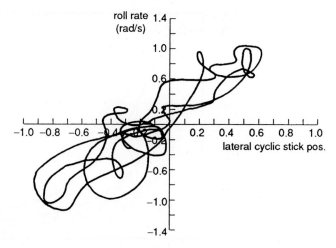

Fig. 6.9 Lateral cyclic-roll rate cross-plot

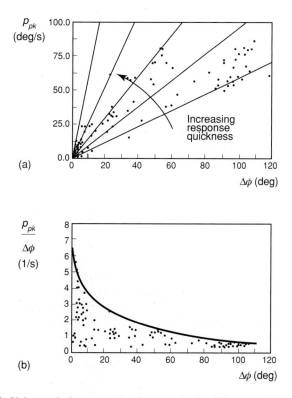

Fig. 6.10 Slalom task signature: (a) roll rate peaks for different attitude changes; (b) roll attitude quickness points

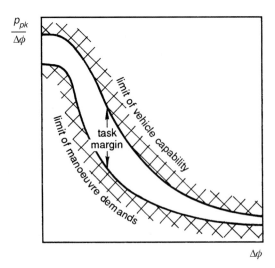

Fig. 6.11 Generalized response quickness diagram (Ref. 6.15)

of the boundary lines against the attitude changes, we arrive at Fig. 6.11, showing the characteristic hyperbolic shape with amplitude of Fig. 6.5. For a given roll attitude change, there will be a maximum value of achievable quickness defined by the limit of the vehicle capability. When the manoeuvre amplitude is high enough, the limiting function in Fig. 6.11 will genuinely be hyperbolic, as the maximum rate is achieved and the limit is inversely proportional to the attitude change. This trend is confirmed in Fig. 6.10 (b), which shows the envelope of maximum quickness derived in the Lynx slalom flight trials described in Ref. 6.16. The highest roll attitude changes of more than 100° were experienced during the roll reversal phases of the MTE. Values of quickness up to 1 rad/s were measured during these reversals, indicating that pilots were using at least 100°/s of roll rate at the highest levels of aggression. In the low amplitude range the quickness rises to more than 5 rad/s, although with the small values of roll attitude change here, the extraction of accurate values of quickness is difficult. The quickness parameter has gained acceptance as one of the innovations of ADS-33, applicable to the moderate amplitude range of manoeuvres. We shall return to this discrimination later in this section but first we examine some of the theoretical aspects of quickness, applied to a simple model of roll control.

The first-order approximation to roll response has been discussed in Chapters 2, 4 and 5, and the reader needs to be aware of the limited range of validity when applied to helicopters. Nevertheless, this simple model can be used to gain useful insight into the theoretical properties of quickness. We consider the first-order, differential equation of motion of a rate command response-type helicopter, written in the form

$$\dot{p} - L_p p = L_{\theta_{1c}} \theta_{1c} = -L_p p_s \theta_{1c} \tag{6.2}$$

where p is the roll rate and θ_{1c} the lateral cyclic control; we have neglected any rotor or actuation dynamics in this model. The damping and control derivatives have been discussed in detail in Chapter 4. The response to a step input in lateral cyclic is an exponential growth to a steady-state roll rate p_s. To derive a value for attitude

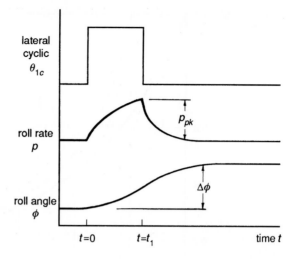

Fig. 6.12 Simple rate response to pulse lateral cyclic input

quickness we need to consider the response to a pulse input of duration t_1, which leads to a discrete attitude change $\Delta\phi$ (Fig. 6.12). Analytic expressions for the roll rate and attitude response expressions then have two forms, one during the application of the pulse and the second after the pulse:

$$t \le t_1 : p = p_s\left(1 - e^{L_pt}\right)\theta_{1c}, \qquad \phi = \frac{p_s}{L_p}\left(1 + L_pt - e^{L_pt}\right)\theta_{1c}$$

$$\tag{6.3}$$

$$t > t_1 : \quad p = p_s e^{L_pt}\left(e^{-L_pt_1} - 1\right)\theta_{1c}, \qquad \phi = \phi(t_1) - \frac{p_s}{L_p}\left(e^{L_pt_1} - 1\right)$$

$$\left(e^{L_p(t - t_1)} - 1\right)\theta_{1c} \tag{6.4}$$

From these expressions the attitude quickness can be formed and, after some reduction, we obtain the simple expression

$$\frac{p_{pk}}{\Delta\phi} = -\frac{L_p}{\hat{t}_1}(1 - e^{-\hat{t}_1}) \tag{6.5}$$

where

$$\hat{t}_1 = -L_p t_1 \tag{6.6}$$

The normalized time $\hat{t}_1$ given by eqn 6.6 can be thought of as the ratio of the manoeuvre duration to the time constant of the aircraft. The quickness, normalized by the roll damping, is shown plotted against $\hat{t}_1$ in Fig. 6.13. One important result of this analysis is that the quickness is independent of control input size. For a 2-s pulse, the quickness will be the same from a small and large input; this is essentially a property of the linear system described by eqn 6.2 and may no longer be true when nonlinearities are

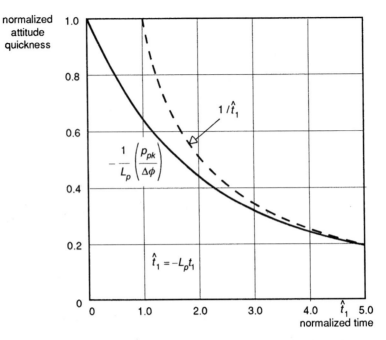

Fig. 6.13 Variation of normalized quickness with manoeuvre time ratio

present. For small values of $\hat{t}_1$, corresponding to short-duration manoeuvres relative to the aircraft time constant, the quickness approximates to the roll damping itself, i.e.,

$$\hat{t}_1 \to 0, \qquad \frac{p_{pk}}{\Delta\phi} \to -L_p \tag{6.7}$$

As $\hat{t}_1$ becomes large, the quickness decreases inversely with $\hat{t}_1$, i.e.,

$$\hat{t}_1 \to \infty, \qquad \frac{p_{pk}}{\Delta\phi} \to \frac{1}{t_1} \tag{6.8}$$

Equation 6.8 tells us that when the manoeuvre is slow relative to the aircraft time constant, then the latter plays a small part in the quickness and the attitude change is practically equal to the roll rate times the pulse time. Equation 6.7 describes the limit for small-duration control inputs, when the roll transient response is still evolving. This case requires closer examination because of its deeper significance which should become apparent. The inverse of the roll damping is equal to the time to reach 63% of p_s following a step input, but the parameter has another related interpretation in the frequency domain. Heuristically, frequency would appear to be more significant than amplitude in view of the insensitivity of quickness to control input size. The phase angle between the roll rate as output and the lateral cyclic as input as a function of frequency is given by the relationship

$$\Phi = \left(\frac{\bar{p}}{\bar{\theta}_{1c}}(\omega)\right) = \tan^{-1}\left(\frac{\omega}{L_p}\right) \tag{6.9}$$

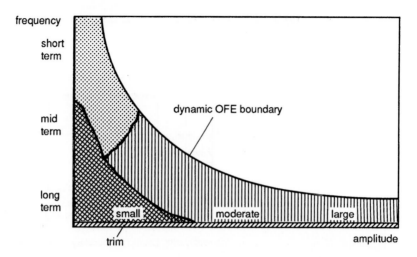

Fig. 6.14 Characterization of aircraft response in four regions

When the phase between p and θ_{1c} is 45°, then the frequency is numerically equal to the damping L_p. This corresponds to the case when the attitude response is 135° out of phase with the control input. We shall see later in this section that the frequency when the attitude response lags the control by 135° is defined by a fundamental handling parameter – the (open-loop) attitude bandwidth. For non-classical response types we shall show that the attitude bandwidth is a more significant parameter than the roll damping and conforms more closely with many of the CACTUS rules. Bandwidth is one of the central parameters in ADS-33.

Returning to our framework diagram, we are now in a better position to examine quality criteria and the associated flight test measurement techniques; we divide the diagram into three 'dynamic' regions as shown (Fig. 6.14). We broadly follow the ADS-33 definition of the amplitude ranges:

(1) small, $\phi < 10°$, continuous closed loop, compensatory tracking;
(2) moderate, $10° < \phi < 60°$, pursuit tracking, terrain avoidance, repositioning;
(3) large, $\phi > 60°$, maximum manoeuvres;

and review selected military and civil criteria. On Fig. 6.14 we have included the narrow range of zero to very low frequency to classify trim and quasi-static behaviour. The first of the 'manoeuvre' regions combines moderate and large amplitude roll attitude criteria.

6.3.2   Moderate to large amplitude/low to moderate frequency: quickness and control power

The most appropriate parameter for defining the quality of flying large amplitude manoeuvres is the control power available, i.e., the maximum response achievable by applying full control from the trim condition. For rate command systems this will be measured in deg/s, while for attitude command response, the control power is measured in degrees. This 'new' definition found in ADS-33 contrasts with the earlier MIL-H-8501 and fixed-wing criteria where the control power related to the maximum control

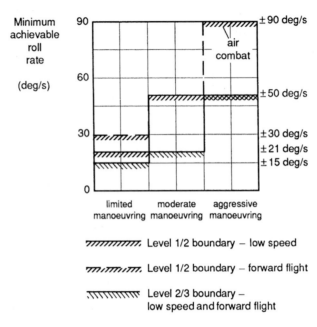

Fig. 6.15 Minimum roll control power requirements – rate response type (Ref. 6.5)

moment available. To avoid confusion we conform with the ADS-33 definition. Perhaps more than any other handling qualities parameter, the control power is strongly task dependent. Figure 6.15 illustrates this with the minimum control power requirements for Level 1 handling qualities (according to ADS-33) corresponding to MTEs that require limited, moderate and aggressive manoeuvring. The figure shows requirements for rate response types in low speed and forward flight MTEs. The minimum rate control power requirements vary from 15°/s in forward flight IMC through to 90°/s in air combat.

Ground-based simulations conducted at the RAE in the late 1970s (Ref. 6.17) were aimed at defining the agility requirements for battlefield helicopters and roll control power was given particular attention. Figure 6.16 shows the maximum roll rates used in the roll reversal phase of a triple bend manoeuvre as a function of roll angle change for various flight speeds. The dashed manoeuvre line corresponds to the theoretical case when the reversal is accomplished in just 2 s. For the cases shown, the control power was set at a high level (> 120°/s) to give the test pilots freedom to exploit as much as they needed. For speeds up to about 70 knots, the pilot control strategy followed the theoretical line, but as the speed increased to 90 and 100 knots there was a marked increase in the maximum roll rates used. This rapid change in control strategy at some critical point as task demands increase is significant and will be discussed later in Chapter 7. The spread of data points corresponds to different rotor configurations, resulting in different roll time constants as shown in the figure. With the larger time constants, corresponding to articulated or teetering rotor heads, the pilots typically used 30–40% more roll rate than with the shorter time constants typical of hingeless rotors. It appears that with 'soft' rotors, the pilot will use more control to quicken the manoeuvre. This more complex control strategy leads to an increase in

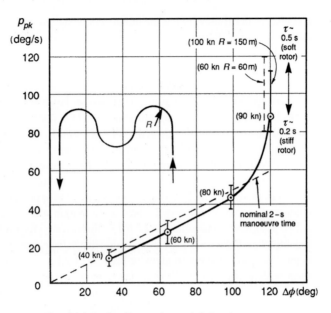

Fig. 6.16 Peak roll rates from triple bend manoeuvre

workload and a degradation in the pilot opinion of handling qualities, a topic we shall return to in Chapter 7. The study of Ref. 6.17 concluded that for rapid nap-of-the-earth (NoE) manoeuvring in the mid-speed range, a minimum control power requirement of 100°/s was necessary for helicopters with moderately stiff rotors, typical of today's hingeless rotor configurations.

The measurement of control power is, in theory, quite straightforward; establish trimmed flight and apply maximum control until the response reaches its steady state. In practice, unless great care is taken, this is likely to result in large excursions in roll, taking the aircraft to potentially unsafe conditions, especially for rate command response types. A safer technique is to establish a trimmed bank angle ($< 60°$) and apply a moderate step input in lateral cyclic, recovering before the aircraft has rolled to the same bank in the opposite direction. The manoeuvre can now be repeated with increasing control input sizes and several data points collected to establish the functional relationship between the roll response and control step size. Applying this incremental technique it will usually be unnecessary to test at the extremes of control input size. Either the minimum requirements will have been met at lower input sizes or the response will be clearly linear and extrapolation to full control is permissible. For the cases requiring the highest control powers, e.g., air combat MTEs, it may be necessary (safer) to capture the data in a closed-loop flight test, e.g., with the aircraft flown in an air combat MTE. In both open- and closed-loop testing, two additional considerations need to be taken into account. First, care should be taken to avoid the use of pedals to augment the roll rate. In operational situations, the pilot may choose to do this to increase performance, but it can obscure the measurement of roll control power. Second, at high rotor thrust, the rotor blades can stall during aggressive manoeuvring, with two effects. The loss of lift can reduce the roll control effectiveness and the increase in drag can cause torque increases that lead to increased power demands. These are real effects and if they inhibit the attainment of the minimum requirements then the design is lacking. Because of the potentially damaging nature of the test manoeuvres for control power estimation, online monitoring of critical rotor and airframe stresses is desirable, if not essential.

Neither the civil handling requirements (Ref. 6.10), nor Def Stan 00970 (Ref. 6.9), refer to criteria for control power *per se*, but instead set minimum limits on control margin in terms of aircraft response. This normally relates to the ability to manoeuvre from trimmed flight at the edge of the OFE. Def Stan 00970 defines control margin in terms of the control available to generate a response of 15°/s in 1.5 s. The old MIL-H-8501A required that at the flight envelope boundary, cyclic control margins were enough to produce at least 10% of the maximum attainable hover roll (or pitch) moment. The FAAs adopt a more flexible approach on the basis that some configurations have been tested where a 5% margin was sufficient and others where a 20% margin was inadequate (Ref. 6.13). Specifically, for FAA certification, what is required from flight tests is a demonstration that at the never-exceed airspeed, 'a lateral control margin sufficient to allow at least 30° banked turns at reasonable roll rates' must be demonstrated. ADS-33 is clearly more performance oriented when it comes to setting minimum control power requirements, and this philosophy extends to the moderate amplitude criteria, where the introduction of a new parameter, the manoeuvre quickness, has taken flying qualities well and truly into the nonlinear domain.

Moderate amplitude roll requirements broadly apply to manoeuvres within the range $-60° < \phi < 60°$ that include military NoE MTEs, such as quick-stop, slalom, and civil helicopter operations in harsh conditions, e.g., recovery to confined areas in gusty conditions and recovery following failed engine or SCAS. The development of attitude quickness has already been discussed in some detail. The definition of quickness used in ADS-33 has been developed to relate to non-pure response types and includes a subtlety to account for oscillatory responses. Figure 6.17 shows the roll axis quickness criteria boundaries from ADS-33, including the definition of the attitude parameters required to derive quickness. Once again the task-oriented nature of flying qualities is highlighted by the fact that there are different boundary lines for different MTE classes, even within the low-speed regime (see Ref. 6.5 for full details). Figure 6.18 shows the quickness envelope for the Lynx flying a lateral sidestep compared with the two Level 1 ADS-33 boundaries. The companion Fig. 6.19 illustrates the phase plane portraits for the sidestep flown at three levels of aggressiveness. Even in this relatively small-scale MTE, roll rates of nearly 70°/s are being used during the reversals (cf. Fig. 6.8). The sidestep task, flown in low wind conditions, strictly relates to the 'general MTE' class, indicating that the Lynx has at least a 60% task margin when flying this particular MTE. For the track boundary, the margin appears less, but it should be noted that the sidestep task in low wind is not the most demanding of MTEs and the Lynx will have a higher task margin than shown. To highlight this, Fig. 6.20 shows the roll quickness envelope for Lynx flying the lateral slalom MTE at 60 knots, with the ADS-33 Level 1/2 boundaries for forward flight MTEs. The rise in quickness to limiting values occurred for this aircraft at the highest aspect ratio when the pilot reached the lateral cyclic stops in the roll reversals. Lynx is particularly agile in roll and we see in Fig. 6.20, possibly some of the highest values of quickness achievable with a modern helicopter. The additional points on Fig. 6.18 will be discussed below.

The compliance testing for quickness depends on roll response type; with rate command, a pulse-type input in lateral cyclic from the trim condition will produce a

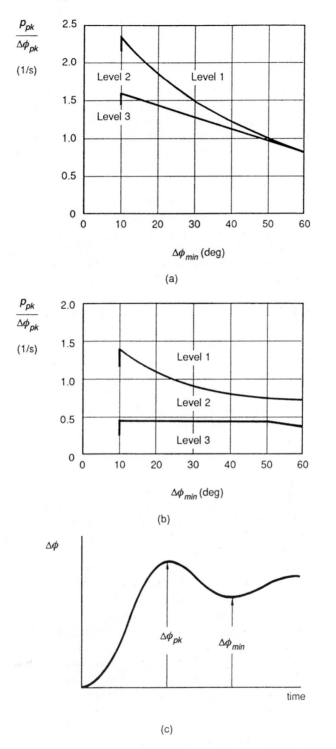

Fig. 6.17 Roll attitude quickness criteria for hover and low-speed MTEs (Ref. 6.5): (a) target acquisition and tracking (roll); (b) general MTEs; (c) definition of attitude parameters

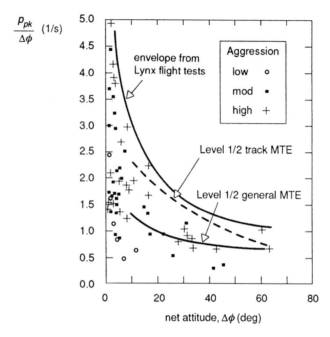

Fig. 6.18 Roll quickness results for lateral sidestep manoeuvre (Ref. 6.18)

100' right sidesteps

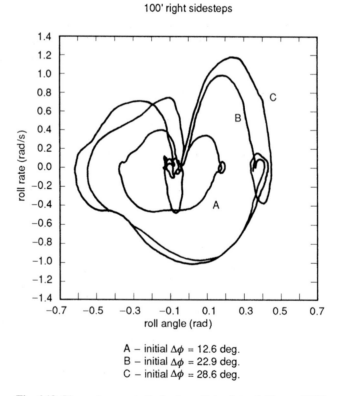

A – initial $\Delta\phi$ = 12.6 deg.
B – initial $\Delta\phi$ = 22.9 deg.
C – initial $\Delta\phi$ = 28.6 deg.

Fig. 6.19 Phase plane portraits for Lynx flying lateral sidestep MTE

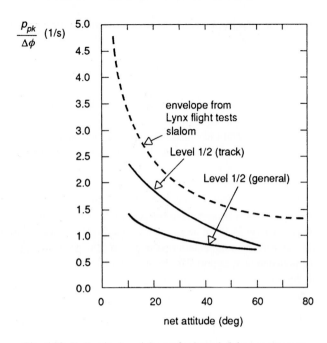

Fig. 6.20 Roll attitude quickness for lateral slalom manoeuvre

discrete attitude change. As noted above, for a helicopter with linear response characteristics, the quickness will be independent of input size and what is required is a decrease in the cyclic pulse duration until the required level of response is attained. In practice, the shorter the pulse, the larger the pulse amplitude has to be in order to achieve a measurable response. Figure 6.21 illustrates flight test results from the DLR's Bo105, showing different values of roll quickness achieved between 10° and 20° attitude change (Ref. 6.19). A maximum quickness of 4 rad/s was measured for this aircraft at the lower limit of the moderate amplitude range. For longer duration inputs, quickness values only just above the ADS-33 boundary were measured, as shown, highlighting the importance of applying sharp enough inputs to establish the

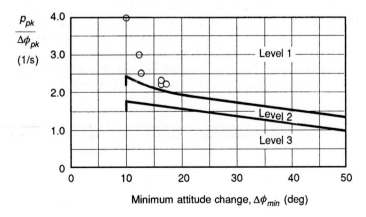

Fig. 6.21 Roll attitude quickness measured on Bo105 at 80 knots (Ref. 6.19)

quickness margins. It should be remembered that for compliance demonstration with ADS-33, all that is required is to achieve values in the Level 1 region.

For attitude response types, it may be necessary to overdrive initially the control input, followed by a return to steady state consistent with the commanded attitude. The control reversal is recommended to overcome the natural stability associated with the attitude command response type, and a moderate amount is allowable to achieve the maximum quickness.

The use of attitude quickness has some appeal in establishing control effectiveness requirements for civil helicopter handling qualities. As with the military requirements, however, establishing a test database from civil MTEs is essential before boundaries can be set. One area that could be well served by quickness is the response characteristics required to recover from SCAS failures; another could be the recovery from upsets due to vortex wakes of fixed-wing aircraft or other helicopters. Establishing quantitative criteria in these areas could have direct impact on the integrity requirements of the stability augmentation systems on the one hand and the operating procedures of helicopters at airports on the other. One of the obvious benefits of robust handling criteria is that they can help to quantify such aspects at the design stage.

In Fig. 6.18, the additional data points shown are computed from the time histories of sidestep data taken from tests conducted on the DRA advanced flight simulator (Ref. 6.3). These and other MTE tests will be discussed in more detail in Chapter 7, but a point worth highlighting here is the spread of quickness values in relation to the level of aggression adopted by the pilot. The level of aggression was defined by the initial roll angle, and hence translational acceleration, flown. High aggression corresponded to roll angles of about $30°$, with hover thrust margins around 15%. At the lower end of the moderate amplitude range (between $10°$ and $20°$), the maximum quickness achieved at low aggression was about 0.7, rad/s, which would correspond to the level of performance necessary to fly in UCE 2 or 3; ADS-33 states that meeting the quickness requirements is not mandatory for these cases (Ref. 6.5). The moderate aggression case would be typical of normal manoeuvring, and the maximum quickness achieved, around 1.5 rad/s, conforms well with the ADS-33 minimum requirement boundary. At the higher levels of aggression, quickness values correlate closely with those achieved with the Lynx in flight test at around 2.5 rad/s, a not-too-surprising result, as the simulation trials were designed to explore the maximum achievable task performance at similar thrust margins to the flight trials.

The large and moderate amplitude criteria extend down to $10°$ roll attitude. Below this, in the small amplitude range, we can see from Fig. 6.18 that the quickness measurements increase to values well beyond the moderate amplitude boundary. But quickness is no longer an appropriate parameter in this region and we have to look at a different formulation to measure flying quality here.

6.3.3   Small amplitude/moderate to high frequency: bandwidth

Early efforts in the time domain

It would not be an exaggeration to say that handling qualities research, for both fixed- and rotary-wing aircraft, has concentrated on the short-term response to control inputs, and the roll and pitch axes have absorbed most of the efforts within this research. The primary piloting task under consideration in this region is attitude regulation,

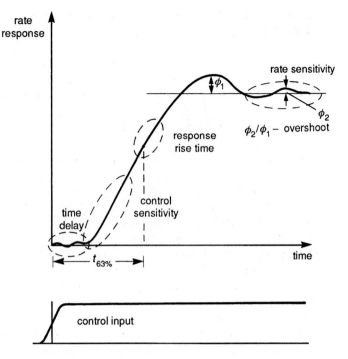

Fig. 6.22 Handling qualities parameters from step response

appropriate to tracking tasks when fine attitude corrections are required to maintain a precise flight trajectory or position. Most of the early work aimed at defining quality in this region focused on the response to step control inputs. Figure 6.22 illustrates the key characteristics associated with the step response:

(1) control sensitivity – the initial angular acceleration per unit of control input (e.g., inch of stick travel);
(2) rate sensitivity – the final steady-state rate per unit of control input; more generally this would be called response sensitivity to account for other response types;
(3) response time constant(s) – the time(s) to some proportion of the steady-state response; for simple first-order systems, the time constant $t_{63\%}$ is sufficient to characterize the simple exponential growth;
(4) time delay – delay before there is a measurable aircraft response; this can be grouped into category (3) but we separate it because of its special meaning;
(5) overshoot – ratio of successive peaks in oscillatory time response.

This list suggests that there needs to be at least five, and perhaps even more, handling qualities parameters that characterize the rise times, sensitivities and damping of the step control response in the time domain. Before we discuss the appropriateness of this further, it is worth reviewing one particular criterion that had gained widespread acceptance, prior to the publication of ADS-33 – the so-called damping/sensitivity diagram (Fig. 6.23). The damping derivative $L_p(1/s)$ is plotted against the control sensitivity derivative $L_{\delta y}$ (rad/(s^2 in)), where δy is the lateral cyclic stick displacement. The use of derivatives in this criteria format stems from the assumption of a linear

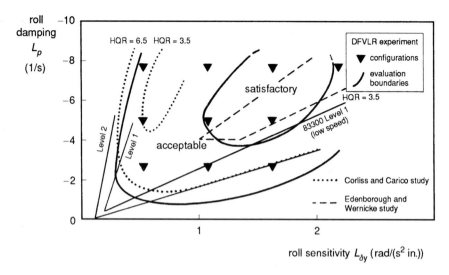

Fig. 6.23 Short-term roll handling qualities – damping/sensitivity boundaries (Ref. 6.20)

first-order type response where these two parameters completely describe the time response characteristics (L_p is the inverse of $t_{63\%}$, and $L_{\delta y}$ is the initial acceleration of the step response function; the ratio of control sensitivity to damping gives the rate sensitivity). On Fig. 6.23, taken from Ref. 6.20, we have drawn the various boundaries set by data from previous experiments. While we have to recognize that boundary lines on this two-parameter handling qualities chart will be task dependent, there is good reason to believe that the wide spread of quality boundaries in Fig. 6.23 is actually a signal that the criteria are not appropriate to short-term response criteria in general. For simple first-order systems, or classical roll rate response types easily modelled in simulation, the contour shapes in the figure will be entirely appropriate. However, there are two principal reasons why Fig. 6.23 is not generally applicable and hence the situation does not meet the CACTUS rules:

(1) Short-term helicopter roll response is typically non-classical, with higher order dynamics distorting the first-order contributions.
(2) For tasks requiring the pilot to perform attitude regulation, there is strong evidence that the pilot's impression of handling qualities is not particularly sensitive to the shape of the response to a step input.

With regard to the first point, in the early unpublished versions of the revision to MIL-H-8501, time domain step response criteria were proposed, based on fixed-wing experience, which included three rise time parameters and one overshoot parameter as shown in Table 6.2.

This criteria set was based entirely on flight data, largely gathered on the variable stability Bell 205 operated by the Flight Research Laboratory in Ottawa; ground-based simulator results were discounted because there were too many unresolved questions about data from rate response types obtained from simulation. Reference 6.21, in summarizing the contribution of the Canadian Flight Research Laboratory to the ADS-33 effort, presents results showing the discrepancy between ground-based simulation and in-flight simulation results; the ground-based tests showed no significant Level 1

Table 6.2 Limiting values of time response parameters for roll rate response type in hover and low speed MTEs

Parameter	Level 1	Levels 2 and 3
t_{r10} (s) max	0.14	0.27
t_{r50} (s) max	0.35	0.69
t_{r90} (s) max	1.15	2.30
ϕ_2/ϕ_1 max	0.3	0.44

achievement. The proposed time domain parameters in Table 6.2 were derived from the limited flight test results available at the time. The criteria proposed for attitude command systems required the rise times to be a function of the effective damping ratio of the response, and were therefore even more complicated. The emphasis on finding suitable *time domain criteria* for both roll and pitch attitude short-term response was partly driven by the helicopter community's familiarity with this format, stemming partly from the history of usage of MIL-H-8501A and the traditional damping/sensitivity two-parameter handling qualities diagram. Had there not been a potentially very effective alternative being developed in parallel with the time domain parameters, the ADS-33 development may well have persisted with this kind of format.

Measurement problems aside (and these are potentially significant), the reality is that, although the step response function may be a simple clinical concept, pilots rarely use step response control strategies in attitude regulation and tracking tasks, and it was only a matter of time before the community became convincingly won over to the *frequency domain* and the alternate proposal – the bandwidth criterion. Before discussing bandwidth in some detail it is worth saying a few words about the archetypal frequency domain approach – low-order equivalent systems (LOES), used extensively in fixed-wing handling criteria (Ref. 6.6). The argument goes that the higher frequency ranges of vehicle dynamics characterizing the short-term responses are dominated by the roll subsidence and short-period pitch modes. Obtaining frequency response data, by exciting the aircraft around the natural frequencies, provides amplitude and phase characteristics to which low-order equivalent systems can be fitted numerically to estimate natural frequency and damping, for which quality boundaries can be defined on two-parameter diagrams. The fixed-wing handling requirements state that this approach is not applicable to non-classical response types and offer the bandwidth criterion as an alternate approach in these cases. It was recognized early on in the development effort that the naturally non-classical behaviour of helicopters would exclude LOES as a general candidate for roll and pitch response (Ref. 6.22), particularly for Level 2 and 3 configurations, and little research has been conducted in this area, at least for roll response.

Bandwidth

In Ref. 6.23, Hoh describes results from a simulator assessment of attitude command response types for a recovery to a ship MTE. The tests were conducted on the NASA vertical motion simulator (VMS) specifically to evaluate the effectiveness of rise time criteria. The step response characteristics of the different configurations tested are shown in Fig. 6.24. An important result of the tests was that the three evaluation pilots rated all the configurations within a fairly tight HQR spread. The pilots were

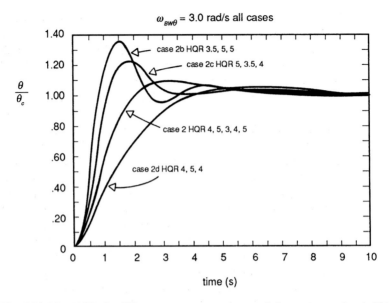

Fig. 6.24 Pilot HQRs for different step response characteristics at constant bandwidth
(Ref. 6.23)

almost unaware of the different time domain characteristics for this precision landing
manoeuvre. What the configurations in Fig. 6.24 do have in common is the *attitude
bandwidth*, even though the damping ratio varies from 0.5 to 1.3. This is a very compelling result and calls for a definition and description of this unique new handling
qualities parameter.

The bandwidth parameter is defined in Fig. 6.25 as the lesser of two frequencies,
the phase-limited or gain-limited bandwidth, derived from the phase and gain of the

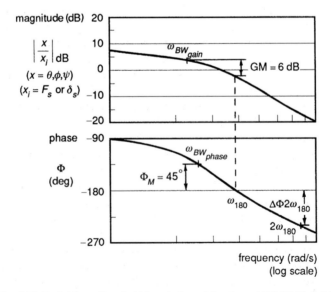

Fig. 6.25 Definition of bandwidth and phase delay from ADS-33 (Ref. 6.5)

frequency response of attitude to pilot's cyclic control. The phase bandwidth is given by the frequency at which the phase is 135°, i.e., the attitude lags behind the control by 135°. The gain bandwidth is given by the frequency at which the gain function has increased by 6 dB relative to the gain when the phase is 180°. The 180° phase reference is significant because it represents a potential stability boundary for closed-loop tracking control by the pilot. If a pilot is required to track a manoeuvring target or to maintain tight flight path control during turbulent conditions, then there are three related problems that hinder control effectiveness. First, at high enough frequencies, the aircraft response will lag behind the pilot's control input by 180°, requiring the pilot to apply significant control lead to anticipate the tracked or disturbed motion (as the aircraft rolls to port, the pilot must also apply lateral cyclic to port to cancel the motion). Second, at higher frequencies the response becomes attenuated and, to achieve the same tracking performance, the pilot has to increase his control gain. Third, and most significant, any natural lags in the feedback loop between an attitude error developing and the pilot applying corrective cyclic control can result in the pilot–aircraft combination becoming weakly damped or even unstable. The combination of these three effects will make all air vehicles prone to PIOs above some disturbance frequency, and one of the aims of the bandwidth criteria is to ensure that this frequency is well outside the range required to fly the specified MTEs with the required precision. Thus, a high phase–bandwidth will ensure that the phase margin of 45°, relative to the 180° phase lag, is sufficient to allow the pilot to operate as a pure gain controller, accepting his own natural phase lags, without threatening stability. The gain bandwidth limit protects against instability should the pilot elect to increase his gain or his level of aggression at high frequency. Of course, a skilful pilot can operate effectively well beyond the limits defined by simple theory, by applying more complex control strategies. This always leads to an increase in workload and hence less spare capacity for the pilot to give attention to secondary tasks, any of which could become primary at any time in consideration of overall safety or survivability. For a wide range of systems, the phase bandwidth is equal to or less than the gain bandwidth.

The bandwidth criteria apply in ADS-33 to both rate and attitude response-types, except that for attitude response types, only the phase bandwidth applies. This nuance leads us to examine the gain-limited bandwidth in a little more detail, following the discussion in Ref. 6.24, where Hoh reports that the '. . . gain bandwidth is included because a low value of gain margin tends to result in a configuration that is PIO prone. Low gain margin is a good predictor of PIO prone configurations because small changes in the pilot gain result in a rapid reduction in phase margin'. An example of a gain-limited bandwidth system is given in Ref. 6.24 and reproduced here as Fig. 6.26. Here, there is a modest value of phase bandwidth, but the gain margin available to the pilot when operating around this frequency, for example, to 'tighten-up' in an effort to improve performance, is limited and considerably less than the 6 dB available at the gain bandwidth. Hoh describes the problem succinctly when he states that, 'The phenomenon is insidious because it depends on pilot technique. A smooth, non-aggressive pilot may never encounter the problem, whereas a more aggressive pilot could encounter a severe PIO'. Hoh goes on to discuss the rationale for not including the gain margin limit for attitude command systems. Basically, because the attitude stabilization task is accomplished by the augmentation system, the pilot should not be required to work at high gains with inner-loop attitude control. If he does, and experiences a PIO tendency, then simply backing off from the tight control strategy

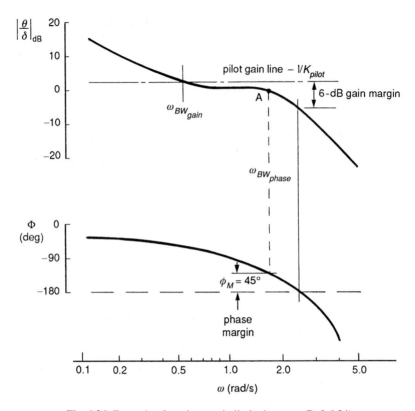

Fig. 6.26 Example of a gain-margin limited system (Ref. 6.24)

will solve the problem. To conclude this discussion we quote further from Hoh (Ref. 6.24):

> *We are faced with a dilemma: on the one hand gain-margin-limited ACAH response types lead to PIO for super precision tasks, and on the other, disallowing such config- urations robs the pilot of workload relief for many other, less aggressive, MTEs. The approach taken herein (in the ADS-33C spec) has been to eliminate gain margin from the definition of bandwidth for ACAH response types, but to recommend avoidance of ACAH systems where the gain bandwidth is less than the phase bandwidth, especially if super precision tasks are required. Additional motivation for not including gain bandwidth as a formal requirement for ACAH was that the PIO due to gain margin limiting has not been found to be as violent for ACAH response types. It should be emphasised that this is not expected to be the case for rate or RCAH response types, where the pilot attitude closure is necessary to maintain the stable hover, and consequently, it is not possible to completely 'back out' of the loop. Therefore gain bandwidth is retained for these response types.*
>
>   *Gain-margin-limited systems result from a large phase delay, combined with flat amplitude characteristics such as shown in Fig. 6.27. Large phase delays usually result from inherent rotor system time delay (65 to 130 ms), combined with computer throughput delays, actuator lags, filters, etc. The flat amplitude characteristic is, of*

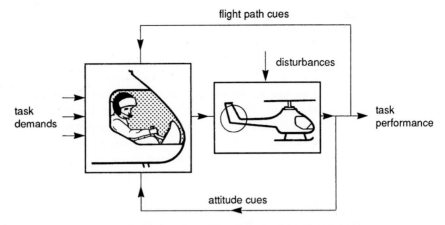

Fig. 6.27 Pilot as sensor and motivator in a task feedback loop

*course, inherent to ACAH, and can occur in RCAH response types due to the nature
of feedforward equalization.*

Another interpretation of bandwidth can be gleaned from its origins out of the development of the so-called crossover model of human pilot behaviour (Ref. 6.25), which treats the pilot action in performing tracking control tasks as an element in a feedback loop (Fig. 6.27). In single-axis tasks, for a wide variety of aircraft dynamic characteristics, the pilot adapts his control strategy so that the product of the pilot and aircraft dynamics take the simple transfer function form

$$Y_p(s)Y_a(s) \approx \frac{\omega_c e^{-\tau s}}{s} \tag{6.10}$$

Therefore, for example, if the rate response is a simple first-order lag, then the pilot will compensate by applying a simple lead with approximately the same time constant as the response lag. This form of overall open-loop characteristic will be applicable over a range of frequencies depending on the application. A key property of this form of model is highlighted by the root locus diagram of its stability characteristics when in a closed-loop system (Fig. 6.28). The pilot can increase the overall gain ω_c to regulate the

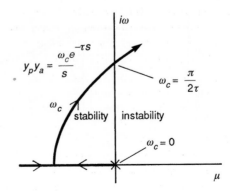

Fig. 6.28 Root locus of crossover model eigenvalues as pilot gain is increased

performance of the tracking task, but doing so will degrade the stability of the closed-loop system. The pure time delay, caused by mental processing and neuromuscular lags, is represented by the exponential function in the complex plane (i.e., Laplace transform) that has an infinite number of poles in the left-hand transfer function plane. The smallest of these moves to the right as the pilot gain is increased in the crossover model links up with the left moving rate-like pole, and the pair eventually become neutrally stable, with 180° phase shift, at the so-called crossover frequency as shown in Fig. 6.28. This simple but very effective model of human pilot behaviour has been shown to work well for small amplitude single-axis tracking tasks and leads to the concept of a natural stability margin defined by the gain or phase margin from the point of neutral stability. This is the origin of the bandwidth criteria.

Phase delay

The quality of flying in the small amplitude–high frequency range of our framework diagram was initially encapsulated in a single-parameter bandwidth. Unfortunately, the situation turned out more complicated and it was not long before cases of equi-bandwidth configurations with widely varying handling qualities were found. Again, Hoh sheds light on this in Refs 6.23 and 6.24. For a wide variety of systems there is a unique relationship between the bandwidth frequency and the shape of the phase curve in the frequency domain beyond the bandwidth frequency. The steeper the phase 'roll-off', then the lower the bandwidth and, with relatively simple high-order effects like transport delays and actuator lags, the increasing phase slope correlates directly with bandwidth. However, for more general high-order dynamics, the phase delay has to be computed as an independent measure of handling, since configurations with markedly different phase slope can have the same bandwidth. A case is cited in Ref. 6.24 where two configurations with the same bandwidth were rated as Level 2 and 3, simply because the phase slopes were very different (Fig. 6.29). Pilots are particularly sensitive to the slope of the phase at high frequency, well beyond the bandwidth frequency but still within the range of piloting, e.g., >10 rad/s. In a closed-loop tracking task, when high precision is required, pilots will find that high values of phase slope make it very difficult for them to adapt their control strategy to even small changes in frequency,

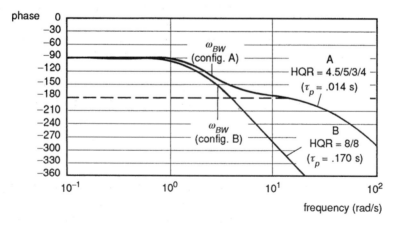

Fig. 6.29 Sensitivity of HQRs to phase characteristics at frequencies beyond ω_{bw} (Ref. 6.24)

hence task disturbance. This and related effects reinforce the point that for tracking tasks the pilot is very sensitive to effects easily described in the frequency domain but hardly noticeable as delays following step inputs. The actual parameter selected to represent the shape of the phase is the phase delay, τ_p, defined as

$$\tau_p = \frac{\Delta\Phi_{2\omega_{180}}}{57.3 \times 2\omega_{180}} \tag{6.11}$$

where $\Delta\Phi_{2\omega_{180}}$ is the phase change between ω_{180} and $2\omega_{180}$. The phase delay is therefore related to the slope of the phase between the crossover frequency and $2\omega_{180}$. Reference 6.5 notes that if the phase is nonlinear in this region, then the phase delay can be determined from a linear least-squares curve fit, in a similar way to the computation of equivalent time delay in LOES analysis.

Bandwidth/phase delay boundaries

The ADS-33C quality boundaries for bandwidth and phase delay are presented on two-parameter handling qualities diagrams as shown in Figs 6.30(a)–(c), corresponding to the different MTE classes shown; the roll axis boundaries are applicable both to low speed and to forward flight regimes. The references provided in the legend to each figure record the supporting data from which the boundaries were developed. It is probably true that more effort has been applied, and continues to be applied, to defining these boundaries than any other. The criteria are novel and considerable effort was required to convince the manufacturing community in particular that the frequency domain criteria were more appropriate than the time domain parameters. The lower, vertical portions of each boundary indicate the minimum acceptable bandwidths, with tracking and air-combat MTEs demanding the highest at 3.5 rad/s for Level 1. The curved and upper portions of the boundaries indicate the general principle that the higher the bandwidth, the higher is the acceptable phase delay, the one compensating for the other.

It may seem surprising that Level 1 handling qualities are possible with phase delays of more than 300 ms. Two points need to be made about this feature. First, a study of the references will indicate that, although the data in these areas is very sparse, they genuinely indicate the trends shown. Second, it would be practically impossible to build a helicopter with a bandwidth of, say, 3 rad/s and with a phase delay of 300 ms (Fig. 6.30(b)); the latter would almost certainly drive the bandwidth down to less than 1 rad/s. This dependence of bandwidth on the same parameters that have a first-order effect on phase delay is perhaps the only weakness of this two-parameter handling qualities diagram. The point is illustrated in Fig. 6.31, which shows the contours of equi-damping and time delay overlaid on the UCE 1 roll bandwidth boundaries. The contours are derived from a simple, rate response 'conceptual handling qualities model' (Refs 6.3, 6.35), which can be written in transfer function form

$$\frac{\bar{p}}{\bar{\eta}_{1c}}(s) = \frac{Ke^{-\tau s}}{\left(\frac{s}{\omega_m} + 1\right)} \tag{6.12}$$

where K is the overall gain or, in this case, the rate sensitivity, τ is a pure time delay; ω_m can be considered to be equivalent to the roll damping, $-L_p$. The results in Fig. 6.31 show that the addition of a pure time delay can have a dramatic effect on both bandwidth and phase delay. With τ set to zero, the bandwidth would be equal to ω_m.

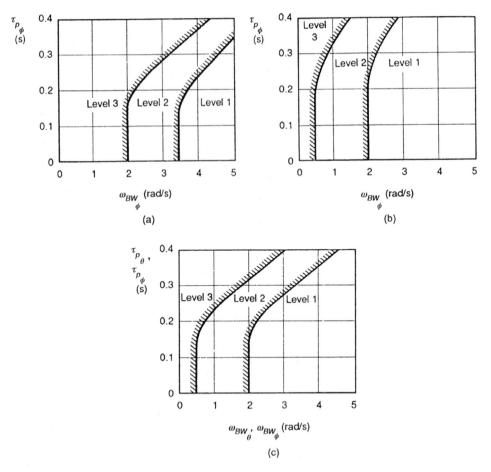

Fig. 6.30 ADS-33C requirements for small amplitude roll attitude changes – hover/low-speed and forward flight MTEs (Ref. 6.5): (a) target acquisition and tracking (roll) (Refs 6.26–6.29); (b) all other MTEs – UCE = 1, VMC and fully attended operations (roll) (Refs 6.30–6.32); (c) all other MTEs – UCE > 1, IMC and/or divided attention operations (pitch and roll) (Refs 6.33, 6.34)

Therefore, a 70-ms pure time delay can reduce the bandwidth of a 12 rad/s aircraft (e.g., with hingeless rotor) down to 4 rad/s. The bandwidth reduction is much less significant on helicopters with low roll damping (e.g., with teetering rotors); the same lags reduce the bandwidth of a 3 rad/s aircraft to about 1.9 rad/s. Note that, according to Fig. 6.30(a), defining the roll bandwidth requirements for tracking tasks, a bandwidth of 4 rad/s corresponds to Level 1 while a bandwidth of 1.9 rad/s corresponds to Level 3.

The model similar to that described by eqn 6.12 was used to investigate the effects of different levels of pilot aggression, or task bandwidth, on the position of the handling qualities boundaries in Fig. 6.31, using the DRA advanced flight simulator (AFS) (Ref. 6.3). The results will be presented later in the discussion on subjective measurement of quality in Chapter 7, but the test results confirmed the ADS-33C boundaries to within 0.5 HQR, up to moderate levels of aggression. The research reported in Ref. 6.3 was part of a larger European ACT programme aimed at

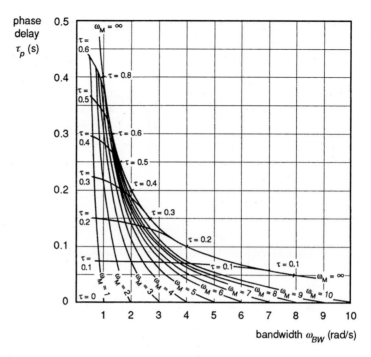

Fig. 6.31 Equi-damping and time delay contours overlaid on ADS-33C handling qualities chart (Ref. 6.8)

establishing guidelines for the handling characteristics of future ACT helicopters (Ref. 6.36). This international programme made complementary use of ground-based and in-flight simulation facilities. Of particular concern was the effect of transport delays introduced by the digital computing associated with ACT, and tests were conducted to try to establish whether the curved boundaries on the ADS-33C criteria would still be appropriate. In a similar time frame a new series of flight and simulator tests was conducted under the US Army/German MoU to check the location of the upper phase delay boundaries (Ref. 6.37). A new lateral slalom task was derived that contained tight tracking elements that could potentially discern PIO tendencies. Both the EuroACT and US/GE research derived results that suggested a levelling of the phase delay boundary between 100 and 150 ms would be required. Figure 6.32 summarizes the results, showing the recommended phase delay caps from the two evaluations. At the time of writing, these recommendations are regarded as tentative, although they have led to a revision of the 'official' requirements, appearing in the latest version of ADS-33 (ADS-33D, Ref. 6.38), as shown in Fig. 6.33. The reduction in phase delay is accompanied by a relaxation in the bandwidth requirement for roll tracking tasks. The evolution of these criteria illustrates once again the powerful effect of *task* on handling qualities and the strong design driver that handling qualities will be for future ACT helicopters.

Civil applications
The bandwidth criterion aims to discern handling qualities that avoid or exacerbate the problems that some pilots experience when 'tightening-up' in a closed-loop compensatory tracking task. For obvious reasons this high precision/performance criterion has

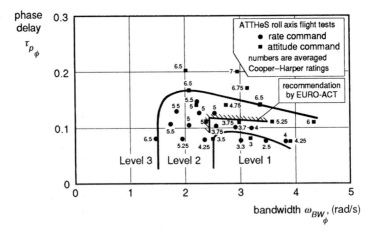

Fig. 6.32 Proposed roll axis bandwidth criteria from European tests (Refs 6.36, 6.37)

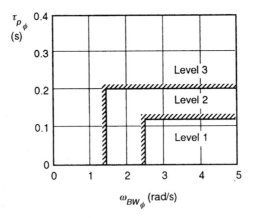

Fig. 6.33 Bandwidth/phase delay criteria for roll axis tracking task according to ADS-33D
(Ref. 6.38)

broad application across military helicopter uses. Considering civil helicopter design, certification and operations, there are several application areas that could potentially benefit from bandwidth (Ref. 6.39). Precise positioning of an underslung load is a good example, although even the military requirements are, at the time of writing, fairly immature in this area. All-weather operations requiring recoveries to moving decks is another example. The whole area of search and rescue is one where civil (and military) helicopters can be flown close to the pilot's limits, with the requirements for precise positioning in confined spaces. With safety as an emphasis in civil helicopter operations, the case for introducing civil MTEs that include high-precision elements into the certification process is considered to be strong. The future application of ACT to civil helicopters, with the potential for increased phase delay, will strengthen this case; it is far better to highlight potential problems in certification than to experience them for the first time in operation. Of course, one of the great strengths of substantiated criteria, like bandwidth, is that they can be used in the design process to ensure satisfactory flying qualities are built in, with the aim of making the certification process a formality.

The measurement of bandwidth

One of the failings of time domain criteria arises when trying to make accurate measurements of the rise times in the step response. While it is relatively easy and economical to apply a step input, the shapes of the rate and attitude response are very sensitive to the detailed form of the control input, and aircraft initial conditions. Errors in rise time computation, particularly for the smaller values ($O(0.1$ s$)$), can be large. Since the significant handling qualities parameter is actually the slope of the phase, any errors in rise time calculation will reflect in a poor estimation of the high frequency phase. On the other hand, the frequency response function is fairly robust to analysis, although considerably more calculation effort is required, and frequency response data are more difficult and far more time consuming to capture in flight test. Since the first publication of ADS-33 in the mid-1980s, considerable experience has been gained in the measurement of bandwidth and phase delay in flight (Refs 6.40–6.49). The recommended test input is a sine wave form with gradually increasing frequency applied at the pilot's controls. Figure 6.34 illustrates a roll axis frequency 'sweep'

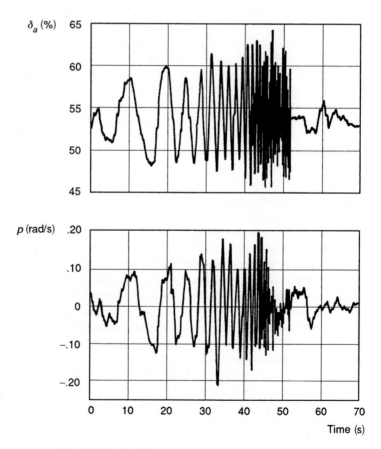

Fig. 6.34 Roll axis frequency sweep for Bo105 (Ref. 6.43)

showing the pilot's control position and aircraft roll rate response for the Bo105 from a test speed of 80 knots. The test manoeuvre is complete in about 1 min, the pilot uses about 10–15% of the control range and the roll rate is contained within the range ±20°/s.

From the accumulated knowledge of frequency sweeping, a number of rules of thumb can be applied when designing and conducting a flight test. These concern both safety and performance aspects and are now addressed in turn.

(1) *Frequency range*   The range of frequencies covered in the sweep need only be high enough to capture the phase characteristics up to twice the 180° phase lag frequency. Unfortunately, this latter frequency may not be known precisely prior to the test, and experience has shown that it can vary widely across different types, e.g., 22 rad/s on the Bo105 (Ref. 6.43) and 12 rad/s on the OH-58D (Refs 6.41, 6.42). Also, there may be only limited data available on the airframe/rotor structural modes within this frequency range. It is therefore very important to establish the upper frequency limit and the influence on structural modes very carefully with exploratory test inputs before applying a full frequency sweep. Reference 6.42 recommends a frequency range of 0.1–2 Hz, but clearly this is inadequate for higher bandwidth helicopters like the Bo105.

(2) *Maintaining trim – instability*   One of the principal problems with the frequency sweeping of helicopters without stability augmentation is their natural tendency to diverge from the trim condition, particularly during the low-frequency portion of the sweep input. If this is allowed to happen, then clearly the validity of the data is questionable; the engineer can no longer relate the computed bandwidth to a particular flight condition, and nonlinear effects are likely to spoil the overall quality of the data. The pilot needs to apply 'uncorrelated' corrective inputs, superimposed on the sweep, to keep the aircraft manoeuvring about the trim condition. This can sometimes be very difficult if not impossible to accomplish satisfactorily, particularly close to hover or for pitch axis sweeps at high speed. In cases where the natural stability of the aircraft is so poor that frequency sweeping is not practicable, then it may well be necessary to deduce the open-loop, bare-airframe, characteristics from sweep results with the SCAS engaged.

(3) *Cross-coupling*   While cross-coupling, in itself, is not a problem during frequency sweeping, it has become a practice for pilots to negate the cross-coupled motion with control inputs, primarily to preserve the mean trim condition. For example, the pitch and yaw moments generated during a roll sweep can soon give rise to large flight-path excursions. Even assuming the pilot is able to apply perfect cancelling inputs there are two data contamination effects that will need to be taken into account. First, the roll response will no longer be due to the lateral cyclic only, but there will be components at various frequencies due to the cross-control inputs. These can, in principle, be extracted using conditional frequency analysis (see below), effectively deriving the secondary frequency characteristics as well as the primary. Second, any correlation between the primary and secondary control inputs will make it very difficult to separate out the primary frequency response, even using the conditional techniques. As noted above, one solution is to apply corrective cross-coupled control inputs that are uncorrelated with the primary axis, e.g., occasional pulse-type inputs to recover airspeed, pitch attitude or sideslip.

(4) *Control amplitude*   The magnitude of the control input is a compromise between achieving the highest signal-to-noise ratio to maximize the information content, and minimizing the excursions from trim and the potential for exciting dangerous loads. It has been found that there is a natural tendency for the inexperienced pilot to increase the control amplitudes as the frequency increases, in order to maintain the same overall amplitude of aircraft response. This should reduce with training (see point (7)). A general rule of thumb is that the control inputs should be kept to within ±10% of full control throw.

(5) *Repeats*   Repeat runs are always required in flying qualities tests to ensure that at least one good data point is captured; for frequency sweeping, the recommended minimum number of repeats is two, to provide at least two quality runs for averaging in the frequency analysis.

(6) *Duration*   The time duration of a frequency sweep depends on the frequency range to be covered, primarily the lower limit, and the rate of change of frequency. Assuming the latter to be about 0.1 Hz, experience has shown that sweep durations between 50 and 100 s are typical. Constraining factors will be the natural stability of the aircraft, influencing the time spent at low frequencies, and the rotor speed, largely influencing the upper limit.

(7) *Training and practice*   One of the most important safety factors that can be included in a frequency sweep test programme is an adequate level of training for pilots. First, there is the simple matter of training pilots to apply a slowly varying sine wave with an amplitude of perhaps ±1 cm. Experience has shown that pilots new to sweeping tend to increase the amplitude of the controls as the frequency increases and they are not always aware that this is happening. Also, it is very difficult for a pilot to judge what is a 2-, 3- or 4-Hz input without experience. Ideally, the pilot would initiate his or her training by following through on the controls while the instructor applied the sweep with the aircraft on the ground. Sweeps on all controls could be taught this way to give the pilot a feel for the kind of hand and feet motion required. The pilot could then practice with a display providing feedback on the frequency and amplitude. After the trainee pilot is confident in his or her ability to apply the input shape, the training can continue in the simulator and eventually in flight, where the pilot needs to practice before the definitive inputs are made. Practising gives the pilot knowledge about what corrective inputs are required in other axes to maintain the aircraft close to the reference flight condition. Practising also allows the engineers conducting the structural loads monitoring to guide the trial better. But pilot-applied frequency sweeps are best done with two crew, one applying the input, the other calling the tune.

(8) *Manual or automated inputs – it takes two to sweep*   In theory, it should be possible to design a frequency sweep for application through an automatic control input device that has superior properties to a manual input, e.g., repeatability, and better defined frequency content. However, the sweeping experience to date, especially with unstabilized helicopters, indicates that manual inputs are to be preferred, because of the increased flexibility in uncertain situations; and the more irregular shapes to the manual inputs actually have a richer information content. This situation has to improve in favour of the auto-inputs with time, but it should be remembered that the bandwidth frequency relates to the attitude response to the pilot's stick input and not to the control servo input. With auto-inputs applied at the

servo actuator, the additional transfer function between stick and servo would still need to be determined. Experience at the DRA with manual frequency sweeping has emphasized the value of the second crew member providing timing assistance to the pilot by counting out with rhythm, particularly at the lower frequencies. Counting out periods of 20, 16, 12, 8, 4 and 2 s helps the pilot to concentrate on applying a series of sine waves at increasing frequencies. At higher frequencies the pilot needs to rely on a learned technique, the counting then being a significant distraction.

(9) *Load monitoring for structural resonances*   Frequency sweeping can damage a helicopter's health and it is important to take this warning seriously. However, with the right preparations and precautions, the damage can be controlled and quantified. Some of the precautions have already been discussed under the headings of frequency range and amplitude, but it is important to know as much as possible about potential structural resonances before embarking on this kind of test input. In the case of a new aircraft, it is prudent to establish the rotor/fuselage coupled modes using the structural test development aircraft prior to making the bandwidth measurements. However, most of the testing carried out in the late 1980s was conducted on experimental aircraft, sometimes without a thorough analysis of potential resonances. Tests conducted by the US Army on the AH-64A and OH-58D (Ref. 6.42) revealed several potential problems. A divergent vertical bounce was experienced during longitudinal cyclic hover sweeps in the AH-64 at about 5 Hz. Damaged tail rotor support components were found following yaw axis sweeps, again on the AH-64. On the OH-58D, sweep tests excited an oscillation in the mast mounted sight, which was not felt by the crew, but only detected visually by the crew of the chase aircraft and through the telemetry at the ground station.

Earlier, at the Royal Aircraft Establishment, the first UK sweep tests were conducted with the research Puma fitted with a full fatigue usage monitoring strain gauge fit (Ref. 6.45). Higher fatigue usage was encountered in pitch axis sweeps in forward flight, and although we are still discussing the roll axis, the results are of such general significance in understanding the role of load monitoring that they are presented here. The tests were conducted to derive equivalent low-order system models for pitch axis dynamics (reported in Refs 6.28 and 6.45), but the test inputs were essentially the same as for the bandwidth measurement. Figure 6.35(a) illustrates two longitudinal cyclic frequency sweeps, one with the SCAS engaged, the other disengaged, captured at 60 knots airspeed. The additional data are the normal acceleration at the fuselage floor and the stress in the forward gearbox strut, derived from the component strain, which transpired to be the most critical for the pitch manoeuvre. The control input is maintained within the recommended range and the control frequency spectrum is primarily below 2 Hz, the required test upper limit. The larger response at the lower frequencies with the SCAS disengaged is noted. Figure 6.35(b) shows results at 100 knots, for two cases, one where the frequency range was limited to 2 Hz, the second where it was extended to 4 Hz. In the second case, the crew experienced significant vertical bounce at the higher end of the range. The normal acceleration record shows amplitude excursions of ± 0.25 g at high frequency. A combination of real-time monitoring through a telemetry link to a ground station coupled with post-flight fatigue life accumulation analysis revealed the extent of the damage done during these tests. Figure 6.36 shows data for one flight (Flt No 728) comprising nine sweeps over the speed range 60–120 knots. The

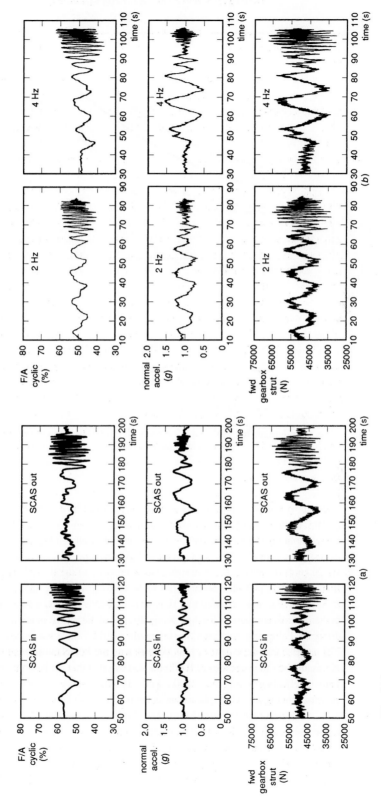

Fig. 6.35 Longitudinal cyclic frequency sweeps on DRA research Puma: (a) 60 knots: SCAS on and off; (b) 100 knots: 2 Hz and 4 Hz

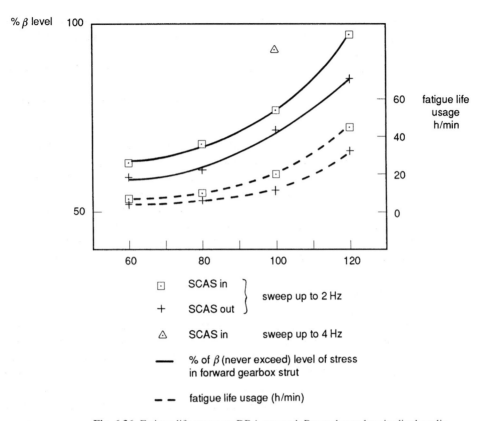

Fig. 6.36 Fatigue life usage on DRA research Puma due to longitudinal cyclic frequency sweeps

figure shows the percentage of the never-exceed fatigue load level, the so-called β-level, in the forward gear box strut and the fatigue life usage across the speed range, for both SCAS-in and SCAS-out. A striking result is that the SCAS-out manoeuvres were less damaging than the SCAS-in manoeuvres. The SCAS-in sweep at 120 knots resulted in gearbox strut loads within 5% of the β-level. The single triangle point at 100 knots corresponds to the case shown in Fig. 6.35(b), when the frequency range was extended to 4 Hz, again taking the load close to the limit. At the higher speeds, component life was being fatigued at the rate of more than 40 h/min. Following these tests, the calculation of the fatigue life used during flight 728 revealed that more than 11 h of life had been used in just nine sweeps. Accumulated life over the period of the tests indicated that the gearbox mounts were prematurely approaching their 2000-h limit. The aircraft was grounded while the gearbox mounts and other related components were replaced.

The lessons learned during these first few years of sweeping suggest that load monitoring, preferably in real time, is the safest precaution against undue structural damage. This requirement elevates the risk level associated with flying qualities testing, placing increased instrumentation demands on prototype aircraft. However, as will be discussed in the sections on subjective assessment in Chapter 7, flying

MTEs can also incur larger than usual fatigue life usage. The new approach to flying qualities criteria and test, epitomized by ADS-33, is clearly much more than just a new cookbook.

(10) *Incrementally safe* This final rule attempts to sum up the approach by emphasizing the importance to safety of engaging with frequency sweeping in an incremental manner, increasing frequency and amplitude only when confidence has been gained at lower values.

These ten rules have been laboured because of the novelty and the safety implications associated with frequency sweeping and because of the lack of guidance in the open literature. A frequency sweep is a flying qualities test but it shares many of the same characteristics as a loads test. Indeed, it might be argued that a rotor/airframe loads scan using the same test technique would yield valuable data for the stress engineers to conduct their part of the airframe qualification. It seems only natural, and certainly economical, to combine the activities in one test programme for a new aircraft, hence elevating the safety issues involved in bandwidth testing to the proper level.

Estimating ω_{bw} and τ_p

Having measured the frequency response of the aircraft to control excitation, the remaining task is to estimate the bandwidth and phase delay from graphical representations of the amplitude and phase of the response as shown in Fig. 6.25. But how do we ensure that the estimated frequency response functions are as accurate as possible or even valid? The frequency response analysis, whereby the time response data are converted into the frequency domain using a Fast Fourier Transform (FFT) technique (Ref. 6.50), assumes that the input–output relationship is approximately linear and that any 'noise' on the signals is random and uncorrelated with the response. Both of these assumptions break down to some degree in practice and it is important to process the time histories systematically to calibrate the data quality. The linear FFT converts a sweep time history of, say, roll rate of duration T, into a complex function of frequency (with in-phase and quadrature components) given by the relation

$$\overline{p}(\omega,\, T) = \int_0^T p(t)\,e^{-j\omega t}\,dt \tag{6.13}$$

The minimum frequency in the transformed function is related to the time duration of the sweep by the simple function

$$\omega_{min} = \frac{2\pi}{T} \tag{6.14}$$

In practice, with digitized data, the transformation is conducted discretely, over the time response samples p_n, measured every Δt, in the form

$$\overline{P}_k(\omega_k) = \Delta t \sum_{n=0}^{N-1} p_n \exp\left(-j2\pi\frac{kn}{N}\right), \quad k = 0,\, 1,\, 2,\, \ldots,\, N-1 \tag{6.15}$$

The frequency response functions H for all required input–output pairs (e.g., η_{1c}, p) can be assembled from the spectral density functions G (Ref. 6.50) as

$$H_{\eta_{1c}p}(\omega) = \frac{G_{\eta_{1c}p}(\omega)}{G_{\eta_{1c}\eta_{1c}}(\omega)} \qquad (6.16)$$

A measure of accuracy of the derived frequency response function in terms of the linear correlation between output and input is given by the coherence function

$$\gamma_{\eta_{1c}p}^{2} = \frac{|G_{\eta_{1c}p}(\omega)|}{G_{\eta_{1c}\eta_{1c}}(\omega)G_{pp}(\omega)} \qquad (6.17)$$

Any coherence less than unity signifies the presence of nonlinearities or correlated noise on the response. In close-to-ideal conditions, the computations given by eqns 6.15 and 6.16 will generate frequency response data from which good estimates of bandwidth and phase delay can be derived. In practice, further and more detailed processing is often required to ensure that the handling qualities parameter estimates are the best obtainable. In Ref. 6.51, Tischler and Cauffman discuss the details as implemented in the US Army's CIPHER analysis software, involving concatenation of multiple sweeps in the time domain and windowing to derive the best estimates of the individual power spectra. A second stage involves the derivation of the conditional frequency responses to take account of the effects of corrective control inputs in secondary axes. The associated partial coherence functions serve as a guide to the accuracy of the results and the linearity of the input–output relationships. The third stage in the data quality improvement ensures that the degrading effects of noise on the data are minimized. Effectively, composite frequency responses are derived from averaging with different data 'window' sizes in the frequency domain – small for the high-frequency range and large for the lower frequencies. A rough rule of thumb for data validity is given when the coherence function exceeds 0.8.

The calculation of bandwidth and phase delay follows according to the procedure given in Fig. 6.25. Most of the data improvement process described above is actually aimed at raising the coherence in the critical frequency range between ω_{180} and $2\omega_{180}$, where the phase delay is computed. An accurate estimate of phase delay is clearly important to define the handling qualities, but measuring the slope from the phase roll-off is not always straightforward. Reference 6.41 describes how the least-squares fit of the phase line had to be restricted to avoid being distorted by a high-frequency phase drop due to a rotor structural mode.

Bandwidth and phase delay have emerged as two key parameters reflecting attitude handling qualities in the small amplitude regime. The supporting test and analysis methodologies have received considerable attention since the initial debate on the merits of time and frequency domain methods, and the extensive, and more general, coverage given to the topic in this roll control section reflects the level of effort and importance given to the bandwidth concept by the rotorcraft community.

Table 6.3 gives the roll axis bandwidth and phase delay estimates for a number of current operational helicopters in hover, together with the relevant data sources.

In the characterization of helicopter response portrayed by the framework diagram, Fig. 6.5, there is no reference to a handling quality that enjoyed centre stage prior to the publication of ADS-33 – the control sensitivity, and before leaving

Table 6.3 Roll attitude bandwidth results for current helicopters

Test aircraft	Bandwidth (rad/s)	Phase delay (ms)	Data source
Bo105	5.72	62	Refs 6.18, 6.43
Bell OH-58D	3.4	120	Refs 6.40–6.42
Bell 214ST	2.4	85	Ref. 6.44
UH-60A ADOCS	2.33	181	Ref. 6.52

small amplitude dynamics, it is important to discuss the apparent demise of this parameter.

Control sensitivity

Control sensitivity is a measure of the initial angular acceleration of the aircraft following a step input command, is traditionally measured in rad/s^2 in, and is recognized as a primary parameter affecting pilot opinion of aircraft handling. ADS-33 does not dispute this but says that, 'all controller sensitivities shall be consistent with the aircraft dynamic response characteristics in each axis at all flight conditions'; no criteria for the acceleration sensitivity are given. This is not difficult to live with for simple first-order-type responses where the control sensitivity is given by the product of the bandwidth and control power. In simple derivative language, the sensitivity would then be related to the control derivative through the control gearing, i.e.,

$$(\theta_{1c\,max})L_{\theta_{1c}} = -p_s L_p \tag{6.18}$$

For simple response types, the requirements on sensitivity are therefore defined by those for response characteristics already discussed. The most obvious interpretation of this relationship was given by Edenborough and Wernicke (Ref. 6.53) who first attempted to define requirements for roll control characteristics for combat helicopters. The boundary lines are shown in the earlier Fig. 6.23, with a minimum sensitivity level of 1 rad/(s^2 in) and an increasing range of acceptable sensitivities for increasing roll damping. The upper limits on sensitivity reflect the fact that the initial response can be too jerky as well as too sluggish. In Fig. 6.23, the boundaries from a variety of different studies, conducted over the last few decades, illustrate the wide range of sensitivity that appears to be acceptable. In Ref. 6.54, Pausder and Von Grunhagen map quality boundaries onto a similar diagram, based on flight data from the DLR ACT Bo105, replacing roll damping with bandwidth (Fig. 6.37). This would seem to be the most suitable format for relating the short-term response to the sensitivity but, like all the other criteria we have discussed, will almost certainly have different boundary lines for different types of MTEs. Note that the minimum bandwidths for Level 1 and 2 handling qualities do not conform to the ADS-33C boundaries. A second series of in-flight experiments to explore sensitivity boundaries is reported in Ref. 6.55, based on tests with the Canadian ACT Bell 205. The authors of this work argue that a more meaningful measure of sensitivity is the rate sensitivity (measured in °/(s in)), rather than the control sensitivity. Note that on Figs 6.23 and 6.37, the rate sensitivity is constant along radial lines. The results presented in Ref. 6.55 confirm that there is a range of acceptable sensitivities for given bandwidth, but with much sharper

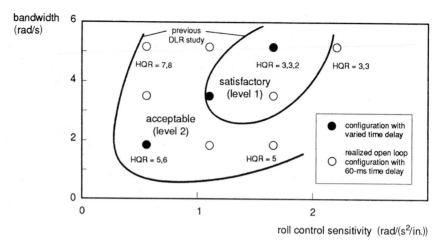

Fig. 6.37 Handling qualities boundaries for bandwidth versus control sensitivity (Ref. 6.54)

degradations for reducing sensitivity than increasing. This may reflect the difference between configurations that are becoming impossible to fly and those that are merely difficult. One thing that seems clear from all studies on sensitivity is that the lower the bandwidth, the narrower the range of acceptable sensitivities. The results in Fig. 6.37 suggest that the Level 1 boundary may be a closed contour, as postulated by Edenborough and Wernicke (Ref. 6.53), reflecting the potential for having an over-responsive aircraft. To date, insufficient attention has been given to this topic to give clear guidance, but there are parallels with fixed-wing flying qualities where very high-performance fighter aircraft do have prescribed upper limits on sensitivity and bandwidth (Ref. 6.6).

But there is another major influence on the quality of control or rate sensitivity – the characteristics of the pilot's controller or inceptor. Even conventional centre sticks can vary in shape and size and, given the control power requirements, the sensitivity is dependent on the size of the control throw. With the advent of sidestick controllers, the sensitivity requirements have become even more complicated. Early research into sidesticks for helicopters soon established the need for nonlinear shaping of the response/control deflexion relationship (Refs 6.56–6.58). For small displacement controllers with linear response gradients, pilots find that the sensitivity for small amplitude inputs is too high to allow smooth control actions, and a much reduced gradient near centre is required. To allow a high control power to be achieved at maximum control throw, the gradient will then typically have to increase several-fold with the possibility of too high a sensitivity at larger displacements. There are many flying qualities issues that are accentuated with sidesticks, and we devote more discussion to these in Chapter 7. So we leave sensitivity, a vital influence but still something of a mystery, with very little data to substantiate well-defined quality boundaries. Perhaps it is as well that such a critical parameter is left for the engineer and pilot to optimize in the design phase.

Staying with small amplitude motions we now increase the timescale to discuss a facet of helicopter flying qualities that is perhaps the most notorious, if not the most critical, of all – stability.

6.3.4 Small amplitude/low to moderate frequency: dynamic stability

Stability is important in any dynamic system, and for helicopters this is reflected in the need for the aircraft to not diverge from its trim condition if the pilot's controls are left momentarily unattended. The theoretical foundation for dynamic stability has already been covered in detail in Chapter 4, and the reader is referred there for discussions on modes of motion and associated eigenvalues. Stability was discussed in terms of the character of the response to small disturbances and the tendency of the aircraft to return to or depart from equilibrium. One of the problems encountered when discussing stability criteria in separated axes form is that the natural modes of motion are generally coupled and the roll DoF actually appears in most. However, there is often, but not always, a single dominant axis per mode and this appears the most logical manner by which to approach the discussion. With this rationale we discuss the lateral/directional oscillatory mode under the yaw axis stability and the pitch–roll long period oscillation under pitch axis stability, although both have implications on roll stability. The remaining mode for which there are stability concerns is the roll–yaw spiral and we choose to discuss relevant criteria in this section.

The characteristics of the spiral mode will determine the tendency of the aircraft to return to or depart from a level trim condition following a perturbation in roll. Spiral and Dutch roll stability are naturally at variance with one another so that a strongly stable spiral mode will result in an attitude command response type in roll, accompanied by a strong excitation of a weakly stable, or even unstable, roll–sideslip oscillation during simple uncoordinated turns. Criteria relating to the roll–sideslip coupling are discussed in Section 6.7 and, of course, the Dutch roll stability itself in Section 6.5. ADS-33 sets the handling boundaries on the time-to-double amplitude of the roll angle following a pulse input in lateral cyclic; i.e.

Level 1: $t_d > 20$ s
Level 2: $t_d > 12$ s
Level 3: $t_d > 4$ s

The degree of spiral stability can be demonstrated qualitatively by the 'turns on one control' technique. Having established a trim condition, lateral cyclic is used to roll the aircraft to a small bank angle. Speed is held constant with longitudinal cyclic and the lateral cyclic retrimmed to hold the new bank angle and turn rate; pedal and collective are held fixed. The manoeuvre is repeated in the opposite direction and for a range of increasing bank angles. Similar tests can be performed using yaw pedals to initiate and trim in the turn. For both tests, the control deflexion required to maintain the steady turn gives a direct indication of the spiral stability. If out-of-turn control is required then the aircraft exhibits spiral instability; conversely, if into-turn control is required then the aircraft is spirally stable. Recalling the linearized derivative theory in Chapter 4 and combining terms in the rolling and yawing equations of motion in a steady turn, the control perturbations can be written as

$$\delta\eta_{1c} = \frac{\left(L_v N_r - N_v L_r\right)}{L_{\eta_{1c}} N_v} r \qquad (6.19)$$

$$\delta\eta_p = \frac{\left(L_v N_r - N_v L_r\right)}{\left(L_{\eta_p} N_v - N_{\eta_p} L_v\right)} r \qquad (6.20)$$

Here r is the yaw rate in the turn and an additional assumption is made that rolling moments generated by the helicopter's pitch rate in the turn can be neglected. The numerator in the above equations is the spiral stability parameter derived in Chapter 4. From the test results, only the ratio of this parameter with the control derivatives can be obtained as a function of flight condition, and the inclusion of the rolling moment due to pedal complicates the analysis. The spiral stability test technique recommended by the FAA (Refs 6.10, 6.13) involves establishing an out-of-balance trim, returning controls to the level trim position and measuring the bank angle response. Reference 6.13 states that the time for the bank angle to pass $20°$ should not be so short as to cause the aircraft to have objectionable flight characteristics in the IFR environment (UCE >1). For unstable aircraft, the time-to-double amplitude should be at least 9 s.

As we examine handling qualities boundaries based on stability for other axes, we shall see that pilots can tolerate some degree of instability in the long period modes of helicopter motion, particularly during attentive flight phases. But before the aircraft even moves, the pilot will be concerned about the ability to establish and maintain trim. We now come to the final area on our response diagram, encompassing trim and classical quasi-static stability.

6.3.5 Trim and quasi-static stability

A key flying qualities concern relates to the ability to trim a helicopter with adequate control margins remaining for manoeuvring, throughout the OFE. We have already discussed aspects of control adequacy in the section on control power but this can now be expanded in an examination of the roll axis. Just as with dynamic stability, however, it is difficult to discuss roll motion in isolation. The ease with which a pilot can coordinate entry to a turn, maintain trim in asymmetric flight or point the fuselage away from the direction of flight depends critically upon the ratio of two static stability effects, the yawing moment (N_v) and rolling moment (L_v) due to sideslip, i.e., directional and dihedral stability, respectively. ADS-33 requires the dihedral to be positive and essentially linear for Level 1 flying qualities. To protect against control limits being reached in sideslipping or sideways flight, upper limits on dihedral effects in the required MTEs are defined in terms of amount of control used, as follows:

> Level 1: <75% control/49 N (11 lbf) control force;
> Level 2: <90% control/60 N (13.5 lbf) control force.

Estimates of both the rolling and the yawing moments can be derived from steady heading sideslip flight tests at a range of forward speeds from climbing through to autorotative flight. Such tests will also highlight any control problems within the sideslip envelope which is usually defined from fuselage stress considerations as a piecewise linear function of airspeed. At each test point, control angles to trim and aircraft attitudes are recorded. Figures 6.38(a)–(c) illustrate trim control results for the Puma; Fig. 6.38(a) illustrates how the control gradients vary with forward speed while Figs 6.38(b) and 6.38(c) show results at three different flight states descent, level and climb at 100 knots, with the slopes of the curves again indicating directional and dihedral stability. The calculation of derivative ratios can be demonstrated using the analysis of Chapter 4. The following ratios can be derived from the steady moment balance in a

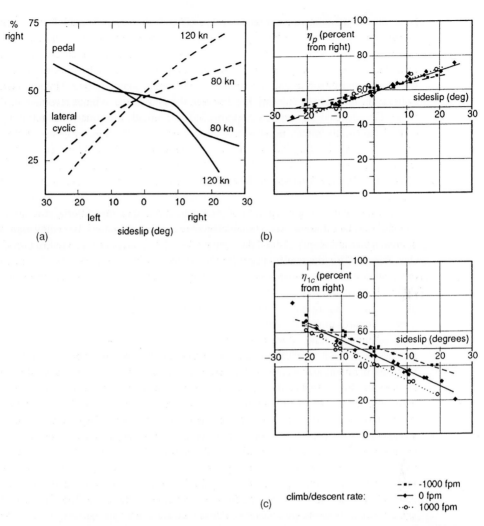

Fig. 6.38 Puma lateral cyclic and pedal positions in sideslip tests: (a) control variations with sideslip at different flight speeds; (b) pedal variations with sideslip in climb/level/descent flight conditions at 100 *kn*; (c) lateral cyclic variations with sideslip in climb/level/descent flight conditions at 100 *kn*

sideslip manoeuvre:

$$\frac{\delta \eta_{1c}}{v} = -\frac{\left(L_v - L_{\eta_p} N_v / N_{\eta_p}\right)}{L_{\eta_{1c}}} \tag{6.21}$$

$$\frac{\delta \eta_p}{v} = -\frac{N_v}{N_{\eta_p}} \tag{6.22}$$

where $\delta \eta_{1c}$ and $\delta \eta_p$ are the pilot's control deflections from level trim and v is the sideslip velocity. Provided that the variation of the control derivatives with speed can be neglected, the trends, though not absolute variations, in dihedral and directional stability can therefore be derived. For helicopter configurations with a high set tail rotor, the rolling moment from the tail rotor will contribute significantly to the lateral cyclic required in steady sideslipping flight. When the dihedral effect is small the trimmed cyclic may be in the same direction as the pedal trim, leading to Level 2 qualities according to ADS-33. An overriding pilot consideration when testing for directional and dihedral stability should be that clear unambiguous sideforce cues indicate the direction of sideslip. In particular, the pilot needs to be clearly alerted by these cues when sideslip limits are approached, as normally information on sideslip is not available to the pilot.

6.4 PITCH AXIS RESPONSE CRITERIA

Before we embark on a discussion of pitch axis flying qualities criteria, it is useful to reflect on the different axis pairings that arise in flying qualities. The conventional approach places roll and yaw together, and pitch and heave together, in classical lateral-directional and longitudinal motions. This is certainly the approach we took in Chapter 4 when analysing flight dynamics in terms of the natural modes. But when it comes to flying qualities criteria, we typically find roll and pitch having much in common. Just as with fixed-wing aircraft, in high-speed flight the pilot has the most powerful control over an aircraft's flight path with his centre-stick, through ailerons and elevator or lateral and longitudinal cyclic. At hover, and in low-speed flight, the cyclic is used to redirect the helicopter rotor's thrust, and harmony between roll and pitch flying qualities is particularly important because mixed pilot commands are a regular occurrence. In this context we should expect similar formats for roll and pitch flying qualities criteria. While this is the case in low-speed manoeuvres, the requirements on the pitch axis in forward flight are quite different from roll in many details. Pitch cyclic is the primary speed control, provides the mechanism for pulling g in manoeuvres, enables fuselage pointing and is a powerful motivator for the control of flight path angle in high-speed flight. Pilots of conventional helicopters are familiar with an impure response type in the pitch axis in forward flight rate in the short term, washing off quite rapidly as speed and incidence change, to give an attitude change in the mid term. The longitudinal stick position therefore provides a powerful cue to the pilot of the forward airspeed and pitch attitude of the helicopter. In response-type terms, however, this most closely resembles a rate response type and hence the related criteria apply.

As we examine the criteria for different areas on the response diagram, we shall find many similarities with the roll axis, but we will also see differences, especially in the areas of dynamic and quasi-static stability. Comparison with equivalent criteria for fixed-wing aircraft will provide interesting points for discussion when comparing the different roles and associated task bandwidths associated with the two types of aircraft.

6.4.1 Moderate to large amplitude/low to moderate frequency: quickness and control power

For low speed and hover MTEs, criteria for moderate and large amplitude pitch axis handling qualities mirror the roll axis very closely. The pilot's ability to manoeuvre is determined by the same performance or agility parameters – control power and attitude quickness. Figure 6.39 illustrates flight results from the DRA research Lynx (Ref. 6.45) performing a quickhop MTE; the aircraft is repositioned from one hover

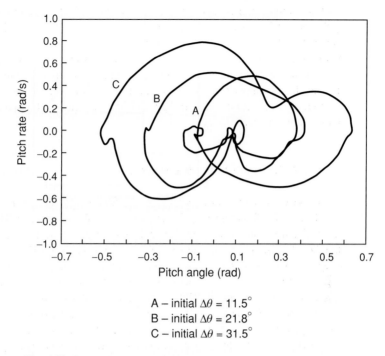

A – initial $\Delta\theta = 11.5°$
B – initial $\Delta\theta = 21.8°$
C – initial $\Delta\theta = 31.5°$

Fig. 6.39 Phase plane portraits for Lynx quickhop manoeuvres (Ref. 6.45)

position to another, across a 50-m (150 ft) clearing. The results are displayed as a phase plane portrait with pitch rate plotted against attitude for three different levels of pilot aggressiveness – low, moderate and high, defined by the initial pitch angle and the rate of application of cyclic control. At the highest level of aggressiveness, the pilot is nominally attempting to fly the manoeuvre as quickly as possible, achieving pitch angles of over 30° during the acceleration phase and corresponding rates of 40°/s. Pitch rates of 50°/s were used in the reversal phase of the manoeuvre to initiate the deceleration. In many respects the quickhop is similar to the lateral sidestep described earlier and illustrated in Fig. 6.19. Moreover, in hover, the control power in the pitch axis is essentially the same as in the roll axis, scaled by the ratio of control ranges. This can mean that the pitch axis control power is actually higher than the corresponding roll axis control power. However, there are two handling aspects that serve to differentiate between pitch and roll requirements for control power and quickness. First, the field-of-view constraints resulting from large positive and negative pitch excursions tend to make pilots less willing to use the full agility in the pitch axis. This is coupled with pilot concern of where the tail of the aircraft is; during the quickhop reversal at maximum aggression, for example, the Lynx's tail rotor descends 10 m (30 ft) closer to the ground. While the same is true for the blade tips in the sidestep, the pilot can more easily monitor the aircraft's safety as he or she looks into the manoeuvre. Second, to achieve similar quickness levels in pitch and roll, the pilot needs to apply larger control inputs to quicken the pitch response effectively, since the higher inertia in the pitch axis reduces the achievable angular acceleration and hence bandwidth, for the same applied control moments. This can lead to overcontrolling and reduced safety margins

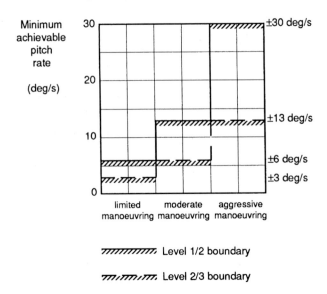

Fig. 6.40 Minimum pitch control power requirements – rate response type (Ref. 6.5)

when manoeuvring close to the ground. The result of these effects is that requirements for pitch axis control power and quickness tend to be lower than for roll. Figures 6.40 and 6.41 show the ADS-33 minimum control power and quickness required for rate response types, for the different classes of MTE. Unlike the roll axis criteria shown in Figs 6.15 and 6.17, the pitch criteria are defined only for hover/low-speed MTEs. In forward flight MTEs, ADS-33 is much more qualitative, requiring the pitch authority to be sufficient to accelerate between defined speeds at constant altitude, with no levels of aggressiveness defined.

The minimum control power levels in Fig. 6.40 were developed from flight and simulation experiments conducted on ground-based and in-flight simulators and apply to the cases of aircraft manoeuvring from the hover at the most critical wind state for pitch manoeuvring. They represent the minimum manoeuvre margins for successfully accomplishing battlefield helicopter operations. For moderate amplitude manouevres, the quickness minima in Fig. 6.41 apply. Compared with the roll boundaries, we immediately see the levels are reduced across the range by significant amounts, for the reasons given above. While the rationale for the mismatch between pitch and roll requirements is understandable, when these are realized in practice, the pilot does not have fully harmonized cyclic control; if he or she pushes the cyclic 45° to the right, the aircraft might accelerate away at 70°, simply because the roll quickness is higher than the pitch. The author believes that there is a strong case for full harmonization with rate or attitude response types for low-speed MTEs, as would be found, for example, with TRC (see Section 6.8).

The use of quickness and response to large control inputs to quantify attitude flying qualities at moderate to large amplitude is an innovation of ADS-33 and replaces the earlier measures adopted in MIL-H-8501A and the UK's Def Stan, based on the attitude response in a defined time, independent of response type. Earlier versions of the ADS, in the original draft Mil Spec 8501B version, did adopt the 'attitude change in one second' criterion, but the very compelling and more intuitive quickness, which

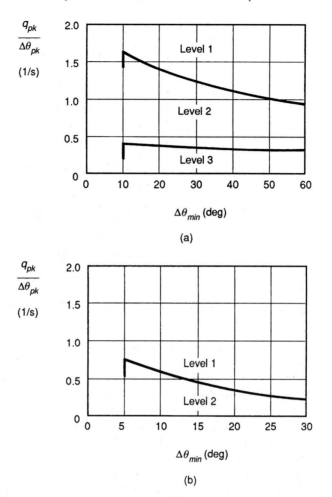

Fig. 6.41 Pitch attitude quickness criteria (Ref. 6.5): (a) target acquisition and tracking (pitch); (b) general MTEs (pitch)

had emerged as a natural roll axis handling and agility parameter, soon replaced this for moderate pitch attitude manoeuvres, with excursions between 10° and 30°.

Figure 6.42 shows the Lynx quickness envelope from the DRA Quickhop tests, overlaid with the ADS-33 Level 1/2 boundaries for tracking and general MTEs. The attitude change has been extended out to beyond 60° to include the excursions during the pitch reversal. The quickness, corresponding to the ADS-33 minimum control power requirement at this end of the manoeuvre range, would be about 0.5 rad/s, or approximately half that achieved by the Lynx. A similar result was found with the roll axis sidesteps. The Lynx is a very agile airframe of course, empowered by its hingeless rotor and it does raise the question as to what are the desirable levels, rather than the minimum levels, of quickness for different MTEs. We shall return to this subject under the special topic of agility in Chapter 7. At the lower end of the amplitude range in Fig. 6.42 the measured quickness values rise up to well beyond the minimum requirements; here we are in the domain of the tracking phase of the MTE and we have to look again to the bandwidth criterion to set the standards.

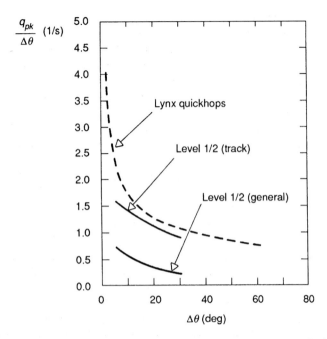

Fig. 6.42 Pitch attitude quickness – envelope from Lynx quickhop tests (Ref. 6.18)

6.4.2 Small amplitude/moderate to high frequency: bandwidth

In the development of pitch handling qualities for fixed-wing aircraft (Ref. 6.6) there has been a history of controversy over the most suitable format for the primary criteria. Most unaugmented or partially augmented aircraft have a characteristic short period pitch mode that dominates the short-term response to elevator, with a frequency that increases with airspeed. The natural parameters associated with this mode are its frequency and damping (ζ and ω), and the response is also shaped by the zero in the numerator (μ_0) of the pitch attitude (θ) to elevator (η) transfer function given by

$$\frac{\theta}{\eta}(s) = \frac{M_\eta(s + \mu_0)e^{-\tau_e s}}{s\left(s^2 + 2\zeta_{sp}\omega_{sp}s + \omega_{sp}^2\right)} \qquad (6.23)$$

The exponential function has been added to account for any unmodelled time delays or high frequency lags in the aircraft, e.g., actuators with time constant τ. Fixed-wing aircraft short-term pitch handling qualities can be established on the basis of the parameter set in the model structure for the short-period mode given above. As discussed in Ref. 6.59, the parameters are used to derive the control anticipation parameter, which is the fundamental manoeuvre margin parameter for fixed-wing aircraft. This so-called LOES approach (Ref. 6.60), whereby the parameters are derived from a model matched to frequency response flight test data, currently enjoys the role of primary criterion for classical response types or essentially where the fit error is small, implying second-order dynamic characteristics. For conventional fixed-wing aircraft, without stability and control augmentation in the pitch axis, the phugoid mode is normally well separated from the short period in frequency terms and the approximation has a wide range of application. For non-classical response types, or when the fit error is too large to trust the estimated frequency and damping, one of the proposed alternate criteria is bandwidth. The bandwidth and phase delay parameter pair were, in fact, born out of the difficulties encountered in achieving satisfactory equivalent system matching for fixed-wing aircraft with complex, high-order control systems that completely changed the shape of the frequency response and replaced the classical short-period mode with a combination of others.

A discussion of the bandwidth concept formed part of the treatment of roll axis handling qualities in the previous section and this reads across directly to the pitch axis, where for helicopters, bandwidth is no longer the alternate, but primary parameter. Indeed, the need for an alternate to LOES for pitch axis handling of helicopters is even stronger, with typical phugoid and short-period modes much closer together in the frequency range. Research results presented by Houston and Horton (Ref. 6.28) showed that second-order equivalent systems for pitch–heave dynamics in forward flight do have potential, and can be used to simulate the response to limited bandwidth inputs, although not all of the estimated handling parameters reported were physically plausible. The character of the longitudinal modes was discussed in Chapter 4 along with the theoretical framework for linearized models of pitch dynamics.

Comparison of the bandwidth/phase delay handling qualities boundaries for fixed- and rotary-wing aircraft are shown in Figs 6.43 and 6.44. Figure 6.43 compares the boundaries for air combat and hover/low-speed tracking tasks, while Fig. 6.44 compares the boundaries for more general rotorcraft MTEs with fixed-wing aircraft in Category C flight phases, including landing. Two points are immediately apparent. The first is that the fixed-wing Level 1/2 boundaries are typically set at bandwidths two to four times those for helicopters. Second, that the phase delay boundaries are set much lower for fixed-wing aircraft. Both of these differences serve to reflect the different character of the rotary- and fixed-wing aircraft MTEs, which in turn is a reflection

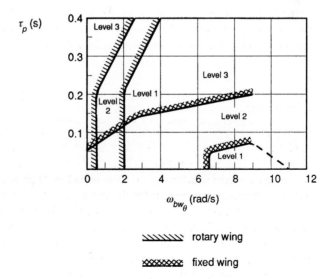

Fig. 6.43 Pitch attitude bandwidth boundaries – comparison of rotary- and fixed-wing aircraft (Category A flight phases) for air combat and tracking tasks (Refs 6.5, 6.6)

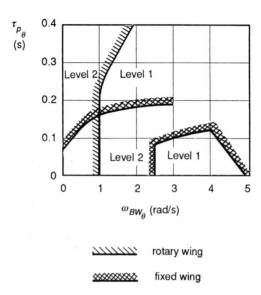

Fig. 6.44 Pitch attitude bandwidth boundaries – comparison of rotary- and fixed-wing aircraft for general MTEs and Category C flight phases (Refs 6.5, 6.6)

of the different speed ranges over which the aircraft operate. It is no coincidence that fixed-wing air combat typically takes place at speeds three to four times those envisaged for rotary-wing aircraft with similar differences in target closure range and rate. Not only is the higher bandwidth required to enable the pilot to track effectively, but the higher speeds in fixed-wing combat provide the aerodynamic forces to achieve the higher bandwidth. It would be very difficult, if not impossible, to engineer the 6 rad/s capability in rotorcraft manoeuvring at 100 knots. The much greater allowed phase delay for rotorcraft is still somewhat controversial, for similar reasons to those discussed for the roll axis. Some research findings have indicated (Refs 6.45, 6.51) that capping of the phase delay boundary down to 200 ms, or even lower, is warranted, and the author has supported this view. However, until a more substantial handling qualities database for pitch axis MTEs is available that clearly demonstrates degraded handling or even PIO tendencies for the higher bandwidth/phase delay configurations, the ADS-33 boundaries will probably be preserved. It should be emphasized that to achieve a phase delay of 300 ms with a 4 rad/s bandwidth, the manufacturer would be working very hard and incorporating very unusual features in the design; in fact, it seems a highly unlikely, if not impossible, practical combination. With the application of digital flight controls to helicopters, however, the controversial issue of phase delay limits for pitch axis dynamics may well re-emerge.

6.4.3 Small amplitude/low to moderate frequency: dynamic stability

The lack of natural longitudinal stability in helicopters was highlighted in Chapter 2 as one of the significant differences between fixed- and rotary-wing aircraft; this particular aspect was also discussed in some detail in Chapter 4, where approximate theoretical models provided some insight into the physical mechanisms – the pitching moments due to incidence and speed – that cause the unstable behaviour. The unstable mode

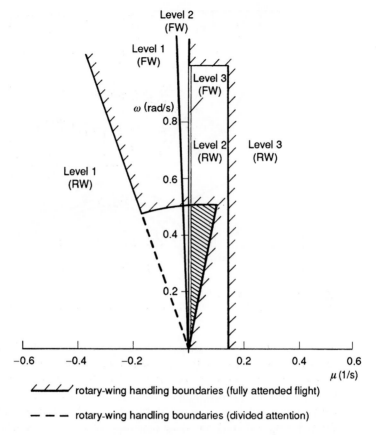

Fig. 6.45 Stability of long-period pitch oscillations – comparison of rotary- and fixed-wing requirements (Refs 6.5, 6.6)

of an unaugmented helicopter is often referred to as the phugoid, for both hover and forward flight, even though the character of the mode is significantly different in the two speed regimes. As discussed in Chapter 4, for some configurations the 'phugoid' frequency reduces to zero in high-speed flight and the motion can become so divergent that the main influence is on short-term control response rather than long-term stability *per se*.

We reproduce Fig. 2.39 here for the reader's convenience as Fig. 6.45, showing a comparison of fixed- and rotary-wing handling qualities boundaries on the frequency/damping plane for the long-period mode. The rotary-wing requirements are taken from ADS-33 and strictly apply only to RC response types, but criteria in Def Stan 970 and the civil standards are very similar. The dashed boundary, corresponding to a damping ratio of 0.35, applies to cases where the pilot is required to divide his or her attention between tasks and for flight in degraded visual environments. For fully attended operations, a small amount of instability is allowed, but this is curtailed abruptly for frequencies above 0.5 rad/s. An interesting comparison with the fixed-wing boundaries is the presence of the shaded region where a Level 1 rotary-wing aircraft handling corresponds to worse than Level 3 handling for a fixed-wing

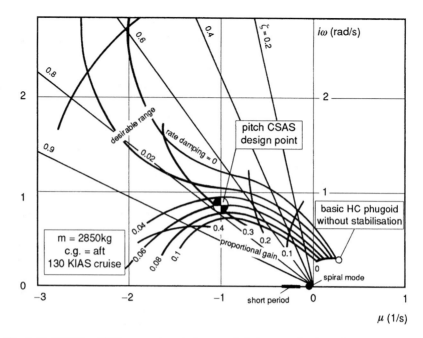

Fig. 6.46 BK117 at 130 knots cruise – influence of pitch rate and attitude feedback gains on phugoid mode (Ref. 6.61)

aircraft. Also, for frequencies above 0.5 rad/s, there is a large region where Level 1 rotary-wing handling coincides with Level 2 fixed-wing handling. We shall discuss the first of these observations only, which appears to be quite anomalous. It has to be recognized that in the development of new flying qualities requirements, any new criterion should not immediately exclude existing operational aircraft (unless there is very good reason to), and the allowance of the region of instability on Fig. 6.45 conforms with this philosophy. On the other hand, the requirement for the 0.35 damping indicates that if any 'serious' mission-related flying is to be conducted, then some form of artificial stability and control augmentation is mandatory. This is exactly how designs evolve in practice, and typically a significant proportion of the development flying on a new type will be dedicated to the refinement of the stability and control augmentation system, with particular emphasis on longitudinal handling. Figure 6.46 shows how the phugoid mode of the BK117 helicopter at 130 knots was stabilized with a combination of rate and attitude feedback with the relatively low gain values $-0.06°/°$s and $0.3°/°$, with the attitude stabilization providing by far the strongest contribution (Ref. 6.61).

A pulse input in longitudinal cyclic is usually sufficient to excite the pitch long period oscillation; the period can lie between 10 and 30 s, or even higher, hence the motion will have to be allowed to develop over a long time to obtain good estimates of both damping and frequency. As discussed in Chapter 2, this can lead to large amplitude motions from which the recovery can be even more dramatic than the test manoeuvre itself. The lesson is to apply a small amplitude exciting pulse, and to ensure that the motion does not exceed the normal linear range (e.g., attitude excursions $<10°$, speed excursions <10 knots).

6.4.4   Trim and quasi-static stability

A pilot flying under IFR in turbulent conditions will have his or her workload significantly increased if, in attempting to control speed errors with cyclic, the new stick position to trim is in the opposite sense to that initially required to cancel the perturbation. Likewise, when manoeuvring to avoid obstacles, a pilot will need to work harder if having rolled into a turn and pulled back on cyclic to increase turn rate, the pilot finds that he or she needs to push forward to avoid 'digging-in'. Both of these handling characteristics, are, generally speaking, unacceptable by any military or civil requirements standards, and flight tests need to be performed to establish if they are present within the OFE. They represent negative margins of speed and manoeuvre stability, respectively, that, together with their close companion flight path stability, form the topic of this section. Requirements tend to be very qualitative for trim and static/manoeuvre stability and therefore the emphasis below is on the required flight test techniques.

Figure 6.47 illustrates the consequences of positive and negative speed stability for cyclic control — in both cases the speed excursion is the same, but, with negative speed stability, the cyclic retrims the 'wrong way'. There are two concepts traditionally associated with this characteristic, namely, apparent and true speed stability. The apparent speed stability is determined by the slope of the longitudinal cyclic trim variation with speed, i.e., with collective varying to maintain level flight or a defined rate of climb or descent. True speed stability, on the other hand, usually of more concern to the pilot, is determined at a given speed by noting the new trim stick position for speed increments at constant collective pitch. The two results are sketched in Fig. 6.48 where, for illustration, the true speed stability is shown to be negative and contrary to the apparent speed stability at the lower speed trim condition.

The test technique to investigate true speed stability is fairly straightforward. Having established trimmed flight at a defined airspeed and power setting, the helicopter is retrimmed in a series of speed increments, below and above the test airspeed, with cyclic. Alternation between positive and negative increments allows the aircraft to remain within a sensible altitude band (e.g., 1000 ft) for level flight airspeed tests.

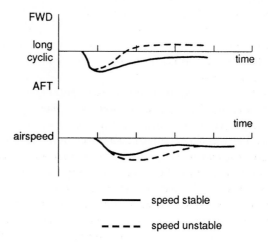

Fig. 6.47 Effects of speed stability; impact on cyclic trim (Ref. 6.62)

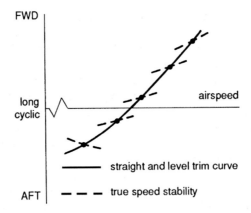

Fig. 6.48 Effects of speed stability; true and apparent speed stability (Ref. 6.62)

For climb and descent conditions, two passes through the required altitude band are typically required. While conducting these tests, the pilot will also be concerned with any related 'ease of trimming' issues, e.g., controller breakout forces and force gradients. Particular attention will be paid to identifying strong nonlinearities, for example, discontinuities in the speed stability and to distinguishing these from any adverse controller force characteristics or the effects of atmospheric disturbances.

Most certification requirements allow a limited degree of speed instability at low speeds, on the basis that the effect is not so critical here with the pilot normally controlling both speed and flight path angle with a combined cyclic/collective control strategy. At higher speeds, particularly for cold weather operations, adverse speed stability can limit the safe maximum flight speed, and careful testing is required to highlight any advancing blade Mach number effects. One such problem arises when a forward speed increment results in the centre of pressure moving further aft on the outboard sections of the advancing blade. This compressibility effect twists the blade cyclically to give a nose-down pitching moment on the aircraft, which needs to be counteracted with aft cyclic.

Within the framework of the linearized stability theory discussed in Chapter 4, the speed stability of a helicopter is determined by the value of the effective derivative

$$M_u^* = M_u - \frac{Z_u}{Z_w} M_w \qquad (6.24)$$

obtained from the equations for the initial pitching moment due to a speed disturbance and the final steady-state cyclic increment. The effect is usually dominated by the pitching moment derivative M_u which has a stabilizing contribution from the main rotor. The fuselage and tailplane contributions will depend upon the trimmed incidence of these components. Tailplane effects can dominate in some situations. Reference 6.63 describes the adverse effect on speed stability caused by tailplane stall during climbing flight in the SA 365N helicopter. Fitting small trailing edge strips on the tailplane attenuated this effect, but to guarantee speed stability for steep climbs in the range 80–100 knots, an additional speed hold function was incorporated into the autopilot. Similar small design modifications to the tailplane leading and trailing edges were required to achieve speed stability for the BK117 helicopter (Ref. 6.61).

In addition to the speed stability testing described above, further tests are required to explore the cyclic trim changes with power settings at different speeds from autorotation to max power climb. These tests are required largely to check that adequate control margins are available in these conditions but will also highlight the essential features of flight path stability. Although there are no general requirements concerned with helicopter flight path stability, for aircraft roles that demand precise flight path control, e.g., guided approaches, testing will need to be carried out to establish the optimum pilot control strategy for the various flight phases. Such tests are likely to be carried out in conjunction with the development of the associated displays and stability augmentation. Collective is, of course, the natural control to counteract flight path errors, but above the minimum power speed the use of cyclic can achieve a similar effect. If the aircraft has, for example, fallen below the glide path and is flying too fast, pulling back on the stick will eventually cancel both errors. Problems arise below minimum power speed where, although the initial effect of pulling back on the stick is to climb the aircraft, the new equilibrium state will be an increased rate of descent. Although normal control strategy should preclude such problems under 'controlled' approach conditions, for unguided steep approaches or emergency situations the pilot needs to be aware of the potential problems. At very steep descent angles the problem can be exacerbated by power settling effects (Ref. 6.64) and ultimately the vortex-ring condition (see Section 6.5), where static stability characteristics are overshadowed by dynamic effects.

While speed and flight path stability are concerned essentially with cyclic to trim requirements in 1 *g* flight, manoeuvre stability is related to cyclic changes required in manoeuvres involving a change in normal acceleration, or the stick displacement (or force) per *g*. All handling requirements specify that this should be positive, i.e., aft stick is required to hold an increased load factor, and as a consequence, there should be no tendency to 'dig in' during turning flight. The manoeuvre stability can be determined in flight from either symmetric pull-up and push-over manoeuvres or steady turns, and needs to be measured across the full range of operational conditions, i.e., speeds, atmospheric conditions, aircraft loading. For the pull-up tests, the aircraft is trimmed in level flight at the test airspeed. With collective fixed, the aircraft is then decelerated with cyclic and then dived to accelerate back to the test airspeed. As the test speed is approached, an aft cyclic step is applied to achieve the desired load factor and airspeed as the aircraft passes through a level attitude. The test is repeated with increasing increments of aft cyclic until the maximum permitted load factor is achieved. Similar tests are performed to establish the manoeuvre margin for load factors less than 1, using the push-over technique. For steady turn tests, the aircraft is again trimmed in level flight at the test airspeed. Load factor is applied incrementally by increasing bank angle at constant collective and airspeed, and maintaining balance with pedals. Cyclic is retrimmed at each test condition and the tests conducted for both left and right turns. Rotorspeed should be adjusted only to remain within power-on limits, and since high rates of descent may be achieved, care should be taken to remain within a defined altitude band (e.g., +1000 ft of test condition).

Figure 6.49 illustrates results that may be derived from these tests; the manoeuvre stability is deliberately shown to be negative (and therefore not Level 1) at the higher speed. The cyclic to trim variation with load factor in the steady turn will typically be steeper than the corresponding pull-up result on account of the increased pitch rate in a turn for a given load factor. The relationship between cyclic to trim and pitch

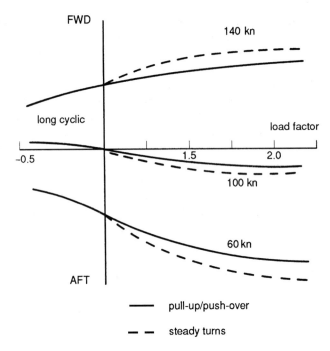

Fig. 6.49 Effects of manoeuvre stability (Ref. 6.62)

rate or load factor can be derived from linearized theory (see Chapter 4) in the form, neglecting flight path angle effects,

For pull-ups:

$$\delta\theta_{1s} = -\left(\frac{M_q Z_w - M_w V}{Z_w M_{\theta_{1s}} - M_w Z_{\theta_{1s}}}\right) q \qquad (6.25)$$

$$q = \frac{g}{V}(n-1) \qquad (6.26)$$

For turns:

$$\delta\theta_{1s} = -\left(\frac{M_q Z_w - M_w V\left(\frac{n}{n+1}\right)}{Z_w M_{\theta_{1s}} - M_w Z_{\theta_{1s}}}\right) q \qquad (6.27)$$

$$q = \frac{g}{V}\left(n - \frac{1}{n}\right) \qquad (6.28)$$

Here θ_{1s} is the applied cyclic pitch (positive aft), q the pitch rate, V the flight speed and n the load factor. The stability and control derivatives will themselves vary with rotor thrust and rotor disc incidence, and a more exact analysis will certainly be required for higher values of n. Nevertheless, eqns 6.25 and 6.27 are valid representations of manoeuvre stability parameters. The numerator in eqn 6.25 is the classical manoeuvre margin parameter that should be positive for 'stability' and acceptable handling characteristics. Typically, an increasingly positive M_w variation with speed will lead to a deterioration in manoeuvre stability to the point where the margin can change sign. The load factor parameter in the manoeuvre margin for steady turns arises from the inclination of the weight component from the fuselage normal and, at low bank angles, will serve to reduce any undesirable effects of a positive M_w. At higher bank angles, however, any unstable tendencies are likely to re-emerge.

The tests described above to establish the manoeuvre margin are carried out at constant collective pitch settings. In many practical situations, however, the pilot will use collective in conjunction with cyclic to maintain height. The pitching moment generated by collective application will be nose up and hence the cyclic position to trim will be further forward than indicated by the tests at constant collective unless control interlinks have been built in. This effect can be compounded by an increased download on the tail from the main rotor downwash. On other occasions, the pilot may choose to decelerate the aircraft in the turn, hence requiring increased aft cyclic displacement. This variability of stick position with load factor, depending on the type of manoeuvre flown, does not provide the pilot with a reliable tactile cue in manoeuvres. In any case, stick force per g is of more concern to the pilot, particularly in the mid–high speed band, and several current operational helicopters (e.g., AH-64, SH-60) have force feel systems that provide a positive and reliable cue to the pilot of manoeuvre margin.

6.5 HEAVE AXIS RESPONSE CRITERIA

Heave, or vertical, axis handling qualities criteria are concerned principally with the response of the aircraft to collective pitch application. In hover and low-speed flight, collective provides the pilot with direct lift control, a feature that clearly makes the helicopter almost unique. In forward flight, control of the aircraft's flight path can be achieved through a combination of collective and cyclic, but in this section we shall restrict the discussion to the response to collective control. The extent to which the pilot is able to exercise this degree of freedom depends on a number of factors, which we shall discuss, but is often dominated by the thrust margin available before transmission torque, rotorspeed or engine limits are exceeded. The thrust margin is a strong function of airspeed through the variation of the power required in trimmed flight; this point has been discussed earlier in the modelling chapters of this book, but it is worth recalling the shape of the power required curve given in Fig. 4.11. Typically, for a fully laden helicopter at its mission weight, the power margin at hover is very low, of the order of 5–10%, giving thrust margins between 3 and 7%. At minimum power speed, the same helicopter may have more than a 100% thrust margin, enabling the aircraft to sustain a 2 g turn. Other fundamental response parameters are the heave damping and control sensitivity derivatives. Again, both vary significantly with forward speed. These aerodynamic effects are a reflection of the increasing efficiency of the rotor as a lifting device through to the mid-speed range. As speed is further increased, the power required increases again and the response derivatives level off to their maximum values as the aerodynamic lift becomes dissipated in higher harmonic loadings that contribute nothing to flight path response. Unlike roll response characteristics, the heave dynamic characteristics therefore vary significantly with forward speed. The pilot is able to exploit these varying characteristics in different ways, and we must expect the associated handling criteria to reflect this.

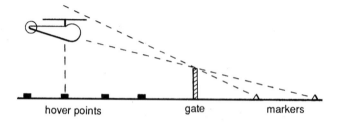

Fig. 6.50 Bob-up MTE

The low-speed vertical axis response characteristics are highlighted in the bob-up task, a vertical unmask manoeuvre illustrated in Fig. 6.50. Results from DRA tests with the research Puma are shown in Figs. 6.51 and 6.52; the pilot's task was to climb with maximum power from the low hover position and to re-establish a hover when the ground markers were lined up with the top of the gate (Ref. 6.65). Height responses are shown in Fig. 6.51 for bob-up heights from 25 to 80 ft together with the case of a maximum power vertical climb, when the climb rate exceeded 30 ft/s (10 m/s). In comparison, the maximum rate achieved during the 25-ft bob-up was only 14 ft/s. This result is a function of the vertical damping of the rotor, which gives an effective time

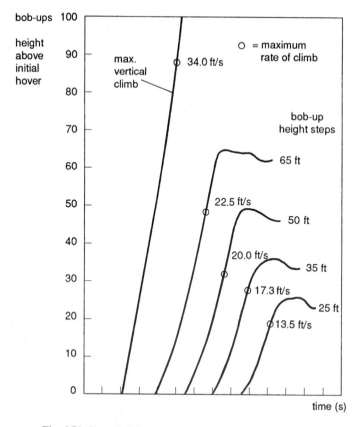

Fig. 6.51 Puma height responses in bob-up MTE (Ref. 6.62)

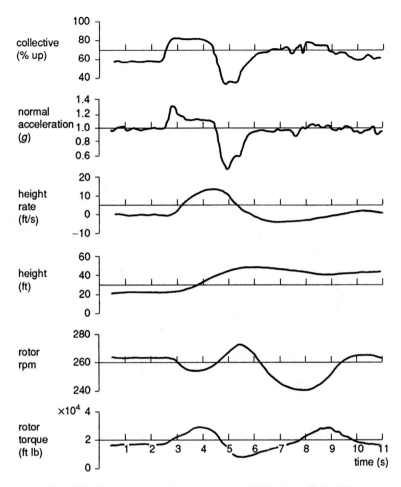

Fig. 6.52 Puma response characteristics in 25-ft bob-up (Ref. 6.62)

constant of several seconds, together with the constraint on the pilot to respect the Puma's collective pitch limits. Figure 6.52 shows the variation with time of selected variables during a 25-ft (7.5 m) bob-up. The pilot pulls in a 20% collective input, causing a sharp rise in normal acceleration. The overshoot in the time history of normal *g* is explained by the delay in build up of the induced inflow, described in Chapters 3 and 5. A thrust margin of about 15% is sustained for about 1 s before the pilot lowers the collective by more than 40% and almost immediately pulls in power again to arrest the deceleration and level out at the top of the bob-up. The manoeuvre is relatively simple but has required the pilot to apply large control inputs in three phases. The lower traces in Fig. 6.52 show the excursions of rotor rpm and torque. The lower rotor rpm limit of 240 rpm is actually reached during the settling phase at the top of the bob-up. The need to respect rotor collective, torque and rpm limits plays a significant role in pilot subjective opinion of vertical axis handling qualities. This has been exposed in most of the flight and ground-based simulation work supporting the development of associated criteria, and we shall return to this aspect later in this section and in Chapter 7.

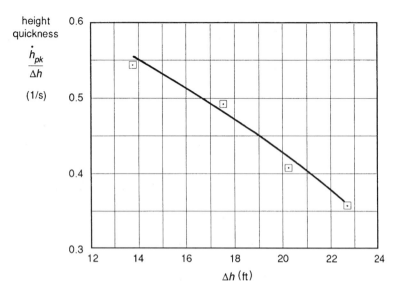

Fig. 6.53 Puma height quickness in bob-up task

Height quickness results derived from the Puma tests are shown in Fig. 6.53 and highlight the much lower values than those found for attitude response, even though the bob-up has been quickened by the pilot through the application of a collective doublet. Control of helicopter vertical motion is generally regarded as a relatively low-gain task for pilots, and criteria developments have been limited to fairly simple formats that apply across the frequency range. The RAE Puma tests referred to above were one of a series conducted during the early- to mid-1980s to develop new heave handling criteria for rotorcraft (Refs 6.65–6.68), building on previous work applied to VSTOL aircraft, with particular emphasis on hover and low speed. Flight path control in forward flight will be discussed later in this section.

6.5.1   Criteria for hover and low-speed flight

The work reported in Refs 6.65 and 6.66 generally supported the use of VSTOL aircraft formats and an early version of the revision to MIL-H-8501 placed the boundaries on time domain parameters – rise time, response shape and control sensitivity, based on the height rate response to a step collective input. The lower boundary for Level 1/2 handling corresponded to a vertical damping Z_w of -0.25 / s centred around a collective sensitivity of 0.4 g/in. Later tests conducted on the NASA VMS (Ref. 6.68) and at the Canadian Flight Research Laboratories (Ref. 6.69) demonstrated the importance of thrust to weight (T/W). A new format, based on T/W, was proposed with revised heave damping boundaries (Fig. 6.54). The Canadian trials provided a range of new insights into rotorcraft vertical axis handling. First, for the bob-up task, the required heave damping for Level 1 handling qualities appeared to be independent of T/W down to the boundary at 1.08 (i.e., 8% thrust margin), provided that the damping was above the minimum required value of 0.25; Level 2 handling could be achieved with thrust margins greater than 4%, for any value of damping, down to zero. Second, the response shape criteria discussed earlier appeared to have little significance in determining Level 1

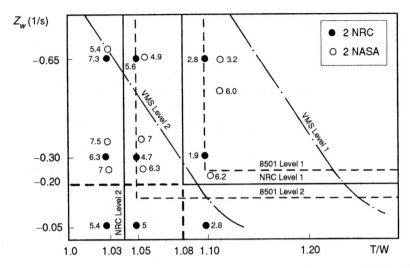

Fig. 6.54 Heave handling qualities boundaries on damping versus T/W diagram (Ref. 6.69)

handling qualities for low values of T/W typical of rotorcraft loaded to their mission gross weights. Third, the dynamics of the torque response, and particularly the dynamic response of the cockpit-displayed torque, significantly affected pilot control strategy and hence handling qualities, emphasizing the need for criteria relating to this effect. Fourth, pilots preferred a collective control sensitivity linearly matched to the heave damping such that the ratio was a constant. Fifth, the boundaries on the damping/thrust margin charts, suggested by the earlier NASA VMS trials (Ref. 6.66), actually sloped a different way; the results from NRC flight trials suggested that as the damping increased, the thrust margin should at least be held constant and possibly increase, to give the pilot a similar level of climb performance. There is a definite trade-off involved here – performance versus stability – and the flight data favours the former, at least for the bob-up task. Reference 6.69 argues that in ground-based simulation experiments, pilots have greater difficulty with the stabilization task than in real flight due to deficient visual cues, hence biasing their preference towards greater stability. Finally, the level of augmentation in pitch and roll had a significant effect on the workload capacity available for pilots to concentrate on the primary vertical axis task. These results fed into the development of the US Army's military handling requirements, and a major revision to the height response criteria eventually appeared in ADS-33 (Ref. 6.5).

The current requirements on vertical axis response characteristics in Ref. 6.5 are based on the premise that the height rate response to a collective step input should have a qualitative first-order shape as shown in Fig. 6.55. Handling qualities parameters can be derived from a response model structure in the first-order form

$$\frac{\dot{h}}{\delta_c} = \frac{K e^{-\tau_{h_{eq}} s}}{T_{h_{eq}} s + 1} \tag{6.29}$$

where h is the height, δ_c the pilot's collective lever and the estimated (time domain) handling qualities parameters should have values less than those given in Table 6.4.

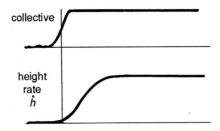

Fig. 6.55 First-order shape of height rate response

Table 6.4 Maximum values for time parameters in height response to collective (Ref. 6.5)

Level	$T_{\dot{h}_{eq}}$ (s)	$\tau_{\dot{h}_{eq}}$ (s)
1	5.0	0.20
2	∞	0.30

The maximum acceptable value of the time constant $T_{\dot{h}_{eq}}$ for Level 1 handling corresponds to a minimum value of heave damping of -0.2. The time delay is included in the model structure to account for actuation and rotor dynamic lags. The gain or control power K is determined from the steady-state response to the step input and, for the simple first-order representation, is given by the ratio of control sensitivity to damping. In ADS-33, the limits on vertical axis control power are expressed in terms of the achievable height rate in 1.5 s and are given in Table 6.5.

The requirements of Table 6.5 can be interpreted in terms of required hover thrust margins if it is assumed that the initial 1.5 s of height response takes the form of a first-order exponential function. The Level 1 requirements then correspond to a thrust margin of 5.5% while the Level 2 boundary lies at 1.9% at zero damping; both values are far lower than any previous results obtained in clinical flying qualities tests with the bob-up task. In this requirement we see a degree of conflict between current capabilities and future requirements. Most helicopters, whether civil or military, will carry the maximum allowed payload on a mission, and this generally leaves little margin for manoeuvring at low speed. As fuel burns off, the available excess thrust and power margins increase, but to insist on 8%, 10% or even higher margins at take-off can significantly reduce the payload and hence mission effectiveness from a productivity perspective. Thus, a compromise has been made in the performance requirements of vertical axis handling, in recognition of current operational practices. For increasing heave damping, the T/W has to increase to achieve the same vertical velocity, and the control power requirements of Table 6.5 can be interpreted as boundaries on Fig. 6.54. The lines again indicate the apparent preference for performance, rather than stability,

Table 6.5 Vertical axis minimum control power requirements (Ref. 6.5)

Level	Achievable vertical rate in 1.5 s, m/s (ft/min)
1	0.81 (160)
2	0.28 (55)
3	0.20 (40)

in MTEs like the bob-up. In contrast, preliminary results from DRA piloted simulation trials of ship landings in poor weather (Ref. 6.70) indicate that increased stability (damping) is preferred as T/W is reduced. The tentative results published in Ref. 6.70 suggest that in sea state 5 (typical worst operating conditions with deck motion ± 2 m) a heave damping of -0.4 would be required with a T/W of 1.08.

The handling parameters in eqn 6.29 can be derived from a curve-fitting procedure defined in Ref. 6.5 and summarized below. This is the only criterion in ADS-33 that requires explicit parameter estimation from a model-fitting process. The technique and some examples have already been discussed in Chapter 5 of this book. In the present case the fitting process is classed as least squares, output error and is accomplished as follows:

(1) The helicopter is trimmed in hover and a step input in collective applied; measurements of height rate are obtained at 0.05-s intervals for a 5-s duration.
(2) Setting initial values for the parameters in eqn 6.29, based on a priori knowledge, an estimate of the height rate is obtained from the solution to eqn 6.29

$$\dot{h}_{est} = K \left[1 - e^{-\frac{1}{T_{heq}}\left(t - \tau_{heq}\right)} \right] \delta_c, \quad t \geq \tau \tag{6.30}$$

$$\dot{h}_{est} = 0, \quad t < \tau \tag{6.31}$$

The $t \geq \tau$ requirement in eqn 6.30 is made to ensure that the response is causal, a point noted in Ref. 6.71.
(3) The difference between the flight measurement and the estimated height rate is constructed as an error function $\varepsilon(t)$ given by

$$\varepsilon^2 = \sum_{i=1}^{101} \left(\dot{h}_i - \dot{h}_{est_i} \right)^2 \tag{6.32}$$

and the sum of squares of this error function is minimized by varying the parameter set K, T_{heq} and τ_{heq}.
(4) The goodness or quality of fit can be derived from the coefficient of determination given by

$$r^2 = \frac{\sum_{i=1}^{101} \left(\dot{h}_{est_i} - \bar{\dot{h}} \right)^2}{\sum_{i=1}^{101} \left(\dot{h}_i - \bar{\dot{h}} \right)^2} \tag{6.33}$$

where the mean value of measured height rate is given by

$$\bar{\dot{h}} = \frac{\sum_{i=1}^{101} \dot{h}_i}{101} \tag{6.34}$$

(5) For a satisfactory fit, the coefficient of determination should exceed 0.97 and be less than 1.03.

The LOES for low-speed heave axis handling qualities described above has evolved from a number of attempts to model important handling effects during the development of ADS-33. It appears to capture, with a reasonably high degree of fidelity, the natural characteristics of unaugmented helicopter heave motion in the frequency range of pilot closed-loop control in manoeuvres like the bob-up and precision landing.

One final aspect of the response to collective concerns the shape of the normal acceleration following a very sharp control input. This subject was discussed in Chapter 5 and is also given some attention in Ref. 6.67. The delay in the build up of the rotor inflow causes the acceleration response to peak at much higher values than the 'steady state' (see Fig. 6.52). This 'high order' effect will be reflected in the height rate response and will 'spoil' the simple first-order character, with the potential consequence that the model parameters will be distorted, in trying to match the more complex response shape. One solution to this potential difficulty is to ensure that the pilot applies a ramp collective input over, say, a 1-s period, thus allowing the inflow time to develop during the input. This is an expedient measure to satisfy the low–moderate frequency requirements of the handling effects accommodated by eqn 6.29, but obscures any additional handling effects at higher pilot gains. We have already stated that control of vertical motion is largely a low–moderate gain task for the pilot, but automatic height-keeping controllers will typically have to work at much higher frequencies where the simple model structure given above will be inadequate. Raising this issue here highlights the different modelling requirements for handling qualities and control law design, a topic given some attention in Chapter 5 of this book.

6.5.2   Criteria for torque and rotorspeed during vertical axis manoeuvres

The vertical handling qualities research exercises at NASA Ames and the Canadian FRL, described in Refs 6.67 and 6.69, both highlighted the importance of monitoring rotor rpm and torque during MTEs like the bob-up. In some cases the monitoring requirements on the pilot dominated the workload and hence the pilot HQRs. This was particularly true for configurations with low T/W or with slow engine/rotorspeed governor systems leading to large excursions in rotorspeed. Various attempts were made to develop supplementary criteria relating to the response of these variables, but the findings from the available database were not entirely consistent. The eventual formats that were settled on for ADS-33 were of a qualitative form for rotorspeed governing, requiring transients to remain within limits for all tasks flown within the OFE, and a quantitative form for displayed torque as shown in Fig. 6.56. The Level 1/2 and 2/3 boundaries are based on the character of the displayed torque in terms of time to first peak and overshoot ratio. To the author's knowledge, there are very few data in the open literature to validate the criterion set down in Fig. 6.56, and the topic has to be one of the weak areas of current handling criteria and, in view of its importance to pilot workload, in need of further research.

6.5.3   Heave response criteria in forward flight

In the field of fixed-wing aircraft flying qualities, the subject of flight path response has received considerable attention and remains one of the areas of ongoing research and even controversy (Ref. 6.23). Two reasons explain why this level of interest has not carried over to rotary-wing aircraft handling. First, a critical flight phase and MTE for fixed-wing aircraft is the approach and precision flare and landing. The flight path response during the flare is very different for classical aircraft than for highly augmented

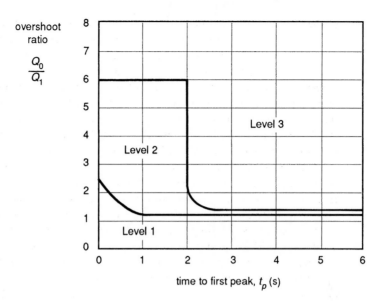

Fig. 6.56 ADS-33C requirements on displayed torque in terms of overshoot rates and time to first peak (Ref. 6.5)

aircraft, and different criteria are required for each, accommodating both attitude and path angle bandwidth requirements. For helicopters, there is no real equivalent MTE, for while guided approaches are common, by the time the helicopter is close to touchdown, the speed will have been reduced to the point that pitch and flight path are independent degrees of freedom. Second, through collective pitch, the helicopter pilot has direct lift or direct flight path control, and can normally use a combination of collective and cyclic to achieve a combination of pitch and flight path angle in forward flight to suit the mission requirements. These two reasons go some way towards explaining why the same level of attention has not been given to rotorcraft; the problems are not the same and the low-speed criteria for rotorcraft are more important. In the absence of a substantial test database, ADS-33 proposed an identical criterion to the one derived for hover and low-speed flight, based on an equivalent first-order system response. In the years since ADS-33 was published, several attempts at using this format have been published. For unaugmented aircraft, the important coupling parameter between pitch and incidence motion is the static stability derivative M_w. For aircraft with close to neutral stability, the pitch and heave motions are uncoupled in the short term and the flight path has a distinct first-order shape. An example of this case is presented for the AH-64A at 130 knots in Ref. 6.72; the time domain fit is shown in Fig. 6.57 and the handling parameters estimated from the first-order fit compare well with stability and control derivatives estimated from test data using six degrees of freedom models. Much poorer results are presented in Refs 6.19 and 6.71. In Ref. 6.19, the corresponding results for the Bo105 at 80 knots are presented. Figure 6.58 shows the height rate response to a step collective input, indicating a non-first-order-like shape in the 5-s window. For the Bo105, the pitch response to a collective step is very strong, causing the speed to reduce and the aircraft flight path to change as the nose pitches up. Applying cyclic to minimize pitch excursions resulted in a first-order height rate response (see full

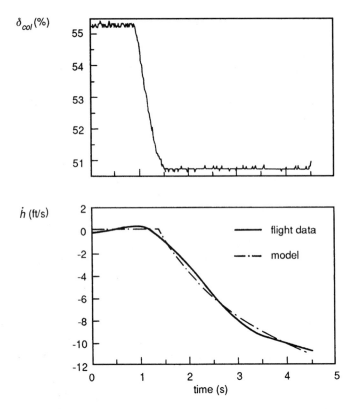

Fig. 6.57 Fit of handling qualities model to step collective response – AH-64 (Ref. 6.72)

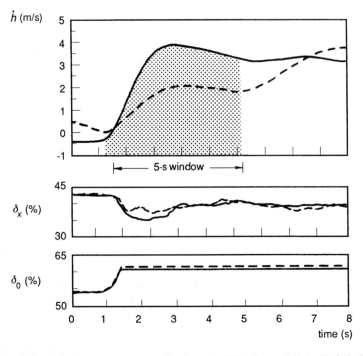

Fig. 6.58 Vertical rate response to collective – Bo105 in forward flight (Ref. 6.19)

curve on Fig. 6.58), but the estimated parameters were dependent on the cyclic control strategy. The Bo105 results show clearly that the simple first-order equivalent system is not a consistently good approximation to heave dynamics and needs to be applied with considerable caution. Heave axis handling qualities in forward flight is an area clearly needing more attention.

6.5.4   Heave response characteristics in steep descent

Flight in conditions of steep descent presents particular dangers for helicopters. It is generally avoided by pilots but the continuing occurrence of accidents and incidents in this flight regime indicates that it is both operationally useful (e.g., positioning underslung loads) and easy to encroach inadvertently. In Chapter 5 we saw that the response to collective pitch at steep angles of descent can reverse, and so increased collective is required to descend more rapidly. Under controlled or directed conditions, this feature can be contained by a pilot, but in other conditions it can easily give rise to serious handling deficiencies. In Chapters 4 and 5 we discussed the problem of power settling, when the flight path steepens in response to an increase in collective. In Ref. 6.64 Heyson presents a useful theoretical analysis of the problem; the reader is referred to Fig. 4.12 where the power requirements in steep descent are illustrated. Heyson also comments on the operational implications of power settling – to quote from Ref. 6.64.

> *Operationally, the appearance of the phenomenon is rapid and usually unexpected. Pilots sometimes refer to it as 'stepping in the sinkhole'. The particular problem is that the pilot has no means of determining his aerodynamic flightpath. He may successfully negotiate a combination of geometric glide slope and speed so many times that he is confident of its safety; however, the next approach may encounter winds that produce disastrous consequences.*
>
> *A similar sequence of events can be encountered even without a tailwind. If any disturbance increases the speed along the glide slope, the instinctive reaction of a pilot is to correct the airspeed by pulling back on the cyclic-pitch stick to increase the rotor inclination. If the original stabilised glide slope was near a minimum power condition, comparison of the various parts of Fig. 12 shows that such a rearward stick movement may result in a power requirement far in excess of that available in the helicopter.*
>
> *The operational significance of this effect is that pilots should be specifically cautioned against any large or rapid rearward stick motions while in steep descents.*
>
> *Any recovery from power settling is likely to result in a significant loss in altitude. Thus, the safest procedure is to fly so as to avoid power settling at all times.*

Power settling and the associated nonlinear flight path response to controls occurs in steep descent ($>60°$) in the speed range of about one to two times the hover induced velocity (20–50 knots, depending on rotor disc loading). At much lower rates of descent and in near-vertical descent, a helicopter can enter a potentially hazardous flight state where high rates of descent can build up rapidly and erratic pitch and roll oscillations can develop. In addition, control effectiveness can change markedly, particularly collective control, with normal recovery techniques seeming only to exacerbate the situation. Analogous to the stall in fixed-wing aircraft, at least in terms of the consequences to the flight path trajectory, but quite dissimilar in aerodynamic origin, this so-called vortex-ring condition is definitely a state to avoid, especially at low altitude. Flying qualities in vortex ring become severely degraded and a pilot's first consideration should be to fly out of the condition.

The phenomenon has its origin in the peculiar flow characteristics that develop through the rotor in the intermediate range between the helicopter and windmill working states (see Fig. 2.8). At very low flight speeds (<15 knots) and rates of descent between 500 and 1500 ft/min, depending on the rotor disc.loading, the flow becomes entrained in a toroidal-shaped vortex ring that leads to extensive recirculation in the outer regions of the rotor disc. The vortex ring is very sensitive to small changes in flow direction, and rapid fluctuating asymmetric development of the ring can lead to fierce moments being applied to the fuselage.

The standard recovery technique involves lowering the nose of the aircraft until sufficient speed is gained that the vortex is 'washed' away, and then applying collective pitch to cancel the rate of descent. Different aircraft types have their own peculiar characteristics in the vortex-ring state. Early tests conducted at the RAE (Ref. 6.73) produced results that varied from loss of control to mild wallowing instability. In general, the aircrew manual for a type will contain an entry describing any particular features and advising the best recovery procedures. One such manual notes that rates of descent can build up to 6000 ft/min if vortex ring becomes fully established and that 'the aircraft pitches sharply nose down if rearward flight is attained'. Another refers to 'an uncontrollable yaw in either direction' eventually occurring. This same manual adds that 'any increase in collective pitch during established vortex ring state creates a marked pitching moment and should be avoided'. All such references make it clear that considerable height will be lost if the vortex-ring state is allowed to develop fully before recovery action is taken.

Interest in the effectiveness of collective control during recovery prompted a series of trials being carried out by the author at RAE Bedford using Wessex 2 and Puma helicopters. The tests were qualitative in nature and aimed at exploring the behaviour of these two aircraft in the vortex-ring state and establishing the benefits to recovery profile of increasing collective pitch before the aircraft nose is lowered to gain air-speed. The test technique options for approaching the vortex-ring condition were somewhat constrained by the need to operate well above the ground (minimum height for initiating recovery action, 3000 ft above ground level) and the lack of reliable low airspeed measurement on both aircraft. The procedure adopted involved a deceleration from 50 knots to the hover, maintaining a constant pre-established (hover) attitude and rate of descent. The rate of descent was then increased incrementally until the vortex region was encountered (Fig. 6.59). For both test aircraft the vortex region was quite difficult to find and apparently limited to a range of very low airspeed. With the Wessex, the region was first encountered with the entry profile at 800 ft/min rate of descent. To quote from the pilot's report (Ref. 6.62)

> ... with the rate of descent at about 800 ft/min we settled into the vortex ring; the rate of descent increased through 2000 ft/min in spite of increasing power to 3000 ft lb (hover torque reading). The vibration level was marked and a considerable amount of control activity was required to hold the attitude, though the cyclic controls always responded normally. Applying full power produced a rapid reduction of the rate of descent as soon as the rotor moved into clear air.

A major result of the tests was that applying collective prior to lowering the nose resulted in a height loss of about 150 ft during recovery, whereas if the collective was lowered

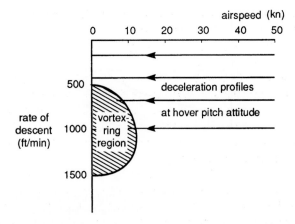

Fig. 6.59 Decelerating profiles into the vortex-ring region (Ref. 6.62)

first and then increased when airspeed developed, the height loss was about 500 ft. Similar results were found with the Puma, except that the pitching and rolling moments were of higher amplitude and frequency and became more intense as the collective lever was raised during recovery. It is emphasized here that the results discussed above are particular to type, and the beneficial use of collective during recovery may not read across to other aircraft. The difference in height loss during recovery for the two techniques is, however, quite marked and is operationally significant, particularly for low-level sorties. Vortex ring is a real hazard area and can be encountered in a variety of situations, some less obvious than near-vertical descents into restricted landing areas. If a pilot misjudges the wind direction, for example, and inadvertently turns and descends downwind into a landing area, concentrating perhaps more on ground speed than airspeed, then he may fly dangerously close to the vortex-ring condition. The final stages of a quick stop manoeuvre can also take the rotor through the vortex-ring condition as the pilot pulls in power. Such manoeuvres are typically carried out close to the ground and the consequences of a delayed or inappropriate recovery procedure could be serious.

Specific flying qualities criteria for the response characteristics in flight at steep descent angles do not exist, but perhaps the emphasis should be on deriving methods to assist the pilot in respecting the very real limits to safe flight that exist in this flight regime, conferring carefree handling, a topic returned to in Chapter 7.

6.6 YAW AXIS RESPONSE CRITERIA

As we turn our attention to the fourth and final axis of control, the reader may find it useful to reflect on the fact that of all the 'control' axes available to the pilot, yaw is, arguably, the most complex and the one that defines the greatest extent of the flight envelope boundary, both directly or indirectly. Figures 6.60 (a) and (b), for example, show the SA330 Puma control limits for the forward flight sideslip envelope, bounding the envelope at higher speeds, and for hovering in a wind from the starboard side, bounding the low-speed envelope. Excursions beyond these boundaries can lead to

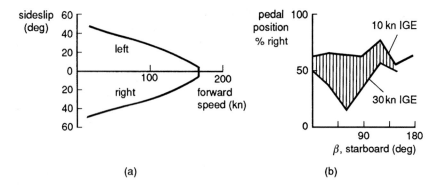

Fig. 6.60 Puma sideslip and sideways flight limits: (a) sideslip envelope in forward flight; pedal margin for hover in wind

loss of control or structural damage. Within these constraints, the pilot may feel able to command yaw motion in a relatively carefree manner. However, the pilot is not provided with a cue as to the magnitude of the loads in the tail rotor critical components. The tail rotor can absorb up to 30% of the total engine power, and in some flight conditions, tail rotor torque transients can lead to damaging loads. The pilot is also not provided with precise knowledge of sideways velocity or sideslip angle, but will typically fly at low level with primary reference to ground cues, oblivious to velocities relative to the air mass, and relying on tactile cues through control position for information on the proximity to aerodynamic limits. Our discussion suggests that yaw control is far from carefree and any handling deficiencies can contribute significantly to pilot workload for both civil and military operations.

Yaw control functions can be grouped into the following categories:

(1) balance of powerplant torque reaction on the fuselage, in steady state and manoeuvring flight;
(2) control of heading and yaw rate in hover and low-speed flight, giving all-aspect flight capability;
(3) sideslip control in forward flight, giving fuselage pointing capability;
(4) balancing or unbalancing manoeuvres, to increase or decrease turn rate.

Of these, the control functions in category (2) have probably accounted for by far the greatest range of yaw handling problems, stemming largely from the effects of main rotor wake–tail rotor–rear fuselage–empennage interactions (Refs 6.74–6.76).

In ADS-33, criteria for yaw handling are defined in much the same formats as for roll and pitch. These will be reviewed briefly.

6.6.1 Moderate to large amplitude/low to moderate frequency: quickness and control power

Following the formats adopted for the roll and pitch handling criteria, Fig. 6.61 shows the heading quickness boundaries for hover and low-speed MTEs, and Fig. 6.62 shows the minimum control power requirements for rate response types. It can be seen that the requirements for quickness are as demanding as for roll response, placing a particularly strong emphasis on yaw moment capability. For example, the ability to

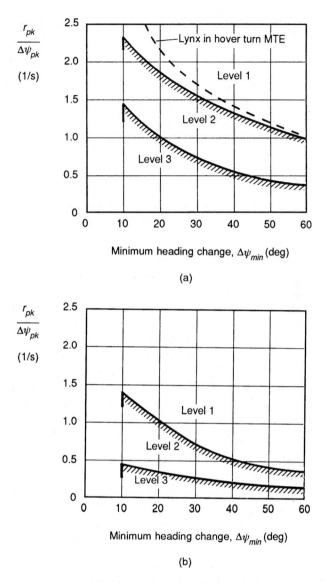

Fig. 6.61 Yaw axis quickness – hover and low-speed flight (Ref. 6.5): (a) target acquisition and tracking; (b) general MTEs

achieve a yaw rate of 40°/s in a discrete 20° heading change requires a maximum acceleration of about 2 rad/s². For an aircraft like Lynx this corresponds to generating a tail rotor thrust perturbation of about 1000 lbf. Overlaid on Fig. 6.61 is the boundary of maximum quickness values measured on Lynx performing precision hover turns, with heading changes from 30° through to 180°. The ADS-33 requirements for target acquisition and tracking are fairly demanding and call for a powerful tail rotor, or fantail in the case of the aircraft for which ADS-33 was developed, the RAH-66. The high quickness levels were partly established through simulation trials conducted on the VMS. Reference 6.77 presents results from simulation trials that included target

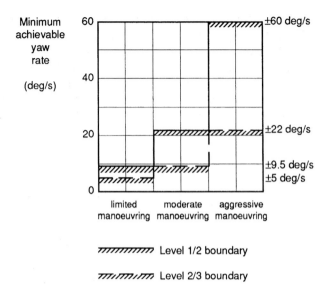

Fig. 6.62 Minimum yaw control power requirements – rate response type (Ref. 6.5)

acquisition and tracking MTEs. Experimental variables under investigation included the yaw damping, weathercock stability and response shape. Figure 6.63, taken from Ref. 6.77, shows the apparently very limited region fit for Level 1 handling in air-to-air target engagement, on a damping/response-shape diagram. The high levels of yaw damping required to achieve Level 1 for this kind of operation could not normally be produced without significant artificial response augmentation. Similar results were reported in Ref. 6.78 for forward flight MTEs. The tracking phase of an aerial combat engagement is more concerned with the higher frequency, small amplitude behaviour, and once again, the authors of ADS-33 turned to bandwidth to discern quality.

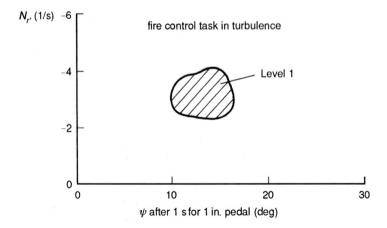

Fig. 6.63 Short-term yaw response requirements in air-to-air tracking task (Ref. 6.77)

6.6.2 Small amplitude/moderate to high frequency: bandwidth

The heading response bandwidth requirements are presented in Figs 6.64(a) and (b). The higher performance required for tracking tasks is common to both hover/low-speed and forward flight MTEs, e.g., Level 1 boundary at 3.5 rad/s. Such high values of bandwidth do not occur naturally in helicopters; typically, the yaw axis has very low damping, particularly at low speed, with rise times of the order of 2 s (see Chapter 4). The results of Refs 6.77 and 6.78 have already indicated the levels of damping that pilots feel are appropriate for aggressive yaw tasks. Bandwidths of 3.5 rad/s and higher are more consistent with rise times of the order 0.5 s and hence require some form of response quickening control augmentation.

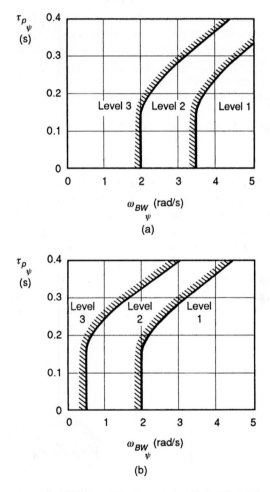

Fig. 6.64 Yaw axis bandwidth/phase delay boundaries (Ref. 6.5): (a) (low speed) target acquisition and tracking – (forward flight) air combat (yaw); (b) general MTEs

6.6.3 Small amplitude/low to moderate frequency: dynamic stability

At high forward speed, helicopters typically suffer from the same, so-called, nuisance mode as fixed-wing aircraft – the Dutch roll, exacerbated by weak weathercock stability and strong dihedral effect. The theory for this coupled mode has been presented in Chapter 4. In response to a doublet pedal input, the aircraft motion will soon be dominated by a weakly damped oscillation, comprising strongly coupled yaw, roll, sideslip and, for helicopters, pitch motions. The two fundamental parameters are the natural frequency and damping, and it is not surprising that efforts to define handling qualities related to the stability characteristics of this mode should have been focused on the corresponding two-parameter chart, or classical frequency/damping plane. The quality boundaries defined in ADS-33 are derived largely from the considerable database for fixed-wing aircraft (Ref. 6.6), with slightly relaxed stability requirements. A comparison of military and civil requirements (for single pilot IFR) is shown in Figs 6.65(a) and (b). The variety of boundaries drawn in Fig. 6.65(a) once again reflects the mission orientation of the military requirements. The comparison between civil and military requirements highlights several aspects already met in previous criteria – chiefly the greater demands made on designers of military aircraft. Another noticeable difference is shown in the low-frequency damping requirements. Both are based on minimum total damping at zero frequency. The more stringent military Level 1/2 boundary is set at time to half amplitude of 0.69 s, while the civil boundary is set at time to double amplitude of 9 s. The evidence supporting the minimum ω_n boundary for military

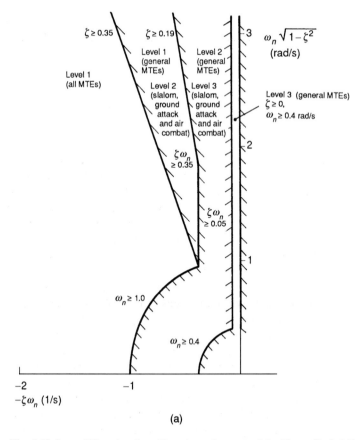

(a)

Fig. 6.65 Lateral/directional oscillatory requirements: (a) military (Ref. 6.5)

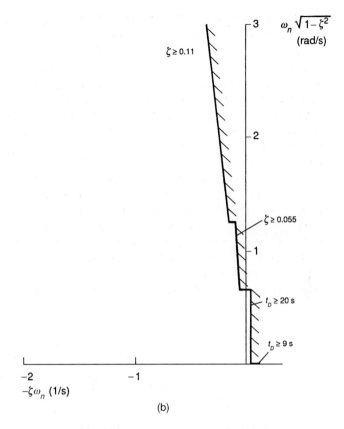

Fig. 6.65 (*continued*) (b) civil (Ref. 6.13)

helicopters in Fig. 6.65(a) is thought to be fairly limited; to the author's knowledge, no supporting data for these boundaries relevant to helicopters have appeared in the open literature since the publication of ADS-33. It is interesting to note that the criterion for the stability of long-period pitch and roll modes (frequencies less than 1 rad/s) lies within the $\omega_n > 1.0$ contour (see Fig. 6.45). One aspect raised here is the importance to handling qualities of the separation of frequencies between the modes with low stability. Modes with overlapping frequencies can cause additional pilot workload, especially when strong cross-couplings are present.

The reader is referred to the analysis of Dutch roll stability and response in Chapters 4 and 5 where the results of the AGARD WG18 study on Dutch roll stability are discussed (Ref. 6.79). Most military and civil helicopters have autostabilization in the yaw axis to improve the Dutch roll damping and, generally, augmented rate damping is sufficient to achieve Level 1. In some cases, design efforts are successful in improving the natural aerodynamic stability in yaw. Reference 6.63, for example, describes how modifications to the fin of the SA332 Super Puma significantly improved the Dutch roll characteristics of this aircraft compared with the original Puma. Figure 6.66, taken from Ref. 6.61, illustrates the marked improvement in Dutch roll damping through the fitting of end-plates on the BK117 helicopter; the new design met the FAA requirement without autostabilization.

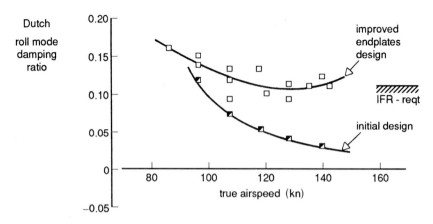

Fig. 6.66 Variation of Dutch roll damping with airspeed – BK117 (Ref. 6.61)

Two final points need to be made about the lateral/directional oscillatory mode. First, at high speed, the frequencies are encroaching on the range appropriate to small amplitude tracking, and the requirements on damping should be seen as supplementing the bandwidth criterion. Second, a very important handling characteristic associated with Dutch roll motion is the phase and relative amplitude between roll and yaw; separate criteria address these issues under the heading yaw–sideslip response to lateral cyclic, and we shall address these when discussing cross-coupling, in Section 6.7.

6.6.4 Trim and quasi-static stability

We have already discussed many of the issues relevant to lateral/directional trim and quasi-static stability in the section on roll axis response characteristics, particularly the need for positive trim control gradients. One additional handling criterion that fits best in this category is the requirement on heading (or roll and pitch attitude) hold functions as defined in ADS-33. With heading hold engaged and activated by the release of the yaw control device, the reference heading should be captured within 10% of the yaw rate at release. In addition, following a disturbance in yaw, the heading should return to within 10% of the peak excursion within 20 s for UCE 1, and within 10 s for UCE > 1.

As discussed in the introduction to this section, at hover and low speed, particularly close to the ground, the helicopter creates a disturbed aerodynamic environment in which the tail rotor is required to work. When the powerful main rotor vortex wake strikes the tail, particularly from the port side, or in the form of the ground vortex in rearward flight (see discussion on interactional aerodynamics in Chapter 3), large yawing moment disturbances can make it difficult for the pilot and even for the simple automatic hold functions to perform well. The problem is intimately associated with the open tail rotor; while generally more efficient than an enclosed fan or jet thrust device in clean aerodynamic conditions, the tail rotor tends to be more sensitive to wind strength and direction, particularly when positioned close to the vertical stabilizer. This sensitivity manifests itself as a non-uniform distribution of lift over the tail rotor, giving rise to large collective and power requirements in critical flight conditions. Reference 6.76 discusses the merits of tail rotor cyclic control in this context, which could be scheduled with collective to provide the optimum lift distribution in all flight conditions.

6.7 Cross-Coupling Criteria

Helicopters are characterized by cross-couplings in practically every axis-pairing, and the ubiquitous nature of cross-coupling constitutes one of the chief reasons why piloting this type of aircraft requires such high skill levels developed through long training programmes. Satisfying the direct, or 'on-axis', response characteristics, described in previous sections for roll, pitch, heave and yaw, is necessary but not sufficient to guarantee good helicopter flying qualities. Any helicopter test pilot would be quick to confirm this and might even advise that fixing the off-axis, cross-coupled response, was a higher priority for conferring Level 1 on-axis handling. Ideally, a designer would like to eliminate all sources of coupling. This is not only impossible (with only four controls), but probably also unnecessary, and one focus of the efforts in handling research has been to establish the maximum level of tolerable coupling. As with on-axis response criteria, this has proved to be task specific and particularly task-gain, or task-bandwidth, dependent. In very general terms, the low frequency/trim coupling effects are driven by the velocity couplings; the moderate frequency effects are reflected in the angular rate couplings and the higher frequency effects are dominated by the control couplings, in either sustained or washed-out form. Pilot subjective opinion of the degrading influence of coupling will therefore depend on the task, e.g., precise positioning, rapid slalom or target tracking. Many of the physical sources of cross-coupling have been described and discussed in Chapter 4 of this book. Here, we shall review the major types of couplings, and the database of results relating to handling qualities criteria and discuss what more needs to be done to set quality requirements. In the following subsections, the use of the condensed descriptor, e.g., pitch to roll, refers to the roll response due to pitch; any distinctions between control and motion couplings will be made as appropriate.

6.7.1 Pitch-to-roll and roll-to-pitch couplings

Pitch–roll and roll–pitch cross-couplings can be powerful and insidious. The natural sources of both are the gyroscopic and aerodynamic moments developed by the main rotor and, in dynamic manoeuvres with large attitude excursions, the uncommanded and sometimes unpredictable off-axis motion can require continuous attention by the pilot. Generally, the magnitude of the pitch-to-roll couplings are more severe than roll to pitch, due to the large ratio of pitch to roll moment of inertia, but are, arguably, more easily contained by the pilot, at least at low to moderate frequencies. Roll-to-pitch coupling effects can have a much stronger impact on flight path and speed control and hence handling qualities in moderate to large manoeuvres. From the results of a piloted simulation study on the NASA Flight Simulator for Advanced Aircraft (FSAA), Chen and Talbot (Ref. 6.80) hypothesized that the critical cross-coupling handling qualities parameters were the ratios of short-term steady-state roll (pitch) to pitch (roll) angular rates, approximated by the ratios of aerodynamic derivatives

$$\left(\frac{L_q}{L_p}\right) \quad \text{and} \quad \left(\frac{M_p}{M_q}\right)$$

Pilot HQRs were consistently awarded in the Level 2 area for values of the ratios greater than about 0.35. When the revision to MIL-H-8501 was initiated in the early 1980s, the

NASA results were initially used as the basis of new pitch–roll criteria. The derivative ratios clearly took no account of the control couplings, however, and were also difficult to measure with accuracy. After some refinement, the criteria adopted in ADS-33 were based on a time domain formulation, in terms of the ratio of the peak off-axis response to the desired on-axis response after 4 s following an abrupt step input, in the form

$$\text{roll step}\left(\frac{\theta_{pk}}{\phi}\right), \quad \text{pitch step}\left(\frac{\phi_{pk}}{\theta}\right) \quad \leq \pm 0.25 \,(\text{Level 1}), \quad \pm 0.6 \,(\text{Level 2}) \quad (6.35)$$

A series of additional piloted simulation and flight trials, conducted in the late 1980s at the Ames Research Center (Refs 6.81, 6.82), confirmed the importance of the derivative ratios, but argued that the new ADS-33 criteria did not cater for the higher frequency control coupling effects, or the interaction with on-axis characteristics. With regard to control coupling, the data in Refs 6.81 and 6.82 suggested that equivalent rotor control phase angles of about 30° would lead to Level 3 handling, confirming the RAE results reported earlier in Ref. 6.17. The relationship between the ADS-33 criteria and the equivalent linear system parameters can be illustrated using the simple first-order rate response formulation (Ref. 6.83), given by the equations

$$\dot{p} - L_p\left(p + \left(\frac{L_q}{L_p}\right)q\right) = L_{\eta_{1c}}\left(\eta_{1c} + \left(\frac{L_{\eta_{1s}}}{L_p}\right)\left(\frac{L_p}{L_{\eta_{1c}}}\right)\eta_{1s}\right) \quad (6.36)$$

$$\dot{q} - M_q\left(q + \left(\frac{M_p}{M_q}\right)p\right) = M_{\eta_{1s}}\left(\eta_{1s} + \left(\frac{M_{\eta_{1c}}}{M_q}\right)\left(\frac{M_q}{M_{\eta_{1s}}}\right)\eta_{1c}\right) \quad (6.37)$$

The relationship between the ADS-33 criteria and the parameters in eqns 6.36 and 6.37 can be reduced to the form

$$\left(\frac{\theta_{pk}}{\phi}\right) = \left(\left(\frac{L_p}{M_q}\right)\left(\frac{M_{\eta_{1c}}}{L_{\eta_{1c}}}\right) - \left(\frac{M_p}{M_q}\right)\right)L_{\eta_{1c}} \quad (6.38)$$

indicating that the pitch attitude coupling in a roll manoeuvre is dependent on both the cross-damping ratio of Ref. 6.80 (M_p/M_q) and the control sensitivity ratio scaled by the ratio of roll to pitch damping. Even with a zero value for the rate coupling M_p, control couplings can give rise to similar levels of pitch attitude excursion. The ADS time domain parameter in this simple model is therefore linearly related to the derivative ratios M_p/M_q and $M_{\eta_{1c}}/L_{\eta_{1c}}$. Given the roll axis control sensitivity and bandwidth, the importance of the control coupling is therefore inversely proportional to the pitch attitude bandwidth, M_q, hence emphasizing the importance of pitch axis effectiveness in cancelling coupling effects. Contours of equal ADS response are therefore given as shown in Fig. 6.67.

Our understanding of the handling qualities effects of roll–pitch coupling has been significantly extended by the series of flight/simulation experiments conducted by the US Army/DLR in the early 1990s, with support from DRA. The work to date is reported in Refs 6.84 and 6.85 and has focused on evaluating handling qualities in forward flight MTEs typified by the lateral slalom. In Ref. 6.84, couplings are classified into three types – those due to rate and control effects and the so-called washed-out coupling effects, more typical of augmented rotorcraft. Reference 6.85 concludes that the current ADS format is adequate for discriminating against unacceptable characteristics in the

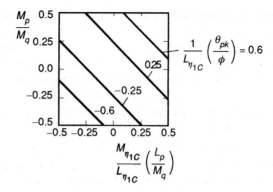

Fig. 6.67 Contours of equi-response on cross-coupling chart

first two categories, but not the washed-out effects, which appear to be frequency dependent. However, data are presented in Ref. 6.85 which suggest a modification to the ADS Level 1/2 boundary as shown in Fig. 6.68(a), where the acceptable level of coupling has been reduced to 0.1. A new frequency domain criterion is proposed in Ref. 6.85 which appears to give a more consistent picture for all three types of coupling. The general form of the criterion is presented in Fig. 6.68(b), where the key parameters are the magnitudes of the frequency response functions between pitch (roll) and roll (pitch) rates, evaluated at the bandwidth of the off-axis attitude response. This format, therefore, again reflects the importance of the response characteristics in the coupled axis. Strictly, the data from Ref. 6.85 will define only the vertical portions of the boundaries in Fig. 6.68(b): the author has hypothesized the upper, horizontal boundaries, which would be defined for pitch axis tasks, and the curved boundary between, reflecting the additional degradation in multi-axis tasks, when couplings in both axes are present.

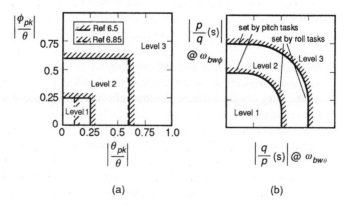

Fig. 6.68 Comparison of ADS-33C and Pausder–Blanken criteria for roll–pitch coupling requirements (Ref. 6.85); (b) proposed frequency domain format for roll–pitch–roll coupling (based on Ref. 6.85)

6.7.2 Collective to pitch coupling

At high speed, the application of main rotor collective pitch can generate powerful pitch and roll moments on the fuselage. Experiments to quantify the effects of collective to pitch coupling on handling qualities were reported in Ref. 6.86. The results were not conclusive but did indicate the powerful degrading effect, sending HQRs across the full span of the Level 2 range as the coupling parameter increased. ADS-33 reflects the limited dataset for collective to attitude couplings and sets limits on the pitch attitude change occurring within 3 s of an abrupt collective input. The limits are set as the ratio of pitch attitude change to the corresponding change in normal acceleration and take the form

$$\eta_c < 0.2\eta_{c_{max}}, \qquad \left| \frac{\theta_{pk}}{n_{z_{pk}}} \right| < 3.0°/(\text{m/s}^2) \qquad (6.39)$$

$$\eta_c \geq 0.2\eta_{c_{max}}, \qquad \left| \frac{\theta_{pk}}{n_{z_{pk}}} \right| < +1.5\,(-0.76)°/(\text{m/s}^2) \qquad (6.40)$$

where the negative value in eqn 6.40 corresponds to down collective inputs. The above criteria apply to forward flight. In low-speed flight the emphasis is more on the collective to yaw couplings.

6.7.3 Collective to yaw coupling

The application of collective pitch causes the rotor to slow down (or speed up) and the governor to increase (or decrease) the fuel flow, hence to increase (or decrease) the engine torque, which in turn results in a yawing moment reaction on the fuselage. Helicopter pilots learn to compensate for this effect early in their training and need to allocate a certain level of compensatory workload for harmonious inputs in pedal when applying collective. Most helicopters are built with a mechanical interlink between tail rotor collective pitch and main rotor collective lever, hence nullifying the gross effects at one particular flight condition. ADS-33 sets a limit of maximum yaw rate excursions of 5°/s following abrupt collective inputs, and also sets more complex limits on the ratio of yaw rate to vertical velocity, for which no substantiating data have appeared in the open literature; readers are referred to Ref. 6.5 for details.

6.7.4 Sideslip to pitch and roll coupling

The remaining cross-coupling effect to which we give some attention is the attitude response to sideslip. Pitch response to sideslip is a peculiar helicopter phenomenon that can lead to control problems in uncoordinated manoeuvres when it is required to point the fuselage off the flight path. The rotor downwash field can affect the horizontal stabilizer giving a powerful nose-up pitching moment in zero sideslip conditions. As sideslip builds up, the wake washes off to one side, exposing the tail to free air and leading to pitch down moments in both port and starboard manoeuvres. Figure 6.69, taken from Ref. 6.87, illustrates the various contributions to the pitching moment due to sideslip on the UH-60 helicopter. The canted tail rotor contributes 50% of the strong cross-coupling and the horizontal stabilizer, contributes 25%. The overall value for the derivative M_β is about 2000 ft lb/° and is equally as powerful as main rotor cyclic

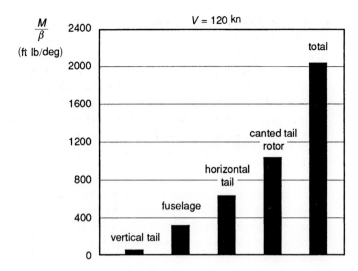

Fig. 6.69 Contribution of aircraft components to the pitching moment due to sideslip – UH-60 (Ref. 6.87)

control, therefore requiring significant pilot compensation. A strong pitch response to sideslip can exacerbate pilot disorientation problems following tail rotor failures. To the author's knowledge there are no published data defining handling qualities boundaries for sideslip to pitch effects; it remains a topic for future research.

Roll–sideslip coupling is defined in ADS-33 as part of the forward flight criteria for lateral/directional oscillatory characteristics (Ref. 6.5). It follows the fixed-wing format and is expressed in terms of the ratio of the oscillatory to the average component of roll attitude response following a lateral cyclic control input. The assumption with this type of format is that the roll oscillations are caused by sideslip excursions in a roll manoeuvre. The Level 1/2 boundary for this parameter depends on the phase angle between roll and sideslip.

Two final points need to be made on cross-coupling in general. First, it should be stated that most helicopters are designed with mechanical interlinks, or control couplings, that minimize the initial coupled motions following abrupt control inputs. In both the Sikorsky CH-53E and UH-60A, for example, application of collective lever couples to all the remaining controls through mechanical interlinks. This is a relatively simple and effective way of reducing some of the primary effects, but does nothing about the rate and velocity couplings. Second, there is evidence that the maximum level of acceptable coupling is a strong function of on-axis response characteristics – the poorer the on-axis handling qualities, the less tolerant pilots are of coupling. This is consistent with the intuitive rule that the presence of more than one degrading handling influence will lead to a combined handling worse than the average of the individual characteristics. This is bad news of course, but shouldn't come as a surprise to any pilot who has tried to 'tighten-up' on the controls with an unstabilized or partially stabilized helicopter, or tried to define a Level 1 roll axis response boundary, with a configuration having Level 2 pitch or cross-coupling characteristics. But future helicopters with active control technology will have low levels of couplings by design. What does need consideration with these highly augmented aircraft is the level of mission criticality and even flight criticality of the coupling augmentation. Future pilots may not be as well trained to fly cross-coupled helicopters, and loss of augmentation may be analogous to engine or tail rotor failure in today's helicopter operations. The central issue then becomes one of system integrity, particularly relating to sensors, and sufficient integrity/redundancy needs to be incorporated so that the risk of loss of coupling augmentation is remote.

6.8 MULTI-AXIS RESPONSE CRITERIA AND NOVEL-RESPONSE TYPES

This section covers two areas that are relatively immature in terms of the existence of any underlying flying qualities database. The primary emphasis of all flying qualities requirements has been the division of criteria into axes over which the pilot has control. In practice, most MTEs require coordinated control inputs in all axes, and the question arises as to whether the combination of single axis criteria is sufficient to ensure pilot acceptance in multi-axis tasks. Practically all the material in the earlier sections of this chapter deals with the most conventional, rate or attitude command, response types. With the advent of fly-by-wire/light and the attendant active controls technologies, the scope for changing the way pilots fly helicopters is very broad indeed. The term novel response types is coined to classify non-attitude-based systems, and some discussion on the current status and thinking in this area constitutes the final topic in this section.

6.8.1 Multi-axis response criteria

Most of the test MTEs in ADS-33 are primarily single axis tasks, e.g., accel–decel (pitch), bob-up (heave), sidestep (roll) and hover turn (yaw). For these, at least in theory, off-axis control inputs are required only to compensate for cross-couplings. Flying qualities requirements on couplings (see Section 6.7), at least when fully developed, should ensure that aircraft are built that demand minimum compensation only. Other MTEs are in their nature multi-axis and require the pilot to apply coordinated controls to achieve satisfactory task performance, e.g., pirouette, angled approach to hover, yo-yo combat manoeuvres and roll reversals at reduced and elevated load factors. Very little research has been done, at least in recent years and hence related to modern missions, on flying qualities criteria specifically suited for combined-axis helicopter manoeuvres. ADS-33 refers only to the requirement that control sensitivities should be compatible and responses should be harmonious. Control harmony is arguably one of the most important aspects of flying qualities, but finding any formal quantification has proved difficult. An intuitive definition seems to be that harmony is a quality achieved by having similar levels of characteristic response parameters, at least in the interacting axes. At a fundamental level, harmony then implies the same response types in the different axes, e.g., rate command in pitch combined with attitude in roll would not be harmonious, perhaps even leading to degraded ratings. Harmony applies most of all to pitch and roll, normally commanded through the same right-hand controller. Manoeuvring at low speed and close to the ground, the pilot directs the rotor thrust with the right-hand controller. The author is of the view that harmony in this mode of control should, as far as possible, encompass response type, bandwidth and control power (particularly for AC response types). Then if the pilot wants to fly at 45° to the right, he initiates and terminates the manoeuvre by moving his controller in the desired direction. This requirement is naturally met in TRC response types discussed below, but would not be for AC or RC types if the ratio of the minimum requirements of ADS-33 were maintained (e.g., $\pm 30°$ pitch, $\pm 60°$ roll for aggressive manoeuvring with attitude response types in UCE 1).

In forward flight, one of the important multi-axis criteria that has received attention is the requirement for turn coordination. As a pilot rolls into a turn, two compensating controls have to be applied. Aft cyclic is required, for helicopters with manoeuvre stability, to compensate for the pitch damping moment in the turn. Into-turn pedal is required to compensate for the yaw damping in the turn. Additional compensation will be needed for any steady-state incidence or sideslip required to augment the turn performance. The requirements for manoeuvre stability have already been discussed in Section 6.3. The requirements on yaw control harmony, and on the attendant sideslip response, are more complicated as they depend on the phase between roll and sideslip in the Dutch roll lateral/directional oscillation. The turn coordination requirements in ADS-33, for example, focus on the amount of sideslip resulting from an abrupt lateral cyclic control input; the criterion also highlights the point that sideslip response is more tolerable when it obviously lags the roll response.

The requirement for cyclic control harmony in manoeuvring flight at moderate to high speed translates into the need for similar time constants for roll attitude and normal acceleration response. Fortunately, this is normally the case, with the pitch bandwidth and control power being harmonized with the correspondingly higher parameters for roll. For example, as a pilot rolls into a turn with rate command in both pitch and roll, a bank angle of $60°$ and load factor of 2 can be achieved in similar times (about 1.5 s for an agile helicopter). One potential problem can arise with a pure RC response type during a roll reversal manoeuvre. Flying a steady turn the pilot will be pulling back to maintain the pitch rate in the turn. As the pilot executes the roll reversal, he has to judge his control strategy carefully, making sure that the cyclic passes through the centre with zero pitch input, to avoid making a discrete change in pitch attitude. Reference 6.35 reports on a study to evaluate the relative benefits of rate and attitude command response types. One of the workload problems with RC highlighted by pilots was the care required when reversing a roll to avoid making a pitch change that inevitably led to a speed decrease or increase as the manoeuvre progressed. To overcome this problem, speed hold functions were proposed. Also, automatic turn coordination was generally preferred by pilots, at least up to a moderate level of agility, obviating the need to apply any compensating pitch or yaw inputs.

One final point on multi-axis tasks, and to make it we assume that an aircraft that has been demonstrated as Level 1 in all axes according to clinical objective criteria will also consistently achieve 'desired performance' in practice. It is recognized that this is a contentious issue, but for the moment we assume that the individual criteria are robust enough that failure to comply will guarantee bad flying qualities. It is well recognized that at high levels of pilot aggression or in degraded environmental conditions, an otherwise Level 1 aircraft can degrade to Level 2. This is generally accepted as being an inevitable consequence of operating helicopters in harsh environments and can apply to both military and civil operations. But this raises the question as to whether a helicopter that has degraded to Level 2 in two or more axes will still be able to meet adequate performance levels in multi-axis manoeuvres. There is some evidence to suggest that the answer is negative. In discussing combined axis handling qualities,

Hoh (Ref. 6.60) advances an advisory 'product rule' that predicts that an aircraft with two axes both receiving ratings of 5 on the Cooper–Harper scale will actually work out as a 7 in practice, i.e., Level 3. We shall return to this and other related issues concerning subjective pilot opinion in the next chapter.

6.8.2 Novel response types

The unique capability of helicopter flight in three dimensions is typified by low-speed manoeuvring close to the ground. The pilot's task can be conceptually divided into three subtasks – navigation, guidance and stabilization. Navigation, or generally where the pilot wants to go in the long term, requires low workload and intermittent attention by the pilot to make course corrections. Guidance relates more to where the pilot wants to go in the shorter term and requires moderate levels of workload that depend principally on the speed of flight and on the level of visibility, or how many flight seconds the pilot can see ahead. In poor visibility and at low level the guidance workload can become very high. Stabilization relates to the continuous activity and workload to maintain the required aircraft attitudes. With unaugmented helicopters, the stabilization workload can be high and requires continuous pilot compensation as the helicopter is disturbed and deviates from the intended flight path. Most of the effort into augmentation for supporting precise flight path control close to the ground and obstacles has been directed towards reducing the stabilization workload by incorporating attitude hold functions, in combination with rate command, or even attitude command response types. For example, both Puma and Lynx have short-term attitude hold as part of their limited authority SCAS, triggered when the attitude falls within a small range close to zero (Lynx), or the pilot's cyclic is stationary (Puma). With the advent of high authority digital flight control, the capability now exists for providing response types that not only remove the stabilization workload, but also directly support the guidance task. Conceptually, the pilot requires control over the magnitude and direction of the aircraft velocity vector. To date the only criterion developed for flying qualities in the guidance task has been the TRC response type required for operations in the DVE according to ADS-33 (Ref. 6.5). TRC refers to a response characteristic where constant pilot controller input leads to a proportional earth-referenced translational velocity response. Level 1 flying qualities are defined by a TRC response having a qualitative first-order shape and equivalent rise time of between 2.5 and 5 s. The lower limit is set to avoid abrupt attitude changes during TRC manoeuvres. Equivalent rise time and the Level 1 TRC control power and sensitivity boundaries are defined as shown in Fig. 6.70. The limited supporting data for TRC response characteristics are published in Ref. 6.32.

Two notable examples of the implementation of novel response types are worth highlighting – the advanced digital flight control system (ADFCS) implemented in the McDonnell Douglas experimental AH-64 Apache AV05 (Ref. 6.88) and the Velstab System designed for the production Boeing/Sikorsky RAH-66 Comanche (Refs 6.89, 6.90). The philosophy behind the ADFCS experimental system, designed and flown in the mid–late 1980s, was to provide low workload management of aircraft control for single-pilot operations. The provision of automatic moding was part of this philosophy, hence not 'trading flight path with button management'. The flight path control logic implemented in AV05 comprised two selectable modes – the flight path vector system (FPVS) and aerobatic system. The control logic for the FPVS contained many

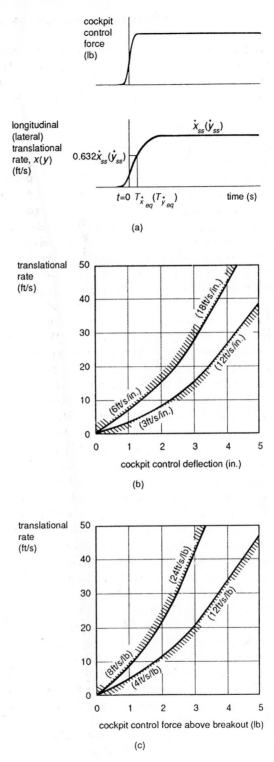

Fig. 6.70 TRC response sensitivity boundaries (Ref. 6.5): (a) Definition of equivalent rise time, $T_{\dot{x}_{eq}}(T_{\dot{y}_{eq}})$; (b) control/response requirement for centre-stick controllers; (c) control/response requirement for sidestick controllers

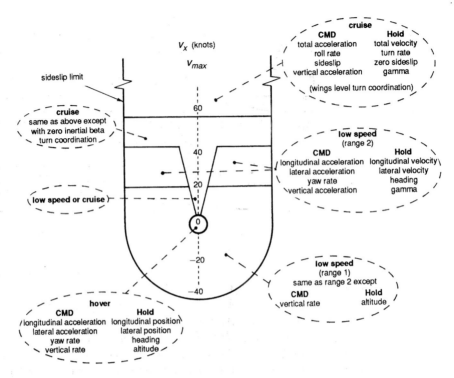

Fig. 6.71 Polar plot of speed/azimuth control logic for FPVS AH-64 (Ref. 6.88)

innovative features as summarized in Fig. 6.71, showing three auto-transition modes for hover, low speed and cruise. In the low-speed mode, (translational) inertial acceleration is commanded with the right-hand controller, with inertial velocity hold. This response type was selected for low-speed NoE flight to ensure that pilots would not be required to hold stick forces for long periods, as would occur with a pure TRC system. In the cruise mode, the turn rate hold feature gives the pilot the ability to maintain relative flight paths while changing speed. In both low-speed and cruise modes, the vertical axis response type of acceleration command/flight path hold simplified the task of terrain flight – the pilot could place the flight path vector symbol in the helmet-mounted display on the desired point on the terrain, e.g., hill top, to ensure clearance of vertical obstacles. In the author's view, AV05 represented, in its day, the state of the art in a full flight envelope ACT system with novel response types. To quote from Ref. 6.88, '... non-pilots could command near envelope limit performance from the aircraft in the course of a one hour demonstration flight'; the present author can testify to this, as he was one of the privileged engineers to fly AV05 in exactly this fashion.

At the time of writing the first edition of this book, the core active flight control system (AFCS) of the RAH-66 Comanche had been demonstrated in piloted simulation to confer Level 1/ good Level 2 flying qualities for the UCE 1 ADS-33 MTEs (Ref. 6.89). In addition, the selectable control modes, giving the pilot hybrid ACVH (attitude command, velocity hold) for flight in DVE, were evaluated as solid Level 1 for the ADS-33 DVE MTEs (Ref. 6.90). On the RAH-66, the DVE control system is described as the VELSTAB mode, and the characteristics relative to inertial groundspeed (V <60

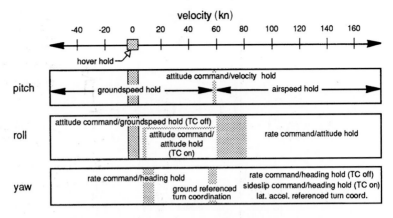

Fig. 6.72 Comanche VELSTAB characteristics (Ref. 6.90)

knots)/airspeed ($V > 60$ knots) are illustrated in Fig. 6.72 (from Ref. 6.90). At very low speed (groundspeed within ± 5 knots), in the shaded hover-hold region on Fig. 6.72, TRCPH is provided, giving the pilot a precise positioning aid. Hover-hold break-out is enabled when the pilot demands a velocity outside the threshold or applies a large cyclic demand. Below 60 knots groundspeed, the pilot flies with ACVH, with wind compensation to eliminate, as far as possible, non-uniformities in the required pilot control strategies in windy conditions, and to ensure a smooth blend between ground and airspeed at 60 knots. Low-speed turn coordination combined with altitude hold allows the pilot to fly single handed with the aircraft body axis always aligned with the flight path.

ADFCS on the Apache and AFCS with VELSTAB on the Comanche are visions of things to come in helicopter flight control and flying qualities, which have been realized successfully in flight and simulation. Flight with novel ground-referenced response types is being enabled by advances in sensor and digital flight control system technologies and clearly offers the potential for significant reductions in pilot workload, particularly for flight in DVE. The military driver is to provide capabilities previously not possible, but significant safety improvements in civil operations in poor visibility or congested and/or confined areas are also likely to be realized with this technology.

6.9 Objective Criteria Revisited

In this chapter a great play has been made of the concept of dynamic response criteria, used to form the predicted handling qualities, fitting into their place on the frequency–amplitude chart, conceptualized earlier in Figs 6.5 and 6.14. This can be summarized more holistically using Fig. 6.73 where the different criteria can be further classified into two groups – those determining the aircraft's agility and those determining the aircraft's stability (Ref. 6.91).

As before, the manoeuvre envelope line is shown to restrict criteria to practical manoeuvres, the achievable manoeuvre amplitude reducing as frequency increases. Within this overall envelope, four areas can be distinguished, two dealing with stability and two dealing with agility. The dynamic response requirements relating to agility and

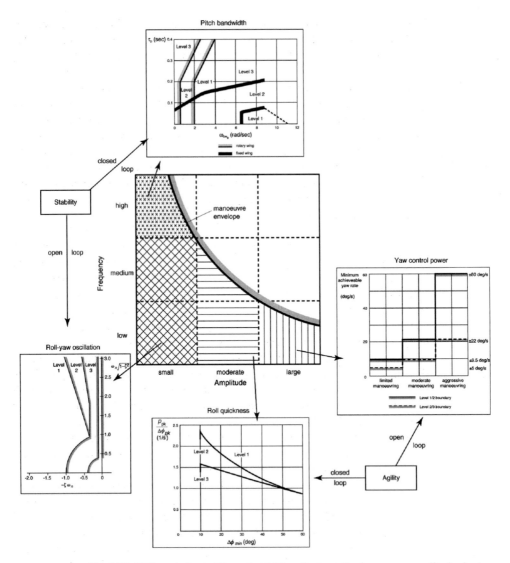

Fig. 6.73 Differentiating agility and stability criteria on the frequency–amplitude chart

stability cannot easily be 'divorced', since too much stability can degrade agility and vice versa. Designing to achieve the right balance requires careful optimization and, as with the design of fixed-wing aircraft, digital fly-by-wire/light flight control technology has provided the designer with considerably more freedom than hitherto in this trade-off. However, it is significantly not essential for control augmentation systems to have full authority over the control actuation to be able to deal with this. The essence of this challenging compromise can be seen in the designs of the augmentation systems on two of the aircraft featured in this book – the Lynx and the Puma. The Lynx system features both pitch–roll attitude and rate feedback signals, but the gain on the attitude signal has two values – a high value for small perturbations and a much reduced value when the attitude increases above certain values. The response type is therefore ACAH for small amplitude inputs and RC for large amplitude inputs. The increased stability conferred by the attitude feedback also improves the aircraft flying qualities in turbulent conditions where the pilot does not have to apply continuous corrective actions to maintain a desired attitude and speed. In the Puma system, a different approach is used. A pseudo-attitude is derived by integrating the signal from the rate gyro and used to provide short-term attitude stabilization. When the pilot moves the cyclic stick outside a prescribed range from the trim position, the pseudo-attitude component is switched out, providing the pilot with full RC response. Both the Lynx and Puma designs were innovative 40 years ago and specifically designed to address the stability–agility trade-off; the augmentation on both aircraft acts through limited authority ($\approx \pm 10\%$) series actuators. Nowadays, task-tailored control and flying qualities are commonplace concepts, although the implementation of such strongly nonlinear design functionality seems to be much less common.

In Fig. 6.73, examples of agility and stability criteria from ADS-33 are shown. The moderate amplitude quickness criteria (shown for the roll axis) provide a direct link between closed-loop stability, encapsulated in the bandwidth criteria (shown for the pitch axis), and the maximum agility, encapsulated in the control power criteria (shown for the yaw axis). The basic stability is defined by the position of the eigenvalues on the complex plane, shown in the figure for the Dutch roll–yaw oscillation. Flying qualities requirements extend far beyond those summarized in Fig. 6.73, of course, including trim and static stability, flight path response, cross-coupling behaviour and controller characteristics, and most of these aspects have been covered to varying extent in this chapter. During the period since the publication of the first edition of *Helicopter Flight Dynamics*, various basic research and application studies have refined the understanding of helicopter flying/handling qualities and some of this has been embodied in the performance specification version of ADS-33E (Ref. 6.92). The changes from the C-version, used throughout Chapter 6, have been numerous and no attempt has been made to fully revise the material presented earlier. Rather, a number of key developments are highlighted here to draw specific attention to them.

ADS-33 contains the first truly mission-oriented set of requirements, embodied in the fact that the location of criteria boundaries are related to the types of 'mission-task-element' to be flown, rather than the aircraft size or weight. In ADS-33E-PRF the aircraft role (i.e., attack, scout, utility, cargo) is then described by a subset of recommended MTEs to be flown and level of agility used (i.e., limited, moderate, aggressive, tracking). The objective criteria then link with the MTEs in two ways – through the Response Type table (see Table 7.4) and also the level of agility associated with an MTE that, in turn, defines the handling qualities boundary to be used. An example from Ref. 6.92 is the set of requirements for large amplitude attitude changes in hover and low-speed flight, i.e., the control power, shown in Table 6.6. The normal levels of agility for the 13 different MTEs that are used to define the control power requirements are defined. For example, slaloms would normally be flown at moderate levels of agility while the acceleration–deceleration would require aggressive agility.

A change to the closed-loop stability requirements appears in the bandwidth for the general or 'all other' MTEs in a UCE > 1 and/or divided attention. Figure 6.74 should be compared with Fig. 6.30(c), showing the shift of the Level 2/3 boundary from 0.5 rad/s to 1 rad/s; the boundary is also raised for forward flight. Generally speaking, less research has been conducted to define the Level 2/3 handling qualities boundary than that conducted for the Level 1/2 boundary; it is known that pilot perception of

Table 6.6 Criteria for large amplitude response in hover

	Rate response types						Attitude command response types			
	Achievable angular rate (deg/s)						Achievable angle (deg)			
	Level 1			Levels 2 and 3			Level 1		Levels 2 and 3	
AGILITY CATEGORY MTE	Pitch	Roll	Yaw	Pitch	Roll	Yaw	Pitch	Roll	Pitch	Roll
Limited agility										
Hover										
Landing	±6	±21	±9.5	±3	±15	±5	±15	±15	±7	±10
Slope landing										
Moderate agility										
Hovering turn										
Pirouette										
Vertical manoeuvre	±13	±50	±22	±6	±21	±9.5	+20 −30	±60	±13	±30
Depart/abort										
Lateral reposition										
Slalom										
Aggressive agility										
Vertical remask										
Acceleration/Deceleration	±30	±50	±60	±13	±50	±22	±30	±60	+20 −30	±30
Sidestep										
Target acquisition and track										
Turn to target										

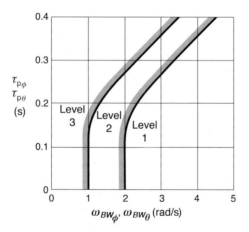

Fig. 6.74 Requirements for small amplitude (roll–pitch) attitude changes in hover and low-speed flight

handling qualities at the Level 2/3 boundary is very dependent on skill and training. Considering the simple first-order roll response (analysis from eqn 6.2 and Fig. 6.3), a bandwidth of 0.5 corresponds to a roll time constant of 2 s, and it is not surprising that a pilot would experience major difficulties flying the precision elements of a slalom or a sidestep, faced with such response lags.

The attitude quickness criteria remain as documented in ADS-33C, apart from clarification on the testing requirements, particularly relating to the need for the pilot to change attitude 'as rapidly as possible without significant reversals in the sign of the cockpit control input'. Control overshoots are known to result in increased quickness and give a 'false' sense of agility, even though the pilot might commonly use this technique to change attitude. The point is also emphasized that the full range of moderate amplitudes should be covered. Figures 6.10 and 6.18 showed what we described as closed-loop quickness (pilot using overshoot technique) for the Lynx flying slalom and sidestep MTEs, measured across a wide range of attitude changes. A significant change is that aircraft with response types appropriate to UCE = 2 and 3 no longer have to meet the quickness requirements, on the basis that operations in DVEs require only limited agility.

In Section 6.5.3, discussion centred around the complexities of helicopter flight path control in forward flight and the shortcomings of the requirements of ADS-33C were highlighted, particularly relating to testing difficulties, and the distinctly non-first-order response of some types (e.g., see Fig. 6.58). In ADS-33E-PRF, new criteria have been introduced for flight path behaviour in response to a pitch attitude change through cyclic pitch with collective fixed. Criteria are distinguished for front-side and back-side (of the power curve) operations. If $\Delta\gamma_{ss}$ is the change in flight path and ΔV_{ss} is the change in speed resulting from a step change in pitch attitude, then

$$\text{Front-side operation} \qquad \Delta\gamma_{ss}/\Delta V_{ss} < 0 \qquad\qquad (6.41)$$

$$\text{Back-side operation} \qquad \Delta\gamma_{ss}/\Delta V_{ss} \geq 0 \qquad\qquad (6.42)$$

For back-side operation, the flight path handling requirements are essentially the same as the low-speed height response to collective requirements discussed in Section 6.51, except that the maximum value of the time constant, $T_{h_{eq}}$, for Level 2 handling, is reduced to 10 s. For front-side operation, the criteria is based on the lag between flight path and pitch attitude (equivalent to the heave time constant, or inverse of the derivative $-Z_w$, at low frequency), expressed in the frequency domain as follows: the lag should be <45° at all frequencies below 0.4 rad/s for Level 1, and below 0.25 rad/s for Level 2. It is considered that these criteria are still open to development, particularly for complex precision approach trajectories envisaged to enable the expansion of simultaneous, non-interfering operations of helicopter at busy hubs.

Finally, we turn to cross-coupling criteria and Fig. 6.75 summarizes the various important cross-coupling effects found in helicopters. The starred boxes denote response couplings for which no handling criteria exist.

Requirements in ADS-33 include detailed quantitative ratio criteria, for example, the roll to pitch or collective to yaw, and also qualitative 'not objectionable' type statements, although the flight path response to pitch attitude changes has been developed in Ref. 6.92 into the quantitative criteria described in the previous paragraph. Cross-couplings emerge as a serious impediment to task performance for manoeuvres where higher levels of precision and aggressiveness are required, and this has been taken into account in Ref. 6.92 by requiring that the coupling requirements on yaw from collective and pitch–roll and roll–pitch be applicable to aircraft that need to meet the aggressive and acquisition/tracking levels of agility. New requirements have been developed for the tracking level of agility based on the research reported in Refs 6.93 and 6.94, which

	PITCH	ROLL	HEAVE	YAW
PITCH		$\Delta\theta_{pk/\Delta\theta_4}$ (hover and fwd flight)	flight path response not objectionable in forward flight	★ yaw response due to rotor torque changes in aggressive pitch manoeuvres
ROLL	$\Delta\theta_{pk/\Delta\phi_4}$ (hover and fwd flight)		★ thrust/torque spikes in rapid roll reversals	$\Delta\beta_{/\Delta\phi}$ ratios in forward flight
HEAVE	$\Delta\theta_{pk/\Delta n_{zpk}}$ (forward flight)	★ $\Delta\phi_{pk/\Delta n_{zpk}}$ (hover and fwd flight)		$r/|h|$ ratios in hover
YAW	★ pitching moments due to sideslip in forward flight	dihedral effect on roll control power	not objectionable in hover	

★ no current requirements

Fig. 6.75 Dynamic response criteria for cross-couplings or off-axis response

identified the frequency dependence of the handling qualities for small amplitude tracking. In Fig. 6.76 the response ratios (average q/p (dB), average p/q (dB)) are derived from the amplitudes of the frequency response functions q/δ_{lat} divided by p/δ_{lat} and p/δ_{long} divided by q/δ_{long}, averaged between the attitude bandwidth frequency and frequency at which the attitude response phase is $-180°$. The requirements focus on the pitch due to roll requirements, derived from tests where pitch control was disrupted by varying levels of coupling during a roll tracking task (Ref. 6.93).

The methodology expressed in ADS-33 has been applied extensively since the publication of the first edition of this book. Some guidance for the tailoring of the requirements to specific roles was given in Ref. 6.95. In the United Kingdom an initial emphasis was placed on the application to the attack helicopter procurement competition (Ref. 6.96), while in mainland Europe to the design and development of the NH90 (Ref. 6.97). A continuing theme in the research community has been the development of a maritime version of ADS-33, with particular application to operations to and from ships. A series of flight and flight simulation trials have been conducted and results reported (Refs 6.98–6.101), which have guided specific applications, but no generally accepted and conclusive product has emerged. What seems to be universally agreed, however, is that (Ref. 6.101) 'Level 1 handling qualities are not achievable at high sea

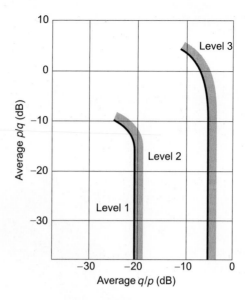

Fig. 6.76 Pitch–roll cross-coupling requirements for target acquisition and tracking MTEs

states using current landing practices with standard levels of aircraft augmentation'. A somewhat similar situation has arisen with regard to extending the scout/attack helicopter requirements – originally the focus of ADS-33 – to cargo helicopters, particularly for operations with external, underslung loads. References 6.102–6.104 document part of the story, but ultimately ADS-33E-PRF, in its discussion on the flight and simulation tests conducted to develop criteria, concludes that, 'The outcome of this testing has been overwhelming evidence that quantitative criteria will be extremely difficult to derive'. In Ref. 6.105 a comprehensive analysis of the handling qualities of the UH-60M prior to first flight is reported, demonstrating the utility of the methodology to vehicle upgrades.

The author's own research has also taken the methodology into new directions, particularly the relationship between handling qualities and loads. In Ref. 6.106, for example, some of the structural fatigue issues in helicopter flight testing are addressed, with particular emphasis to flying qualities tests. In Ref. 6.107 an approach to integrating the handling qualities/agility and load alleviation design processes is presented, taking advantage of and extending the ADS-33 metrics to embrace both disciplines.

Finally, it seems appropriate to briefly mention the main intended recipient of the ADS-33 development efforts, the RAH-66 Comanche. The programme was cancelled on 23 February 2004, but not before the aircraft had demonstrated the fruits of the effort to design and build the first helicopter with Level 1 handling qualities throughout the OFE, at least in the good visual environment (Ref. 6.108). A salutary lesson lies in the conclusions of Ref. 6.108, however, that '... while the analytical requirements of ADS-33D, Section 3, are an indispensable resource for control law development, they do not obviate the requirement for a vigorous flight test programme with active engagement between pilot and engineers, without which many of the critical improvements ... would not have been possible'. With these words, the authors of Ref. 6.108 lead us naturally to the subject of subjective pilot assessment and the assigned handling qualities.

*The MDHC variable stability (fly-by-wire) Apache AV05 during a
handling qualities evaluation over the Arizona desert
(Photograph from the author's collection)*

7 Flying qualities: subjective assessment and other topics

> *If test manoeuvres are too dangerous for a skilled test pilot to perform in a tightly controlled environment, it is unreasonable to expect the user to fly such manoeuvres in an unfamiliar, unfriendly environment in the fog of war.*
>
> *(Key, 1993)*

7.1 INTRODUCTION AND SCOPE

While objective measurements and assessment are necessary for demonstrating compliance with quality standards, they are still not sufficient to ensure that a new helicopter will be safe in achieving its operational goals. Gaps in the criteria due to limited test data, and the drive to extend operations to new areas, continue to make it vital that additional piloted tests, with a subjective orientation, are conducted prior to certification. A further issue relates to the robustness of the criteria and an aircraft's flying qualities at higher levels of performance. The point has been made on several occasions that criteria in standards like ADS-33 represent the minimum levels to ensure Level 1 in normal operation. A good design will do better than just meet the objective Level 1 requirements, and the absence of upper limits on most of the handling parameters means that there is practically no guidance as to when or whether handling might degrade again. An aircraft will need to be flight tested to assess its flying qualities in a range of mission task elements MTEs, throughout its intended operational flight envelope (OFE), and including operations at the performance limits to expose any potential handling cliff edges. During such testing, measurements will be made of aircraft task performance and control activity, but there is, as yet, no practical substitute for pilot subjective opinion. The measurement and interpretation of pilot opinion is a continuing theme throughout this chapter but is exclusively the subject of Section 7.2, where a range of topics are covered, including handling qualities ratings (HQRs), MTEs and the design and conduct of a handling qualities experiment.

Section 7.3 deals with a selection of what we have described as special flying qualities, including agility, carefree handling and flight in poor visual conditions.

One of the areas omitted from the comprehensive treatment of objective assessment in Chapter 6 was the requirements for pilot's inceptors or controllers. The issues surrounding the assessment of quality for inceptors, particularly sidesticks, are so pilot centred that coverage in this chapter was considered more appropriate; Section 7.4 deals with this topic.

For both military and civil helicopters, the potential improvements in flying qualities offered by active control technology (ACT) through the almost infinite variety of response shaping, where computers take on the 'compensation', have prompted a more serious examination of the benefits of improved flying qualities to safety and performance. Helicopters' accidents and incidents due to so-called pilot error are still far too high and many can be attributed to poor flying qualities in a broad sense. Even those accidents caused by system failures can ultimately be attributed to degraded flying qualities, as the pilot struggles to fly a disabled aircraft to a safe landing. With these ideas in mind, Section 7.5 examines the contribution of flying qualities to performance and safety by viewing the pilot as a system element, with the potential of failing when under 'stress', and outlines a new approach to quantifying the risk of failure.

7.2   The Subjective Assessment of Flying Quality

Opinion on what constitutes quality when it comes to flying has been demonstrated over the years to be wide and varied amongst pilots, and will undoubtedly continue to be so. Individuals can have different preferences and achieving universal agreement over all aspects of quality is probably unrealistic, and perhaps even undesirable. Fortunately, pilots, like most of the human race, are exceptionally adaptable and can learn to use someone else's favourite flight vehicle very effectively. If we consider flying quality to be valued in both aesthetic and functional terms by pilots, then by far the most effort has been expended by flying qualities engineers on trying to establish a consensus regarding functional quality. This effort has received considerable leverage through the development of mission-oriented or functional flying qualities criteria. Prior to this, over several decades, the merging of functional values with aesthetic values has led to flying quality being a 'nice to have' attribute rather than essential for achieving safety and performance. The importance of aesthetic quality is recognized, but treating this aspect is beyond the scope of this engineering text. The emphasis with mission-oriented flying qualities is the ability to perform a defined set of tasks with temporal and spatial constraints, and what better test of quality than flying the tasks themselves. When flying a task, or to use the parlance of ADS-33 (Ref. 7.1), an MTE, the pilot will adopt a control strategy to maximize performance and minimize workload. Control strategy may vary from pilot to pilot, reflecting the complex network of influences on how different pilots elect to use their controls. Figure 7.1, taken from Ref. 7.2, illustrates the point. The task requirements in a given environment will determine the accuracy and spare workload capacity required; McRuer has described this through the dual concepts of attentional demand and spare control capacity (Ref. 7.3). Landing a helicopter on the deck of a small ship may require considerable accuracy, and a pilot may well take his time to achieve a safe landing. Evading a threat may place greater demands on acting quickly than flying precisely. Whatever the drivers, the combination of vehicle response characteristics and task cues will determine the control strategy adopted by the pilot, which in turn will be reflected in the realized task performance and actual pilot workload. In making a subjective assessment of the flying quality, a pilot will need to take into account these interacting influences and then articulate his or her thoughts to the flying qualities engineer, whose job at this stage is to make changes for the better, if at all possible. It is the articulation and the associated interpretation of the pilot's subjective opinion that underpins any successful development of flying qualities,

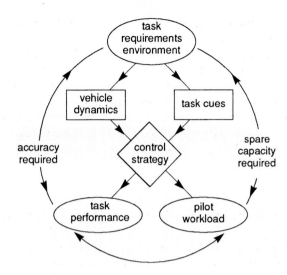

Fig. 7.1 The influences on pilot control strategy

and it is hardly surprising that this activity has been 'assisted' by a wide range of different support tools including rating scales and questionnaires. One rating scale has achieved more universal acceptance than any other since it was first proposed in the late 1960s – the Cooper and Harper handling qualities rating scale (Ref. 7.4). In view of its importance to the subject and partly to highlight potential misuses, the next subsection will give exclusive coverage to this scale and the associated pilot handling qualities rating (HQR).

7.2.1 Pilot handling qualities ratings – HQRs

Figure 7.2 illustrates the decision tree format of the Cooper–Harper handling qualities rating scale. Test pilots and flying qualities engineers need to be intimately familiar with its format, its intended uses and potential misuses. Before we begin a discussion on the rating scale, we refer the reader back to Fig. 7.1 and to the key influences on control strategy which should be reflected in pilot opinion; we could look even further back to Fig. 6.1, highlighting the internal attributes and external factors as influences on flying qualities. The pilot judges quality in terms of his or her ability to perform a task, usually requiring closed-loop control action. A key point in both figures is that the handling qualities and pilot control strategy are a result of the combined quality of the aircraft characteristics and the task cues. The same aircraft can be Level 1 flying routine operations at day and then Level 3 at night, or when the wind blows hard, or when the pilot tries to accomplish a landing in a confined area. An aircraft may be improved from Level 3 to 2 by providing the pilot with a night vision aid or from Level 2 to 1 by including appropriate symbology on a helmet-mounted display. Handling qualities are task dependent and that includes the natural environmental conditions in which the task is to be performed, and the pilot will be rating the situation as much as the aircraft. We will discuss the scale and HQRs in the form of a set of rules of thumb for their application.

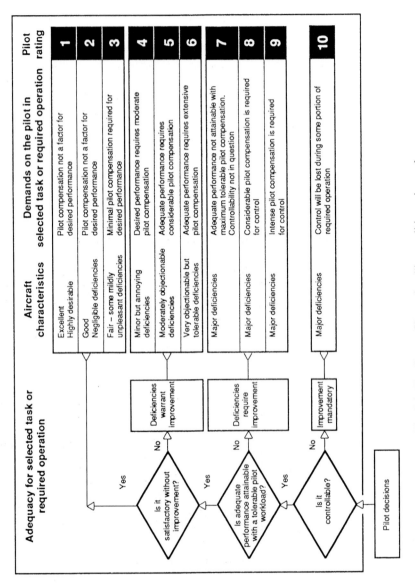

Fig. 7.2 The Cooper–Harper handling qualities rating scale

(1) *Follow the decision tree from left to right.* Pilots should arrive at their ratings by working through the decision tree systematically. This is rule number 1 because it helps the pilot to address the critical issue of whether the aircraft is Level 1, 2 or 3 in the intended task or subtask. The decision tree solicits from the pilot his opinion of the aircraft's ability to achieve defined performance levels at perceived levels of workload.

(2) *An HQR is a summary of pilot subjective opinion on the workload required to fly a task with a defined level of performance.* An HQR can be meaningless without back-up pilot comment. It is the recorded pilot opinion which will be used to make technical decisions, not the HQR, because the HQR does not tell the engineer or his manager what the problems are. Often a structured approach to qualitative assessment will draw on a questionnaire that ensures that all the subject pilots address at least a common set of issues. We will consider the ingredients of a questionnaire in more detail later.

(3) *Pilot HQRs should be a reflection of an aircraft's ability to perform an operational role.* The MTEs should be designed with realistic performance requirements and realistic task constraints. The pilot then needs to base his rating on his judgement of how an 'average' pilot with normal additional tactical duties could be expected to perform in a similar real-world task.

(4) *Task performance and workload come together to make up the rating, but workload should be the driver.* This is most important. To highlight the emphasis we refer to Fig. 7.3 where workload and task performance are shown as two dimensions in the piloting trade-off. Task performance is shown in three categories – desired, adequate and inadequate. Workload is also shown in three categories – low, moderate to extensive and maximum tolerable, reflecting the rating scale parlance. HQRs on the Cooper – Harper scale fall into the areas shown shaded. Figure 7.3 acknowledges that a pilot may be able to work very hard (e.g., using maximum tolerable

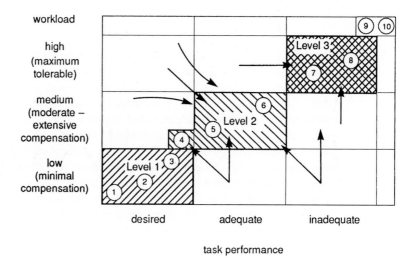

Fig. 7.3 The contributions of workload and task performance to the HQR

compensation) and achieve desired performance, but it is not appropriate then to return a Level 1 rating. Instead, he should aim for adequate performance with some spare workload capacity. Similarly, a pilot should not be satisfied with achieving adequate performance at low workload; he should strive to do better. A common target for pilots in this situation would be to try to achieve desired performance at the lower end of Level 2, i.e., HQR 4. Level 1 characteristics should be reserved for the very best, those aircraft that are fit for operational service. An HQR 4 means that the aircraft is almost good enough, but deficiencies still warrant improvement.

(5) *Two wrongs can make a disaster.* Handling qualities experiments often focus on one response axis at a time, two at the most. We have seen in Chapter 6 how much it takes to be a Level 1 helicopter, and a question often arises as to how much one or two deficiencies, among other superb qualities, can degrade an aircraft. The answer is that any Level 2 or 3 deficiency will degrade the whole vehicle. A second point is that several Level 2 deficiencies can accumulate into a Level 3 aircraft. Unfortunately, there seems to be very little rotorcraft data on this topic, but Hoh has given a hint of the potential degradations in the advisory 'product rule' (Ref. 7.5)

$$R_m = 10 + \frac{-1^{(m+1)}}{8.3^{(m-1)}} \prod^m (R_i - 10) \qquad (7.1)$$

where R_m is the predicted overall rating and R_i are the predicted ratings in the individual m axes. According to the above, two individual HQR 5s would lead to a multi-axis rating of 7. The fragile nature of such prediction algorithms emphasizes the critical role of the pilot in judging overall handling qualities and the importance of tasks that properly exercise the aircraft in its multi-axis roles. So while sidesteps and quickhops might be appropriate MTEs for establishing roll and pitch control power requirements, the evaluations should culminate with tasks that require the pilot to check the harmony when flying a mixed roll–pitch manoeuvre. Ultimately, the combined handling should be evaluated in real missions with the attendant mission duties, before being passed as fit for duty.

(6) *The HQR scale is an ordinal one, and the intervals are far from uniform.* For example, a pilot returning an HQR of 6 is not necessarily working twice as hard as when he returns an HQR of 3; Cooper and Harper, in discussing this topic, suggest that '. . . the change in pilot rating per unit quality should be the same throughout the rating scale'. The implied workload nonlinearity has not hindered the almost universal practice of averaging ratings and analysing their statistical significance. Many examples in this book present HQRs with a mean and outer ratings shown, so the author clearly supports simple arithmetic operations with HQRs. However, this practice should be undertaken with great care, particularly paying attention to the extent of the rating spread. If this is large, with ratings for one configuration appearing in all three levels for example, then averaging would seem to be inappropriate. If the rating spread is only 1 or 2 points, then it is likely that the pilots are 'experiencing' the same handling qualities. Of course, if the ratings still cross a boundary, and the mean value works out at close to 3.5 or 6.5, then it may be necessary to put more pilots through the evaluation or explore some task variations. The whole issue of averaging, which can make data presentation so appealing, has to be undertaken in the light of the pilots' subjective comments. Clearly, it would be

inappropriate to average a group of ratings when the perceived handling deficiencies recorded in each of the pilot's notes were different.

(7) *Are non-whole ratings legal?* There appears to be universal agreement that pilots should not give ratings of 3.5 or 6.5; there is no space here to sit on the fence and the trial engineer should always reinforce this point. Beyond this restriction, there seems to be no good reason to limit pilots to the whole numbers, provided they can explain why they need to award ratings at the finer detail. A good example is the 'distance' between HQR 4 and 5. It is one of the most important in the rating scale and pilots should be particularly careful not to get stuck in the handling qualities 'potential well' syndrome of the HQR 4. In many ways the step from HQR 4 to HQR 5 is a bigger workload step than from 3 to 4 and pilots may feel the need to return HQRs between 4 and 5; equally, pilots may prefer to distinguish between good and very good configurations in the region between HQR 2 and 3.

(8) *How many pilots make a good rating?* This question is always raised when designing a handling qualities experiment. The obvious trade-off involving the data value is expressed in terms of authenticity versus economy. Three pilots seems to be the bare minimum with four or five likely to lead to a more reliable result and six being optimal for establishing confidence in the average HQR (Ref. 7.6). For a well-designed handling qualities experiment, there will inevitably be variations in pilot ratings as a result of different pilot backgrounds, skill level, pilots' perception of cues, their natural piloting techniques and standards to which they are accustomed (e.g., one pilot's HQR 4 might be another's 5). Measuring this 'scatter' is an important part of the process of understanding how well the aircraft will work in practice. But if the scatter is greater than about two ratings, the engineer may need to consider redesigning the experiment.

(9) *How to know when things are going wrong.* A wide variation of ratings for the same MTE should ring alarm bells for the trial engineer. There are many legitimate reasons for a spread in HQRs but also some illegitimate ones. One reason could be that the pilots are not flying the same task. Part of the task definition are the standards for desired and adequate performance. These will be based on some realistic scenario, e.g., sidestep from one cover point to another, 100 m distant, and establish a hover within a defined world-referenced box, with permitted errors up to, say, 2 m. The task definition might also add that the pilot should maintain his flight path below 10 m above the ground and accomplish the task in a defined time. The more detail that is added to the task definition, the more likely it is that each pilot will try to fly the same MTE and the more the HQR scatter will be left to pilot differences, which is what is required. Conversely, the less detail there is, the more likely is the chance of different pilots interpreting the task differently; one may fly the task in 15 s, another in 20 s and the different demands will drive the workload and hence ratings. Next to the need for complete and coherent task definition comes the need to provide the pilot with sufficient cues to enable him to judge his task performance. This is a critical issue. In real-world scenarios, pilots will judge their own task performance requirements and they will usually do this on the basis of requiring low to moderate pilot compensation. Pilots do not usually choose to fly at high levels of workload, unless they have to, and will normally set performance requirements based on task cues that they can clearly perceive. Unless a pilot has made an error of judgement, he or she will not normally fly into a condition where the task cues are insufficient for guidance and stabilization. In clinical flying

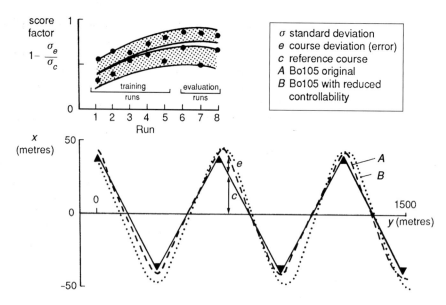

Fig. 7.4 The DLR score factor (Ref. 7.7)

qualities tests, it is important for the trials engineer to work closely with the 'work-up' pilot to define realistic performance goals that an average pilot would be expected to perceive in operations. Then, when the pilot returns an HQR, the actual task performance achieved should correlate well with that perceived by the pilot. Unless properly addressed, this issue can devalue results of handling qualities experiments. A third factor in HQR scatter deserves a rule all on its own.

(10) *How long before a pilot is ready to give a rating?* There is no simple answer to this question, but pilots and engineers should be sensitive to the effects of learning with a new configuration. Briefly, pilots should be allowed enough time to familiarize with a configuration, for general flying and in the test MTEs, before they are ready to fly the evaluation runs. The DLR test technique, adopted during in-flight simulation trials, involves computing the 'score factor' of the MTE, i.e., the ratio of successive performance measures (Fig. 7.4, Ref. 7.7). When the score factor rises above a pre-defined level, then the pilot is at least achieving repeatable task performance, if not workload. Ultimately, the pilot should judge when he is ready to give a formal evaluation and the trial engineer should resist forcing a 'half-baked' HQR. Something for both the test pilot and the engineer to bear in mind is that the subjective comments recorded during the learning phase are very important for understanding the basis for the eventual HQR; a communicative pilot will usually have a lot of very useful things to say at this stage. This brings us to the subject of communication between the pilot and engineer and flying qualities jargon.

(11) *Interpreting test pilot talk.* In handling qualities evaluations, test pilots will use a variety of descriptors within their subjective comment to explain the impact of good and bad characteristics. To simplify this discussion, we will relate two categories – the classical pilot qualitative language, e.g., sluggish, crisp, smooth and predictable, and the engineering parlance, e.g., control power, damping and bandwidth. HQRs are the summary of pilot comment, and it is important that the pilot comment is

consistent and understandable; once again, it is the pilot comment that directs the engineer towards improvement. Two observations on pilot comment are worth highlighting here. First, any classical parlance should be defined in terms relating to task; e.g., the roll response of this aircraft is sluggish because it takes too long to achieve the required bank angle. There is no universal dictionary for classical parlance so it is a good idea to establish agreed meanings early in a trial; the HQRs will then be more valuable. Second, it is the author's considered opinion that test pilots should be strongly discouraged from using engineering parlance during evaluations, unless they are conversant with the engineering background. Sometimes quite different engineering parameters can lead to similar effects and if pilots try to associate effects with causes, they run the risk of making predictive judgements based on what they think will be the case. Engineers need test pilots to tell them what aspects are good or bad and not try to diagnose why. Ironically, it is the very skill that test pilots are valued for – the ability to think about and interpret their response – that can spoil their ratings. When it comes to the evaluation, deeply learned and instinctive skills are being exercised and, to an extent, thinking can intrude on this process. It is far better for test pilots to describe their workload in terms that are subjective but unambiguous.

(12) *When is an HQR not an HQR?* During the assembly of a handling qualities database, configurations will be evaluated that span the range from good to bad, and pilots need not think that it is their fault if they cannot achieve the performance targets. During the development of a new product, flying qualities deficiencies may appear and the test pilots need to present their findings in a detached manner. Above all, test pilots that participate in such evaluations need to be free from commercial constraints or programme commitments that might influence their ratings. This point is stressed by Hoh in Ref. 7.8. Eventually it will be in both the user's and manufacturer's interest to establish the best level of flying quality.

(13) *Pilot fatigue – when does an HQR lose its freshness?* This will certainly vary from pilot to pilot and task to task, but evaluation periods between 45 and 90 min seem from experience to cover an acceptable range. The pilot fatigue level, and to an extent this can be influenced by their attitude to the evaluation, can be a primary cause for spread in HQRs. The pilot is usually the best judge of when his performance is being impaired by fatigue, but a useful practice is to introduce a reference configuration into the test matrix on occasions as a means of pilot calibration.

(14) *HQRs are absolute, not relative.* This is an important rule, but perhaps the most difficult to apply or live by, especially if several different aircraft are being compared in an experiment; there will always be the temptation for the pilot to compare an aircraft or configuration with another that has already achieved a particular standard and been awarded a rating. Disciplined use of the Cooper–Harper decision tree should help the pilots resist this temptation, and appropriate training and good early practice would seem to be the best preventative medicine for this particular bad habit.

(15) *The HQR is for the aircraft, not for the pilot.* Piloting workload determines the rating but the rating needs to be attributed to characteristics of the aircraft and task cues as defined in the Cooper–Harper rating scale. Emphasis on HQR, rather than pilot rating, can help with this important distinction.

(16) *An HQR does not tell the whole story.* In this last point we reiterate rule number 2 that every HQR should be accompanied by a sheet of pilot comments to give the full story. This can often be derived as a series of answers on a questionnaire addressing

the various aspects covered in the Cooper–Harper decision tree – vehicle characteristics, workload (compensation) and task performance; task cues also need to be addressed, and the absence of any reference to task cues in the Cooper–Harper decision tree is explained by the assumption that sufficient task cues exist for flying the task. The subject of task cues and the need for pilot subjective impressions of the quality of task cues have received more prominence with the introduction of vision aids to support flying at night and in poor weather. This topic is addressed further in Section 7.3.3.

These 16 rules represent this flying qualities engineer's assessment of the important facets of the subjective measurement scale and the HQR. Put together, the issues raised above highlight the importance of the special skills required of test pilots, enhanced by extensive training programmes. To examine how these work in practice, we need to discuss the design, conduct and test results from handling qualities evaluations.

7.2.2   Conducting a handling qualities experiment

Depending on the objectives, a handling qualities experiment can commit flight or ground-based simulation facilities and a trials team for periods from several days to several weeks or even months. The subjective and objective data gathered may take months or even years to be analysed fully. The success, and hence value, of such an endeavour rests heavily on the experimental design and trials planning. There are a multitude of issues involved here, most of which would be inappropriate for discussion in this book. One of the critical elements is the design of the MTEs in which the handling qualities are to be evaluated. This has already been raised as an important issue in the discussion on HQRs above; the task performance drives the workload, which drives the pilot rating. Before we examine results from a handling experiment, it is worth looking more closely at the design of an MTE.

Designing a mission task element

The concept of the MTE was introduced in Chapter 2, the Introductory Tour to this book. Any mission can be analysed in terms of mission phases and MTEs and sample manoeuvres. An MTE is identifiable by its clearly defined start and end conditions. To be viable as a test for handling qualities, an MTE also needs to be defined in terms of spatial and temporal constraints. Above all, the constraints need to be related to real operational needs, or the data will be of questionable value and test pilots will quickly lose interest. During the 10 years between 1984 and 1994, the MTE has become central to the development of military handling qualities criteria and work reported at conferences, specialist meetings and in journals abound with examples of different MTEs and related HQR diagrams. At the core of these activities, the ADS-33 MTEs have evolved into a set of mature test manoeuvres, aimed at providing the acid test for new military helicopters. In the early 1990s, a major refinement exercise was undertaken on these manoeuvres, as reported in Refs 7.9–7.11. The emphases of the refinements were (Ref. 7.10) ease of understanding, mission-oriented performance standards for good and degraded visual environments (DVEs), simple task cueing and affordable instrumentation. In this programme, several current operational helicopters were used in a flight test activity that served to concentrate attention on flight safety issues. Handling qualities testing, by its very nature, carries risk as the boundaries to

Table 7.1 ADS-33 flight test manoeuvres (Ref. 7.12)

Good visual environment		Degraded visual environment	
Precision tasks	Aggressive tasks	Precision tasks	Agressive tasks

transition to hover
hovering turn
landing
pirouette
slope landing
	turn to target
	bob-up/down
	vertical remask
	accel–decel
	sidestep
	slalom
	decel to dash
	transient turn
	pull-up/push-over
	roll reversal at reduced
	  and elevated load
	  factors
	high/low yo-yo
		decel in IMC
		transition to hover
		hovering turn
		landing
		pirouette
			bob-up/down
			accel–decel
			sidestep
			slalom

safe operation are mapped out. The new ADS-33 MTEs were designed to highlight any deficiencies in a pseudo-operational context and test programmes will certainly need to give a higher level of attention to safety than previously. The importance and justification for this approach is well summarized by Key in Ref. 7.10 when referring to the AH-64 ADS-33 flight tests: 'Some of the aggressive manoeuvres, especially in DVE, were quite thrilling...if they are too dangerous for a skilled test pilot to perform in a tightly controlled environment, it is unreasonable to expect the user to fly such manoeuvres in an unfamiliar, unfriendly environment in the fog of war'.

The test manoeuvres proposed for the new Military Standard (Ref. 7.12) are summarized in Table 7.1, and include both GVE (good visual environment) and DVE cases.

Figure 7.5, taken from Ref. 7.9, illustrates results from handling qualities tests during development of the refined MTEs using three test aircraft – the NRC variable stability Bell 205, the AH-64A and the UH-60A; the 205 was tested with both Level 1 and 2 response characteristics, according to the objective ADS-33 criteria. ADS-33 was targeted at a new design of course, the RAH-66 Comanche, so it is hardly surprising that current operational aircraft appear as good Level 2 on average.

To illustrate an MTE in more detail we have chosen the lateral sidestep repositioning manoeuvre and also to compare the ADS-33 task description and performance standards with those developed for the DRA's ACT research programme (Ref. 7.13). The layout of the sidestep ground markers for the DRA ACT simulations is sketched in Fig. 7.6 and quantified in Table 7.2. The pilot is required to initiate the MTE from a hover point with the triangle and square aligned, sidestepping to a new hover position again aligning the triangle and square. There is a close comparison between the DRA and ADS-33 manoeuvres, but the DRA requirements are slightly more demanding, reflecting the expected improvements conferred by full authority ACT. Important differences appear in the temporal and spatial constraints, with the DRA placing more emphasis on measuring the effects of piloting aggressiveness (three levels specified) and repositioning at a defined point, to introduce a realistic spatial constraint. In contrast,

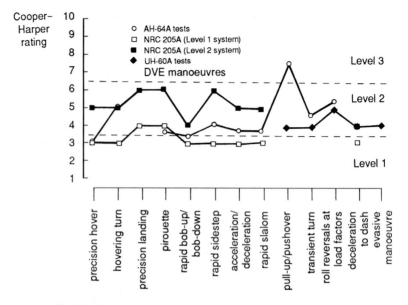

Fig. 7.5 HQRs for various aircraft flying ADS-33 tasks (Ref. 7.9)

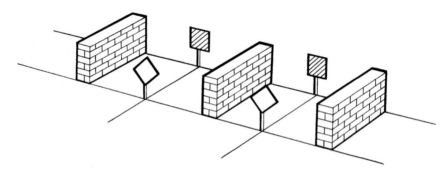

Fig. 7.6 Layout of the DRA sidestep MTE (Ref. 7.13)

the ADS-33 standards place more emphasis on pilots' achieving close to maximum lateral velocities with aggressiveness defined by the times to accelerate and decelerate.

Evaluating roll axis handling characteristics

Roll axis handling characteristics have figured prominently in Chapter 6, where many of the new concepts associated with modern mission-oriented response criteria were introduced and the development processes described. As we continue the discussion of subjective measurement and assessment we return to this reference topic and present results of tests conducted at the DRA in the early 1990s utilizing the ground-based flying qualities facility – the advanced flight simulator (Ref. 7.13). Before discussing details of this work it is appropriate to review the original data that contributed to the definition of satisfactory roll axis characteristics, in particular the small amplitude bandwidth criteria. In Ref. 7.14, Condon highlighted the point that during the early 1980s the fidelity of ground-based simulators was considered inadequate for defining

Table 7.2 Comparison of task description and performance standards for ADS-33 and DRA ACT sidestep mission task element

	Sidestep task performance requirements			
	Desired		Adequate	
	ADS-33	DRA ACT	ADS-33	DRA ACT
Height	9.14 ± 3.05 m	8 ± 2.5 m	9.14 ± 4.57 m	8 ± 5.0 m
Track	±3.05 m	±3.0 m	±4.57 m	±3.0 m
Heading	±10°	±5°	±15°	±10°
Hover	not specified	±3.0 m	not specified	±6 m
T_{stab}	5 s	not specified	10 s	not specified
ϕ_{accel}	25° (in 1.5 s)	10°, 20°, 30°	25° (in 3 s)	10°, 20°, 30°
ϕ_{decel}	30° (in 1.5 s)	not specified	30° (in 3 s)	not specified
V_{max}	V_{limit} − 5 kn	not specified	V_{limit} − 5 kn	not specified
S_{ss}	not specified	50 m	not specified	50 m

Task description

Both sidesteps are intended to assess lateral directional handling qualities for aggressive manoeuvring near the rotorcraft limits of performance. The flight path constraints reflect operations close to the ground and obstacles. Secondary objectives are to check for any objectionable cross-couplings and to evaluate the ability to coordinate bank angle and collective to hold constant altitude.

Both sidesteps require the pilot to reposition from hover to hover with a lateral manoeuvre maintaining task performance requirements as shown above. The ADS-33 sidestep puts emphasis on achieving close to limiting lateral velocity without step size constraints. The DRA ACT sidestep places emphasis on repositioning to a particular location, hence requiring the pilot to judge the acceleration and deceleration phases carefully, on the basis that an operational sidestep is likely to have relatively tight terminal position constraints. The ADS-33 step requires the pilot to achieve a minimum bank angle in a maximum time during the accel and decel phases. The DRA ACT sidestep requires the pilot to fly at three initial bank angles to quantify the effects of pilot aggressiveness.

the Level 1/2 boundaries for rate command rotorcraft. Figure 7.7, from Ref. 7.11, shows how HQRs derived from the NRC Bell 205 compare with those derived from the NASA VMS facility, for equivalent MTEs. As a result of this kind of comparison, ground-based simulation data were considered unreliable and were not used in the early development of ADS-33 for rate command systems. Problems were attributed to a number of different areas including poor visual cueing, particularly fine detail and field-of-view, the harmony between visual and motion cues and time delays in the cue development, all areas where there were no equivalent flight problems. During the late 1980s and early 1990s, simulation technology improved significantly and a number of studies were reported with varying degrees of success, but all acknowledged

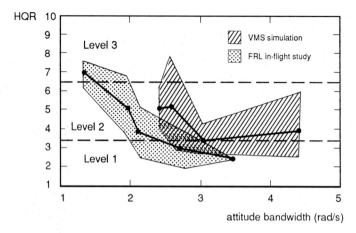

Fig. 7.7 Comparison of flight and simulation results for rate command aircraft in sidestep MTE (Ref. 7.11)

continuing limitations compared with in-flight simulation (Refs 7.14–7.16). With the commissioning of the DRA's AFS for helicopter research in 1991, it was considered important to calibrate handling qualities results for rate command systems to determine whether the fidelity of the AFS was good enough for definitive flying qualities research. The study, reported in Ref. 7.13, explored roll and pitch axes with a primary objective of establishing at what values of attitude bandwidth pilots would start returning Level 1 ratings consistently, if at all. The trial was configured with a futuristic ACT helicopter, with a two-axis sidestick, conventional collective and pedals, with primary flight instruments displayed head-up and pure rate response characteristics.

The key elements of the experiment are summarized in Fig. 7.8, showing the large motion system, computer-generated imagery (CGI) visuals, generic helicopter cockpit and the conceptual simulation model (CSM). A number of MTEs were developed on the CGI database that included sufficient textural detail for the evaluation pilots to perceive the desired and adequate task performance standards clearly. The photographs in Fig. 7.9 show the layout of four of the critical MTEs for evaluating roll and pitch handling at low speed and in forward flight – the low-speed sidestep and quickhop and the forward flight lateral jinking and hurdles. The task definitions included specification of the level of task aggressiveness to be flown by the pilots, as illustrated in the sidestep in Table 7.2 where initial bank angle was used as the defining parameter. We will return to the results for the sidestep later, but first we will consider the lateral jinking MTE in more detail, shown in plan form in Fig. 7.10, with the task performance standards defined in Table 7.3.

The lateral jinking or slalom manoeuvre is essentially a forward flight roll axis task comprising a sequence of 'S' turn manoeuvres followed by line tracking elements, as pilots attempt to fly through the gates shown in Fig. 7.10. Secondary handling qualities considerations include the ability to coordinate turns with pitch and yaw control and the harmonized use of collective and roll to maintain height. Task aggression is defined in terms of the maximum roll attitude used during the turning phases; values of 15°, 30° and 45° were found appropriate for designating low, moderate and high aggressiveness. These levels correspond to relaxed flying, normal operations with a

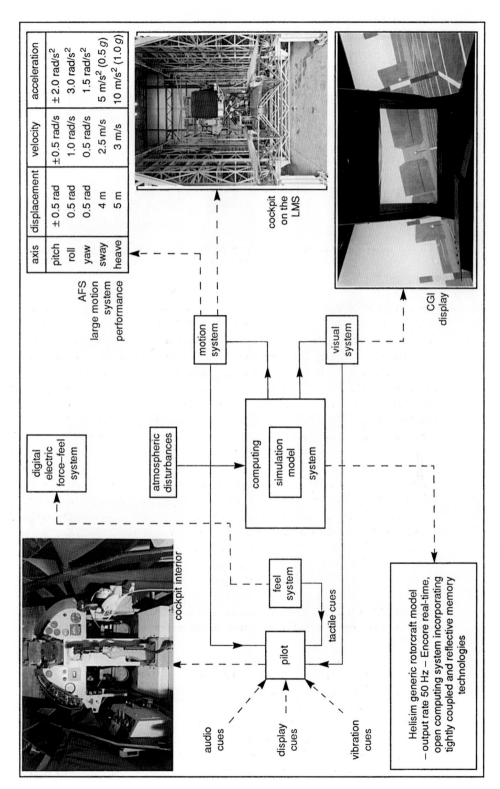

Fig. 7.8 Elements of DRA simulation trials

axis	displacement	velocity	acceleration
pitch	± 0.5 rad	± 0.5 rad/s	± 2.0 rad/s²
roll	0.5 rad	1.0 rad/s	3.0 rad/s²
yaw	0.5 rad	0.5 rad/s	1.5 rad/s²
sway	4 m	2.5 m/s	5 m/s² (0.5 g)
heave	5 m	3 m/s	10 m/s² (1.0 g)

AFS
large motion
system
performance

cockpit on the
LMS

CGI
display

motion
system

visual
system

digital
electric
force–feel
system

atmospheric
disturbances

computing
system

simulation
model

cockpit interior

feel
system

tactile cues

pilot

audio
cues

display
cues

vibration
cues

Helisim generic rotorcraft model
– output rate 50 Hz – Encore real-time,
open computing system incorporating
tightly coupled and reflective memory
technologies

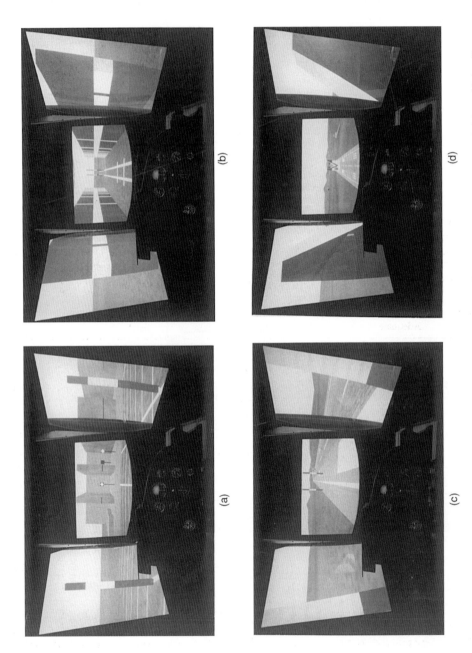

Fig. 7.9 The MTEs flown in the DRA simulation trials (Ref. 7.13): (a) sidestep; (b) quickhop; (c) lateral slalom; (d) hurdles

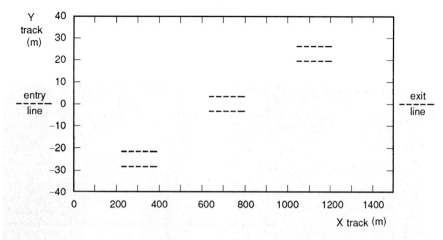

Fig. 7.10 Plan view of lateral slalom MTE (Ref. 7.13)

Table 7.3 Task performance requirements for lateral jinking MTE

MTE phase	Performance	Speed (kn)	Height (m)	Track (m)	Heading (deg)	End gate (m)
Translation	Desired	60 ± 5	8 ± 2.5	–	–	±3
	Adequate	60 ± 7.5	8 ± 5	–	–	±6
Tracking	Desired	60 ± 5	8 ± 2.5	±3	±5	–
	Adequate	60 ± 7.5	8 ± 5	±6	±10	–

degree of urgency and emergency or other life-threatening situations, respectively. The task objective is to fly through the course whilst maintaining a height of 8 m and a speed of 60 knots, turning at the designated gates to acquire the new tracking line as quickly as possible, within the constraints of the set level of aggression. The turning gates are represented by adjacent vertical posts, which also provide height cueing – the white band on the posts delineating the desired performance margin. The intermediate gates were added to give enhanced tracking cues supplementing the runway lines. The width of the gates was determined by the adequate margin of performance for the tracking task (±20 ft/6 m).

The helicopter model used in the trial was the equivalent system CSM (Ref. 7.17). The roll axis characteristics are described by the simple second-order system

$$\frac{\overline{p}}{\overline{\eta}_{1c}}(s) = \frac{K\,e^{-\tau s}}{\left(\frac{s}{\omega_m}+1\right)\left(\frac{s}{\omega_a}+1\right)} \tag{7.2}$$

where K is the overall gain or in this case the rate sensitivity (deg/s per unit control), τ is a pure time delay and ω_m can be considered to be equivalent to the roll damping aircraft L_p; ω_a is the bandwidth of a pseudo-actuator lag. The actuator effectively reduces the transient acceleration jerk following a step control input, to realistic values. The value of ω_a was set to 20 rad/s for the tests. Variations in bandwidth and phase delay, the principal handling parameters of interest, can be achieved through the CSM parameters given in eqn 7.2. Figure 7.11 illustrates contours of constant damping

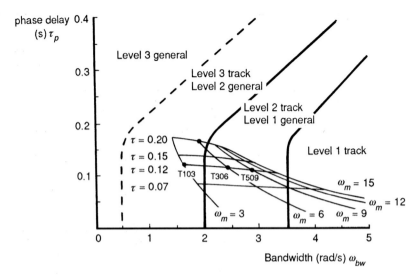

Fig. 7.11 CSM configurations overlaid on ADS-33 roll bandwidth chart (Ref. 7.13)

and time delay for the CSM overlaid on the bandwidth–phase delay diagram with the ADS-33 handling qualities boundaries. The four configurations to be discussed are spotted on the figure, with the designations of Ref. 7.13 – T103, 306 and 509. The last two digits denote the value of roll damping, as shown in the figure (e.g., T509 has a roll damping of 9 rad/s). The first digit assigns the control sensitivity (T1 = 0.1, T3 = 0.2, T5 = 0.3 rad/s² per %). All configurations share the same roll control power, 96 deg/s, hence the control sensitivity increased in proportion with the damping. Contributions to the approximate 110-ms phase delay for all three configurations include the actuator lag and pure time delay from the AFS system computing and image generation. It is interesting to note that the bandwidth of configuration T509, with a natural damping of 9 rad/s, is reduced to 3 rad/s by the time delays. Configuration T306 + 80 included 80-ms additional pure time delay. The configurations spanned the ADS-33C Level 1/2 handling qualities boundary for general MTEs situated at 2 rad/s.

The trial was flown by six test pilots whose HQRs are shown as a function of roll bandwidth in Fig. 7.12. The ratings are shown with the mean, maximum and minimum values. For each configuration, ratings are shown for the three levels of pilot aggression. The maximum spread of the HQRs for each configuration/aggression level is about 2. If the spread had been much greater than this, then there would have been cause for concern, but a spread of 2 is regarded as acceptable. Several observations can be made about these subjective results, drawn from the pilot comments gathered in the tests, as follows:

(1) First, we address the primary objective of the trial – to establish the Level 1/2 boundary for rate response types on the AFS. The result depends on which level of aggression is taken, and we will return to this particular issue later. On the basis that pilots can be expected to fly with moderate levels of aggression on a regular basis to accomplish tasks with some urgency, we would argue that Level 1 qualities are achievable with configuration T306 (≈2.5 rad/s) and higher bandwidths. This is a little higher than the ADS-33 value at 2 rad/s but actually agrees with the NRC mean

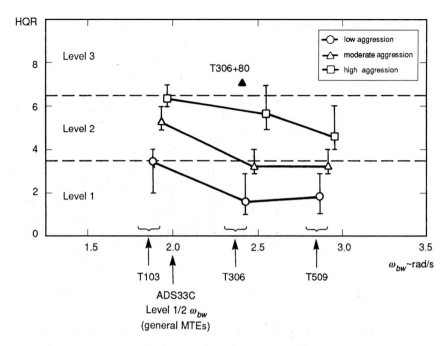

Fig. 7.12 HQRs for lateral slalom MTE versus roll attitude bandwidth (Ref. 7.13)

rating line shown in Fig. 7.7. This result was a clear indication that the AFS was able to predict Level 1 handling qualities with good accuracy.

(2) At the high aggression level, Level 1 ratings were not achievable, and ratings strayed into the Level 3 region for the lower bandwidth configurations T103 and T306. Pilots complained of insufficient control and a sluggish response in negotiating the tighter turns for the lower bandwidth configurations. Configuration T509 was solid Level 2, and it could be speculated that even higher bandwidths would confer better flying qualities still.

(3) The spread of ratings for each configuration gives an indication of the powerful effect of task demands on pilot workload. Pilots rated the same aircraft, configuration T103, between a 2 and a 7 as the urgency level increased from low to high. This emphasizes the importance of defining the level of pilot aggression required; it is one of the parameters that workload is most sensitive to, even more so than bandwidth over the range considered. At the higher urgency level, several new handling qualities issues come to light, including flight envelope limit monitoring, task cue deficiencies at high-bank angles and the need for improved pilot judgement of flight path trajectory. We will address these in more detail in later sections of this chapter.

(4) One of the classic problems experienced by pilots flying low-bandwidth aircraft in moderately demanding manoeuvres is the need to command a high roll rate to compensate for the long rise time, combined with the need to arrest the rate quickly to stabilize on a new attitude. This can lead to overcontrolling and difficulties with flight path control. Figure 7.13 shows the attitude quickness values (see Chapter 6) for configurations T306 and T509 flown up to moderate levels of aggression – both achieved borderline Level 1/2 ratings. Pilots flew the lower bandwidth configuration,

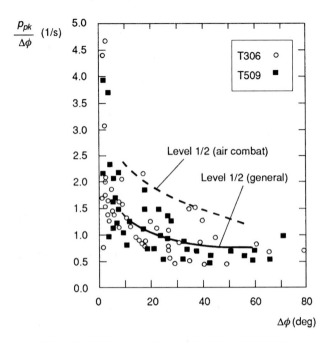

Fig. 7.13 Roll attitude quickness for slalom (Ref. 7.13)

T306, with significantly higher levels of quickness than T509, compensating for the lower bandwidth by using more of the control power. There is a usable limit to this trade-off between bandwidth and control power; the reader might note that the achieved roll quickness for T306 rises to the ADS-33 Level 1/2 track boundary.

(5) The single HQR 7 for configuration T306 + 80 at the moderate aggression level is also shown in Fig. 7.12. According to Fig. 7.11, the addition of 80-ms time delay to T306 should lead to Level 2 flying qualities for general MTEs and Level 3 flying qualities for tracking MTEs; in the event, a Level 3 rating was returned, indicating the significant tracking content of the lateral jinking task at moderate to high levels of aggression. For this case the pilot actually experienced a roll pilot-induced oscillation (PIO) while trying to tighten up the flight path to negotiate the gates. Figure 7.14 illustrates the plan view of the task showing the aircraft ground track for the same pilot flying T306 (upper figure) and T306 + 80 (lower figure). The (roll) PIO on the approach to, and flying through, the third gate is quite pronounced, and this experience highlights the real dangers of operating with low-bandwidth aircraft with large values of phase delay, in this kind of task. It carries a particularly strong message to the designers of ACT helicopters with digital flight control systems where high values of phase delay can be introduced by digital system transport delays and filters.

(6) The deterioration from borderline Level 1/2 to Level 3 with the addition of 80-ms time delay is an important result, suggesting that the pilots are more sensitive to increases in phase delay than the boundaries in Fig. 7.11 would suggest. We have already presented results in Chapter 6, which suggest that the phase delay boundaries should be capped rather than extend linearly out above 150 ms. The AFS slalom data tend to confirm this, although the pilot was forced to use a more aggressive control strategy for T306 + 80 than for the standard T306, often hitting the control stops during the roll reversals. We

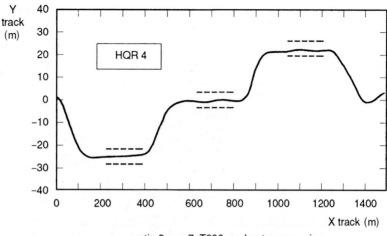

sortie 8, run 7 T306 moderate aggression

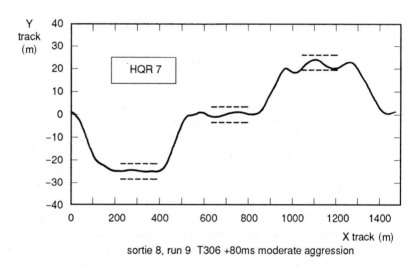

sortie 8, run 9 T306 +80ms moderate aggression

Fig. 7.14 Comparison of ground tracks in slalom MTE (Ref. 7.13)

have made the point on several occasions in this book that handling deficiencies can emerge as cliff edges, developing rapidly as some detail of the task or configuration is changed. The slalom PIO is a classic example of this and serves as a reminder of the importance of testing through moderate and up to high levels of aggression.

The pilot ratings for the DRA sidestep are shown in Fig. 7.15, comparing well with the ADS-33 flight test data; the results show a similar trend to the slalom data, except that the degradation with level of aggression does not appear nearly as strong for the higher bandwidth configurations. This is partly explained by the pilot comments that the sidestep task is considerably easier and more natural to fly with a more aggressive control strategy than the slalom, provided the attitude bandwidth and control power are available.

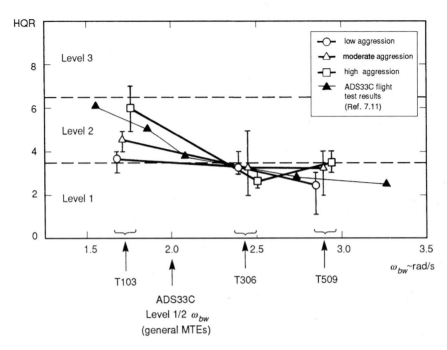

Fig. 7.15 HQRs for lateral sidestep MTE versus roll attitude bandwidth (Ref. 7.13)

In Ref. 7.18, results are reported from trials on the DRA's AFS for a maritime mission – the recovery of large helicopters to non-aviation ships in various sea states. The landing MTE was quickly identified as by far the most critical for handling qualities for all sea states. Figure 7.16 illustrates the task performance requirements for the landing, based on the need to touch down on the deck grid within defined velocity constraints. The pilot is required to bring the helicopter to the hover alongside the ship, wait for a quiescent period in the ship motion, exercise a lateral sidestep towards, and land onto, the deck. As often happens in practice, as the aircraft arrives over the deck, the pilot is unable to execute a successful landing immediately and has to maintain station over the deck grid, waiting for another quiescent period. The study reported in Ref. 7.18 examined roll, pitch and heave handling qualities, again using the CSM, but now configured as a much larger helicopter. Figure 7.17 shows the pilot HQRs for the landing MTE as a function of roll attitude bandwidth for various sea states. Figure 7.18 shows the achieved task performance in terms of landing scatter and touchdown velocities. For comparison, flight test results from a Sea King helicopter are included. The pilot was able to achieve adequate performance for the points shown and the HQRs were driven by the extreme levels of workload required at the higher sea states. For this MTE, sea state is the principal task driver, just as urgency level was for the battlefield sidestep and slalom MTEs, and the pilot's ability to achieve desired performance levels at low workload is a strong function of the deck motion induced by the sea state (SS). While the data indicate that SS3 can be achieved with relatively low-bandwidth configurations ($\approx$1.5 rad/s), SS5 will require considerably higher values, perhaps as high as the 3.5 rad/s boundary of the ADS-33 air combat/tracking tasks, as indicated by the suggested performance requirements shaded on Fig. 7.17. This will be more

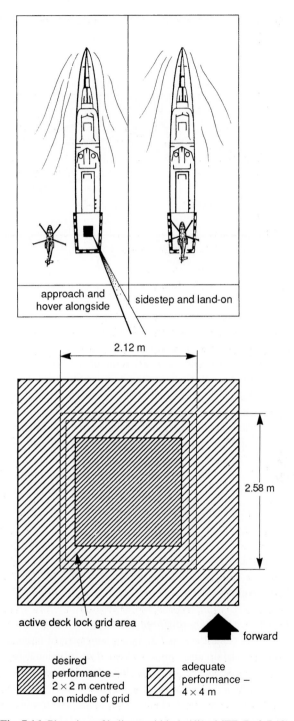

approach and hover alongside   sidestep and land-on

2.12 m

2.58 m

active deck lock grid area

forward

desired performance – 2 × 2 m centred on middle of grid

adequate performance – 4 × 4 m

Fig. 7.16 Plan view of helicopter/ship landing MTE (Ref. 7.18)

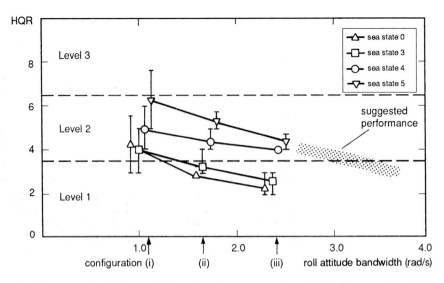

Fig. 7.17 HQRs for helicopter/ship landing MTE versus roll attitude bandwidth (Ref. 7.18)

difficult to achieve with large helicopters, and high gain/high authority active control may be required to guarantee consistent performance in poor weather conditions.

This section has discussed some of the important issues associated with pilot subjective opinion of aircraft handling qualities and the practical use of the Cooper–Harper HQR scale. The reader will be able to find many examples of handling qualities experiment reported in the open literature during the 1980s and 1990s; it has been a rich and productive period for this subject, spurred on to a large extent by the new handling qualities specifications on the one hand and the advent of active control technology on the other. These broad and concerted efforts to define and improve handling qualities have exposed and highlighted many new areas and facets of handling, where previous work had not been definitive. We now turn to examine some of these under the heading – Special flying qualities.

7.3   SPECIAL FLYING QUALITIES

7.3.1   Agility

Agility as a military attribute
In Chapter 6 and Section 7.2, the measurement of flying qualities from objective and subjective standpoints was discussed. Two additional issues arise out of the quality scale and assessment. First, the boundaries are defined for minimum requirements that reflect and exercise moderate levels of the dynamic OFE only, rather than high or extreme levels. Second, the assessments are usually made in 'clean' or clinical conditions, uncluttered by secondary tasks, degraded visual cues or the stress of real combat. Beyond the minimum quality levels there remains the question of the value of good flying qualities to the overall mission effectiveness. For example, how much more effective is an aircraft that has, say, double the minimum required (Level 1) roll control

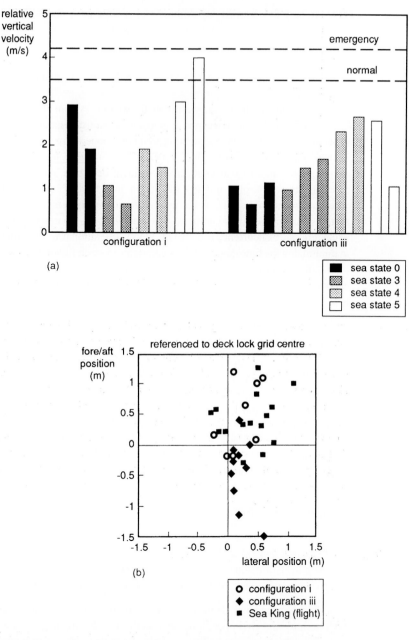

Fig. 7.18 Task performance in helicopter/ship landing MTE (Ref. 7.18): (a) touchdown velocity; (b) landing scatter

power? A second question asks whether there are any upper limits to the flying qualities parameters, making quality boundaries closed contours. The answers to these questions cannot generally be found in flying qualities requirements like ADS-33. At higher performance levels, very little data are available on flying qualities and, consequently, there are very few defined upper limits on handling parameters. Regular and safe, or carefree, use of high levels of transient performance has come to be synonymous with agility. The relationship between flying qualities and agility is important because it potentially quantifies the value of flying qualities to operational effectiveness. We will return to this question of value in Section 7.5, but first we examine agility in a more general context.

Operational agility is a key attribute for weapon system effectiveness. Within the broader context of the total weapon system, the mission task naturally extends to include the actions of the different cooperating, and non-cooperating, subsystems, each having its own associated time delay (Ref. 7.19). We can imagine, for example, the sequence of actions for an air-to-air engagement – threat detection, engagement, combat and disengagement; the pilot initiates the action and stays in command throughout, but a key to operational agility is the automation of subsystems – the sensors, mission systems, airframe/engine/control systems and weapon – to maximize the concurrency in the process. Concurrency is one of the keys to operational agility. Another key relates to minimizing the time delays of the subsystems to reach full operational capability and hence effectiveness in the MTE. Extensions to the MTE concept are required which encompass the functions and operations of the subsystems and so provide an approach to assessing system operational agility. Working Group 19, set up by AGARD in 1990, was tasked to address these issues (Ref. 7.19). In this study, addressing both fixed- and rotary-wing aircraft, flying qualities were a major concern. Minimizing time delays is crucial for the airframe, but flying qualities can suffer if the accelerations are too high or time constants too short, leading to jerky motion. The following discussion is based on the author's contribution to AGARD WG19.

We need to examine how well existing flying qualities requirements address agility, but to set the scene we first reflect on the WG19 generalized definition of agility:

> *the ability to adapt and respond rapidly and precisely with safety and with poise, to maximise mission effectiveness.*

To place this definition in context it is useful to list the four mission phases where agility might be important:

(1) stealthy flying, in particular terrain-masked, to avoid detection;
(2) threat avoidance once detected;
(3) the primary mission (e.g., threat engagement);
(4) recovery and launch from confined, or otherwise demanding, areas.

In addition, we can include the need for agility in response to emergency situations for both military and civil operations, such as those following major system failures. The key attributes of airframe agility, as contained in the above definition, are as follows:

(1) *Rapid.* Emphasizing speed of response, including both transient and steady-state phases in the manoeuvre change; the pilot is concerned to complete the manoeuvre change in the *shortest possible time*; what is possible will be bounded by a number of different aspects.
(2) *Precise.* Accuracy is the driver here, with the motivation that the greater the task precision, e.g., pointing, flight path achievable, the greater the chance of a successful outcome.

(*Note:* the combination of speed and precision emphasizes the special nature of agility; one would normally conduct a process slowly to achieve precision, but agility requires both.)

(3) *Safety.* This reflects the need to reduce piloting workload, making flying easy and freeing the pilot from unnecessary concerns relating to safety of flight, e.g., respecting flight envelope limits.

(4) *Poise.* This relates to the ability of the pilot to establish new steady-state conditions quickly and to be free to attend to the next task; it relates to precision in the last moments of the manoeuvre change but is also a key driver for ride qualities that enhance steadiness in the presence of disturbances.
(Poise can be thought of as an efficiency factor, or measure of the unused energy potential.)

(5) *Adapt.* The special emphasis here relates to the requirements on the pilot and aircraft systems to be continuously updating awareness of the operational situation; the possibility of rapid changes in the external factors, discussed earlier in this chapter (e.g., threats, UCE, wind shear/vortex wakes), or the internals, through failed or damaged systems, makes it important that agility is considered, not just in relation to set-piece manoeuvres and classical engagements, but also for initial conditions of low energy and/or high vulnerability or uncertainty.

Existing flying qualities requirements address some of these agility attributes implicitly, through the use of the HQRs, which relate the pilot workload to task performance achieved, and explicitly through criteria on response performance, e.g., control power, bandwidth, stability. A new parameter, the agility factor, makes a direct link between inherent vehicle performance and handling.

The agility factor
One of the most common causes of dispersion in pilot HQRs stems from poor or imprecise definition of the performance requirements in an MTE, leading to variations in interpretation and hence perception of achieved task performance and associated workload. We have already illustrated this with the controlled experiment data from the AFS slalom and sidestep MTEs. In operational situations, this translates into the variability and uncertainty of task drivers, commonly expressed in terms of precision, but the temporal demands are equally important. The effects of task time constraints on perceived handling have been well documented (Refs 7.20–7.24) and represent one of the most important external factors that impact pilot workload. Flight results gathered on Puma and Lynx test aircraft at DRA (Refs 7.20, 7.23, 7.24) showed that a critical parameter was the ratio of the task performance achieved to the maximum available from the aircraft; this ratio gives an indirect measure of the spare capacity or performance margin and was consequently named the *agility factor*. The notion developed that if a pilot could use the full performance safely, while achieving desired task precision requirements, then the aircraft could be described as agile. If not, then no matter how much performance margin was built into the helicopter, it could not be described as agile. The DRA agility trials were conducted with Lynx and Puma operating at light weights to simulate the higher levels of performance margin expected to be readily available, even at mission gross weights, in future types (e.g., up to 20–30% hover thrust margin). A convenient method of computing the agility factor was developed as the ratio of ideal task time to actual task time. The task was

deemed to commence at the first pilot control input and to complete when the aircraft motion decayed to within prescribed limits (e.g., position within a prescribed cube, rates <5°/s) for repositioning tasks, or when the accuracy/time requirements were met for tracking or pursuit tasks. The ideal task time is calculated by assuming that the maximum acceleration is achieved instantaneously, in much the same way that some aircraft models work in combat games. So, for example, in a sidestep repositioning manoeuvre, the ideal task time is derived with the assumption that the maximum translational acceleration (hence aircraft roll angle) is achieved instantaneously and sustained for half the manoeuvre, when it is reversed and sustained until the velocity is again zero.

The ideal task time is then simply given by

$$T_i = \sqrt{(4S/a_{max})} \tag{7.3}$$

where S is the sidestep length and a_{max} is the maximum translational acceleration. With a 15% hover thrust margin, the corresponding maximum bank angle is about 30°, with a_{max} equal to 0.58 g. For a 100-ft sidestep, T_i then equals 4.6 s. Factors that increase the achieved task time, beyond the ideal, include

(1) delays in achieving the maximum acceleration (e.g., due to low roll attitude bandwidth/control power);
(2) pilot reluctance to use the maximum performance (e.g., no carefree handling capability, fear of hitting ground);
(3) inability to sustain the maximum acceleration due to drag effects and sideways velocity limits;
(4) pilot errors of judgement leading to terminal repositioning problems (e.g., caused by poor task cues, strong cross-coupling).

To establish the kinds of agility factors that could be achieved in flight test, pilots were required to fly the Lynx and Puma with various levels of aggressiveness or manoeuvre 'attack', defined by the maximum attitude angles used and rate of control application. For the low speed, repositioning sidestep and quickhop MTEs, data were gathered at roll and pitch angles of 10°, 20° and 30° corresponding to low, moderate and high levels of attack, respectively. Figure 7.19 illustrates the variation of HQRs with agility factor for the two aircraft (Ref. 7.24). The higher agility factors achieved with Lynx are principally attributed to the hingeless rotor system and faster engine/governor response. Even so, maximum values of only 0.6–0.7 were recorded compared with 0.5–0.6 for the Puma. For both aircraft, the highest agility factors were achieved at marginal Level 2/3 handling. In these conditions, the pilot is either working with little or no spare capacity, or not able to achieve the flight path precision requirements. According to Fig. 7.19, the situation rapidly deteriorates from Level 1 to Level 3 as the pilot attempts to exploit the full performance, emphasizing the 'cliff edge' nature of the effects of handling deficiencies. The Lynx and Puma are typical of current operational types with low authority stability and control augmentation. While they may be adequate for their current roles, flying qualities deficiencies emerge when simulating the higher performance required in future combat helicopters.

The different possibilities are illustrated in Fig. 7.20. All three configurations are assumed to have the same performance margin and hence ideal task time. Configuration

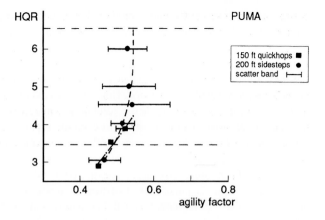

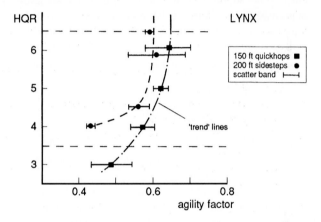

Fig. 7.19 Variation of HQRs with A_f showing the cliff edge of handling deficiencies (Ref. 7.24)

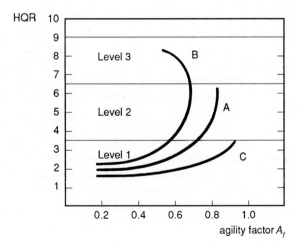

Fig. 7.20 Variation of HQR with A_f for different notional configurations (Ref. 7.25)

A can achieve the task performance requirements at high agility factors but only at the expense of maximum pilot effort (poor Level 2 HQRs); the aircraft cannot be described as agile. Configuration B cannot achieve the task performance when the pilot increases his or her attack and Level 3 ratings are returned; in addition, the attempts to improve task performance by increasing manoeuvre attack have led to a decrease in agility factor, hence a waste of performance. This situation can arise when an aircraft is PIO prone, is difficult to re-trim or when control or airframe limits are easily exceeded in the transient response. Configuration B is certainly not agile and the proverb 'more haste, less speed' sums the situation up. With configuration C, the pilot is able to exploit the full performance at low workload. The pilot has spare capacity for situation awareness and being prepared for the unexpected. Configuration C can be described as truly agile. The inclusion of such attributes as safeness and poise within the concept of agility emphasizes its nature as a flying quality and suggests a correspondence with the quality levels. These conceptual findings are significant because the flying qualities boundaries, which separate different quality levels, now become boundaries of available agility. Although good flying qualities are sometimes thought to be merely 'nice to have', with this interpretation they can actually delineate a vehicle's achievable performance. This lends a much greater urgency to defining where those boundaries should be. Put simply, if high performance is dangerous to use, then most pilots will avoid using it.

In agility factor experiments the definition of the level of manoeuvre attack needs to be related to the key manoeuvre parameter, e.g., aircraft speed, attitude, turn rate or target motion. By increasing attack in an experiment, we are trying to reduce the time constant of the task, or increasing the task bandwidth. It is adequate to define three levels – low, moderate and high, the lower corresponding to normal manoeuvring, the upper to emergency manoeuvres.

There are also potential misuses of the agility factor when comparing aircraft. The primary use of the A_f is in measuring the characteristics of a particular aircraft performing different MTEs with different performance requirements. However, A_f also compares different aircraft flying the same MTE. Clearly, a low-performance aircraft will take longer to complete a task than one with high performance, all else being equal. The normalizing ideal time will therefore be greater for the lower than the higher performer, and if the agility factors are compared, this will bias in favour of the poor performer. Also, the ratio of time in the steady state to time in the transients may well be higher for the low performer. To ensure that such potential anomalies are not encountered, when comparing aircraft using the agility factor it is important to use the same normalizing factor – defined by the ideal time computed from a performance requirement.

Relating agility to handling qualities parameters

Conferring operational agility on future rotary-wing aircraft, emulating configuration C above in Fig. 7.20, requires significant improvements in handling, but research into criteria at high-performance levels are needed to lead the way. A natural agility parameter has developed as one of the ADS-33 innovations – the response quickness. We have already discussed the properties of this parameter in Chapter 6 but it is useful to take a closer look at the effect of this and other handling parameters on the equi-response charts shown in Fig. 7.21. For a simple illustration we refer back to the CSM

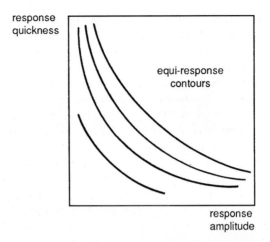

Fig. 7.21 Response characteristics on the frequency–amplitude plane: equi-response contours

model structure for roll rate command response type

$$\frac{\overline{p}}{\eta_{1c}}(s) = \frac{K\,e^{-\tau s}}{\left(\frac{s}{\omega_m} + 1\right)\left(\frac{s}{\omega_a} + 1\right)} \tag{7.4}$$

If we interpret the frequency axis as roll response quickness as shown in Fig. 7.21, the effect of independent variation of the different parameters in eqn 7.4 can be illustrated as in Fig. 7.22. The sensitivity of agility factor with the parameters of the CSM is relatively easy to establish. For example, if we consider a bank and stop MTE (Fig. 7.23), some useful insight can be gained. A pulse-type control input will be assumed, although, in practice, pilots would adopt a more complex strategy to increase the agility factor. To illustrate the primary effect, we consider the case where the 'secondary' time delays are set to zero (i.e., $\tau = 0$, $\omega_a = \infty$). For a roll angle change of $\Delta\phi$, the ideal

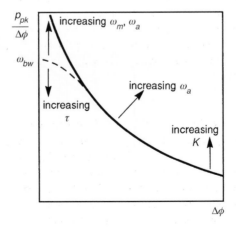

Fig. 7.22 CSM parameters on frequency–amplitude diagram (Ref. 7.25)

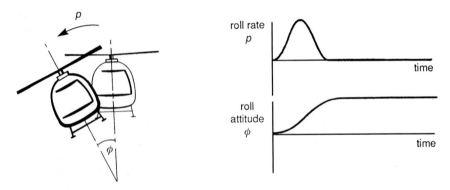

Fig. 7.23 Bank and stop MTE

time (assuming the time to achieve maximum rate is zero) is then given by

$$T_i = \Delta\phi/K = \Delta t \tag{7.5}$$

where Δt is the control pulse duration.

The time to reduce the bank angle to within 5% of the peak value achieved is given by

$$T_a = \Delta t - \ln(0.05)/\omega_m \tag{7.6}$$

The agility factor is then given by the expression

$$A_f = \frac{T_i}{T_a} = \frac{\omega_m \Delta t}{\omega_m \Delta t - \ln(0.05)} \tag{7.7}$$

Figure 7.24 illustrates the variation of A_f with $\omega_m \Delta t$. The bandwidth ω_m is the maximum achievable value of quickness for this simple case and hence the function shows

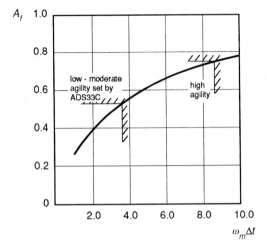

Fig. 7.24 Variation of A_f with normalized bandwidth for bank and stop MTE (Ref. 7.25)

the sensitivity of A_f to both bandwidth and quickness. The normalized bandwidth is a useful parameter as it represents the ratio of aircraft bandwidth to control input bandwidth, albeit rather approximately. For short, sharp control inputs, typical in tracking corrections, high aircraft bandwidths are required to achieve reasonable agility factors. For example, at the ADS-33C minimum required roll attitude bandwidth of 3.5 rad/s and with 1-s pulses, the pilot can expect to achieve agility factors of 0.5 using simple control strategies in the bank and stop manoeuvre. To achieve the same agility factor with a 0.5-s pulse would require double the bandwidth. This is entirely consistent with the argument that the ADS-33C boundaries are set for low to moderate levels of attack. If values of agility factor up to 0.75 are to be achieved, it is suggested as in Fig. 7.24, that bandwidths up to 8 rad/s will be required. Whether the 30% reduction in task time is worth the additional effort and cost to develop the higher bandwidth can be judged only in an overall operational context. Such high values of roll bandwidth may be achievable in very high performance fixed-wing aircraft and Fig. 7.24 serves to illustrate and underline the different operational requirements of the two vehicle classes, and also, to a large extent, the different expectations of the operators.

This simple example has many questionable assumptions, but the underlying point that increasing the key flying qualities parameters above the Level 1/2 boundary has a first-order effect on task performance still holds. But it provides no clues to possible upper performance boundaries set by flying qualities considerations. Existing requirements do not address upper limits directly, and more research with high-performance variable stability helicopters is required to address this issue. Intuitively, we might expect upper limits to be related to the acceleration capability of the aircraft (Ref. 7.25). This is largely the case with fixed-wing aircraft but there are also tentative upper limits on pitch attitude bandwidth (see Figs 6.43, 6.44). However, it is suspected that these are actually a reflection of the high control sensitivity required to achieve a defined level of control power, rather than the high values of bandwidth *per se*. Upper limits on control sensitivity are typically set to reduce the jerkiness or abruptness for small amplitude precision control, but the numerical values depend very much on the inceptor characteristics. Regarding the moderate and large amplitude motions, the best we can say at the moment is that the parameters on the quickness–amplitude charts are likely to have upper bounds beyond which agility would deteriorate.

Agility is a special flying quality catering for extreme operational requirements and a key technology driver for military functions. At the other end of the spectrum we find another, equally demanding, requirement for flight in very poor visibility. Here the pilot is not so much interested in agility as increased stabilization and the enhancement of his visual cues for the guidance task. Flight in degraded visual conditions exemplifies the tension and contrast between stability and agility requirements and is pressing hard on cockpit-related technologies that support pilotage; it is also the next topic of investigation.

7.3.2 The integration of controls and displays for flight in degraded visual environments

Flight in DVE

With fixed-wing aircraft, pilots can be flying under either visual or instrument flight rules (VFR or IFR), corresponding to defined levels of outside visual cues or meteorological conditions (VMC or IMC). If aircraft have to operate in IMC then typically there will be two crew, one flying while the other keeps an eye open for any hazards appearing in the visual scene. Except for the important case of military fixed-wing aircraft flying low level to avoid radar and other detection systems, nearly all fixed-wing IMC flying is conducted at altitude, well away from obstacles, and means little more than flying on instruments while in cloud or at night. Key instruments that the pilot would scan include the attitude indicator, heading gyro, airspeed indicator, ball and slip and rate of climb/descent indicator. A guided approach to a landing site would, in addition, require the pilot to follow a flight path as directed by special guidance instruments until the aircraft emerges into VMC below cloud to carry out a normal touchdown. Airports are equipped with various levels of guidance facilities enabling up to fully automatic landings in poor visibility or IMC. With fixed-wing aircraft, the characterization of visibility conditions is therefore fairly simple and the associated operational decision making, e.g., whether to initiate a sortie, can be based on relatively simple criteria, e.g., how much of the airfield can be seen. Rotary-wing aircraft operations have also been constrained by the same physical conditions but the ability to operate at low speed, combined with the military need to operate at very low level to avoid detection, has led to the development of a considerably more detailed and structured approach to the characterization of outside visual cues (OVCs). The general term adopted in the rotary-wing technical community for characterizing poor visibility conditions is the DVE – degraded visual environment, and this section examines some of the special considerations that accompany helicopter operations and flying qualities in the DVE.

Pilotage functions

To initiate the discussion it is useful to reflect on the piloting task and to review our previous classification into three subtasks – navigation (and general situation awareness), guidance and stabilization. The discussion will be aided by Fig. 7.25 showing the pilot performing as the feedback element in the closed-loop task. Navigation is concerned with knowing where you are and where you are going – the very outer loop in Fig. 7.25, with typical time/space scales measured in min/km. Most of the time, the pilot will not

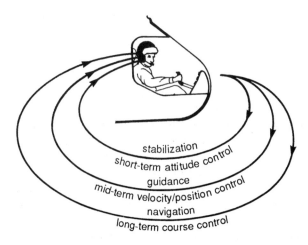

Fig. 7.25 The three piloting activities

be concerned with applying control actions to support the navigation function. Typically, he will be following maps that lead him from one 'way-point' to the next, when he applies control to direct the aircraft on to the next heading and perhaps speed and height. Flying at low level and in the nap of the earth (NoE) makes considerably more demands on piloting, particularly on the inner-loop functions. The guidance function works within time/space scales of a few seconds and tens of metres and is concerned with avoiding obstacles and the ground. The closer the pilot has to fly to obstacles, the more arduous the guidance task becomes and, typically, the slower the groundspeed selected. A general rule of thumb is that pilots will select a speed that gives them a perception – action timeframe of between 3 and 5 s. Hence, a 3-s pilot may elect to fly at 40 knots to give a 60-m straight line see-ahead distance for avoiding obstacles. At 80 knots the same pilot would need to be able to see 120 m ahead. These are minimum distances and pilots normally fly with much greater safety margins. In the same terrain, for example, a different pilot may prefer to fly with less urgency, making more spare capacity available for observation, and chose to fly at 20 knots, giving a 3-s to fly 30 m, or 6-s see-ahead time in a DVE with a 60-m visibility. Similarly, the vertical flight path excursions caused by disturbances will increase with increasing forward speed, forcing the pilot to fly higher to maintain the same level of safety. In the NoE, a general rule is to trade feet for knots – 10 knots at a height of 10 ft, 60 knots at a height of 60 ft. The selected overall piloting strategy for guidance will depend on a number of factors – pilot familiarity with the terrain and experience of NoE flying, the aircraft response characteristics, the level of task urgency and last, but perhaps most important of all, the quality of the OVCs. This is the cue to the main subject of this section but before discussing flying qualities in the DVE in more detail, we need to examine the third, and perhaps most, distinguishing feature of helicopter pilotage – stabilization.

Automobile drivers are generally unconcerned with stability, except in tight curves or on slippery surfaces, or perhaps with faulty steering/uneven tyre wear or balance. However, without some form of artificial stability augmentation, helicopter pilots need to make continual corrections with their controls to ensure that the aircraft does not depart from a prescribed flight path. Different helicopters have their own particular stability characteristics and problems, but most suffer from natural instabilities in both longitudinal and lateral motions that are difficult to cure completely with limited authority artificial stability augmentation. Control of instabilities is primarily achieved through attitude as illustrated in Fig. 7.25 and often requires the pilot to be continuously attentive to flight path control, hence contributing significantly to pilot workload.

Flying in DVE

Military helicopter operations require pilots to fly at low level in the NoE at night and in bad weather, and clearly the DVE has a major impact on all three pilotage functions – navigation, guidance and stabilization. To a lesser extent, recovery of civil transport helicopters in poor weather to confined landing sites, such as ships and building tops, also makes additional demands on flying qualities. Pilots need support for all three functions described above. Fear of getting lost may well be a primary concern but navigation is not directly a flying qualities issue. We are more concerned with guidance and stabilization. As the OVCs degrade, pilots will have two related concerns. First, they will need to supplement the disappearing outside world position and velocity cues to enable them to continue low-level flight without risk of bumping into things, with potentially catastrophic consequences. Second, they will need to fixate on their attitude instruments, particularly in gusty conditions, to prevent the aircraft departing from trim or level flight. Without any artificial guidance and stabilization aids, these requirements are clearly incompatible (the one requiring the pilot to keep eyes out, the other to fix gaze on displays) and a pilot will sensibly climb out of the unsafe flight condition. To enable helicopters to continue operations in low-level DVE, special guidance and control technologies are being developed, and requirements on these have been clarified in the new parlance of ADS-33.

It is recognized that the guidance function can really be augmented only through the provision to the pilot of augmented visual cues projected either onto the visor of his helmet or onto cockpit panels, either head-up or -down. The first generation of such displays can be found in systems like the AH-64A Apache helicopter with the integrated helmet and display system (IHADS), which provides a thermal image from a forward-looking infra-red sensor (FLIR) onto a monocular display, overlaid with flight path symbology and integrated and slaved with the pilot's helmet (Ref. 7.26). We will discuss this as representative of current operational technology later in this section. It is also recognized that the stabilization task can be augmented properly only through feedback control functions, augmenting the poor natural damping and aerodynamic stiffness of the helicopter (which is practically absent at low speed). Two outstanding questions arise from this simple analysis – how best to ensure a harmonious integration of the guidance and stabilization augmentations for flight in DVE and what trade-offs exist in the design of the related display and SCAS technologies.

The usable cue environment

ADS-33 addresses these issues through the requirement for different response types (effectively the stabilization level) in different usable cue environments (UCE) to ensure Level 1 flying qualities. Of course, the quality levels still depend upon what the pilot is trying to achieve in terms of MTE, and Fig. 7.26 illustrates conceptually the three dimensions associated with this problem. We have briefly reviewed some of the issues involved previously in Chapters 2 and 6, but with this discussion of flying qualities in the DVE we are at the heart of the UCE concept and can give it more attention. One of the first discussions on the need for a more elaborate structure to define the quality of visual cues for rotorcraft operations appeared in Ref. 7.27. Hoh introduced the concept of the OVC scale in the form of Fig. 7.27, to establish the quality of task cues for the

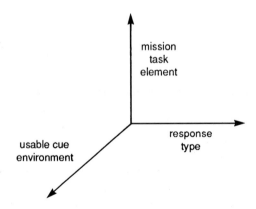

Fig. 7.26 The three dimensions of flight in DVE

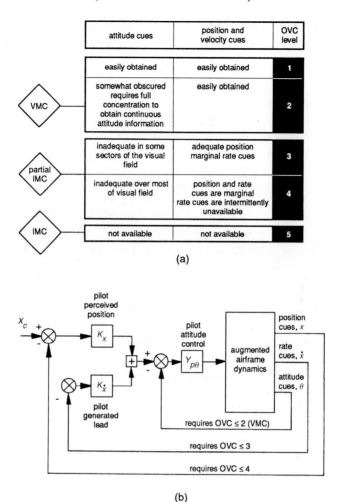

Fig. 7.27 The outside visual cue scale (Ref. 7.27): (a) quantification of outside visual cues (OVC); (b) required outside visual cues for control

control of attitude and velocity or translational rate. In a flight test study to define the relative importance of such attributes as texture and field of view, Hoh developed the OVC concept and gathered pilot ratings for visual cues – the so-called VCRs (Ref. 7.28). A conclusion of this study was that the stabilization function can be performed well with only a narrow field of view but fine-grain texture is vital, and also that the guidance function, and more general situation awareness, requires a wide field of view with macro texture. The OVC scale was further developed into the UCE (Refs 7.28, 7.29) to measure the usefulness and quality of artificial vision aids. The UCE scale along with the adjectival meanings of the different subjective VCRs are shown in Fig. 7.28. Pilots must rate the visual cues based on their ability to perform various low-speed/hover MTEs (with the DVE, rather than day, task performance requirements). The method for deriving the UCE is described in Refs 7.29 and 7.30. VCRs from at least three test pilots must be used to derive the UCE. Once derived for a given task,

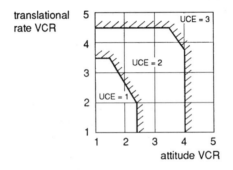

	attitude	horizontal translational rate	vertical translational rate

definitions of cues

X = pitch or roll attitude and lateral, longitudinal, or vertical translational rate

good X cues — can make aggressive and precise X corrections with confidence and precision is good

fair X cues — can make limited X corrections with confidence and precision is only fair

poor X cues — only small and gentle corrections in X are possible, and consistent precision is not attainable

Fig. 7.28 Usable cue environment (Ref. 7.1)

the individual VCRs are then processed according to the following rules derived from Ref. 7.30.

(1) Choose the worst attitude VCR and worse translational VCR for each pilot in each task.
(2) Average all the pilots' worst attitude VCRs and average all the pilots' worst translational VCRs.
(3) Calculate the standard deviation for the VCRs found in step (2).
(4) Check that the standard deviation is less than 0.75.
(5) Plot the two average VCRs on the UCE 2D scale to derive an overall task UCE (Fig. 7.28).

An important point in the derivation of the UCE is that the baseline aircraft used should exhibit Level 1 rate response characteristics in day visual conditions, i.e., before significant stability or guidance augmentation functions are added. This is to ensure that the VCRs are not corrupted by an aircraft's poor handling qualities. The UCE innovation is used in ADS-33 to identify the requirement for enhanced stability augmentation and/or display augmentation. The requirement is summarized in Table 7.4, illustrating the response types needed to confer Level 1 flying qualities in different UCEs. Thus, if an aircraft with its vision aids is required to operate in a UCE 3 then only the full augmentation provided by translational rate command/position hold (TRCPH) will confer Level 1. Alternatively, if the vision aids could be enhanced to improve the UCE from 3 to 2, then attitude command/attitude hold (ACAH) will be sufficient. If a UCE 3 could be upgraded to 1, through the provision of high-quality world scene cues inside the cockpit, then the lowest level of augmentation provided by rate command (RC) will be sufficient. We can now see the trade-off between vision, or guidance, aids and control, or stability, augmentation, and with both technologies advancing rapidly in the 1990s it is likely that a wide variety of options will be available on future types depending on the character of the tasks. For example, recovering civil or military helicopters to small ships in poor visibility will certainly require both improved guidance and stabilization aids if the operational risks are to be significantly reduced.

The basic substantiating data for the requirements of Table 7.4 came originally from flight test data on the Canadian Bell 205 in-flight simulator (Ref. 7.31). Tests were conducted with varying levels of stability augmentation while the pilot flew with night vision goggles, fogged to vary the UCE. Later, a more systematic piloted simulation investigation was performed on the NASA vertical motion simulator (VMS), designed specifically to explore the need for enhanced stability augmentation in DVE (Ref. 7.30). In Fig. 7.29, taken from Ref. 7.30, HQRs are shown plotted against response type for five low-speed MTEs flown in the sequence – hover, hover, vertical landing, pirouette, 2 accel/decels, sidesteps, hover, hover and landing – as shown in Fig. 7.30. The CGI visual scene was degraded to UCE 3 by fogging the far field and reducing the micro- and macrotexture in the near- to mid-field. The HQRs illustrate clearly how the workload reduces as the augmentation is increased, poor Level 2 ratings characterizing the RC response type in all MTEs. Level 1 ratings were given for the TRC in most of the MTEs, with the ACAH system generally lying in the good Level 2 region. The results were obtained without any visual display augmentation. In a series of similar, more recent, trials using the advanced flight simulator at the DRA, pilots viewed the UCE 3 world scene through a monochrome, bi-ocular HMD (Ref. 7.32) shown previously in Fig. 2.49. The outside world scene, with an allround field of regard, but only a $48 \times 36°$ field of view, was overlaid with different symbology sets to aid the pilots' stabilization and guidance tasks. HQR data for two DVE tasks from ADS-33, the recovery to hover and sidestep, are superimposed on the results in Fig. 7.29 for comparison. In the AFS trials, the UCE 3 was obtained with a combination of a sparse outside world scene and superimposed symbology; strictly speaking, the data cannot be directly compared with the VMS data, where the pilots flew with the outside world scene alone. Nevertheless, the data correlate very well and confirm the marked change in performance and workload with level of stability augmentation. One of the symbology sets evaluated in Ref. 7.32 is illustrated in Fig. 7.31 and is based on the horizontal situation display featured in the current generation AH-64A helicopter. We shall discuss this type of format in more detail later in this section. Results from the AFS trial have highlighted the importance of the height hold facility to ensure Level 1 ratings with the ACAH and TRC response types. Another result of the AFS trial questioned

Table 7.4 Response types for Level 1/2 handling qualities in different UCEs (Ref. 7.1)

	UCE =1		UCE =2		UCE =3	
	Level 1	Level 2	Level 1	Level 2	Level 1	Level 2
vertical takeoff and transition to forward flight – clear of earth	rate	rate	rate	rate	rate	rate
precision hover / slung load pickup and delivery / slung load carrying / shipboard landing including RAST recovery			ACAH + RCDH + RCHH →	rate + RCDH →	TRC + RCDH + RCHH + PH	ACAH + RCDH + RCHH
vertical takeoff and transition to near – earth flight / hover – taxi/NoE – travelling / rapid slalom[a]		→				
slope landing / precision vertical landing / pull-up/push-over[a]			ACAH + RCDH →			ACAH + RCDH
rapid bob-up and bob-down[a] / rapid hovering turn			ACAH + RCDH + PH →			rate + RCDH + PH
tasks involving divided attention operation / sonar dunking[b] / mine sweeping[b]	rate + RCDH + PH →					
rapid transition to precision hover[a] / rapid sidestep[a] / rapid accel. and decel.[a] / target acquisition and tracking[a,c]	rate →					

[a] high levels of aggressiveness may not be achieveable for UCE = 2 and 3.

[b] these tasks are normally accomplished in an environment where visual cueing may be consistent with UCE = 2 or 3 even in 'day VFR conditions'.

[c] increase in rank to TRC not recommended for pitch pointing tasks.

rate = rate or rate command attitude hold (RCAH) response type; TC = turn coordination; ACAH = attitude command attitude hold response type; RCHH = vertical-rate command with altitude (height) hold response type; RCDH = rate-command with heading (direction) hold response-type; PH = position hold response type; TRC = translational rate command response type.

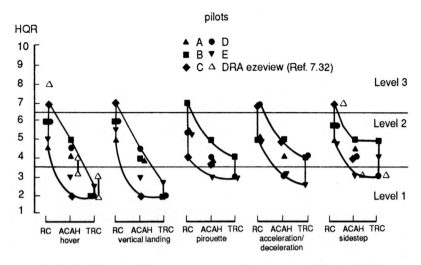

Fig. 7.29 HQRs for different response types flying various MTEs (Ref. 7.30)

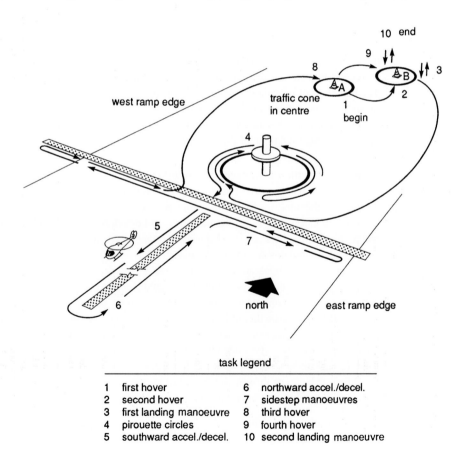

Fig. 7.30 Ten contiguous MTEs (Ref. 7.30)

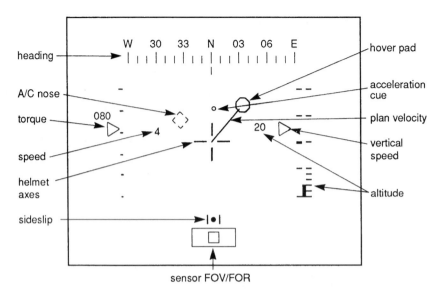

Fig. 7.31 Low-speed display symbology format used in the AH-64A Apache (Ref. 7.26)

the value of attitude bars on displays during very-low-speed MTEs, particularly when attitude stabilization is provided artificially, as in the ACAH and TRC response types. The bars did not provide significant improvements with the RC response type. Also, the dynamic cues provided by the moving bars could be distracting while not fulfilling a useful function. This introduction to the use of symbology to supplement the OVC brings us to the final topic of this section.

UCE augmentation with overlaid symbology

It should be obvious to the reader that NoE flight in the DVE makes considerable demands on the piloting task. The pilot needs support with both the stabilization and the guidance functions that he or she performs, and developments in display and control technologies have been so rapid during the late 1980s and early 1990s that flying qualities requirements for their effective use have been slow to catch up. Even as the chapters of this book are laid down the subject is expanding in several directions embracing and mixing the traditional flying qualities and human factors disciplines. It is therefore not appropriate to be definitive at this stage and as one of the contributors at a TTCP workshop on the subject (Ref. 7.33) remarked, 'the more we know about the subject the less we seem to understand'. This old adage seems particularly germane to the present topic and is enhanced by the apparent notion that every pilot would prefer to design the display in his own image. Stepping back from the detail we can review the different kinds of information required by the crew during a DVE mission, and which are candidates for superposition on an outside world scene:

(1) primary flight path information for guidance and stabilization, including speed, height, attitude, heading, etc;

(2) guidance information related to special tasks, e.g., recovery to ship, pathway in the sky for flying NoE, target acquisition and weapon aiming;

(3) flight envelope and carefree handling cueing;

(4) aircraft system status, e.g., engine torque, AFCS modes;
(5) situational (tactical) awareness data, e.g., navigation – Where am I?, hazards – Where are the obstacles or the threats?

Any attempt to cram all the above information on to one display will quickly lead to crew overload. One of the basic messages in display design is to provide information only as and when it fulfils a useful function, hence increasing situational awareness or task performance. Unfortunately, the achievement of maximum situational awareness seems at times to be at odds with achieving a specific task performance. Put another way, well-designed displays can help to recover the full OFE, previously reduced for flight in the DVE, but only at the expense of high pilot workload caused by poor fields of view, poor resolution and the increased potential for spatial disorientation. There are times when the crew need to gaze down a narrow field-of-view 'soda straw' and see fine detail with precision symbology, and others when they need to scan continuously a 220° field-of-view scene overlaid with guidance symbology cues. Each makes different demands on the display technology but, ultimately, field of view and symbology content, like many display attributes, need provide only the functionality required for a given task.

A good example of how symbology can be designed to aid specific tasks is provided by the AH-64A format shown previously in Fig. 7.31, in the so-called hover 'pad capture' or 'bob-up' mode. We supplement Fig. 7.31 with Figs 7.32 and 7.33, taken from Ref. 7.26, showing how the pilot uses this particular display to position the aircraft in the very-low-speed regime. The display is intended to aid the pilot maintain an accurate hover in a DVE. By 'flying' the acceleration cue into the hover pad and 'flying' both the cue and pad to the fixed aircraft reticle, the pilot is able to achieve a hover at a prescribed location, defined by the hover pad. Other flight data on the display include the heading, height and rate of climb and airspeed. The velocity vector indicates the aircraft's horizontal inertial velocity. The box in the lower portion of the display shows the pilot where the FLIR image on the monocular display is positioned in the field of regard of the FLIR sensor. The three plan features that dominate the pilot's attention during the pad capture MTE are shown in Fig. 7.33 and consist of the velocity vector, the acceleration cue and the hover pad. The 8-ft hover pad is driven on the display relative to the fixed reticle. The velocity vector has a full scale deflection of 12 ft/s, hence the display is intended to support very-low-speed manoeuvring to recover to the hover over a pilot-selected plan position. This display format provides

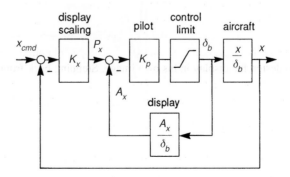

Fig. 7.32 Pilot–vehicle display block diagram (Ref. 7.26)

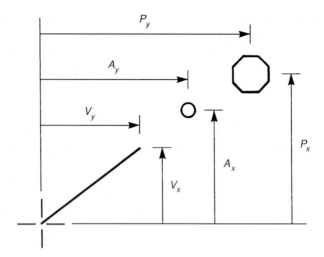

Fig. 7.33 Central symbology variables in AH-64A display format (Ref. 7.26)

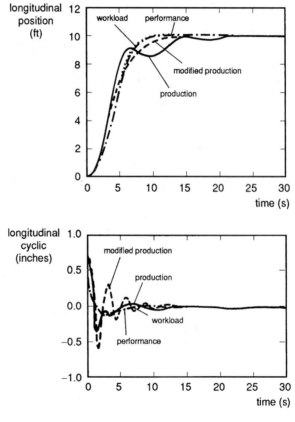

Fig. 7.34 Comparison of control inputs and aircraft responses for various display
dynamics (Ref. 7.26)

insight into how display dynamics can play a critical role in the overall achievement of task performance. Display dynamics are represented by the A_x/δ_b (translational acceleration to stick) transfer function in Fig. 7.32, defining the dynamic relationship between the acceleration cue (A_x) and the pilot cyclic stick (δ_b). As pointed out in Ref. 7.26, a pure gain display law would be the easiest to control but would be likely to lead to poor hover performance. At the other extreme, driving the acceleration cue with the trajectory demand would enable an improved task performance but would certainly require the pilot to work a lot harder. In the production version of the Apache display laws, a compromise is struck with a blend of the two. Reference 7.26 discusses three alternative drive law designs, designated the modified-production, performance and workload designs, that appear to offer significant improvements over the production version, principally by tailoring the response characteristics at frequencies above 2 rad/s, where the pilot will be working to make small and precise position corrections. Figure 7.34 shows the time responses of the aircraft horizontal position and longitudinal cyclic derived from an analytic simulation of Fig. 7.32. The theoretical predictions of improved performance and reduced workload are apparent. The results of an extensive piloted simulation exercise are also reported in Ref. 7.26, where ten pilots evaluated the different designs and concluded that the performance and workload designs were far superior. Figure 7.35 shows the HQRs for the different designs, indicating an improvement from poor to good Level 2 ratings for the pad capture task, with very high statistical significance.

The design of the display laws are clearly dependent on the response characteristics of the aircraft, as indicated conceptually in Fig. 7.32. For example, in the simulation trials conducted at DRA and reported in Ref. 7.32, the Ames display law

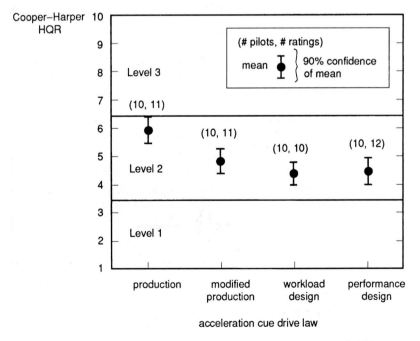

Fig. 7.35 Comparison of HQRs with various display dynamics (Ref. 7.26)

design methodology (Ref. 7.26), using the performance variant, was applied to derive laws that were compatible with the three response types evaluated, RC, ACAH and TRC. The general principle is to tailor the acceleration cue response to aircraft position *x*, attitude θ and cyclic δ, through transfer functions given by the general form

$$\frac{A_x}{\delta_b}(s) = f_{\dot{x}}(s)\frac{\dot{x}}{\delta_b}(s) + f_{\theta}(s)\frac{\theta}{\delta_b}(s) + f_{\delta_b}(s) \qquad (7.8)$$

by matching to desired response and cancelling unwanted vehicle dynamics from the cue motion. The principles for achieving different design goals are described in Ref. 7.26. As with control law design for tailoring the aircraft response characteristics, display law design can be partially completed off-line using linear control techniques, but the final optimization still requires piloted evaluation. This example highlights some of the important integration aspects between displays that support pilotage and the response characteristics of the aircraft, hence any automatic guidance and control augmentation. Several other examples of emerging display formats for supplementing OVCs are discussed in Ref. 7.33 and indicate the potential of things to come, but are perhaps only stepping stones towards the tenuous aviation concepts of virtual reality and the computerized Pilot's Associate that performs all the mechanical aviation functions, including pilotage, leaving the mission manager to direct the operation.

Flight in the DVE and agility represent the extremes of operation and obviously have a significant, although not exclusive, military relevance. Our third topic, carefree handling, is applicable to both civil and military operations, although once again the leading edge research has been forged by military requirements.

7.3.3   Carefree flying qualities

A survey conducted with UK operational military pilots from all three services during the 1980s concluded that some 40% of the piloting workload derived from the need to monitor aircraft and flight envelope limits (Ref. 7.34). Some 70 pilots completed questionnaires in the survey and seven different aircraft types were addressed. One of the questions enquired as to which limits were the most demanding on pilot workload. From the response it was clear that the top two limits were engine/gearbox torque, selected by 75% of pilots, and rotorspeed, selected by about 60% of pilots. Some of the limits considered were actual limits, i.e., the pilot refers to an instrument showing the critical flight parameter with appropriate green and red zones, e.g., torque, engine temperature, rotorspeed. Others were derived limits, with parameters displayed on instruments giving essentially kinematic information about the aircraft state. Examples in this category are airspeed (reflecting rotor and fuselage loads), bank angle in steady turns and normal acceleration (reflecting rotor fatigue loads and static strength). Some limits are not normally presented to the pilot at all, e.g., sideslip and lateral velocity (reflecting rear fuselage strength) and yaw rate (reflecting tail rotor gearbox torque). The study reported in Ref. 7.34 also solicited pilot opinion of the potential value of different types of system that might assist in the monitoring and respecting of limits. The class of such systems was described as carefree handling systems and included head-up/down visual cues, audio cues, tactile cues and direct intervention control systems, with and without pilot override. The majority of pilots believed that the display of flight envelope limits on a helmet-mounted device would satisfy most of their concerns and would be effective in reducing the monitoring workload. Equally, the majority of pilots considered that direct intervention control systems without pilot override would not be acceptable – about half of the pilots interviewed rated the potential effectiveness of such systems as zero. This last point should be placed in the context of fixed-wing aircraft experience, where most of the carefree handling features are in the direct intervention class; the only way the pilot has to override them is to turn them off. The results of this review of current perception and practice spawned a UK research activity into the functional attributes of helicopter carefree handling systems, which is ongoing at the time of writing. Selected results from the study are presented below.

Carefree handling is a concept very familiar in the fixed-wing world, with Tornado, F-16, F-18 and Airbus A320 all featuring some form of system that protects the aircraft from exceeding limits; in general, as noted above, such systems cannot be overridden by the pilot unless he turns them off. The principal reason for this is that the protected limits are bounding regions where there is a high risk of loss of control, e.g., deep stall on the F-16, stall and spin departure on Tornado. In contrast, with helicopters, most of the limits are associated with structural considerations and, generally speaking, overstressing is preferable to hitting the ground. It is convenient to classify helicopter limits into four categories, as shown in Fig. 7.36, and related to structural/aerodynamic loads on the rotor or fuselage, engine speed and temperature, transmission loads and loss of control. For all but the last category (which has dominated the fixed-wing experience), the limits can be described as soft or with varying degrees of hardness. For example, a gearbox transient torque limit can be exceeded as the pilot pulls up to avoid an obstacle; permanent damage may have been done and the gearbox may need replacing, but the aircraft and crew have survived. It is for this reason that helicopter pilots, almost unanimously, are unwilling to accept carefree handling without an override capability. Before discussing the research efforts in this area it is important to give brief attention to

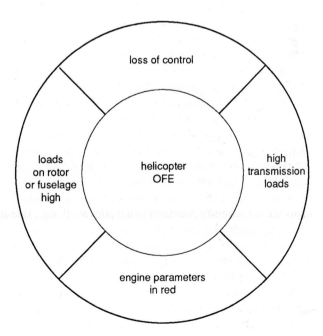

Fig. 7.36 Sources of flight envelope limits

Table 7.5 Carefree handling features evaluated in Ref. 7.34

Configuration	Carefree handling features
1	none
2	visual warnings on HUD master warning light on coaming collective shaker for torque limits pedal shakers for lateral airspeed
3	monitoring of hub moment audible warnings for all limits
4	direct intervention protecting all limits manoeuvre limiting based on hub moment and g tactile cueing via hard-stops on sticks
5	torque demand protecting torque and rotorspeed direct intervention to protect limits manoeuvre limiting based on g tactile cueing via feel forces

potential loss of control regimes in helicopters. Vortex ring can be as severe for helicopter pilots as stall is for fixed-wing pilots, and is a definite inhibition to manoeuvring vertically at low speed. Similarly, loss of tail rotor control can lead to a period of uncontrolled yaw motion which can be disastrous in confined areas. Both these examples require good knowledge of the aircraft's velocity relative to the air, which is notoriously inaccurate at low speed. A third example where the helicopter's flight envelope is limited by control problems is at low normal 'g', which is a particular concern for teetering rotors; control power can reduce to zero or even reverse at negative 'g'. Helicopters have been lost because of excursions into unsafe control regions, and these corners of the flight envelope should not be neglected in the striving for safe and carefree handling.

In the study reported in Ref. 7.34, four combinations of different carefree handling features were trialled on the ground-based flight simulator at RAE Bedford. Table 7.5 lists the features evaluated. Configurations 4 and 5 featured direct intervention carefree handling. The error between the aircraft flight state and flight envelope limit was continuously estimated from measurements and triggered high-gain feedback control as the limit was approached. Warning systems included visual cues on a head-up display, audio tones and tactile cues fed through the variable force–feel control system. The vertical axis included a torque command system (TCS), as an alternative to the direct drive collective.

Six test pilots participated in the trial which included eight MTEs, designed to exercise the limits in both single and combined ways. The results of this study were quite illuminating. As predicted by the pilot opinion survey, protection of rotor torque and rotorspeed was valued the most. Contrary to the results of the pilot opinion reviews, the presentation of visual warning (flashing) cues on the head-up display (HUD) did not improve performance in the selected MTEs (configuration 2). Typically, pilots could be distracted by the visual warnings or even ignore them in high workload situations, hence limit transgressions were typically as numerous and high as without any carefree handling features (configuration 1). Of the warning systems, both audio and tactile were judged to be useful, because they reduced pilot workload although they still demanded pilot corrective action following the approach to, or exceedance of, the limit. The direct intervention systems scored the highest in terms of performance improvement and workload reduction, even when tactile cues, in the form of hard stops on the controls, inhibited the pilots from pulling through. On balance, the configuration with soft stops and stiffening control forces was preferred because pilots were more confident that the excess performance was available, if required.

Figure 7.37 shows results for the 100-m sidestep manoeuvre. Mean HQRs are plotted against task time for the five configurations tested, showing the marked

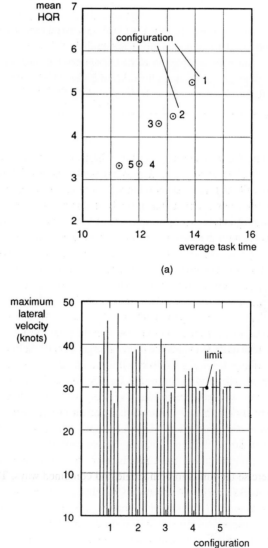

(a)

(b)

Fig. 7.37 Comparison of simulation results with different carefree handling systems (Ref. 7.34): (a) mean HQRs; (b) peak lateral velocity excursions

improvement in task time and reduction in workload as the direct intervention systems are introduced (Fig. 7.37(a)). The principal flight limit of interest in this MTE was the lateral velocity, set at 30 knots, shown in Fig. 7.37(b). With only the warning systems, 30% limit exceedances were typical, while the direct intervention system held the limits to within 10%. The achievement of marginal Level 1/2 HQRs for the MTEs flown with high levels of aggression on a ground-based simulator was a significant achievement when these trials were conducted and was attributed to the truly carefree manner in which the pilots were able to fly the tasks. The TCS was also well received by pilots and few cases of overtorquing occurred, even in multi-axis manoeuvres like the slalom and accel–decel. However, pilots did complain that the TCS appeared to reduce vertical axis performance, compared with the direct drive. The aircraft model used in the simulation was the RAE's CSM, discussed earlier in this chapter, that possessed good Level 1 rate command flying qualities.

Since the completion of the UK conceptual studies into carefree handling, a number of extensions and applications to different aircraft types have been accomplished. In Ref. 7.35, the problem of torque control received primary attention, applied to a simulation of the Bo105 helicopter. The sluggish vertical response characteristic of a TCS was demonstrated to be an inherent feature of the linear deadbeat torque response. The key to resolving the conflict, and thus conferring both crisp height rate response and torque command, lay in an innovative control law design technique that effectively varied the control law gains and structure as the limit was approached. Figure 7.38 illustrates the torque and height rate responses to small and large collective pulse inputs in the improved TCS design. For the small, 10% pulse, the crisp height response is accompanied by a 50% torque overshoot, which would, of course, be unacceptable if the test input had been applied closer to the maximum transient torque limit. For the larger input, demanding 50% torque, the height rate is constrained as the torque is held at the limit. For hands-off collective operation, a height hold, trim follow-up function automatically backdrove the collective to give the desired torque demand. This design was successfully trialled in a simulation on the DRA's advanced flight simulator (AFS) with pilots flying air-to-air (ATA) target tracking and terrain following MTEs. The ATA MTEs are illustrated in Fig. 7.39, with the pilot's task being to turn, climb and accelerate to acquire and track the moving target aircraft. Selected results from the simulation are illustrated in Fig. 7.40, showing a comparison of task time (a), transmission torque (b) and pilot HQRs (c) for the Bo105 with and without the torque carefree handling system. The carefree handling system enabled the target to be acquired 20% sooner, virtually eliminated unintentional limit transgressions and conferred Level 1 flying qualities on an otherwise Level 2 aircraft. The baseline aircraft simulated in this study was a Bo105 with full-authority active control system having solid Level 1 handling according to the ADS-33 criteria.

The results of the UK simulation programme appear quite convincing regarding the benefits of carefree handling qualities, at least for military operations, where the requirement to use the full performance potential of the helicopter on a regular basis is clear. In comparison, for civil operations there is no requirement for pilots to fly close to envelope limits, except in emergencies. It is therefore likely that the military application will continue to drive the enabling active control technologies; improved safety in civil operations will almost certainly be a fallout however. One of the findings of the results to date is that, given the safety of operations at the limits, pilots can be expected to fly there more often and, hence, aircraft incorporating carefree handling may well be

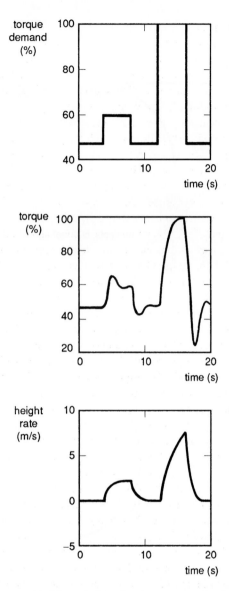

Fig. 7.38 Torque and height variations showing response shaping (Ref. 7.35)

exposed to more damaging fatigue usage. In this context, carefree handling will almost certainly need to be integrated with a fatigue usage monitoring system. The positive side to this additional complexity is that pilots, together with the carefree handling associate, can learn to fly with less damaging control strategies, if required. There will always be trade-offs involved, this time between performance and structural integrity, but the pilot should be able to make the decision which way to play the weightings, in any given situation.

One aircraft where a degree of carefree handling has been incorporated into the development programme is the Boeing-Bell V22 Osprey tiltrotor. Reference 7.36

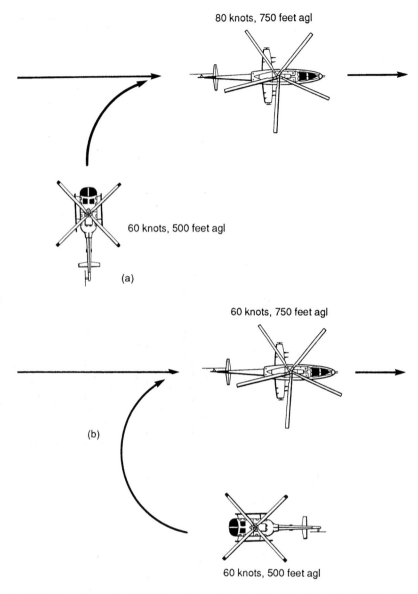

Fig. 7.39 Air-to-air combat MTEs flown in carefree handling simulation (Ref. 7.35): (a) 90° turn, climb and accelerate to acquire target; (b) 180° turn and climb to acquire target

describes a number of innovative features aimed at protecting the aircraft from the effects of structural load exceedances, including

(1) limiting the rotor disc angle of attack during high load factor manoeuvres, using elevator to reduce the blade stall on the high disc loading rotor (helicopter mode);
(2) reduction of transient rotor flapping and yoke chord bending loads during aggressive pitch manoeuvres through limiting of high-frequency rotor commands (helicopter mode);

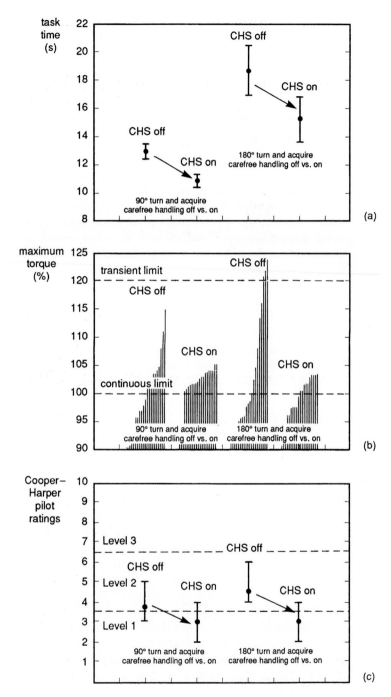

Fig. 7.40 Comparison of results with torque command carefree handling system on and off (Ref. 7.35): (a) task time; (b) maximum torque; (c) handling qualities ratings

(3) reduction of trim rotor flapping with elevator control (helicopter mode);
(4) reduction of transient mast and driveshaft torques in roll manoeuvres through roll rate feedback to differential rotor collective (airplane mode);
(5) prevention of nacelles from lifting off the downstop during aggressive roll manoeuvres through roll acceleration limiter (airplane mode);
(6) reduction of oscillatory yoke chord bending in pitch manoeuvres through tailoring pitch response characteristics.

The functionality of these design control law shaping features has been verified in simulation as reported in Ref. 7.36, highlighting that the loads in worst cases have been contained within the design limit loads, with an almost insignificant effect on handling qualities when flying the V-22 MTEs.

It could be argued that helicopters should be designed so that the flight limits are outside the capability of the aircraft, providing it remains within the OFE. Then the flying qualities engineer would not need to be concerned with artificial aids, and the aircraft would possess natural carefree handling. The pilot could never overtorque the gearbox, droop the rotor, pull too much g or exceed the sideslip or sideways velocity limits. The problem is that achieving this multi-objective design goal is actually very difficult, if not impossible, and with the classical helicopter design, the large control ranges to trim throughout the speed range provide sufficient control power at most flight conditions to inadvertently exceed one or other limit. With the tilt rotor, achieving a balanced design appears to be even more difficult and this aircraft has demonstrated that true carefree handling, where the full performance is not inhibited for safety reasons, will come only through the application of active flight control.

7.4 PILOT'S CONTROLLERS

Evaluation of a helicopter's flying qualities for a particular role will include an assessment of the mechanical characteristics of the pilot's controls. No matter how good the response characteristics of the aircraft are, the overall flying qualities will be judged by the quality of the operation of the pilot's controls reflected in a range of design features including cyclic self-centring, breakout forces and force gradients, deadbands and trimming actuators. Breakout forces that are too high, for example, can inhibit the pilot from making small, precise changes in flight path, and sluggish hydraulic systems can impede manoeuvrability. Slow trim motors can increase pilot workload and control force gradients that are too light or too strong can spoil the use of smooth control action by the pilot. Nowadays there should be little controversy about what constitutes good controller characteristics when discussing central cyclic control sticks with large displacement; most medium to large helicopters have featured such devices with fixed-stiffness centring springs and operating through hydraulic actuation systems, for several decades. The range of acceptable characteristics needs to be fairly broad as the control forces generally have to be harmonized with the response characteristics. Figure 7.41 illustrates the general form of the control force/displacement relationship showing maximum and minimum values for the breakout force levels and gradients. The accompanying table gives values for Level 1 qualities as defined by ADS-33 (Ref. 7.1) and Def Stan 00970 (Ref. 7.37). Notable differences are in the roll force gradients and the maximum tolerable breakout forces, which appear to reflect traditional preferences in the different countries' armed services. Unlike fixed-wing aircraft, helicopters

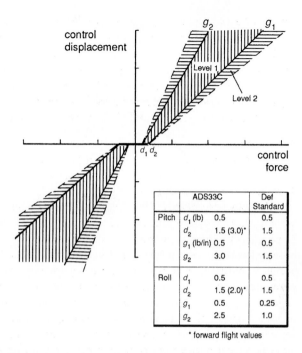

Fig. 7.41 Control force versus control displacement for centre-sticks (Refs 7.1, 7.37)

		ADS33C	Def Standard
Pitch	d_1 (lb)	0.5	0.5
	d_2	1.5 (3.0)*	1.5
	g_1 (lb/in)	0.5	0.5
	g_2	3.0	1.5
Roll	d_1	0.5	0.5
	d_2	1.5 (2.0)*	1.5
	g_1	0.5	0.25
	g_2	2.5	1.0

* forward flight values

do not usually include any artificial feel augmentation to cue the pilot in manoeuvres. However, a new requirement in ADS-33 for achieving Level 1 handling is for the stick force per *g* to lie between 3 lb (13 N) and 15 lb (67 N) per *g* (Ref. 7.1).

All current operational helicopters are fitted with a conventional cyclic centre-stick, collective lever and pedals, with a wide variety of different mechanical characteristics, reflecting the varying design preferences and, ultimately, pilot adaptability. Future helicopters with fly-by-wire control systems are likely to feature integrated side-stick controllers, and during the period between the late 1970s and late 1980s much of the basic research was undertaken to explore the potential of such devices (Refs 7.38–7.41). A key initial concern was whether equivalent handling qualities and performance could be achieved with sidesticks, considering the high levels of cross-coupling in helicopters. Other issues related to the required level of stability and control augmentation to enable satisfactory performance with sidesticks, the trim mechanisms, grip designs and force–feel characteristics. The review paper of in-flight simulation activities at the Canadian Flight Research Laboratory (FRL), Ref. 7.41, offers the most comprehensive and coherent insight into what is currently known about passive sidesticks for helicopters and forms the basic material for this short discussion. Sidestick control is now generally recognized as being a step in the right direction for helicopters and entirely commensurate with the development of fly-by-wire control. Significant improvements in cockpit ergonomics, including a dramatic influence on seating posture, relieving pilot fatigue and lower spinal damage, coupled with the potential for greater precision through integrated hand movements, are powerful reasons for their application.

One of the first issues to be encountered with helicopter sidesticks was concerned with the need for command shaping to tailor the control sensitivity. Basically, to provide

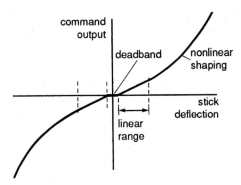

Fig. 7.42 Typical nonlinear shaping function for sidestick controllers

the same levels of control power as with conventional controllers, the stick to response gearing would need to be nonlinear with a steepening gradient (Fig. 7.42). Linear gearing results in the control sensitivity being too high for small amplitude control inputs, giving rise to a strong tendency to overcontrol. A major question regarding the use of sidesticks has been how many control functions should be included in a single inceptor – 2, 3 or 4. Another question is related to whether there is a strong preference for displacement or force sensing. These two issues are not unrelated. Reference 7.41 reports pilot preference for a separate collective with a displacement sidestick, compared with no preference between 4 + 0 and 3 + 1 (collective) with a force sensing stick. The Canadian studies have demonstrated that four-axis control is entirely feasible for low–moderate gain tasks. For higher levels of aggression and higher bandwidth helicopters than the FRL Bell 205, little flight data have been published in the open literature, but it is suspected that 3 + 1 (collective) will be the extent of the integration. The Canadian research confidently proclaims that, to quote from Ref. 7.41:

> *The studies. . .consistently suggest that there is no evidence that the use of integrated side-mounted controllers in a helicopter. . .is detrimental to the overall handling qualities of the vehicle, nor that they demand of the pilot any unusual or exceptional skills in their use. Neither handling qualities nor pilot performance should control decisions as to the use of integrated sidesticks in helicopters.*

For any specific application, however, there is an insufficient database to draw firm guidelines on the many design issues involved – the optimum force characteristics, grip shape and orientation, etc. ADS-33 reserves the section on sidesticks for future requirements. Two future projects are already committed to sidesticks. In Ref. 7.42, the design for the RAH-66 sidestick is referred to in passing as a three-axis (roll, pitch and yaw) sidestick with limited vertical axis capability, used in conjunction with the altitude hold function. In Ref. 7.43, the design of the NH-90 is shown to include a more conservative two-axis sidestick. Both will be passive, in the sense that the force characteristics will be fixed. New research into active sidesticks is underway at the time of writing. These should enable tactile cueing and more general, variable force–feel characteristics, tailored to the changing response types in an actively controlled helicopter.

7.5   THE CONTRIBUTION OF FLYING QUALITIES TO OPERATIONAL EFFECTIVENESS AND THE SAFETY OF FLIGHT

The two overriding considerations for both civil and military rotorcraft operations are to achieve good performance at low workload. Like stability and agility, these dual aims can often conflict. Typically, military operations are characterized by achieving performance goals as a priority, while civil operations are biased towards safety. In any mission or operational situation, the pilot will make the tactical decision of which to favour but the tension between performance and safety is ever present. In considering the contribution of flying qualities to effectiveness and safety, this tension forms a centrepoint of the discussion. The HQR scale measures pilot workload required to achieve a defined performance, hence giving an indication of the safety margins available. HQRs are explicit measures of pilot workload and implicit measures of aircraft stability and control characteristics. However, there appears to have been very little work done on the operational benefits using the HQR approach. For example, how much more mission effective is a Level 1 than a Level 2 aircraft when, for example, the pilot is stressed due to poor weather or the need for rapid action? Generally, and in objective terms, the value of good flying qualities should be reflected in three principal areas:

(1) productivity – how many missions or sorties can be accomplished;
(2) performance – how well can each sortie be accomplished;
(3) attrition – how many losses can be expected.

We will examine these issues within the framework of a probabilistic approach along the lines first put forward in Refs 7.44 and 7.45 and later developed in Ref. 7.25. The basic notion is that flying qualities deficiencies increase the chance of pilot error, hence can lead to accidents, incidents or MTE failures. This is a controversial concept. A significant proportion of accidents and incidents are attributed to human error, but there is often a counter-argument put forward that suggests some deficiency in the aircraft's handling qualities. In this context, Refs 7.44 and 7.45 considered the benefits to flight safety using the Cooper–Harper pilot rating scale as a metric (Fig. 7.2). These references considered the pilot as a vital system component who can fail (i.e., be stressed to failure) in an operational context, just like any mechanical or electrical component. Pilot failure can be manifested in MTE failure, corresponding to HQRs > 6.5 or, in the extreme, a loss of control, corresponding to a HQR > 9.5. We have already discussed on several occasions the variability of flying qualities with both internal attributes and external factors. In the life of an aircraft, there is a finite probability that 'virtual' ratings across the whole range will be experienced. We refer to these as virtual ratings because in reality they are not awarded; one can imagine, however, an HQR meter, sampling workload and pilot-set performance targets. For every distinguishable MTE that is flown, the HQR meter takes a recording. Examples might be as given in Table 7.6.

The next assumption we make is that over a long period of time the distribution of the virtual ratings takes a normal form as shown conceptually in Fig. 7.43. The regions of desired, adequate and inadequate performance are clearly identified. The desired and adequate regions can be considered as reflecting varying degrees of MTE success, while the inadequate level corresponds to MTE failure. Effectively, each mission is composed of a number of contiguous MTEs, each having its own virtual HQR. If a particular MTE

Table 7.6 Possible HQRs for same aircraft in different MTEs

Handling qualities rating	Mission task element
1	cruise on autopilot
3	landing in confined area in GVE
4	sidestep during NoE training flight
5	landing on ship in gusty conditions
6	landing in confined area in DVE
7	turn to acquire target, but target out of range
9	deliver underslung load in gusty DVE
10	crash following tail rotor failure

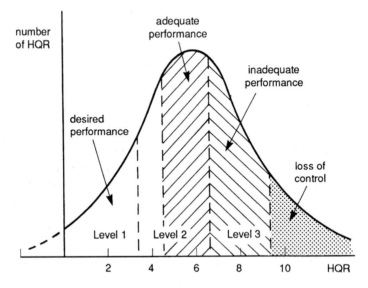

Fig. 7.43 Notional distribution of pilot HQRs for a given aircraft (Ref. 7.25)

was assigned a Level 3 rating, then the pilot would have to either try again or give up on that particular MTE. Loss of control has obvious ramifications on mission success. For certain types of operation, loss of control will almost certainly result in a crash. The probability of obtaining a rating in one of the regions is proportional to the area under the distribution in that region. Note that, as discussed in Ref. 7.25, we include ratings greater than 10 and less than 1 in the analysis. The rationale is that there are especially bad and good aircraft or situations, whose qualities correspond to ratings like 13 or −2. However, the scale enforces recording them as 10 or 1.

Note too that the scatter produces, even with a good mean rating, a large probability of merely adequate performance and even a finite probability of total loss of control and, in some cases, a crash. We have said elsewhere in this book that flying qualities are determined by the synergy between internal attributes and external influences. It follows then that sources of scatter originate both internally and externally. Internals include divided attention, stress and fatigue, pilot skill and experience. Externals include atmospheric disturbances, changing operational requirements and timelines, threats, etc. The flying qualities community has done much to minimize scatter by careful

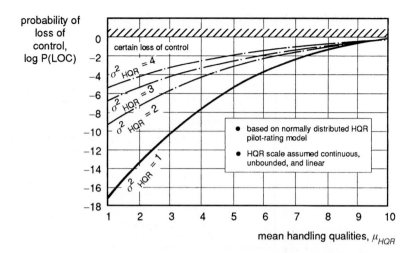

Fig. 7.44 Relationship between mean HQR and P(LOC) (Ref. 7.44)

attention to experimental protocol (Ref. 7.6), but here we emphasize that in operational environments, the effective pilot rating scatter is omnipresent.

With the assumed normal distribution of ratings, the probability of control loss, P_{loc}, can be calculated for various mean ratings and dispersions; these are plotted in Fig. 7.44. P_{loc} is the probability of obtaining a rating greater or worse than 9.5, which in turn is simply proportional to the area under the distribution to the right of the 9.5 rating. Thus, the probability of loss of control (i.e., flight failure) due to flying qualities deficiencies can be estimated. For the cases studied in Refs 7.44 and 7.45 and depicted in Fig. 7.44, operating a Level 1 aircraft can be seen to reduce the probability of a loss of control by an order of magnitude relative to a Level 2 aircraft. Interestingly, the P_{loc} of an aircraft with a mean HQR of 3.5, on the Level 1/2 boundary, is 1 in 10^9, the value quoted for flight critical component reliability in civil transports.

If we now consider the same approach applied to the full extent of the rating scale, the effectiveness in terms of MTE success or failure can be estimated. Figure 7.45 shows the probability of obtaining ratings in the various regions when the standard deviation of the ratings is unity. This curve has some interesting characteristics. First, the intersections of the lines fall on the ratings 4.5, 6.5 and 9.5, as expected. Also it turns out that for a mean rating of 7, the probability of achieving inadequate performance is, of course, high, and we can also see that the probability of achieving desired performance is about the same as that for loss of control – about one in a hundred. Improving the mean HQR to 2 lowers the probability of loss to 10^{-13} (for our purposes zero) and ensures that performance is mostly at desired levels. Degrading the mean rating from 2 to 5 will increase the chances of mission failure by three orders of magnitude.

If we consider the above results applying to a fleet of 100 of the same aircraft type, some interesting statistics begin to emerge. We assume that each aircraft in the fleet flies one mission per day, and each mission comprises 20 MTEs. Over a 20-year period the fleet will fly about 15×10^6 MTEs. If we assume that loss of control equates to loss of an aircraft, then Fig. 7.45 provides information on the expected losses due to flying qualities deficiencies over the life of the fleet. For an aircraft with a mean

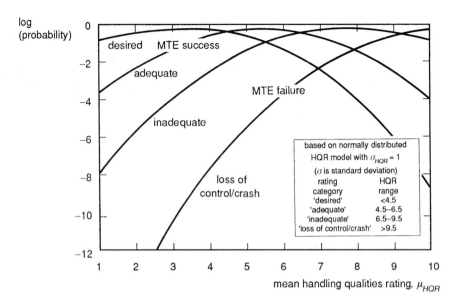

Fig. 7.45 Relationship between mean HQR and probability of mission success, failure and loss of control (Ref. 7.25)

HQR of 5, the fleet can expect to lose one aircraft per year (i.e., 20% of the fleet over the fleet life). With a mean HQR of 3, no aircraft will be lost from poor handling qualities during the life of the fleet. It is likely that most operational aircraft in service today do not have mean HQRs in the Level 1 region, because of limited stability and control augmentation, the poor cueing of flight envelope limits and the degraded flying qualities associated with failures and in emergency situations. It would be inappropriate to discuss data on particular types in this book, but these preliminary results give some cause for concern; on the other hand, they also offer a methodology for quantifying the value of good flying qualities.

We describe these results as preliminary because we assume that there is a rational continuum between desired performance, adequate performance and control loss. For example, desired and adequate performance may be represented by discrete touchdown zones/velocities on the back of a ship, and loss of control might be represented by, say, landing on the edge of the ship or hanger door. On a smaller ship (or bigger helicopter, for example), the desired and adequate zones may be the same size, which puts the deck-edge closer to the adequate boundary, or represent a similar fraction of the deck size, hence tightening up the whole continuum. This raises some fundamental questions about the underlying linearity of the scale. Assuming these issues can be resolved satisfactorily, there is also the question of how to establish the mean rating for a particular aircraft type. This could, in principle, be estimated from a series of tests as outlined in Chapters 6 and 7 of this book, but data in the most degraded conditions in which the aircraft is expected to operate will need to be captured.

Though these questions remain, pilot rating and mission success or failure are powerfully related through the preliminary data in Fig. 7.45. Put simply, flying qualities alone can determine whether operational agility and flight safety are flawless or whether control is lost. Flying qualities are at the heart of the subject of Flight Dynamics and, through a holistic approach, this book has attempted to establish the intimate connection between the theoretical foundations of modelling and handling criteria and the operational world of flight safety and performance.

Nowhere is safety more prominent an issue as when handling qualities degrade, as a result of loss of visual cues, loss of some flight control function or when the aircraft is exposed to the effect of severe atmospheric disturbances; this is the subject of the last chapter of this book.

*A Merlin Mk 3 in the desert creating its own
Degraded Visual Environment*

8   Flying qualities: forms of degradation

The future of the helicopter is immense and later the craft will be a very familiar sight in the air to everyone. It will also be capable of rendering a great number of services which no other craft can render, and can be described as the greatest friend in need in the case of an emergency.
Igor Sikorsky at the end of his lecture 'Sikorsky Helicopter Development' presented to the Helicopter Association of Great Britain at Mansion House, London, on Saturday 8 September 1947

8.1   INTRODUCTION AND SCOPE

As Sikorsky foretold with confidence and optimism 60 years ago in Ref. 8.1, the helicopter would indeed serve mankind as a 'friend in need', but as often happens, its unique capability would be usable only by pilots exercising very high levels of flying skills, and, in dangerous and emergency situations, by pushing both safety and performance to the limits. Sikorsky talked in his lecture about the significance of the helicopter in rescue service. He recounted a recent occurrence with 'satisfaction and great encouragement', to quote:

> *The police rang up the factory to say that an oil carrying barge with two men on board was in distress and was starting to disintegrate, water sweeping over the surface of the barge. We immediately dispatched a helicopter with a hoisting sling and in spite of a wind of 60 m.p.h and gusty, the helicopter quickly reached the barge and was able to hover 20–25 ft. over it, lower the hoisting sling and take the men off, one after another. The rescue was made as the end of the day was approaching and the general consensus of opinion was that these two would certainly not have been able to stay on the damaged barge overnight.*

Igor Sikorsky presented the lecture published as Ref. 8.1 just a few years after the birth of the practical helicopter. He talked about '. . . absolute accuracy of the control' and '. . . control as perfect as any other system of control.' Today, we can only try to imagine the motivation, the courage and the optimism of the early pioneers as they shaped the first vertical flight machines with four axes of control. A few months later, on 19 April 1948, as reported in Ref. 8.2 and discussed in the Introductory Tour to this book (Chapter 2), a Sikorsky S-51 during a test flight at the Royal Aircraft Establishment would almost crash as the pilot momentarily lost control during a high-speed (4 g) pull-out and inadvertent rapid roll to 90° of bank. The other side of the coin, so to speak, was experienced with the consequences of degraded handling qualities. Helicopter control, while qualitatively precise, would always require close pilot attentiveness and relatively high workload.

Chapter 7 ended with a discussion on the impact of flying qualities on safety and mission effectiveness. The twin goals of safety and performance, with the consequent tension between them, have pervaded the whole business of aviation since the Wright brothers' first flight in December 1903. In the helicopter world, the performance–safety tension is perhaps strongest when flying close to the surface with what is sometimes described as mission imperative, or at the edges of the operational envelope, in harsh environments, or when the pilot has to deal with flight system failures. When flying close to the surface, the first priority for the pilot is to maintain a sufficient margin of 'spatial awareness' to guarantee safe flight. This spatial awareness also has a temporal dimension; the pilot is actually trying to predict and control the future flight trajectory. We can imagine a pilot flying to maintain a safe time margin, avoiding obstacles and the ground, with a relaxed control strategy allowing plenty of time for navigation and monitoring aircraft systems. The pilot will want to maintain a sufficiently long 'time to encounter' between the aircraft and any potential hazard, so that there is ample time to manoeuvre around, climb over or even stop, if required. But external pressures can make things more difficult for the pilot, increasing the workload. Imagine that the task is to transit, within tight time constraints, to deliver an underslung load to a confined forest clearing at night, with the threat of enemy action. Under relentless time pressures, the pilot has some scope for trading off performance and workload, depending on the requirements of the moment. He or she will be forced to fly low to avoid detection by the enemy. Increasing the tempo at low level reduces the safety margin; more precision or more agility requires higher levels of concentration on flight path guidance and attitude stabilization. The more the pilot concentrates on flight management, the more the global situation awareness is compromised with increased risk of getting lost or becoming disconnected with the military situation. Flying qualities affect and are powerfully affected by these demands and nowadays can be sensibly discussed only in terms of mission-oriented requirements and criteria, hence the considerable emphasis on the development of handling qualities engineering and the standards, particularly Aeronautical Design Standard-33 (Ref. 8.3).

Military standards have wholly embraced the concept of handling qualities levels and pilot assessment through the Cooper–Harper handling qualities rating scale, discussed extensively in Chapter 7 of this book. For an aircraft to be fit for service (i.e., according to ADS-33 '... *no limitations on flight safety or on the capability to perform intended missions will result from deficiencies in flying qualities*'), it has to exhibit Level 1 handling qualities throughout the normal operational flight envelope (OFE). Degradation to Level 2 is 'acceptable' following the failure of some flight functions, in emergency situations or when the aircraft strays outside the OFE. Some operators may also allow Level 2 handling qualities in parts of the OFE, provided exposure is limited, e.g., deck landings in high sea states. Even though guided and constrained by their own experience and standard operational procedures, pilots need to make judgements all the time as to whether a particular manoeuvre is achievable or not. Sometimes they make the wrong judgement but the usual outcome is that the pilot gets a second chance at the landing or to position the load or pick up the survivor. Failing a mission task element (MTE) might push the aircraft into Level 3, but provided the degradation is not too severe the situation is recoverable. A more sudden or rapid degradation can push the aircraft towards the Level 4 condition however, where there is a high risk of loss of control. Chapter 7 closed with a statistical interpretation of the consequences on flight safety of an aircraft exhibiting different handling qualities (see Fig. 7.45). Acknowledging the assumptions of the analysis adopted, we drew the tentative conclusion that for an aircraft exhibiting a mean HQR at the Level 1/2 borderline, the probability of loss of control would be approximately 1 in 10^9 MTEs across the fleet. In comparison, an aircraft that exhibited a mean HQR in the middle of the Level 2 range would have a probability of loss of control across the fleet of about 1 in 10^5 MTEs, a massive increase in risk to safety.

These conclusions are borne out by the accident data. For example, in Ref. 8.4, Key pointed out that 54% of all accidents on the H-60 Blackhawk in the 10-year period up to 1996 involved deficiencies in handling qualities or situation awareness. The data also revealed that marginal handling was much more of a problem for low-time pilots. In a complementary study on US civil helicopter accidents, Ref. 8.5 reports that of the 547 accidents that occurred between 1993 and 2004, 23% could be '... attributed to loss of control by the pilot – caused or aggravated by inadequate or deficient handling qualities'. The relationship between handling and safety is an important link to make, even more so because in the drive to 'weather-proof' flight operations future rotorcraft will be required to perform roles in more degraded conditions than is currently possible with safety, hence an understanding of the ways degradation can occur, and some of the consequences, can assist in forming the requirements for day–night, all-weather augmentation systems. This chapter addresses these issues and material is drawn from the author's own research over the 10 years since the publication of the first edition of this book, e.g., Refs 8.6–8.9. During the second half of this period, the author relocated to The University of Liverpool, creating and building a research group focused on all aspects of *Flight Science and Technology*, and with a strong emphasis on flight safety. Central to the research at Liverpool is the Bibby Flight Simulator and, within this chapter, research results using this facility are presented liberally; the simulation facility is described in some detail in Appendix 8A.

To create a framework for the chapter, handling qualities degradation is described in four categories:

(a) degradation resulting from flight in degraded visual conditions;
(b) degradation resulting from flight system failures, both transient and steady state;
(c) degradation resulting from flight in severe atmospheric disturbances;
(d) degradation resulting from loss of control effectiveness.

Strictly speaking, category (d) should not occur, almost by definition, within the OFE and usually results from pilots inadvertently straying outside this, as a result of degradations in categories (a)–(c). Discussion on category (d) situations, for example, loss of heave control following entry into vortex ring, loss of tail rotor effectiveness in quartering flight or loss of pitch/roll control power in high-speed stall, will not be included.

8.2 FLIGHT IN DEGRADED VISUAL ENVIRONMENTS

Imagine a bird flying through a cluttered environment; a sparrow hawk is a good example. It is so successful at avoiding bumping into things and eventually catching prey on the wing that we can assume that the bird has very accurate knowledge of where it is heading, its rates of closure with objects in its path, its orientation and, more generally, its flight trajectory. How does it pick up the required information from the 'visual flow' of the world around it, projected onto its visual sensors? We might ask the same question of a fell-runner who successfully tracks over rough terrain without stumbling, or indeed an athlete who somersaults and lands, precisely balanced, on two feet, or a pigeon landing gently on the ledge rather than overshooting and crashing into the window. Motion control is ubiquitous in the natural world, and without completely reliable and precise functioning life would be very vulnerable. When the visual world is obscured, so too are the stimuli to the perception system and again life becomes vulnerable; most life sleeps at night, with the visual sensors, the eyes, shut, although there are some notable exceptions, of course. However, in the world of man-made flying machines, we regularly practice flight at night and in poor visual conditions, and technology even allows us to land fixed-wing aircraft on narrow runways, or bring a helicopter to hover, in zero visibility. Without precise control augmentation however, such manoeuvres would not be possible and such precision approaches are only really possible in tightly controlled airspace. Inadvertent flight into a degraded visual environment (DVE) is extremely hazardous with a high risk of loss of control through a loss of awareness of spatial orientation. Looking to the future, technology is under development that will provide pilots with a sufficiently reliable 'synthetic' world in which they have confidence to manoeuvre, to exercise motion control, in a cluttered environment with no natural outside world information. Until then, flight will be risky in poor visibility. We can gain valuable information on motion control by studying flight in good visual environments (GVE). By doing this we can also attempt to build an engineering framework for motion control using visual stimuli, which can inform the development of vision augmentation systems. This is the theme of this section. Through his research into motion control in the natural world, this author has observed that the subject is still in development with different 'schools of thought' existing on the key stimuli and mechanisms involved. I have had to be selective in attempting to build the bridge between the engineering and psycho-physics approaches to flight control, and my foundation has, naturally, been the work of James Gibson and the developments of his theory of optical flow. In the quest for solutions to how to design for completely autonomous flight in a cluttered, undulating environment however, I believe strongly that there is much still to be learned and understood, and that much of the contemporary, seemingly contradictory, research will have helped to inform progress.

8.2.1 Recapping the usable cue environment

Just like the sparrow hawk, a pilot flying a helicopter close to the surface and near obstacles requires clear visual information for attitude stabilization and flight path guidance, tasks not too dissimilar to cycling or walking over uneven or rough terrain. Although critical for short-term stabilization, vestibular motion cues are generally unreliable for guidance; turn the lights off or shut the eyes, and the cyclist or walker would soon fall over. Attitude stabilization cues for helicopter flight are derived from knowledge of the horizon, an awareness of spatial orientation and rotational motion. The requirements of ADS-33 are quite clear about the importance of stability augmentation when the usable cue environment (UCE) degrades below 1. Figure 8.1 summarizes the material presented in Chapter 7 – increased attitude stabilization (attitude command) as the

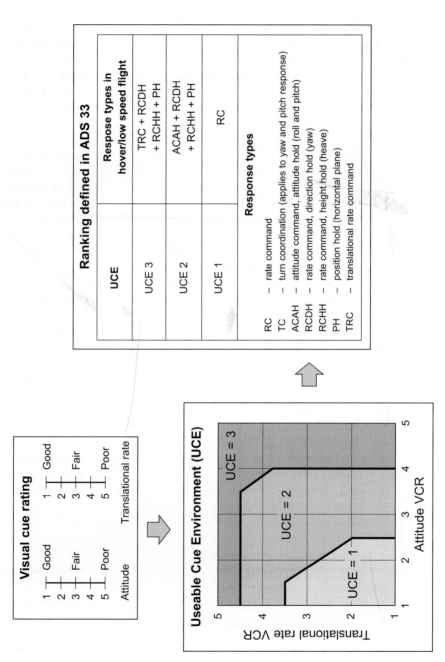

Fig. 8.1 The usable cue environment process summarized.

UCE degrades to 2 and increased velocity stabilization (translational rate command) as the UCE degrades to 3.

In Ref. 8.10, Hoh applied the UCE/VCR approach to quantifying the risk of spatial disorientation when flying in the DVE, using the original ADS-33 flight test database. The work reported addresses the wide class of ground/obstacle collisions that occur when aircrew are unaware that they have an inaccurate perception of their position, altitude or their motion. Hoh's analysis models situations where the overall pilot workload is a combination of the attentional demands (AD) of flight control and the effort required to maintain situation awareness (SA). The greater the requirements for control attention, the less capacity remains for SA and Hoh hypothesizes a relationship. To quote from Ref. 8.10

> *The risk of a spatial disorientation accident is linked to the attentional demand required for control as follows. High risk is defined when attentional demand exceeds 42% of the total available workload capacity. Extreme risk is defined when the AD exceeds 66% of the available workload capacity. The attentional demand for rotorcraft control in the DVE depends on two factors, 1) the basic handling qualities in the GVE and 2) the Response Type (Rate or ACAH + HH). The relationship between these factors is summarised in Fig. 8 where the attitude VCR and translational VCR are assumed to be equal to simplify the presentation of the effects. These results indicate that as the visual environment is degraded: 1) the use of ACAH+HH is highly effective in minimising the increase in AD, and 2) helicopters with a rate response type (conventional) suffer a rapid increase in AD. Any factor that degrades the HQR in the GVE (e.g. marginal basic handling qualities or turbulence) exacerbates the second result.*

The presentation in Fig. 8.2 is compelling but also conceptual, since it is acknowledged that the relationships between handling qualities, control workload and UCE proposed

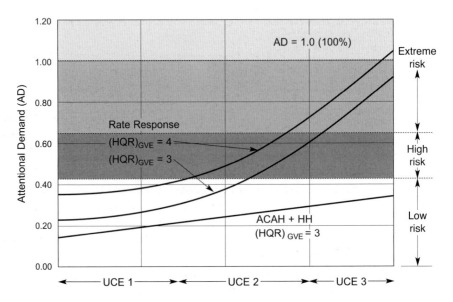

Fig. 8.2 Summary of the effect of the DVE on attentional demand (from Ref. 8.10)

are approximate and have not been fully quantified or validated. Nevertheless, they represent an intuitive and compelling argument for the importance of providing the pilot with augmented attitude control in the DVE. Moreover, Hoh concludes that providing additional instruments or displayed information to 'cue' the pilot can actually increase, rather than decrease, the AD, further increasing the risk of disorientation.

With this line of thinking, research into improving the UCE becomes focused on improving spatial awareness for the pilot. This research needs to establish relationships between the pilot's VCRs, features in the visual scene and the pilot's control strategy. The two components of a pilot's VCR can be thought of the adequacy of cues for flight guidance (translational rate) on the one hand, and the adequacy of cues for flight stabilization (attitude) on the other, i.e., the two dimensions of spatial awareness. While the previous discussions in this book on flying qualities have centred on the vehicle and the associated response characteristics, when addressing spatial awareness, we have to face the most adaptable and least well-understood element of the system and, indeed, the whole flying qualities discipline – the pilot and his or her perception system. To understand more about what makes up the UCE/VCR, we need to develop an engineer's appreciation of how the pilot organizes visual information and the human factors of flight control. Improved understanding here can lead to the development of more efficient pilot aids that function harmoniously with the natural systems, and are hence more effective at helping pilots maintain spatial awareness. Unlike the aircraft motion however, pilot action is not simply governed by Newton's laws and the rules of continuum mechanics; the perception–action system is far from completely understood and behaviour is often confused by misjudgements and malfunctions that are difficult to describe, let alone model. The following sections should be read in the light of this uncertainty.

8.2.2 Visual perception in flight control – optical flow and motion parallax

One of the earliest published works on visual perception in flight control presented a mathematical analysis of 'motion perspective' as used by pilots when landing aircraft (Ref. 8.11). The first author of this work, James Gibson, introduced the concept of the optical flow and the centre of expansion when considering locomotion relative to, and particularly approaching, a surface. Gibson suggested that the 'psychology of aircraft landing does not consist of the classical problems of space perception and the cues to depth'. In making this suggestion, Gibson was challenging conventional wisdom that piloting ability was determined by the sufficiency of linear/aerial perspective and parallax cues. Gibson had earlier introduced the concept of motion perspective in Ref. 8.12, but in applying it to flight control he laid the foundation for a new understanding of, what we might generally call, spatial awareness. To quote from Ref. 8.11:

> *Speaking in terms of visual sensations, there might be said to exist two distinct characteristics of flow in the visual field, one being the gradients of 'amount' of flow and the other being the radial patterns of 'directions' of flow. The former may be considered a cue for the perception of distance and the latter a cue for the perception of direction of locomotion relative to the surface.*

Gibson focused mainly on fixed-wing landings but he also presented an example of the optical flow-field generated by motion perspective for the case of a helicopter landing vertically, as shown in Fig. 8.3. 'For the case of a helicopter landing, the apparent

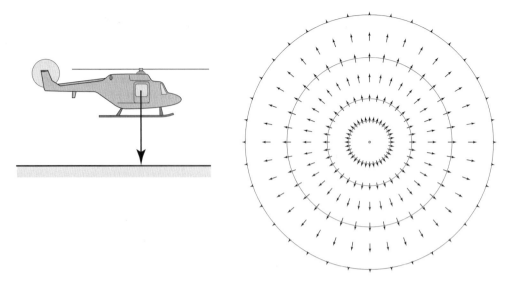

Fig. 8.3 Projected differential velocities (optical flow-field) on the ground in a helicopter vertical landing

velocity of points in the plane below first increases to a maximum and then decreases again.' The optical flow-field concept clearly has relevance to a helicopter landing at a heliport, on a moving deck or in a clearing, and raises questions as to how pilots reconstruct a sufficiently coherent motion picture from within the confines of a closed-in cockpit to allow efficient use of such cues.

Gibson's ecological approach (Ref. 8.13) is a 'direct' theory of visual perception, in contrast with the 'indirect' theories which deal more with the reconstruction and organization of components in the visual scene by the visual system and associated mental processes (Ref. 8.14). The direct theory can be related to the engineering theory of handling qualities. The flight variables of interest when flying close to obstacles and the surface are encapsulated in the definition of performance requirements in the ADS-33 flight manoeuvres – speed, heading, height above surface, flight path accuracies, etc. In visual perception parlance these have been described as ego-motion attributes (Ref. 8.14) and key questions concern the relationship between these and the optical variables, like Gibson's motion perspective. If the relationships are not one-to-one then there is a risk of uncertainty when controlling the ego motion. Also, are the relationships consistent and hence predictable? The framework for the discussion is a set of three optical variables considered critical to recovering a safe UCE for helicopter nap-of-the-earth (NoE) flight – optical flow, differential motion parallax and the temporal variable tau, the time to contact or close a gap.

Figure 8.4, from Ref. 8.15 (contained within Ref. 8.16), illustrates the optical flow-field when flying over a surface at 3 eye-heights per second (corresponds to fast NoE flight – about 50 knots at 30-ft height – or the flow-field observed by a running person). The eye-height scale has been used in human sciences because of its value to deriving body-scaled information about the environment during motion. Each flow vector represents the angular change of a point on the ground during a 0.25-s snapshot. Inter-point distance is 1 eye-height. The scene is shown for a limited field-of-view

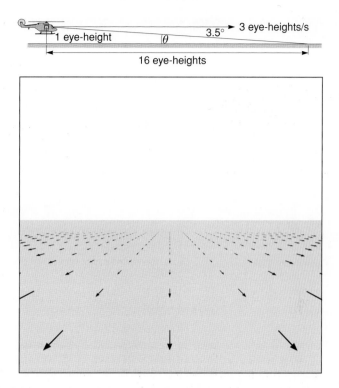

Fig. 8.4 Optical flow-field for motion over a flat surface (speed 3 eye-heights/s, snapshot 0.25 s)

window, typical of current helmet-mounted display formats. A 360° perspective would show flow vectors curving around the sides and to the rear of the aircraft (see Gibson, Ref. 8.12). The centre of optical expansion is on the horizon, although the flow vectors are shown to 'disappear' well before that, to indicate the consequent 'disappearance' of motion information to an observer with normal eyesight. If the pilot were to descend, the centre of optical expansion would move closer to the aircraft, in theory giving the pilot a cue that his or her flight trajectory has changed.

The length of the flow vectors gives an indication of the motion cues available to a pilot; they appear to decrease rapidly with distance. In the figure, the 'flow' is shown to disappear after 16 eye-heights.

The velocity in eye-heights per second is given by

$$\dot{x}_e = \frac{dx}{dt}\frac{1}{z} \tag{8.1}$$

In terms of the optical flow, or rate of change of elevation angle θ (Fig. 8.5), we can write

$$\frac{d\theta}{dt} = \frac{\dot{x}_e}{1 + x_e^2} \tag{8.2}$$

where x_e is the pilot's viewpoint distance ahead of the aircraft scaled in eye-heights.

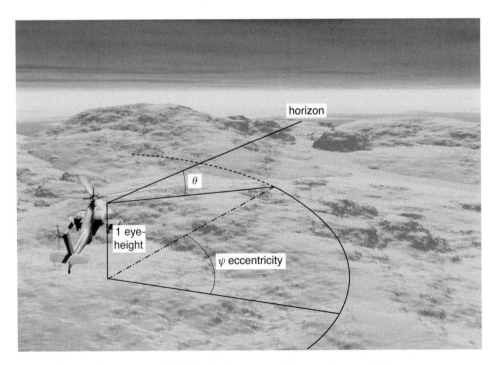

Fig. 8.5 Viewing eccentricity and elevation angles

When the eye-height velocity, $\dot{x}_e$, is constant, then the optical flow is also constant; they are in effect measures of the same quantity. However, the simple linear relationship between $\dot{x}_e$ and ground speed given by eqn 8.2 is disrupted by changes in altitude. If the pilot descends while keeping forward speed constant, $\dot{x}_e$ increases; if he climbs, $\dot{x}_e$ decreases. A similar effect is brought about by changes in surface layout, e.g., if the ground ahead of the aircraft rises or falls away. Generalizing eqn 8.2 to the case where the aircraft has a climb or descent rate ($\frac{dz}{dt}$) relative to the ground, we obtain

$$\frac{d\theta}{dt} = -\frac{\frac{dx}{dt}z - \frac{dz}{dt}x}{x^2 + z^2} \tag{8.3}$$

The relationship between optical flow rate and the motion variables is no longer straightforward. Flow rate and ground speed are uniquely linked only when flying at constant altitude.

A related optical variable comes in the form of a discrete version of that given by eqn 8.2 and occurs when optically specified edges within the surface texture pass some reference in the pilot's field of vision, e.g., the cockpit frame usually serves as such a reference. This optical edge rate is defined as

$$e_r = \frac{dx}{dt}\frac{1}{T_x} \tag{8.4}$$

where T_x is the spacing between the surface edges. A pilot flying at 50 ft/s over a network of 50-ft square grids would therefore experience an edge rate of 1/s. Unlike

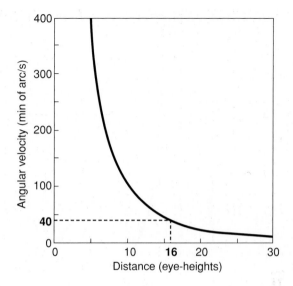

Fig. 8.6 Angular velocity versus distance along ground plane (Ref. 8.15)

optical flow rate, edge rate is invariant as altitude changes. However, when ground speed is constant, edge rate increases as the edges in the ground texture become denser, and decreases as they becomes sparser.

From eqn 8.2, it can be seen that flow rate falls off as the square of the distance from the observer. Figure 8.6, from Ref. 8.15, shows how the velocity, in minutes-of-arc/second, varies with distance for an eye-point moving at 3 eye-heights/s.

Perrone suggests that a realistic value for the threshold of velocity perception in complex situations would be about 40 min arc/s. In Fig. 8.6, this corresponds to information being subthreshold at about 15–16 eye-heights distant from the observer. To quote from Ref. 8.15, '*This is the length of the "headlight beam" defined by motion information alone. At a speed of 3 eye-heights/sec, this only gives about 5 seconds to respond to features on the ground that are revealed by the motion process*'. The value of optical streaming for the detection and control of speed and altitude has been discussed in a series of papers by Johnson and co-workers (Refs 8.16–8.20). Flow rate and texture/edge rate are identified as primary cues. These velocity cues can be picked up from both foveal (information detected by the central retinal fovea) and ambient (information detected by the peripheral retina) vision. An issue with ambient information, however, is the significant degradation in visual acuity as a function of eccentricity. The fovea of the human eye, where there is a massive concentration of visual sensors, has a field of regard of less than $1°$ (approximately a thumb's width at arm's length). The visual acuity at $20°$ eccentricity is about 15% as good as the fovea for resolution, although Cutting points out that this increases to 30% for motion detection (Ref. 8.21). Cutting also observes that the product of motion sensitivity and motion flow (magnitude of flow vectors) when moving over a surface is such that 'the thresholds for detecting motion resulting from linear movement over a plane are roughly the same across a horizontal meridian of the retina'.

In Fig. 8.4, the centre of optical expansion or outflow is on the horizon. If the pilot is looking directly at this point then the information seen will be 'filtered' by the

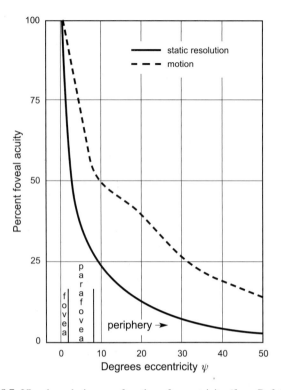

Fig. 8.7 Visual resolution as a function of eccentricity (from Ref. 8.21)

variable sensitivity across the retina. The motion acuity gradient or visual resolution takes the form illustrated in Fig. 8.7 (Ref. 8.21), based on the eccentricity and elevation angles defined in Fig. 8.5.

In Fig. 8.7, data are shown for static resolution and motion resolution referenced to the fovea performance of 100. The results show that the strength of visual inputs 20° off-centre reduce to about 40% of those picked up by the fovea when in motion compared with 15% statically. However, the magnitude of the motion flow vectors depicted in Fig. 8.4 increases away from the line of sight in a normalized manner shown in Fig. 8.8 (also from Ref. 8.21). The sensitivity of the retina to motion is therefore the resultant product of the two effects and is actually fairly uniform across a horizontal meridian. Figure 8.9 shows the case for viewing at 8° below the horizon, corresponding to about 7 eye-heights ahead of the aircraft.

A more irregular surface will give rise to deformations in the sensitivity but the same underlying effect will be present, leading to the conjecture that the pilot's gaze will naturally be drawn to the direction of flight, i.e., that direction which, on average, gives uniform stimulation across the retina. This is good news for pilots, and a determining factor on piloting skill is how well this capability is 'programmed' into an individual's perceptual system.

The subject of way-finding, or establishing the direction of flight, has also been addressed in some detail in Ref. 8.21, where the notion of directed perception was introduced. Cutting developed the optical flow-field concept, arguing that people and animals make more use of the retinal flow-field, fixating with the fovea on specific

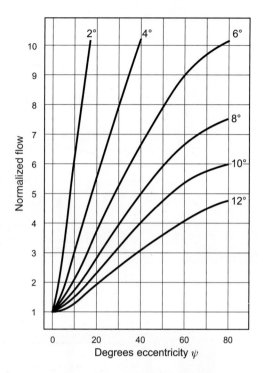

Fig. 8.8 Normalized flow vectors as a function of eccentricity (from Ref. 8.21)

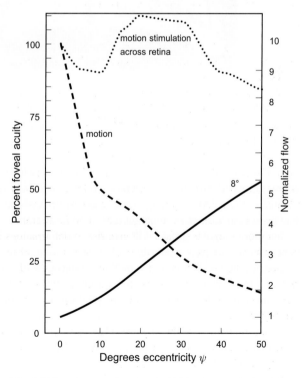

Fig. 8.9 Resultant motion threshold function across the retina

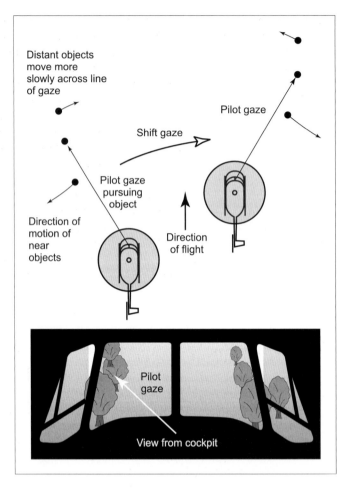

Fig. 8.10 Differential motion parallax as an optical invariant to aid way finding (from Ref. 8.21)

parts of the environment and deriving information from the way in which surrounding features move relative to that point on the retina. In this way the concept of differential motion parallax (DMP) was hypothesized as the principal optical variable used for way-finding in a cluttered environment. Figure 8.10 illustrates how motion and direction of motion can be derived from DMP. The helicopter is being flown through a cluttered environment. The pilot fixates his or her gaze on one of the obstacles (to the left of motion heading) and observes the motion parallax effects on objects closer and farther away. Objects farther away move to the right and those closer move to the left of the gaze (as seen on the retinal array). The pilot can judge which objects are closer and further away by the relative velocities. Figure 8.10 indicates that closer objects move more quickly across the line of gaze. There is no requirement to know the actual size or distance of any of the objects in the clutter. The pilot can judge from this motion perception that the direction of motion is to the right of the fixated point. He or she can now fixate on a different object. If objects further away (slower movements) move to the left and those closer (faster movements) move to the right, then the pilot will perceive that motion is to the left of the fixated object. By applying a series of such fixations the pilot will be able to keep updating his or her information about direction of motion, and home in on the true direction with potentially great accuracy (the point where there is no flow across the line of gaze). Cutting observed that in high-performance situations, for example, deck landings of fixed- and rotary-wing aircraft, required heading accuracies might need to be 0.5° or better. DMP does not always work however, e.g., in the direction of motion itself or in the far field, where there is no DMP, or in the near field, where DMP will fail if there are no objects nearer than half the distance to the point of gaze (Ref. 8.21).

In Ref. 8.15, Perrone also discusses the question of how pilots might infer surface layout, or the slants of surfaces, ahead of the aircraft. This is particularly relevant to flight in a DVE where controlled flight into terrain is a major hazard and still all too common. The correct perception of slope is critical for achieving 'desired' height safety margins for flight over undulating terrain, and hence for providing good visual cue ratings for vertical translational rate, for example. Figure 8.11 illustrates the flow-field when approaching a 60° slope hill about 8 eye-heights away. The centre of optical

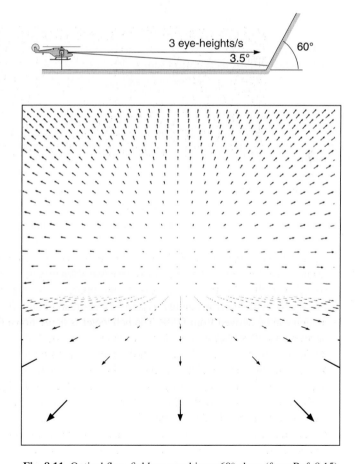

Fig. 8.11 Optical flow-field approaching a 60° slope (from Ref. 8.15)

expansion has now moved up the slope and the motion cues over a significant area around this are very sparse. If the pilot wants to maintain gaze at a point where the motion threshold cuts in (e.g., 5 s ahead) he or she will have to lift his or her gaze, and pilots will tend to do this as they approach a hill. This aspect is discussed again later in this chapter when results are presented from simulation research into terrain flight, where the question – how long do pilots look forward? – is addressed.

Any vision augmentation system that tries to infer slope based on flow vectors around the centre of expansion is likely be fairly ineffective. In Ref. 8.22, a novel vision augmentation system was proposed for aiding flight over featureless terrain at night. An obstacle detector system was evaluated in simulation, consisting of a set of cueing lights, each with a different look-ahead time, presenting a cluster of spots to the pilot of the light beams on the terrain ahead of the aircraft. As altitude or the terrain layout ahead changed, so the cluster changed shape, providing the pilot with an 'intuitive spatial motion cue' to climb or descend.

In a cluttered environment, the optical variables – flow/edge rate and DMP – appear to provide primary cues to pilots for judging the direction in which they are heading. The question as to how they judge their speed and distance takes us onto the third optical flow variable in this discussion, the results of which suggest that pilots do not actually need to know speed and distance for safe flight control; rather the prospective control is temporally based within an ordered spatial environment.

8.2.3   Time to contact; optical tau, τ

When $x_e \gg 1$ (or $x \gg z$), we can simplify eqns 8.1 and 8.3 to the form

$$\dot{\theta} = \frac{\dot{x}_e}{x_e}\theta = \frac{\dot{x}}{x}\theta \tag{8.5}$$

The ratio of distance to velocity is the instantaneous time to reach the viewpoint, which we designate as $\tau(t)$,

$$\tau(t) = \frac{x}{\dot{x}} \tag{8.6}$$

This temporal optical variable is considered to be important in flight control. A clear requirement for pilots to maintain safe flight is that they are able to predict the future trajectory of their aircraft far enough ahead so that they can stop, turn or climb to avoid a hazard or follow a required track. This requirement can be interpreted in terms of the pilot's ability to detect motion ahead of the aircraft. In his explorations of temporal optical variables in nature (Refs 8.23–8.28), David Lee makes the fundamental point that an animal's ability to determine the time to pass or contact an obstacle or piece of ground does not depend on explicit knowledge of the size of the obstacle, its distance away or relative velocity. The ratio of the size to rate of growth of the image of an obstacle on the pilot's retina is equal to the ratio of distance to rate of closure, as conceptualized in Fig. 8.12, and given in angular form by eqn 8.7.

$$\tau(t) = \frac{x}{\dot{x}} = \frac{\theta}{\dot{\theta}} \tag{8.7}$$

Lee hypothesized that this 'looming' is a fundamental optical variable that has evolved in nature, featuring properties of simplicity and robustness. The brain does not have to

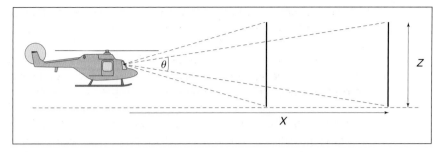

Fig. 8.12 Optical looming when approaching an object: (a) τ of horizontal velocity in a deceleration manoeuvre; (b) τ of flight path angle in a climb manoeuvre; (c) τ of heading angle in a turn manoeuvre

apply computations on the more primitive variables of distance or speed, thus avoiding the associated lags and noise contamination. The time-to-contact information can readily be body scaled in terms of eye-heights, using a combination of surface and obstacle $\tau(t)$'s, thus affording animals with knowledge of, for example, obstacle heights relative to themselves.

While making these assertions, it is recognized that much spatial information is available to a pilot and will provide critical cues to position and orientation and perhaps even motion. For example, familiar objects clearly provide a scale reference and can guide a pilot's judgement about clearances or manoeuvre options. However, the temporal view of motion perception purports that the spatial information is not essential to the primitive, instinctive processes involved in the control of motion.

Tau research has led to an improved understanding of how animals and humans control their motion and humans control vehicles. A particular interest is how a driver or pilot might use τ to avoid a crash state, or how τ might help animals alight on objects. A driver approaching an obstacle needs to apply a braking (deceleration) strategy that will avoid collision. One collision-avoid strategy is to control directly the rate of change of optical tau, which can be written in terms of the instantaneous distance to stop (x), velocity $(\dot{x})$ and acceleration $(\ddot{x})$ in the form:

$$\dot{\tau} = 1 - \frac{x\,\ddot{x}}{\dot{x}^2} \tag{8.8}$$

The system used here for defining the kinematics of motion is based on a negative gap x being closed. Hence, with $x < 0$ and $\dot{x} > 0$, $\dot{\tau} > 1$ implies accelerating flight, $\dot{\tau} = 1$ implies constant velocity and $\dot{\tau}<1$ corresponds to deceleration. In the special case of a constant deceleration, the stopping distance from a velocity $\dot{x}$ is given by

$$x = -\frac{\dot{x}^2}{2\ddot{x}} \tag{8.9}$$

Hence, a decelerating helicopter will stop short of the intended hover point if at any point in the manoeuvre

$$\frac{-\dot{x}^2}{2\ddot{x}} < -x \quad \text{or} \quad \frac{x\ddot{x}}{\dot{x}^2} > 0.5 \tag{8.10}$$

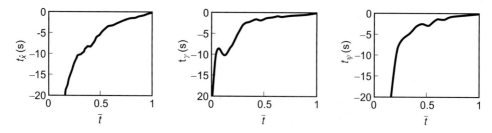

Fig. 8.13 Motion τ's in helicopter manoeuvres as a function of normalized time

Using eqns 8.7 and 8.8, this condition can be written more concisely as

$$\frac{d\tau}{dt} < 0.5 \qquad (8.11)$$

A constant deceleration results in $\dot{\tau}$ progressively decreasing with time and the pilot stopping short of the obstacle, unless $\dot{\tau} = 0.5$ when the pilot just reaches the destination.

The hypothesis that optical τ and $\dot{\tau}$ are the variables that evolution has provided the animal world with to detect and rapidly process visual information, suggests that these should be key variables in flight guidance. In Ref. 8.26, Lee extends the concept to the control of rotations, related to how athletes ensure that they land on their feet after a somersault. For helicopter manoeuvring, this can be applied to control in turns, connecting with the heading component of flight motion, or in vertical manoeuvres, with the flight path angle component of the motion. For example, with heading angle ψ and turn rate $\dot{\psi}$, we can write angular τ as

$$\tau(t) = \frac{\psi}{\dot{\psi}} \qquad (8.12)$$

A combination of angular and translational τ's, associated with physical gaps, needs to be successfully picked up by pilots to ensure flight safety. Figure 8.13 illustrates three examples of motion τ variations as a function of normalized manoeuvre time. The results are derived from flight simulation tests undertaken on the Liverpool Flight Simulator in, nominally, good visual conditions. In all three cases the final stages of the manoeuvre ($\bar{t}$ approaches 1) are characterized by a roughly constant $\dot{\tau}$, implying, as noted above, a constant deceleration to the goal.

Reaching a goal with a constant $\dot{\tau}$ can be achieved without a constant deceleration of course, and we shall see later in this chapter what the different deceleration profiles look like. For example, if the maximum deceleration towards the goal occurs late in the manoeuvre, then $0.5 < \dot{\tau} < 1.0$, while an earlier peak deceleration corresponds to $0.0 < \dot{\tau} < 0.5$. An interesting case occurs when $\dot{\tau} = 0$, so that however close to the goal τ remains a constant c, i.e.,

$$\tau = \frac{x}{\dot{x}} = c \qquad (8.13)$$

The only motion that satisfies this relationship is an exponential one, with the goal approached asymptotically.

An interesting discovery of τ control, although unbeknown at the time, is described in Ref. 8.29. In the early 1970s, researchers at NASA Langley conducted flight

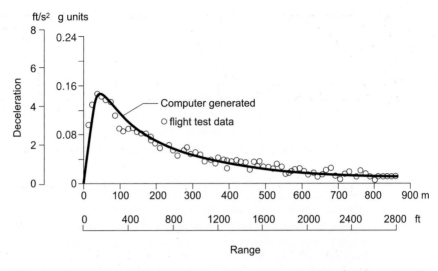

Fig. 8.14 Deceleration profile for helicopter descending to a landing pad (from Ref. 8.29)

trials using several different helicopter types in support of the development of instrument flight procedures and the design of flight director displays to aid pilots during the approach and landing phases in poor visibility conditions. The engineers had postulated particular deceleration profiles as functions of height and distance from the landing zone, and the pilots were asked to evaluate the systems based on workload and performance. Several different design philosophies were evaluated and the pilots commented that none felt intuitive and that they would use different control strategies in manual landings, particularly during the final stages of the approach. The linear deceleration profiles resulted in pilots concerned that they were being commanded to hover well short of the touchdown point. The constant deceleration profile was equally undesirable and led to a high pitch/low power condition as the hover was approached. The pilots were asked to fly the approaches manually in good visual conditions, from which the deceleration profiles would be derived and then used to drive the flight director. Figure 8.14 shows a typical variation of deceleration during the approach with 50 knots initial velocity at 500 ft above the ground.

Also shown in the figure is the computer-generated profile showing a gradual reduction in speed until the peak deceleration of about 0.15 g is reached 70 m from the landing pad. The 'computer-generated' relationship between acceleration, velocity and distance took the form

$$\ddot{x} = \frac{k\,\dot{x}^2}{x^n} \tag{8.14}$$

where k is a constant derived from the initial conditions. Recalling the formulae for $\dot{\tau}$ in eqn 8.8, the relationship given by eqn 8.14 can be written in the form

$$1 - \dot{\tau} = k\,x^{1-n} \tag{8.15}$$

The parameters k and n were computed as constants in any single deceleration but varied with initial condition and across the different pilots. The range power parameter

n varied between 1.2 and 1.7. Note that a value of unity corresponds to a constant $\dot{\tau}$ for the whole manoeuvre. Equation 8.15 suggests that, at long range, the pilot is maintaining constant velocity ($\dot{\tau}=1$), consistent, of course, with the steady initial flight condition. As range is reduced, $\dot{\tau}$ reduces until $x = k^{\frac{1}{n-1}}$ when $\dot{\tau} = 0$ and the approach becomes exponential. Beyond this, the representation in eqn 8.15 breaks down, as it predicts a negative $\dot{\tau}$, i.e., the helicopter backs away from the landing pad, although this can happen in practice. Approaches at constant τ may seem ineffective because the goal is never reached, but there is evidence that pilots sometimes use this strategy during the landing flare in fixed-wing aircraft, perhaps as a 'holding' strategy as the flight path lines up with the desired trajectory (see Ref. 8.30).

The concept of τ in motion control has significance for helicopter flight in degraded visual conditions. If the critical issue for sufficiency of visual cues is that they afford the information to allow the pilot to pick up τ's of objects and surfaces, then it follows that the τ's are measures of spatial awareness. It would also follow that they would be appropriate measures to use to judge the quality of artificial vision aids and form the underlying basis for the design and the information content of vision aids. This author has conducted a number of experiments to address the question – how do pilots know when to stop, or to turn or pull up to avoid collision? In the following sections some results from this research will be presented.

8.2.4 τ control in the deceleration-to-stop manoeuvre
Figure 8.15 shows a schematic of an acceleration–deceleration manoeuvre, showing the distance to go to stop at $x = 0$. Figure 8.16 shows the kinematic profile of a helicopter

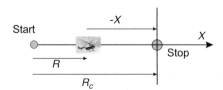

Fig. 8.15 Kinematics of the acceleration–deceleration manoeuvre

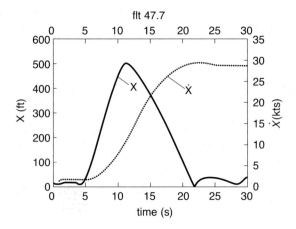

Fig. 8.16 Kinematics of the acceleration–deceleration manoeuvre

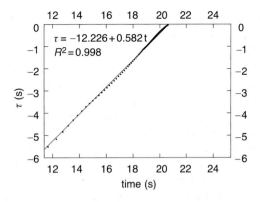

Fig. 8.17 Time to stop as a function of time in the deceleration phase

flying a 500 ft accel–decel in a flight simulator trial, and Fig. 8.17 shows the variation of time to stop, τ_x, during the deceleration phase of the manoeuvre. The data are taken from Ref. 8.31, where the author and his colleagues introduced the concept of τ-control in helicopter flight, also described as prospective control in recognition of the temporal nature of flight control. The deceleration is seen to extend from about 11 s into the manoeuvre, when the peak velocity is about 30 knots, for about 10 s, when the time to stop is nearly 6 s.

Figure 8.17 shows that the correlation of τ_x with time is very strong, with a correlation coefficient R^2 of 0.998. The slope of the fit, i.e., $\dot{\tau}$, is 0.58, indicating a non-constant deceleration with peak during the second half of the manoeuvre. So the pilot initiates the deceleration when the time to stop is about 6 s and holds an approximately constant $\dot{\tau}$ strategy through to the stop.

As referred to above, the use of τ in motion control has been the subject of research in the natural world for some time. In Ref. 8.32, Lee and colleagues have measured the τ control strategy of pigeons approaching a perch to land. Figure 8.18 shows a sequence of stills taken during the final 0.5 s of the manoeuvre. The analysis of the photographic data shows that the pigeon controls braking in the last few moments of flight by maintaining a constant $\dot{\tau}$ for the gap between its feet and the perch, as shown in Fig. 8.19. The τ of the pigeon's feet to the landing position is given by $\tau(x_{feet}, lp)$. The feet are moving forward and the head is moving back, so the visual 'cues' are far from simple. The average slope of the lines in Fig. 8.19 ($\dot{\tau}$) is about 0.8, indicating that

Fig. 8.18 Pigeon approaching a landing perch (Ref. 8.32)

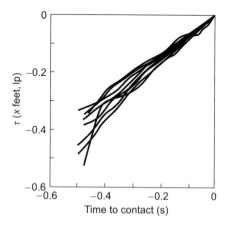

Fig. 8.19 Time to land for pigeon approaching a perch (Ref. 8.32)

the maximum braking occurs very late in the manoeuvre – the pigeon almost crash lands, or at least experiences a hard touchdown, which ensures positive contact and is probably quite deliberate.

So how does a pilot or a pigeon manage to maintain a constant $\dot{\tau}$, or indeed any other τ variation, as they approach a goal? In addressing this question for action in the natural world, Lee gives a new interpretation to the whole process of motion control, which has significant implications for helicopter flight control and the design of augmentation systems. We now turn to the general theory of τ-coupling, which also addresses the need for controlling several motion τ's in more complex manoeuvres and introduces the concept of the motion guide and its associated τ.

8.2.5 Tau-Coupling – a paradigm for safety in action

General τ theory hypothesizes that the closure of any motion gap is guided by sensing and adjusting the τ of the associated physical gap (Ref. 8.24). The theory reinforces the evidence presented in the previous section that information solely about $\dot{\tau}_x$ is sufficient to enable the gap x to be closed in a controlled manner, as when making a gentle landing or coming to hover next to an obstacle. According to the theory, and contrary to what might be expected, information about the distance to the landing surface or about the speed and deceleration of approach is not necessary for precise control of the approach and landing. The theory further suggests that a pilot might perceive τ of a motion gap by virtue of its proportionality to the τ of a gap in a 'sensory flow-field' within the visual perception system. In helicopter flight dynamics, the example of decelerating a helicopter to hover over a landing point on the ground serves to illustrate the point. The τ of the gap in the optic flow-field between the image of the landing point and the centre of optical outflow (which specifies the instantaneous direction of travel, see Fig. 8.4) is equal to the τ of the motion gap between the pilot and the vertical plane through the landing point. This is always so, despite the actual sizes of the optical and motion gaps being quite different; the same applies to stopping at a point adjacent to an obstacle – see Fig. 8.20.

Often movements have to be rapidly coordinated, as when simultaneously making a turn and decelerating to stop, or descending and stopping, or performing a bob-up

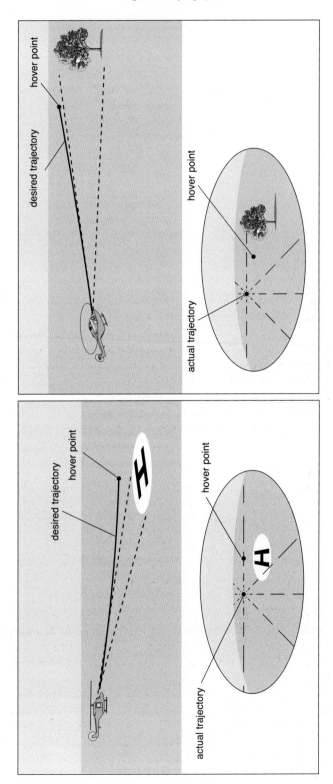

Fig. 8.20 τ gaps for a helicopter approaching a hover (from Ref. 8.31)

and, simultaneously, a 90° turn. This requires accurate synchronizing and sequencing of the closure of different gaps. To achieve this, visual cues have to be picked up rapidly and continuously and used to guide the action. τ theory shows how such closed-loop control might be accomplished by keeping the τ's of gaps in constant ratio during the movement, i.e., by τ-coupling. Evidence of τ-coupling in nature is presented in Refs 8.27 and 8.28 for experiments with echo-locating bats landing on a perch and infants feeding. In the present context, if a helicopter pilot, descending (along z) and decelerating (along x), follows the τ-coupling law

$$\tau_x = k\tau_z \tag{8.16}$$

then the desired height will automatically be attained just as the aircraft comes to a stop at the landing pad. The kinematics of the motion can be regulated by appropriate choice of the value of the coupling constant k. There is evidence that such coupling can be exploited successfully in vision aids. For example, the system reported in Ref. 8.22 functioned on the principle of the matching of a cluster of forward-directed light beams with different look-ahead distances, which translated into times at a given speed. Such a system was designed as an aid in situations where the natural optical flow was obscured.

In many manoeuvres such as a hover turn or bob-up, there is essentially only one gap to be closed, yet the feedback actions must in principle be similar, whether there are two coupled motion gaps or just one. When a pilot is able to perceive the motion gaps associated with both the displacement and velocity, then τ-coupling takes a special form,

$$\tau_x = \frac{x}{\dot{x}}, \qquad \tau_{\dot{x}} = \frac{\dot{x}}{\ddot{x}} \tag{8.17}$$

with

$$\tau_x = k\tau_{\dot{x}} \tag{8.18}$$

Combining eqns 8.17 and 8.18, we can write

$$\dot{\tau}_x = 1 - \frac{\ddot{x}x}{\dot{x}^2} = 1 - k \tag{8.19}$$

Hence, the $\dot{\tau}$ constant strategy can be expressed as the pilot maintaining the τ's of the displacement and velocity in a constant ratio. The more general hypothesis is that the closure of a single motion gap is controlled by keeping the τ of the motion gap coupled onto what has been described as an intrinsically generated τ guide, τ_g (Ref. 8.24). One form of such a guide is a constant deceleration motion, from an initial condition $x_{g0}(<0)$, v_{g0} (>0) given by ($c < 0$)

$$a_g = c, \qquad v_g = v_{g0} + ct, \qquad x_g = x_{g0} + v_{g0}t + \frac{c}{2}t^2 \tag{8.20}$$

At $t = T$, the manoeuvre duration, we can write

$$x_{g0} = \frac{cT^2}{2}, \qquad v_{g0} = -cT \tag{8.21}$$

Substitution into the kinematic relationships in eqn 8.20 results in

$$\dot{\tau}_g = 1 - \frac{cx_g}{v_g^2} = \frac{1}{2} \qquad (8.22)$$

The constant deceleration τ guide has a $\dot{\tau} = 0.5$, a result obtained earlier in this section, and coupling onto this guide with coupling constant k implies

$$\dot{\tau}_x = \frac{k}{2} \qquad (8.23)$$

For motions that start at rest, $v_g = 0$, and end at rest, we need to find a different form of guide. In Ref. 8.24, Lee argues that for natural motions such as reaching, which involve simple phases of acceleration followed by deceleration, it is reasonable to hypothesize that a simple form of intrinsic τ guide will have evolved that is adequate for guiding such fundamental movements. In the context of helicopter NoE flight, any of the classic hover-to-hover repositioning manoeuvres fit into this category of motions. Indeed, in any manoeuvre that takes the aircraft from one state to another, the pilot is, in essence, closing a gap of one kind or another. The hypothesized intrinsic tau guide corresponds to a time-varying quantity, perhaps a triggered pattern within the perception system, which changes from one state to another with a constant acceleration. Surprisingly, coupling onto this 'constant acceleration' intrinsic guide does not, however, generate a motion of constant acceleration. The resultant motion is, rather, one with an accelerating phase followed by a decelerating phase. From a similar analyses to the case of the constant deceleration guide, the equations describing the changing τ_g can be derived and written in the form

$$\tau_g = \frac{1}{2}\left(t - \frac{T^2}{t}\right) \qquad \dot{\tau}_g = \frac{1}{2}\left(1 + \left(\frac{T}{t}\right)^2\right) \qquad (8.24)$$

where T is the duration of the aircraft or body movement and t is the time from the start of the movement. Coupling the τ of a motion gap, τ_x, onto such an intrinsic tauguide, τ_g, is then described by the equation

$$\tau_x = k\tau_g \qquad (8.25)$$

for some coupling constant k. The intrinsic tau guide, τ_g, has a single adjustable parameter, T, i.e., its duration. The value of T is assumed to be set to fit the movement either into a defined temporal structure, as when coming to a stop in a confined space, or in a relatively free way, as in the simple movement of reaching for an object. In the case of a helicopter flying from hover to hover across a clearing, we can hypothesize that time constraints are mission related and the pilot can adjust the urgency, within limits, through the level of aggressiveness applied to the controls. The kinematics of a movement can be regulated by setting both T and the coupling constant, k in eqn 8.18, to appropriate values. For example, the higher the value of k, the longer will be the acceleration period of the movement, the shorter the deceleration period and the more abruptly will the movement end. We describe situations with k values >0.5 as hard stops (i.e., k close to unity corresponds to a situation where the peak velocity is pushed close to the end of the manoeuvre) and situations with $k < 0.5$ as soft stops, similar to the constant $\dot{\tau}$ strategy.

The following of a constant acceleration guide in an accel–decel manoeuvre is conceptualized in Fig. 8.21. Both aircraft and guide, shown as a ball, start at the same

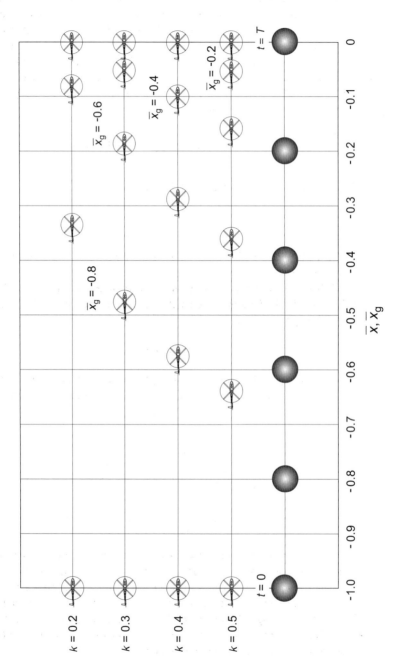

Fig. 8.21 Following a constant acceleration τ guide in an accel–decel manoeuvre

point (normalized distance $\bar{x}_g = -1.0$) and time ($t = 0$) and reach the end of the manoeuvre at the same time, $t = T$. The ball is continuing to accelerate at this point of course, while the helicopter has come to the hover. For the case $k = 0.5$, the aircraft has covered about 35% of the manoeuvre distance when $\bar{x}_g = -0.8$ and about 85% of the distance when $\bar{x}_g = -0.4$. With $k = 0.2$, when $\bar{x}_g = -0.8$, the aircraft has covered two-thirds of the manoeuvre and when $\bar{x}_g = -0.6$, the aircraft is within 10% of the stopping point. As k increases, so does the point in the manoeuvre when the reversal from acceleration to deceleration occurs. The time in the manoeuvre when the reversal occurs, t_r, can be derived as a function of the coupling coefficient k by noting that, at this point, $\dot{t}_x = 1$; from eqns 8.24 and 8.25, we can write

$$\dot{t}_x = \frac{k}{2}\left(1 + \left(\frac{T}{t_r}\right)^2\right) = 1 \tag{8.26}$$

This equation can be rearranged into the form

$$t_r = \sqrt{\frac{k}{2 - k}}\, T \tag{8.27}$$

Thus, when $k = 0.2$, $t_r = 0.333T$, when $k = 0.4$, $t_r = 0.5T$ and when $k = 0.6$, $t_r = 0.67T$, etc.

When two variables, i.e., the motion x and the motion guide x_g, are related through their τ-coupling in the form of eqn 8.25, it can be shown (Ref. 8.23) that they are also related through a power law

$$x \propto x_g^{1/k} \tag{8.28}$$

This relationship is ubiquitous in nature, governing the relationships between stimuli and sensory responses. Normalizing the distance and time by the manoeuvre length and duration respectively, the motion kinematics (for negative initial x) can be written as

$$\bar{x} = -(1 - \bar{t}^2)^{\left(\frac{1}{k}\right)} \tag{8.29}$$

$$\bar{x}' = \frac{2}{k}\bar{t}(1 - \bar{t}^2)^{\left(\frac{1}{k}-1\right)} \tag{8.30}$$

$$\bar{x}'' = -\frac{2}{k}\left[\left(\frac{2}{k} - 1\right)\bar{t}^2 - 1\right](1 - \bar{t}^2)^{\left(\frac{1}{k}-2\right)} \tag{8.31}$$

The coupling parameter k determines exactly how the closed-loop control functions, e.g., proportional as k approaches 1 or according to a square law when $k = 0.5$. The manoeuvre kinematics are presented in Fig. 8.22 of a motion that perfectly tracks the constant acceleration τ guide for various values of coupling constant k. The motion t is shown in Fig. 8.22(a), the motion gap, $\bar{x}$, in Fig. 8.22(b), the gap closure rate (normalized velocity) in Fig. 8.22(c) and the normalized acceleration in Fig. 8.22(d); all are shown plotted against normalized time.

The closure rate, shown in Fig. 8.22(c), illustrates a typical accel-decel-type velocity profile (cf. $\dot{x}$ in Fig. 8.16). For $k = 0.2$ the maximum velocity occurs about 30% into the manoeuvre, while for $k = 0.8$ the peak occurs close to the end of the manoeuvre.

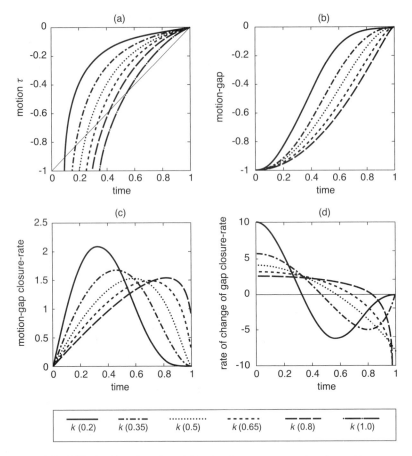

Fig. 8.22 Profiles of motions following the constant acceleration guide

An example of the success of this more general strategy is shown in Fig. 8.23, showing the same case from Ref. 8.31, illustrated previously in Fig. 8.16, but now for the helicopter flying the complete accel–decel manoeuvre. The coupling coefficient is 0.28, giving a power factor of 3.5, with a correlation coefficient of 0.98.

In the analysis of the Ref. 8.31 data, the start and end of the manoeuvres were cropped at 10% of the peak velocity. The tests were flown on the DERA/QinetiQ large motion simulator as part of a series of tests with a Lynx-like helicopter examining the effect of levels of aggressiveness on handling qualities and simulation fidelity. Considering all 15 accel–decels that were flown, the mean values of k follow the trends expected based on Fig. 8.22 (low aggression, $k = 0.381$; moderate aggression, $k = 0.324$; high aggression, $k = 0.317$). As the aggression level increases, the pilot elects to initiate the deceleration earlier in the manoeuvre: low aggression $0.5T$ into manoeuvre when $\tau \approx 6$ s; high aggression $0.4T$ into manoeuvre when $\tau \approx 4.5$ s. The pilot is more constrained during the deceleration phase, with the pilot limiting the nose-up attitude to about $20°$ to avoid a complete loss of visual cues in the vertical field of view.

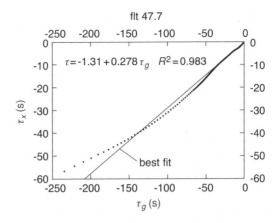

Fig. 8.23 Correlation of τ_x and τ_g for helicopter in accel–decel manoeuvre (Ref. 8.31)

An intrinsic τ guide is effectively a mental model, created by the nervous system, which directs the motion. It clearly has to be well informed (by visual cues in the present case) to be safe. Constant acceleration is one of the few natural motions, created in the short term by the gravitational field, so it is not surprising that the perception system might well have developed to exploit such motions. But the nature of the coupling, the chosen manoeuvre time T and profile parameter k must depend on the performance capability of the aircraft (or the animal performing a purposeful action), and pilots (or birds) need to train to 'programme' these patterns into their repertoire of flying skills. When the visual environment degrades so does the ability of the pilot's perception system to pick up the required information and hence to track the error between actual motions and intrinsic guides. The UCE is, in a sense, a measure of this ability, suggesting that there should be a relationship between UCE and the τ of the motion when initiating a manoeuvre to stop, turn or pull-up. In the low-aggression case of Ref. 8.31, the pilot returned Level 1 HQRs and initiated the deceleration when $\tau \approx 6$ s, taking about 10 s to come to the hover. The pilot could clearly pick up the visual 'cues' of the trees at the stopping point throughout.

The question of how far, or more appropriately how long, into the future the pilot needs to be able to see is critical to flight safety and the design of vision augmentation systems. The research reported in Ref. 8.31 has been extended to address this question specifically, and preliminary results are reported in Ref. 8.33. The focus of attention in this study was terrain following in the presence of degraded visibility, in particular fog, and we continue this chapter with a review of the results of this work and analysis of low-speed terrain following in the DVE.

8.2.6 Terrain-following flight in degraded visibility

As a pilot approaches rising ground, the point at which the climb is initiated depends on the forward speed and also the dynamic characteristics of the aircraft, reflected in the vertical performance capability and the time constant in response to collective pitch inputs; at speeds below minimum power, height control is exercised almost solely through collective inputs. A matched manoeuvre could be postulated as one where the pilot applies the required amount of collective at the last possible moment so that the climb rate reaches steady state, with the aircraft flying parallel to the surface of the hill. For low-speed flight, vertical manoeuvres can be approximately described by a first-order differential equation (see Chapter 5, eqn 5.52), with its solution to a step input in the pilot's collective lever given by

$$\dot{w} - Z_w w = Z_{\theta 0}\theta_0$$
$$w = w_{ss}(1 - e^{Z_w t}) \tag{8.32}$$

In the usual notation, w is the aircraft normal velocity (positive down), w_{ss} the steady state value of w and θ_0 is the collective pitch angle. Z_w is the heave damping derivative (see Chapter 4), or the negative inverse of the aircraft time constant in the heave axis, t_w. Writing $\delta_w = \frac{w}{w_{ss}}$, the time to achieve δ_w can be written in the form

$$\frac{t_{\delta w}}{t_w} = -\log_e(1 - \delta_w) \tag{8.33}$$

When $\delta_w = 0.63$, $t_{\delta w} = t_w$, the heave time constant. To reach 90% of the final steady state would take 2.3 time constants and to reach 99% would take nearly 5 time constants. In ADS-33, Level 1 handling qualities are achieved if $t_w < 5$ s, and for many aircraft types, values of 3–4 s are typical. In reality, because of the exponential nature of the response, the aircraft never reaches the steady-state climb following a step input. In fact, following a step collective, the aircraft approaches its steady-state in a particular manner. The instantaneous time to reach steady-state rate of climb $-w$, τ_w, varies with time and is defined as the ratio of the instantaneous differential (negative) velocity to the acceleration; hence

$$\dot{w} = -w_{ss} Z_w e^{Z_w t}$$
$$\tau_w = \frac{w - w_{ss}}{\dot{w}} = \frac{1}{Z_w} = -t_w \tag{8.34}$$

The instantaneous time to reach steady state, τ_w, is therefore a constant and equal to the negative of the time constant of the aircraft t_w – a somewhat novel interpretation of the heave time constant. The step input requires no compensatory workload but, theoretically, the aircraft never reaches its destination. To achieve the steady-state goal, the pilot needs to adopt a more complex control strategy and will use the available visual cues to ensure that τ_w reaches zero when the aircraft has reached the appropriate climb rate; the complexity of this strategy determines the pilot workload.

In the simulation trial reported in Ref. 8.33, the pilot was launched in a low hover and requested to accelerate forward and climb to a level flight trim condition that he or she considered suited the environment. To ensure that all the visual information for stabilization and guidance was derived from the outside world, head-down instruments were turned off. After establishing the trim condition, the pilot was required to negotiate a hill with 5° slope rising 60 m above the terrain. The terrain was textured with a rich, relatively unstructured surface, and to explore the effects of degraded visual conditions fog was located at distances of 80, 240, 480 and 720 m ahead of the aircraft. The fog was simulated as a shell of abrupt obscuration surrounding a sphere of 'clear air' centred on the pilot.

Table 8.1 HQRs and UCEs for terrain-hugging manoeuvres (Ref. 8.33)

Fog-line	80 m	240 m	480 m	720 m
HQR	6	5	4	4
UCE	3	2/3	2	1
VCRs				
Pitch	4.0	3.5	2.5	2.0
Roll	3.5	3.0	2.0	2.0
Yaw	3.5	3.0	2.0	2.0
Longitudinal	4.0	3.5	2.5	2.0
Lateral	3.0	3.0	2.0	1.5
Vertical	4.0	4.0	3.0	2.5

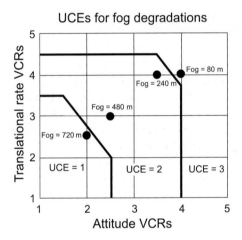

Fig. 8.24 Terrain following – UCE chart for different fog cases

The visual cue ratings (VCRs) and associated UCEs and handling qualities ratings (HQRs) for the different cases are given in Table 8.1. The methodology adopted was an adaptation of that in ADS-33, where the UCE is derived from VCRs given by three pilots flying a set of low-speed manoeuvres. The concept of UCE >3 is also an adaptation to reflect visual conditions where the pilot was not prepared to award a VCR within the defined scale (1–5). The VCRs are also plotted on the UCE chart in Fig. 8.24. As expected, the increased workload in the DVE led the pilot to award poorer HQRs, and the UCE degraded from 1 to 3. For the HQRs, the adequate performance boundary was set at 50% of nominal height and the desired boundary at 25% of nominal height. No numerical constraints were placed on speed but the pilot was requested to maintain a reasonably constant speed.

Key questions addressed in Ref. 8.33 were 'would the pilot elect to fly at different heights and speeds in the different conditions,' and 'how would these relate to the body-scaled measure, the eye-height?' Would the pilot use intrinsic τ guides to successfully transition into the climb, and what form would these take? Could the pilot control strategy be modelled based on τ following principles? Earlier in this chapter we discussed a pilot's ability to pick up visual information from the surface over which

Table 8.2 Average flight parameters for the terrain-hugging manoeuvres

Distance and time to fog (m, s)	Height (m)	Velocity (m/s, x_e/s)	Distance and time to 12 x_e point (m, s)
720, 13.0	24.4	55, 2.25	292, 5.3
480, 12.0	14.8	40, 2.70	176, 4.4
240, 7.7	13.8	31, 2.24	166, 5.36
80, 8.0	5.6	10, 1.78	67.4, 6.74

the aircraft was flying and drew attention to Ref. 8.15, where Perrone had hypothesized that the looming of patterns on a rough surface would become detectable at about 16 eye-heights (x_e) ahead of the aircraft. In the various simulation exercises conducted at Liverpool there is some evidence that this reduces to about $12x_e$ for the textured surfaces used; hence, we reference results to this metric in the following discussion.

The average distances and times to the fog-lines, along with the velocity and time to the 12 eye-height point ahead of the aircraft, are given in Table 8.2. The results indicate that as the distance to the fog-line reduces the pilot flies lower and slower, while maintaining eye-height speed relatively constant. Comparing the 720-m fog-line case with the 240-m case, the average eye-height velocity is almost identical, while the actual speed and height has almost doubled. For the UCE = 3 case, the aircraft has slowed to below 2 x_e per second, as the distance to the fog-line has reduced to within 20% of the 12 eye-height point. It is worth noting that the test pilot, who has extensive military and civil piloting experience, declared that the UCE = 3 case would not be acceptable unless urgent operational requirements prevailed; it simply would not be safe in an undulating, cluttered environment, and where the navigational demands would strongly interfere with guidance.

Figure 8.25 shows the vertical flight path (height in metres) and flight velocity (in metres/second and x_e per second) plotted against range (metres) for the different fog cases. The pilots were requested to fly along the top of the hill for a further 2000 m to complete the run.

The distances along the flight path to the terrain surface, as the sloping ground is approached, are shown in Fig. 8.26. While the actual distances vary significantly, the distances in eye-heights to the surface reduce to between 12 and 16 eye-heights, as the hill is approached and during the initial climb phase. The times to contact the terrain, $\tau_{surface}$, are shown in Fig. 8.27. Typically, the pilot allowed $\tau_{surface}$ to reduce to between 6 and 8 s before initiating the climb. These results are consistent with those derived in the acceleration–deceleration manoeuvre. In the UCE = 3 case (fog at 80 m) the pilot is flying at 10 m/s, giving about 8 s look-ahead time to the fog-line and only 2–3 s margin from the 12 x_e point. The results suggest a relationship between the UCE and the margin between the postulated, 12 x_e, look-ahead point and any obscuration; this point will be revisited towards the end of this section, but prior to this the variation of flight path angle during the climb will be analyzed to investigate the degree of τ guide following during the manoeuvre.

τ on the rising curve

For the τ analysis, the aircraft flight path angle γ is converted to γ_a, the negative perturbation in γ from the final state, as illustrated in Fig. 8.28.

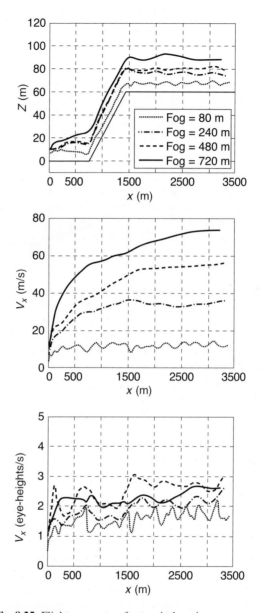

Fig. 8.25 Flight parameters for terrain-hugging manoeuvre

If the aircraft's normal velocity w is small relative to the forward velocity V, the flight path angle can be approximated as

$$\gamma \approx -\frac{w}{V} \tag{8.35}$$

If the final flight path angle is γ_f, then the γ-to-go, γ_a can be written as

$$\gamma_a = \gamma - \gamma_f \tag{8.36}$$

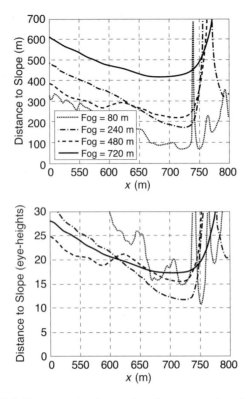

Fig. 8.26 Distance to the slope surface (in meters and eye-heights)

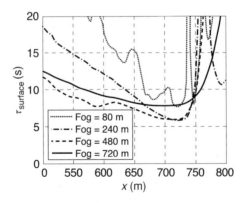

Fig. 8.27 Times to the slope surface (τ to surface)

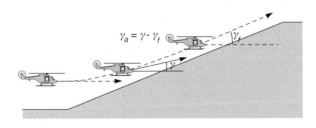

Fig. 8.28 Flight path angle

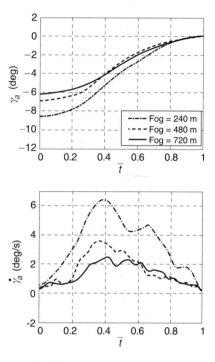

Fig. 8.29 γ_a, $\dot{\gamma}_a$ during the climb

The γ's-to-go and associated time rates of change are plotted in Fig. 8.29 for the UCE = 1 and UCE = 2 fog-line cases, although it should be noted that the 240-m case is actually borderline UCE = 2/UCE = 3. In Fig. 8.29, time has been normalized by the duration of the climb transient (fog = 240 m, $T = 2.5$ s; fog = 480 m, $T = 4.3$ s; fog = 720 m, $T = 4.7$ s). The final value of the flight path angle was chosen as the value when $\dot{\gamma}$ first became zero (thus defining T and γ_f); hence, although the hill had a 5° slope, the values used correspond to the first overshoot peak value. From that time on, the pilot closes the loop on a new τ gap, related to the flight path error from above. As can be seen from the initial conditions, the pilot tends to overshoot the 5° hill slope with increasing extent as the UCE degrades – flight path angle of 8.5° for the 240-m fog-line (UCE 2/3), 7° for the 480-m fog-line case (UCE 2) and 6° for the 720-m case (UCE 1).

Figure 8.30 shows the variation of τ_γ with normalized time. The fluctuations reflect the higher frequency content in the $\dot{\gamma}$ function. As the goal is approached the curves straighten out and develop a slope of between 0.6 and 0.7, corresponding to the pilot following the $\dot{\tau}$ constant guide with peak deceleration close to the goal.

As with the accel–decel manoeuvre, the pilot is changing from one state to another (horizontal position for the accel–decel, flight path angle for the climb), and as discussed earlier, the natural guide for ensuring that such changes of state are achieved successfully is the constant acceleration guide, with the relationship, $\tau_\gamma = k\tau_g$.

Figure 8.31 shows results for the τ_γ versus τ_g correlation for the terrain climb in the three fog conditions. In the UCE = 1 case (720 m fog), apart from a slight departure at the end of the manoeuvre, the fit is tight for the full 5 s ($R^2 = 0.99$). The departures from the close fit at both the beginning and end of such state change

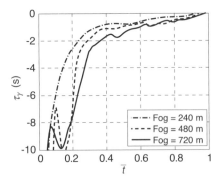

Fig. 8.30 Variation of τ_γ for climb transient

manoeuvres are considered to be transient effects, partly due to the need for the pilot to 'organize' the visual information so that the required gaps are clearly perceived and partly due to the contaminating effects of the pitch changes that disrupt the visual cues for flight path changes picked up from the optic flow. Table 8.3 gives the coupling coefficients, k, for all cases flown. The lower the k value, the earlier in the manoeuvre the maximum motion gap closure rate occurs (e.g., a value of 0.5 corresponds to a symmetric manoeuvre). There is a suggestion that k reduces as UCE degrades, confirmed by the results in Fig. 8.31. The pilot has commanded a flight path angle rate of about 6°/s less than 40% into the manoeuvre in the UCE = 2–3 case compared with about 2.5°/s about 50% into the manoeuvre in the UCE = 1 case.

Typical correlations between τ_γ and τ_g are shown in Fig. 8.32, plotted against normalized time. The relatively constant slope during the second half of the manoeuvre indicates that the pilot has adopted a constant $\dot{\tau}$ strategy. As expected, the test data track the constant acceleration guide fairly closely over the whole manoeuvre.

The results reveal a strong level of coupling with the τ guide. This was not unexpected. In a complementary study, τ analysis has been conducted on data from approach and landing manoeuvres for fixed-wing aircraft (Ref. 8.30). During the landing flare, the pilot follows the τ guide to the touchdown. Instrument approaches where the visibility was reduced to the equivalent of Cat IIIb (cloud base 50 ft, runway visual range 150 ft) were investigated, and in some cases the coupling reached the limiting case of constant τ, the pilot effectively levelling off just above the runway. The results presented are consistent with those presented in Ref. 8.30, unsurprisingly as the flare and terrain climb tasks make very similar demands on the pilot in terms of visual information.

From eqns 8.34 and 8.35, we can write the equation for flight path perturbation dynamics as

$$\dot{\gamma}_a - Z_w \gamma_a = Z_w \gamma_f - \frac{Z_{\theta 0}}{V} \theta_0 \tag{8.37}$$

with

$$\gamma_a(t = 0) = -\gamma_f \tag{8.38}$$

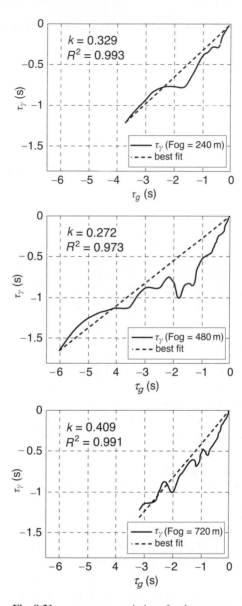

Fig. 8.31 τ_γ versus τ_g variations for three cases

Table 8.3 Correlation constants and fit coefficients – following the constant acceleration guide

Case	k	R^2
Fog = 720 m (run 1)	0.3930	0.9820
Fog = 720 m (run 2)	0.3490	0.9130
Fog = 720 m (run 3)	0.4090	0.9910
Fog = 480 m (run 1)	0.2720	0.9730
Fog = 480 m (run 2)	0.3670	0.9960
Fog = 240 m (run 1)	0.3490	0.9960
Fog = 240 m (run 2)	0.2060	0.8850
Fog = 240 m (run 3)	0.3290	0.9930

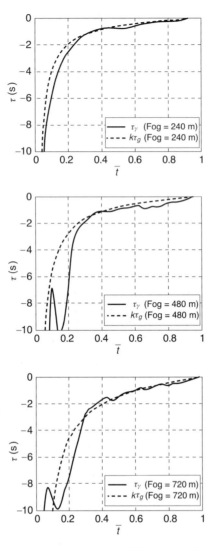

Fig. 8.32 Correlation between times to close the flight path gap for the motion and guides

Normalizing with the final values and manoeuvre time gives

$$\overline{\gamma}'_a + \frac{1}{\overline{t}_w}\overline{\gamma}_a = -\frac{1}{\overline{t}_w}(1 - \overline{\theta}_0) \tag{8.39}$$

where

$$\overline{\gamma}_a = \frac{\gamma_a}{\gamma_f}, \qquad \overline{t}_w = \frac{t_w}{T}$$

$$\gamma' = \frac{d\gamma}{d\overline{t}} = T\,\dot{\gamma} \tag{8.40}$$

$$\overline{\theta}_0 = \frac{\theta_0}{\theta_{0f}}, \qquad \theta_{0f} = \frac{V\,Z_w\,\gamma_f}{Z_{\theta 0}}$$

The instantaneous time to reach the goal of $\gamma = \gamma_f$ is defined as

$$\tau_{\gamma a} = \frac{\gamma_a}{\dot{\gamma}_a} \tag{8.41}$$

Following a τ guide such that $\tau_x = k\tau_g$ results in motion that follows the guided motion as a power law

$$x = C\,x_g^{1/k} \tag{8.42}$$

where C is a constant.

Recapping from earlier in the chapter, the constant acceleration guide has the forms given by

$$\tau_x = k\tau_g = \frac{k}{2}\left(t - \frac{T^2}{t}\right) = \frac{kT}{2}\left(\bar{t} - \frac{1}{\bar{t}}\right) \tag{8.43}$$

$$x_g = \frac{a_g}{2}\,T^2\,(\bar{t}^2 - 1) \tag{8.44}$$

a_g is the constant acceleration of the guide. As discussed earlier and shown in Fig. 8.21, the motion begins and ends hand-in-hand with the guide, but initially overtakes before being caught up and passed by the guide at the goal.

The equations for the flight path motion and its derivatives can then be developed (see Ref. 8.24) and written in the normalized form

$$\overline{\gamma}_a = -(1 - \bar{t}^2)^{(1/k)} \tag{8.45}$$

$$\overline{\gamma}'_a = \frac{2\bar{t}}{k}(1 - \bar{t}^2)^{(\frac{1}{k}-1)} \tag{8.46}$$

From eqn 8.39, the collective control can then be written in the general form

$$\overline{\theta}_0 = 1 + \bar{t}_a\,\overline{\gamma}'_a + \overline{\gamma}_a \tag{8.47}$$

Combining with eqns 8.45 and 8.46, the normalized collective pitch is given by the expressions

$$\theta_0 = \left(\frac{V\,Z_w\,\gamma_f}{Z_{\theta 0}}\right)\overline{\theta}_0$$

$$\overline{\theta}_0 = \left(1 - (1 - \bar{t}^2)^{1/k}\left[1 - \frac{2\bar{t}_w\,\bar{t}}{k(1 - \bar{t}^2)}\right]\right) \tag{8.48}$$

The normalized functions $\overline{\theta}_0$ and $\overline{\gamma}_a$ (independent of $\bar{t}_w$) are plotted in Figs 8.33 and 8.34, as a function of normalized time for different values of coupling constant k. The three cases correspond to the parameter $\bar{t}_w$ set at 0.5, 1.0 and 1.5. The strategy involves increasing the collective gradually, and well beyond the steady-state value, and then decreasing as the target rate of climb is approached. For $\bar{t}_w = 0.5$, when the heave time constant is half the manoeuvre duration, the overdriving of the control is limited to about 50% of the steady-state value. As the ratio increases to 1.5, so too does the overdriving to as much as 250% at the lower k values. This overshoot is unlikely to be achievable even when operating with a low-power margin. As k increases, the peak

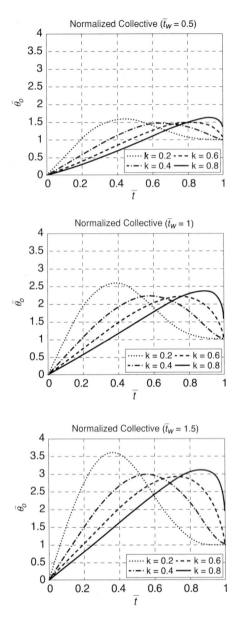

Fig. 8.33 Normalized collective pitch for a flight path angle change following a constant acceleration τ guide – variations with k and t_w

collective lever position occurs later in the manoeuvre until the limiting case where it is reduced as a down-step in the final instant to bring $\dot{\bar{\gamma}}_a$ to zero. When approaching a slope, the pilot has scope to select T, hence $\bar{t}_w$, and k, to ensure that the control and hence the manoeuvre trajectory are within the capability of the aircraft. Whatever values are selected, the control strategy is far removed from the abrupt, open-loop character associated with a step input.

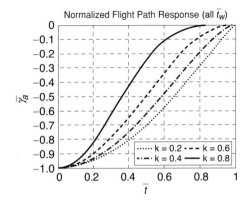

Fig. 8.34 Normalized flight path response for a flight path angle change following a constant acceleration τ guide–variations with k

A comparison of normalized collective inputs and flight path angles with the τ-coupled predictions for representative cases is shown in Fig. 8.35. The large peak for the 240-m fog-line case resulted in the overshoot to $10°$ flight path angle discussed earlier and could be argued in a case where the pilot has lost track of the cues that enable the τ-coupling to remain coherent. The actual pilot control inputs appear more abrupt than the predicted values but it again could be argued that the pilot needs to stimulate the flow-field initially with such inputs. There is good agreement for the flight path angle variations.

In Ref. 8.31, the notion was put forward that '*the overall pilot's goal is to overlay the optic flow-field over the required flight trajectory – the chosen path between the trees, over the hill or through the valley – thus matching the optical and required flight motion*'. This concept can be extended to embrace the idea that the overlay technique can happen within a temporal as well as spatial context. The results presented above convey a compelling impression of pilots coupling onto a natural τ guide during the 3–5 s of the climb phase of the terrain-hugging manoeuvre. As in the accel–decel manoeuvre, pilots appear to pick up their visual information from about 12 to 16 eye-heights ahead of the aircraft and establish a flight speed that gives a corresponding look-ahead time of about 6–8 s. As height is reduced, the pilot slows down to maintain velocity in eye-heights, and corresponding look-ahead time, relatively constant. The manoeuvre is typically initiated when the $\tau_{surface}$ reduces to about 6 s and takes between 4 and 5 s to complete. Of course, the manoeuvre time must depend on the heave time constant of the aircraft being flown. For the FLIGHTLAB Generic Rotorcraft simulation model used in the trials, t_w varies between 3 s in hover and 1.3 s at 60 knots, reducing to below 1 s above 100 knots. Much stronger interference between the aircraft and task dynamics would be expected with aircraft that exhibited much slower heave response (e.g., aircraft featuring rotors with high-disc loadings), as illustrated in Fig. 8.32.

The temporal framework of flight control offers the potential for developing more quantitative UCE metrics. For example, in the terrain manoeuvres described above, the UCE = 3 case was characterized by the distance to visual obscuration coming within about 20% of the 12 eye-height point (i.e., $\approx$ 1 s). For the UCE = 1 case, the margin was more than 6 s. Sufficiency of visual information for the task is the essence of safe flight and τ envelopes can be imagined which relate to the terrain contouring and

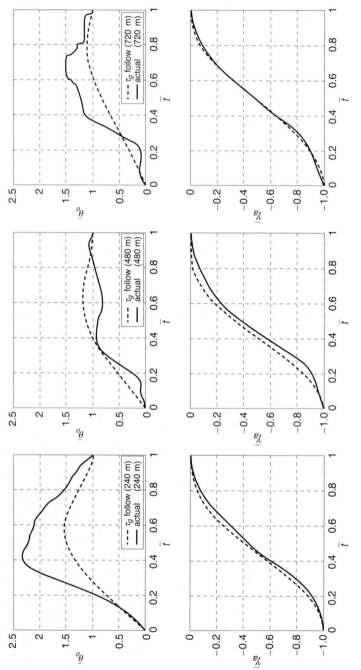

Fig. 8.35 Comparisons of normalized pilot control activity and flight path with τ-following strategy

associated MTEs. These constructs might then be used to define the appropriate speed and height to be flown in given terrain and visual environment. Visual information that provides clear cues to the pilot to enable coupling onto the natural τ guides could then form the basis for quantitative requirements for artificial aids to visual guidance that improve both attitude and translational rate contributions to the UCE.

The goal of a pilotage augmentation system, designed to extend operational capability in a DVE, must be to achieve performance without compromising safety, reducing fatigue by reducing cognitive workload and increasing confidence to allow aggressive manoeuvring. The designers of such synthetic vision systems can utilize the natural, reflexive pilot skills, and several pathway-in-the-sky type formats are currently under development or being explored in research (e.g., Refs 8.34, 8.35) that exhibit such properties. Designers also have the freedom to combine such formats with more detailed display structures for precision tracking, e.g., the pad-capture mode on the AH-64A (Ref. 8.36). This type of format requires the pilot to apply cognitive attention, closing the control loop using detailed individual features to achieve the desired precision, hence risking a loss of situation awareness with respect to the outside world. Achieving a balance between precision and situation awareness is the pilot's task and what is appropriate will change with different circumstances. Quite generally however, when equipped with an adequate sensor suite, there seems no good reason why a large part of the stabilization and tracking tasks should not be accomplished by the automatic flight control system.

Improving flying qualities for flight in a DVE is about the integration of vision and control augmentation. The UCE describes the utility and adequacy of visual cues for guidance and stabilization. The pilot rates the visual cues based on how aggressively and precisely corrections to attitude and velocity can be made. An assumption in this approach is that the aircraft has Level 1, rate command handling qualities in a GVE. In a DVE, the handling qualities of the aircraft degrade because of the impoverishment

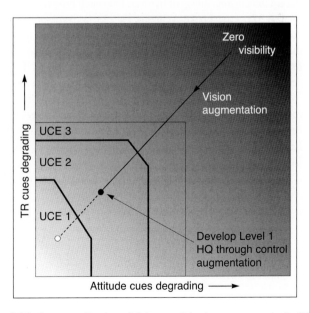

Fig. 8.36 Conceptualization of flying qualities improvements in the DVE

of the visual cues. According to the UCE methodology of ADS-33, provided the DVE is no worse than UCE = 3, Level 1 handling qualities can be 'recovered' by control augmentation. The augmentation process therefore appears straightforward, at least in principle: recover to UCE = 3 or better via vision augmentation, and then use control augmentation, to recover Level 1 flying qualities – Fig. 8.36 conceptualizes this idea.

Helicopter flight in degraded visibility will remain dangerous, with a consequent higher risk to flight safety, without vision augmentation. A key characteristic of a good vision aid is that it should provide the pilot with clear and coherent cues for judging operationally relevant, desired and adequate performance standards for flight path and attitude control. The ADS-33 handling qualities requirements then provide the design criteria for control augmentation. The higher the levels of augmentation and the stronger the control feedback gains, required an increased level of safety monitoring and redundancy to protect against the negative effects of failure, which leads us to the second topic in this chapter.

8.3 HANDLING QUALITIES DEGRADATION THROUGH FLIGHT SYSTEM FAILURES

The second issue of the *Journal of the Helicopter Society of Great Britain*, published in 1947, featured just two papers. The first was by Sikorsky and has already been referred to in the Introduction to this chapter (Ref. 8.1). The second was by O.L.L. Fitzwilliams, or 'Fitz' as he was affectionately known to his colleagues at Westland Helicopters, where he worked at the time of writing the paper in late 1947. Fitz had previously worked at the Airborne Forces Experimental Establishment at Beaulieu, near Southampton, England, during the Second World War and his paper partly covered his experiences there, including tests conducted on German rotorcraft acquired during the closing stages of the war. One such type was the first production helicopter (Ref. 8.37) – the Focke-Achgelis Fa 223, a development of the first practical helicopter, the FW.61. The Fa 223 aircraft was flown 'by its German crew, via Paris, to the A.F.E.E., at Beaulieu, where it arrived in September 1945, having performed the first crossing of the English Channel by a helicopter'. The Fa 223 was a twin rotor configuration with longitudinal cyclic control for pitch and differential collective for roll. Differential longitudinal cyclic gave yaw control in hover, supplemented by the rudder in forward flight. All these functions are nowadays to be found on a modern tilt rotor aircraft. More details of the flight control system on the Fa 223 are reported in Refs 8.38 and 8.39. Figure 8.37, from Ref. 8.39, shows a photograph of the aircraft at Beaulieu.

The handling qualities problems of the Fa 223 largely stemmed from the mechanism for lift control – essentially throttle and rotorspeed, which resulted in major deficiencies. To quote from Ref. 8.39,

> In hovering or in low speed flight, the control of the lift by means of the throttle is extremely sluggish and has contributed to the destruction of at least one aircraft following a downwind turn after take off. Moreover, the sluggishness of the lift control necessitates a high approach for landing and a protracted landing manoeuvre, during which the aircraft is exposed to the dangers consequent on operation of the change mechanism.

Fig. 8.37 The Fa 223 twin rotor helicopter at Beaulieu in British markings

The 'change mechanism' allowed the pilot, via a two-position lever, to change the mean blade pitch to its helicopter position (up) or its autorotation position (down). Lowering the lever caused the engine clutch to be disengaged, and the rotor blades rotate at a controlled rate (via a hydraulic ram and spring) to the autorotative pitch setting. This mechanism operated automatically in the event of engine failure, transmission failure and a number of other 'failure modes', some of which appear not to have been fully taken into account during design (Ref. 8.39). The failure mechanism was also irreversible and Fitz recounts his experience during an early flight test with the aircraft, when an auxiliary drive failure caused an automatic change to the autorotative condition (Ref. 8.37).

Once the mechanism had operated, even voluntarily, it was impossible to regain the helicopter condition in flight and a glide landing was necessary. In fact, with the high disc loading of this aircraft (author's note; 5.9 lb/ft^2 at 9,500 lb) and the absence of any control over the blade pitch, a glide landing was essential and if there was not enough height for this purpose the operation of this so-called safety mechanism would dump the aircraft as a heap of wreckage on to the ground. This actually happened, at about 60–70 ft above the ground, shortly after the machine arrived at Beaulieu, and I was among those who were sitting in it at the time. In consequence, I have a strong prejudice against trick gadgets in helicopter control systems and also a rooted objection to helicopters, however light their disc loadings, which do not allow the pilot direct manual control over the blade pitch in order to cushion a forced landing.

Although the Fa 223 first flew in August 1940, at the cessation of hostilities only three aircraft existed and the loss of the aircraft at Beaulieu brought to a premature end to the testing of what was undoubtedly a remarkable aircraft with a number of ingenious design features, notwithstanding Fitz's prejudices.

Nowadays, the safety assessment of this design through a failure modes and effects analysis (FMEA) would have deemed the consequences of this failure mode close to the ground 'catastrophic', and a greater reliability would be required in the basic design. An engine failure at low altitude would have been equally catastrophic of course, without control of collective pitch, as Fitz implied, but this is no justification for having a safety device that itself had a hazardous failure mode. In handling qualities terms, the failure, at least while the aircraft was in hover close to the ground, resulted in degradation to Level 4 conditions, the pilot effectively losing control of the aircraft. In the Introductory Tour to this book in Chapter 2, the author cited another example of a helicopter being flown in severely degraded handling qualities. The cases of the S.51 and the Fa 223 are highlighted not to demonstrate poor design features of early types (hindsight offers some clarity but usually fails to show the complete picture), but rather to draw to the reader's attention to the way in which the helicopter brought new experiences to the world of aviation, 40 years after the Wright brothers' first flight, at a time when 'flying qualities' was in its infancy and still a very immature discipline. But a holistic discipline it would become, spurred by the need for pilots and engineers to define a framework within which the performance increases pursued by operators, for commercial or military advantage, could be accommodated with safety. How to deal with failures has always been an important part of this framework and we continue this chapter with a discussion of current practices for quantifying flying qualities degradation following failures of flight system functions.

8.3.1 Methodology for quantifying flying qualities following flight function failures

The structure of Flying Qualities Levels provides the framework for analysing and quantifying the effects in the event of a flight system failure. Failures can be described under three headings – loss, malfunction or degradation – as described below:

(a) *Loss of function*: for example, when a control becomes locked at a particular value or some default status, hence where the control surface does not respond at all to a control input;

(b) *Malfunction*: for example, when the control surface does not move consistently with the input, as in a hard-over, slow-over or oscillatory movement;

(c) *Degradation of function*: in this case the function is still operating but with degraded performance, e.g., low-voltage power supply or reduced hydraulic pressure.

The first stage in a flying qualities degradation assessment involves drawing up a failure hazard analysis table, whereby every possible control function (e.g., pitch through longitudinal cyclic, yaw through tail rotor collective, trim switch) is examined for the effects of loss, malfunction and degradation. This assessment is normally conducted by an experienced team of engineers and pilots to establish the failure effect as minor, major, hazardous or catastrophic. Table 8.4 summarizes the definitions of these hazard categories in terms of the effects of the failure and the associated allowable maximum probability of occurrence per flight hour (Ref. 8.40). The table refers to the system safety requirements for civil aircraft.

In the military standard ADS-33, the approach taken is defined in the following steps (Ref. 8.3) (*author's italics for emphasis*):

(a) tabulate all rotorcraft failure states (*loss, malfunction, degradation*),

(b) determine the degree of handling qualities degradation associated with the *transient* for each rotorcraft failure state,

(c) determine the degree of handling qualities degradation associated with the subsequent *steady* rotorcraft failure state,

Table 8.4 Failure classification

Failure severity	Maximum probability of occurrence per flight hour	Failure condition effect
Catastrophic	Extremely improbable $<10^{-9}$	All failure conditions that prevent continued safe flight and landing.
Hazardous	Extremely remote $<10^{-7}$	Large reductions in safety margins or functional capabilities. Higher workload or physical distress such that the crew could not be relied upon to perform tasks accurately or completely. Adverse effect upon occupants.
Major	Remote $<10^{-5}$	Significant reductions on safety margins or functional capabilities. Significant increases in crew workload or in conditions impairing crew efficiency. Some discomfort to occupants.
Minor	Probable $<10^{-3}$	Slight reduction in safety margins. Slight increases in workload. Some inconvenience to occupants.

(d) calculate the probability of encountering each identified rotorcraft failure state *per flight hour*,

(e) compute the total probabilities of encountering Level 2 and Level 3 flying qualities in the *Operational and Service Flight Envelopes*. This total is the sum of the rate of each failure only if the failures are statistically independent.

Degradation in the handling qualities level, due to a failure, is permitted only if the probability of encountering the degraded level is sufficiently small. These probabilities shall be less than the values shown in Table 8.5. The probabilities used in ADS-33 are based on the fixed-wing requirements in Ref. 8.41, but converted from the probability per flight to the probability per flight hour, with the premise that a typical fixed-wing mission lasts 4 h. The requirements are not nearly as demanding as the civil requirements of Table 8.4 where the probabilities are typically two orders of magnitude lower.

In contrast, the UK Defence Standard (Ref. 8.42) defines safety criteria for failures of automatic flight control systems (AFCS) according to Table 8.6. The effect is defined

Table 8.5 Levels for rotorcraft failure states

Probability of encountering failure	
Within operational flight envelope	**Within service flight envelope**
Level 2 after failure $<2.5 \times 10^{-3}$ per flight hour	
Level 3 after failure $<2.5 \times 10^{-5}$ per flight hour	$<2.5 \times 10^{-3}$ per flight hour
Loss of control $<2.5 \times 10^{-7}$ per flight hour	

Table 8.6 AFCS failure criteria (Def Stan 00970, Ref. 8.42)

effect on FQ ->	minor	major	hazardous	catastrophic
probability of failure having 'effect' within intervention time per flying hour	$<10^{-2}$	$<10^{-4}$	$<10^{-6}$	$<10^{-7}$

within the so-called intervention time, which is a function of the pilot attentive state. With the pilot flying attentive hands-on, for example, the intervention time is 3 s, but in passive hands-on mode, the time increases to 5 s.

In the following sections, examples are given of failures in the three categories along with results from supporting research.

8.3.2   Loss of control function

Loss of control is a most serious event, and huge emphasis on safety in the aviation world is there to ensure that all possible events that might lead to a loss of a flight critical function are thoroughly examined and steps taken in the design process to ensure that such losses are extremely improbable. In military use, when helicopters can be exposed to the hazards of war, steps are sometimes taken to build in additional levels of redundancy in case of battle damage. For example, the AH-64 Apache features a back-up, fly-by-wire control system that can be engaged following a jam or damage in the mechanical control runs. The tail rotor is particularly vulnerable to battle damage, and a study carried out by DERA and Westland for the UK MoD and CAA, during the mid–late 1990s, identified that tail rotor failures occur in training and peace-time operations at a rate significantly higher than the airworthiness requirements demand. In the following section some of the findings of that study are presented and discussed.

Tail rotor failures

We broaden the scope to include both types of tail rotor failure: drive failure, where the drive-train is broken and a complete loss of tail rotor effectiveness results, and control failure, where the drive is maintained but the pilot is no longer able to apply pitch to the tail rotor. Both examples result in a loss of the yaw control function and can occur because of technical faults or operational damage. References 8.43 and 8.44 describe a programme of research aimed at reviewing the whole issue of tail rotor failures and developing improved advice to aircrew on the actions required, following a tail rotor failure in flight. The activity was spurred by the findings of the UK MoD/CAA Tail Rotor Action Committee (TRAC), which in particular were as follows (Ref. 8.9):

(a)   Tail rotor failures occur at an unacceptably high rate. MoD statistics between 1974 and 1993 showed a tail rotor technical failure rate of about 11 per million flying hours; the design standards require the probability of transmission/drive failure that would prevent a subsequent landing to be remote (<1 in a million flying hours, Ref. 8.42); a review of UK civil accident and incident data revealed a similar failure rate.

(b)   Tail rotor drive failures are three times more prevalent than control failures.

(c)   There appear to be significant differences in the handling qualities post-tail rotor failure, between different types (e.g., some designs appeared to be uncontrollable, and the probability of an accident resulting from a failure is greater with some types than others), although there is a dearth of knowledge on individual types.

(d) Improved handling advice would enhance survivability.

TRAC recommended that work should be undertaken to develop validated advice for pilot action in the event of a tail rotor failure for the different types in the UK military fleet, and also that airworthiness requirements should be reviewed and updated to minimize the likelihood of tail rotor failures on future designs. In the study that followed, validation was classified into three types – validation type 1 corresponding to full demonstration in flight, validation type 2 corresponding to demonstration in piloted simulation combined with best analysis and validation type 3 corresponding to engineering judgement based on calculation and also read-across from other types. It was judged that the best advice that could be achieved would be supported by type 1 validation for control failures and type 2 validation for drive failures.

The study, reported fully in Ref. 8.44, drew data from a variety of sources including the MoD and CAA, the US Navy, Marine Corps and Coast Guard and the US National Transport Safety Board (NTSB). The overall failures rates were relatively consistent across all helicopter 'fleets' and occurred in the range 9–16 per million flying hours. Recall that civil transport category aircraft are required to have a failure rate for flight critical components of no more than 1 in 10^9 flying hours. Of the 100 'tailfails' in the UK helicopter fleet between the mid-70s and mid-90s, 30% were caused by drive failure and 16% by control failure or loss of control effectiveness. Tail rotor loss due to collision with obstacle or vice versa accounted for 45% of the failures.

When investigating flying qualities in failed conditions, two different aspects need to be addressed – the characteristics during the failure transient and post-failure flying qualities, including those during any emergency landing. Both are, to some extent, influenced by the flight condition from which the failure has occurred. For example, the failure transients and optimum pilot actions will be quite different when in a low hover compared with those in high-speed cruise, well clear of the ground. The required actions will also be different for drive and control failures. Furthermore, in the case of control failures, the aircraft and appropriate pilot responses will depend on whether the control fails to a high pitch or low pitch, or some intermediate value, perhaps designed in as a fail-safe mechanism to mitigate the adverse effects of a control linkage failure.

Reference 8.44 describes a flight trial, using a Lynx helicopter, where control failures were 'simulated' by the second pilot (P2) applying pedals to the failure condition. P2 held the failed condition, while P1 endeavoured to develop successful recovery strategies using a combination of cyclic and collective. The high-pitch control failure mode results in a nose-left yaw (for anti-clockwise rotors), the severity of which depends on the initial power setting and aircraft speed. For example, the magnitude of control and yaw excursions will be greater from flight at minimum power speed than cruise. Accompanying the yaw will be roll and pitch motions, driven by the increasing sideslip. In the flight trials, a number of different techniques were explored to recover the aircraft to a stable and controllable flight condition. For failures in high-speed cruise, attempts to decelerate through the power bucket to a safe-landing speed were unsuccessful; the right sideslip (left yaw) built up to limiting values, and controlling heading with cyclic demanded a very high workload. A successful strategy was developed as illustrated in Fig. 8.38.

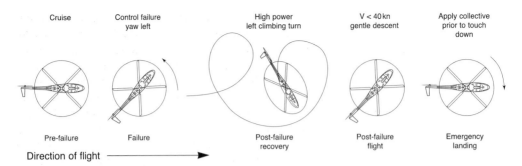

Fig. 8.38 Sequence of events following high-pitch tail rotor failure in cruise

A high-power climbing turn to the left gave a sufficiently stable flight condition so that deceleration could be accomplished without the aircraft diverging in yaw. The aircraft could then be levelled out at about 40 knots and a slow decelerating descent initiated. Gentle turns to both right and left (left preferred) were possible in this condition. The landing was accomplished by lining the aircraft up with the nose well to port and applying collective, and levelling the aircraft, just before touchdown to arrest the rate of descent and align the aircraft with the flight path. Running landings between 20 and 40 knots could be achieved with this strategy. In comparison, low thrust control failures resulted in the aircraft yawing to starboard. Reducing power then arrests the yaw transient and allows the aircraft to be manoeuvred to a new trimmed airspeed. During recovery it was important that the pilot yawed the aircraft with collective to achieve a right sideslip condition, so that collective cushioning prior to landing yawed the aircraft into the flight path.

The drive failures were conducted in the relative safety of the DERA advanced flight simulator (see Chapter 7, Section 7.2.2). The trial was conducted within the broad framework of the flying qualities methodology with task performance judged by the pilot's ability to land within the airframe limits, i.e., touchdown velocities and drift angle. Unlike a control failure, where the tail rotor continues to provide directional stability in forward flight, in a drive failure this stability augmentation reduces to zero as the tail rotor runs down. For failures from both hover and forward flight, survival is critically dependent on the pilot recognizing the failure and reducing the power to zero as quickly as possible. Figure 8.39 shows the sequence of events following a drive failure from a cruise condition. The aircraft will yaw violently to the right as tail

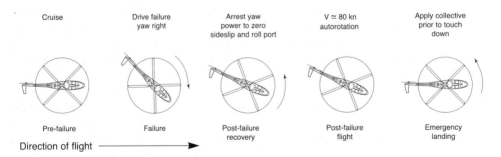

Fig. 8.39 Sequence of events following a tail rotor drive failure in cruise

rotor thrust reduces. The study showed that a short pilot intervention time is critical here to avoid sideslip excursions beyond the structural limits of the aircraft. The pilot should reduce power to zero as quickly as possible by lowering the collective lever. Once the yaw transients have been successfully contained, and the aircraft is in a stable condition, the engines can be shut down and the aircraft retrimmed at an airspeed of about 80 knots. With the Lynx, this gives about a 20% margin above the speed where loss of yaw control is threatened. Any attempt to find a speed–power combination that enabled continued powered flight risked a yaw breakaway which could drive the aircraft into a flat spin. Gentle turns to right and left (more stable) were possible from the 80 knots autorotation. The pilot approaches the landing with the aircraft nose to starboard and, in this case, raising collective to cushion touchdown yaws the nose to port and aligns with the flight path.

Reference 8.44 describes typical examples of tail rotor failure that were in the database investigated. One such example involved a Lynx helicopter taking off on a test flight following the fitting of a new tail rotor gearbox. With the aircraft in a low hover, a 'low power' control failure occurred. To quote from Ref. 8.44,

> *As the aircraft lifted there was a slight yaw to the right which the pilot compensated for, but by the time the aircraft was established in a 10 feet hover, a matter of only 2–3 seconds after launch, the aircraft was continuing to diverge to the right with full left pedal applied. The pilot called out 'full left pedal', and the aircraft accelerated into a right hand spot turn over which the aircrew had no control. The aircrew recalled the AEO's briefing and reduced the MR speed (which also reduced tail rotor speed and thrust), the yaw accelerated further, exacerbated by the fact that they were entering the downwind arc. The words of the briefing were then recalled 'right hand turn equals low power setting, therefore increase NR'. The speed select lever was pushed forward to increase MR speed (and hence tail rotor speed and thrust), the yaw rate slowed down. The aircrew regained control of the aircraft and were able to land without further incident.*

The Aircraft Engineering Officer (AEO) referred to here had actually led the tail rotor flight and simulation programme at DERA and is the first author of Ref. 8.43, and hence was very familiar with tail rotor failures. He had briefed the maintenance flight aircrew on actions to take in the event of a tail rotor failure. The advice proved crucial and the pilot's actions averted a crash; the story is told in Ref. 8.45.

Reference 8.44 also identifies a number of candidate technologies that could mitigate the effects of tail rotor failure, e.g., warning systems, integrated with health and usage monitoring systems, emergency drag parachutes. This is an important line of development in the context of safety. The accident data highlight that drive failures on most types are not very survivable. The two illustrations used to describe the failure types show a straightforward transition from the failure through the recovery to the landing. In practice, however, the pilot is likely to be confused initially by what has happened (note above example where the pilot operated the speed select lever in the wrong direction initially) and can quickly become disoriented as the aircraft not only yaws, but also rolls and pitches, as sideslip builds up. Also, the accident/incident data show that on several occasions the pilot has successfully recovered from the failure but the aircraft has turned over during the landing. Tail rotor failures make undue demands on pilot skill and attention and the way forward has to be to ensure that designs have sufficiently reliable drive and control systems so that the likelihood of component failure is extremely remote in the life of a fleet. Reference 8.44 recommends that the Joint Aviation Requirements be revised to provide a two-path solution to 'closing the regulatory gap' in respect of tail rotor control systems. Firstly, fixed-wing aircraft levels of redundancy of flight critical components are required. Secondly, where redundancy may be impractical, 'the design assessment should include a failure analysis to identify all failure modes that will prevent continued safe flight and landing and identification of the means provided to minimise the likelihood of their occurrence'.

Reference 8.44 also recommended that the ADS-33 approach of specifying failure transients (see next section) be adopted along with the collective to yaw coupling requirements and sideslip excursion limitations as a method of quantifying the effects of failure. Such criteria could also form the basis for evaluating the effectiveness of retrofit technologies, including contributions from the automatic flight control system. Tail rotor failures require the pilot to exercise supreme skill to survive what is, quite simply, a loss of control situation. If flying qualities degradation could be contained within the Level 3 regime, with controllability itself not threatened, then the probability of losing aircraft to such failures would be reduced significantly. Time will tell how effectively the recommendations of Ref. 8.44 are taken up by the Industry.

8.3.3   Malfunction of control – hard-over failures

A control malfunction occurs when the control surface does not move consistently with the input, and in this section the failure corresponding to an actuator moving 'hard-over' to its limit is considered. The effect of such a failure in the longer term is likely to be that the actuator is disengaged, although it does not necessarily follow that the control function is then 'lost'. The failure may be in a limited-authority actuator, feeding augmentation signals to the control surface in series with the pilot's inputs. The loss of this function is unlikely to be flight critical although it may be mission critical. For example, the loss of control augmentation may reduce the handling qualities in degraded visual conditions from Level 1 to Level 3 (e.g., loss of TRC sensor systems degrading response type to RC in a UCE = 3). The aircraft is still controllable but should the pilot attempt any manoeuvring close to the ground, the high risk of loss of spatial awareness would render the operation unsafe. If the actuator forms part of the primary flight control system then it would be normal to have sufficient redundancy so that a back-up system is brought into play to retain the control function following the failure. The question then becomes how much of a failure transient can be tolerated before the back-up system takes over? Similarly, how much failure transient can be tolerated before a runaway augmentation function is made safe? The transient response of the aircraft to failures therefore becomes part of the FMEA. ADS-33 (Ref. 8.3) addresses the consequences of these transients in a threefold context – possible loss of control, exceedance of structural limits and collision with nearby objects. Table 8.7 summarizes the requirements in terms of attitude excursions, translational accelerations and proximity to the OFE. The hover/low-speed requirements are based on the pilot being in a passive, hands-on state, perhaps engaged with other mission-related tasks. The 3-s intervention time then takes account of pilot recognition and diagnosis of the failure, before initiating the correct recovery action. The Level 3 requirements relate to the aircraft having been disturbed about 50 ft from its hover position before the pilot reacts. The assumption is that in such circumstances, the aircraft would have collided with surrounding obstacles or the ground. The Level

Table 8.7 Failure transient requirements (ADS-33)

		Flight condition	
			Forward flight
Level	Hover and low speed	Near earth	Up-and-away
1	3° roll, pitch, yaw $0.05\ g\ n_x, n_y, n_z$ no recovery action for 3 s	both hover and low speed and forward flight up-and-away reqts. apply	stay within the OFE no recovery action for 10 s
2	10° roll, pitch, yaw $0.2\ g\ n_x, n_y, n_z$ no recovery action for 3 s	both hover and low speed and forward flight up-and-away reqts. apply	stay within the OFE no recovery action for 5 s
3	24° roll, pitch, yaw $0.4\ g\ n_x, n_y, n_z$ no recovery action for 3 s	both hover and low speed and forward flight up-and-away reqts. apply	stay within the OFE no recovery action for 3 s

2 and Level 1 requirements then provide increasing margins from this 'loss of control' situation.

References 8.46 and 8.47 both deal with failure transients and degraded flying qualities of tilt rotor aircraft. In Ref. 8.46, the methodology for dealing with loss, malfunction and degradation in the development of the European Civil Tilt Rotor is described. Reference 8.47 is concerned with the V-22 and will be returned to later in this section. In Ref. 8.46, the up-and-away requirements for the civil tilt rotor were expressed in terms of the transient attitude excursions following a failure, shown in Table 8.8, with the assumption that the pilot was hands-off the controls and would require 3.5 s to initiate recovery action (Ref. 8.48).

Degradation into Level 4 handling qualities would result from attitude transients shown with the consequent high risk of loss of spatial awareness and hence control. An analysis was conducted using the civil tilt rotor simulation model to establish the handling qualities boundaries as a function of the parameters of the hard-over as summarized in Fig. 8.40. The control surface is driven at the maximum actuation rate to a value X_1, which is then held for the so-called passivation time, after which the surface returns to an offset value X_2.

Figure 8.41 shows results for the roll angle following a failure of the left aileron to 16° initiated at 0.1 s. For the case shown, the aileron reached the failure limit, driven at the maximum actuation rate, in 0.4 s. The aileron holds the hard-over position for the passivation time of 1.5 s, after which the surface is returned to an offset value of 3° at the reduced rate of the back-up system. The pilot takes control at 3.5 s, applying full right aileron and achieving this in 1 s (reduced actuation rate of 100%/s). In the example

Table 8.8 Failure transient requirements (Ref. 8.46)

Transient attitude excursions; forward flight, up-and-away	
Level 1	20° roll, 10° pitch, 5° yaw
Level 2	30° roll, 15° pitch, 10° yaw
Level 3	60° roll, 30° pitch, 20° yaw

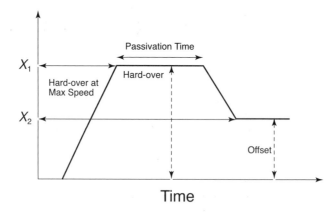

Fig. 8.40 The general form of the control malfunction (Ref. 8.46)

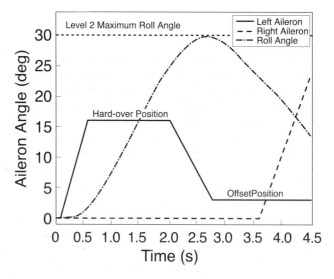

Fig. 8.41 Example of the roll angle response to aileron failure for the tilt rotor in airplane mode (Ref. 8.46)

shown, the maximum roll angle of 30° occurred at about 3 s and the transient response was already reducing by the time the pilot applied corrective action. In the study reported in Ref. 8.46, the failure parameters in Fig. 8.40 were varied to define the handling qualities boundaries according to Table 8.8, using the methodology typified in Fig. 8.41. In this way the designer can use the results to establish the required safety margins in the design that guarantee that the handling stays within the Level 1 or Level 2 regions.

Figure 8.42 shows the handling qualities regions using the two-parameter chart of maximum aileron deflection versus passivation time. The results are shown for the zero offset condition. So, for example, with a passivation time of 1.5 s, the Level 3 boundary is reached with failure amplitude of about 15°. The methodology allows a wide range of different scenarios to be assessed. Cases where the failed actuator is not

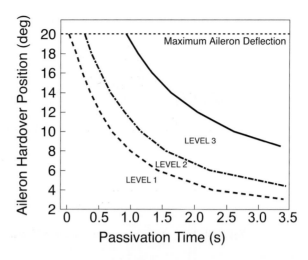

Fig. 8.42 Handling qualities levels for roll response shown as a function of passivation time and aileron hard-over amplitude (Ref. 8.46)

returned to an offset can also be considered, as can cases where the failure magnitude is limited to the authority of the in-series, stability augmentation.

The recovery control action discussed above is formulated clinically as shown in Fig. 8.40 and with the very large number of test cases needing quantification, off-line production of the knowledge contained in charts like Fig. 8.42 is the only realistic approach. The results derived from such analysis provide the 'predicted' handling qualities. But, as with flying qualities testing in normal conditions, piloted tests are required to support and validate the analysis. It has become a normal practice in some qualification standards to require flight testing to be carried out, e.g., SCAS failures in the UK Defence Standard (Ref. 8.42), but in most cases, the risk to flight safety is so high that such testing is actually never carried out, particularly addressing the question – what impact does the degradation have on flying qualities post-failure? In Ref. 8.47, the methodology adopted during qualification of the V-22 flying qualities is described, wherein extensive use of piloted simulation was made to answer this question. Following the recovery from the failure transient, it is expected that this aircraft will need to fly the equivalent of MTEs even in fly-home mode, although some may be impossible to set up. Reference 8.47 highlights the importance of maintaining the same performance standards as when flying operationally without failures. To quote from Ref. 8.47,

> *Relaxing task requirements can open the possibility of a very undesirable dilemma: the severely crippled aircraft could receive HQRs that are not much worse than, or possibly are even better than, those for the unfailed aircraft. For the precision hover example, suppose the performance limits were relaxed from 'hover within an area that is X feet on each side' to 'don't hit the ground'. Precision hover is typically more difficult in the simulator than in flight, so Level 2 HQRs (4, 5, or 6) would not be surprising for the unfailed aircraft performing the tight hover MTE. Artificially opening the performance limits, to accommodate the presence of the failure, could lead a pilot to assign a comparable – or better – HQR for what might be an almost uncontrollable configuration.*

So, the extent of the handling degradation following system failures can be properly measured only through a direct comparison with the unfailed aircraft, using both predictive (off-line) and assignment (pilot assessment) methods.

It is also important to establish the pilot's impressions of the transient effect of the failure and ability to recover, aspects not covered by a handling rating *per se*. The failure rating scale developed by Hindson, Eshow and Schroeder (Ref. 8.49) in support

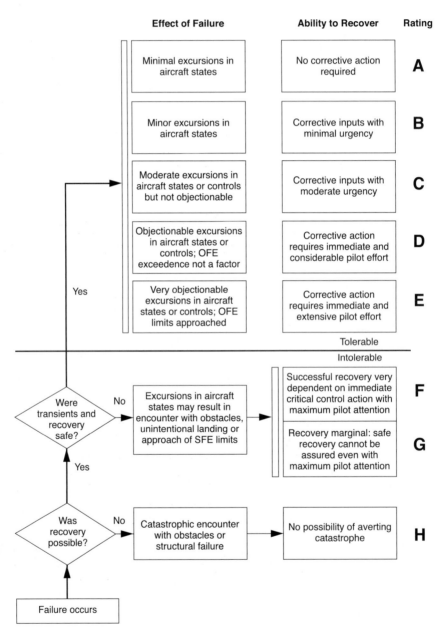

Fig. 8.43 Failure transient and recovery rating scale

of the development of an experimental fly-by-wire helicopter was modified in the V-22 study, and this version is reproduced here as Fig. 8.43. The essential modifications relative to the original Ref. 8.49 scale were firstly the nature of the questions on the left-hand side; positive answers moved up the scale, as in the Cooper–Harper handling qualities scale. Secondly, the exceedances in failure categories A to F were referred to the safe flight envelope (SFE) rather than the OFE, and thus to effectively maintain Level 2 handling qualities.

Pilots rate two aspects of the failure using Fig. 8.43 – the effect of the failure itself and the consequent ability to recover to a safe equilibrium state. Failure ratings (FR) A to E would be regarded as tolerable, F to G as intolerable, with a marginal recovery capability, while a rating of H means there is 'no possibility of averting a catastrophe'. In the programme to develop the European civil tilt rotor this methodology has been extended to produce an integrated classification of failures as illustrated in Fig. 8.44 (Ref. 8.50) and is itself an extension of that adopted in the development and certification of the NH-90 helicopter. The integration brings together the failure category concept (minor-catastrophic), the FR and the HQR. In Fig. 8.44, the OFE exceedance requirements were maintained corresponding to failures A to E rather than the SFE modification in Ref. 8.47. We can see that a 'minor' failure that elicits an FR

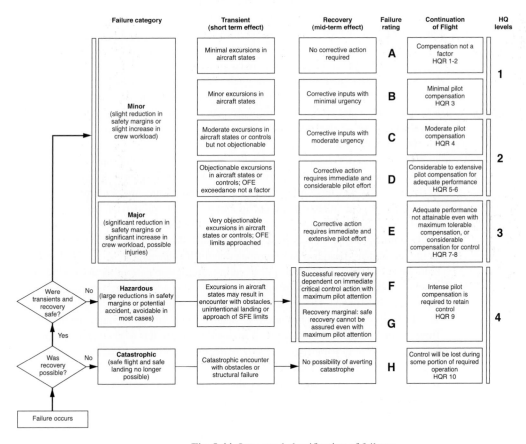

Fig. 8.44 Integrated classification of failures

of A or B results in the aircraft maintaining its Level 1 handling qualities. If the ratings degrade to C or D, the aircraft falls into the Level 2 category. Major failures correspond to degradations to Level 3 handling qualities while Hazardous or Catastrophic failures correspond to the aircraft being 'thrown into' the Level 4 region where loss of control is threatened.

The integration is considered to offer an important new framework for relating the impact of flight system failures on flight handling qualities, within which engineers and pilots can develop and qualify systems that are safe.

As discussed above, a malfunction can often lead to a loss of control function, but we need now to consider the third failure type where the control function is still operating but with degraded performance.

8.3.4   Degradation of control function – actuator rate limiting

Degradation in performance of a control function typically occurs when the power supply to a control surface actuator, mechanical or electrical, fails in some way. With the exception of some very small types, most helicopters feature powered control systems. The pilot moves the cockpit inceptors, and through a system of rods, bell-cranks, levers and pulleys (or computers and electrical signalling in a fly-by-wire system) the movement is transmitted to the input side of a hydraulic (or electro-mechanical) actuator. The output side of the actuator is connected to the non-rotating face of the rotor swashplate, which can be tilted or raised/lowered, transmitting signals through the rotating face of the swashplate and through the mechanical pitch link rods to the rotor blades. Three actuators are required on the main rotor to provide collective pitch, longitudinal cyclic and lateral cyclic pitch. The tail rotor normally requires only a collective pitch actuator. The high-reliability level required for rotor controls is usually achieved, 'below the swashplate', through redundancy of the mechanical or electrical actuation system. A dual-redundant hydraulic actuator would normally be supplied by two hydraulic systems, each providing 50% of the power. Failure of one of the supplies results in a degradation of performance, such that the maximum velocity and acceleration at which the control surface can be moved are reduced. The extent of the reduction depends on the pilot's control inputs, since the same power system is typically driving collective and cyclic, but a straight comparison before and after failure would normally show a corresponding 50% reduction in maximum rate. This rather simple assumption is being used to establish the actuation power requirements in the preliminary design of the European civil tilt rotor aircraft as reported in Ref. 8.46. Degradation in the 'predicted' handling qualities, e.g., attitude bandwidth and quickness, can be derived from off-line analysis of the nonlinear simulation of the aircraft and its systems. The actuation rate at which the predicted handling falls into the Level 3 region establishes a minimum acceptable value corresponding to the transition from minor to major failure category (Fig. 8.44). Because it is likely that the predicted handling for several parameters is likely to degrade at the same time in this scenario, it is especially important to check the predictions by carrying out piloted tests using operational MTEs. The importance of this requirement is emphasized in Ref. 8.47, where the point is made that anything less than the same performance requirements as demanded in the MTEs for normal operational flight would lead to pilots relaxing their control strategy and consequently not really experiencing the adverse effects of degradation. While this might be possible in some or even most conditions, flight

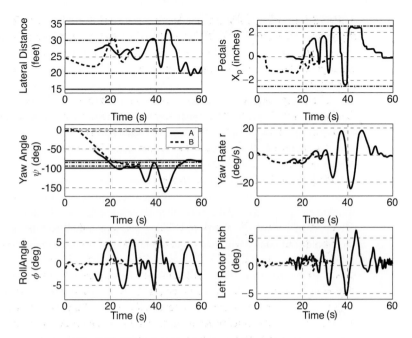

Fig. 8.45 Control and responses in a hover turn manoeuvre

through heavy turbulence, recovering to a moving deck or a confined area in poor visibility may be unavoidable, hence some assessment of the likely consequences needs to be conducted.

Figure 8.45, from Ref. 8.46, captures this point perfectly. The data show results for two test pilots flying a tilt rotor aircraft in a 90° hover-turn manoeuvre to the ADS-33 GVE performance standards. The actuation rate for the yaw control function, in this case provided by differential longitudinal cyclic pitch, was reduced incrementally until Level 3 ratings were awarded. The case shown corresponds to a reduced cyclic pitch rate of 3°/s. It can be seen that Pilot B flew the manoeuvre with an initial step control input, allowing the aircraft to slow gradually to the required heading with minimal closed-loop action. Pilot A, on the other hand, attempted to stop the yaw rate with a more abrupt input but immediately entered a pilot-induced oscillation. Over about 10 s the oscillatory pedal inputs increased to the stops with yaw rate building up to greater than 20°/s. The rate limiting on the cyclic pitch (differential longitudinal cyclic provides yaw control in hover) can be seen clearly in the lower-right figure. Pilot B returned an HQR 4 for this case, the same rating that both pilots had returned for the case with no rate limiting. Pilot B had continued to adapt his strategy as the system performance degraded to fly the task with moderate levels of compensation. Pilot A, with each new configuration, initially attempted the manoeuvre with a similar strategy, deliberately to focus on changes in handling. For the case shown, the pilot returned an HQR of 10 on the basis that he had to stop flying the aircraft (get 'out-of-the-loop') for several seconds. The aircraft had not, however, drifted outside the adequate performance limits on position and bank, only yaw angle, and arguably, a Level 3 rating would have been more appropriate. However, the pilot was holding right pedal on the stops for more than 5 s to bring the aircraft onto heading before coming out of the loop and his impression was that he had, indeed, lost control. The rate limit on longitudinal cyclic was set at $4°/$s, a very similar value to that defined for pitch control in the accel–decel manoeuvre. The full power actuation rate authority was actually set at $10°/$s in this preliminary design study, 2.5 times the rate at which the Level 2/3 boundary had been predicted, hence giving an additional margin of safety for compensation in harsher environmental conditions.

Studying the effects of system failure brings home the importance and associated cost of safety in aviation, by far the safest sector in the transport industry. Ensuring that the effects of flight system loss, malfunction or degradation do not lead to a worsening of handling beyond Level 2 provides a major contribution to this safety. Players in this aspect of safety include regulatory bodies, requirements writers, design engineers, manufacturers, certification agencies, operators, maintenance engineers, training organizations and the pilots themselves; in other words, practically the whole aerospace community is involved. Nothing less than a total commitment to safety by the whole community will lead to an eradication of accidents resulting from system failures.

This brings us to the third situation where an otherwise Level 1 aircraft can be literally thrown into a degraded condition – by encountering severe atmospheric disturbances.

8.4 ENCOUNTERS WITH ATMOSPHERIC DISTURBANCES

To a first, albeit rather crude, approximation, the response of a helicopter to an atmospheric disturbance can be measured in terms of the force and moment derivatives discussed in Chapters 1, 4 and 5 of this book. In Chapter 1, the heave response to a vertical gust was touched on, and expressions for the contributing derivative Z_w were then developed in Chapter 4. The discussion was extended in Section 5.4 to the modelling of atmospheric disturbances and the subsequent ride qualities. Heave response tends to dominate the concern because the rotor is the dominant lifting component on a helicopter. As the forward velocity increases, the energy of the 'gust response' is absorbed more and more by the vibratory loading, since this dominates the component of lift proportional to forward speed. The heave response derivative, Z_w, becomes asymptotic to the expression $-\frac{\rho a_0 (\Omega R)}{4 \ell_b}$ as velocity increases (see eqn 5.79). This represents an approximation to the initial vertical bump when flying into a vertical gust and is proportional to rotor blade tip speed and inversely proportional to blade loading (ℓ_b). In comparison, as a fixed-wing aircraft flies faster, the product of dynamic pressure and incidence leads to a heave response proportional to forward velocity V ($-\frac{\rho a_0 V}{2 \ell_w}$) and inversely proportional to wing loading (ℓ_w). The charts and tables of derivatives at the end of Chapter 4 give a 'feel' for the magnitude of the gust response; a typical helicopter has a value of Z_w of about 1 m/s^2 per m/s at high speed, giving a 1-g bump when entering a vertical gust of magnitude about 10 m/s. On entering such a gust the aircraft would be climbing at 6.3 m/s after 1 s ($t_{63\%} = -\frac{1}{Z_w}$) and would continue climbing, approaching 10 m/s asymptotically. Similarly, the response to flight through a variable gust field can be approximated by the aerodynamic components of the damping derivatives L_p and M_q (the gyroscopic components in expressions like eqns 4.86–4.89 are not included, only the aerodynamic terms), assuming the gust field can be approximated by a linear variation across the rotor disc. Linear approximations have been used extensively by the fixed-wing community to analyze and quantify the gust response of aircraft, and similar methods are available and used in helicopter design.

Very strong atmospheric disturbances, where linear models are questionable, should be avoided in operations where possible. However, there are some situations where a helicopter has to be flown through a vortex-infested, swirling flow-field to reach its landing site. The helicopter, recovering to a ship or helideck, having to fly through the airwake from the superstructure presents such an example. In Ref. 8.7 (see also Refs 8.51 and 8.52), the ship airwake was described as the 'invisible enemy' by virtue of the fact that a pilot is very vulnerable to the degraded handling qualities arising from the effects of the unseen, unsteady and swirling vortical flow structures in the lee of a ship's superstructure, which commonly is where the helicopter landing deck is situated. Over the decade since the publication of the first edition of this book a considerable amount of research, typified by that reported in Refs 8.51 and 8.52, has been conducted to develop modelling and simulation capabilities able to predict the ship–helicopter operating limits (SHOLs) in the presence of the ship's airwake. The kind of problem faced by operators is shown in the example of the Royal Navy's Royal Fleet Auxiliary (RFA), which has two landing spots, spot 1 on the port side close to the hangar and spot 2 on the starboard side to the aft of the flight deck. The difference in the SHOLs for spots 1 and 2 is compared with the original requirement during procurement in Fig. 8.46 (Ref. 8.51). The SHOL is the shaded area on the polar

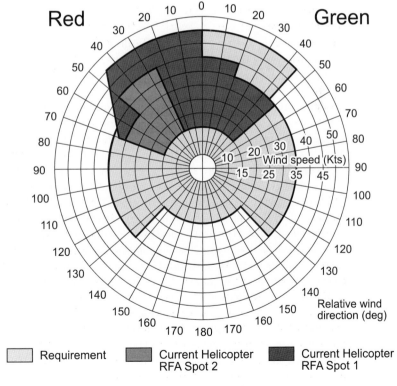

Fig. 8.46 Comparison of SHOLs for front and aft spots on RFA (from Ref. 8.51)

plot of relative wind speed and direction. Although both are restricted, the SHOL for spot 1 makes it almost unusable in most wind-over-deck conditions. The problem is caused by the combination of the helicopter operating close to the hangar face at low heights and the airwake created by the geometry of the ship, the hangar and the flight deck. Mean and unsteady downwash velocity components are so high in this region that the pilot has difficulty taking off and hovering.

The SHOL is defined to allow operations to be conducted safely in the presence of disturbed atmospheric conditions and the additional difficulties associated with ship motion and degraded visibility. For the main discussion in this section of the chapter, we turn to a situation where operating limits are more difficult to define – the effect of fixed-wing aircraft trailing vortices on helicopter handling qualities.

8.4.1   Helicopter response to aircraft vortex wakes

A key part in the process of assessing the response of helicopters to the vortices of fixed-wing aircraft is the development of severity criteria for the encounters. Severity criteria based on handling qualities analysis link directly with the central thrust of this book. The results presented here are drawn from the author's research with colleagues (Ref. 8.53–8.56), which was initially aimed at developing safety cases for the positioning of final approach and take-off areas (FATO) at airports. The work has since expanded to inform the development of operating procedures for runway-independent aircraft (Ref. 8.56), hence assisting the timely expansion of vertical flight aircraft operations, both helicopters and tilt rotor aircraft, to and from busy hubs.

The wake vortex

Wake vortices are an extension of the so-called bound vorticity of a lifting surface, shed from the wing tips as distinct vortex structures and rolling up with the span-wise shed vorticity into a counter-rotating pair. The resulting flow structure descends in the mean downwash from the wing, moves laterally with any horizontal wind and eventually breaks up as the inner core, kept together almost as a solid body by strong viscous forces, becomes unstable. Both fixed- and rotary-wing aircraft leave a vortex wake behind, dissipating the energy required to maintain the aircraft aloft. Figure 8.47 illustrates the flow topology in the tip vortex.

The velocity in the vortex core increases linearly with radial location from the centre, the fluid rotating effectively as a solid body. The flow here is 'rotational' so that elements of fluid rotate as they are drawn around in a circular pattern (see the white triangular fluid elements in Fig. 8.47). Outside the core, the flow is largely irrotational and the velocity decreases with distance from the core centre. An element of fluid would be drawn into the vortex from the surroundings, and would move toward the centre along a spiralling streamline, without rotation (the decreasing velocity with radius allows this to happen – see the grey fluid elements in Fig. 8.47) until reaching the outer edge of the core. The manner in which the rotational core is fed with irrotational fluid and the 3-dimensional development of the vortex, both radially and streamwise, has been the subject of aviation research for decades (see Ref. 8.57 for '. . . a consolidated European view on the current status of knowledge of the nature and characteristics of aircraft wakes . . .'). For the purposes of this analysis a rather simple model of the vortex structure is used and will be described following an appraisal of the severity criteria in handling qualities terms.

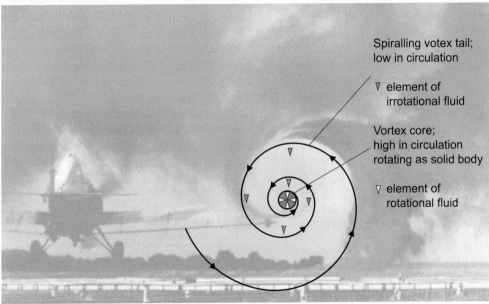

Fig. 8.47 The wake vortex structure

Hazard severity criteria

The definition of a hazard set out in the SAE's Aerospace Recommended Practice ARP4761 (Ref. 8.40) is 'a potentially unsafe condition resulting from failures, malfunctions, external events, errors or a combination thereof'. In busy airspace, aircraft are regularly exposed to the risk of experiencing unsafe conditions through wake-vortex encounters (Ref. 8.58). Separation is designed to minimize this risk, but the risk is ever present, and its 'acceptability' is a function of the severity of the disturbance and the probability of occurrence. Generally, severe disturbances must be improbable and as the level of severity decreases, the frequency of occurrence can increase for the same risk. This critical relationship underpins aviation safety and system design.

There are two main concerns and related questions regarding disturbance severity:

(a)   Does the disturbed aircraft have sufficient control margin for the pilot to overcome the disturbance?
(b)   Can the disturbance transient lead to an unsafe flight condition if not checked within a reasonable pilot intervention time, in terms of collision with surfaces, exceedance of flight envelope, risk of pilot disorientation or loss of control?

The detailed answers to these questions lie in understanding the nature of the response of the aircraft to a vortex disturbance. In handling qualities terms, the response characteristics of immediate interest relate to the moderate to large amplitude criteria – quickness and control power (see Chapter 6). To recap, the control power is the amount of response achievable with the available control margin; the response quickness is the ratio of peak attitude rate to attitude change in a discrete attitude change manoeuvre. Quickness is inversely related to the time to change attitude and will be affected by roll/pitch damping, actuator limits and, to an extent, static stability effects, e.g., how much and in what sense sideslip or incidence changes occur during the manoeuvre.

The sufficiency of attitude control margins in terms of quickness (for pitch manoeuvres up to $30°$, roll up to $60°$) and control power (for pitch manoeuvres $>30°$, roll $>60°$) is of primary concern. Reference 8.53 highlighted that the initial disturbance to an encounter with a vortex, aligned in the same direction as the helicopter (parallel encounter), will be in pitch. The non-uniform (lateral) incidence distribution imposed on the rotor disc by the vortex in a parallel encounter has a similar effect to the application of longitudinal cyclic pitch, the flapping response occurring $90°$ later to give pitch up/down moments. This is in contrast to the rolling moment disturbance experienced by fixed-wing aircraft following a parallel encounter. Figures 8.48 and 8.49 show the MTE-dependent pitch axis quickness and control power criteria boundaries for low-speed/hover tasks according to ADS-33 (Ref. 8.3, also Section 6.4).

A helicopter flying into the irrotational 'tail' of the vortex wake will experience a more uniform incidence distribution across the rotor disc, leading to thrust and power changes. In the heave axis, the corresponding Level 1 response criteria are defined in terms of control power (minimum of 160 ft/min, 1.5 s after initiation of rapid displacement of collective control from trim) and vertical rate time constant ($t_{63\%} < 5$ s). These correspond approximately to a hover rate of climb performance of 650 ft/min with a 5% thrust margin. Level 2 performance is obtained with a minimum climb rate of 55 ft/min and Level 3 with 40 ft/min. These values are relatively low and it can be appreciated that a general downwash of magnitude about 10 ft/s would nearly swamp the Level 1 performance margin. An aircraft should possess at least the Level 1 performance standards described above for the pilot to be able to fly moderately aggressive low-speed manoeuvres with precision and low compensation. The question arises as to whether an aircraft designed to meet the ADS-33 performance standards will have sufficient margin for the pilot to overcome the effects of a vortex encounter. The second issue listed above concerns the aircraft motion transients in response to the vortex encounter, and this will be addressed using the same methodology described for

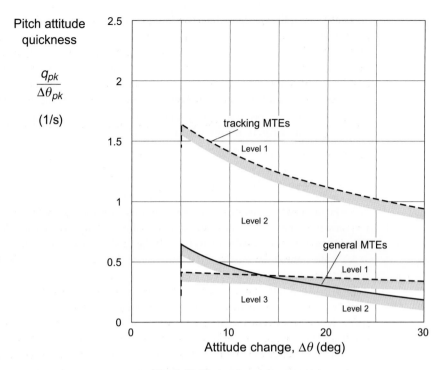

Fig. 8.48 Pitch axis quickness

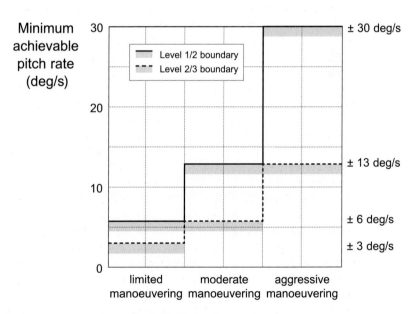

Fig. 8.49 Pitch axis control power

control system failures. ADS-33 sets requirements for the response to system failure transients in the form of Table 8.7 presented earlier in this chapter.

The focus in the study described in Ref. 8.55 was the severity of encounters for rotorcraft flying in hover and low speed (<45 knots), including low-speed climbs. The main concern is with the first column in Table 8.7, and particularly the Level 2/3 boundary, on the basis that this differentiates between safe and unsafe conditions. The Level 2/3 boundary corresponds to a transient that would result in a displacement of the aircraft of about 20 ft (6.1 m), with a velocity of about 20 ft/s (6.1 m/s) and angular rate of $10°/s$, after 3 s. A question that arises when expressing the encounter transients using these criteria is – what should the pilot intervention time be? The 3 s in ADS-33 corresponds to a scenario of a single pilot attending to other mission duties while in hover with auto-hover engaged. In the UK Defence Standard (Ref. 8.42) this would correspond to passive hands-on operation. For attentive hands-on operation, the pilot response time is 1.5 s according to Ref. 8.42, following control system failures. In the US civil certification standards (Ref. 8.59) the response time (for hover operations) is set at the normal pilot recognition time (0.5 s). However, a strong argument could be made for increasing this to 1.5 s in divided-attention situations or when operating with auto-hover engaged.

To address the two questions posed above, a series of piloted simulation trials using the facility at The University of Liverpool (see Appendix 8A) was carried out. The modelling and simulation environment used in the studies was FLIGHTLAB and the HELIFLIGHT motion simulator (Ref. 8.60). The two aircraft featured in the study were the Westland Lynx and the FLIGHTLAB Generic Rotorcraft (FGR), configured as a UH-60 type helicopter. Key configuration parameters of the two aircraft are given in Table 8.9.

The aeromechanics modelling features are summarized in the following:

- blade element rotor with look-up tables of quasi-steady, nonlinear lift, drag and pitching moment as functions of incidence and Mach number (five equi-annulus segments),
- FGR – four rigid blades with offset flap hinge; Lynx – four elastic blades with first three coupled modes,
- three-state dynamic inflow model,
- Bailey disc tail rotor with δ_3 coupling,
- three-state turbo-shaft engine/rotorspeed governor (rotorspeed, torque, fuel flow),
- look-up tables of fuselage and empennage forces and moments as nonlinear functions of incidence and sideslip,
- rudimentary quasi-steady interference between rotor wake and fuselage/empennage,

Table 8.9 Helicopter parameters in the vortex encounter study

	Lynx	FGR
rotor radius	21 ft (6.4 m)	27 ft (8.2 m)
weight	11000 lbf (4911 kgf)	16300 lbf (7277 kgf)
disc loading	7.9 lbf/ft^2 (38.2 kgf/m^2)	7 lbf/ft^2 (34.4 kgf/m^2)
flap hinge offset	12% (equivalent)	5% (actual)
rotorspeed	35 rad/s	27 rad/s
nominal hover power margin	21%	34%

- basic mechanical control system with mixing unit and actuators plus limited-authority stability and control augmentation system (SCAS – rate damping with attitude control characteristics at small attitudes in Lynx),
- rudimentary three-point undercarriage.

This level of modelling is generally regarded as medium fidelity, capable of capturing the primary trim and on-axis responses within about 10% of test data. Handling qualities parameters are also reasonably well predicted by this modelling standard. A variety of empirical models have been used to describe the tangential velocity profile of a tip vortex. Two commonly used examples are the 'Dispersion' model (Ref. 8.61) and the 'Burnham' model (Refs 8.62, 8.63); the Dispersion model takes the form

$$V_T(r) = \frac{\Gamma \, r}{2 \, \pi \, (r^2 + r_c^2)} \tag{8.49}$$

where $V_T(r)$ is the tangential velocity at a distance r from the vortex core, r_c is the core radius (defined as the distance from the centre of the vortex to the peak of the tangential velocity) and Γ is the total circulation around the vortex (with units of m^2-s^{-1}).

The Burnham model takes the form

$$V_T(r) = \frac{V_c \, (1 + \ln(r/r_c))}{r/r_c}, \quad |r| > r_c$$

$$V_T(r) = V_c \, (r/r_c), \quad |r| \le r_c \tag{8.50}$$

where $V_T(r)$ and r_c are as defined previously, and V_c is the peak velocity, i.e., the value of $V_T(r)$ at the edge of the rotational core, $r = r_c$. These vortex models are compared to LIDAR (Coherent Laser Radar) measurements of the tangential velocities in the (young) vortex wake of a Boeing 747 in Fig. 8.50 (Ref. 8.54). A best fit was obtained for the velocity profiles of several aircraft types, and the resulting parameters are given in Table 8.10.

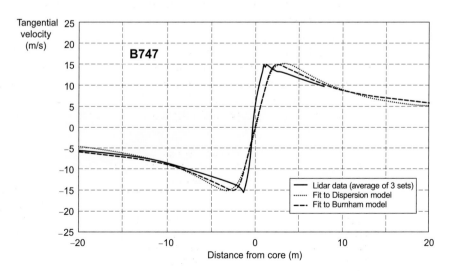

Fig. 8.50 Velocity distribution in Boeing 747 vortex wake

Table 8.10 Best fit parameter values to LIDAR velocity profiles for the Burnham and dispersion models (Ref. 8.54)

Aircraft type	'Burnham' model		'Dispersion' model		
	r_c (m)	V_c (m s^{-1})	Γ(m^2 s^{-1})	r_c (m)	V_c (m s^{-1})
B747	2.4	14.9	612	3.2	15.2
B757	<0.8	>21.2	251	<0.9	>22
A340	2.0	11.4	385	2.5	12.2
A310	<1.0	>20	283	<1.0	>22

As discussed in Ref. 8.54, the parameter values for the larger aircraft (Boeing 747, Airbus A340) should be reliable, but the maximum velocities for the medium twin engine aircraft (Boeing 757, Airbus A310) are estimates, which will be equal to or less than the true value, as the LIDAR sensitivity was insufficient to detect the peak.

In the study, the encounters occur when the vortex is at the (full) strength. Vortices do decay with time and the decay rate is a function of prevailing wind, humidity and wing flap configuration. The results presented therefore probably represent worse-case scenarios and the encounter effects in a real scenario may differ considerably. Key assumptions are that the vortex flow-field is unaffected by the rotorcraft and is superimposed on the quasi-steady incidence changes on the rotor. These assumptions are clearly open to question, but there is little reliable information on the interactional effects and they are likely to be very complex, particularly if the rotor blade cuts through the vortex core.

The velocity field of a Boeing 747 vortex when centred at the rotor hub is sketched in Fig. 8.51. Note that, at the rotor tips, the downwash/upwash is still considerable

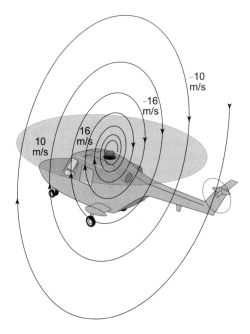

Fig. 8.51 Velocity flow-field of Boeing 747 vortex around Lynx rotor

(12 m/s, 39 ft/s), and with a rotor tip speed of about 220 m/s (720 ft/s) the perturbation in incidence is approximately 3°. This cyclic variation in incidence will result in longitudinal, forward, flapping of the rotor blades and a nose-down pitch moment for the anti-clockwise rotors on the Lynx and FGR.

A similar rationale can be applied to the perturbations in heave velocity. In this case the greatest disturbances are experienced when the rotorcraft is in the vortex tail, close to the core. The cyclic stick and collective lever margins available to the pilot to negate the effects of the vortex depend on trim position of the controls.

The technique of constrained simulation was used extensively in the study described in Refs 8.53–8.55 to ensure that the rotorcraft–vortex encounters have predictable initial conditions. Also, it proved more convenient and tractable to fix the position of the vortex in space and to move the aircraft laterally at different encounter velocities through the tails and core. As described in Ref. 8.53, with unconstrained simulations it was found that as the vortex approached the aircraft at the same height the aircraft would be lifted up in the approaching tail of the vortex and carried over the top and down in the following tail, as shown in Fig. 8.52.

In contrast, Fig. 8.53 illustrates the case when the initial position of the helicopter was such that an encounter with the vortex core was forced to occur. This scenario is not unrealistic as the vortex wakes tend to remain at about a semi-span (of the fixed-wing aircraft) above the ground. Hence, to avoid the complications of having to set different initial conditions for the different helicopters, vortex wakes and encounter speeds and to ensure that worst-case scenarios are explored, the constrained simulation approach was adopted. The initial condition was with the rotorcraft positioned 100 ft (~30 m)

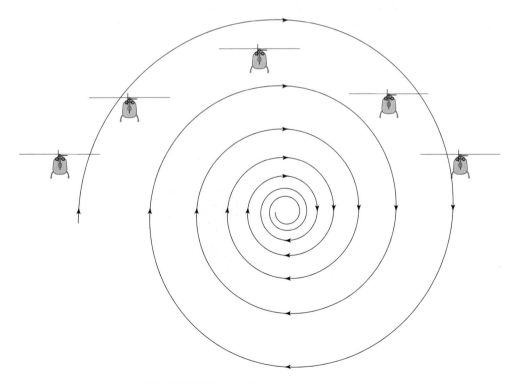

Fig. 8.52 Helicopter lifted above vortex core during encounter

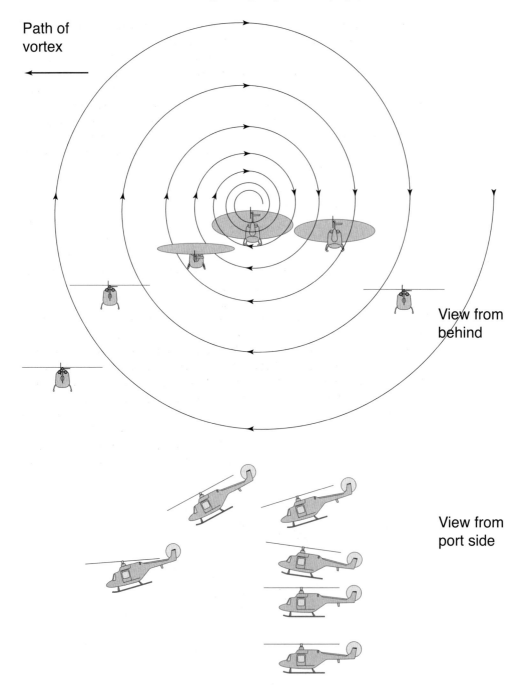

Path of vortex

View from behind

View from port side

Fig. 8.53 Helicopter encountering vortex core

to the port side of the (port wing) vortex, simulating an encounter with a vortex shed from an aircraft taking off to starboard of the rotorcraft.

In the following section, results are presented for two cases: (i) with constrained vertical/forward motion and heading to explore pitch attitude perturbations as the core is encountered and (ii) with constrained attitude, heading and forward motion to explore heave perturbations as the tails are traversed. Both Lynx and FGR have been investigated, with and without their SCAS engaged, although only results with SCAS-on are presented as this will be the normal configuration. Encounters with the vortex wake of the Boeing 747 are shown; Ref. 8.54 presents comparisons of encounters with the different aircraft mentioned in Table 8.10.

Analysis of encounters – attitude response

Figures 8.54–8.57 show aspects of the helicopter attitude response to a vortex encounter with no pilot recovery inputs. Figures 8.54 and 8.55 show the pitch attitude and rate response of the rotorcraft for three vortex encounter speeds: 5, 10 and 20 ft/s ($\sim$1.5, 3 and 6 m/s). The attitude transients increase as vortex-passing speed decreases as expected, since the aircraft is exposed to the vortex flow-field for longer. Note that the attitude hold system in the Lynx SCAS returns the aircraft to the hover attitude after the passage of the vortex, contrasting with the rate-damping SCAS in the FGR, which leaves the aircraft in a disturbed attitude state. Both rotorcraft initially pitch up as they pass through the advancing tail of the vortex induced by the lateral distribution of inflow through the rotor disc. As the rotor hub encounters the vortex core, the lateral inflow distribution reverses, leading to a much larger flapping and nose-down pitching moment. The attitude perturbations for the 10 and 20 ft/s encounters are approximately 30 and 20°, respectively, in 3–4 s, similar for both aircraft, while the slower encounter results in a pitch of nearly 40° in 5 s for the Lynx and more than 50° in 10 s for the FGR.

The pitching moment and corresponding accelerations are much higher on the Lynx with its hingeless rotor system, but the FGR is pitched to the larger attitude

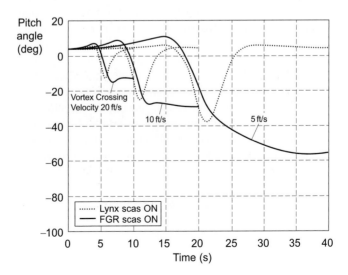

Fig. 8.54 Pitch attitude response

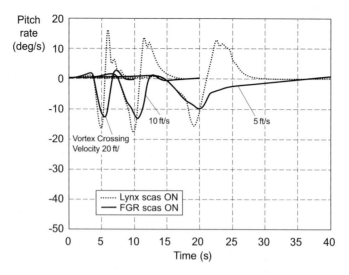

Fig. 8.55 Pitch rate response

because the increased diameter rotor is in the vortex for about 30% longer. The pitch response would be reversed for clockwise rotors (e.g., Eurocopter Super Puma).

The attitude responses are plotted on the pitch quickness charts in Figs 8.55 and 8.56. On each chart the maximum quickness is also plotted as a function of attitude derived from applying high-amplitude pulse inputs with varying duration. The ADS-33 handling qualities boundaries are also included.

Both aircraft have significant quickness margin (40–100%) to overcome the vortex, up to the 30° pitch attitude change (limit set in the ADS manoeuvre performance). Also, just meeting the ADS-33 minimum quickness requirements for tracking tasks (Level 1/2 boundary) gives a significant response margin (50–100%) for attitude changes up to 30°. It should be recognized that the pitch rates are transient and the nature of encounters is such that the pilot should need to apply compensatory control inputs only momentarily. This is not to say that the transient disturbance is not a serious handling 'problem' for the pilot. Such encounters are most likely to occur close to airports, and Ref. 8.56 highlights the result that pilots would most likely abort an approach following such upsets. The results also indicate that an aircraft that just met the minimum Level 1/2 quickness requirements for general MTEs would have wholly inadequate control for counteracting the effects of a vortex encounter. As with control system failures, quantifying response in terms of flight-handling qualities parameters provides a description of severity that links with safety and provides the basis for safety cases. While the attitude response is important, the vertical disturbance can be even more serious.

Analysis of encounters – vertical response

The vertical motions of the rotorcraft during the vortex encounters are illustrated in Fig. 8.58 (height), Fig. 8.59 (height rate) and Fig. 8.60 (vertical acceleration). The effects of SCAS are negligible in most cases; hence only SCAS-off results are presented. An exception is the vertical acceleration response of the Lynx, which has a feedback loop from acceleration to collective to improve high-speed stability characteristics.

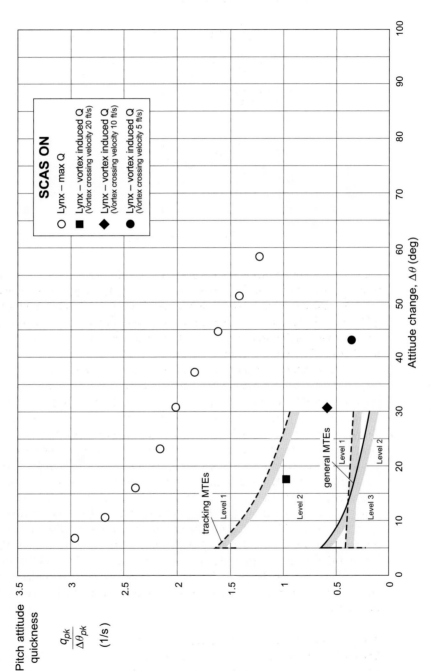

Fig. 8.56 Pitch quickness (Lynx)

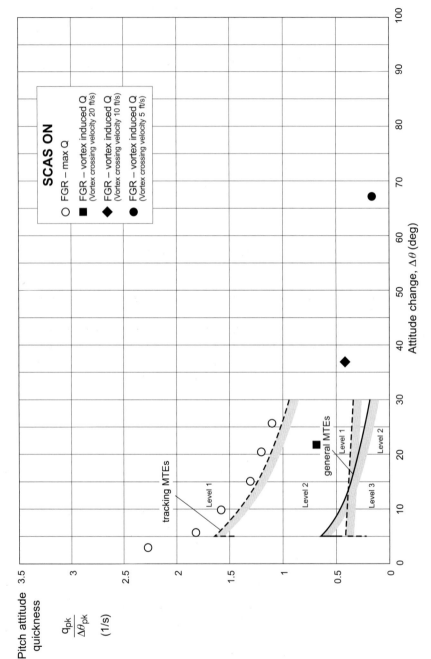

Fig. 8.57 Pitch quickness (FGR)

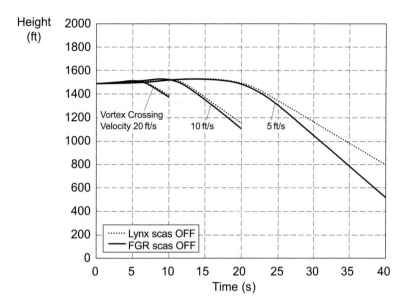

Fig. 8.58 Height response in encounter

The acceleration peaks in Fig. 8.60 would be reduced by 20% with SCAS engaged on the Lynx.

An important point to take into account when interpreting these data is that the initial trim of both aircraft is 100 ft to port of the clockwise-rotating vortex. The collective pitch is therefore lower than the hover value by an amount depending on the rotorspeed and rotor solidity. The fixed collective setting then results in a descent rate in

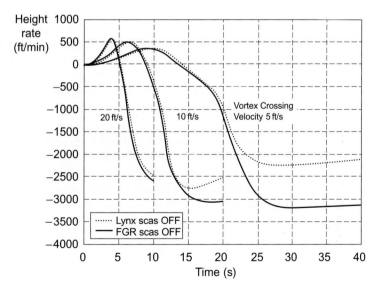

Fig. 8.59 Height rate response in encounter

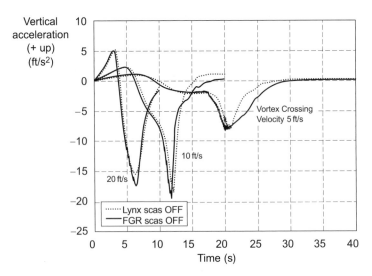

Fig. 8.60 Vertical acceleration response

the receding vortex tail, which is higher than that corresponding to the hover collective setting. For the Lynx, the reference rate of descent (i.e., the descent rate corresponding to the decreased collective at the initial condition) is about 900 ft/min and for the FGR about 1200 ft/min.

Approaching the vortex core the rotorcraft are lifted up to a maximum rate of climb of about 500 ft/min, followed by a reversal to a rapid descent rate of more than 2500 ft/min (more than 1500 ft/min relative to reference rate of descent noted in the previous paragraph). The slower the encounter, the greater time is spent in the vortex wakes and the larger height loss. At 20 ft/s encounter, 100 ft is lost in about 3 s; at 10 ft/s, 100 ft in 5 s; at 5 ft/s, 100 ft in about 8 s. The lower disc loading on the FGR results in larger peak accelerations and higher descent rates; at the fastest encounter of 20 ft/s, a bump of about -0.5 g is experienced within 3 s of a small positive bump. The descent rates induced in the vortex tail (1100 ft/min – Lynx; 2000 ft/min – FGR, again, relative to reference) are significantly higher than the 650 ft/min minimum requirement for Level 1 performance defined in ADS-33. This suggests that thrust margins of 10–15% would be required to enable a pilot to completely counteract the effects of a vortex encounter.

Handling qualities criteria provide a natural framework within which to set performance margins and quantify severity during upsets caused by vortex encounters. The preceding analysis has demonstrated that an aircraft satisfying minimum Level 1 (tracking) attitude quickness and (aggressive manoeuvring) control power performance should have sufficient control margin to overcome the effects of a full-strength vortex. Satisfying the normal minimum performance requirements for general MTEs will not provide an adequate margin, however. A rate SCAS significantly reduces the disturbance, while the addition of the attitude hold function (Lynx) returns the aircraft to the hover attitude, further reducing the upset. In terms of vertical performance, the minimum Level 1 standard, when translated into a margin for climb performance, is insufficient by a significant margin to overcome the effects of the downwash in the vortex tail.

Table 8.11 Transient pitch attitudes following the vortex encounter

| Aircraft | Encounter velocity | | | ADS-33 Level 3 |
| | 5ft/s | 10ft/s | 20ft/s | |
	Pitch attitude in 3 s (deg)			
Lynx SCAS on	15	*30*	16	
Lynx SCAS off	*40*	*50*	*45*	$10 < \theta < 24$
FGR SCAS on	10	*25*	22	
FGR SCAS off	*30*	*35*	*50*	

The performance criteria indicate what is ultimately achievable, but further insight can be gained by comparing the severity of the disturbance against the ADS-33 criteria for the transient response following failures.

8.4.2 Severity of transient response

Referring back to Table 8.7, showing the limits on attitudes and accelerations following a failure, the questions asked are – can this approach also apply to the response caused by external disturbances and are the same standards applicable? Table 8.11 shows the approximate pitch attitude transients at 3 s following the maximum pitch-up attitude. The values represent the changes in attitude from the maximum pitch-up rather than the initial pitch. This method leads to significantly greater transients in some cases but is justified because although the pilot would not be expected to allow the aircraft to pitch, he/she would have to apply forward cyclic to maintain the hover, which would exacerbate the pitch-down as the vortex core was crossed. The italicized numbers in Table 8.11 correspond to the cases where the Level 3 boundary is exceeded. SCAS disengaged results are also shown to illustrate the power of the SCAS and its positive impact on safety.

Similarly, Table 8.12 lists the 3-s perturbations in vertical acceleration. Only SCAS-off data are included; the SCAS does not change the level. In this case the reference conditions are the points where the larger negative bump begins (e.g., at 17.5 s for the FGR with the 5 ft/s crossing, Fig. 8.60).

If the pilot intervention time had been set at 1.5 s, the perturbations would have reduced to less than 50% of those in Tables 8.11 and 8.12 (with the possible exception of some SCAS-off cases); the italic cases would then be within the Level 3 boundary

Table 8.12 Transient vertical acceleration following the vortex encounter

| Aircraft | Encounter velocity | | | ADS-33 Level 3 |
| | 5ft/s | 10ft/s | 20ft/s | |
	Vertical acceleration in 3 seconds (g)			
Lynx SCAS off	0.16	0.31	*0.47*	
				$0.2 < n_z < 0.4$
FGR SCAS off	0.19	0.38	*0.53*	

and most other cases would be Level 2. Combining the ADS-33 approach with the hazard categories in Fig. 8.44 leads to the following relationships:

- Handling qualities Level 1,2 – hazard category MINOR (safety of flight not compromised; slight reduction in safety margin or increase in pilot workload)
- Handling qualities Level 3 – hazard category MAJOR (safety of flight compromised; significant reduction in safety margins or increase in crew workload)
- Handling qualities Level >3 – hazard category HAZARDOUS (safety of flight compromised; large reduction in safety margin)

From this classification, and without considering control margins, it can be deduced that with a 3-s pilot intervention time the vortex encounter is HAZARDOUS, and with a 1.5-s intervention time the hazard category of the encounter is MAJOR. Both relate to the disturbance-induced flight path variations and the resulting risk of disorientation or loss of control. It should be noted that the largest attitude and acceleration changes occur after the initial pitch-up or negative bump. It could be argued that the pilot may at this stage be aware of the vortex encounter and the normal, full-attention, 0.5-s intervention time is more appropriate. The hazard category may then reduce to MINOR.

The effect of intervention time on the severity of the response can be investigated with the aid of the upset severity rating (USR) scale shown in Fig. 8.61. This scale is based on the pilot rating scale for failure transients described in Ref. 8.49 and already presented in modified form in Fig. 8.43. In summary, ratings A to E indicate tolerable severity and are awarded for cases where the disturbed excursions range from minimal, requiring no corrective action, to very objectionable, requiring immediate and intense pilot effort. For cases A through E, safety of flight is judged not to be compromised, and the hazard category is MINOR. Safety of flight is compromised with ratings of F through G, with excursions leading to possible encounter with obstacles, unintentional landing or exceedance of flight envelope limits; recovery is marginal and the hazard category is MAJOR (F) or HAZARDOUS (G). A rating of H means that the pilot judged recovery to be impossible with the hazard category CATASTROPHIC.

Figure 8.62 shows results from simulation trials at Liverpool with the FGR, with power, collective and vertical motion changes as the vortex is traversed at a nominal 10 ft/s. Also shown is the lateral track as a function of time with the core and outer boundaries indicated. The pilot reduces collective as the rotor enters the upwash of the advancing tail. The pilot is able to maintain height within ±10 ft during this phase of flight and reduces collective to command a very low engine torque, less than 20% of the hover setting. At about 25 s the vortex core is passed and as the helicopter moves into the downwash of the retreating tail, a descent rate of more than 1000 ft/min builds up in about 5 s, arrested by the pilot applying significantly more than the 106% transient torque limit. This transient overtorque limited the height loss to about 50 ft. Height and collective excursions when the helicopter experiences the downwash in the vortex tail are double those during the 'upwash' phase. The effect of the helicopter being rolled and accelerated to starboard during the core encounter, i.e., pushed out of the vortex, can be seen in the increased slope of the lateral position trace. An HQR of 7 and USR of F (MAJOR) were awarded for this case on the grounds that the torque limit was exceeded and the height excursion was beyond the adequate boundary of ±30 ft.

The ability to counteract the vertical motion induced by the vortex clearly depends on the available power and thrust margin. As shown in Fig. 8.62, the FGR was being flown with a power margin of more than 30%, reinforcing the point made earlier that the

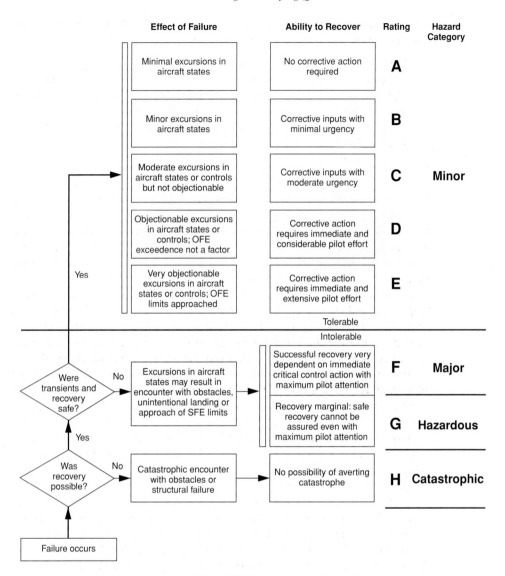

Fig. 8.61 Upset severity rating scale

ADS-33 minimum standards for Level 1 performance margins in hover are insufficient in this respect. Another observation made by pilots during the trials related to the large changes in roll and yaw attitude during the encounter. Yaw motion can be induced by the lateral velocities in the lower and upper portions of the vortex. If the aircraft yaws by 90° then the pitch effects described earlier would transform into roll.

The solution to the wake vortex problem for runway-dependent aircraft approaching and departing along similar trajectories is to define minimum longitudinal separation distances. The severity of encounters can be catastrophic close to the ground, but the risk is lowered to an acceptable level by reducing the probability of occurrence through separation. When considering runway-independent aircraft and the associated

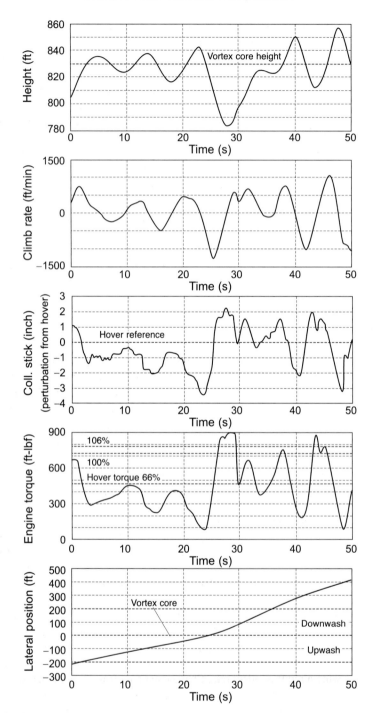

Fig. 8.62 Vertical response during encounter (FGR – SCAS on)

concept of simultaneous, non-interfering operations (SNIOps), the problem is more complex and lateral separation of approach and departure flight paths also becomes a major issue. At any particular location, the positioning of a helicopter final approach and landing area can be optimized on the basis of prevailing winds and atmospheric conditions, fixed-wing aircraft landing and take-off patterns and the nature of the traffic at any particular time. Whether it will ever be acceptable to operate with this flexibility is another question, but the risk certainly needs to be carefully managed through flight path constraints and positioning of the terminal area ground operations. The most concerning result is the potential loss of height due to encounters with the downwash side of a vortex. Reference 8.64 documents an accident following a suspected encounter of a light helicopter with a vortex, and it was the vertical motion of the aircraft that most disturbed the crew prior to the loss of control and crash. Helicopters typically operate with fairly low power/thrust margins in hover ($<10\%$). Although these may satisfy the handling standards for vertical performance, the results of both off-line and piloted simulations show that they may be wholly inadequate to overcome the effects of a vortex encounter. The situation will improve when some forward velocity has been gained and also when the helicopter has a rate of climb.

8.5 CHAPTER REVIEW

In this chapter some of the ways in which handling qualities can degrade have been discussed, and methodologies for taking them into account are outlined. What should be clear to the reader from the ideas and results presented is that the pilot's task can become very difficult if the visual cues degrade, if flight systems fail or when strong atmospheric disturbances are encountered; the risk to safety and the likelihood of an accident increase in such situations. If such degradations happen quickly and are unforeseen, taking the pilot by surprise, then the risk further increases. With a good understanding of the degrading mechanisms, appropriate design criteria, more stringent operational procedures and the availability of safety-related technologies, for both new and old aircraft, there seems to be no good reason, apart from cost, why all existing and new helicopters cannot be made more 'accident proof'. At the time of writing, this goal is being pursued in an international initiative, stimulated partly by the revelations of a comprehensive analysis of US civil helicopter accidents over a 40-year period by Harris *et al.* (Ref. 8.65), summarized in the 2006 AHS Nikolsky Lecture (Ref. 8.66) – *No Accidents – That's the Objective*. In the 40-year period from 1964, the US civil helicopter accident rate per 100 000 flying hours decreased from 65 in 1966 to 11 in 2004; these data relate to a total number of accidents i.e., 10 410 in the period, where nearly 2700 people lost their lives. The accident rate per 1000 aircraft has also decreased substantially over this period – 120 in 1964 to 12 in 2005. However, the helicopter accident rate is still about an order of magnitude greater than the fixed-wing aircraft accident rate. Challenging the oft-made point that safety improvements are uneconomical, Harris presents data showing that the cost of one accident is about 1 million US dollars, so that the total cost to the Industry over the 40-year period has been about 11 billion US dollars. More than three quarters of this relates to insurance claims.

Figure 8.63 summarizes the distribution of US civil rotorcraft accidents presented by Harris (Ref. 8.66). The data show that loss of control is a growing problem; with 1114 in total, less than 10% were in this category in 1964 but greater than 20% in 2005,

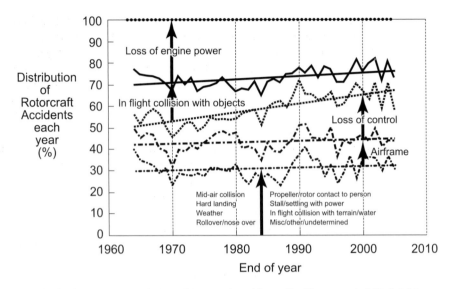

Fig. 8.63 Distribution of US civil rotorcraft accidents of a 40-year period (Ref. 8.66)

making loss of control equal the number one contributor to accidents, alongside loss of engine power. In-flight collision with objects (including wires, poles, trees, but not the surface) has reduced as a cause, but a total of 1322 accidents in this category over the period highlights the problem that pilots too often bump into things and the helicopter is very unforgiving to such. The surface collision accidents are contained in the lower 30% 'mixed-bag' in Fig. 8.63, so details are unclear, but it is likely, based on more dedicated studies (e.g., Ref. 8.4, 8.67), that the proportion in this category, where loss of visual cues is a major factor, will be significant.

The safety initiative referred to above, led by the International Helicopter Safety Team (IHST), was launched at the AHS Safety Symposium in Montreal in September 2005 (Ref. 8.68), where a commitment was made to reduce the helicopter accident rate by 80%, from 8.09 to 1.62 per 100 000 flying hours, within 10 years. The activity is being modelled on the US Commercial Aircraft Safety Team (CAST) programme, which had set a similar goal for fixed-wing aircraft accidents in the mid-1990s. While it might seem unusual to describe such contemporary initiatives in a textbook, it is considered by the author that this activity is vitally important to the helicopter industry and to the theme of this chapter. IHST has defined a three-stage process summarized as follows: conduct data analysis, set safety priorities and integrate safety enhancements. A substantial number of operators, regulators and manufacturers worldwide have signed up to the master plan summarized in three components (Ref. 8.69):

> *'IHST Mission: To provide government, industry and operator leadership to develop*
>   *and focus implementation of an integrated, data-driven strategy to improve*
>   *helicopter aviation safety worldwide, both military and civil*
> *IHST Vision: To achieve the highest levels of safety in the international helicopter*
>   *communities by focusing on appropriate initiatives prioritized to result in the*
>   *greatest improvement in helicopter aviation safety*
> *IHST Goal: To reduce the helicopter aviation accident rate by 80 percent by 2016'*

The IHST was consolidated at the 62nd Annual Forum of the AHS in May 2006 (Ref. 8.69), where a number of participants reported on analysis conducted to date. Cross, in Ref. 8.70, examined the potential impact on safety of various mitigating technologies, noting that the airline industry had made significant improvements in its safety record through the introduction of, for example, damage-tolerant/fail-safe designs, extensive use of simulators in flight training, safety management systems and quality assurance to reduce human errors, flight data monitoring programs, disciplined take-off and landing profiles (e.g., stabilized approach), digital flight management systems to reduce pilot workload, improved situational awareness, help to cope with emergencies, improved one-engine-inoperative performance and various terrain/collision avoidance systems. Specifically, Cross drew the conclusion that more than 50% of accidents were preventable with a combination of enhanced handling, conferred by meeting modern FAR standards, and improved pilot training.

Handling qualities are central to both flight performance and flight safety and much has been made of the trade-off between these twin goals in the design of aircraft. Nowadays, the environmental impact and the economics of a system's life cycle introduce further constraints in the management of this trade-off. There is much to be done, and much will no doubt be accomplished, in the relentless pursuit of perfection in flight.

APPENDIX 8A   HELIFLIGHT AND FLIGHTLAB AT THE UNIVERSITY OF LIVERPOOL

In the production of this second edition, research results derived from the University of Liverpool's motion simulator, HELIFLIGHT, and its simulation environment, FLIGHTLAB, have been used extensively. This appendix provides an overview of the facility; the material is derived largely from Ref. 8A.1.

The HELIFLIGHT facility can be described as a reconfigurable flight simulator, with six key components that are combined to produce a relatively high-fidelity system, including

(a)   interchangeable flight dynamics modelling software (FLIGHTLAB), featuring 'selective fidelity', e.g., different types of rotor wake model, with a real-time interface, PilotStation;

(b)   6-DoF motion platform;

(c)   four-axis dynamic control loading;

(d)   three-channel collimated visual display system ($135 \times 40°$) plus two flat panel chin windows ($60°$), each channel running its own visual database;

(e)   reconfigurable, computer-generated instrument display panel and heads-up-display (HUD)

(f)   data record and time history capture facility.

A schematic of the HELIFLIGHT configuration is shown in Fig. 8A.1.

The main host computer is a dual processor PC running Linux. One processor runs FLIGHTLAB and PilotStation, whilst the second processor drives the control loaders. In addition, this machine acts as both a file server and a server for other hosts. The use of two Ethernet cards (one to access the Internet and the other to access the HELIFLIGHT network via a hub) enables isolation of the local area network from the

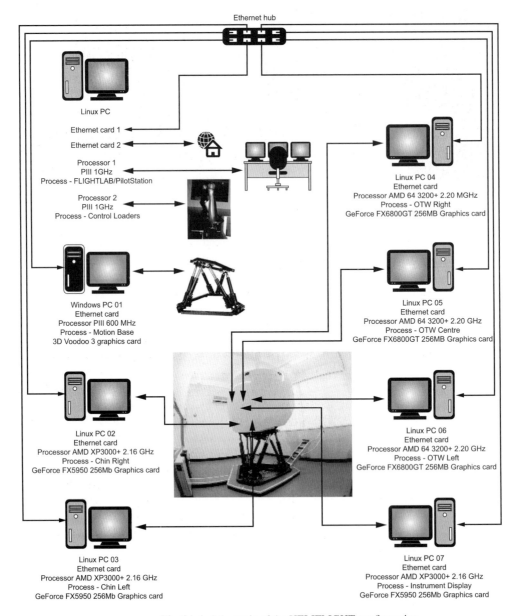

Fig. 8A.1 Schematic of the HELIFLIGHT configuration

Internet, maximizing throughput and security. There are seven other Windows-based host computers running the motion base, the two chin windows, the three forward Out the Window (OTW) displays and the instrument display. The HUD on the OTW centre can be toggled on/off. All the Windows computers are equipped with graphics cards that send signals to the cockpit displays, asynchronously. The keyboard and mouse of each computer are also multiplexed, allowing each Windows computer to be controlled from a single station.

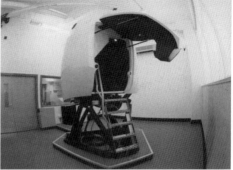

Fig. 8A.2 Flight simulation laboratory at The University of Liverpool

The simulation laboratory has two main areas: the simulator control room and the cockpit pod room. An authorized simulator operator controls the real-time operation of the simulator from the main host running PilotStation in the control room and interacts with the pilot in the cockpit room using a two-way communication system. From this viewpoint, the operator can observe both the motion of the cockpit and also the displays which are replicas of those present in the cockpit pod (Fig. 8A.2).

During a real-time session or 'sortie', the operator is responsible for ensuring the safe operation of the motion base and can override a pilot's inputs in the event of loss of pilot control. A lap belt is worn by the pilot during a sortie and is part of the safety interlock system that incorporates electromagnetic door releases on the gull wing capsule door and a cockpit room door interlock. Emergency stop buttons are available to both the pilot and the operator. In the case of an emergency or power failure, the simulator parks, returning the capsule safely to its down position and the cockpit pod door opens.

Throughout a sortie, a video/DVD record is taken of OTW centre, generating both a visual and audio log of the mission for use in post-trial analysis. PilotStation also has a data-logging function, allowing a range of aircraft performance parameters, flight model outputs and pilot control inputs to be captured for subsequent processing. Using a computer image of the aircraft being flown, flights can later be reconstructed from any viewing point.

Flightlab

The software at the centre of operation of the facility is FLIGHTLAB, a multi-body modelling environment, providing a modular approach to the creation of flight dynamics models, including enabling the user to produce a complete vehicle system from a library of pre-defined components. In particular, FLIGHTLAB provides a range of tools to assist in the rapid generation of nonlinear, multi-body models, significantly reducing the effort required for computer coding. Although FLIGHTLAB was originally developed for rotorcraft using blade element models, it can readily be used as a simulation tool for fixed-wing aircraft. For example, within the FLIGHTLAB library at Liverpool are models of the Wright Flyers, Grob 115, X-29, Boeing 707/747, Jetstream and Space Shuttle Orbiter.

To aid the generation and analysis of flight models, three graphical user interfaces (GUIs) are available: GSCOPE, FLIGHTLAB Model Editor (FLME) and Xanalysis. A schematic representation of the desired model can be generated using a component-level editor called GSCOPE. Components are selected from a menu of icons, which are then interconnected to produce the desired architecture and data are assigned to the component fields. When the representation is complete, the user selects the script generation option and a simulation script in FLIGHTLAB's *Scope* language is automatically generated from the schematic. *Scope* is an interpretive language that uses MATLAB syntax, together with new language constructs, for building and solving nonlinear dynamic models.

FLME is a subsystem model editor allowing a user to create models from higher level primitives such as rotors and airframes. Typically, a user will select and configure the subsystem of interest by inputting data values and selecting options that determine the required level of sophistication. This approach provides a selective-fidelity modelling capability while maximizing computational efficiency. Models are created hierarchically, with a complete vehicle model consisting of lower level subsystem models, which in turn are collections of primitive components. This is the Model Editor Tree, which puts all the pre-defined aircraft subsystems into a logical 'tree' structure. This tool facilitates configuration management by keeping all models in a pre-defined structure, while at the same time allowing the user flexibility in defining the individual aircraft structure and subsystems.

Prior to running a real-time simulation, the model generated using the above tools can be analysed using Xanalysis. This GUI has a number of tools allowing a user to change model parameters and examine the dynamic response, static stability, performance and handling qualities of design alternatives. Additional tools are available to generate linear models with prescribed perturbation sizes, perform eigen-analyses, time and frequency response analyses and control system design. The nonlinear model may also be directly evaluated through utilities that support trim and time and frequency response.

The real-time simulation is coordinated using PilotStation, which controls and interfaces image generation for the OTW displays, instruments and the HUD with the control loaders, motion base and flight dynamics models generated using FLIGHT-LAB, in real time. Typically, a helicopter simulation, with a four-bladed rotor and five elements per blade, runs at 200 Hz. The frame time can be increased or decreased to ensure optimized performance, taking account of model complexity (number of operations per second) and the highest frequency modes (numerical stability). During a simulation, a circular buffer is continuously updated containing pre-defined output variables. Selecting the History option makes the buffer accessible to the operator, which can be plotted or saved for off-line analysis. The operator console can be used to modify the vehicle configuration and flight condition and initiate faults or inputs on-line, e.g., SCAS on/off, tail rotor failure, gusts.

Immersive cockpit environment

The flight dynamics models are an important part of a flight simulator and ultimately define the fidelity level of the simulation. Of equal importance is the environment into which a pilot is immersed. HELIFLIGHT uses six-axis motion cueing together with

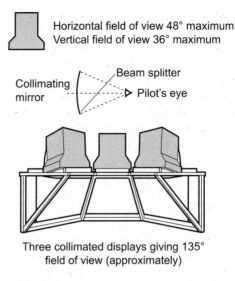

Horizontal field of view 48° maximum
Vertical field of view 36° maximum

Beam splitter

Collimating mirror

Pilot's eye

Three collimated displays giving 135° field of view (approximately)

Fig. 8A.3 Collimated display system in the HELIFLIGHT

collimated displays and pilot control loaders to create a virtual flying experience. A pilot will derive information about the vehicle behaviour from a number of sources. The basic mechanisms are visual perception, perception through the vestibular system of the inner ears and perception through the proprioceptors distributed throughout the body. Each of these mechanisms provides important information or 'cues' to the pilot.

Three collimated visual displays (Fig. 8A.3) are used to provide infinity optics for enhanced depth perception, which is particularly important for hovering and low-speed flying tasks. The displays provide 135° horizontal by 40° vertical field of view, extended to 60° vertical field of view using two flat-screen displays in the foot-well chin windows (Fig. 8A.4). The displays have a 1024 × 768 pixel resolution, refreshing at 60 Hz giving good visual cues when displaying a texture-rich visual database (Fig. 8A.5).

The capsule has a main instrument panel that can be reconfigured to represent displays from different aircraft presented on a flat screen monitor. The 'standard' HUD is displayed in OTW centre and contains an attitude indicator, vertical speed indicator, airspeed and altitude indicator and has a 'hover box' to aid helicopter control at low speed.

The sensation of motion is generated using the six-axis motion platform, with movement envelope as given in Table 8A.1.

The electrically actuated motion platform has a position resolution of 0.6 μm. The human visual system is relatively slow to detect changes in speed, compared with the vestibular system, which is much quicker to react to accelerations. As a result, certain tasks may be difficult to perform without motion cues, in particular helicopter hovering. To ensure that the pilot does not receive 'false' cues, the motion cueing algorithms can be tuned to correspond with the desired vehicle performance and MTE requirements. The parameters are accessible in a configuration file, which can be made aircraft specific. A major limitation with motion platforms is the stroke available.

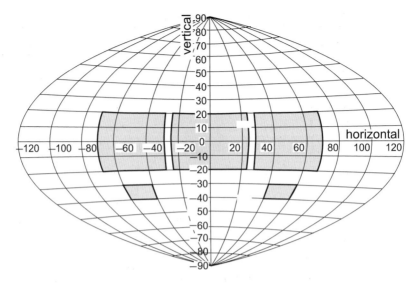

Fig. 8A.4 Outside world field of view in HELIFLIGHT simulator

Fig. 8A.5 Typical pilot's eye view in HELIFLIGHT capsule

To maximize the usable motion envelope, the drive algorithms features conventional washout filters that return the simulator to its neutral position after a period of simulator motion at low enough acceleration rates to minimize false cues.

Pilots can gain information about the behaviour of the aircraft by the feel and position of the controls. HELIFLIGHT uses electric control loaders for the three primary pilot inceptors: cyclic, collective and pedals. The collective lever and cyclic stick host several switches for various functions that can be reprogrammed or are directly

Table 8A.1 HELIFLIGHT motion envelope

Motion parameter	Range
Heave range	500 mm[a]
Peak heave velocity	±0.6 m/s
Peak heave acceleration	±0.6 g[b]
Surge range	930 mm[a]
Peak surge velocity	±0.7 m/s
Peak surge acceleration	±0.6 g
Sway range	860 mm[a]
Peak sway velocity	±0.7 m/s
Peak sway acceleration	±0.6 g
Roll range	±28°
Peak roll rate	40°/s
Pitch range	+34°/−32°
Peak pitch rate	40°/s
Yaw range	±44°
Peak yaw rate	60°/s

[a] All motions are stated from mid heave with all other axes neutral. By coupling one or more motions, a larger range may be obtained.

[b] Measured over whole motion envelope. Heave accelerations of +1 g, −2 g may be produced near the centre of the motion envelope.

associated with PilotStation (e.g., run/pause, trim release). The HELIFLIGHT capsule also contains two secondary controls – a joystick and a throttle lever. All the controls, buttons and switches are configurable, e.g., the hat button on the cyclic controls nacelle tilt in the FLIGHTLABXV-15 and the collective button configured as a brake for the undercarriage wheels on fixed-wing aircraft models. Digital control of the stick gradient and control position is carried out with a resolution of 2.5 μm. Such accuracy allows a pilot to utilize the force trim release feature to zero the control forces at the trim position. The force feel characteristics are also reconfigurable through software to represent an aircraft-specific control system.

Vibration and audio cues contribute to the realism of the simulation. Aircraft-specific noise is played through two loudspeakers in the HELIFLIGHT cockpit to provide audio cues to the pilot. Vibration can be detected directly through the motion platform driven by variables in the model. A 'low' frequency audio actuator is mounted under the floor of the capsule, directly beneath the pilot. This can transmit sounds of frequency 20–100 Hz into the floor of the capsule to provide vibration or impact cues.

An important aspect of the overall fidelity of the system is the amount of latency present. The latency is produced by the transport delays in the transfer of information between the various components of the simulator, from the control inputs to the flight model outputs through the motion base and the visual system to the pilot and back through to the flight model via the pilot's controls. If the degree of latency is high, the pilot is likely to notice a lag between an input control command and perceived response of the system. This can seriously affect handling, particularly for precision-tracking tasks. In HELIFLIGHT, the flight dynamics model, running typically at 200 Hz, produces a 5-ms delay. A delay of less than 16 ms occurs as the output from the flight model is converted to produce a corresponding change in the simulator motion system. The graphics cards receive a signal broadcast across the HELIFLIGHT network near the start of each time frame. However, variable latency in the visuals occurs due to the terrain texture density being displayed, which also varies with the specification of the graphics card. Currently, this causes delays of between 16 and 30 ms in the redrawing of the terrain. In addition to this, the monitors are refreshing at 60 Hz. Finally, the Loadcue feel system introduces a potential 5-ms delay into the system. With all these contributions, the overall transport delay between pilot stick and motion base and visual response is estimated to be below 50 ms.

The scope of activity undertaken on HELIFLIGHT over the first 5 years of operation is documented in the wide range of journal and conference papers (Refs 8A.1–8A.28). These publications give something of the flavour of what can be achieved with a research quality flight simulator in the continuing development of flight handling qualities.

References

CHAPTER 1

1.1 Duncan, W.J., The Principles of Control and Stability of Aircraft, Cambridge University Press, 1952

1.2 Crawford, C.C., Jr., Rotorcraft Analytical Improvement Needed to Reduce Development Risk – The 1989 Alexander A. Nikolsky Lecture, J. Am. Helicopter Soc., Vol 35, No 1, 1990

1.3 Padfield, G.D., Helicopter Handling Qualities and Flight Control – Is the Helicopter Community Prepared for Change? Overview Paper, Helicopter Handling Qualities and Control, RAeSoc International Conference, London, November 1988

1.4 Anonymous, Handling Qualities Requirements for Military Rotorcraft, TTCP Achievement Award, 1994

1.5 Pirsig, R.M., Zen and the Art of Motorcycle Maintenance, Bodley Head, 1974

1.6 Cooper, G.E. and Harper, R.P., Jr., The Use of Pilot Ratings in the Evaluation of Aircraft Handling Qualities, NASA TM D-5133, 1969

1.7 AGARD, Rotorcraft System Identification, AGARD AR 280, LS 178, 1991

1.8 Padfield, G.D., Pavel, M.D., Casoralo, D., Hamers, M., Roth, G. and Taghizard, A., Simulation Fidelity of Real-Time Helicopter Simulation Models, 61st Annual Forum of the American Helicopter Society, Grapevine, Tex., June 2005

CHAPTER 2

2.1 AVSCOM, Aeronautical Design Standard (ADS) 33C – Handling Qualities for Military Helicopters, US Army AVSCOM, 1989

2.2 Anonymous, General Requirements for Helicopter Flying and Ground Handling Qualities, Mil-H-8501A, 1961

2.3 USAF, Flying Qualities of Piloted Airplanes, Mil-F-8785C, USAF, 1980

2.4 Hoh, R., New Developments in Flying Qualities Criteria with Application to Rotary Wing Aircraft, Helicopter Handling Qualities, NASA CP 2219, April 1982

2.5 Mouille, R. and d'Ambra, F., The 'Fenestron' a Shrouded Tail Rotor Concept for Helicopters, 42nd Annual Forum of the American Helicopter Society, Washington, D.C., June 1986

2.6 Morse, C.S., ADFCS and Notar; Two Ways to Fix Flying Qualities, AGARD CP 508, Flying Qualities, Quebec, October 1990

2.7 Drees, J.M., A Theory of Airflow Through Rotors and Its Application to Some Helicopter Problems, J. Helicopter Assoc. Great Brit., Vol 3, 1949

2.8 Johnson, W., Helicopter Theory, Princeton University Press, 1980

2.9   Chen, R.T.N. and Hindson, W.S., Influence of High Order Dynamics on Helicopter Flight Control System Bandwidth, Paper No 83, 11th European Rotorcraft Forum, London, September 1985

2.10 Silverio, F.J., SH-60B Test Program, Paper No 26, 7th European Rotorcraft Forum, Garmisch Partenkirchen, Germany, September 1981

2.11 Schrage, D., The Impact of TQM and Concurrent Engineering on the Aircraft Design Process, AHS Conference on Vertical Lift Technology, San Francisco, 1989

2.12 Lock, C.N.H., The Application of Goldstein's Theory to the Practical Design of Airscrews, Aeronautical Research Council R&M 1377, 1931

2.13 Bryan, G.H., Stability in Aviation, Macmillan, 1911

2.14 Milne, R.D., The Analysis of Weakly Coupled Dynamical Systems, Int. J. Control, Vol 2, No 2, 1965

2.15 Padfield, G.D., A Theoretical Model of Helicopter Flight Mechanics for Application to Piloted Simulation, RAE TR 81048, 1981

2.16 Padfield, G.D., On the Use of Approximate Models in Helicopter Flight Mechanics, Vertica, Vol 5, 1981

2.17 Bramwell, A.R.S., Helicopter Dynamics, Arnold Publishers, 1976

2.18 Duff, G.F.D. and Naylor, D., Differential Equations of Applied Mathematics, John Wiley, 1966

2.19 Padfield, G.D. and DuVal, R.W., Application Areas for Rotorcraft System Identification, Rotorcraft System Identification, AGARD LS 178, 1991

2.20 Padfield, G.D. (ed.), Applications of System Identification in Rotorcraft Flight Dynamics, Vertica Special Edition, Vol 13, No 3, 1989

2.21 AGARD, Rotorcraft System Identification, AGARD AR 280, 1991

2.22 AGARD, Rotorcraft System Identification, AGARD LS 178, 1991

2.23 Shinbrot, M., A Least Squares Curve Fitting Method with Application to the Calculation of Stability Coefficients from Transient Response Data, NACA TN 2341, 1951

2.24 Klein, V., Aircraft Parameter Identification Methods, AGARD LS 104, November 1979

2.25 McCallum, A., Padfield, G.D. and Simpson, A., Current Studies in Rotorcraft Simulation Validation at DRA Bedford, RAeSoc Conference on Rotorcraft Simulation, London, May 1994

2.26 Gray, G.J. and von Grunhagen, W., An Investigation of Open-Loop and Inverse Simulation as Nonlinear Model Validation Tools for Helicopter Flight Mechanics, RAeSoc International Conference, Rotorcraft Simulation, May 1994

2.27 Bradley, R., Padfield, G.D., Murray-Smith, D.J. and Thomson, D.G., Validation of Helicopter Mathematical Models, Trans. Inst. Measure. Control, Vol 12, No 4, 1990

2.28 Thomson, D.G. and Bradley, R., Validation of Helicopter Mathematical Models by Comparison of Data from Nap-of-the-Earth Flight Tests and Inverse Simulation, 14th European Rotorcraft Forum, Milan, September 1988

2.29 Thomson, D.G., Evaluation of Helicopter Agility Through Inverse Solution of the Equations of Motion, PhD Thesis, University of Glasgow, 1987

2.30 Cooper, G.E. and Harper R.P., Jr., The Use of Pilot Ratings in the Evaluation of Aircraft Handling Qualities, NASA TM D-5133, 1969

2.31 Stewart, W. and Zbrozek, J.K., Loss of Control Incident on S-51 Helicopter VW 209, RAE Technical Report Aero 2270, 1948

2.32 Hoh, R.H., Concepts and Criteria for a Mission Oriented Flying Qualities Specification; Advances in Flying Qualities, AGARD LS 157, 1988

2.33 McRuer, D.T., Pilot Modelling, AGARD LS 157, 1988

2.34 Buckingham, S.L. and Padfield, G.D., Piloted Simulations to Explore Helicopter Advanced Control Systems, RAE Technical Report 86022, April 1986

2.35 Padfield, G.D., Helicopter Handling Qualities and Flight Control – Is the Helicopter Community Prepared for Change? Overview Paper, Helicopter Handling Qualities and Control, RAeSoc Conference, London, November 1988

2.36 Charlton, M.T., Padfield, G.D. and Horton, R., Helicopter Agility in Low Speed Manoeuvres, Proceedings of the 13th European Rotorcraft Forum, Arles, France, September 1987 (also RAE TM FM22, 1989)

2.37 Heffley, R.K., Curtiss, H.C., Hindson, W.S. and Hess, R.A., Study of Helicopter Roll Control Effectiveness Criteria, NASA CR 177404, April 1986

2.38 Hodgkinson, J., Page, M., Preston, J. and Gillette, D., Continuous Flying Quality Improvement – The Measure and the Payoff, AIAA Paper 92-4327, 1992 Guidance, Navigation and Control Conference, Hilton Head Island, SC, August 1992

2.39 Padfield, G.D. and Hodgkinson, J., The Influence of Flying Qualities on Operational Agility, Technologies for Highly Manoeuvrable Aircraft, AGARD CP 548, Maryland, 1993

2.40 Sissingh, G.J., Response Characteristics of the Gyro-Controlled Lockheed Rotor System, Proceedings of the 23rd Annual Forum of the American Helicopter Society, May 1967

2.41 Potthast, A.J. and Kerr, A.W., Rotor Moment Control with Flap Moment Feedback, 30th Annual Forum of the American Helicopter Society, May 1974

2.42 Ellin, A.D.S., An In-Flight Experimental Investigation of Helicopter Main Rotor/Tail Rotor Interactions, PhD Thesis, Aerospace Engineering Department, University of Glasgow, April 1993

2.43 Prouty, R. and Amer, K., The YAH-64 Empennage and Tail Rotor – A Technical History, 38th Annual Forum of the American Helicopter Society, Washington, D.C., May 1982

2.44 Roesch, P. and Vuillet, A., New Designs for Improved Aerodynamic Stability on Recent Aerospatiale Helicopters, 37th Annual Forum of the American Helicopter Society, New Orleans, La., May 1981

2.45 Prouty, R., Helicopter Performance, Stability and Control, PWS Publishers, 1986

CHAPTER 3

3.1 Duncan, W.J., The Principles of Control and Stability of Aircraft, Cambridge University Press, 1952

3.2 Gessow, A. and Myers, G.C., Jr., Aerodynamics of the Helicopter, Frederick Ungar, 1952

3.3 AGARD, Aerodynamics and Aeroacoustics of Rotorcraft, Proceedings of the 75th AGARD FDP Panel Meeting, Berlin, October 1994

3.4 Padfield, G.D., A Theoretical Model of Helicopter Flight Mechanics for Application to Piloted Simulation, RAE Technical Report 81048, April 1981

3.5 Bisplinghoff, R.L., Ashley, H. and Halfman, R.L., Aeroelasticity, Addison-Wesley, 1955

3.6 Bramwell, A.R.S., Helicopter Dynamics, Edward Arnold, 1976

3.7 Johnson, W., Helicopter Theory, Princeton University Press, 1980

3.8 Hohenemser, K.H. and Yin, S.-K., Some Applications of the Method of Multi-Blade Coordinates, J. Am. Helicopter Soc., Vol 17, No 3, 1972

3.9 Gaonker, G.H. and Peters, D.A., Review of Dynamic Inflow Modelling for Rotorcraft Flight Dynamics, Vertica, Vol 12, No 3, 1988

3.10 Chen, R.T.N., A Survey of Nonuniform Inflow Models for Rotorcraft Flight Dynamics and Control Applications, 15th European Rotorcraft Forum, Amsterdam, September 1989

3.11 Young, C., A Note on the Velocity Induced by a Helicopter Rotor in the Vortex Ring State, RAE Technical Report 78125, 1978

3.12 Castles, W., Jr. and Gray, R.B., Empirical Relation Between Induced Velocity, Thrust and Rate of Descent of a Helicopter Rotor as Determined by Wind Tunnel Tests on Four Model Rotors, NACA TN 2474, October 1951

3.13 Duncan, W.J., Thom, A.S. and Young, A.D., An Elementary Treatise on the Mechanics of Fluids, Edward Arnold, 1960

3.14 Glauert, H., A General Theory of the Autogyro, Aeronautical Research Council R&M 1111, 1926

3.15 Coleman, R.P., Feingold, A.M. and Stempin, C.W., Evaluation of the Induced Velocity Field of an Idealised Helicopter Rotor, NACA WR L-126, 1945

3.16 Mangler, K.W., Fourier Coefficients for Downwash of a Helicopter Rotor, RAE Report No Aero 1958, 1948

3.17 Padfield, G.D., Flight Testing for Performance and Flying Qualities, Helicopter Aero-mechanics, AGARD LS 139, 1985

3.18 Heyson, H.H., A Momentum Analysis of Helicopters and Autogyros in Inclined Descent, with Comments on Operational Restrictions, NASA TN D-7917, 1975

3.19 Wolkovitch, J., Analytic Prediction of Vortex Ring Boundaries, J. Am. Helicopter Soc., Vol 17, No 3, 1972

3.20 Wang, Shi-cun, Analytical Approach to the Induced Velocity of a Helicopter Rotor in Vertical Descent, J. Am. Helicopter Soc., Vol 35, No 1, 1990

3.21 Drees, J.M., A Theory of Airflow Through Rotors and Its Application to Some Helicopter Problems, J. Helicopter Assoc. Great Brit., 1949

3.22 Sissingh, G.J., The Effect of Induced Velocity Variation on Helicopter Rotor Damping in Pitch and Roll, RAE TN Aero 2132, November 1951

3.23 Ormiston, R.A. and Peters, D.A., Hingeless Rotor Response with Nonuniform Inflow and Elastic Blade Bending – Theory and Experiment, AIAA Paper No 72-65, 1972

3.24 Azuma, A. and Nakamura, Y., Pitch Damping of Helicopter Rotor with Nonuniform Inflow, J. Aircr., Vol 11, No 10, 1974

3.25 Peters, D.A., Hingeless Rotor Frequency Response with Unsteady Inflow, AHS/NASA Specialists' Meeting on Rotorcraft Dynamics, NASA SP 352, Ames Research Center, February 1974

3.26 Pitt, D.M. and Peters, D.A., Theoretical Prediction of Dynamic-Inflow Derivatives, Vertica, Vol 5, No 1, 1981

3.27 Peters, D.A. and Ninh, HaQuang, Dynamic Inflow for Practical Application, J. Am. Helicopter Soc., Vol 33, No 4, 1988

3.28 Peters, D.A., Boyd, D.D. and He, Chengjian, Finite State Induced Flow Model for Rotors in Hover and Forward Flight, J. Am. Helicopter Soc., Vol 34, No 4, 1989

3.29 Peters, D.A. and He, C.J., Correlation of Measured Induced Velocities with a Finite State Wake Model, J. Am. Helicopter Soc., Vol 36, No 3, 1991

3.30 Carpenter, P.J. and Fridovitch, B., Effect of a Rapid Blade Pitch Increase on the Thrust and Induced Velocity Response of a Full Scale Helicopter Rotor, NACA TN 3044, 1953

3.31 Lowey, R.G., A Two Dimensional Approach to the Unsteady Aerodynamics of Rotary Wings, J. Aeronaut. Sci., Vol 24, No 2, 1957

3.32 Sissingh, G.J., Response Characteristics of the Gyro-Controlled Lockheed Rotor System, Proceedings of the 23rd Annual Forum of the AHS, May 1967

3.33 Young, M.I., A Simplified Theory of Hingeless Rotors with Application to Tandem Rotors, Proceedings of the 18th Annual National Forum of the AHS, May 1962

3.34 Bramwell, A.R.S., A Method for Calculating the Stability and Control Derivatives of Helicopters with Hingeless Rotors, The City University, Research Memorandum Aero 69/4, 1969

3.35 Reichert, G., The Influence of Aeroelasticity on the Stability and Control of a Helicopter with a Hingeless Rotor, Paper No 10, Aeroelastic Effects from a Flight Mechanics Standpoint, AGARD CP-46, Marseilles, France, April 1969

3.36 Shupe, N.K., A Study of the Dynamic Motions of Hingeless Rotored Helicopters, US Army Electronics Command TR ECOM-3323, Fort Monmouth, NJ, August 1970

3.37 Curtiss, H.C., Jr. and Shupe, N.K., A Stability and Control Theory for Hingeless Rotors, 27th Annual Forum of the American Helicopter Society, Washington, D.C., May 1971

3.38 Anderson, W.D. and Johnston, J.F., Comparison of Flight Data and Analysis for Hingeless Rotor Regressive In-plane Mode Stability, Rotorcraft Dynamics, NASA SP 352, 1974

3.39 Young, M.I., Bailey, D.J. and Hirschbein, M.S., Open and Closed Loop Stability of Hingeless Rotor Helicopter Air and Ground Resonance, Rotorcraft Dynamics, NASA SP 352, 1974

3.40 Curtiss H.C., Jr., Stability and Control Modelling, 12th European Rotorcraft Forum, Garmisch-Partenkirchen, Germany, September 1986

3.41 Tischler, M.B., System Identification Requirements for High Bandwidth Rotorcraft Flight Control System Design, Rotorcraft System Identification, AGARD LS-178, 1991

3.42 Simons, I.A., The Effect of Gyroscopic Feathering Moments on Helicopter Rotor Manoeuvre Behaviour, Westland Helicopters RP 404, August 1971 (see also J. Am. Helicopter Soc., Vol 52, No 1, pp 69–74, 2007)

3.43 Cheeseman, I.C. and Bennett, W.E., The Effect of the Ground on a Helicopter Rotor in Forward Flight, Aeronautical Research Council R&M No 3021, 1957

3.44 Prouty, R.W., Helicopter Performance, Stability and Control, PWS Publishers, 1986

3.45 Curtiss, H.C., Jr., Erdman, W. and Sun, M., Ground Effect Aerodynamics, Vertica, Vol 11, No 1, 1987

3.46 Lynn, R.R., Robinson, F., Batra, N.N. and Duhon, J.M., Tail Rotor Design – Part 1: Aerodynamics, J. Am. Helicopter Soc., Vol 15, No 4, 1970

3.47 Wilson, J.C. and Mineck, R.E., Wind Tunnel Investigation of Helicopter-Rotor Wake Effects on Three Helicopter Fuselage Models, NASA TM-X-3185, 1975

3.48 Biggers, J.C., McCloud, J.L. and Patterakis, P., Wind Tunnel Tests on Two Full Scale Helicopter Fuselages, NASA TN-D-154-8, 1962

3.49 Loftin, L.K., Jr., Airfoil Section Characteristics at High Angle of Attack, NACA TN 3421, 1954

3.50 Cooper, D.E., YUH-60A Stability and Control, J. Am. Helicopter Soc., Vol 23, No 3, 1978

3.51 Curtiss, H.C. and McKillip, R.M., Studies in Interactive System Identification on Helicopter Rotor/Body Dynamics Using an Analytically Based Linear Model, RAeSoc Conference on Helicopter Handling Qualities and Control, London, November 1988

3.52 Hoerner, S.F. and Borst, H.V., Fluid Dynamic Lift, Hoerner Fluid Dynamics, New Jersey, 1975

3.53 Samoni, G., Simulation Helicoptére Characteristique du SA 330, Aerospatiale Note Technique, 330.05.0080, 1975

3.54 Anonymous, Mathematical Model for the Simulation of Naval and Utility Variants of the Lynx Helicopter, Westland Helicopters Ltd, Tech Note FM/L/031, Issue 2, May 1981

3.55 Sweeting, D., Some Design Aspects of the Stability Augmentation System for the WG13 Rigid Rotor Helicopter, Helicopter Guidance and Control Systems, AGARD CP 86, 1971

3.56 Lanczos, C., Applied Analysis, Pitman Press, London, 1957

3.57 Padfield, G.D., Theoretical Modelling for Helicopter Flight Dynamics; Development and Validation, Proceedings of 16th Congress of ICAS, Jerusalem, August 1988 (RAE TM FM25, April 1989)

3.58 Prouty, R.W., A State-of-the-Art Survey of Two-Dimensional Airfoil Data, J. Am. Helicopter Soc., Vol 20, No 4, 1975

3.59 Pearcey, H.H., Wilby, P.G., Riley, M.J. and Brotherhood, P., The Derivation and Verification of a New Rotor Profile on the Basis of Flow Phenomena; Aerofoil Research and Flight Tests, AGARD Specialists Meeting – The Aerodynamics of Rotary Wings, Marseilles, France, September 1972, RAE Technical Memo Aero 1440, August 1972

3.60 Leishman, J.G., Modelling Sweep Effects on Dynamic Stall, J. Am. Helicopter Soc., Vol 34, No 3, 1989

3.61 Beddoes, T.S., A Synthesis of Unsteady Aerodynamic Effects Including Stall Hysteresis, 1st European Rotorcraft Forum, Southampton, England, September 1975

3.62 Beddoes, T.S., Representation of Airfoil Behaviour, Vertica, Vol 7, No 2, 1983

3.63 Leishman, J.G. and Beddoes, T.S., A Semi-Empirical Model For Dynamic Stall, J. Am. Helicopter Soc., Vol 34, No 3, 1989

3.64 Leishman, J.G., Modelling of Subsonic Unsteady Aerodynamics for Rotary Wing Applications, J. Am. Helicopter Soc., Vol 35, No 1, 1990

3.65 Johnson, W. and Ham, N.D., On the Mechanism of Dynamic Stall, J. Am. Helicopter Soc., Vol 17, No 4, 1972

3.66 DuVal, R.W., A Real-Time Blade Element Helicopter Simulation for Handling Qualities Analysis, 15th European Rotorcraft Forum, Amsterdam, September 1989

3.67 He, Chengjian and Lewis, W.D., A Parametric Study of Real Time Mathematical Modelling Incorporating Dynamic Wake and Elastic Blades, 48th Annual Forum of the American Helicopter Society, Washington, D.C., June 1992

3.68 Turnour, S.R. and Celi, R., Effects of Blade Flexibility on Helicopter Stability and Frequency Response, 19th European Rotorcraft Forum, Cernobbio, Italy, September 1993

3.69 Hansford, R.E. and Simons, I.A., Torsion–Flap–Lag Coupling on Helicopter Rotor Blades, J. Am. Helicopter Soc., Vol 18, No 4, 1973

3.70 Sheridan, P.F. and Smith, R.P., Interactional Aerodynamics – A New Challenge to Helicopter Technology, J. Am. Helicopter Soc., Vol 25, No 1, 1980

3.71 Brocklehurst, A., A Significant Improvement to the Low Speed Yaw Control of the Sea King Using a Tail Boom Strake, 11th European Rotorcraft Forum, London, September 1985

3.72 Wilson, J.C., Kelley, H.L., Donahue, C.C. and Yenni, K.R., Development in Helicopter Tail Boom Strake Applications in the United States, RAeSoc International Conference on Helicopter Handling Qualities and Control, London, November 1988

3.73 Ellin, A.D.S., An In-Flight Experimental Investigation of Helicopter Main Rotor/Tail Rotor Interactions, PhD Thesis, Aerospace Engineering Department, University of Glasgow, April 1993

3.74 Ellin, A.D.S., An In-Flight Investigation of Lynx AH Mk 5 Main Rotor/Tail Rotor Interactions, 19th European Rotorcraft Forum, Cernobbio, Italy, September 1993

3.75 Beddoes, T.S., A Wake Model for High Resolution Airloads, US Army/AHS Conference on Rotorcraft Basic Research, North Carolina, 1985

3.76 Srinivas, V., Chopra, I., Haas, D. and McCool, K., Prediction of Yaw Control Effectiveness and Tail Rotor Loads, 19th European Rotorcraft Forum, Cernobbio, Italy, September 1993

3.77 Xin, Z. and Curtiss, H.C., A Linearized Model of Helicopter Dynamics Including Correlation with Flight Test, Proceedings of 2nd International Conference on Rotorcraft Basic Research, Maryland, February 1988

3.78 Curtiss, H.C., Jr. and Quackenbush, T.R., The Influence of the Rotor Wake on Rotorcraft Stability and Control, 15th European Rotorcraft Forum, Amsterdam, September 1989

3.79 Baskin, V.E., Theory of the Lifting Airscrew, NASA TT F-823, 1976

3.80 Quackenbush, T.R. and Bliss, D.B., Free Wake Prediction of Rotor Flow Fields for Interactional Aerodynamics, 44th Annual Forum of the American Helicopter Society, Washington, D.C., June 1988

3.81 Crawford, C.C., Jr., Rotorcraft Analytical Improvement Needed to Reduce Development Risk – The 1989 Alexander A. Nikolsky Lecture, J. Am. Helicopter Soc., Vol 35, No 1, 1990

3A.1 Bramwell, A.R.S., Helicopter Theory, Edward Arnold, 1976

CHAPTER 4

4.1   Meriam, J.L., Dynamics, John Wiley, 1966

4.2   Duncan, W.J., The Principles of the Control and Stability of Aircraft, Cambridge University Press, 1952

4.3   Heyson, H.H., A Momentum Analysis of Helicopters and Autogyros in Inclined Descent, with Comments on Operational Restrictions, NASA TN D-7917, October 1975

4.4   Bryan, G.H., Stability in Aviation, Macmillan, London, 1911

4.5   Gessow, A. and Myers, G.C., Aerodynamics of the Helicopter, Macmillan, New York, 1952

4.6   AGARD, Dynamic Stability Parameters, AGARD FDP CP 235, 1978

4.7   Blake, B. and Alansky, I., Stability and Control of the YUH-61A, J. Am. Helicopter Soc., Vol 21, No 3, 1976

4.8   Cooper, D.E., YUH-60A Stability and Control, J. Am. Helicopter Soc., Vol 23, No 3, 1978

4.9   Prouty, R., Helicopter Performance, Stability and Control, PWS Publishers. 1986

4.10 Curtiss, H.C., Jr. and McKillip, R.M., Jr., Studies in Interactive System Identification of Helicopter Rotor/Body Dynamics Using an Analytically Based Linear Model, RAeSoc Conference on Helicopter Handling Qualities and Control, London, November 1988

4.11 Hoerner, S.F. and Borst, H.V., Fluid Dynamic Lift, Hoerner Fluid Dynamics, New Jersey, 1975

4.12 Padfield, G.D. and DuVal, R.W., Applications of System Identification Methods to the Prediction of Helicopter Stability. Control and Handling Characteristics, NASA/AHS Specialist Meeting on Helicopter Handling Qualities, NASA CP 2219, 1982

4.13 Amer, K.B., Theory of Helicopter Damping in Pitch or Roll and a Comparison with Flight Measurements, NACA TN 2136, 1950

4.14 Bramwell, A.R.S., Helicopter Dynamics, Edward Arnold, 1976

4.15 Sissingh, G.J., The Effect of Induced Velocity Variation on Helicopter Rotor Damping in Pitch and Roll, RAE TN 2132, 1951

4.16 Wilson, J.C., Kelley, H.L., Donahue, C.C. and Yenni, K.R., Development in Helicopter Tail Boom Strake Application in the United States, RAeSoc International Conference on Helicopter Handling Qualities and Control, London, November 1988

4.17 Brocklehurst, A., A Significant Improvement to the Low Speed Yaw Control of the Sea King Using a Tail Boom Strake, 11th European Rotorcraft Forum, London, September 1985

4.18 Livingston, C.L. and Murphy, M.R., Flying Qualities Considerations in the Design and Development of the HueyCobra, J. Am. Helicopter Soc., Vol 14, No 1, 1969

4.19 Driscoll, J.T. and Sweet, D.H., RAH-66 Comanche/T800 Engine Integration Features and Their Effects on Vehicle Handling Qualities, 50th Annual Forum of the American Helicopter Society, Washington, D.C., May 1994

4.20 Shupe, N.K., A Study of the Dynamic Motions of Hingeless Rotored Helicopters, R&D Technical Report ECOM-3323, US Army Electronics Command, August 1970

4.21 Azuma, A. and Nakamura, Y., Pitch Damping of Helicopter Rotor with Non-uniform Inflow, J. Aircr., Vol 11, No 10, 1974

4.22 AGARD, Rotorcraft System Identification, Final Report of WG18, AGARD AR 280, 1991

4.23 AGARD, Rotorcraft System Identification, AGARD LS 178, 1991

4.24 Milne, R.D., The Analysis of Weakly Coupled Dynamical Systems, Int. J. Control, Vol 2, No 2, 1965

4.25 Padfield, G.D., On the Use of Approximate Models in Helicopter Flight Mechanics, Vertica, Vol 5, pp 243–259, 1981

4.26 Padfield, G.D. and DuVal, R.W., Application Areas for Rotorcraft System Identification; Simulation Model Validation, Rotorcraft System Identification, AGARD LS 178, October 1991

4A.1 Collar, R. and Simpson, A., Matrices and Engineering Dynamics, Ellis Horwood Series in Engineering Science, 1987

4A.2 Duff, G.F.D. and Naylor, D., Differential Equations of Applied Mathematics, John Wiley, 1966

4A.3 Korevvar, J., Mathematical Methods – Vol 1, Academic Press, 1968

4A.4 Milne, R.D., The Analysis of Weakly Coupled Dynamical Systems, Int. J. Control, Vol 2, No 2, 1965

4A.5 Duncan, W.J., The Principles of the Control and Stability of Aircraft, Cambridge University Press, 1952

4A.6 Milne, R.D. and Padfield, G.D., The Strongly Controlled Aircraft, Aeronaut. Q., Vol XXII, Pt 2, May 1971

4B.1 AGARD, Rotorcraft System Identification, AGARD Advisory Report AR 280, 1991

CHAPTER 5

5.1 Neumark, S., Problems of Longitudinal Stability Below Minimum Drag Speed and Theory of Stability Under Constraint, ARC R&M 2983, 1957

5.2 Pinsker. W.J.G., Directional Stability in Flight with Bank Angle Constraint as a Condition Defining a Minimum Acceptable Value for N_v, RAE Technical Report 67 127, June 1967

5.3 Milne, R.D., The Analysis of Weakly Coupled Dynamical Systems, Int. J. Control, Vol 2, No 2, 1965

5.4 Milne, R.D. and Padfield, G.D., The Strongly Controlled Aircraft, Aeronaut. Q., Vol XXII, Pt 2, May 1971

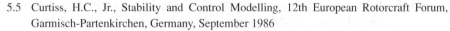 5.5 Curtiss, H.C., Jr., Stability and Control Modelling, 12th European Rotorcraft Forum, Garmisch-Partenkirchen, Germany, September 1986

 5.6 Duncan, W.J., The Principles of the Control and Stability of Aircraft, Cambridge University Press, 1952

5.7 Tischler, M.B., System Identification Requirements for High Bandwidth Rotorcraft Flight Control System Design, Rotorcraft System Identification, AGARD LS 178, 1991

5.8 Diftler, M.A., UH-60A Helicopter Stability Augmentation Study, 14th European Rotorcraft Forum, Paper No 74, Milano, Italy, September 1988

5.9 Pausder, H.J. and Jordan, D., Handling Qualities Evaluation of Helicopters with Different Stability and Control Characteristics, Vertica, Vol 1, 1976

5.10 Wolkovitch, J. and Hoffman, J.A., Stability and Control of Helicopters in Steep Approaches, Vol 1, Summary Report USAVLABS Technical Report 70-74A, 1971

5.11 Heyson, H.H., A Momentum Analysis of Helicopters and Autogyros in Inclined Descent, with Comments on Operational Restrictions, NASA TN D-7917, 1975

5.12 Padfield, G.D., Jones, J.P., Charlton, M.T., Howell, S.E. and Bradley, R., Where Does the Workload Go When Pilots Attack Manoeuvres? An Analysis of Results from Flying Qualities Theory and Experiment, 20th European Rotorcraft Forum, Amsterdam, October 1994

5.13 Thomson, D.G. and Bradley, R., An Investigation of Flight Path Constrained Helicopter Manoeuvres by Inverse Simulation, Proceedings of 13th European Rotorcraft Forum, Arles, France, 1987

5.14 AGARD, Rotorcraft System Identification, AGARD AR 280, LS 178, 1991

5.15 Lanczos, C., Applied Analysis, Issac Pitman, London, 1957

5.16 Houston, S.S., Identification of Factors Influencing Heave Axis Damping and Control Sensitivity, RAE TR 88067, 1989

5.17 Carpenter, P.J. and Fridovitch, B., Effect of a Rapid Blade Pitch Increase on the Thrust and Induced Velocity Response of a Full Scale Helicopter Rotor, NACA TN 3044, 1953

5.18 Ormiston, R.A., An Actuator Disc Theory for Rotor Wake Induced Velocities, AGARD Specialists Meeting on The Aerodynamics of Rotary Wings, Marseilles, France, 1972

5.19 Pitt, D.M. and Peters, D.A., Theoretical Prediction of Dynamic Inflow Derivatives, Vertica, Vol 5, No 1, 1981

5.20 Chen, R.T.N. and Hindson, W.S., Influence of Dynamic Inflow on the Helicopter Vertical Response, Vertica, Vol 11, No 1, 1987

5.21 Houston, S.S., Identification of a Coupled Body/Coning/Inflow Model of Puma Vertical Response in the Hover, Vertica, Vol 13, No 3, 1989

5.22 Houston, S.S. and Tarttelin, P.C., Validation of Mathematical Simulations of Helicopter Vertical Response Characteristics in Hover, J. Am. Helicopter Soc., Vol 36, No 1, 1991

5.23 Houston, S.S. and Black, C.G., Identifiability of Helicopter Models Incorporating Higher Order Dynamics, J. Guid. Control Dyn., Vol 14, No 4, 1991

5.24 Feik, R.A. and Perrin, R.H., Identification of an Adequate Model for Collective Response Dynamics of a Sea King Helicopter, Vertica, Vol 13, No 3, 1989

5.25 Ballin, M.G. and Dalang-Secretan, M.A., Validation of the Dynamic Response of a Blade-Element UH-60 Simulation Model in Hovering Flight, J. Am. Helicopter Soc., Vol 36, No 4, 1991

5.26 Fu, K.H. and Kaletka, J., Frequency Domain Identification of Bo105 Derivative Models with Rotor Degrees of Freedom, J. Am. Helicopter Soc., Vol 38, No 1, 1993

5.27 Tischler, M.B. and Cauffman, M.G., Frequency Response Method for Rotorcraft System Identification: Flight Applications to Bo105 Coupled Rotor/Fuselage Dynamics, J. Am. Helicopter Soc., Vol 37, No 3, 1992

5.28 Hanson, R.S., Towards a Better Understanding of Helicopter Stability Derivatives, J. Am. Helicopter Soc., Vol 28, No 4, 1983

5.29 Black, C.G., Murray-Smith, D.J. and Padfield, G.D., Experience with Frequency Domain Methods in Helicopter System Identification, Proceedings of the 12th European Rotorcraft Forum, Garmisch Partenkirchen, Germany, September 1986

5.30 Padfield, G.D., Theoretical Modelling for Helicopter Flight Dynamics: Development and Validation, Proceedings of the 16th ICAS Congress, Jerusalem, September 1988 (also RAE Tech Memo TM 25, April 1989)

5.31 Xin, Z. and Curtiss, H.C., A Linearised Model of Helicopter Dynamics Including Correlation with Flight Test, 2nd International Conference on Rotorcraft Basic Research, Maryland, February 1988

5.32 He, Chengjian and Lewis, W.D., A Parametric Study of Real-Time Mathematical Modelling Incorporating Dynamic Wake and Elastic Blades, 48th Annual Forum of the American Helicopter Society, Washington, D.C., June 1992

5.33 Padfield, G.D. and DuVal R.W., Application Areas for Rotorcraft System Identification; Simulation Model Validation, Rotorcraft System Identification, AGARD LS 178, 1991

5.34 Padfield, G.D., Thorne, R.T., Murray-Smith, D.J., Black, C. and Caldwell, A.E., UK Research into System Identification for Helicopter Flight Mechanics, Vertica, Vol 11, No 4, 1987

5.35 Padfield, G.D., Integrated System Identification Methodology for Helicopter Flight Dynamics, Proceedings of the 42nd Annual Forum of the American Helicopter Society, Washington, D.C., June 1986

5.36 Gaonker, G.H. and Hohenemser, K.H., Flapping Response of Lifting Rotor Blades to Atmospheric Turbulence, J. Aircr., Vol 6, No 6, 1969

5.37 Drees, J.M. and Harvey, K.W., Helicopter Gust Response at High Forward Speed, J. Aircr., Vol 7, No 3, 1970

5.38 Arcidiaconon, P.J., Bergqist, R.R. and Alexander, W.T., Helicopter Gust Response Characteristics Including Unsteady Aerodynamic Stall Effects, J. Am. Helicopter Soc., Vol 34, No 4, 1974

5.39 Judd, M. and Newman, S.J., An Analysis of Helicopter Rotor Response Due to Gusts and Turbulence, Vertica, Vol 1, No 3, 1977

5.40 Reichert, G. and Rade, M., Influence of Turbulence on Helicopter Design and Operation, AGARD FMP Symposium on Flight in Turbulence, Woburn, England, May 1973

5.41 Dahl, H.J. and Faulkner, A.J., Helicopter Simulation in Atmospheric Turbulence, Vertica, Vol 3, 1979

5.42 Costello, M.F., Prasad, J.V.R., Schrage, D.P. and Gaonker, G.H., Some Issues on Modeling Atmospheric Turbulence Experienced by Helicopter Rotor Blades, J. Am. Helicopter Soc., Vol 37, No 2, 1992

5.43 Riaz, J., Prasad, J.V.R., Schrage, D.P. and Gaonker, G.H., Atmospheric Turbulence Simulation for Rotorcraft Applications, J. Am. Helicopter Soc., Vol 38, No 1, 1992

5.44 Bradley, R., Sinclair, M., Turner, G.P. and Jones, J.G., Wavelet Analysis of Helicopter Response to Atmospheric Turbulence in Ride Quality Assessment, 20th European Rotorcraft Forum, Amsterdam, October 1994

5.45 Anonymous, UK Defence Standard 00970; Design and Airworthiness Requirements for Service Aircraft, Vol 2, Rotorcraft – Book 1, 1988

5.46 Lawson, T.V., Wind Effects on Buildings, Applied Science Publishers, London, 1980

5.47 von Karman, T., Progress in the Statistical Theory of Turbulence, Turbulence: Classical Papers in Statistical Theory, Interscience Publications, New York, 1961

5.48 Jones, J.G., Models of Atmospheric Turbulence for Helicopter Response Studies, DRA Working Paper FDS WP 94052/1, March 1994

5.49 Jones, J.G., Statistical Discrete Gust Theory for Aircraft Loads, RAE Technical Report TR 73167, November 1973

5.50 Jones, J.G., The Statistical Discrete Gust Method for Predicting Aircraft Loads and Dynamic Response, J. Aircr., Vol 26, No 4, 1989

5.51 Jones, J.G., Foster, G.W. and Earwicker, P.G., Wavelet Analysis of Gust Structure in Measured Atmospheric Turbulence Data, J. Aircr., Vol 30, No 1, 1993

5.52   Bir, G.S. and Chopra, I., Gust Response of Hingeless Rotors, J. Am. Helicopter Soc., Vol 31, No 2, 1986

5.53   Elliot, A.S. and Chopra, I., Helicopter Response to Atmospheric Disturbances in Forward Flight, J. Am. Helicopter Soc., Vol 35, No 2, 1990

5.54   Gaonker, G.H., A Perspective on Modelling Rotorcraft in Turbulence, Probab. Eng. Mech. J., Vol 3, No 1, 1988

5.55   Costello, M.F., A Theory for the Analysis of Rotorcraft Operating in Atmospheric Turbulence, 40th Annual Forum of the American Helicopter Society, Washington, D.C., May 1990

5.56   Prasad, J.V.R., Gaonker, G.H. and Yingyi, D., Real Time Implementation Aspects of a Rotorcraft Turbulence Simulation Method, 49th Annual Forum of the American Helicopter Society, St Louis, Mo., May 1993

5.57   Jones, J.G., Ride-Bumpiness and the Influence of Active Control Systems, RAE Technical Memo FS 268, September 1979

5.58   AVSCOM, Aeronautical Design Standard (ADS-33C) – Handling Qualities for Military Helicopters, US Army AVSCOM, 1989

CHAPTER 6

6.1   Stewart, W. and Zbrozek, J.K., Loss of Control Incident on S-51 Helicopter VW 209, RAE Technical Report Aero 2270, 1948

6.2   Cooper, G.E. and Harper, R.P., Jr., The Use of Pilot Ratings in the Evaluation of Aircraft Handling Qualities, NASA TM D-5133, 1969

6.3   Padfield, G.D., Charlton, M.T. and Kimberley, A.K., Helicopter Flying Qualities in Critical Mission Task Elements; Initial Experience with the DRA Bedford Large Motion Simulator, 18th European Rotorcraft Forum, Avignon, France, September 1992

6.4   Andrews, H., Technical Evaluation Report on the Flight Mechanics Panel Symposium on Flying Qualities, AGARD Advisory Report 311, April 1992

6.5   AVSCOM, Aeronautical Design Standard (ADS-33C) – Handling Qualities for Military Helicopters, US Army AVSCOM, 1989

6.6   USAF, Flying Qualities of Piloted Airplanes, MIL-F-8785C (USAF), 1980

6.7   McRuer, D., Estimation of Pilot Ratings via Pilot Modelling, Flying Qualities, AGARD CP 508, 1991

6.8   Key, D.L., A New Handling Qualities Specification for US Military Rotorcraft, Proceedings of International Conference, Helicopter Handling Qualities and Flight Control, RAeSoc, London, 1988

6.9   Anonymous. UK Defence Standard 00970; Design and Airworthiness Requirements for Service Aircraft, Vol 2, Rotorcraft – Book 1, 1988

6.10   FAA, Federal Aviation Regulations Part 29 – Airworthiness Standards for Transport Category Rotorcraft, FAA, 1983

6.11   CAA, British Civil Airworthiness Requirements, BCAR 29, Rotorcraft, CAA, Issue 83. 1986

6.12   Charlton, M.T. and Padfield, G.D., A Review of Handling Requirements for Civil Helicopters, DRA Working Paper FSB WP(91) 040, July 1991

6.13   Anonymous, Certification of Transport Category Rotorcraft. FAA Advisory Circular AC No 29-2(A), 1983

6.14 Padfield, G.D., Helicopter Handling Qualities and Flight Control – Is the Helicopter Community Prepared for Change? Overview Paper, Helicopter Handling Qualities and Flight Control, RAeSoc Conference, London, November 1988

6.15 Heffley, R.K., Curtiss, H.C., Hindson, W.S. and Hess, R.A., Study of Helicopter Roll Control Effectiveness Criteria, NASA CR 177404, April 1986

6.16 Padfield, G.D., Jones, J.P., Charlton, M.T., Howell, S. and Bradley, R., Where Does the Workload Go When Pilots Attack Manoeuvres? – An Analysis of Results from Flying Qualities Theory and Experiment, Proceedings of the 20th European Rotorcraft Forum, Amsterdam, October 1994

6.17 Tomlinson, B.N. and Padfield, G.D., Simulation Studies of Helicopter Agility, Vertica, Vol 4, 1980

6.18 Charlton, M.T., Padfield, G.D. and Horton, R., Helicopter Agility in Low Speed Manoeuvres, Proceedings of the 13th European Rotorcraft Forum, Arles, France, September 1987 (also RAE TM FM22, 1989)

6.19 Ockier, C.J., Flight Evaluation of the New Handling Qualities Criteria Using the Bo105, 49th Annual Forum of the American Helicopter Society, St Louis, Mo., May 1993

6.20 Gmelin, B. and Pausder, H.-J., Test Techniques for Helicopter Handling Qualities Evaluation and Design, Proceedings of the RAeSoc International Conference on Helicopter Handling Qualities and Flight Control, London, 1988

6.21 Morgan, J.M. and Baillie, S.W., ADS-33C Related Handling Qualities Research Performed Using the NRC Bell 205 Airborne Simulator, Proceedings of NASA/AHS Conference, Piloting Vertical Flight Aircraft – A Conference on Flying Qualities and Human Factors, San Francisco, January 1993

6.22 Hoh, R.H., New Developments in Flying Qualities Criteria with Application to Rotary Wing Aircraft, Helicopter Handling Qualities, NASA CP 2219, April 1982

6.23 Hoh, R.H., Concepts and Criteria for a Mission Oriented Flying Qualities Specification, Advances in Flying Qualities, AGARD LS 157, 1988

6.24 Hoh, R.H., Dynamic Requirements in the New Handling Qualities Specification for US Military Rotorcraft, Helicopter Handling Qualities and Flight Control, RAeSoc International Conference, London, 1988

6.25 McRuer, D.T. and Krendal, E.S., Mathematical Models of Human Pilot Behaviour, AGARD AG 188, 1974

6.26 Decker, W.A., A Piloted Simulation Investigation of Yaw Dynamic Requirements for Turreted Gun Use in Low Level Helicopter Air Combat, 44th Annual Forum of the American Helicopter Society, Washington, D.C., June 1988

6.27 Whalley, M.S. and Carpenter, W.R., A Piloted Simulation Investigation of Pitch and Roll Handling Qualities Requirements for Air-to-Air Combat, 48th Annual Forum of the American Helicopter Society. Washington, D.C., June 1992

6.28 Houston, S.S. and Horton, R.I., The Identification of Reduced Order Models of Helicopter Behaviour for Handling Qualities Studies, 13th European Rotorcraft Forum, Arles, France, September 1987

6.29 Aponso, B.L., Mitchell, D.G. and Hoh, R.H., Simulation Investigation of the Effects of Helicopter Hovering Dynamics on Pilot Performance. J. Guid. Control Dyn., Vol 13, No 1. 1990

6.30 Corliss, L.D. and Carico, D.G., A Flight Investigation of Roll Control Sensitivity, Damping and Cross Coupling in a Low Altitude Lateral Manoeuvring Task, NASA TM 84376, December 1983

6.31 Pausder, H.-J. and Gerdes, R.M., The Effects of Pilot Stress Factors on Handling Qualities

Assessment During US/German Helicopter Agility Flight Tests, 8th European Rotorcraft Forum, Aix-en-Provence, France, September 1982

6.32 Mitchell, D.G., Hoh, R.H. and Morgan, J.M., Flight Investigation of Helicopter Low Speed Response Requirements, J. Guid. Control Dyn., Vol 12, No 5, 1989

6.33 Hoh, R.H., Handling Qualities Criterion for Very Low Visibility Rotorcraft NoE Operations, AGARD CP 423, October 1986

6.34 Baillie, S.W. and Hoh, R.H., The Effect of Reduced Usable Cue Environment on Helicopter Handling Qualities, CASI J., Vol 34, No 3, 1988

6.35 Buckingham, S.L. and Padfield, G.D., Piloted Simulations to Explore Helicopter Advanced Control Systems, RAE Technical Report 86022, April 1986

6.36 Charlton, M.T., Guyomard, C., Lane, R., Modden, H., Schimke, D. and Taghizad, A., The European ACT Programme: A Collaborative Approach to Handling Qualities Evaluation and Control Law Design, 49th Annual Forum of the American Helicopter Society, St Louis, Mo., May 1993

6.37 Pausder, H.-J. and Blanken, C.L., Investigation of the Effects of Bandwidth and Time Delay on Helicopter Roll Axis Handling Qualities, NASA/AHS Specialists Meeting on Piloting Vertical Flight Aircraft – A Conference on Flying Qualities and Human Factors, San Francisco, January 1993

6.38 AVSCOM, Aeronautical Design Standard (ADS-33D) – Handling Qualities for Military Helicopters, US Army AVSCOM, 1994

6.39 Talbot, N., Helicopter Handling Qualities and Control – Civil Certification Aspects, RAeSoc Conference, Helicopter Handling Qualities and Control, London, 1988

6.40 Ham, J.A. and Butler, C.P., Flight Testing the Handling Qualities Requirements of ADS-33C – Lessons Learned at ATTC, 47th Annual Forum of the American Helicopter Society, May 1991

6.41 Ham, J. and Tischler, M.B., Flight Testing and Frequency Domain Analysis for Rotorcraft Handling Qualities Characteristics, NASA/AHS Specialists Meeting on Piloting Vertical Flight Aircraft – A Conference on Flying Qualities and Human Factors, San Francisco, January 1993

6.42 Ham, J.A., Frequency Domain Flight Testing and Analysis of an OH-58D Helicopter, J. Am. Helicopter Soc., Vol 37, No 4, 1992

6.43 Tischler, M.B., System Identification Requirements for High Bandwidth Rotorcraft Flight Control System Design, Rotorcraft System Identification, AGARD LS 178, 1991

6.44 Tischler, M.B., Demonstration of Frequency Domain Testing Technique Using a Bell 214ST Helicopter, NASA TM 89422, ARMY TM 87-A-1, April 1987

6.45 Charlton, M.T. and Houston, S.S., Flight Test and Analysis for New Handling Criteria, Proceedings of the RAeSoc Conference on Handling Qualities and Control, London, 1988

6.46 Ballin. M.B. and Dalang-Secretan, M.-A., Validation of the Dynamic Response of a Blade-Element UH-60 Simulation Model in Hovering Flight, J. Am. Helicopter Soc., Vol 36, No 4, 1991

6.47 Houston, S.S. and Horton, R.I., An Examination of Bandwidth Criteria for Helicopter Pitch Axis Handling Qualities, RAE Technical Memo FS(B) 681, October 1987

6.48 Tischler, M.B., Leung, J.G.M. and Dugan, D.C., Frequency Domain Identification of XV-15 Tilt Rotor Dynamics in Hovering Flight, J. Am. Helicopter Soc., Vol 30, No 2, 1985

6.49 Harding, J.W., Frequency Domain Identification of Coupled Rotor/Body Models of an Advanced Attack Helicopter, 48th Annual Forum of the American Helicopter Society, Washington, D.C., June 1992

6.50 Otnes, R.K. and Enochson, I., Applied Time Series Analysis, John Wiley, New York, 1978

6.51 Tischler, M.B. and Cauffman, M.G., Frequency Response Method for Rotorcraft System Identification: Flight Applications to BO105 Coupled Rotor/Fuselage Dynamics, J. Am. Helicopter Soc., Vol 37, No 3, 1992

6.52 Tischler, M.B., Flight Control System Methods for Advanced Combat Rotorcraft, RAeSoc Conference on Helicopter Handling Qualities and Control, London, November 1988

6.53 Edenborough, H.K. and Wernicke, K.G., Control and Manoeuvre Requirements for Armed Helicopters, 20th Annual Forum of the American Helicopter Society, Washington, D.C., May 1964

6.54 Pausder, H.-J. and Von Grunhagen, W., Handling Qualities Evaluation for Highly Augmented Helicopters, Flying Qualities, AGARD CP 508, 1990

6.55 Baillie, S.W. and Morgan, J.M., Control Sensitivity, Bandwidth and Disturbance Rejection Concerns for Advanced Rotorcraft, 45th Annual Forum of the American Helicopter Society, Boston, May 1989

6.56 Padfield, G.D., Tomlinson, B.N. and Wells, P.M., Simulation Studies of Helicopter Agility and other Topics, RAE Technical Memo FS197, 1978

6.57 Landis, K.H. and Aiken, E.W., An Assessment of Various Sidestick Controller/Stability and Control Augmentation Systems for Night Nap-of-the-Earth Flight Using Piloted Simulation, Helicopter Handling Qualities, NASA CP 2219, 1982

6.58 Morgan, J.M., Some Piloting Experiences with Multi-Function Isometric Side-Arm Controllers in a Helicopter, Helicopter Handling Qualities, NASA CP 2219, 1982

6.59 Hodgkinson, J. and LaManna, W.J., Equivalent Systems Approaches to Handling Qualities and Design Problems of Augmented Aircraft, AIAA Atmospheric Flight Conference, Hollywood, Fla., August 1977

6.60 Hoh, R.H., Advances in Flying Qualities; Concepts and Criteria for a Mission Oriented Flying Qualities Specification, AGARD LS 157, 1988

6.61 Huber, H. and Masue, T., Flight Characteristics, Design and Development of the MBB/KHI BK117, 7th European Rotorcraft Forum, Garmisch Partenkirchen, Germany, 1981

6.62 Padfield, G.D., Flight Testing for Performance and Flying Qualities, Helicopter Aeromechanics, AGARD LS 139, 1985

6.63 Roesch, P. and Vuillet, A., New Designs for Improved Aerodynamic Stability on Recent Aerospatiale Helicopters, 37th Annual Forum of the American Helicopter Society, New Orleans, La., 1981

6.64 Heyson, H.H., A Momentum Analysis of Helicopters and Autogyros in Inclined Descent, with Comments on Operational Restrictions, NASA TN D-7917, 1975

6.65 Padfield, G.D. and Charlton, M.T., Aspects of RAE Flight Research into Helicopter Agility and Pilot Control Strategies, Proceedings of the Handling Qualities Specialists Meeting, Ames Research Center, Moffett Field, Calif., June 1986

6.66 Corliss, L.D., The Effects of Engine and Height Control Characteristics on Helicopter Handling Qualities, J. Am. Helicopter Soc., Vol 28, No 3, 1983

6.67 Hindson, W.S., Tucker, G.E., Lebacqz, J.V. and Hilbert, K.B., Flight Evaluation of Height Response Characteristics for the Hover Bob-up Task and Comparison with Proposed Criteria, 42nd Annual National Forum of the American Helicopter Society, Washington, D.C., June 1986

6.68 Blanken, C.L., Whalley, M.S., Heffley, R.K. and Bourne, S.M., Helicopter Thrust Response Study, Proceedings of the Handling Qualities Specialists Meeting, Ames Research Center, Moffett Field, Calif., June 1986

6.69   Baillie, S.W. and Morgan, J.M., The Impact of Vertical Axis Characteristics on Helicopter Handling Qualities, NRC Aeronautical Report LR-619, Ottawa, August 1987

6.70   Tate, S. and Padfield, G.D., Simulating Flying Qualities at the Helicopter/Ship Dynamic Interface, Proceedings of the 50th Annual Forum of the American Helicopter Society, Washington, D.C., May 1994

6.71   Howitt, J., Comments on the Proposed MIL-H-8501 Update Criterion on Height Response Characteristics, RAE Working Paper WP FM 041, 1990

6.72   Tischler, M.B., System Identification Methods for Handling Qualities Evaluation, Rotorcraft System Identification, AGARD LS 178, October 1991

6.73   Stewart, W., Helicopter Behaviour in the Vortex Ring State, ARC R&M 3117, 1951

6.74   Smith, A.C., Review of Helicopter Low Speed Handling Problems Associated with the Tail Rotor, RAE TM FS(B) 683, November 1987

6.75   Brocklehurst, A., A Significant Improvement to the Low Speed Yaw Control of the Sea King Using a Tail Boom Strake, 11th European Rotorcraft Forum, London, September 1985

6.76   Ellin, A.D.S., An In-Flight Experimental Investigation of Helicopter Main Rotor/Tail Rotor Interactions, PhD Thesis, Aerospace Engineering Department, University of Glasgow, April 1993

6.77   Bivens, C.C., Directional Handling Qualities Requirements for Nap-of-the-Earth Tasks, 41st Annual Forum of the American Helicopter Society, May 1985

6.78   Decker, W.A., Morris, P.M. and Williams, J.W., A Piloted Simulation Investigation of Yaw Dynamics Requirements for Turretted Gun Use in Low Level Helicopter Air Combat, 44th Annual Forum of the American Helicopter Society, Washington, D.C., June 1988

6.79   Padfield, G.D. and DuVal, R.W., Application Areas for Rotorcraft System Identification – Simulation Model Validation, Rotorcraft System Identification, AGARD LS 178, 1991

6.80   Chen, R.T.N. and Talbot, P.D., An Exploratory Investigation of the Effects of Large Variations in Rotor System Dynamics Design Parameters on Helicopter Handling Characteristics in NoE Flight, 33rd Annual Forum of the American Helicopter Society, Washington, D.C., May 1977

6.81   Watson, D.C. and Aiken, E.W., An Investigation of the Effects of Pitch-Roll Cross Coupling on Helicopter Handling Qualities for Terrain Flight, AIAA-87-2534-CP, 1987

6.82   Watson, D.C. and Hindson, W.S., In-Flight Simulation Investigation of Rotorcraft Pitch-Roll Cross Coupling, RAeSoc Conference on Handling Qualities and Control, London, 1988

6.83   Charlton, M.T., Padfield, G.D. and Howitt, J., A Piloted Simulation Investigation of Helicopter Roll-Pitch Cross-Coupling Requirements, DRA Working Paper FS(B) WP(93) 049, 1993

6.84   Pausder, H.-J. and Blanken, C.L., Rotorcraft Pitch-Roll Cross Coupling Evaluation for Aggressive Tracking Manoeuvring, 19th European Rotorcraft Forum, Cernobbio, Italy, September 1993

6.85   Blanken, C. and Pausder, J.-H., Rotorcraft Pitch-Roll Decoupling Requirements from a Roll Tracking Manoeuvre, 50th Annual Forum of the American Helicopter Society, Washington, D.C., May 1994

6.86   Chen, R.T.N., Talbot, P.D., Gerdes, R.M. and Dugan, D.C., A Piloted Simulator Investigation of Augmentation Systems to Improve Helicopter Nap-of-the-Earth Handling Qualities, 34th Annual Forum of the American Helicopter Society, Washington, D.C., May 1978

6.87   Hansen, K.C., Handling Qualities Design and Development of the CH-53E, UH-60A and S-76, RAeSoc Conference on Helicopter Handling Qualities and Control, London, November 1988

6.88 Morsc, C.S., ADFCS and NOTOR^{TM}: Two Ways to Fix Flying Qualities, Flying Qualities, AGARD CP 508, February 1991

6.89 Fogler, D.L., Jr. and Keller, J.F., Design and Pilot Evaluation of the RAH-66 Comanche Core AFCS, NASA/AHS Specialists Meeting on Piloting Vertical Flight Aircraft, A Conference on Flying Qualities and Human Factors, San Francisco, January 1993

6.90 Gold, P.J. and Dryfoos, J.B., Design and Pilot Evaluation of the RAH-66 Comanche Selectable Control Modes, NASA/AHS Specialists Meeting on Piloting Vertical Flight Aircraft, a NASA/AHS Conference on Flying Qualities and Human Factors, San Francisco, January 1993

6.91 Padfield, G.D., The Making of Helicopter Flying Qualities: A Requirements Perspective, The Aeronaut. J. R. Aeronaut. Soc., Vol 102, No 1018, pp 409–443, 1998

6.92 Anonymous, Aeronautical Design Standard-33E-PRF, Performance Specification, Handling Qualities Requirements for Military Rotorcraft, US Army AMCOM, Redstone, Alabama, March 21, 2000

6.93 Blanken, C.L., Ockier, C.J., Pausder, H-J. and Simmons, R.C., Rotorcraft Pitch-Roll Decoupling Requirements from a Roll Tracking Maneuver, J. Am. Helicopter Soc., Vol 42, 1997

6.94 Ockier, C.J., Evaluation of the ADS-33D Handling Qualities Criteria Using the BO 105 Helicopter, DLR Report Forschungsbericht 98-07, 1998

6.95 Key, D.L., Hoh, R.H. and Blanken, C.L., Tailoring ADS-33 for a Specific End Item, Proceedings of the 54th Annual Forum of the American Helicopter Society, Washington, D.C., May 1998

6.96 Padfield, G.D., Charlton, M.T., Mace, T. and Morton, R., Handling Qualities Evaluation of the UK Attack Helicopters Contenders Using ADS-33, 21st European Rotorcraft Forum, St Petersburg, Russia, September 1995

6.97 Benquet, P., Rollet, P., Pausder, H.J. and Gollnick, V., Tailoring of ADS-33 for the NH90 Programme, Proceedings of the 52nd American Helicopter Society Annual Forum, Washington, D.C., June 1996

6.98 Tate, S., Padfield, G.D. and Tailby, A.J., Handling Qualities Criteria for Maritime Helicopter Operations – Can ADS-33 Meet the Need? 21st European Rotorcraft Forum, St Petersburg, Russia, September 1995

6.99 Padfield, G.D. and Wilkinson, C.H., Handling Qualities Criteria for Maritime Helicopter Operations, 53rd Annual Forum of the American Helicopter Society, Virginia Beach, Va., April 1997

6.100 Carignan, J.R.P.S. and Gubbels, A.W., Assessment of Vertical Axis Handling Qualities for the Shipborne Recovery Task – ADS 33 (Maritime), Proceedings of the 54th Annual Forum of the American Helicopter Society, Washington, D.C., May 20–22, 1998

6.101 Carignan, S.J.R.P., Gubbels, A.W. and Ellis, K. Assessment of Handling Qualities for the Shipborne Recovery Task – ADS 33 (Maritime), Proceedings of the 56th Annual Forum of the American Helicopter Society, Washington, D.C., May 2000

6.102 Strachan, M.W., Shubert, A.W. and Wilson, A.W., Development and Validation of ADS-33C Handling Qualities Flight Test Manoeuvres for Cargo Helicopters, Proceedings of the 50th American Helicopter Society Annual Forum, Washington, D.C., May 1994

6.103 Hoh, R.H. and Heffley, R.K., Development of Handling Qualities Criteria for Rotorcraft with Externally Slung Loads, Proceedings of the 58th Annual Forum of the American Helicopter Society, Montreal, Canada, June 2002

6.104 Blanken, C.L., Hoefinger, M.T. and Strecker, G., Evaluation of ADS-33E Cargo Helicopter Requirements Using a CH-53G, Proceedings of the 62nd Annual forum of the American Helicopter Society, Phoenix, Ariz., May 2006

6.105 Chokesy, F. and Gonzalez, P., Simulation Evaluation of the UH-60M Helicopter Against the ADS-33 Handling Qualities Specification, 60th Annual Forum of the American Helicopter Society, Baltimore, Md., May 2004

6.106 Martyn, A.W., Charlton, M.T. and Padfield, G.D., Vehicle Structural Fatigue Issues in Rotorcraft Flying Qualities Testing, 21st European Rotorcraft Forum, St Petersburg, Russia, September 1995

6.107 Pavel, M.D. and Padfield, G.D., The Extension of ADS-33 Metrics for Agility Enhancement and Structural Load Alleviation, J. Am. Helicopter Soc., Vol 51, No 4, 2006

6.108 Kothmann, B.D. and Armbrust, J., RAH-66 Comanche Core AFCS Control Law Development, Proceedings of the 58th Annual Forum of the American Helicopter Society, Montreal, Canada, June 2002

CHAPTER 7

7.1 AVSCOM, Aeronautical Design Standard ADS-33C – Handling Qualities for Military Helicopters, US Army AVSCOM, St Louis, Mo., 1989

7.2 Padfield, G.D., Charlton, M.T., Houston, S.S., Pausder, H.J. and Hummes, D., Observations of Pilot Control Strategy in Low Level Helicopter Flying Tasks, Vertica, Vol 12, No 3, 1988

7.3 McRuer, D., Estimation of Pilot Ratings via Pilot Modelling, Flying Qualities, AGARD CP 508, 1991

7.4 Cooper, G.E. and Harper, R.P., Jr., The Use of Pilot Ratings in the Evaluation of Aircraft Handling Qualities, NASA TM D-5133, 1969

7.5 Hoh, R.H., Advances in Flying Qualities – Concepts and Criteria for a Mission Oriented Flying Qualities Specification, AGARD LS 157, 1988

7.6 Wilson, D.J. and Riley, D.R., Cooper–Harper Rating Variability, Atmospheric Flight Mechanics Conference, AIAA Paper 89-3358, Boston, August 1989

7.7 Pausder, H.J. and von Grunhagen, W., Handling Qualities Evaluation for Highly Augmented Helicopters, Flying Qualities, AGARD CP 508, 1990

7.8 Hoh, R.H., Lessons Learned Concerning the Interpretation of Subjective Handling Qualities Pilot Rating Data, AIAA Atmospheric Flight Mechanics Conference, AIAA Paper 902824, Portland, August 1990

7.9 Ham, J.A., Metzger, M. and Hoh, R.H., Handling Qualities Testing Using the Mission Oriented Requirements of ADS-33C, 48th Annual Forum of the American Helicopter Society, Washington, D.C., June 1992

7.10 Key, D.L., Blanken, C.L. and Hoh, R.H., Some Lessons Learned in Three Years with ADS-33C, Proceedings of the AHS/NASA Conference on Piloting Vertical Flight Aircraft, San Francisco, January 1993

7.11 Morgan, J.M., and Baillie, S.W., ADS-33C Related Handling Qualities Research Performed Using the NRC Bell 205 Airborne Simulator, Proceedings of the AHS/NASA Conference on Piloting Vertical Flight Aircraft, San Francisco, January 1993

7.12 Anonymous, Draft MIL-STD, 8501B, Rotorcraft Flight and Ground Handling Qualities – General Requirements for, US AATC, November 1993

7.13 Padfield, G.D., Charlton, M.T. and Kimberley, A.M., Helicopter Flying Qualities in Critical Mission Task Elements, 18th European Rotorcraft Forum, Avignon, France, September 1992

7.14 Condon, G.W., Simulation of Nap-of-the-Earth Flight in Helicopters, Computer Aided System Design and Simulation, AGARD CP 473, 1990

7.15 Mayo, J.R., Occhiato, J.J. and Hong, S.W., Helicopter Modeling Requirements for Full Mission Simulation and Handling Qualities Assessment, Proceedings of the 47th Annual Forum of the American Helicopter Society, May 1991

7.16 Braun, D., Kampa, K. and Schimke, D., Mission Oriented Investigation of Handling Qualities Through Simulation, Proceedings of the 17th European Rotorcraft Forum, Berlin, September 1991

7.17 Buckingham, S.L. and Padfield, G.D., Piloted Simulations to Explore Helicopter Advanced Control Systems, RAE Technical Report 86022, April 1986

7.18 Tate, S. and Padfield, G.D., Simulating Flying Qualities at the Helicopter/Ship Dynamic Interface, Proceedings of the 50th Annual Forum of the American Helicopter Society, Washington, D.C., May 1994

7.19 AGARD, Operational Agility, Final Report of AGARD Working Group 19, AGARD AR 314, 1994

7.20 Brotherhood, P. and Charlton, M.T., An Assessment of Helicopter Turning Performance During NoE Flight, RAE TM FS(B) 534, January 1984

7.21 Heffley, R.E., A Review of Roll Control Effectiveness Study, Handling Qualities Specialists Meeting, NASA Ames Research Center, June 1986

7.22 Blanken, C.L. and Whalley, M.S., Helicopter Thrust Response Study, Handling Qualities Specialists Meeting, NASA-Ames Research Center, June 1986

7.23 Padfield, G.D. and Charlton, M.T., Aspects of RAE Flight Research into Helicopter Agility and Pilot Control Strategy, Handling Qualities Specialists Meeting, NASA Ames Research Center, June 1986

7.24 Charlton, M.T., Padfield, G.D. and Horton, R.I., Helicopter Agility in Low Speed Manoeuvres, 13th European Rotorcraft Forum, Aries, France, September 1987 (also RAE TM FM 22, April 1989)

7.25 Padfield, G.D. and Hodgkinson, J., The Influence of Flying Qualities on Operational Agility, AGARD FMP Symposium on Technologies for Highly Manoeuvrable Aircraft, (AGARD CP 548), Annapolis, Maryland, October 1993

7.26 Schroeder, J.A. and Eshow, M.E., Improvements in Hover Display Dynamics for a Combat Helicopter, Piloting Vertical Flight Aircraft – A NASA/AHS Conference on Flying Qualities and Human Factors, San Francisco, January 1993

7.27 Hoh, R.H., New Developments in Flying Qualities Criteria with Application to Rotary-Wing Aircraft, Helicopter Handling Qualities, NASA CP 2219, April 1982

7.28 Hoh, R.H., Investigation of Outside Visual Cues Required for Low Speed and Hover, AIAA Paper 85-1808 CP, 1985

7.29 Hoh, R.H., Handling Qualities for Very Low Visibility Rotorcraft NoE Operations, AGARD CP 423, October 1986

7.30 Blanken, C.L.. Hart, D.C. and Hoh, R.H., Helicopter Control Response Types for Hover and Low Speed Near-Earth Tasks in Degraded Visual Conditions, Proceedings of the 47th Annual Forum of the American Helicopter Society, Phoenix, May 1991

7.31 Hoh, R.H., Bailey, S.W. and Morgan, J.M., Flight Investigation of the Tradeoff Between Augmentation and Displays for NoE Flight in Low Visibility, Rotorcraft Flight Controls and Avionics, AHS National Specialists Meeting, Cherry Hill, NJ, October 1987

7.32 Kimberley, A.M. and Padfield, G.D., An Evaluation of Flying Qualities in UCE 3, Workshop on Helmet Mounted Displays, NASA Ames Research Center, February 1994

7.33 Bucher, N.M. (ed.), Proceedings of a TTCP Workshop on Flight-worthy Helmet Mounted Displays and Symbology for Helicopters, US Army ATCOM, NASA Ames Research Center, February 1994

7.34 Massey, C.P. and Wells, P.M., Helicopter Carefree Handling Systems, RAeSoc Conference on Handling Qualities and Control, London, November 1988

7.35 Howitt, J., Carefree Manoeuvring in Helicopter Flight Control, 51st Annual Forum of the American Helicopter Society, Fort Worth, Tex., May 1995

7.36 King, D.W., Dabundo, C., Kisor, R.L. and Agnihotri, A., V-22 Load Limiting Control Law Development, 49th Annual Forum of the American Helicopter Society, St Louis, Mo., May 1993

7.37 Anonymous, UK Defence Standard 00970; Design and Airworthiness Requirements for Service Aircraft. Vol 2, Rotorcraft – Book 1, 1988

7.38 Padfield, G.D., Tomlinson, B.N. and Wells. P.M., Simulation Studies of Helicopter Agility and Other Topics, RAE Technical Memo FS197, July 1978

7.39 Massey, C.P., Pilot/Control System Interfaces for Future ACT Equipped Helicopters, Proceedings of the 12th European Rotorcraft Forum, Garmisch Partenkirchen, Germany, September 1986

7.40 Landis, K.H. and Aiken, E.W., An Assessment of Various Sidestick Controller/Stability and Control Augmentation Systems for Night Nap-of-the-Earth Flight Using Piloted Simulation, NASA CP 2219, 1982

7.41 Morgan, J.M., In-Flight Research into the Use of Integrated Side-Stick Controllers in a Variable Stability Helicopter, RAeSoc Conference on Helicopter Handling Qualities and Control, London, November 1988

7.42 Fogler, D.L., Jr. and Keller, J.F., Design and Pilot Evaluation of the RAH-66 Comanche Core AFCS. Piloting Vertical Flight Aircraft, A Conference on Flying Qualities and Human Factors, NASA/AHS Conference, San Francisco, January 1993

7.43 Huber, H. and Hamel, P., Helicopter Flight Control; State of the Art and Future Directions, 19th European Rotorcraft Forum, Como, Italy, September 1993

7.44 Hodgkinson, J., Page, M., Preston, J. and Gillette, D., Continuous Flying Quality Improvement – The Measure and the Payoff, AIAA Paper 92-4327, Guidance, Navigation and Control Conference, Hilton Head Island, SC, August 1992

7.45 Page, M., Gillette, D., Hodgkinson, J. and Preston, J., Quantifying the Pilot's Contribution to Flight Safety, International Air Safety Seminar, MDC Paper 92K0377, Flight Safety Foundation, Long Beach, Calif., November 1992

CHAPTER 8

8.1 Sikorsky, I., Sikorsky Helicopter Development, J. Helicopter Soc. Great Brit., Vol 1, No 2, pp 5–12, 1947 (5th Lecture to the Helicopter Association of Great Britain Delivered September 8, 1947)

8.2 Stewart, W. and Zbrozek, J.K., Loss of Control Incident on S-51 Helicopter VW209, Royal Aircraft Establishment Report No Aero 2270, June 1948

8.3 Anonymous, Aeronautical Design Standard-33E-PRF, Performance Specification, Handling Qualities Requirements for Military Rotorcraft, US Army AMCOM, Redstone, Alabama, March 21, 2000

8.4 Key, D.L., Analysis of Army Helicopter Pilot Error Mishap Data and the Implications for Handling Qualities, Proceedings of the 25th European Rotorcraft Forum, Rome, Italy, September 1999

8.5 Dugan, D.C. and Delamer, K.J., The Implications of Handling Qualities in Civil Helicopter Accidents Involving Hover and Low Speed Flight, Proceedings of the American Helicopter

Society 'International Helicopter Safety Symposium', Montreal, Canada, September 2005

8.6   Padfield, G.D., Controlling the Tension Between Performance and Safety in Helicopter Operations, A Perspective on Flying Qualities, Proceedings of the 24th European Rotorcraft Forum, Marseilles, France, September 1998

8.7   Lumsden, R.B. and Padfield, G.D., Challenges at the Helicopter–Ship Dynamic Interface, Proceedings of the 24th European Rotorcraft Forum, Marseilles, France, September 1998

8.8   Padfield, G.D., The Contribution of Handling Qualities to Flight Safety, Royal Aeronautical Society Conference of Rotorcraft Flight Safety, London, November 1998

8.9   Padfield, G.D., The Making of Helicopter Flying Qualities, A Requirements Perspective, J. R. Aero. Soc., Vol 102, No 1018, pp 409–437 1998

8.10   Hoh, R., ACAH Augmentation as a Means to Alleviate Spatial Disorientation for Low Speed and Hover in Helicopters, American Helicopter Society International Conference on Advanced Rotorcraft Technology and Disaster Relief, Heli Japan, Gifu City, Japan, April 1998

8.11   Gibson, J., Olum, P. and Rosenblatt, F., Parallax and Perspective During Aircraft Landings, Am. J. Psychol., Vol 68, 1955

8.12   Gibson, J.J., Perception of the Visual World, The Riverside Press, Houghton Mifflen Company, Boston, 1950

8.13   Gibson, J.J., The Ecological Approach to Visual Perception, LEA Inc. Publishers, New Jersey, 1986

8.14   Bruce, V., Green, P.R. and Georgeson, M.A., Visual Perception Physiology, Psychology and Ecology, 3rd edn, Psychology Press, Hove, UK, 1996

8.15   Perrone, J.A., The Perception of Surface Layout During Low Level Flight, NASA CP 3118, 1991

8.16   Johnson, W.W. and Kaiser, M.K. (eds), Visually Guided Control of Movement, NASA CP 3118, 1991

8.17   Johnson, W.W. and Awe, C.A., The Selective Use of Functional Optical Variables in the Control of Forward Speed, NASA TM 108849, September 1994

8.18   Johnson, W.W., *et al.*, Optical Variables Useful in the Active Control of Altitude, 23rd Manual Control Conference, Cambridge, Mass., June 1988

8.19   Johnson, W.W. and Phatak, A.V., Optical Variables and Control Strategy Used in a Visual Hover Task, IEEE Conference on Systems, Man and Cybernetics, Cambridge, Mass., November 1989

8.20   Johnson, W.W. and Andre, A.D., Visual Cueing Aids for Rotorcraft Landings, Piloting Vertical Flight Aircraft: A NASA/AHS Conference on Flying Qualities and Human Factors, San Francisco, 1993

8.21   Cutting, J.E., Perception with an Eye for Motion, A Bradford Book, The MIT Press, Cambridge, Mass., 1986

8.22   Kaiser, M.K., Johnson, W.W., Mowafy, L., Hennessy, R.T. and Matsumoto, J.A., Visual Augmentation for Night Flight over Featureless Terrain, 48th Annual Forum of the American Helicopter Society, Washington, D.C., June 1992

8.23   Lee, D.N., The Optic Flow-Field: The Foundation of Vision, Phil. Trans. R. Soc. Lond., B, Vol 290, pp 169–179, 1980

8.24   Lee, D.N., Guiding Movement by Coupling Taus, Ecol. Psychol., Vol 10, pp 221–250, 1998

8.25   Lee, D.N. and Young, D.S., Visual Timing of Interceptive Action, Brain Mechanisms and Spatial Vision, Martinus Nijhoff, Dordretch pp 1–30, 1985

8.26   Lee, D.N., Young, D.S. and Rewt, D., How Do Somersaulters Land on Their Feet? J. Exp. Psychol. Human Percept. Perform., Vol 18, pp 1195–1202, 1992

8.27 Lee, D.N., Simmons, J.A., Saillant, P.A. and Bouffard, F., Steering by Echolocation: A Paradigm of Ecological Acoustics. J. Comp. Physiol. A, Vol 176, pp 347–354, 1995

8.28 Lee, D.N., Craig, C.M. and Grealy, M.A., Sensory and Intrinsic Guidance of Movement. Proc. R. Soc. Lond. B, Vol 266, pp 2029–2035, 1999

8.29 Moen, G.C., DiCarlo, D.J. and Yenni, K.R., A Parametric Analysis of Visual Approaches for Helicopters, NASA TN-8275, December 1976

8.30 Jump, M. and Padfield, G.D., Investigation of Flare Manoeuvres Using Optical Tau, AIAA J. Guid. Control Dyn., Vol 29, No 3, 2006

8.31 Padfield, G.D., Lee, D. and Bradley, R., How Do Pilots Know When to Stop, Turn or Pull-Up? Developing Guidelines for the Design of Vision Aids, J. Am. Helicopter Soc., Vol 48, No 2, pp 108–119, 2003

8.32 Lee, D.N., Davies, M.N.O., Green P.R. and vander Weel, F.R., Visual Control of Velocity of Approach by Pigeons When Landing, J. Exp. Biol., Vol 180, pp 85–104, 1993

8.33 Padfield, G.D., Clark, G. and Taghizard, A., How Long Do Pilots Look Forward? Prospective Visual Guidance in Terrain Hugging Flight, 31st European Rotorcraft Forum, Florence, Italy, September 2005 (see also J. Am. Helicopter Soc., Vol 52, No 2, 2007)

8.34 Wilkins, R.R., Jr., Precision Pathway Guidance, 57th Annual Forum of the American Helicopter Society, Washington, D.C., May 2001

8.35 Jukes, M., Aircraft Display Systems, Professional Engineering Publishing, London, 2004

8.36 Schroeder, J.A. and Eshow, M.E., Improvements in Hover Display Dynamics for a Combat Helicopter, Proceedings of Piloting Vertical Flight Aircraft – A NASA/AHS Conference on Flying Qualities and Human Factors, San Francisco, January 1993

8.37 Fitzwilliams, O.L.L., Some Work with Rotating-Wing Aircraft, J. Helicopter Soc. Great Brit., Vol 1, No 2, pp 13–148, 1947 (7th Lecture to the Helicopter Association of Great Britain Delivered September 8, 1947)

8.38 Gibbs-Smith, C. H., The Aeroplane; An Historical Survey of Its Origins and Development, Her Majesty's Stationery Office, London, 1960

8.39 Liptrot, R.N., Rotary Wing Activities in Germany During the Period 1939–1945, British Intelligence Objectives Sub-Committee Overall Report No 8, His Majesty's Stationary Office, London, 1948

8.40 Anonymous, Guidelines and Methods for Conducting the Safety Assessment Process on Civil Airborne Systems and Equipment, SAE Aerospace Recommended Practice, ARP 4761, Society of Automotive Engineers, Warrendale, Pa., December, 1996

8.41 Anonymous, Flying Qualities of Piloted Airplanes, Mil-F-8785C, USAF, 1980

8.42 Anonymous, UK Defence Standard 00970, Design and Airworthiness Requirements for Service Aircraft, Vol 2, Rotorcraft – Book 1, MoD, Defence Procurement Agency, 1988

8.43 Martyn, A.W., Phipps, P. and Mustard, E., The Use of Simulation to Develop an Improved Understanding of Helicopter Tail Rotor Failures and Aircrew Emergency Advice, Advances in Rotorcraft Technology, AGARD CP 592, April 1997

8.44 Anonymous, Helicopter Tail Rotor Failures, Civil Aviation Authority, CAA Paper 2003/1, November 2003

8.45 Hulme, T.M., 40 Seconds Airborne and One Spot Turn! Roy. Navy Safety J. Cockpit, Vol 162, pp 16–17, 1998

8.46 Cameron, N. and Padfield, G.D., Handling Qualities Degradation in Tilt Rotor Aircraft Following Flight Control System Failures, 30th European Rotorcraft Forum, Marseilles, France, September 2004

8.47 Weakley, J.M., Kleinhesselink, K.M., Mason, D. and Mitchell, D., Simulation Evaluation of V-22 Degraded Mode Flying Qualities, Proceeding of the 59th Annual Forum of the American Helicopter Society, Phoenix, Ariz., May 6–8, 2003

8.48 Anonymous, Joint Aviation Requirements – 29, Transport Category Rotorcraft, JAA Hoofddorp, November 2004

8.49 Hindson, W.S., Eshow, M.M. and Schroeder, J.A., A Pilot Rating Scale for Evaluating Failure Transients in Electronic Flight Control Systems, AIAA Atmospheric Flight Mechanics Conference Proceedings, Portland, Ore, AIAA-90-2827-CP, pp 270–284, August 1990

8.50 Rollet, P., Guidelines for Failure Transients, Private Communication, July 2006 (also Minutes of the 1st WP1/WP2 Task 2.3 Meeting of ACT-TILT Project, June 6–7, 2002, The University of Liverpool, ACT-TILT/EC/WP1,2/TMM01/1.0)

8.51 Lumsden, B. and Padfield, G.D., Challenges at the Helicopter–Ship Dynamic Interface, Military Aerospace Technologies, FITEC '98, IMechE Conference Transactions, pp 89–122, 1998

8.52 Roper, D.M., Owen, I., Padfield, G.D. and Hodge, S.J., Integrating CFD and Piloted Simulation to Quantify Ship–Helicopter Operating Limits, Aeronaut. J. R. Aeronaut Soc., Vol 110, No 1109, pp 419–428, 2006

8.53 Padfield, G.D. and Turner, G.P., Helicopter Encounters with Aircraft Vortex Wakes, Aeronaut. J. R. Aeronaut. Soc., Vol 105, No 1043, pp 1–8, 2001

8.54 Turner, G.P. and Padfield, G.D., Encounters with Aircraft Vortex Wakes: The Impact on Helicopter Handling Qualities, AIAA J. Aircr., Vol 39, No 5, pp 839–850, 2002

8.55 Padfield, G.D. and Manimala, B. and Turner, G.P., A Severity Analysis for Rotorcraft Encounters with Vortex Wakes, J. Am. Helicopter Soc., Vol 49, No 4, pp 445–457, 2004

8.56 Lawrence, B. and Padfield, G.D., Vortex Wake Encounter Severity for Rotorcraft in Approach and Landing, 31st European Rotorcraft Forum, Florence, Italy, September 2005

8.57 Gerza, T., Holzapfela, F. and Darracq, D., Commercial Aircraft Wake Vortices, Prog. Aerosp. Sci., Vol 38, pp 181–208, 2002

8.58 Kershaw, A., Wake Encounters in the London Heathrow Area and Planned S-Wake Studies, Proceedings of the 4th WAKENET Workshop – 'Wake Vortex Encounter', NLR, Amsterdam, October 2000

8.59 Anonymous, FAA FAR 29, Certification of Transport Rotorcraft, Advisory Circular AC 29-2A, 1994

8.60 Padfield, G.D. and White, M.D., Flight Simulation in Academia; HELIFLIGHT in Its First Year of Operation, Aeronaut. J. R. Aeronaut. Soc., Vol 107, No 1076, 2003

8.61 Constant, G., Foord, R., Forrester, P.A. and Vaughan, J.M., Coherent Laser Radar and the Problem of Aircraft Wake Vortices, J. Mod. Opt., Vol 41, 1994

8.62 Burnham, D.C., B747 Vortex Alleviation Flight Tests: Ground Based sensor Measurements, US DOT/FAA report DOT-FAA-RD-81-99, 1982

8.63 Hallock, J.M. and Burnham, D.C., Decay Characteristics of Wake Vortices from Jet Transport Aircraft, 35th Aerospace Sciences Meeting, AIAA 97-0060, Reno, Nev., 1997

8.64 Anonymous, Accident to EC-135, National Transportation Safety Board Ref. NYC99FA032, December 1998, www.nstb.gov.

8.65 Harris, F.D., Kasper, E.F. and Iseler, L.E., U.S. Civil Rotorcraft Accidents, 1963 Through 1997, NASA/TM-2000-209597, 2000

8.66 Harris, F., No Accidents – That's the Objective, Alexander A Nikolsky Honorary Lecture, Proceedings of the 62nd Annual Forum of the American Helicopter Society, Phoenix, Ariz., May 2006 (see also J. Am. Helicopter Soc., Vol 52, No 1, pp 3–14, 2007)

8.67 Anonymous, Investigation and Review of Helicopter Accidents Involving Surface Collision, CAA Paper 97004, London, May 1997

8.68 Anonymous, Proceedings of the American Helicopter Society 'International Helicopter Safety Symposium,' Montreal, Canada, September 2005

8.69 Anonymous, Proceedings of the American Helicopter Society 'Special Session on Helicopter Safety' at the 62nd Annual AHS Forum, Phoenix, Ariz., May 2006

8.70 Cross, T., Helicopter Safety – Potential Mitigation Measures, International Helicopter Safety Team (IHST) Meeting, American Helicopter Society Forum, Phoenix, Ariz., May 6, 2006

8A.1 Padfield, G.D. and White, M.D., Flight Simulation in Academia; HELIFLIGHT in Its First Year of Operation, Aeronaut. J. R. Aeronaut. Soc., Vol 107, No 1075, pp 529–538, 2003

8A.2 Padfield, G.D., Lee, D. and Bradley, R., How Do Pilots Know When to Stop, Turn or Pull-Up? Developing Guidelines for the Design of Vision Aids, J. Am. Helicopter Soc., Vol 48, No 2, pp 108–119, 2003

8A.3 Padfield, G.D. and Lawrence, B., The Birth of Flight Control; with the Wright Brothers in 1902, Aeronaut. J. R. Aeronaut. Soc., Vol 107, No 1078, pp 697–718, 2003

8A.4 Manimala, B., Padfield, G.D., Walker, D., Naddei, M., Verde, L., Ciniglio, U., Rollet, P. and Sandri, F., Load Alleviation in Tilt Rotor Aircraft Through Active Control; Modelling and Control Concepts, Aeronaut J. R. Aeronaut. Soc., Vol 108, No 1082, pp 169–185, 2004

8A.5 White, M.D. and Padfield, G.D., Flight Simulation in Academia; Progress with HELIFLIGHT at The University of Liverpool, Flight Simulation 1929–2029: A Centennial Perspective, The Royal Aeronautical Society Flight Simulation Conference, London, May 2004

8A.6 White, M.D. and Padfield, G.D., Progress with the Adaptive Pilot Model in Simulation Fidelity, 50th Annual Forum of the American Helicopter Society, Baltimore, Md., June 2004

8A.7 Cameron, N. and Padfield, G.D., Handling Qualities Degradation in Tilt Rotor Aircraft Following Flight Control System Failures, 30th European Rotorcraft Forum, Marseilles, France, September 2004

8A.8 Padfield, G.D., Manimala, B. and Turner, G.P., A Severity Analysis for Rotorcraft Encounters with Vortex Wakes, J. Am. Helicopter Soc., Vol 49, No 4, pp 445–457, 2004

8A.9 Meyer, M. and Padfield, G.D., First Steps in the Development of Handling Qualities Criteria for a Civil Tilt Rotor, J. Am. Helicopter Soc., Vol 50, No 1, pp 33–46, 2005

8A.10 Lawrence, B. and Padfield, G.D., A Handling Qualities Analysis of the Wright Brothers' 1902 Glider, AIAA J. Aircr., Vol 42, No 1, pp 224–237, 2005

8A.11 Padfield, G.D. and White, M.D., Measuring Simulation Fidelity Through an Adaptive Pilot Model, Aerosp. Sci. Technol., Vol 9, pp 400–408, 2005

8A.12 Padfield, G.D., Pavel, M.D., Casoralo, D., Hamers, M., Roth, G. and Taghizard, A., Simulation Fidelity of Real-Time Helicopter Simulation Models, 61st annual forum of the American Helicopter Society, Grapevine, Tex., June 2005

8A.13 Manimala, B., Padfield, G.D., Walker, D. and Childs, S., Synthesis and Analysis of a Multi-Objective Controller for Tilt-Rotor Structural Load Alleviation, 1st International Conference on Innovation and Integration in Aerospace Sciences, Belfast, August 2005

8A.14 Padfield, G.D., Clark, G. and Taghizard, A., How Long Do Pilots Look Forward? Prospective Visual Guidance in Terrain Hugging Flight, 31st European Rotorcraft Forum, Florence, Italy, September 2005 (see also J. Am. Helicopter Soc., Vol 52, No 2, 2007)

8A.15 Lawrence, B. and Padfield, G.D., Vortex Wake Encounter Severity for Rotorcraft in Approach and Landing, 31st European Rotorcraft Forum, Florence, Italy, September 2005

8A.16 Manimala, B., Walker, D. and Padfield, G.D., Rotorcraft Simulation Modelling and Validation for Control Design and Load Prediction, 31st European Rotorcraft Forum, Florence, Italy, September 2005

8A.17 Padfield, G.D. and Lawrence, B., The Birth of the Practical Aeroplane; An Appraisal of the Wright Brothers' Achievements in 1905, Aeronaut. J. R. Aeronaut. Soc., Vol 109, No 110, pp 421–438, 2005

8A.18 Manimala, B., Padfield, G.D. and Walker, D., Load Alleviation for a Tiltrotor Aircraft in Airplane Mode, AIAA J. Aircr., Vol 43, No 1, 2006

8A.19 Padfield, G.D., Brookes, V. and Meyer, M., Progress in Civil Tilt Rotor Handling Qualities, J. Am. Helicopter Soc., Vol 50, No 1, 50th Anniversary issue, pp 80–91, 2006

8A.20 Jump, M. and Padfield, G.D., Progress in the Development of Guidance Strategies for the Landing Flare Manoeuvre Using Tau-Based Parameters, Aircr. Eng. Aerosp. Technol., Vol 78, No 1, pp 4–12, 2006

8A.21 Padfield, G.D., Flight Handling Qualities; A Problem Based Learning Module for Final Year Aerospace Engineering Students, Aeronaut. J. R. Aeronaut. Soc., Vol 110, No 1104, pp 73–84, 2006

8A.22 Pavel, M.D. and Padfield, G.D., The Extension of ADS-33 Metrics for Agility Enhancement and Structural Load Alleviation, J. Am. Helicopter Soc., Vol 50, 2006

8A.23 Lawrence, B. and Padfield, G.D., Flight Handling Qualities of the Wright Flyer III, AIAA J. Aircr., Vol 43, No 5, pp 1307–1316, 2006

8A.24 Jump, M. and Padfield, G.D., Investigation of Flare Manoeuvres Using Optical Tau, AIAA J. Guid. Control, Vol 29, No 5, pp 1189–1200, 2006

8A.25 Kendrick, S. and Walker, D., Modelling, Simulation and Control of Helicopter Operating with External Loads, Proceedings of the 62nd AHS International Forum, Phoenix, Ariz., May 2006

8A.26 Roper, D., Padfield, G.D. and Owen, I., CFD and Flight Simulation Applied to the Helicopter–Ship Dynamic Interface, Aeronaut. J. R. Aeronaut. Soc., Vol 110, No 1109, pp 419–428, 2006

8A.27 Lawrence, B., Padfield, G.D. and Perfect, P., Flexible Uses of Simulation Tools in an Academic Environment, AIAA Atmospheric Flight Mechanics Conference, Keystone, Colorado, August 2006

8A.28 White, M.D. and Padfield, G.D., The Use of Flight Simulation for Research and Teaching in Academia, AIAA Atmospheric Flight Mechanics Conference, Keystone, Colorado, August 2006

Index

A

accel–decel manoeuvre, constant acceleration
guide in, 541–5
acceleration–deceleration manoeuvre,
536–8
accident data, 519, 567
active control, 58, 69, 441–2, 478, 504
active control technology (ACT), 69, 75,
83–4, 389–90, 441, 455, 478
actuator disc, 46, 116–7, 119–20, 128, 234
ADS-33, 3, 5, 14, 24, 61–2, 65–6, 68, 77,
188, 355, 359–62, 364, 368, 371–2, 391,
397–411, 421–5, 430–46, 449–56,
464–7, 472–6, 490, 509–10, 546–7
ADS-33D, 389–90
advance ratio, 21–2, 214–5
advanced flight simulator, 73, 378, 388,
466, 493, 504, 566
aeroelasticity, 89, 91, 135
aerofoil, 33, 150–51, 164–6, 223
Aeronautical Design Standard-33 (ADS-33),
518
agility, 26, 67, 70, 72–3, 77, 81, 297,
331–2, 362, 372, 405, 407–8, 384–7,
443, 447–55, 452–61
    definition, 478
    factor, 72–3, 481–2, 484–7
    operational, 480, 484, 514
AHS Safety Symposium, 598–9
aircraft engineering officer (AEO), 567
angle of attack, 150, 166, 506
angular momentum, 38–9, 173
Apache, McDonnell Douglas AH-64, 394,
417, 425, 444, 446, 465, 496–8, 564
atmospheric disturbance *see* turbulence
attentional demands (AD) of flight control,
522
attitude command *see* response type

attitude stabilization, 520
augmentation system
    actuator inputs, 157–8
    displays, 496–7, 600–03
    failed, 78
    feel, 576
    stability and control, 78, 81–3, 154, 156,
    159, 161, 298–9, 302–6, 374, 378, 392,
    394–6, 412, 444, 490, 571, 583, 587–9,
    591–3, 596
automatic-flight control system (AFCS),
563–4, *see* flight control system
autorotation, 15, 22, 63, 118–9, 152, 154,
206, 226, 415, 561, 567
average distances and times to the fog-lines,
548
axes systems, 26–8, 32, 91–2, 96, 110,
112, 146, 175–6, 180–85, 203, 209,
252, 293
azimuth angle, 31, 96–7, 99, 102, 169, 182,
305

B

bandwidth
    attitude response, 67–8, 325–6, 329–31,
    362, 369, 371, 378, 383, 391, 393, 406,
    420, 439–41, 587–8
    control system, 371
    estimation of, 397–9
    fixed vs rotary wing, 378–9
    flight path, 425
    gain limited, 382–3
    handling criteria, 68, 378, 381, 417, 420,
    424, 430, 451
    measurement of, 391–7
    phase limited, 382–6
    pitch attitude, 409–10, 438, 487
    ratio aircraft/task, 73–4, 314, 388

with roll damping/sensitivity, 399, 487
task, 73–4, 314, 388, 404, 437, 484
yaw axis, 433
bending of blades, 169
in-plane (lag), 135–6
modes, 135
moment, 95
out-of-plane (flap), 31, 93–4, 170
blade element theory, 125, 142, 144
blade loading, 125, 142, 164, 170, 190–91,
220–21, 231, 317, 345, 576
blockage, fin, 143, 171
Bo105, ECD
derivatives, 227, 230, 233–4, 238,
268–92
eigenvalues/stability, 240–49
handling qualities, 355–62, 372–3,
378–82, 398–400, 408–11
hinge sequence, 139
hingeless rotor, 31, 36, 42, 100–01, 128,
244
lag frequency, 136–7, 392
response, 297, 305–52
rotorspeed, 35
sideforce, 145
simulation parameters, 70, 310–18
tail rotor, 142
trim, 197, 201–21
bob-up mission-task-element (MTE),
356–61, 373–4
Boeing 747 vortex, velocity field of,
584–5

C

carefree handling, 18, 22, 24, 71, 78, 84,
455, 482, 500–08
centre of gravity, 26–7, 87, 91, 145–6, 149,
201, 246, 338
centrifugal force, 31, 138, 182, 204
characteristic equation, values *see*
eigenvalues
chord, blade, 20, 34, 100, 170
civil mission, 9, 12–13, 66, 371
clear visual information, 520
climb, climbing flight, 22, 63, 116–8, 152,
187, 191, 194, 200, 204, 206, 226, 294,
324, 342, 402–3, 412, 414–5, 421, 488,
497, 504

collective pitch, lever, 16, 35, 98, 123, 144,
158, 189–91, 204, 206, 226, 231–32,
234, 317, 319, 322–3, 413, 417, 419,
425, 427–8, 440, 545–6, 555
collimated visual displays, 603
Comanche, RAH-66, Boeing-Sikorsky, 3,
444–7, 453, 465
compressibility, 90, 97, 163, 170, 414
computer-generated profile, 535
conceptual simulation model, DRA 70,
169, 378, 423, 465
coning, rotor, 16, 30, 35, 40, 49, 52, 90,
102, 223, 232, 318–22, 325
constant deceleration profile, 535
control axes, 184–5
control power, 67, 69, 101, 121–2, 208,
359, 371–4, 399–402, 404–7, 422,
430–32, 442–4, 449, 460, 472, 474–5,
481–2, 502, 508, 510, 519, 580–81
controllers, pilot's, 455, 508–10
Coriolis force, 136, 138, 184
cross-coupling, 15, 36–7, 46, 62, 78–80,
233, 437–40
collective to pitch, 440
collective to yaw, 440
pitch to roll, 167, 233, 437–9
roll to pitch, 447–9, 451
Cutting, 527
cyclic pitch, stick, 16, 26, 33, 35–7, 82,
100–01, 113, 137, 139, 144, 154–7, 167,
201, 216, 225, 227–8, 235, 299, 307,
309, 325–38, 379, 416, 427, 449, 451,
499, 527, 574–5, 580, 585, 604

D

damping (*see also* derivatives)
ADS-33, 65
aerodynamic, 33–36, 101, 135
damping /sensitivity, 379–81
Dutch roll, 151, 247–8, 343, 435–6
flap, 34, 37–8, 108, 227, 234
gyroscopic, 330
mechanical lag, 137
mode, 240, 249
natural, 79, 472, 490
phugoid, 259
pitch, 228, 243, 308, 331, 337
ratio, 255, 309, 381–2

roll,   247, 311, 336, 369–71, 387–8
trim factor,   194
vertical/heave,   46, 48, 190–91, 220–22, 230–31, 418, 420
yaw,   147, 410–12
deceleration towards goal,   534–6
degradation of control function,   574–6
degraded visibility
  helicopter flight in,   559
  terrain-following flight,   545–8
degraded visual environment (DVE),   520
degraded visual environment,   62, 361, 411, 464–5, 487–500
derivatives,   41, 45, 55, 189, 211–29, 238–9, 244, 246, 248–9, 306, 319–22, 329, 340, 343–5, 368, 379, 402–3, 417, 425
  control,   15, 50, 56–57, 133–4, 191–2, 231–6, 269–89, 301, 338–9, 359, 382
  flapping,   36, 108, 133
  hub moment,   131–4
  inflow,   123
  stability,   42–6, 54, 66, 188, 269–89, 317, 338, 417
descent,   15, 19–20, 43, 80, 116–21, 187, 206–7, 294, 307, 310, 342, 402–3, 413–5
differential coning,   102, 105–6
differential motion parallax (DMP),   530
dihedral effect,   222–3, 249, 331, 341, 403, 434
'direct' theory of visual perception,   524
direction of flight,   528
disc loading,   47, 80, 120, 191
displays,   75, 364, 415, 487–500, 535
downwash,   19, 87–9, 99, 105, 116, 119–20, 147, 150–51, 167, 171–4, 192, 196, 294
  effect on derivatives,   214, 219–20
  effect on fuselage/empennage,   146–52, 169–71, 176–81, 204, 220–22, 228, 396, 420
  inflow roll,   18
  nonuniform inflow,   234–5
  rotor induced inflow,   34, 46–8, 89, 99, 105, 112, 115–6, 141, 146–51, 167, 171–4, 196–8, 201–3, 219–20, 230, 235, 417, 440
  tail rotor,   203

drag
  divergence,   97, 164
  fuselage, empennage,   146, 171, 200, 205
  induced,   98, 189, 205
  profile,   98–9, 112
  rotor, rotorblade,   33, 96–9, 109, 119, 136, 145, 163–5, 205
drive failure from a cruise condition,   566–7
dual-redundant hydraulic actuator,   574
Dutch roll mode,   247–50, 331, 334, 338, 342–43
dynamic stability, *see* stability
dynamic stall,   165–6

E
eigenvalue/vector,   28–9, 40–41, 50–51, 105–6, 137, 190–92, 237–60, 277–92, 300–3, 305, 308, 316, 322, 338, 449
empennage,   79–81, 84, 91, 146–52, 163, 171, 173, 196–7, 204, 207, 209, 217, 219–20, 340, 364
engine,   69–71, 148, 152–5, 161, 188, 203, 230, 261–3, 310, 424
equation(s) of motion,   27, 30, 42–3, 45, 50–51, 90–91, 104, 176, 178–80, 191–2, 207–9, 304, 319
  integrated,   159–62
  linearized,   51, 209–14, 253
equilibrium *see* trim
European civil tilt rotor,   569
eye-height scale,   524
eye-height velocity,   526, 548

F
FAA,   374, 402, 435
failure modes and effects analysis (FMEA),   561–2
Failure ratings, pilot's perspective,   573–4
fatigue, structural,   21, 26, 453
feedback *see* flight control system
fidelity
  functional,   53
  physical,   53
fin, vertical (*see also* empennage),   92, 149, 151, 172, 194
flapping, rotor,   30, 35, 39, 124, 128–35, 188, 205, 207, 215, 239, 318, 326, 328, 506, 508

equation,   99

multi-blade coordinates,   16, 101, 113,
   283, 286

flight control system,   26, 70, 91, 154–5,
   297, 334, 444–7, 559–60

AFCS,   70, 154, 156, 446–7, 497, 563

flight function failures,   562

flight simulation laboratory,   601

FLIGHTLAB,   338, 601–2

flying qualities degradation assessment,
   562–3

flying qualities for flight in a DVE,   559

flying qualities in failed conditions,   565

flying qualities levels,   562

flying qualities,   1, 3–4, 58–9

   ADS-33 requirements,   62, 186, 359,
      364

   agility, relationship with,   484–7

   carefree,   500–08

   civil helicopter,   66, 360, 378

   clinical tests,   461–2

   criteria,   3, 11, 64, 344, 363–4, 397, 429

   cross-coupling requirements,   453

   deficiencies,   62, 66, 70–72, 75, 80–81,
      322, 460–1, 463, 465, 482–3

   degraded,   456, 514

   in degraded visual environments,   411,
      487–500, 519–59

   design challenge,   13

   effects of interactions,   143, 165

   engineers 187,   209, 356, 393, 446, 453

   experiment,   358

   functional criteria,   456

   general introduction,   5, 59, 355

   in guidance task,   444

   handling and ride qualities,   59–62, 77–8,
      346, 355

   heave/vertical axis,   69

   helicopter vs fixed-wing,   65, 364, 381,
      409

   levels,   482, 484, 509–10, 517–8, 562

   military helicopter,   2–3, 355, 390, 464

   mission/task-oriented,   4, 11–12, 59,
      356

   in MTE/UCE,   356

   objective assessment,   355

   operational benefits,   75, 511

   parameters,   358

parameter prediction,   56

pilot-centred attributes,   84

pilot's impression/opinion,   297

PIO onset,   74

pitch-axis,   404

process,   23

RAE/DRA research,   86, 261, 263, 319,
   323, 396, 404

reference points,   10

roll-axis,   364

special,   455

in steep descent/vortex ring,   427, 429

subjective assessment,   455

super-safe,   1

synergy,   356, 512

test database,   522

vs handling qualities,   356

workload metrics for,   314

Focke-Achgelis Fa 223,   560–61

free wake analysis,   167

frequency response,   37, 256–7, 305,
   320–22, 335–6, 383, 391–92, 397–98,
   408

fuselage,   15–16, 26–8, 35–36, 87, 90

   aerodynamics,   264–8

   derivatives,   43

   dynamics,   39-42, 87, 90–92, 101, 105,
      146–8, 293–4, 303–12

   gust response,   49, 345

   trim,   192–207

G

gain

   in AFCS,   70, 156, 158, 446–7, 497,
      563–4

   bandwidth, *see* bandwidth

   display,   499

   engine/rotorspeed,   152, 230

   frequency response,   381–2

   gust response,   345

   inflow,   127

   pilot,   68, 299, 383, 385–6, 424

   transfer function,   69, 152–3, 256–7,
      319–20, 385–6

γ, flight path angle,   548–59

Genhel,   170, 323

Gibson, James,   523

glide slope 206–7,   477

good visual environments (GVE), 520
governor, rotorspeed, 70–71, 230
graphical user interfaces (GUIs), 602
ground effect, 139-42
gust, tuned, 349-50
gyroscopic motion
  aircraft, 32
  flap, 30, 37–8, 139
  pitch, 437

H
handling and safety, relationship between, 519
handling qualities, 24
  attitude control considerations, 305
  chart, 389f
  civil, 378, 389-91
  combined axis, 443
  conceptual model, 359
  cross-coupling, 435
  definition by Cooper-Harper, 60, 357f
  in degraded visual conditions, 14
  derivatives, 226
  design driver for ACT, 389
  experiment, 455, 464–78
  fixed-wing aircraft, 408–10
  flying qualities, 356
  heave/vertical axis, 69, 323, 417
  lateral-directional, 404
  levels, 78, 358, 372, 387, 399
  military, 9, 11, 61, 438
  parameters, 39, 56, 71, 379, 387, 409, 421, 484
  pilot opinion of, 373, 399
  pitch axis, 404, 409
  ratings (HQR), 358, 381, 424, 436, 457–64, 493
  roll axis, 364
  short-term, 39
  synergy, 77
  task, effects of, 481
  turbulence effects, 349
  two-parameter diagram, 381
  workload increase, 297
  yaw axis, 449
handling qualities degradation, 519
hazard severity criteria, 579–87
heave response, 576

helicopter attitude response to vortex encounter, 587–8
helicopter control, 517
helicopter manoeuvring, 534
HELIFLIGHT facility, 599–600
high-pitch control failure mode, 565
hinge
  delta, 3, 232
  flap, 31
  offset/effective offset, 42, 81, 96, 112, 134, 139, 226, 312
hover
  cross coupling, 80
  cyclic response, 317–30, 394
  derivatives, 215–45
  flap response, 36, 81, 108, 201, 227, 232–33
  flight regime, 17
  ground effect, 139
  gust response, 46–50, 345
  handling (*see also* handling qualities), 80
  hub moments, 126
  manoeuvres, 66, 70, 81
  mode shape, 50, 95
  momentum theory/induced velocity, 116–27, 206, 317–18
  precision, 12, 431, 494
  speed stability, 45
  stability, 243, 413–15
  thrust margin, 17, 70, 378, 422, 481, 482
  trim, 187, 196
  vertical response, 317–25
hover/low-speed requirements, 568–9
HQR *see* handling qualities
hub moment, 31–3, 37–8, 93–7, 101, 112–15, 128–35, 167–70, 218, 228–29, 232, 234, 305, 327

I
immersive cockpit environment, 602–6
inceptors (*see also* controllers) 75, 455
incidence (*see also* angle of attack)
  aircraft, 43–4, 215–19, 246, 330
  empennage, 146–9, 171, 219
  fuselage, 146–54
  rotor/rotorblade, 20, 36, 63, 87, 97, 131, 162–63, 222, 325
induced power, 47, 120, 139

induced velocity *see* downwash

in-flight collision with objects, statistics, 598

inflow *see* downwash

inflow angle, 109

in-plane motion *see* lead-lag

instantaneous time to reach steady state, 546

interactional aerodynamics 146, 171–75, 340

International Helicopter Safety Team (IHST), 598–9

intrinsic τ guide, 545, 547

inverse simulation, 57–8, 162, 311–14

irrotational 'tail' of vortex wake, helicopter flight into, 580

K

kinematics of a movement, 541

L

latency, 605–6

lead-lag (*see also* bending of blades), 42, 135–7, 304, 334

level 4 conditions, 518

lift coefficient, 150, 164

lift curve slope
main rotor, 34, 97, 100
tail rotor, 141

linear deceleration profiles, 535

Liverpool flight simulator, 534

load factor, 21–22, 415–17

Lock number, 34, 38–40, 99, 108, 130, 133–35, 216, 222, 225–9, 232, 234, 305, 320–21, 329–30

looming, 532–3

loss of control, 564–76

low-order-equivalent system (LOES), 67–69, 381, 387, 408–09, 423

Lynx, Westland
aerofoil section, 165, 261
control system, 170, 280, 424
dangleberry, 154
derivatives, 47, 217–18, 225–9, 268–75
eigenvalues/stability, 50, 65, 82–3, 237–46, 249–54, 257–59, 300–03
handling qualities, 71, 217, 345–8, 354–8, 384–6, 410, 424, 438, 455–8

hinge sequence, 128

hingeless rotor, 31, 100–101, 128, 136, 139, 159, 170, 217, 244, 249, 261, 314, 407, 482

hover downwash, 46

lag frequency, 136

response, 47, 53, 71, 295–7

simulation parameters, 200, 261–64, 314

tail rotor, 18, 143, 172, 222

trim, 200–03

M

Mach number, 20, 164–66

malfunction of control, 568–74

matched manoeuvre, 545–6

matrix formulation, 45, 50, 55, 102–4, 226, 242, 246

maximum aileron deflection vs passivation time, 570–71

military mission, 12–13, 84, 355, 434, 464, 478

mission task element (MTE), 518

mission-task-element (MTE), 11–14, 24–25, 62, 70-5, 356-58, 361-62, 372-84, 406–10, 428–33, 442, 449–53, 455, 460–86

modes, modal
6 DoF, 28, 41, 90, 191, 209-11, 214, 236-37, 252-60
constrained, 298–314
elastic, 26, 90, 96, 129, 138–39, 169–70
multi-blade coordinates, 102–03

moment of inertia
rotorblade, 36, 101, 131
whole aircraft, 26, 209, 218

momentum theory, 89–90, 116–27, 206, 318

motion control, 520

motion flow vectors, magnitude of, 528

MTE *see* mission-task-element

multi-blade coordinates *see* flapping

N

nap-of-the-Earth, 12–14, 171, 348, 352, 373, 446, 489, 496, 524, 541

normalized collective inputs vs. flight path angles, 557–8f

normalized collective pitch for a flight path angle change, 555–6f
normalized manoeuvre time, 533

O

obstacle detection system, 532
offset flap hinge *see* hinge
operational flight envelope (OFE), 15, 19–22, 24–26, 60, 62, 80, 83–84, 90, 158, 188, 222, 253, 222, 253, 336, 344, 356, 363, 424, 455, 478, 482, 518–19
optical edge rate, 526
optical expansion, 531–2
optical flow-field concept, 524
optical flow-field, 524
outside visual cues (OVC), 14, 488–91, 496, 500
overall pilot's goal, 557

P

performance-safety tension, 518
perturbations in heave velocity, 585
phase angle, rotor, 37, 134, 156, 228, 233
phase delay, 37, 68, 382, 384, 386–91, 398, 409–10, 471–74
phugoid, 46, 63–4, 238–47, 258–60, 302, 408–12
pilot's VCR, 523
pilotage augmentation system, 559
piloted simulation trials, 582
pilot-induced-oscillation (PIO), 26, 74, 299, 310, 361, 383–84, 389, 410, 474–75, 484, 575
pitch response, 52–53, 79–80, 329–30, 336–37, 405, 425, 441, 508, 588
pitch, rotor *see* collective and cyclic pitch
planes, rotor reference, 184–85
power, 17–22, 47, 52, 80, 98, 139–42, 145, 152–3, 171–5, 205–8, 230, 543–5
power setting, 413, 415, 565, 567
powerplant *see* engine
precise control augmentation, 520
'predicted' handling qualities, degradation in, 574–5
prescribed wake, 167, 173
Puma, ECF
    articulated rotor, 42, 305
    control system, 83, 444

derivatives, 50, 217–9, 228–30, 236, 243, 268, 322
eigenvalues/stability, 50, 65, 237–41, 244, 249–58
handling qualities, 217–8, 420, 449–53
response, 299–305, 317–23, 394–6, 408, 416
rotor incidence, 162
rotorspeed, 35
sideforce, 151
simulation parameters, 162, 187, 263–4
trim, 200, 392, 413
vortex ring, 428

Q

quartering flight, 163, 171–3, 519
quasi steady (*see also* stability), 90, 101, 106, 110, 123–4, 127, 132, 154, 164, 192, 221, 226, 238, 251, 310, 317, 327, 582, 584
quickhop, 70, 72, 404–8, 460, 468, 482
quickness, response, 66–8, 351, 364–71, 374–8, 404–08, 430–31, 451, 473–4, 484–7, 574, 580, 588, 592

R

rate command (*see* response type)
relaxing task requirements, 570
rescue service, role of helicopter, 517
response criteria
    heave, 344
    pitch, 381
    roll, 368
    yaw, 432
response type, 38, 62, 68, 358–62, 374, 399, 404, 442, 490, 500, 522
    acceleration command, 17, 363, 446
    attitude command, 24, 62, 361–2, 371, 378, 381, 401, 442–4, 450
    novel, 442, 444–7
    rate command, 62, 79, 361–2, 368, 373–4, 442–4, 485
    translational rate command, 24, 62, 361–2, 493
reverse flow region, 98
ride qualities, 25, 47, 59, 345, 350–52, 481
roll response, 343, 368, 373, 392, 571
root cut-out, 98

rotor systems
  articulated, 36, 37, 40, 42, 81, 95, 101,
    108, 112, 135, 170, 185, 201, 218, 229,
    239, 259, 300, 304
  hingeless, 31, 36–44, 50, 63, 79–81, 93,
    100–01, 108, 113–14, 129, 135–7, 139,
    156, 159, 170, 217–18, 227, 244, 249,
    261, 300, 314, 331
  teetering, 36, 37, 100–01, 112, 222, 224,
    228, 232–3, 313–14, 502
rotorspeed, 22, 35, 71, 145, 152, 167, 203,
  415, 502, 582
runway-dependent aircraft, wake vortex
  problem for, 595–7

S
SAE's aerospace recommended practice
  ARP4761, 579
safe flight envelope, 15, 19, 80, 90, 188, 573
separation, flow, 147, 166, 230
severe disturbances, 580
severity criteria, 578
shaft tilt, 197
ship–helicopter operating limits (SHOLs),
  577–8
sideslip, 79, 146–9, 151, 173, 183, 187,
  194–5, 203, 207, 215, 222–4, 230,
  249–50, 264, 268, 293, 268, 293, 338,
  343, 402–04, 430, 440–43, 508, 565–68
sidestep, 13, 62, 70–73, 362, 364, 374–76,
  405, 442, 450–51, 460–61, 465–68, 476,
  482, 503
Sikorsky, Igor, 517
slalom, 62, 73–4, 312–14, 365–8, 374, 465,
  468, 470–76
solidity, 21, 47, 110, 143, 189
spatial awareness, 518
spatial disorientation, risk of, 522
stability, 15, 28–9, 40–5, 63, 79–84,
  205–59, 463
  closed loop, 68, 299–301, 308, 424, 449,
    543
  derivatives *see* derivatives
  dynamic, 9, 15, 187, 189, 208, 217, 222,
    401, 410, 433
  manoeuvre, 50, 413, 415–17, 443
  quasi-static, 371, 402–04, 413, 436
  under constraint, 297–52

stabilization, automatic *see* flight control
  system
stabilizer (*see also* empennage)
  horizontal, 43, 79, 173–74, 200, 217,
    224, 332, 440
  vertical, (*see also* fin), 52, 79, 142, 151,
    436
stall, rotor, 20, 125, 164, 209
statistical discrete gust method, 347–50
stiffness number, rotor, 36–7, 101, 216,
  227–8, 233
surface collision accidents, statistics,
  598
swashplate, 15–16, 101, 138–9, 144, 154,
  156, 184, 233, 332–3, 574
sweep effects, 165
symbology, 457, 490, 493, 496–8
system identification, 4, 54–8, 213, 236,
  315

T
tail rotor, 17–19, 52, 79–80, 92, 142–6,
  163, 171–3, 430
tail rotor failures, 564–8
tailplane *see* empennage, stabilizer
task (*see also* mission task element), 11,
  22–5, 59, 62
task bandwidth, 74, 314, 388, 437,
  484
task margin, 364
task portrait, signature, 66, 364
τ in motion control, significance for
  helicopter flight, 536
τ of a motion gap, 538
τ on rising curve, 548–59
τ, optical tau, 532–3
τ-coupled predictions, 557
τ-coupling, 538–45
temporal optical variable, 532
3-s perturbations in vertical acceleration,
  593
3-s pilot intervention time, 594
threshold of velocity perception, 527
thrust margin, 17, 70, 378, 417–22, 481–2,
  580, 592, 597
thrust, rotor, 16, 22, 46–8, 68, 93, 112,
  116–26, 141–2, 185, 196, 201, 222–34,
  311, 330, 416, 442

time constant, 17, 35, 39–49, 55–61, 68–73, 105, 124–9, 151, 190, 249–54, 301, 311–16, 326–37, 349, 363–72, 378–94, 401

time-to-contact information, 533

tip loss, 98

tip path plane, 16, 184–5

tip speed ratio *see* advance ratio

torque, rotor, 50, 98, 114–16, 185, 189, 194, 228–30

transfer function, 69, 152, 334–6, 385–7, 408, 499–500

transient response, severity of, 593–7

translational rate command (TRC) *see* response type

trim, 5, 15, 25–9, 112, 187–94, 392, 402

turbulence, 82, 161, 303, 345–52, 522

twist, rotorblade, 80, 98, 135, 138–9, 170

typical correlations between τ_γ and τ_g, 552

U

U.S. civil rotorcraft accidents, distribution of, 597–8

UCE/VCR, 523

UH-60 type helicopter, key configuration parameters, 582–3

UH-60, SH-60, Sikorsky, 173, 337, 440–41, 582

UK MoD/CAA Tail Rotor Action Committee (TRAC), 564–5

unsteady aerodynamics, 139, 165, 169–75, 191, 322, 349

upset severity rating scale, 595f

usable cue environment (UCE), 14, 24, 62, 341, 358–62, 388, 436, 490–96, 520–22, 547, 551–2, 557, 559

V

V-22, Osprey, Boeing Bell, 508, 569, 571

vertical axis *se* handling/flying qualities

vertical manoeuvres, 546

vertical motion simulator, 381, 420–21, 467, 493

vertical motions of rotorcraft during vortex encounters, 588–93

visual perception in flight control, 523

visual perception in flight control, 523–32

visual sensations, 523

vortex ring state, 20, 122, 209, 428

vortices, 171

W

wake, 19, 46, 53, 79, 89–91, 97–8, 101, 115–22, 146, 150, 167–74, 340, 440, 578, 585

wake vortices, 578–87

wavelets, 347–51

weakly coupled systems, 237–59, 281–97

windmill state, 42, 58, 238, 242, 245, 260, 299

workload, 70–77, 297–315, 358, 444

workload on pilot-external pressures, 518

Y

yaw motion, 595